MICROBIAL PHYSIOLOGY

MICROBIAL PHYSIOLOGY

Third Edition

ALBERT G. MOAT
Marshall University School of Medicine
Huntington, West Virginia

JOHN W. FOSTER
University of South Alabama School of Medicine
Mobile, Alabama

WILEY-LISS

A JOHN WILEY & SONS, INC., PUBLICATION

New York / Chichester / Brisbane / Toronto / Singapore

Address All Inquiries to the Publisher
Wiley-Liss, Inc., 605 Third Avenue, New York, NY 10158-0012

Copyright © 1995 Wiley-Liss, Inc.

Printed in the United States of America

The text of this book is printed on acid-free paper.

Library of Congress Cataloging-in-Publication Data

Moat, Albert G.
    Microbial physiology / Albert G. Moat, John W. Foster.—3rd ed.
      p. cm
    Includes index.
    ISBN 0-471-01295-5 (cloth : alk. paper).—ISBN 0-471-01452-4
(paper : alk. paper)
    1. Microorganisms—Physiology.   I. Foster, John Watkins.
II. Title.
QR84.M64 1995
576'.11—dc20   95–15648

10 9 8 7 6 5 4 3 2 1

To
Irene and Zari

CONTENTS

PREFACE   xv

1 INTRODUCTION TO MICROBIAL PHYSIOLOGY   1

The *Escherichia coli* Paradigm / 1

Cell Structure / 1

   The cell surface / 1

   Cell wall / 2

   Membranes / 3

   Capsules / 4

   Organs of locomotion / 5

   Pili or fimbriae / 5

   Ribosomes / 5

Synthesis of DNA, RNA, and Protein / 6

Metabolic and Genetic Regulation / 9

Microbial Genetics / 10

Chemical Synthesis / 11

   Chemical composition / 11

   Energy / 14

   Oxidation-reduction versus fermentation / 14

   Nitrogen assimilation / 16

Special Topics / 18

   Mitochondria / 18

   Endospores / 19

Growth / 19

   Growth cycle / 20

   Continuous culture / 22

Factors Affecting Growth / 22
  Nutrition / 22
  Oxygen / 24
  Carbon dioxide / 25
  Temperature / 25
  Hydrogen ion concentration / 26
Summary / 27

**2   MACROMOLECULAR SYNTHESIS AND PROCESSING:
    DNA, RNA, AND PROTEIN SYNTHESIS**   **28**
Structure of DNA / 29
  Bacterial nucleoid / 31
  REP elements / 36
DNA Replication / 36
  Initiation of DNA replication / 43
  Termination of DNA replication and chromosome partitioning / 47
RNA Synthesis: Transcription / 47
  RNA synthesis / 47
  Protein-protein interactions with RNA polymerase / 51
  RNA turnover / 52
  RNA processing / 52
Protein Synthesis: Translation / 56
  Transfer RNA / 57
  Charging of tRNA / 60
  Ribosome structure / 61
  Initiation of polypeptide synthesis / 64
  Elongation / 66
  Peptide-bond formation / 70
  Translocation / 71
  Termination / 71
  Ribosome editing / 72
  Coupled transcription and translation / 72
  Protein folding and chaperones / 73
Protein Trafficking / 73
Degradation of Abnormal Proteins / 79
Antibiotics that Affect Nucleic Acid and Protein Synthesis / 80
  Agents affecting DNA metabolism / 80
  Agents affecting transcription / 85
  Agents affecting translation / 86
Selected References / 91

3 BACTERIAL GENETICS **94**

Transfer of Genetic Information in Prokaryotes / 94

Plasmids / 95

   Nonconjugative, mobilizable plasmids / 95

   Resistance (R) plasmids / 96

   Plasmids in other bacterial genera / 96

   Plasmid replication control / 97

Conjugation / 100

Transformation / 110

Transduction / 114

Recombination / 117

   General recombination / 118

   Genetics of recombination / 120

Transposable Elements / 126

   Transposon Tn *10* / 130

   Transposon Tn *3* / 132

   Conjugative transposition / 132

   Evolutionary consideration / 132

Mutagenesis / 133

   Spontaneous mutations / 134

   The nature of mutational events / 134

   Suppressor of mutations / 136

   Directed mutations / 138

DNA Repair Systems / 139

   Photoreactivation / 139

   Nucleotide excision repair / 139

   UvrD (Helicase II) / 141

   Mismatch repair / 142

   DNA glycosylases (Base excision repair) / 142

   Adaptive response to methylating and ethylating agents / 144

   Postreplication daughter strand gap repair / 144

   SOS-inducible repair / 146

Selected References / 149

4 REGULATION OF PROKARYOTIC GENE EXPRESSION **152**

Introduction / 152

Transcriptional Control / 152

   DNA-binding proteins / 153

   The *lac* operon / 155

   Catabolite control / 158

   The *gal* operon / 161

The arabinose operon / 161

The *trp* operon: repression and attenuation controls / 164

Transcriptional attenuation: the *pyrBI* strategy / 167

Translational control: the *pyrC* strategy / 170

Arginine / 171

Membrane-mediated regulation / 171

Recombinational regulation of gene expression (flagellar phase variation) / 171

Translational Repression / 174

Communication with the Environment: Two Component Systems / 177

Osmotic Control of Gene Expression / 179

Global Control Networks: Regulation at the Whole Cell Level / 183

The SOS response / 185

Heat shock responses / 185

Electron Transport: Respiratory Pathways / 187

Stringent Control / 189

Regulation of Nitrogen Assimilation and Nitrogen Fixation / 193

Phosphate Starvation-Controlled Stimulon / 196

Oxidation Stress / 197

The Lon System: Proteolytic Control / 198

Selected References / 199

5   BACTERIOPHAGE GENETICS   **202**

General Characteristics of Bacteriophage / 202

T4 Phage / 205

T4 phage structure / 205

General pattern of gene expression / 209

T4 replication / 212

Regulation of T4 gene expression / 216

Introns in T4 genes / 217

Lambda Phage / 218

The lysis-lysogeny decision / 219

Control of integration and excision / 223

Lambda phage replication / 225

Lambda as a cloning vector / 227

Mu Phage / 228

ϕX174 / 232

Concluding Remarks / 235

Selected References / 235

6   CELL STRUCTURE AND FUNCTION   **237**

The Eukaryotic Nucleus / 237

Bacterial Nucleoid / 239

   Nucleosomes / 243

Mitochondria / 246

Cell Surface of Microorganisms / 247

   Eukaryotic cell surface / 247

   Prokaryotic cell surface / 248

   Structure and synthesis of bacterial peptidoglycan / 249

   Teichoic acids and lipoteichoic acids / 257

   Outer membranes of gram-negative bacteria / 260

   Lipopolysaccharides / 260

   Porins / 263

   Lipopolysaccharide biosynthesis / 264

   Enterobacterial common antigen / 265

Cytoplasmic Membrane and Other Membranous Structures / 266

   Cytoplasmic membrane / 266

   Permeability and transport / 269

   Periplasm / 269

   Other membranous organelles / 270

Capsules / 270

Organs of Locomotion / 278

   Cilia and flagella of eukaryotes / 279

   Bacterial flagella / 279

   Chemotaxis / 284

   Swarming phenomenon / 291

   Motility in spirochetes / 295

   Gliding motility / 296

Pili or Fimbriae / 297

References / 300

7   CARBOHYDRATE METABOLISM AND ENERGY PRODUCTION   305

Glycolytic Pathways / 306

   Fructose bisphosphate aldolase (Embden–Meyerhof–Parnas) pathway / 306

   Hexose monophosphate pathways / 308

   Phosphoketolase pathway / 308

   Oxidative pentose phosphate cycle / 309

   Entner–Doudoroff or ketogluconate pathway / 312

Gluconeogenesis / 314

Tricarboxylic Acid Cycle / 316

Glyoxylate Cycle / 319

Energy Production / 322
  Substrate level phosphorylation / 322
Oxidative Phosphorylation / 325
  Measurement of PMF / 325
Energetics of Chemolithotrophs / 332
pH Homeostasis / 334
Transport / 335
  Specific transport systems / 338
Pathways for Utilization of Sugars Other than Glucose / 344
Pectin and Aldohexuronate Pathways / 348
Cellulose Degradation / 352
Utilization of Starch, Glycogen, and Related Compounds / 355
Metabolism of Aromatic Compounds / 357
Fermentation Pathways in Specific Groups of Microorganisms / 362
  Fermentation balances / 363
  Lactic acid producing fermentations / 367
  Butyric acid and solvent-producing fermentations / 376
  Fermentation of the mixed-acid type / 382
  Propionic acid fermentation / 385
  Acetic acid fermentation / 390
Characteristics and Metabolism of Autotrophs / 390
  Major groups of autotrophs / 390
  Photosynthetic bacteria and cyanobacteria / 391
  Autotrophic CO_2 fixation and mechanisms of photosynthesis / 394
  Hydrogen bacteria / 398
  Nitrifying bacteria / 399
  Sulfur bacteria / 399
  Iron bacteria / 401
  Methylotrophs / 401
  Methanogens / 403
References / 405

8   LIPIDS AND STEROLS   410
Lipid Composition of Microorganisms / 411
  Straight-chain fatty acids / 411
  Branched-chain fatty acids / 414
  Ring-containing fatty acids / 415
  Alk-1-enyl ethers (plasmalogens) / 416
  Alkyl ethers / 416
  Microbial lipids / 418
  Glycolipids / 419
Biosynthesis of Fatty Acids / 421

Biosynthesis of Phospholipids / 426

Degradation of Lipids / 427

Biosynthesis of Mevalonate, squalene, and sterols / 430

References / 434

9   NITROGEN METABOLISM   **436**

Biological Nitrogen Fixation / 436

The Nitrogen Fixation Process / 439

Components of the nitrogenase system / 441

Symbiotic Nitrogen Fixation / 442

Inorganic Nitrogen Metabolism / 446

Assimilation of Inorganic Nitrogen / 450

General Reactions of Amino Acids / 452

Amino acid decarboxylases / 452

Amino acid deaminases / 453

Amino acid transaminases (aminotransferases) / 456

Amino acid racemases / 457

Role of pyridoxal-5′-phosphate in enzymatic reactions with amino
  acids / 458

The Stickland Reaction / 459

References / 459

10   AMINO ACIDS, PURINES, AND PYRIMIDINES   **462**

Amino Acid Biosynthesis / 462

The Glutamate or α-Ketoglutarate Family / 463

Glutamine and glutathione synthesis / 463

The proline pathway / 463

Aminolevulinate synthesis / 463

The arginine pathway / 464

Polyamine biosynthesis / 468

Lysine biosynthesis in fungi / 468

The Aspartate and Pyruvate Families / 471

Asparagine synthesis / 471

The aspartate pathway / 472

The bacterial pathway to lysine / 473

Threonine, isoleucine, and methionine formation / 473

Isoleucine, leucine, and valine biosynthesis / 475

Regulation of the aspartate family / 476

The Serine-Glycine Family / 480

Aminolevulinate and the pathway to tetrapyrroles / 482

The Aromatic Amino Acid Pathway / 484

Phenylalanine, tyrosine, and tryptophan / 484

The common aromatic amino acid pathway / 484

Pathways to tyrosine and phenylalanine / 489

p-Aminobenzoate and folate biosynthesis / 491

Enterobactin biosynthesis / 491

Biosynthetic pathway to ubiquinone / 494

Menaquinone (vitamin K) biosynthesis / 495

Biosynthesis of nicotinamide adenine dinucleotide (NAD) / 496

Histidine biosynthesis / 497

Purines and Pyrimidines / 500

Biosynthesis of purines / 500

Biosynthesis of pyrimidines / 506

Regulation of purine and pyrimidine biosynthesis / 507

Interconversion of nucleotides, nucleosides,
and free bases: salvage pathways / 512

References / 514

11   GROWTH AND ITS REGULATION   **518**

Growth of Gram-Positive Streptococci / 519

Growth of Gram-Negative Rods / 522

Growth of Gram-Positive Bacilli / 533

Effect of Environmental Changes on Microbial Growth / 536

Water / 539

Osmotic pressure and osmoregulation / 539

Hydrogen ion concentration / 541

Temperature / 543

Nutrition-starvation induced proteins / 545

Hydrostatic pressure / 546

References / 546

12   ENDOSPORE FORMATION (DIFFERENTIATION)   **549**

Life Cycle of *Bacillus* / 549

Cytological Aspects of Bacterial Endospore Formation / 550

Physiological and Genetic Aspects of Sporulation / 552

Sporulation genes / 552

Activation, Germination, and Outgrowth of Bacterial Endospores / 558

Activation / 558

Germination / 558

Germination monitoring / 559

Germination loci / 559

Models of spore germination / 559

Metabolic changes during germination / 560

Outgrowth / 561

References / 561

INDEX   **563**

PREFACE

The field of microbial physiology continues to develop at a rapid pace, thanks to the introduction of more and more sophisticated genetic and molecular approaches to the subject. Consequently, we have found it necessary not only to write a new edition to this text but to reorder the text material to present the physiological aspects in their true light. The concepts of microbial genetics and molecular biology have been presented first so the reader might have a thorough grounding in these disciplines before delving into the main aspects of microbial cell structure, intermediary metabolism, and growth. We felt that this order of presentation would help the reader to fully appreciate the current level of our understanding of microbial physiology.

We would like to thank all of those who have provided us with help and encouragement in this undertaking. We especially wish to thank those who granted us permission to use illustrations and/or provided us with original materials for this purpose.

ALBERT G. MOAT
JOHN W. FOSTER

Huntington, West Virginia
Mobile, Alabama

MICROBIAL PHYSIOLOGY

CHAPTER 1

INTRODUCTION TO MICROBIAL PHYSIOLOGY

THE *ESCHERICHIA COLI* PARADIGM

Microbial physiology is an enormous discipline encompassing knowledge culled from the study of thousands of different microorganisms. It is, of course, impossible to convey all that is known within the confines of one book. However, one can build a solid foundation using a limited number of organisms to illustrate key concepts of the field. The goal of this text is to help set the foundation for further inquiry into microbial physiology and genetics. To accomplish this we have taken the gram-negative organism *Escherichia coli* as the paradigm. We will include other organisms that provide significant counterexamples to the paradigm or use alternative strategies to accomplish a similar biochemical goal. In this introductory chapter we would like to paint a broad portrait of the microbial cell with special focus on *E. coli*. Each topic will be covered in depth in a later chapter but this chapter should serve as a point of convergence where the student can see how one aspect of physiology relates to another.

CELL STRUCTURE

As any beginning student of microbiology knows, bacteria come in three basic models; spherical (coccus), rod (bacillus), and spiral (spirillum). Bacteria do not possess a membrane-bound nucleus as do eukaryotic microorganisms and are, therefore, termed prokaryotic. in addition to these basic types of bacteria, there are other more specialized forms described as budding, sheathed, and mycelial bacteria. Figure 1-1 presents a schematic representation of a typical (meaning *E. coli*) bacterial cell. We will briefly tour this cell starting from the exterior.

The Cell Surface

The interface between the microbial cell and its external environment is by definition the cell surface. It is what protects the cell interior from external hazards and maintains the

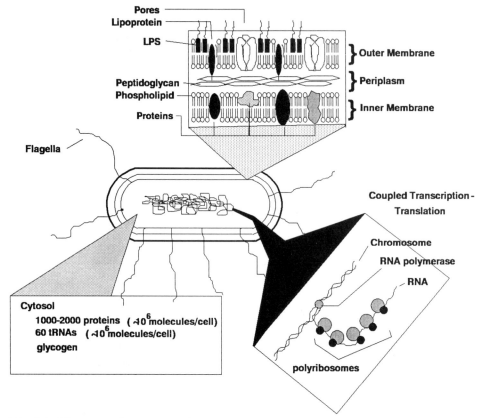

Fig. 1-1. Diagrammatic representation of a "typical" bacterial cell. Portions of the cell are enlarged to show further details.

integrity of the cell as a discrete entity. Although it must be steadfast in fulfilling these functions, it must also enable transport of large molecules into and out of the cell. These large molecules include carbohydrates (e.g., glucose), vitamins (e.g., vitamin B$_{12}$), amino acids, and nucleosides, as well as proteins exported to the exterior to the cell. The structure of the cell surface can vary considerably in its complexity depending on the organism.

Cell Wall. In 1884 the Danish investigator Christian Gram devised a differential stain based on the ability of certain bacterial cells to retain the dye crystal violet after decoloration with 95% ethanol. Those cells that retained the stain were called **gram positive.** Subsequent studies have shown that this fortuitous discovery distinguished two fundamentally different types of bacterial cells. The surface of gram-negative cells is much more complex than that of gram-positive cells. As shown in the schematic drawings in Figure 1-2, the gram-positive cell surface has two major structures: the cell wall and the cell membrane. The cell wall of gram-positive cells is composed of multiple layers of peptidoglycan. Peptidoglycan is a linear polymer of alternating units of *N*-acetylglucosamine (NAG) and *N*-acetylmuramic acid (NAM). A short peptide chain is attached to muramic acid. A common feature in bacterial cell walls is cross-bridging between the peptide

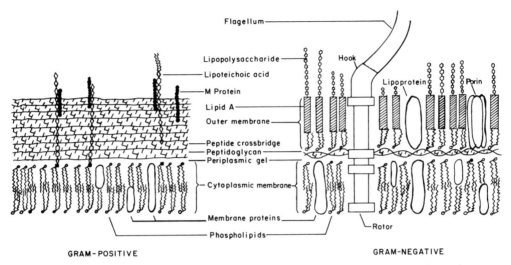

Fig. 1-2. Composition of the cell surfaces of gram-positive and gram-negative bacteria.

chains. In a gram-positive organism, such as *Staphylococcus aureus*, the cross-bridging between adjacent peptides may be close to 100%. By contrast, in *E. coli* (a gram-negative organism) the frequency of cross-bridging may be as low as 30% (Fig. 1-3). Other components, for example, lipoteichoic acid (only present in gram-positive organisms) are synthesized at the membrane surface and may extend through the peptidoglycan layer to the outer surface.

The peptidoglycan layer of gram-negative cells appear to be a single monolayer. An outer membrane surrounding the gram-negative cell is composed of phospholipid, lipopolysaccharide, enzymes, and other proteins, including lipoproteins. The space between this outer membrane and the inner membrane is referred to as the periplasmic space. It may be traversed at several points by various enzymes and other proteins (Fig. 1-2).

Membranes. The cytoplasmic membrane of both gram-positive and gram-negative cells is a lipid bilayer composed of essentially the same basic components: phospholipids, glycolipids, and a variety of proteins. The proteins in the cytoplasmic membrane may extend through the entire thickness of the membrane. Some of these proteins provide structural support to the membrane while others function in the transport of sugars, amino acids, and other metabolites.

The outer membrane of gram-negative cells contains a relatively high content of lipids as lipopolysaccharides. These lipid-containing components represent one of the most important identifying features of gram-negative cells: the O antigens that are formed by the external polysaccharide chains of the lipopolysaccharide. This lipid-containing component also displays endotoxin activity, that is, it is responsible for the shock observed in severe infections caused by gram-negative organisms. Bacterial cell surfaces may also contain specific carbohydrate or protein receptor sites for the attachment of **bacteriophages** (viruses that infect bacteria). Once attached to these receptor sites, the bacteriophage can initiate invasion of the cell.

Gram-positive and gram-negative cells have somewhat different strategies for transporting materials across the membrane and into the cell. The cytoplasmic membrane of

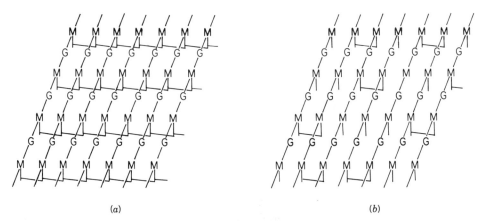

Fig. 1-3. Diagrammatic views of bacterial peptidoglycan in (a) gram-positive and (b) gram-negative cells. G = N-acetylglucosamine; M = N-acetylmuramic acid; verticle lines represent the peptide chains; horizontal lines represent the amino acid cross-bridges between peptide chains. Note the greater number of cross-bridges in the gram-positive peptidoglycan monolayer.

gram-positive organisms has immediate access to media components. However, chemicals and nutrients must first traverse the *outer* membrane of gram-negative organisms before encountering the cytoplasmic membrane. Gram-negative cells have pores formed by protein triplets in their outer membrane that will permit passage of fairly large molecules into the space between the two membranes called the periplasmic space. Subsequent transport across the inner or cytoplasmic membrane is similar in both gram-positive and gram-negative cells.

Capsules. Some bacterial cells produce a layer of material external to the cell referred to as a capsule or a slime layer. Capsules are composed of either polysaccharides, high-molecular-weight polymers of carbohydrates, or polymers of amino acids called poly-peptides (often formed from the D rather than the L isomer of an amino acid). The capsule of *Streptococcus pneumoniae* type III is composed of glucose and glucuronic acid in alternating β-1, 3- and β-1, 4- linkages:

This capsular polysaccharide, sometimes referred to as pneumococcal polysaccharide, is responsible for the virulence of the pneumococcus. *Bacillus anthracis*, the anthrax bacillus, produces a polypeptide capsule composed of D-glutamic acid subunits, which is a virulence factor for this organism.

Organs of Locomotion. Many microorganisms are motile, able to move from place to place in a concerted manner, especially in an aqueous environment. In the case of bacteria, this motility is accomplished by means of simple strands of protein (flagellin) woven into helical organelles called flagella. The bacterial flagellum is attached at the cell surface by means of a basal body (Fig. 1-4). The basal body contains a motor that turns the flagellum, which propels the organism through the liquid environment.

Pili or Fimbriae. Many bacteria possess external structures that are shorter and more rigid than flagella. These structures have been termed **pili** (from the Latin meaning hair) or **fimbriae** (from the Latin meaning fringe). These appendages also appear to arise from a basal body or granule located either within the cytoplasmic membrane or in the cytoplasm immediately beneath the membrane (Fig. 1-5). Generalized or common pili play a role in cellular adhesion to surfaces or to host cells.

Ribosomes. Particles containing approximately 65% ribonucleic acid (RNA) and 35% protein have been isolated from the cytoplasm of all microorganisms including bacteria. These particles, termed **ribosomes**, impart the fine granular appearance to the cytoplasm as observed in many electron micrographs of cells (see Fig. 1-1). It is the ribosome that orchestrates the polymerization of amino acids into proteins. At higher magnification under the electron microscope the ribosome particles can be observed as spheres. In properly prepared specimens the ribosomes are observed as collections or chains held together on a single messenger RNA (mRNA) molecule and referred to as **polyribosomes** or simply **polysomes**.

The more or less spherical ribosome particle, when examined by sucrose gradient sedimentation, has been found to have a Svedberg coefficient of 70S. (A svedberg unit denotes the rate of sedimentation of a macromolecule in a centrifugal field and is related to the molecular size of that macromolecule.) The prokaryotic ribosome may be separated

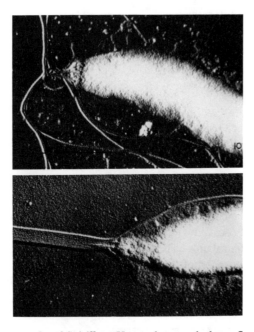

Fig. 1-4. Electron micrographs of *Spirillum*. Upper photograph shows flagella of *Spirillum linum* extending from an aggregated mass of basal granules. Lower photograph shows a terminal body of *Spirillum atlanticum* from which the flagella emerge from the cell. From Williams, M. A. and G. B. Chapman. 1961. *J. Bacteriol.* **81:** 195.

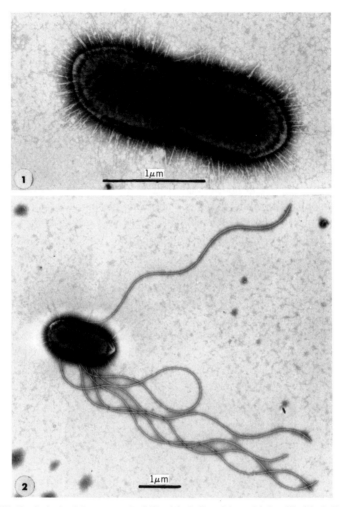

Fig. 1-5. Pili (fimbriae) of *Proteus mirabilis*. (1) Cells with multiple pili. (2) Cells with both pili and flagella. Bar = 1 μm. From Hoeniger, J. F. M. 1965. *J. Gen. Microbiol.* **40**:29.

into two lower molecular weight components, one of 50S and another of 30S. Only the complete 70S particle functions in polypeptide synthesis. By comparison, the ribosomes of eukaryotic cells are associated with the endoplasmic reticulum, are larger (80S), and are composed of 40S and 60S subunits. The function of both 70S and 80S ribosomes in protein synthesis is identical. In support of the concept that mitochondria evolved from endosymbiotic prokaryotic cells, eukaryotic mitochondria characteristically display 70S ribosomes.

SYNTHESIS OF DNA, RNA, AND PROTEIN

The molecular events that affect all aspects of cellular growth and development include DNA **replication**, **transcription**, and **translation**. In bacteria, **replication** involves the accurate duplication of chromosomal DNA enabling the formation of two daughter cells via binary fission. All of the information required for the structure and function of the cell is usually contained in the chromosome (see plasmids). This genetic information is processed in two steps to produce various proteins. The first step is the transcription of the information from DNA to RNA. One genetic information unit (**gene**) can be transcribed into many copies of RNA. Thus, in any one cell of *E. coli*, whose chromosomes consists of a circular, double-stranded DNA molecule, there will usually be one chromosome containing one copy of each gene but many copies of RNA for any specific gene.

A simplified view of DNA replication in *E. coli* is shown in the diagram in Figure 1-6. The double-stranded DNA molecule unwinds from a specific starting point (**origin**). The new DNA is synthesized opposite each strand. The enzyme involved in replication (DNA polymerase) uses a parent strand as a template, placing adenine residues opposite thymine and cytosine residues opposite guanine. New DNA is synthesized in both directions from the origin and continues until both replication forks meet at the terminus 180° from the origin. At this point, cell division proceeds with cross-wall formation occurring between the two newly synthesized chromosomes (Fig. 1-7). Note that the chromosome appears attached to the cell membrane and that growth of this membrane serves to separate the daughter chromosomes. Although this model has been challenged recently, it still serves to illustrate the need for chromosome segregation and a possible model for accomplishing that goal (see Chapter 2, Partitioning).

The synthesis of proteins (**translation**) can be envisioned in an overall process, as depicted in Figure 1-8. The enzyme RNA polymerase (DNA-dependent RNA polymerase) first locates the beginning of a gene. This area of the chromosome then undergoes a localized unwinding allowing RNA polymerase to transcribe RNA from the DNA template. Before the RNA (called **mRNA** or the **message**) is completely transcribed, a ribosome will attach to the beginning of the message. As noted above, the ribosome is composed of two subunits, 30S and 50S, each composed of special ribosomal proteins

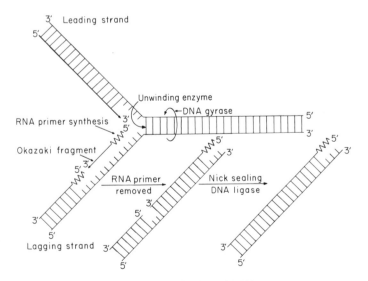

Fig. 1-6. Replication of DNA.

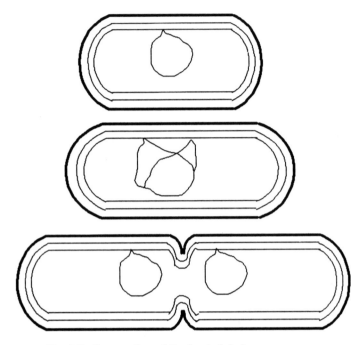

Fig. 1-7. Segregation of the bacterial chromosome.

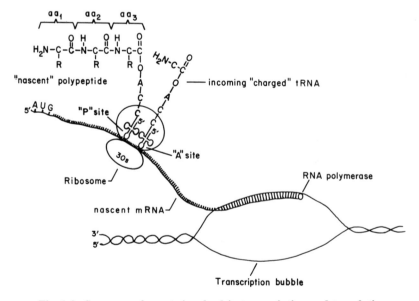

Fig. 1-8. Sequence of events involved in transcription and translation.

and ribosomal ribonucleic acids (**rRNA**). The rRNA molecules do not, by themselves, code for any protein. The ribosome **translates** the RNA message into protein by reading three nucleotides (**triplet codon**) as a specific amino acid. Each amino acid used by the ribosome must first be attached to an *adaptor* or **transfer RNA (tRNA)** molecule specific for that amino acid. A tRNA containing an attached amino acid is referred to as a **charged tRNA** molecule. Part of the tRNA molecule (called the **anticodon**) will base pair (bp) with the codon in a **mRNA**. When two such charged tRNA molecules simultaneously occupy adjacent sites on the ribosome, the ribosome catalyzes the formation of a peptide bond between the two amino acids. At this point, the two amino acids are attached to one tRNA while the other tRNA is uncharged and released from the ribosome. The ribosome is then free to move along the message to the next codon. The process continues until the ribosome reaches the end of the message at which point a complete protein has been formed. Notice that synthesis of the protein begins with the N-terminal amino acid and finishes with the C-terminal amino acid. Also note that the beginning of the message is the 5′ end of the mRNA, while the beginning of the gene is the 3′ end of the template DNA. The processes of replication, transcription, and translation are discussed in detail in Chapter 2.

Metabolic and Genetic Regulation

For a cell to grow efficiently, all the basic building blocks and all the macromolecules derived from them have to be produced in the correct proportions. With complex metabolic pathways, it is important to understand the manner in which a microbial cell regulates the production and concentration of each product. Two very common mechanisms of metabolic and genetic regulation follow:

1. **Feedback inhibition** of enzyme *activity* (**metabolic regulation**).
2. **Repression** of enzyme *synthesis* (**genetic regulation**).

In feedback inhibition, the activity of an enzyme already present in the cell is inhibited by the end-product of the reaction. In repression, the synthesis of an enzyme (see previous discussion of transcription and translation) is inhibited by the end product of the reaction. If, in a hypothetical metabolic pathway (Fig. 1-9A), the formation of an excess concentration of intermediate B results in the inhibition of the activity of enzyme 1, this action is referred to as **feedback or end-product inhibition**. Similarly, an excess of end-product C may inhibit the activity of enzyme 1 by feedback inhibition. Different organisms will display variations in the manner by which they employ feedback inhibition to regulate enzyme activity.

As an alternative to feedback inhibition, an excess concentration of end-product C may cause the cell to stop **synthesizing** enzyme 1 (Fig. 1-9B). This action is referred to as genetic **repression**. Individual organisms may employ quite different combinations of feedback inhibition and repression of enzyme synthesis to regulate a metabolic pathway. In situations where a branched metabolic pathway is involved, a combination of feedback inhibition and repression may be particularly effective, especially if isoenzymes or isozymes (multiple forms of the same enzyme) catalyze reactions just prior to a branch point. Isozymes may differ in their response to end-product repression and feedback inhibition (Fig. 1-9C). In a branched metabolic pathway, each isozyme may respond to only one of the end-products of the branched pathway. End-product C may cause re-

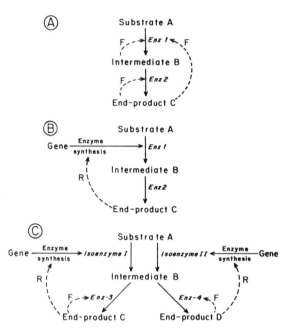

Fig. 1-9. Diagrammatic presentation of feedback inhibition of enzyme activity and end-product repression of enzyme synthesis. Enz = enzyme; F = feedback inhibition; R = repression of enzyme synthesis.

pression of isozyme 1, whereas, end-product D may cause repression of isozyme II. By such mechanisms, the amount of end-products C and D may be kept in balance. In Chapter 4, these and other regulatory mechanisms are discussed in greater detail.

MICROBIAL GENETICS

Having just outlined the processes of transcription, translation, and replication, it is now possible to define several genetic terms. The **gene** may be defined as a heritable unit of function composed of a specific sequence of purine and pyrimidine bases. This sequence of bases determines, by transcription, the sequence of purine and pyrimidine bases in an RNA molecule. By the process of translation, the sequence of bases in an RNA molecule specifies the sequence of amino acids incorporated into a polypeptide chain. The **genotype** of an organism is the sum total of all of the hereditary units of genes. The observed *expression* of the genetic determinants, that is, the structural appearance and physiological properties of an organism, is referred to as its **phenotype**.

An individual gene can exist in different forms as a result of changes in the sequence of nucleotides. These alternate forms of a gene are referred to as **alleles**. Genetic material is not absolutely stable but can change or **mutate**, the process of change being referred to as **mutagenesis**. Altered genes are referred to as **mutant alleles** in contrast to the normal or **wild-type alleles**. **Spontaneous mutations** are considered to arise during replication, repair, and recombination of DNA as a result of errors made by the enzymes involved in DNA metabolism. Mutations may be increased by the activity of a number of environmental influences. Radiation in the form of X-rays, ultraviolet (UV), or cosmic rays may affect the chemical structure of the gene. A variety of chemicals may also give rise to mutations. Physical, chemical, or physicochemical agents capable of increasing the frequency with which mutations occur are referred to as **mutagens**. The resultant alterations are termed **induced mutations** in contrast to those that appear to occur at some constant frequency in the absence of intentionally applied external influence (**spontaneous mutations**). Since bacterial cells are **haploid**, mutants are usually easier to recognize because the altered character is more likely to be expressed, particularly if the environment is favorable to the development of the mutants.

The use of mutants has been a tremendous tool in the study of most if not all biochemical processes. Genes are usually designated by a three letter code based on their function. For example, genes involved in the biosynthesis of the amino acid arginine are called *arg* followed by an uppercase letter to indicate different *arg* genes (e.g., *argA*, *argB*). A gene is always indicated by lower case letters (e.g., *arg*), whereas an uppercase letter in the first position (e.g., ArgA) indicates the gene product. At this point we need to expose a common mistake made by many budding microbial geneticists related to the *interpretation* of mutant phenotypes. Organisms, such as *E. coli*, can grow on a basic minimal medium containing only salts, ammonia as a nitrogen source and a carbon source, such as glucose or lactose, because it can use the carbon skeleton of glucose to synthesize all the building blocks necessary for macromolecular synthesis. The building blocks include amino acids, purines, pyrimidines, cofactors, and so on. A mutant defective in one of the genes necessary to synthesize a building block will *require* that building block as a supplement in the minimal medium (e.g., an *arg* mutant will require arginine in order to grow). Microorganisms also have an amazing capacity to use many different compounds catabolically as carbon sources. But a mutation in a carbon source utilization gene (e.g., *lac*) does not mean it requires that carbon source. It means the mutant will *not grow* on a medium containing that carbon source if it is the only carbon source available (e.g., a *lac* mutant will not grow on lactose).

CHEMICAL SYNTHESIS

Chemical Composition

Our paradigm cell (the gram-negative cell *E. coli*) can reproduce in a minimal glucose medium once every 40 min. As we proceed through a detailed examination of all the processes involved, the amazing nature of this feat will become increasingly obvious. It is useful at this point in our introductory chapter, to discuss the basic chemical composition of our model cell. The total weight of an average cell is 9.5×10^{-13}g with water (at 70% of the cell) contributing 6.7×10^{-13}g. The total dry weight is thus 2.8×10^{-13}g. The components that form the dry weight include protein (55%), ribosomal RNA (16.7%), tRNA (3%), mRNA (0.8%), DNA (3.1%), lipid (9.1%), lipopolysaccharide (3.4%), peptidoglycan (2.5%), building block metabolites, vitamins (2.9%), and inorganic ions (1.0%). It is interesting to note that the periplasmic space forms a full 30% of the cell volume with total cell volume being approximately 9×10^{-13}mL. An appreciation for the dimensions of the cell follows this simple example. One teaspoon of packed *E. coli* weighs approximately 1 g (wet weight). This comprises about 1 trillion cells, more than 100 times the human population of the planet. Another useful number to remember when calculating the concentration of a compound within the cell is that there are 3.3 μl of water/1 mg of dry weight. Our reference cell, although considered haploid, will contain two copies of the chromosome when growing rapidly. It will also contain 18,700 ribosomes and a little over 2 million total molecules of protein of which there are between 1000–2000 different varieties. As you might gather from these figures, the bacterial cell is extremely complex. However, the cell has developed an elegant strategy for molecular economy that we still struggle to understand. Some of what we have learned will be discussed throughout the remaining chapters.

As just noted, in just 40 min an *E. coli* cell can make a perfect copy of itself growing on nothing more than glucose, ammonia, and some salts. How this is accomplished seems almost miraculous! Figure 1-10 illustrates that each of the biosynthetic pathways needed to copy a cell originate from one of just 12 precursor metabolites. To understand microbial physiology one must first discover what the 12 metabolites are and where they come from.

Where they come from is glucose or some other carbohydrate. The catabolic dissimilation of glucose not only produces these core metabolites but also generates the energy needed for all the work carried out by the cell. This work includes biosynthetic reactions as well as movement, transport, and so on. Figure 1-11 is a composite diagram of major pathways for carbohydrate metabolism with the 12 metabolites highlighted. Most of them are produced by the Embden–Meyerhof route and the tricarboxylic acid (TCA) cycle.

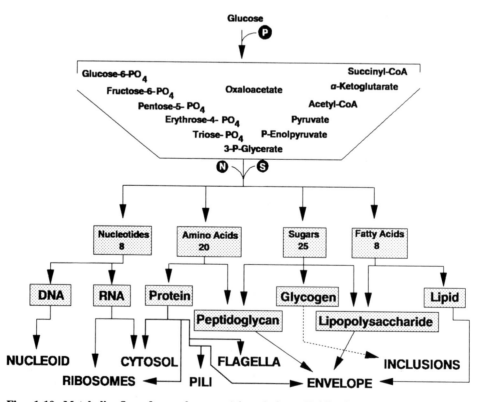

Fig. 1-10. Metabolic flow from glucose. Adapted from Neidhardt, F., J. Ingraham and M. Schaecter. *Physiology of the Bacterial Cell.* Sinauer Associates, Inc. Sunderland, MA.

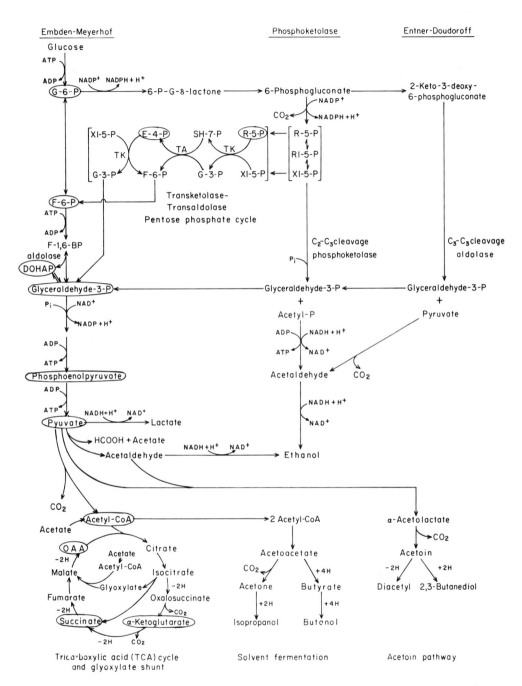

Fig. 1-11. Composite diagram of major pathways of carbohydrate metabolism. The 12 key metabolites are highlighted. G = glucose; E = erythrose; R = ribose; Xl = xylulose; SH = sedoheptulose; DOHAP = dihydroxyacetone phosphate; OAA = oxaloacetate; TK = transketolase; TA = transaldolase.

Two are produced by the pentose phosphate pathway. Figure 1-12 illustrates how these compounds are siphoned off the catabolic pathways and used as starting material for the many amino acids, nucleic acid bases, and cofactors that must be produced. Subsequent chapters will deal with the specifics of each pathway, but these three figures present an integrated picture of cell metabolism.

Energy

Another mission of carbohydrate metabolism is the production of energy. The most universal energy-transfer compound found in living cells is adenosine triphosphate (ATP, Fig. 1-13). The cell can generate ATP in two ways; substrate level phosphorylation in which a high-energy phosphate is transferred from a chemical compound (e.g., phosphoenolpyruvate) to adenosine diphosphate (ADP) during the course of carbohydrate catabolism; or oxidative phosphorylation in which the energy from an electrical and chemical gradient formed across the cell membrane is used to drive a membrane bound ATPase to produce ATP from ADP and inorganic phosphate (P_i).

The generation of an electrical and chemical gradient (collectively called the proton motive force) across the cell membrane requires a complex set of reactions in which H^+ and e^- are transferred from chemical intermediates of the Embden–Myerhof and TCA cycles to a series of membrane-associated proteins called cytochromes. As the e^- is passed from one member of the cytochrome chain to another, the energy released is used to pump H^+ out of the cell. The resulting difference between the inside and outside of the cell in terms of charge (electrical potential) and pH (chemical potential) can be harnessed by the cell to generate ATP. Of course, in order for the cytochrome system to work there must be a terminal electron acceptor. Under aerobic conditions oxygen will serve that function but under anaerobic conditions *E. coli* has a menu of alternate electron acceptors from which it can choose depending on availability (e.g., nitrate). A more detailed accounting of this process is discussed in Chapter 7.

Oxidation–Reduction versus Fermentation

Carbohydrate metabolism is essentially a progressive oxidation of a sugar involving the transfer of hydrogen from intermediates in the pathway to hydrogen-accepting molecules. The most commonly used hydrogen-acceptor compound is nicotinamide adenine dinucleotide (NAD), Fig. 1-14. It is the reduced form of NAD (NADH) that passes the H^+ and e^- to the cytochrome system. A problem can develop for the cell when it is forced to grow in an anaerobic environment without any alternate electron acceptors. This situation could lead to a complete depletion of NAD^+ with all of the NAD pool converted to NADH. Reduced NAD, produced during the early part of glycolysis, would not be able to pass its H atom along, and so the cell could not regenerate NAD^+. If this situation were allowed to develop the cell would stop growing because there would be no NAD^+ to continue glycolysis! To avoid this problem many microorganisms, including *E. coli*, can regenerate NAD^+ by allowing NADH to transfer H to what would otherwise be dead-end intermediates in the glycolytic pathway [e.g., pyruvate or acetyl coenzyme A (CoA)]. The process, known as fermentation, is also depicted in Figure 1-11, as the production of lactic acid, isopropanol, butanol, ethanol, and so on. *Escherichia coli* does not perform all of these fermentation reactions. It is limited to lactate, acetate, formate,

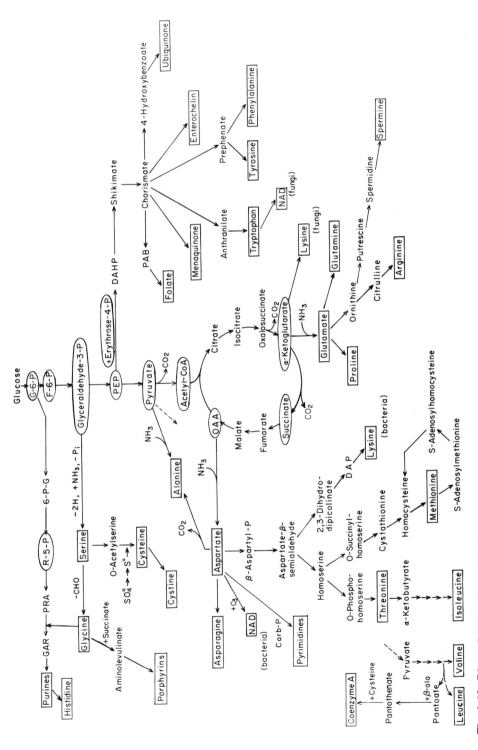

Fig. 1-12. Biosynthetic pathways leading to the major amino acids and related compounds. The highlighted intermediates (ovals) are the 12 key compounds that serve as biosynthetic precursors for a variety of essential end-products.

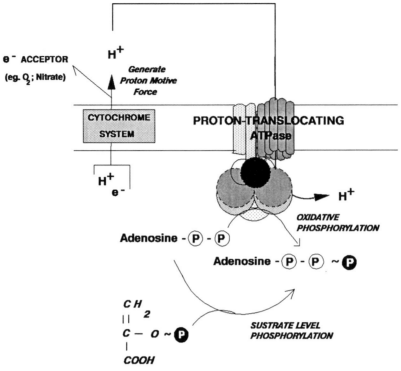

Fig. 1-13. Reactions essential to energy production. *Oxidative phosphorylation.* The energy that comprises the proton motive force can be harnessed and used to generate ATP when protons from outside the cell pass through the membrane-associated proton-translocating ATPase. The energy released will run the ATPase in reverse. It is estimated that passage of three H atoms through the ATPase is required to generate one ATP molecule. *Substate level phosphorylation.* Energy contained within high-energy phosphate bonds of certain glycolytic intermediates can be transferred to ADP forming ATP. The example shows phosphoenolpyruvate.

ethanol, carbon dioxide, and dihydrogen production. Table 1-1 lists the fermentation patterns for some other common organisms.

The cell does not only catabolize glucose via glycolysis. There are alternate metabolic routes available for the dissimilation of glucose. One use for alternate pathways of carbohydrate metabolism (e.g., the phosphoketolase pathway, see Fig. 1-11) is the generation of biosynthetic reducing power. The cofactor NAD is actually divided functionally into two separate pools. The NAD(H) is used primarily for catabolic reactions, whereas a derivative, NAD phosphate (NADP) and its reduced form NADPH, are involved in biosynthetic (anabolic) reactions. The phosphoketolase pathway is necessary for the generation of the NADPH essential for biosynthetic reactions.

Nitrogen Assimilation

A major omission in our discussion to this point involves the considerable amount of nitrogen needed by microorganisms. Every amino acid, purine, pyrimidine, and many other chemicals in the cell include nitrogen in their structures. Since glucose does not

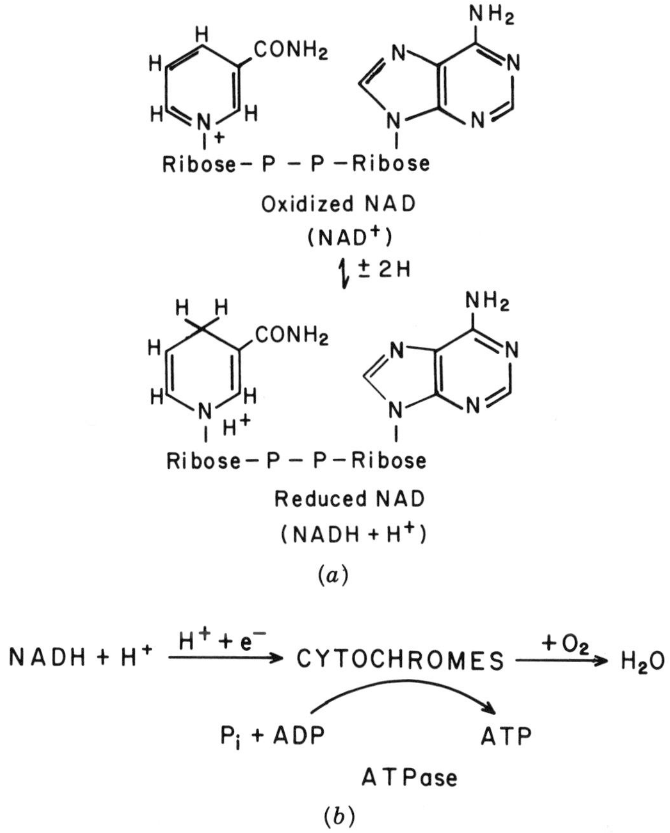

Oxidized NAD
(NAD⁺)

$\uparrow \pm 2H$

Reduced NAD
(NADH + H⁺)

(a)

$$NADH + H^+ \xrightarrow{H^+ + e^-} CYTOCHROMES \xrightarrow{+O_2} H_2O$$

$P_i + ADP \qquad ATP$

ATPase

(b)

Fig. 1-14. Nicotinamide adenine dinucleotide (NAD). Function of NAD in oxidation–reduction reactions. (*a*) Hydrogen atoms removed from a hydrogen donor are transferred to the nicotinamide portion of NAD. (*b*) The hydrogen atoms can be transferred from NAD to an acceptor, such as cytochrome pigments.

TABLE 1-1. Variation in Fermentation Products Formed from Pyruvate

Organism	Product(s)
Saccharomyces (yeast)	Carbon dioxide, ethanol
Streptococcus (bacteria)	Lactic acid
Lactobacillus (bacteria)	Lactic acid
Clostridium (bacteria)	Acetone, butyric acid, isopropanol, butanol
Enterobacter (bacteria)	Acetoin, carbon dioxide, ethanol, lactic acid
E. coli (bacteria)	Lactic acid, H_2, ethanol, formic acid

contain any nitrogen (N) how do cells acquire it? Some microorganisms can fix atmospheric nitrogen via nitrogenase to form ammonia (NH_4^+) and then assimilate the ammonia into amino acids (e.g., *Rhizobium*). Other organisms, such as *E. coli*, must start with NH_4^+. The assimilation of N involves the amidation of one of the 12 key metabolites noted above, α-ketoglutarate, to form glutamic acid (Fig. 1-15). After assimilation into glutamate, the amino nitrogen is passed on to other compounds by transamination reactions. For example, glutamate can pass its amino group to oxaloacetate to form aspartate. From Figure 1-12, one can see that aspartate, like glutamate, is a precursor for several other amino acids. The subject of nitrogen assimilation is covered in depth in Chapter 9.

SPECIAL TOPICS

Mitochondria

All living cells above the level of the prokaryotes contain **mitochondria**, which are highly membranous organelles that contain the respiratory activity. Their structure and function have been studied extensively. Yeast mitochondria (Fig. 1-16) are comparable to those observed in electron photomicrographs of higher organisms. Although bacteria and other prokaryotic cells do not have cytologically distinguishable organelles recognizable as mitochondria, aerobic, and facultative bacteria conduct the same or very similar respiratory activities.

The resemblance of mitochondria to prokaryotic cells has given rise to the popular hypothesis that these organelles may have evolved from bacteria and entered into a symbiotic association in eukaryotic cells. Base sequence analysis of mitochondrial ribosomal RNA and the physicochemical structure of mitochondrial ribosomes suggest a phylogenetic relationship between mitochondria and prokaryotic cells.

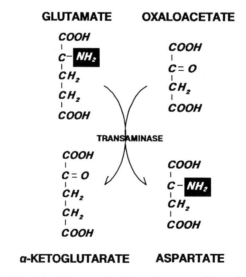

Fig. 1-15. Transamination. In this example, the amine group from glutamic acid is transferred to oxaloacetate forming aspartic acid.

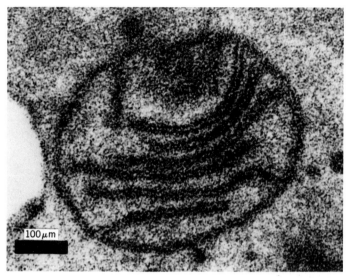

Fig. 1-16. Mitochondrion from a yeast cell. This high-magnification electron micrograph shows prominent cristae (involutions of the inner membrane to which components of the terminal respiratory chain are attached. Bar = 100 nm. From Federman, M. and C. J. Avers. 1967. *J. Bacteriol.* **94**:1236.

Mitochondria contain all of the respiratory enzymes of the cytochrome system (electron-transport system) including those that are concerned with the process of oxidative phosphorylation. They also contain the enzymes of the TCA cycle, which furnish the hydrogen ions and electrons to the cytochrome system. Many other enzymes are also present in mitochondria, but those concerned with fermentation (i.e., the anaerobic breakdown of glucose; glycolysis) are absent and are found in the cytoplasm.

Endospores

A few bacteria, such as *Bacillus* and *Clostridium*, produce specialized structures called endospores. Endospores are bodies that do not stain with ordinary dyes and appear as unstained highly refractile areas when seen under the light microscope. They provide resistance to heat, dessication, radiation, and other environmental factors that may threaten the existence of the organism. Endospores also provide a selective advantage for survival and dissemination of the species that produce them. Under the electron microscope, spores show a well-defined multilayered exosporium, an electron-dense outer coat observed as a much darker area, and a thick inner coat. In the spore interior, the darkly stained ribosomes and the nuclear material may also be visible (Fig. 1-17).

GROWTH

Growth of a cell is the culmination of an ordered interplay between all of the physiological activities of the cell. It is a complex process involving:

1. Entrance of basic nutrients into the cell.

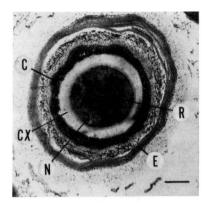

Fig. 1-17. Mature spore of *Clostridium botulinum*. Shown is a well-defined, multilayered exosporium (E), electron dense outer-coat layer (OC), and a thick inner coat (C). The darkly stained ribosomes (R) and nucleoid areas (N) are clearly differentiated in the spore interior. Bar = 0.2 μm. From Stevenson, K. E., R. H. Vaughn, and E. V. Crisan. 1972. *J. Bacteriol.* **109**:1295.

2. Conversion of these compounds into energy and vital cell constituents.
3. Replication of the chromosome.
4. Increase in size and mass of the cell.
5. Division of the cell into two daughter cells, each containing a copy of the genome and other vital components.

Microbiologists usually consider the phenomenon of growth from the viewpoint of population increase. This results from the fact that most current techniques do not allow the detailed study of individual cells. A study of the increase in population implies that each cell, as it is produced, is capable of producing new progeny.

Growth Cycle

Under ideal circumstances in which cell division commences immediately and proceeds in unhampered fashion for a protracted period of time, prokaryotic cell division follows a geometric progression:

$$2^0 \rightarrow 2^1 \rightarrow 2^2 \rightarrow 2^3 \rightarrow 2^4 \rightarrow 2^5 \rightarrow 2^6 \rightarrow 2^7 \rightarrow 2^8 \rightarrow 2^9 \rightarrow \text{etc.}$$

This progression may be expressed as a function of 2, as shown in the second line above. The number of cells (b) present at a given time may be expressed as

$$b = 1 \times 2^n$$

The total number of cells (b) is dependent on the number of generations (n = number of divisions) occurring during a given time period. Starting with an inoculum containing more than one cell, the number of cells in the population can be expressed as

$$b = a \times 2^n$$

where *a* is the number of organisms present in the original inoculum. Since the number of organisms present in the population (*b*) is a function of the number 2, it becomes convenient to plot the logarithmic values rather than the actual numbers. Plotting the number of organisms present as a function of time generates a curvilinear function. By plotting the logarithm of the number, a linear function is obtained as shown in Figure 1-18. For convenience, logarithms to the base 10 are used. This is possible because the logarithm to the base 10 of a number is equal to 0.3010 times the logarithm to the base 2 of a number.

Up to this point we have assumed that the individual generation time (i.e., the time required for a single cell to divide) is the same for all cells in the population. In a given population, the generation times for individual cells vary so that the term **doubling time** is generally invoked in referring to the doubling time for the total population. As shown in Figure 1-18, the cells initially experience a period of adjustment to the new environment and there is a **lag** in the time required for all of the cells to divide. This is referred to as the **lag phase**. Actually, some of the cells in the initial inoculum may not survive and there may be a drop in the number of viable cells. The cells eventually become adjusted to the new environment and begin to divide at a more rapid rate and enter into several rounds of division at a maximum rate. Under favorable environmental conditions there will be a relatively constant doubling time for a protracted period of time. Plotting the logarithm of the number of cells results in a linear function. For this reason this phase of growth is referred to as the **log phase** or, more correctly, the **exponential phase**.

All cultures of microorganisms eventually reach a maximum population density. This is termed the **stationary phase**. Entry into the stationary phase of growth can result from several events. Exhaustion of essential nutrients, accumulation of toxic waste products, depletion of oxygen, or development of an unfavorable pH are the factors most commonly implicated as being responsible for the decline in the growth rate. Although cell division continues during the stationary phase, the number of cells that are able to divide (termed viable cells) are approximately equal to the number that are unable to divide (defined here as nonviable cells). Thus, the stationary phase represents an **equilibrium** between the number of cells able to divide and the number that are unable to divide.

Eventually, the death of organisms in the population results in a decline in the viable population and the **death phase** ensues. The exact shape of the curve during the death phase will depend on the nature of the organism under observation and the many factors that contribute to cell death. The death phase may assume a linear function, such as during heat-induced death where viable cell numbers will decline logarithmically.

Some additional considerations of the growth curve are important in assessing the effect of various internal as well as external factors on growth. Since the number of cells in a population (*b*) is equal to the number of cells in the initial inoculum (*a*) $\times 2^n$:

$$b = a \times 2^n$$

Then,

$$\log_2 b = \log_2 a + n$$
$$\log_{10} b = \log_{10} a + n \log_{10} 2$$
$$\log_{10} b = \log_{10} a + (n \times 0.3010)$$

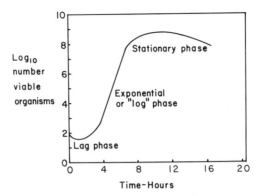

Fig. 1-18. A typical growth curve for a bacterial culture.

Solving the equation for n, the number of generations that occurred between the time of inoculation and the time of sampling is

$$n = \frac{\log_{10} b - \log_{10} a}{\log_{10}} = \frac{\log_{10} b - \log_{10} a}{0.3010}$$

The **generation time** (t_g) or **doubling time** may be determined by dividing the time elapsed (t) by the number of generations (n):

$$t_g = t/n$$

Continuous Culture

Usually bacteria are grown in "batch" culture (or turbidostat) in which a flask containing media is inoculated and growth allowed to occur. This is a closed system that is actually very difficult to control in terms of manipulating such parameters as growth rate. In batch cultures growth rate is determined *internally* by properties of the bacteria themselves. One can use a turbidostat to grow bacteria at different growth rates as long as nutrients are added at a concentration below that which supports maximal growth. But to accomplish this the cell density, and thus cell number, will be too low for certain analyses. To grow bacteria at slow growth rates and at high cell density an apparatus called the *chemostat* is used. In this apparatus, fresh medium containing a limiting nutrient is added from a reservoir to the culture vessel at a set rate. The volume in the culture vessel is kept constant by an overflow device that removes medium and cells at the same rate as fresh medium is added. In a chemostat, growth rate is determined *externally* by altering the rate of flow of fresh medium containing the limiting nutrient to the culture vessel. The more of the limiting nutrient added the faster the growth rate.

FACTORS AFFECTING GROWTH

Nutrition

All living organisms have certain basic nutritional requirements; sources of (a) carbon, (b) nitrogen, (c) energy, and (d) essential growth factors (minerals and vitamins) are needed to support growth. Microorganisms vary widely in their nutritional requirements. Two main groups of organisms are classified on the basis of their ability to gain energy from certain sources and the manner in which they satisfy their carbon and nitrogen requirements for growth:

1. *Lithotrophs.* Utilize CO_2 as the sole source of carbon and gain energy through the oxidation of inorganic compounds (**chemolithotrophs**) or light (**photolithotrophs**). Inorganic nitrogen is utilized for the synthesis of organic compounds.

2. *Organotrophs.* Generally prefer organic substrates as a source of energy and carbon. **Photoorganotrophs** utilize light as a source of energy for assimilation of CO_2 as well as organic compounds. **Chemoorganotrophs** utilize organic compounds for growth.

While their nutritional requirements are remarkably simple, chemolithotrophic bacteria must be metabolically complex since they synthesize all of their cellular components and provide the energy for this activity through the oxidation of inorganic compounds. One fundamental characteristic of **strict chemolithotrophs** is that they are unable to grow on or assimilate exogenous organic compounds. **Facultative chemolithotrophs** can utilize exogenous organic carbon sources. Chemolithotrophs possess unique mechanisms for CO_2 fixation, such as the ribulose bisphosphate (Calvin–Benson) cycle and the reductive carboxylic acid (Campbell–Evans) cycle (see Chapter 7).

Organotrophic organisms may utilize CO_2 as a source of carbon, but usually prefer organic carbon sources and generally cannot subsist on CO_2 as the sole carbon source. Organotrophs may use inorganic nitrogen, but most members of the group grow better when supplied with organic nitrogen compounds. Some organotrophs that have lost (or never achieved) the ability to synthesize many organic nitrogen compounds exhibit a nutritional requirement for them. Some common organotrophic bacteria, such as *E. coli*, *Enterobacter aerogenes*, yeasts, and molds, grow luxuriantly with a carbohydrate, such as glucose, as the only organic nutrient. Streptococci, staphylococci, and a wide variety of other heterotrophs may exhibit specific requirements for one or more nitrogen sources as amino acids, purines, or pyrimidines (see Table 1-2). Many organotrophic organisms require one or more B vitamins for growth. Biotin is required by many yeasts or molds that otherwise require only a carbohydrate as an organic nutrient (e.g., the yeast *Saccharomyces cerevisiae*, and the common bread mold *Neurospora*). Thiamine (B_1), riboflavin (B_2), nicotinic acid (niacin), pyridoxine (B_6), pantothenic acid, and vitamin B_{12} (cobalamin) are required by many fastidious organisms.

The fat-soluble vitamins are not required by many microorganisms. Vitamins A, D, and E are not required. Vitamin K (naphthoquinone) is required by certain members of the genus *Mycobacterium* and genus *Bacteroides* and is known to function as a substitute for coenzyme Q (benzoquinone) in the respiratory electron-transport chain in these organisms. Vitamin C is stimulatory to the growth of a number of organisms but does not represent a true growth factor. It apparently serves to provide the proper oxidation–reduction potential to the medium.

Fatty acids are required by certain organisms, particularly in the absence of certain of the B vitamins. Replacement of a growth factor requirement by the addition of the end-product of a biosynthetic pathway in which the vitamin normally functions is referred to as a *sparing action*. This type of activity has been reported for many growth factors, including amino acids, purines, pyrimidines, and other organic constituents. Insofar as a vitamin can completely replace a particular organic nutrient in a defined medium that

TABLE 1-2. Nutritional Requirements of Some Organotrophs

	Escherichia coli	Salmonella typhi	Staphylococcus aureus[a]	Leuconostoc paramesenteroides[b]
Basic Nutrients				
Glucose				
NH_4^+				
Mn^{2+}				
Mg^{2+}	Required by all for maximum growth in defined medium			
Fe^{2+}				
K^+				
Cl^-				
SO_4^{2-}				
PO_4^{3-}				
Additional requirements				
	None	Tryptophan	Nicotinic acid	Nicotinic acid
			Thiamine	Thiamine
			10 amino acids	Pantothenate
				Pyridoxal
				Riboflavin
				Cobalamin
				Biotin
				p-Aminobenzoate
				Folate
				Guanine
				Adenine
				Uracil
				16 Amino acids
				Sodium acetate
				Tween 80

[a] From Gladstone, G. P. 1937. *Br. J. Exptl. Pathol.* **18**:322.
[b] From Garvie, E. I. 1967. *J. Gen. Microbiol.* **48**:429.

nutrient cannot be regarded as a true growth requirement since it can be synthesized in the presence of the requisite vitamin.

Although most bacterial membranes do not contain sterols, sterols are required in the membranes of some members of the Mycoplasmataceae. (These organisms do not possess a cell wall.) *Mycoplasma* require sterols for growth. *Acholeplasma* do not require sterols; however, they do not synthesize sterols but produce terpenoid compounds that function in the same capacity. Fungi (yeasts and molds) contain sterols in their cell membranes but in most cases appear to be capable of synthesizing them.

Oxygen

Microorganisms that require oxygen for the energy-yielding metabolic processes are called **aerobes**, while those that cannot utilize oxygen for this pupose are called **anaerobes**. Organisms that are capable of using either respiratory or fermentation processes, depending on the availability of oxygen in the cultural environment, are termed **facultative**. Aerobic organisms possess cytochromes and cytochrome oxidase, which are involved in the process of oxidative phosphorylation. Oxygen serves as the terminal electron acceptor in the sequence and water is one of the resultant products of respiration. Some of the oxidation–reduction enzymes interact with molecular oxygen to give rise to superoxide (O_2^-), hydroxyl radicals (OH·), and hydrogen peroxide (H_2O_2), all of which are extremely toxic:

$$O_2 + e^- \xrightarrow{\text{oxidative enzyme}} O_2^-$$
$$O_2 + H_2O_2 \xrightarrow{\text{nonenzymatic}} O_2 + OH\cdot + OH^-$$

The enzyme superoxide dismutase dissipates superoxide:

$$2O_2^- + 2H^+ \longrightarrow H_2O_2 + O_2$$

Superoxide dismutase is present in aerobic organisms and those that are aerotolerant, but not in strict anaerobes. Many, but not all, aerobes also produce catalase, which can eliminate the hydrogen peroxide formed:

$$2H_2O_2 \longrightarrow 2H_2O + O_2$$

Aerotolerant organisms apparently do not produce catalase. Hence, growth of these organisms is frequently enhanced by culture on media containing blood or other natural materials that contain catalase or peroxidase activity. Organisms that do not utilize oxygen may tolerate it because they do not interact in any way with molecular oxygen and do not generate superoxide or peroxide.

Anaerobic bacteria from a variety of genera are present in the normal flora of the animal and human body as well as in a number of natural habitats, such as the soil, marshes, and deep lakes. A number of the more widely known genera of anaerobic organisms are listed in Table 1-3.

Carbon Dioxide

Many organisms are dependent on the fixation of CO_2. On a practical basis, certain organisms thrive better if they are grown in an atmosphere containing increased CO_2. *Haemophilus, Neisseria, Brucella, Campylobacter,* and many other bacteria require at least 5–10% CO_2 in the atmosphere for initiation of growth, particularly on solid media. Also, a variety of intermediates of glycolysis and the TCA cycle are siphoned-off these pathways to serve as biosynthetic precursors. Consequently, CO_2 is required to replenish intermediates in the TCA cycle (**anapleurotic function**) through the action of pyruvate carboxylase, phosphoenolpyruvate carboxylase, or malic enzyme (see Chapter 7).

Temperature

Microorganisms vary widely in regard to their ability to intiate growth over certain ranges of temperature (Table 1-4). Extremes of temperatures probably result in the inactivation

TABLE 1-3. Genera of Anaerobic Bacteria[a]

Bacilli		Cocci	
Gram positive	Gram negative	Gram positive	Gram negative
Clostridium	*Bacteroides*	*Peptococcus*	*Veillonella*
Actinomyces	*Fusbacterium*	*Peptostreptococcus*	
Bifidobacterium	*Vibrio*	*Ruminococcus*	
Eubacterium	*Desulfovibrio*		
Lactobacillus			
Propionibacterium			

[a]Members of the order *Rhodospirillales* are all anaerobic, gram-negative organisms that may be rods, cocci, spirals, or pleomorphic.

of enzymes or other functional cell structures, such as membranes. However, the rather wide range in the optimum temperature at which various organisms successfully conduct their metabolic activities and undergo normal growth requires more careful search for a satisfactory explanation. For example, bacteria found in hot springs and in the thermal vents on the ocean floor appear to function at extremely high temperatures.

Hydrogen Ion Concentration

Bacteria, yeasts, and molds exhibit wide ranges of pH at which growth is most readily initiated (Table 1-5). Some microorganisms prefer to live in an acidic environment (**acidophilic organisms**) while others prefer an alkaline pH as the optimal growth environment (**alkaliphilic organisms**). *Escherichia coli* prefers a neutral pH environment and so is classified as **neutralophilic**. (The older term, neutrophilic, is not consistent with the nomenclature of the other two groups and can be confused with the neutrophiles, a form of white blood cell, and so should no longer be used.) From studies on isolated enzymes, it is obvious the pH exerts a marked control on enzyme activity, each enzyme having an optimal pH range for maximum activity. The pH of the growth medium may affect cell permeability and other physiological activities. For example, certain enzymes are apparently produced only when the pH of the environment reaches a specific level. The pH of the culture medium may alter the fermentation products produced by lactic acid bacteria. At lower, pH ranges, the primary product is lactic acid, while at neutral to alkaline pH, other acid products and ethanol may increase in amount.

TABLE 1-4. Temperature Ranges of Bacterial Growth

Type of Organism	Growth Temperature (°C)		
	Minimum	Optimum	Maximum
Psychrophilic	−5–0	5–15	15–20
Mesophilic	10–20	20–40	40–85
Thermophilic	25–45	45–60	>80

TABLE 1-5. pH Limits for Growth of Various Microorganisms

Organism	Minimum	Optimum	Maximum
Bacteria	2–5	6.5–7.5	8–11
Yeasts	2–3	4.5–5.5	7–8
Molds	1–2	4.5–5.5	7–8

SUMMARY

This chapter was designed to be a highly condensed version of the remainder of this book. This was done in an attempt to build a coherent picture from the start. Very often the student is presented with detailed treatments of one topic after another. The information overload can become so great with this approach that the student is too burdened to design on their own an integrated picture of the cell and what makes it work. Our hope is that the student will take from this chapter the framework needed to appreciate the forthcoming details.

CHAPTER 2

MACROMOLECULAR SYNTHESIS AND PROCESSING: DNA, RNA, AND PROTEIN SYNTHESIS

Macromolecules of the cell include peptidoglycan, which provides the structural integrity of the cell; proteins and nucleic acids (DNA and RNA), which contain the information necessary to coordinate the activities of the cell into a balanced system providing for orderly growth and development. Our discussion of the bacterial cell will begin with what is arguably the core processes of all life: nucleic acid and protein synthesis. Peptidoglycan synthesis will be described later in Chapter 6.

It is obvious that nucleic acid replication is of cardinal importance to the cell. During growth of a population of cells, the continuous synthesis of the various building blocks and their proper assembly into the structures of the cell are both dependent on the fidelity with which the DNA is replicated. The chain of events emanating from DNA follows the general scheme:

$$\text{DNA} \xleftarrow{\text{replication}} \textbf{DNA} \xrightarrow{\text{transcription}} \begin{bmatrix} \text{tRNA} \\ \text{mRNA} \\ \text{rRNA} \end{bmatrix} \xrightarrow{\text{translation}} \text{protein}$$

It should be evident that knowledge of nucleic acid biosynthesis is fundamental to understanding living processes, including energy production, building block biosynthesis, growth, and regulation.

Transcription of the genetic information in one of the strands of DNA into RNA implies that the DNA strand serves as a **template** upon which the new strand of RNA is formed. Three classes of RNA are formed.

1. **Messenger RNA** (mRNA) carries the information coded in DNA to the ribosome on which protein synthesis takes place.
2. **Ribosomal RNA** (rRNA) is a component of the ribosomes, the sites of protein synthesis.
3. **Transfer RNA** (tRNA) guides the specific amino acids into their proper position in the polypeptide chain as it is being formed. The transcription process must be

visualized as a reversible process in certain instances, since the genetic information in some viruses resides in RNA, which can be transposed into DNA by reverse transcription.

Since knowledge of DNA, RNA and protein synthesis is central to understanding the processes of macromolecular synthesis as well as growth and regulation, the ensuing discussion considers these events in some detail. Throughout the next several chapters we will refer to numerous genetic loci in *Escherichia coli*. The review by Bachman (1990) provides a detailed accounting of these genes, their map positions, and the mnemonic basis for their three letter designations.

STRUCTURE OF DNA

In order to appreciate the processes of replication, transcription, and translation of the information in the DNA molecule, it is necessary to describe the structure of DNA and to illustrate its ability to direct the events occurring in the cell. The DNA molecule is composed of the four deoxyribonucleosides [purine (adenine or guanine) and pyrimidine (cytosine or thymine) bases + deoxyribose] arranged in a chain with phosphodiester bonds connecting the 5'-carbon on the deoxyribose of one nucleoside to the 3'-carbon of the deoxyribose of the adjacent nucleoside, as shown in Figure 2-1. The property of **polarity** (illustrated by the lower figure in Fig. 2-1) is maintained by the orientation of the phosphodiester bonds on the 5' and 3' positions of the deoxyribose in the DNA strand. In the bacterial chromosome, DNA is composed of two such strands arranged in apposition; that is, it is **double stranded (dsDNA)**. In dsDNA, each of the two strands has opposite polarity, as shown in Figure 2-2. Because the two chains, when arranged in apposition, have opposite polarity, they are described as being **antiparallel**. One strand will end with a 3'-hydroxyl group (the **3' terminus**), while the other will end with a 5' phosphate (the **5' terminus**). Notice, too, that DNA is a negatively charged molecule under physiological conditions.

The purine and pyrimidine bases are **planar** structures that **pair** with one another through the forces of hydrogen bonding, usually depicted as dashed or dotted lines in

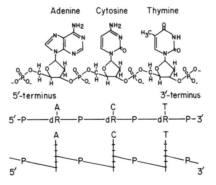

Fig. 2-1. Structure of DNA showing alignment of adjacent nucleotides and some common methods of depicting polarity in abbreviated structures. The crossbars on the vertical lines represent the carbon atoms in deoxyribose.

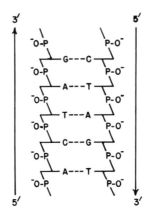

Fig. 2-2. Polarity of the double-stranded DNA molecule. This abbreviated diagrammatic view uses the conventions described in Figure 2-1.

diagrammatic presentations. These **base pairs** are stacked in parallel to one another and lie perpendicular to the phosphodiester backbone of the structure. As a result of the covalent bond angles and the noncovalent forces, native DNA assumes a **helical structure**, which may exist in different forms depending on the environment. In an aqueous environment, DNA exists primarily in a stable conformation (B form) in which there are 3.4 Å between the stacked bases and 34 Å per right-handed helical turn. This means that there are 10 base pairs (bp) per turn, and the bases will be arranged so that they are perpendicular to the helical axis, as shown in Figure 2-3.

Under certain conditions, DNA can shift into either the A conformation (11 rather than 10 bp per turn and tilted bases) or become Z-DNA (left-handed rather than right-handed helix having 12 bp per turn). Native DNA (*in vivo*) very likely exists in a variety

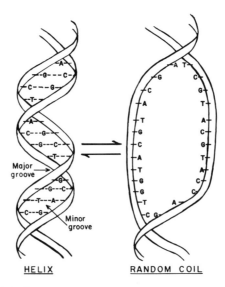

Fig. 2-3. Helical and randomly coiled DNA structures.

of conformations. Different areas of the bacterial chromosome may well differ topologically as a result of variations in the local environment. Structural differences of this kind may be essential to certain functions of DNA, particularly with regard to the binding of regulatory proteins to specific areas of the chromosome.

Over the years there has been some controversy regarding the *in vivo* existence and significance of Z-DNA conformations. The Z-DNA is most often formed as a result of extended runs of deoxyguanosine-deoxycytosine [poly (dG-dC)]. *In vitro* this also requires a fairly high salt environment because poly (dG-dC) will assume a B structure in low salt. Proof that Z-DNA can exist *in vivo* was provided by R. D. Wells and colleagues. They placed segments of poly (dG-dC) immediately next to a sequence of DNA that is specifically methylated only when in the B form. This sequence was dramatically undermethylated *in vivo* when the Z-DNA sequence was included.

Under certain circumstances, DNA may be converted from the double-stranded helix into a **random coil** in which base pairing does not occur. Both of these states may exist in different regions of the chromosome at the same time, and the process of conversion may be freely reversible. This helical–random coil transition (shown in Fig. 2-3) may be essential for the function of DNA in replication and transcription.

One may question how regulatory DNA-binding proteins recognize their proper DNA-binding sequences since DNA duplexes present a fairly uniform surface to potential regulatory proteins (i.e., a uniform phosphodiester backbone that faces outward). There are two potential mechanisms for nucleic acid–protein interaction. The first model predicts that the specificity of interaction relies upon groups present in the major or minor grooves of dsDNA (see Fig. 2-3). Thymine and adenine have groups that will protrude into the grooves. A protein could interact with these groups to bind to specific regions of DNA. The second model is that direct access to bases may occur by a protein interacting with DNA in a randomly melted region. The DNA-binding proteins can have one of several characteristic structures that will aid in sequence-specific binding. These proteins will be discussed in Chapter 4.

Bacterial Nucleoid

In all living cells, DNA takes on a highly compacted tertiary structure. For some time it has been known that for eukaryotic cells the first stage of compaction involves the winding of DNA around an octomeric assembly of histone proteins to form what is referred to as the nucleosome structure. At this stage, a linear stretch of DNA would take on a beadlike appearance. The second stage is thought to be the solenoidal organization of these nucleosomes into a helical network.

The packaging of DNA in prokaryotic cells is understood less completely. The bacterial chromosome is a covalently closed circular DNA molecule. When circular DNA molecules are gently isolated from bacterial cells, the DNA is negatively supercoiled. Supercoiling occurs when a circular dsDNA molecule becomes twisted so that the axis of the helix is itself helical (Fig. 2-4). If DNA is ''overwound'' (twisted in the direction of the helix) or ''underwound'' (twisted in the opposite direction relative to the helix) a significant amount of torsional stress is introduced into the molecule. In order to relieve this stress, the DNA double helix will twist upon itself. Underwound DNA will form negative supercoils, while overwinding results in positive supercoils. The bacterial chromosome is believed to contain about 50 negatively supercoiled loops or domains (Fig. 2-5). Each domain represents a separate topological unit, the boundaries of which may

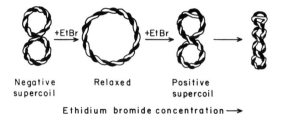

Ethidium bromide concentration →

Fig. 2-4. Superhelicity of a closed circular DNA molecule changing from negative supercoil to positive supercoils with increasing concentrations of ethidium bromide.

be defined by sites on DNA that limit its rotation. Several studies indicate roles for both protein and RNA in the stabilization of the nucleoid.

Topoisomerases are enzymes that can alter the topological form (supercoiling) of a circular DNA molecule (Table 2-1). Type I topoisomerases (e.g., *E. cɔli top A*) relax negatively supercoiled DNA apparently by breaking one of the phosphodiester bonds in dsDNA, allowing the 3'-OH end to "swivel" around the 5'-phosphoryl end and then resealing the "nicked" phosphodiester backbone (Fig. 2-6).

Type II topoisomerases, on the other hand, require energy to *underwind* DNA molecules, thus introducing negative supercoils. Type II topoisomerases, such as DNA gyrase, can also relax negative supercoils in the absence of energy. These enzymes seem to utilize a mechanism as described in Figure 2-7. The model requires the passage of one dsDNA molecule through a second molecule that has a double-stranded break in its phospho-

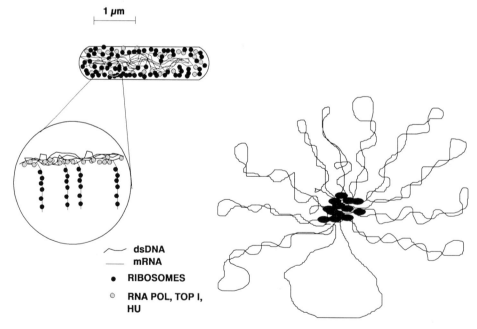

Fig. 2-5. Schematic representation of *E. coli* nucleoids in situ and after isolation and spreading. The insert illustrates an area active in transcription and translation.

TABLE 2-1. DNA Topoisomerases

Enzyme	Type	Relaxation of (−) or (+) Twists	Introduction of (−) Twists	Linking of Duplex DNA Circles	ATP Requirement
E. coli topoisomerase I (omega protein)	1	(−)	No	Yes	No
Eukaryotic nick-closing enzymes	1	(−), (+)	No	Yes?	No
Phage λ *int*	?	(−), (+)	No	Yes	No
Bacterial DNA gyrase	2	(−)	Yes	Yes	Yes
E. coli topoisomerase II′	2	(−), (+)	No	Yes	No
T4 topoisomerase	2	(−), (+)	No	Yes	Yes
Eukaryotic type 2 topoisomerase (*Drosophila* embryos and *Xenopus oocytes*)	2	(−), (?)	No	Yes	Yes

diester backbone. The DNA gyrase is apparently responsible for the negatively super-coiled state of the bacterial chromosome. As will be discussed later, this supercoiling is required for efficient replication and transcription of prokaryotic DNA.

Supercoiling alone cannot account for the level of chromosome compaction. The following exercise illustrates the extent to which the chromosome of *E. coli* must be compacted to fit in the cell. The *E. coli* genome is about 1 mm long (4.3×10^{-15}g). In exponentially growing cells, the DNA comprises 3–4% of the cellular dry mass or about 12–15×10^{-15}g per cell. Thus there are around three genomes in the *average* growing cell. If chromosomal DNA were homogeneously distributed over the entire cell volume (1.4×10^{-12}mL) the packing density would be 10 mg/mL. However, DNA is confined

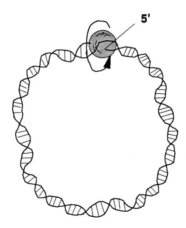

Fig. 2-6. Action of Type I topoisomerases. Type I topoisomerases (e.g., *E. coli top A*) relax negatively supercoiled DNA apparently by breaking one of the phosphodiester bonds in dsDNA, allowing the 3′-OH end to "swivel" around the 5′-phosphoryl end and then resealing the "nicked" phosphodiester backbone.

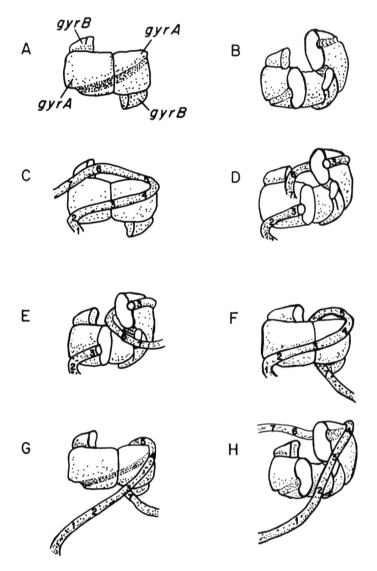

Fig. 2-7. Model for DNA gyrase. (A) Tetrameric gyrase molecule in "closed" conformation. (B) Gyrase molecule in "open" conformation. Illustrations C to H represent steps in a scheme for negative supercoiling by inversion of a right-handed DNA loop (C) into a left-handed one (F). (C) A small section of a circular dsDNA molecule is represented as a tube (numbered). The DNA is bound to gyrase in a right-handed coil. *Three* is the point of DNA cleavage and covalent attachment of subunit A polypeptides. Sections 1-3 and 3-5 are wrapped around the left and right A subunits, respectively, while DNA at 6 contacts a B subunit. (D) Gyrase cleaves DNA at 3 and opens. (E) DNA (point 6) is transferred from the left to the right B subunit through the cleavage point. (F) Gyrase closes and reseals DNA at point 3. The DNA between points 3 and 7 now describes a left-handed coil. (G to H) DNA partially unwraps from gyrase, the gyrase opens releasing the DNA and then recloses. From Wang et al., in Alberts, 1980. Mechanistic studies of DNA replication and genetic recombination. Academic, New York.

to ribosome-free areas of the cell bringing its packing density to 15–30 mg/mL. The inevitable conclusion is that the chromosome is packaged into what is loosely referred to as a ''compactosome.'' As a general term, a compactosome may or may not have a nucleosome structure depending on the organism under discussion. At least four different histone-like proteins have been isolated from *E. coli* that may contribute to compactosome structure (Table 2-2). They are all basic, heat-stable, DNA-binding proteins present in high proportions on the *E. coli* nucleoid. Their similarity to eukaryotic histones was used to extrapolate a role in forming prokaryotic nucleosomes. However, there is some controversy as to the existence of nucleosomes in *E. coli*. For example, the level of histone-like proteins does not appear sufficient to generate a nucleosome structure. The most abundant histone-like protein, HU, is not even associated with the bulk of the DNA. Instead, it is situated where transcription and translation occur and may play a more direct role in those processes.

Another role of histones in eukaryotes is charge neutralization of DNA. Could this also be the case in *E. coli*? The data suggests, No! Because there is so little histone-like protein in *E. coli*, neutralization of DNA charge must be accomplished by other molecules, such as polyamines and Mg^{2+}.

A recent model depicts the nucleoid as a fluid structure. Gentle isolation reveals the presence of between 30 and 100 distinct chromosomal loops (Fig. 2-5). These loops are compacted and centrally located within the cell. The isolated nucleoid shows most of the domain loops supercoiled. A single-stranded nick in a loop will lead to its relaxation without disturbing the other supercoiled loops. The tethering points for each domain are suspected of being DNA gyrase, as is the case in eukaryotes, and perhaps HU. In contrast to eukaryotic nucleosomes, the prokaryotic DNA would be in continuous movement such that transcriptionally active segments are located at the ribosome–nucleoid interface. The inactive parts would be more centrally located. An alternative hypothesis is that the coupled transcription–translation–secretion of exported proteins serves to tether regions

TABLE 2-2. Histone-Like Proteins of *E. coli*

Name	Subunits	Molecular Weight	Function
Hu (*hupA*)	Hu(α)[HLPII$_a$]	9,000	Can form nucleosomelike structures similiar to eukaryotic histone H2B
(*hupB*)	Hu(β)[HLPII$_b$] (associated as heterologous dimers)	9,000	
HLP1		17,000	Affects RNA polymerase transcription (*firA* gene)
H1 (*hns*)		15,000	May selectively modulate *in vivo* transcription
H		28,000	Similar to eukaryotic H2A

of the chromosome to the membrane, thereby defining different chromosomal domains (see Protein Export in this chapter).

What role might DNA supercoiling have besides nucleoid compaction? There is mounting evidence that the expression of many genes can vary with the extent of negative supercoiling. The cell may even regulate certain sets of genes by differentially modulating domain supercoiling. The evidence suggests that this may occur in response to environmental stress, such as an increase in osmolarity or change in pH. The subsequent, coordinated increase in expression of some genes and decreased expression of others would enable the cell to adapt to and survive the stress. Further detail on nucleoid structure of both prokaryotic and eukaryotic microbes is presented in Chapter 6.

REP Elements

One sequence element present in prokaryotic genomes that may play a role in the organization of the nucleoid is called the repetitive extragenic palindrome (REP) or palindromic unit (PU) sequence. It is a 38 bp consensus sequence that is a palindrome capable of producing a stem–loop structure. A consensus sequence is derived by comparing several like-sequences and determining which bases are conserved among them. The consensus for *E. coli* is 5′A(AT)TGCC(TG)GATGCG(GA)CG(CT)NNNN(AG)CG (TC)CTTATC(AC)GGCCTAC(AG). The letter N indicates any base can be found here. Two bases within a parentheses indicate one or the other base can be found at that position. The palindromic region can be observed by starting at the NNNN region and reading in both directions noting that the bases are complementary (see underlined regions). The REP element is located between genes within an operon or at the end of an operon, in different orientations and arrays. Although there are between 500 and 1000 REP copies, they are always found outside of the structural genes. The suggestion that REP or a complex of REP sequences called BIME (bacterial interspersed mosaic elements) play a part in nucleoid structure comes from *in vitro* studies showing specific interactions between REPs and DNA polymerase I and DNA gyrase. However, the actual function for these elements is not known.

DNA REPLICATION

The opposite polarity of the two strands in the DNA molecule provided the suggestion that the replication of DNA must proceed from an initiation point, where the two DNA strands are separated. Isotopic labeling studies have shown that DNA replication proceeds in a **semiconservative** manner, one strand of the parental molecule being contributed to each new (daughter) DNA molecule, as shown in Figure 2-8A. During the first round of DNA replication, each strand is duplicated so that at the end of this step each of the resulting DNA molecules will contain one old and one new strand. Only after the second round of replication will there be two completely new strands formed. Proof of the semiconservative mode of DNA replication was achieved by a classic series of experiments conducted by Meselson and Stahl in 1958. By culturing *E. coli* in a minimal medium in which ^{15}N-labeled ammonium chloride served as the only source of nitrogen, the "heavy" ^{15}N was incorporated into the purine and pyrimidine bases of the DNA. Comparison of the densities of the DNA extracted from *E. coli* grown in ^{15}N-containing medium and from cells grown in a medium with normal light nitrogen (^{14}N) indicated

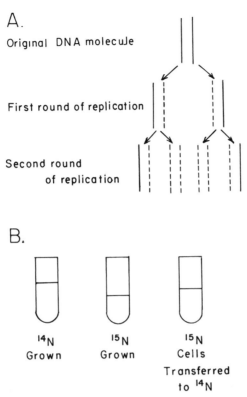

A.

Original DNA molecule

First round of replication

Second round
of replication

B.

^{14}N
Grown

^{15}N
Grown

^{15}N
Cells
Transferred
to ^{14}N

Fig. 2-8. **(A) Semiconservative DNA replication.** Note that after the first round of replication each strand of DNA is composed of one new and one old strand. Only after the second round of replication are two completely new strands formed. **(B) The Meselson–Stahl experiment.** See text for details. Bands within tubes represent bands of DNA extracted from *E. coli* grown on light or heavy nitrogen-containing medium.

the presence of "heavy" (^{15}N) and "light" (^{14}N) bands of DNA in a cesium chloride gradient. Transferring cells grown in a medium containing ^{15}N DNA to a medium containing ^{14}N for one generation and then examining the DNA in cesium chloride gradients revealed only a single DNA band of intermediate density between the "heavy" and "light" bands. This indicated that the DNA formed following one round of replication was a "hybrid molecule" consisting of one heavy strand and one newly synthesized light strand. Autoradiographic studies provided evidence that the bacterial chromosome remained circular throughout replication and that the ring is doubled between the point of initiation and the growing point observed at the time of interruption of the process (Fig. 2-8B). If the replication process is permitted to go to completion, then the entire cyclic chromosome will be duplicated and the two new daughter chromosomes will separate.

Although a number of earlier studies provided investigators with many insights into the potential mechanisms of the polymerization of deoxyribonucleotides to form DNA, the first example of enzymatic DNA synthesis *in vitro* was an enzyme designated DNA polymerase that catalyzed the polymerization of deoxyribonucleotides in a defined sequence dictated by a preexisting DNA template. Polymerase activity required a dsDNA template and magnesium ions. Polymerization of a new strand not only required a template, but also a primer molecule with a 3'-hydroxyl end (Fig. 2-9).

Escherichia coli has since been shown to contain three DNA polymerase activities: Pol I, Pol II, and Pol III. All of these polymerases synthesize DNA only in the 5' $\longrightarrow$ 3' direction and require the presence of a 3'-hydroxyl primer (see Figs. 2-9 and 2-11). Pol I has a bound zinc ion and in addition to the polymerase activity also possesses a 3' $\longrightarrow$ 5' exonuclease activity used for ''proofreading'' (see below) and a 5' $\longrightarrow$ 3' exonuclease activity. Pol II and Pol III also have a 3' $\longrightarrow$ 5' exonuclease activity. In addition, pol III has a single strand-specific 5' $\longrightarrow$ 3' exonuclease activity.

The fact that polymerases synthesize DNA only in the 5' $\longrightarrow$ 3' direction raises a problem when you consider that replication of both strands occurs simultaneously. The *appearance* is that one of the newly synthesized strands elongates in the 3' $\longrightarrow$ 5' direction (Fig. 2-10). This paradox was resolved by finding that DNA synthesis occurs via the formation of small fragments of 1000–2000 nucleotides (Okazaki fragments) at the growing fork, providing a means of explaining how new strand synthesis can take place in the 5' $\longrightarrow$ 3' direction and still appear to proceed in the 3' $\longrightarrow$ 5' direction. Figure 2-11 illustrates that one strand in a replicating fork is synthesized continuously in a 5' $\longrightarrow$ 3' direction, the same direction as the fork movement. The other strand is synthesized discontinuously, 1000 nucleotides at a time, also in a strict 5' $\longrightarrow$ 3' direction, but as the fork moves and more parental DNA is unwound, a new fragment is synthesized. The overall **appearance**, however, is that the strand is elongated in a 3' $\longrightarrow$ 5' direction when, in fact, it is synthesized 5' $\longrightarrow$ 3'.

Thus, one strand is referred to as the leading strand and the other as the lagging strand. Since DNA polymerases cannot initiate new chain synthesis in the absence of an RNA primer, a primer generating polymerase (primase) is required for the initiation of each new Okazaki fragment. The primase in *E. coli* is the product of the *dnaG* locus. The short RNA primers are elongated by DNA polymerase, ''erased'' by the 5' $\longrightarrow$ 3' exonuclease activity of Pol I, and then replaced by DNA as shown in Figure 2-11.

Replication as it occurs *in vivo* represents a complex interaction between replication

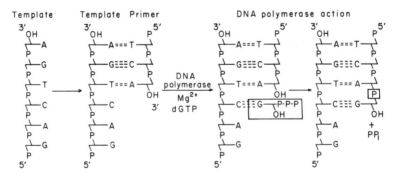

Fig. 2-9. Schematic representation of the action of DNA polymerase in the synthesis of DNA. A DNA template, a complementary RNA primer with a free 3'-OH end, and Mg^{2+} are required. Replication by DNA polymerase takes place in the 5' $\longrightarrow$ 3' direction. dGTP = deoxyguanosine 5'-triphosphate.

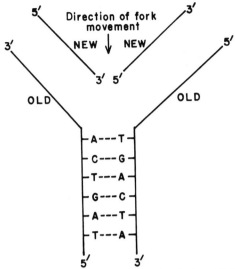

Fig. 2-10. Diagrammatic representation of the replicating fork in DNA synthesis in microbial cells.

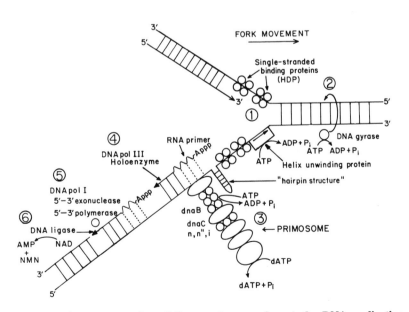

Fig. 2-11. Schematic representation of the events occurring at the DNA replicating fork. Numbers refer to steps outlined in the text. Modified from Kornberg, A. 1980. *DNA Replication.* W. H. Freeman and Co., San Francisco.

proteins and nucleic acid. Difficulties encountered in the analysis of this process include the large number of proteins involved (Tables 2-3 and 2-4) and the fact that some of the components of the replication apparatus dissociate upon purification. Consequently, much of the information concerning chromosomal replication was uncovered using more simplified replicons, such as plasmids and bacteriophages.

Replication of the bacterial chromosome proceeds bidirectionally from a fixed origin (*oriC*) located at 83.5 min on the standard *E. coli* linkage map (Fig. 2-14). The mechanism controlling the initiation of DNA synthesis will be discussed later. Once initiated, however, replication continues until both forks meet at the terminus located approximately 180° from *oriC*. At this point the two daughter chromosomes will form a linked concatemer due to the topological constraints inherent when separating the two strands of a double-stranded helical circle. Along with introducing negative supercoils into dsDNA, DNA gyrase can resolve linked DNA concatemers into two separate chromosomes. Obviously, this process is necessary if the cell is to divide (see the section on Partitioning in this chapter).

It is amazing to consider that the bacterial chromosome ($\sim 3.8 \times 10^5$ bp) replicates at a rate of 800 nucleotides per second, yet the frequency of error only amounts to 1 in 10^{10} bp replicated. Thus, a high degree of fidelity is maintained under normal circumstances.

Once initiated, elongation of the DNA molecule is a complex process requiring a large number of proteins. Figure 2-11 presents a rather detailed account of this process. The following numbered outline describes the numbered events depicted in this figure.

1. Strand separation enables replication to proceed and requries helix destabilizing proteins [single-stranded DNA (ssDNA) binding proteins] that serve to prevent unwound DNA from reannealing and to protect ssDNA from intracellular nucleases; and a helix unwinding protein, either helicase II or the *rep* gene product.

2. Unwinding results in the introduction of positive supercoils in the dsDNA ahead of the fork. The enzyme DNA gyrase removes these positive supercoils by introducing negative supercoils.

3. Prepriming (Primosome). Prior to the synthesis of an RNA primer, a primosome comprising at least six proteins (n,n',n'',i, DnaB, and DnaC) is assembled and functions as a mobile promoter for replication. In the presence of ATP, the primosome will migrate processively in a direction opposite to elongation (however, see the pol III dimer-looping model below). One hypothesis for how the primosome works is that the DnaB protein "engineers" the ssDNA into a hairpin structure. This could serve as a signal for the primase (*dnaG* gene product) to synthesize a RNA primer (beginning with ATP) of about 10 nucleotides.

4. Once the RNA primer is formed, DNA polymerase III can bind to the 3′-OH terminus of the primer and begin to synthesize new DNA. The DNA Pol III holoenzyme is very complex (see Table 2-4). There are at least *10* different proteins that comprise the highly processive holoenzyme form. Table 2-4 summarizes the various forms of Pol III encountered *in vitro* and their effects on DNA synthesis. Once started, synthesis on the lagging strand will continue until the polymerase meets with the 5′ terminus of a previously formed RNA portion of an Okazaki fragment. At this point the polymerase will dissociate from the DNA. Recalling the fidelity with which DNA is replicated in spite of the speed of replication leads one to question how fidelity is maintained. Ex-

TABLE 2-3. Genes and Proteins Involved in DNA Replication

Protein	Gene	Map Location	Molecular Weight	Function
	oriC	83.5		Origin of replication (245 bp)
Protein i *(X)	dnaT	99		Prepriming
Protein n (Z)	priB	95		Prepriming
Protein n' (Y)	priA	89		Prepriming (DNA dependent ATPase)
Protein n"	priC	10		Prepriming
DnaA	dnaA	82	54,000	Initiation, binds oriC, DnaB loading
DnaB	dnaB	91	55,000	Mobile promoter, helicase prepriming, priming, DNA-dependent rNTPase
DnaC	dnaC	99	25,000	Formation of dnaB–dnaC complex
Pol III (α)	dnaE	4	129,000	DNA pol III holoenzyme, elongation
Primase	dnaG (parB, dnaP)	67	60,000	Priming, RNA primer synthesis
γ subunit	dnaZX	10	47,500	Synthesis, part of the gamma complex
β (EFI)	dnaN	82.5	40,600	β-subunit of DNA pol III, processivity
Helix-destabilizing	ssb-1	91	20,000	Single-stranded binding protein
Helix-unwinding	rep	85		Strand separation; not essential for chromosome replication
τ subunit	dnaZX	85	71,000	Promotes dimerization of pol III
δ (EFIII)	holA	14	38,000	Synthesis, part of the gamma complex
δ	holB	26	37,000	Synthesis, part of the gamma complex
ψ	holD		14,000	Part of gamma complex
χ	holC		12,000	Part of gamma complex
θ	holE		10,000	?
	dnaY	12		?
ε subunit	dnaQ(mutD)	5	27,500	Proofreading
DNA pol I	polA	85	109,000	Gap filling, primer degradation
Ligase	lig	52	75,000	Ligation of single-strand nicks in the phosphodiester backbone
DNA gyrase (subunit α)	gyrA(nalA)	48	105,000	Supertwisting
DNA gyrase (subunit β)	gyrB(con)	82	95,000	Relaxation of supercoils
DNA pol II	polB	2	120,000	?
RNA pol, β subunit	rpoB	90	150,000	Initiation(?)
DNA helicase I			180,000	Unwinding
DNA helicase II			75,000	Unwinding

TABLE 2-4. Forms of DNA Polymerase III Peptides Comprising DNA Polymerase III Holoenzyme

Peptide	Gene	Molecular Weight	Functions
Pol III core α	*dnaE*	140,000	Polymerizing activity
ε	*dnaQ*		3' ⟶ 5' proofreading
θ		10,000	?
β (EFI, copol III)	*dnaN*	40,600	Processivity
γ	*dnaZ*	52,000	Needed to add β; enables stimulation by SSB, makes enzyme more processive
δ(EF III)	*holA*	32,000	Enables stimulation by SSB, makes enzymes more processive
Y	*holC*	14,000	Make enzyme more processive
ζ	?	?	10 fold higher processivity than core
δ complex (δ,δ',Y,Ψ)			Transfer of β subunit to primed template, ATP hydrolysis.
Pol III' (α, ε, θ, τ)			Fivefold higher processivity than Pol III
Pol III* (α, ε, θ, τ, γ, δ, Ψ, Y)			?
Pol III holoenzyme (Pol III* + β)			highly processive

periments have shown that polymerases in general do incorporate a relatively large number of incorrect bases while replicating. This, however, sets up a base pair mismatch situation that can be recognized as incorrect by the 3' ⟶ 5' exonuclease proofreading activity of the DNA polymerase. The component of DNA polymerase responsible for proofreading is DnaQ. This subunit allows polymerase to hesitate, excise the incorrect base, and then insert the correct base. However, even with this elegant proofreading system, some mistakes still occur. How the cell deals with these will be discussed in the section concerned with DNA repair.

5. Once pol III has replicated DNA up to the point where a preexisting RNA primer resides (on the lagging strand), the cell must remove that RNA and replace it with DNA. To accomplish this, DNA polymerase I utilizes its 5' ⟶ 3' exonuclease activity to cleave the RNA primer and its 5' ⟶ 3' polymerase activity to replace the cleaved RNA with DNA.

6. When pol I is finished, there remains a phosphodiester break (or nick) between the 3'OH end of the last nucleotide synthesized by pol I and the 5'-phosphoryl end of the adjacent DNA segment. The DNA ligase will "seal" this nick by using NAD as a source of energy (Fig. 2-12).

DNA Polymerase III Functions as a Dimer. At this point we need to readdress the question of how the leading and lagging strands are synthesized simultaneously. *In vivo,*

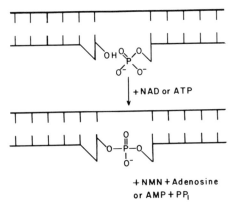

Fig. 2-12. Action of DNA ligase in joining adjacent 3'-hydroxyl and 5'-phosphoryl groups in duplex DNA. After Lehman, I. R. 1974. DNA ligase: structure, mechanism and function. *Science* **186**:790.

an Okazaki fragment is formed each second. This would seem to require a lagging strand DNA polymerase molecule to bind a primer, synthesize a 1000 nucleotide fragment, dissociate, and then reassociate to the next primer each second! *In vitro* it takes DNA polymerase 30-min to dissociate even from termination complexes. A current theory that may resolve this paradox is that pol III holoenzyme is dimeric with one member bound to the leading strand while its partner is associated with the lagging strand (Fig. 2-13). The fact that both halves of the dimer must move in the same direction dictates the formation of a loop in the lagging strand to allow binding of the polymerase to the RNA primer and synthesis of the Okazaki fragment.

INITIATION OF DNA REPLICATION

Genetic and recombinant DNA studies have placed the origin of replication (*oriC*) at 83 min on the circular *E. coli* map (Fig. 2-14) between *uncB* and *asn* in a region of about 240 bp. The region contains extended A + T rich sequences that could result in easier strand separation during initiation (the amount of energy required to separate A + T rich DNA strands is considerably less than that required to separate G + C rich DNA strands because A/T pairs are connected by two hydrogen bonds, whereas G/C pairs are connected by three bonds). A number of potential RNA polymerase binding sites have also been found in this region (see section on RNA synthesis). Some of these sites are required to synthesize an initiator RNA primer molecule. The origin has also been shown to bind to the outer membrane of *E. coli* and to a membrane fraction not characteristic of either inner or outer membrane. Although this would appear to present a geographical paradox, it could provide a solution for a problem inherent with the theory that chromosome segregation in dividing cells is mediated by membrane attachment. Since the inner membrane is too fluid to provide a suitable anchor for segregation, the outer membrane with its covalent attachment to peptidoglycan could serve to define the rigid framework necessary for chromosomal separation. Unfortunately for this model, the origin appears to interact transiently with the membrane, so if the membrane is involved, it is not through attachment at *ori* (see the section on Partitioning, in this chapter).

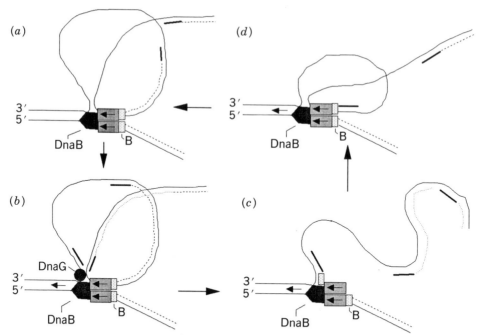

Fig. 2-13. Concurrent synthesis of leading and lagging strands of DNA by a dimer of DNA pol III. The arrows indicate direction of DNA pol III movement (open rectangles). (*a*) DnaB protein (pentagon) is complexed with the pol III dimer and unwinds DNA ahead of the replication fork. (*b*) Synthesis of the Okazaki fragment continues until reaching a previously synthesized fragment. Concurrently, DNA primase (circle) synthesizes a new primer. (*c* + *d*). The β subunit of DNA polymerase (filled rectangle) recruits the new primer for DNA synthesis.

The genetic loci required for initiation of replication include *dnaA*, *dnaB*, *dnaC*, and *dnaG*. One model that attempts to illustrate the events associated with initiation is presented in Figure 2-15. There are two basic hypotheses regarding how a cell determines when to initiate a new round of replication that involve either positive or negative control. Both models take into account that the cell must attain a certain cell mass/chromosome ratio before initiating chromosome replication. The positive control model states that the concentration of a specific effector molecule increases during cell growth and that once a certain threshold concentration is attained, initiation will occur. Alternatively, the negative control theory proposes that upon initiation, a pulse of negative regulator protein is synthesized that must be diluted below a certain threshold limit before a new round of replication can initiate. It appears that both positive and negative controls are involved in coordinating the time of replications. Two proteins, DnaA and IciA, have been implicated, respectively, as positive and negative regulators of initiation. While its precise role *in vivo* is unclear, IciA clearly inhibits initiation *in vitro*. This protein binds to a series of 13-mers at the origin (labeled L,M,R in Fig. 2-15) to prevent initiation and may need to be diluted during cell growth before a new round of replication can begin.

In contrast to IciA, DnaA binds to a series of 9-mers (DnaA boxes), also located within the origin (labeled R1–R4 in Fig. 2-15). When bound to these sites in the presence of ATP, DnaA helps separate the two strands forming an open complex. The histone-like

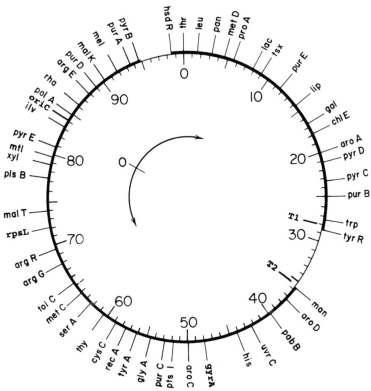

Fig. 2-14. Partial linkage and replication maps of E. coli. Circular reference map of *E. coli* K-12. The large numbers refer to map position in minutes, relative to the *thr* locus. From the complete linkage map 52 loci were chosen on the basis of greatest accuracy of map location, utility in further mapping studies, and/or familiarity as long-standing landmarks of the *E. coli* K-12 genetic map. T1 and T2 are terminators. From Bachmann, B. J., K. B. Low, and A. L. Taylor, *Bacteriol. Rev.*, **40**, 116, 1976.

protein Hu is also required. A prepriming complex then forms as DnaB and DnaC proteins bind, with the helicase activity of DnaB unwinding the template for subsequent priming. The DnaC protein does not remain bound; its role is to facilitate DnaB binding. The polarity of DnaB movement is 5' $\longrightarrow$ 3' on the strand to which it is bound. This places the protein on the lagging strand template *ahead* of the leading strand polymerase (Fig. 2-13). It is not clear whether RNA polymerase or DnaG primase synthesizes the initiator RNA primers although it appears either can accomplish this *in vitro*. Thus, DnaA acts as a positive regulator of initiation. Its accumulation in the cell during growth is controlled by DnaA itself acting as a negative regulator of *dnaA* transcription.

Another intriguing feature of initiation is the role that DNA methylation may play in timing new rounds of replication. Throughout the origin there are many GATC sequences. This sequence is a substrate for DNA adenine methylase (product of the *dam* locus), which adds a methyl group to the N-6 position of adenine. Immediately after initiation these sites will be hemimethylated (one strand methylated, the daughter strand unmethylated). There is evidence that the hemimethylated origin will bind to the membrane. This

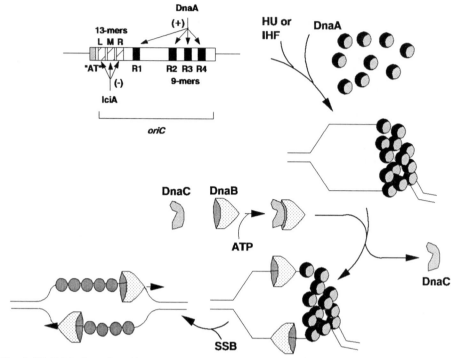

Fig. 2-15. Initiation of replication. The diagram depicts the area of the origin of replication. L, M, R, left, right, and middle 13mers; R1-4, 9 mer DnaA boxes. See text for details. The DnaA proteins bind to the DnaA boxes to initiate replication. In collaboration with the histone-like protein HU and IHF, DnaA helps form an open complex. The DnaC protein guides DnaB to the open complex where it forms a sliding clamp. Its unwinding activity helps elongate the open complex and serves as a loading site for DNA polymerase and primosome components.

complex is a poor substrate for initiation and will also inhibit the methylation of itself. A few minutes after membrane attachment, the origin detaches and can be methylated. Presumably, the negative regulator accumulates during the period of membrane association and then binds to the origin when it detaches. The following summarizes a proposed sequence of events leading to initiation.

1. Dilution of negative regulator, IciA and/or others.
2. DnaA binding to 9-mer DnaA boxes.
3. Open complex formation at 13-mer region.
4. Formation of prepriming complex with DnaB and DnaC.
5. Priming with DnaG primase or RNA polymerase.
6. DNA replication by DNA pol III leads to hemimethylation of origin.
7. Membrane association with hemimethylated origin.
8. Burst of negative regulator synthesis.
9. Detachment of origin from membrane, methylation.
10. Slow dilution of negative regulator during cell growth.

Another role for the *dnaA* gene product is that it is required to protect initiator RNA from RNAse H. This nuclease appears to prevent anomalous initiations by degrading potential initiator RNAs that may arise at various locations around the chromosome.

TERMINATION OF DNA REPLICATION AND CHROMOSOME PARTITIONING

Once intiated, replication proceeds to the terminus (*terA, B, C*) between 30 and 35 min on the linkage map. The time required to complete one round of replication is about 40 min. A question that has puzzled scientists for years is how termination occurs. Recent experiments have shed considerable light on this problem. There are two primary sites, T1 and T2, involved in the inhibition of replication. Each one functions in a polar manner. Terminator T1 (at 28 min) permits clockwise-traveling replication forks to enter the terminus region but inhibits counterclockwise-traveling forks that might exit the region. Terminator T2 (at 35 min) does the opposite, allows counterclockwise forks to enter the region but inhibits clockwise forks from exiting. Another protein essential for termination is the terminus utilization substance, Tus. It maps near T2 and is required for the function of both T1 and T2. At completion, the replicated daughter chromosomes will appear as linked concatemers. Successful partitioning of the two chromosomes into separate daughter cells is essential for successful cell division. Type II topoisomerases are believed to be responsible for resolving the concatemers into individual chromosomes. These include DNA gyrase and the products of the *parC* and *parE* loci (topoisomerase IV). Both ParC and ParE show extensive homology with GyrA and GyrB, respectively, suggesting a common evolutionary origin. However, whereas DNA gyrase *increases* negative supercoiling, top IV *decreases* it. Actual partitioning of the chromosomes into separate regions of a dividing cell requires protein synthesis and a protein called MukB. One hypothesis is that MukB may attach to the chromosome and "walk" along a cytoskeleton-like protein filament. While potential filament proteins have been found, their role in partitioning is primarily conjecture (see further discussion in Chapter 11).

RNA SYNTHESIS: TRANSCRIPTION

RNA Synthesis

The synthesis of all cellular RNAs, including messenger RNA (mRNA), transfer RNA (tRNA), and ribosomal RNA (rRNA) involves the process of transcription. The transcription of DNA to RNA requires a DNA-dependent RNA polymerase and proceeds in a manner similar to DNA synthesis, only using ribonucleic acid triphosphates (rNTP) rather than deoxyribonucleic acid triphosphates (dNTP). RNA polymerase is a complex enzyme consisting of at least four polypeptides: α, β, β', σ. Core polymerase consists of two α subunits plus one β and one β' subunit. The core enzyme can bind to DNA at random sites and will synthesize random lengths of RNA. Holoenzyme, consisting of a core enzyme plus a sigma (σ) subunit (σ-70 is considered the "housekeeping" σ), binds to DNA at specific sites called promoters and transcribes specific lengths of RNA. Thus, the σ subunit plays an important role in promoter recognition by RNA polymerase. The β subunit carries the catalytic site of RNA synthesis as well as the binding sites for substrates and products. The β' subunit appears to play a role in DNA template binding while the two α subunits assemble the two larger subunits into core enzymes ($\alpha_2\beta\beta'$). Rifampicin, a commonly used antibiotic, inhibits transcription by interfering with the β subunit of prokaryotic RNA polymerase. Figure 2-16 presents a current view of RNA polymerase structure and important binding regions for each subunit.

Within a transcribing region, the coding strand (also called sense or designated "+") refers to the DNA strand that is not transcribed but consists of the same sequence as mRNA (with a T instead of U). The anti-sense template strand (designated −) is the strand that is transcribed. Insofar as RNA polymerase binds to a promoter region, then progressively moves along the template strand from the region encoding the N-terminus to the carboxy terminus, the terms **upstream** and **downstream** are used to describe regions of DNA relative to polymerase movement. Thus, the promoter is upstream from the structural gene. It should be noted that RNA polymerase moves along the DNA template strand in the $3' \longrightarrow 5'$ direction while synthesizing RNA $5' \longrightarrow 3'$ (Fig. 2-17). As with DNA replication, transcription can be described in three main steps: **initiation, elongation, and termination**. Initiation involves the binding of polymerase to the promoter with the formation of a stable RNA polymerase–DNA initiation complex and the catalysis of the first $3'-5'$ internucleotide bond. Elongation is the translocation of RNA polymerase along the DNA template with the concomitant elongation of the nascent RNA chain. Obviously, termination involves the dissociation of the complex. The overall process of transcription is illustrated in Figure 2-17. Each step in the transcription process will now be described in more detail.

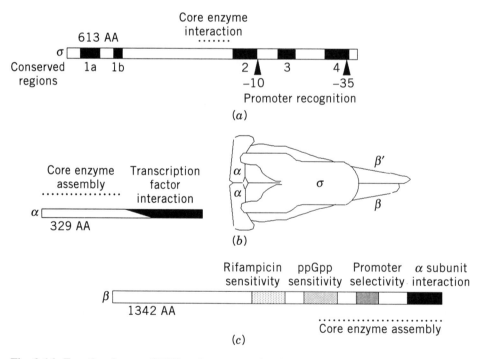

Fig. 2-16. Functional map of RNA polymerase subunits. The molecular model of RNA polymerase holoenzyme is shown in the middle. Areas of each subunit involved with various interactions between subunits, other protein factors, effector molecules and promoter regions are indicated. Derived from Ishihama, A. 1992. *Mol. Microbiol.* **6:**3283.

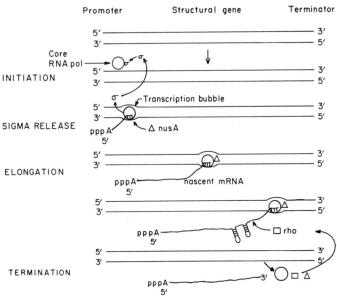

Fig. 2-17. Overview of the transcription process.

Initiation. The recognition site referred to as the promoter includes two DNA sequences, which appear to be conserved in most of the promoters examined. In general, highly conserved sequences are referred to as **consensus sequences**. The promoter consensus sequences are centered at −10 and −35 bp from the transcriptional start point (designated +1) and have been implicated in normal promoter function:

5′ TTGACA   5′ TATAATPu
AACTGT 5′   ATATTAPy 5′
−35   −10

The −10 sequence is referred to as Pribnow's box while the −35 region is called the recognition site. One question that must be addressed, however, is how RNA polymerase can recognize these sequences when it would appear that direct interaction between RNA polymerase and DNA is precluded by the double-stranded nature of DNA. Association is possible for any protein–nucleic acid interaction through base-specific groups that can be recognized in the major or minor grooves. The RNA polymerase can interact with groups in the major groove and recognize the proper sequence upstream (−35 region) from the Pribnow box, then form a stable complex (closed complex) by moving laterally to the −10 region.

Sigma may be involved in recognizing both regions. It is also becoming clear that the superhelical nature of the chromosome may play a role in promoter function. In general, a negatively supercoiled chromosome is a better transcriptional template than a relaxed chromosome. Presumably, the torsional stress imposed by supercoiling makes certain areas of DNA easier to separate by RNA polymerase (i.e., lowers the melting temperature). However, it appears that supercoiling affects the expression of some genes more than others. There is some thought that the cell, through the use of topoisomerases Top

I and DNA gyrase, can affect supercoiling in localized areas of the chromosome, facilitating the transcription of some genes while retarding the transcription of others.

Obviously, σ-70 plays an important role in normal transcription initiation. Alternate σ factors (e.g., σ-32) can change the promoter recognition specificity of core enzyme. Thus, σ-32 factor is utilized when cells undergo heat shock and is believed to be responsible for the production of proteins required to survive that stress (see Chapter 4). Other alternate σ factors are listed in Table 2-5 and in Chapter 12 in the discussion on sporulation. Among the various sigmas there are highly conserved regions designated 1–4, as shown in Figure 2-16.

The closed promoter complex, in which RNA polymerase-bound DNA remains unmelted, must be converted to an open promoter complex. This involves the unwinding of about one helix turn from the middle of Pribnow's box to just beyond the initiation site. The open complex then allows tight binding of RNA polymerase with the subsequent initiation of RNA synthesis. The first triphosphate in a mRNA chain is usually a purine. Initiation ends after the first internucleotide bond is formed (5'pppPupN).

Elongation. After the formation of transcripts eight to nine nucleotides in length, it appears that σ factor is released abruptly from the complex. This suggests that RNA polymerase undergoes a conformational change causing a decrease in the affinity of σ from the RNA polymerase–DNA–nascent RNA complex (ternary complex). The released σ factor can be reused by a free core polymerase for a new initiation.

Once elongation begins, transcription proceeds at a rate of between 30 and 60 nucleotides per second (at 37°C). The elongation reaction as a whole includes the following steps: (a) nucleotide triphosphate binding, (b) bond formation between the bound nucle-

TABLE 2-5. RNA Polymerase Subunits and Other Transcription Factors

Subunit	Structural Gene	Map Position	Molecular Weight	Function
α	rpoA	72	36,000	Initiation
β	rpoB	90	150,000[a]	Initiation, elongation, termination
β'	ropC	90	160,000	Initiation
σ^{70}	rpoD	60	83,000	Initiation
σ^{32}	rpoH	76	32,000	Initiation, heat shock response
σ^{28}(σ^F)	fliA	43	28,000	Flagellar genes, chemotaxis
σ^{24}(σ^E)			24,000	Extreme heat shock
σ^{38}(σ^S)	rpoS(KatF)	59	38,000	Stationary phase
σ^{54}	rpoN	72	54,000	Nitrogen
omega	rpoZ	82		unwinding
CRP	crp	73	22,500	Initiation enhancement
rho(ρ)	rho	84	50,000	Termination
NusA	nusA	69	69,000	Pausing, transcription termination
NusB	nusB	10	14,500	Pausing, transcription termination
NusG	nusG	90	20,000	Pausing, transcription termination
GreA	greA	72	17,000	3' ⟶ 5' RNA hydrolosis, proofreading?
GreB				3' ⟶ 5' RNA hydrolysis, proofreading?

[a]A Zn^{2+} metalloenzyme.

otide and the 3′-OH of the nascent RNA chain with, (c) pyrophosphate release, and (d) translocation of polymerase along the DNA template. Movement of RNA polymerase, by necessity, involves the melting of the DNA template ahead of the transcription bubble (~ 17 nucleotides), as well as reformation of the template behind the bubble. The rate of elongation along a template is not uniform. Regions in a template where elongation rates are very slow are called pausing sites. Pausing sequences in general contain GC-rich regions approximately 10 bp upstream of the 3′-OH end of the paused transcript, as well as regions of dyad symmetry present 16–20 bp upstream of the 3′-OH end. The regions of dyad symmetry involve two closely spaced sequences on a single strand that are capable of base pairing with each other and that cause the formation of RNA–RNA stem–loop structures that in many cases appear to be important for pausing. The mechanism of pausing, however, remains unclear.

The *nusA* and *nusG* proteins are believed to associate with core polymerase some time after σ dissociates and to modulate elongation rates. The Nus interaction with the ternary complex is nonprogressive involving rapid association and disassociation. These proteins enhance pausing at some sites and may be involved with rho-dependent transcription termination (see below).

It should be noted at this point in the discussion that the RNA transcript is normally larger than the structural gene. Most messages contain a promoter-proximal region of mRNA called the leader sequence (or untranslated region) prior to the translation start codon. The leader sequence carries information for ribosome binding (see Shine–Dalgarno sequence) and in some cases is important in determining whether RNA polymerase will proceed into the structural gene(s) proper (see Attenuation in Chapter 4).

There is evidence that RNA polymerase possesses a proofreading activity analogous to the 3′–5′ exonuclease used by DNA polymerase. Two proteins, **GreA** and **GreB**, enable transcriptionally arrested RNA polymerase to back up and cleave 2–3 nucleotides from the 3′ end of the nascent message. This releases RNA polymerase, allowing it to proceed through the arrest point to the 3′ end of the gene. If this model proves accurate, it would help account for the high degree of transcriptional fidelity observed in transcribed messages.

Protein–Protein Interactions with RNA Polymerase. Many regulatory factors appear to directly contact RNA polymerase at the promoter region. These proteins can be divided into two classes. Class I transcriptional activators bind upstream from the promoter (e.g., CRP, AraC, Fnr, and OmpR). Class II transcription factors overlap the promoter region. Many of these regulators contact the C-terminal end of the α subunit (Fig. 2-16) and in some manner activate transcription. Many of these factors will be discussed in Chapter 4.

Termination. Transcriptional termination includes the following events: (a) cessation of elongation, (b) release of the transcript from the ternary complex, and (c) dissociation of polymerase from the template. There are two classes of termination referred to as rho-independent and rho-dependent transcription termination.

The rho-independent termination signal includes a GC-rich region with dyad symmetry allowing formation of a RNA stem–loop structure about 20 bases upstream of the 3′-OH terminus and a stretch of 4–8 consecutive uridine residues. The transcript usually ends within or just distal to the uridine string. The RNA stem–loop structure causes RNA polymerase to pause and disrupts the 5′ portion of the RNA–DNA hybrid helix. The remaining 3′ portion of the RNA–DNA hybrid molecule includes the oligo(rU) sequence. The relative instability of rU–dA bp causes the 3′ end of the hybrid helix to melt, releasing the transcript.

As the term implies, termination of rho-dependent terminators requires the *rho*-gene product. Rho-factor termination only occurs at strong pause sites apparently located at specific distances from a promoter region. In other words, a strong pause site located closer to the promoter will not serve as a termination site. No specific termination sequence has been noted for Rho-action, but Rho, which occurs as a hexamer, requires a single-stranded binding region of RNA (Fig. 2-18). The ssRNA is thought to wind around the outside of rho-factor, facilitated by rho-dependent ATP hydrolysis, which serves to bring Rho in contact with RNA polymerase. The extended dwell-time of RNA polymerase elongation complexes at pause sites allows the Rho protein translocating along the nascent RNA to catch up with the transcription complex. The association of Rho to core enzyme may occur via NusA or NusG proteins. Subsequently, activation of an RNA/ DNA helicase activity causes the mRNA chain and core RNA polymerase to dissociate from the DNA. Free core RNA polymerase can then interact with a σ factor and be used to initiate a new round of transcription.

RNA Turnover

Cellular RNA, thus transcribed, can be classed into two major groups relative to their decay rates *in vivo*: stable and unstable RNA. Stable RNA consists primarily of rRNA and tRNA, while unstable RNA is mRNA. In *E. coli*, 70–80% of all RNA is rRNA, 15–25% is tRNA, and 3–5% is mRNA. Factors that appear to contribute to stability include the association of rRNA in ribonucleoprotein complexes (ribosomes) and the extensive secondary structure exhibited by both rRNA and tRNA. Both of these structural considerations serve to protect stable RNA at the 5′ terminus from ribonucleases.

The average mRNA (~ 1200 nucleotides) has a half-life (time required to reduce the mRNA population by one-half) of approximately 40 s at 37 °C. This figure represents an average as each mRNA species has a unique degradation rate. Degradation occurs overall in the 5′ ⟶ 3′ direction, however, the exoribonucleases present in the cell degrade in the 3′ ⟶ 5′ direction. One theory that attempts to resolve this paradox is that an intitial, random endonucleolytic event occurs in the mRNA near the 5′ end, which would allow for exonucleolytic degradation of the 5′ end. Since no new ribosomes can bind to mRNA that has lost its 5′ end, the remaining message is progressively exposed as the already bound ribosomes move toward the 3′ end. Thus, additional endonucleolytic events followed by 3′ ⟶ 5′ exonucleolytic degradation would be possible. Although no enzyme or series of enzymes have been directly implicated in this degradation, RNase III, which is involved in RNA processing (see below), has been suggested as a candidate. Message stability can be enhanced by the presence of stem–loop structures present at the end or beginning of the mRNA. Stem–loops at the 3′ end of the message will inhibit 3′–5′ exonuclease processivity while the 5′ loops may interfere with endonuclease association.

RNA Processing

All stable RNA species and a few mRNAs of *E. coli* must be processed prior to their use. For example, each of the seven rRNA transcription units is transcribed into a single

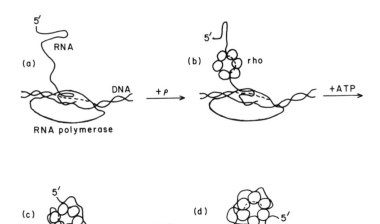

Fig. 2-18. Model for rho factor-mediated release of RNA. (a) An RNA polymerase molecule has reached a transcription termination site on a DNA molecule. The RNA transcript is attached to the transcription complex by a few base pairs with the DNA and by contacts with the RNA polymerase. (b) An exposed segment of the RNA has become bound in the large binding site of rho protein. This step could precede the arrival of RNA polymerase at the termination site. (c) The RNA has become partially wrapped around the outside of rho factor. As a consequence, rho protein has come into contact with the RNA polymerase. This step is dependent on the hydrolysis of nucleoside triphosphates. Continuation of this wrapping causes disruption of the noncovalent bonds that hold the nascent RNA to the DNA and RNA polymerase. (d) The rho protein–RNA complex is released from the RNA polymerase–DNA complex. From Richardson, J. P. (1983). *Microbiology–1983.* American Society of Microbiology, Washington, DC.

message as follows:

5′ leader–16S rRNA–spacer–23S rRNA–5S rRNA–trailer-3′

The spacer always contains some *tRNA* gene. Obviously, this long multigene (*polycistronic*) precursor RNA molecule must be processed to form mature 16S rRNA, 23S rRNA, and 5S rRNA species. Another example of processing involves the *tRNA* genes that are frequently clustered and cotranscribed in multimers of up to seven identical or different tRNAs. These multimers msut be processed to individual tRNAs. Along with the required cutting and trimming, the initial transcripts of rRNA and tRNA lack the modified nucleosides of the mature species. Thus, processing of stable RNAs (i.e., tRNA and rRNA) includes the modification of bases in the initial transcript.

There are four basic types of processing. The first involves the precise separation of polycistronic transcripts into monocistronic precursor tRNAs. Second, there must be a mechanism by which the mature 5′ and 3′ termini are recognized followed by the removal of extraneous nucleotides. The third type of processing involves the addition of terminal residues to RNAs lacking them (e.g., the 3′ CCA end of some bacteriophage tRNAs, such as T4 phage). Finally, the appropriate modification of base or ribose moieties of nucleosides in the RNA chain must be accomplished. Obviously, not every RNA molecule is subject to all four of these processes. The following constitutes a brief description of several enzymes involved with RNA processing in *E. coli*:

RNase P, a ribozyme. This endonucleolytic enzyme is required for the maturation of tRNA (Fig. 2-19). RNase P removes a 41-base long fragment from the 5'- side of a wide variety of tRNA precursor molecules. The enzyme appears to recognize tRNA secondary and tertiary configurations rather than a specific nucleotide sequence. There is an absolute requirement for the presence of the CCA terminal sequence within the precursor. The subunit structure of this enzyme is unusual in that there is a polypeptide component (C5) coded for by the *rnpA* locus and an RNA component (M1/M2) encoded by *rnpB*. The RNA component is integral to the mechanism by which RNase P cleaves tRNA precursor. It is, in fact, an example of catalytic RNA. Hence, the description of RNase P as a ribozyme.

RNase III. Encoded by the *rnc* locus, this enzyme contains two identical polypeptide units of 25,000 molecular weight (MW) with no RNA component. RNase III cleaves dsRNA by making closely spaced single strand (ss) breaks (~ every 15 bases). Other than recognizing perfect double-stranded RNA stems, there does not appear to be a unique recognition sequence for this enzyme.

RNase E. The *rne* locus is the structural gene for RNase E (MW 70,000). The enzyme specifically cleaves p5S precursor rRNA from larger transcripts. The enzyme first cleaves between the ss region of 5S precursor and the ds region. The second cleavage occurs within the ds area (Fig. 2-20). Only two substrates for this enzyme are known, 9S rRNA and RNAI (involved in the control of ColEI plasmid copy number, see Chapter 3).

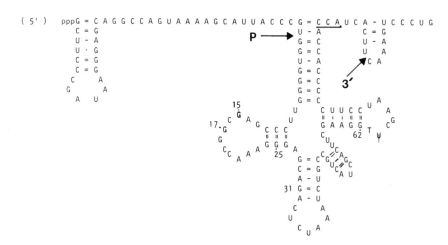

Fig. 2-19. The tRNA precursor structure. The sequence shown is that of the tRNA$_1^{tyr}$ transcript. Transcription *in vivo* proceeds through a second tRNA$_1^{tyr}$ gene. Sites of *in vivo* endonuclease cleavage are shown on the 5' side of the mature tRNA (RNase P) and near the 3' end. The mature 3' terminus is underlined. Numbered nucleotide positions are those at which base changes result in reduced RNase P cleavage *in vivo*. From Gegenheimer, P. and D. Apirion. 1981. Processing of prokaryotic ribonucleic acid. *Microbiol. Rev.* **45**:502–541.

PROCESSING MAP OF RIBOSOMAL RNA

A. SECONDARY STRUCTURE AND CLEAVAGE SITES

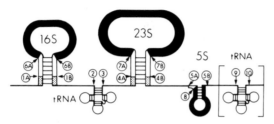

Cuts 1,4. RNase III   2,9. RNase P   3,10. RNase F   5. RNase E
6. RNase M16   7. RNase "M23"   8. RNase "M5"

B. PROCESSING IN WILD-TYPE STRAINS

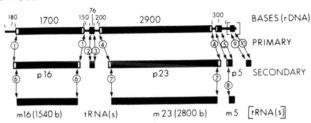

Fig. 2-20. Structure and processing of rRNA transcripts. (A) Structure and cleavage sites of the rRNA primary transcript (not to scale). Distal (trailer) tRNAs are bracketed because not all rDNAs contain them. Transcripts may contain one or two spacer tRNAs, and no, one or two trailer tRNAs. Arrows indicate endonucleolytic cleavage sites. Each cutting event is given a separate number, referring to the enzyme involved; A and B indicate that two (or more) separate cuts may be required. Thick solid segments represent mature rRNA sequences, thick open segments represent precursor-specific sequences removed during secondary processing steps, stippled segments are sequences found only in p16b and p23b of RNase III⁻ cells, and thin lines (except for tRNAs) represent nonconserved sequences discarded during primary processing. Enzymes are discussed in the text. **(B)** Processing in wild-type strains. The first line shows the transcriptional map of a representative rDNA unit, drawn approximately to scale. Distances in bases are between vertical bars above the map. The primary and secondary cuts, numbered as in **(A)**, are shown above the products they generate. Open and solid segments are as in **(A)**. From Gegenheimer, P. and D. Apirion. 1981. Processing of prokaryotic ribonucleic acid. *Microbiol. Rev.* **45**:502–541.

RNase D. Maturation of tRNA precursors requires the removal of extra nucleotides distal to the CCA sequence located within the precursor tRNA destined to become the mature 3'-OH end of the tRNA molecule. The CCA–OH sequence is required for amino acid acceptor activity on all tRNAs. RNase D is a monomeric protein with a molecular weight of 38,000. It is a nonprocessive 3' exonuclease, which releases mononucleotide 5' phosphates from the RNA substrate.

RNase F. Since transcription of tRNA gene clusters and of *tRNA* genes in rRNA transcriptional units continues for a considerable distance past the 3' terminus of the tRNA, a 3' endonuclease is required to expose a 3' end that can be acted upon by RNase D. RNase F is an enzyme proposed to fill such a role.

tRNA nucleotidyltransferase. This enzyme can repair prokaryotic tRNAs in which the CCA sequence is missing. The product of the *cca* locus (67 min), nucleotidyltransferase, will sequentially add a 3' CCA terminal sequence to these tRNAs. Although it is probably not essential, mutants with a reduced level of this enzyme grow at a rate slower than normal.

Modifying enzymes. These enzymes chemically modify nucleotides present within tRNA and rRNA precursor transcripts and include methylases, pseudouridyllating enzymes, thiolases, and others. There are several examples that illustrate the importance of these modified bases. For example, mutants resistant to the antibiotic kasugamycin have lost an enzyme that specifically methylates two adenine residues to form dimethyl adenine in the sequence A-A-C-C-U-G near the 3' end of 16S rRNA. Lack of this modification diminishes the affinity of kasugamycin for 30S ribosomal subunits.

In addition, there are several modifications in tRNA that are important. Modifications in the third base of the anticodon loop can contribute to the "wobble" of anticodon–codon recognition (see below under **Transfer RNA**). Other modifications of tRNA are beginning to be recognized as necessary for signaling various stress conditions placed upon the cell, such as a shift from anaerobic to aerobic growth or limitation of amino acids (see Attenuation in this chapter).

Intervening sequences. Some organisms, such as *Salmonella typhimurium*, do not contain intact 23S rRNA in their ribosomes. The 23S rRNA locus of *S. typhimurium* contains an extra 90 bp sequence that is not present in the *E. coli* homolog. This sequence, called an **intervening sequence** (IVS), is excised from the original transcript much as an intron is excised from within the mRNA of most eukaryotic genes. However, unlike intron processing, the remaining pieces of the *S. typhimurium* 23S rRNA are not spliced together. Since the rRNA does not encode a protein, there is no need to splice the fragments. The pieces function normally in all aspects of ribosomal assembly and protein synthesis. The rRNA superstructure is unaffected because the fragments remain associated through secondary and tertiary contacts. RNase III is the enzyme responsible for processing the IVS as proven by observing normal processing of *S. typhimurium* 23S rRNA in *rnaE*⁺ but not *rnaE*⁻ mutants of *E. coli*. The origins of IVSs are unknown. They may be the "footprints" of transposable elements that inserted and then excised from the 23S rRNA genes (see Chapter 3).

A summary of RNA processing in *E. coli* is shown in Figure 2-20.

Poly (A) tails. A feature thought to be unique to eukaryotic cells is the addition of poly(A) tails to the 3' ends of many of their mRNAs. It is now evident that short stretches of poly(A) can be found in bacterial mRNA especially in mutants lacking 3' exonuclease activity. However, their role in bacteria is unclear.

PROTEIN SYNTHESIS: TRANSLATION

During translation, the information coded in mRNA is specifically read and used to form a polypeptide molecule with a definite function. A given nucleotide sequence in mRNA codes for a given amino acid. Since there are 20 naturally occurring amino acids, this means that there must be at least three nucleotides in the code for each amino acid (**triplet code**). Since there are only four bases, a doublet code would account for only 4^2 or 16 amino acids. With a triplet code, 4^3 or 64 possible amino acids may be encoded. Since there are only 20 amino acids, this means that there is usually more than one triplet that can code for a given amino acid (see Table 2-6). For this reason, the code is termed **degenerate**. On the other hand, a given codon triplet cannot code for any other amino acid. For example, although the triplets UUU or UUC can both code for the amino acid phenylalanine, they cannot code for any other amino acid, indicating that the code is **nonoverlapping**. Of considerable significance is the fact that three codons, UAG, UAA, and UGA do not *usually code* for any amino acid, and tRNA molecules with the corresponding anticodons are uncommon.* These codons (**nonsense** codons) serve as termination signals, stopping the translation process. The code is **commaless**; that is, there is no punctuation. If a single base is deleted or if another base is added in the sequence, then the entire sequence of triplets will be altered from that point on (**frameshift**).

Although the code is considered to be universal, there are some important differences in detail that will be considered later. At this point it will suffice to note just a few specific aspects. Apparently methionine, rather than *n*-formylmethionine, is the initiating amino acid in eukaryotic cells as well as the archebacteria. The termination process is similar in both types of organisms. The process of protein synthesis in mitochondria is more closely analogous to that in bacteria than it is to what occurs in the cytoplasm of eukaryotes. Those differences that do exist between eukaryotic and prokaryotic protein synthesis may form the basis for selective inhibition by antibiotics or toxins on one cell type in the presence of the other. For example, some antibiotics have been shown to preferentially inhibit protein synthesis in eukaryotes. Diphtheria toxin specifically inactivates the elongation factor 2 of eukaryotic cells while having no effect on the analogous factor in prokaryotic cells.

Elucidation of the nature of the code was aided by the discovery that synthetic polynucleotides could substitute for the natural mRNA in the cell-free system obtained from *E. coli*. By using trinucleotides of known base composition, it was possible to demonstrate the specific binding of ^{14}C-aminoacyl–tRNA to ribosomes. Polarity in the recognition system was also demonstrated. The codon triplet for valine proved to be GpUpU, as shown by the fact that this trinucleotide included ^{14}C-valyl–tRNA to bind ribosomes, wheres UpUpG would not. Phenylalanine was bound in the presence of pUpUpU but not UpUpUp. It was also found that some synthetic polynucleotides composed of only one base could replace native mRNA [poly(A) codes for lysine, poly(C) for proline, poly(U) for phenylalanine, and poly(G) for glycine polypeptides].

Transfer RNA

The incorporation of amino acids into polynucleotide chains in a specified order involves the formation of an ''activated complex'' between an amino acid and a specific tRNA molecule. This activation (charging) of specific tRNA molecules is accomplished by aminoacyl tRNA synthetases (aminoacyl tRNA ligases). Each amino acid has its specific synthetase and specific tRNA that can read the code in the mRNA and thereby direct the

---

*There are exceptions, for example, in mitochondria and selenocystine-containing proteins (see below).

TABLE 2-6. The Amino Acid Code[a]

Second →	U	C	A	G	Third ↓
First ↓	Phe	Ser	Tyr	Cys	U
U	Phe	Ser	Tyr	Cys	C
	Leu	Ser	Ochre	Opal	A
	Leu	Ser	Amber	Trp	G
	Leu	Pro	His	Arg	U
C	Leu	Pro	His	Arg	C
	Leu	Pro	Gln	Arg	A
	Leu	Pro	Gln	Arg	G
	Ile	Thr	Asn	Ser	U
A	Ile	Thr	Asn	Ser	C
	Ile	Thr	Lys	Arg	A
	Met	Thr	Lys	Arg	G
	Val	Ala	Asp	Gly	U
G	Val	Ala	Asp	Gly	C
	Val	Ala	Glu	Gly	A
	Val	Ala	Glu	Gly	G

[a] First, second, and third refer to the order of the bases in the triplet sequence. Methionine (Met) has a specific codon (AUG), which functions as an initiation signal. Ochre, amber, and opal are termination codons. In some systems the opal codon (UGA) codes for Cys.

incorporation of the amino acid into the growing peptide chain. Each tRNA has the capacity to recognize a specific triplet code by virtue of a **codon recognition site** and a site that can recognize the specific tRNA synthetase (ligase) that adds the amino acid to the tRNA (**ligase recognition site**). Additional specificities are requried: a specific **amino acid attachment site** and a **ribosome recognition site**. The amino acid attachment site is the 3′-OH of a terminal adenine ribose. All tRNA molecules have a 3′-terminus consisting of a cytosine–cytosine–adenine (CCA) nucleotide sequence and a 5′-terminal guanosine nucleotide, as shown in Figure 2-21.

Transfer RNAs consist of approximately 80 nucleotides. A characteristic feature is the high content of unusual nucleotides (inosine, pseudouridine, and various methylated bases) that occupy specific positions. Figure 2-22 illustrates a number of these unusual bases. These unique bases are formed by modification of the structures by specific **modification enzymes** after the synthesis of the tRNA molecule has been completed (posttranscriptional modification). The secondary structure of tRNA appears as a cloverleaf with three loops containing the unusual bases. Loops result because some bases do not form double-stranded base paired regions. Loop I contains dihydroxyuridine (DHU) and is termed the DHU loop. Loop II contains the **anticodon**, a triplet of bases that recognizes the code in mRNA. Loop III varies in size and, in those tRNAs in which it is large, is double-stranded. Loop IV contains (ribo)thymidylate, pseudouridine, and cytosine (Fig. 2-22), a constant feature of all tRNAs that have been examined. The activity of tRNA is dependent on the secondary and tertiary structure. The first tRNA in which the complex tertiary structure has been elucidated is yeast phenylalanyl-tRNA (Fig. 2-23). The 3′-terminal CCA was thought to be added after the basic tRNA molecule had been synthesized, involving a nucleotidyl transferase. It is now clear that the terminal CCA is present

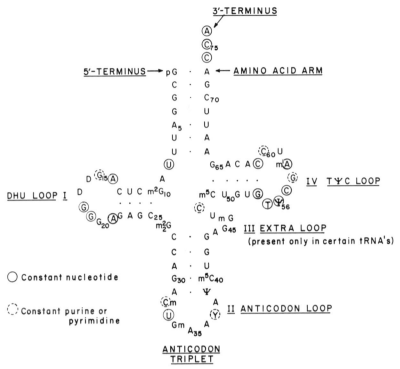

Fig. 2-21. Nucleotide sequence of yeast phenylalanyl tRNA. The structure is shown in the extended cloverleaf configuration. Loop I is termed the DHU loop since it contains dihydroxy-uridine. The anticodon loop (II) contains the triplet bases, which recognize the triplet code words in mRNA. Loop III varies in size and may even be absent from certain tRNAs. Loop IV is termed the TΨC loop since it contains (ribo)thymidylate, pseudouridine (ψ), and cytosine. Circled bases are constant in all tRNAs, and dashed circles indicate positions that are occupied constantly by either purines or pyrimidines. Eukaryotic initiator tRNAs do not have the same constant nucleotides, however. From Kim et al. 1974. *Science* **185**:435.

upon transcription but requires processing (see the section on Processing in this chapter). While it was generally considered that the DHU loop serves as the recognition site for tRNA ligase and loop III as the ribosome recognition site, it is now clear that the unusual tertiary structure of many sites in tRNA is involved in recognition. In the anticodon loop (loop II) each tRNA has a specific anticodon that directs the attachment of the proper amino acid as specified by the codon in the mRNA. Recognition is dependent on base pairing as in DNA. For example.

tRNA anticodon:	AAA	AUG	UUU	GGG	UGA
mRNA codon:	UUU	UAC	AAA	CCC	ACU
Amino acid added:	phe	tyr	lys	pro	thr

Since a number of amino acids have more than one codon, one question that must be addressed is how many tRNAs are required. There are, in a number of instances, isoaccepting species of tRNA. That is, different tRNAs that are charged by the same amino

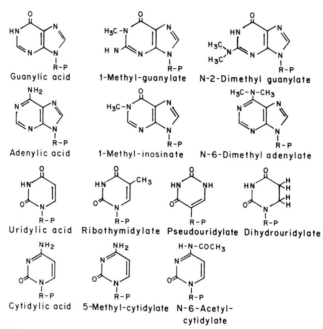

Fig. 2-22. Examples of some of the modified bases occurring in tRNAs.

acid. This accommodates, in part, the degeneracy of the code. However, when there is more than one codon for a single amino acid it is usually only the last nucleotide of the triplet that differs. The tRNA anticodon often contains one of the unusual nucleotides noted in Figure 2-22. These odd bases have what you might call "sloppy" base pairing properties that permit them to bind with several different bases. This capability of recognizing more than one base was termed the "wobble" in codon recognition.

Charging of tRNA

The transfer of an amino acid to tRNA occurs via a two-step process. The amino acid must first be activated by ATP to form an aminoacyl–AMP complex with tRNA synthetase. The aminoacyl–AMP complex is then transferred to tRNA:

$$\text{R--CH--COOH} + \text{ATP} + \text{tRNA synthetase} \longrightarrow \text{R--CH--}\overset{\overset{\text{O}}{\|}}{\text{C}}\text{--O--AMP--synthetase} + \text{PP}_i \quad (1)$$

Amino acid (with NH$_2$)

Aminoacyl–AMP–Synthetase complex (with NH$_2$)

aminoacyl–AMP–synthetase complex + tRNA $\longrightarrow$

Aminoacyl–tRNA + AMP + tRNA synthetase   (2)

Each amino acid has a specific aminoacyl–tRNA synthetase (charging enzyme) and a specific tRNA. Although there may be more than one species of tRNA for a specific

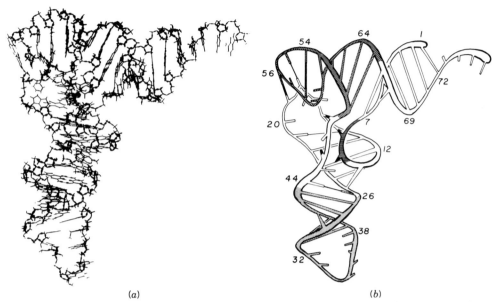

(a)   (b)

Fig. 2-23. Molecular and schematic models of yeast phenylalanyl tRNA. (*a*) Photograph of a molecular model built with Kendrew wire models at a scale of 2 cm/Å. The molecule is oriented with the anticodon loop at the bottom, the acceptor stem at the upper right, and the TψC loop at the upper left. The view is approximately perpendicular to the molecular plane. (*b*) Schematic model drawn from the coordinates of the molecular model shown in *a*. The ribose phosphate backbone is drawn as a continuous cylinder with bars to indicate the hydrogen-bonded base pairs. The positions of single bases are indicated by rods that are intentionally shortened. The TψC arm is heavily stippled, and the anticodon arm is marked by vertical lines. The black segments of the backbone include residues 8 and 9 as well as 26. Tertiary structure interactions are illustrated by black rods. The numbers indicate various nucleotides in the sequence, as shown in Figure 2-21. From Kim et al. 1974. *Science* **185**:435.

amino acid, there is only one charging enzyme for each amino acid. The synthetase must have at least two recognition properties. It must be able to differentiate one amino acid from another and it must also be able to recognize a specific tRNA. These recognition properties are necessary to assure that the specific amino acid is charged on the proper tRNA molecule. Recognition of the codon on mRNA by the specific anticodon on the tRNA is required for the incorporation of an amino acid into the proper position in the polypeptide sequence.

Ribosome Structure

The main component of the protein synthesizing system (PSS) is the ribosome. It is composed of two subunits, 30S (900,000 MW) and 50S (1,600,000 MW), which, when combined, form the 70S ribosome. There are 52 different proteins (r-proteins) and three distinct RNAs (rRNA) that make up an intact ribosome. The small subunit (30S) contains 21 r-proteins (designated S1–S21 for small subunit proteins) and one 16S rRNA. The large subunit (50S) contains 31 r-proteins (designated L1–L34 for large subunit proteins) and one copy each of two rRNA species, 23S rRNA and 5S rRNA. Figure 2-24A presents

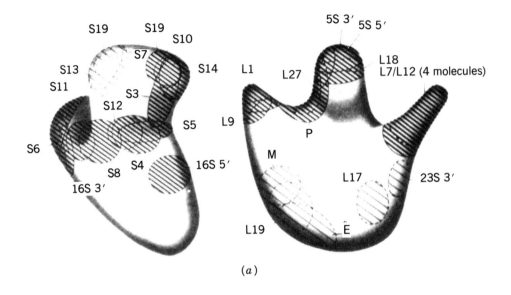

(a)

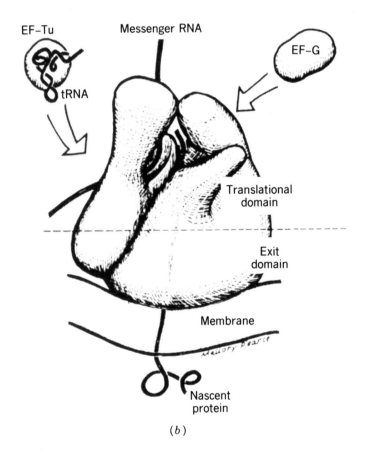

(b)

the three-dimensional structure of the ribosome and the positions of key ribosomal proteins. Figure 2-24B illustrates the association of the two subunits and their relationships with some translational factors and mRNA. Figure 2-25 illustrates in diagrammatic form how the ribosome is constructed using the processed rRNA as scaffolding. Protein S8, encoded by *rpsH*, is a small globular polypeptide component of the *E. coli* ribosome. It plays a critical role in both the assembly of the 30S ribosomal subunit and in the translational regulation of ribosomal proteins encoded by the *spc* operon (see Chapter 4). During 30S assembly, S8 binds to the central domain of the 16S rRNA and interacts cooperatively with several other small-subunit proteins to form a well-defined ribonucleoprotein neighborhood. A similar strategy is employed for 50S subunit assembly.

The r-protein genes of *E. coli* are included among 19 different operons (Table 2-7; Fig. 2-26; see Chapter 4, Translational Control). A considerable amount of redundancy occurs with respect to rRNA genes. There are seven rRNA operons each arranged as

5' leader–16S rRNA–spacer–23S rRNA–5S rRNA–trailer-3'

(see above, RNA Processing). The redundancy may be a protective strategy designed to reduce the potential danger associated with mutations in a single rRNA gene. Note that the various rRNA operons cluster near the origin of replication and that transcription of a given operon occurs in the same direction as replication. This is thought to minimize collisions between the two polymerases.

Although the general structure of ribosomes from different organisms appear similar, detailed studies have shown that ribosomes from eubacterial, archaebacterial, and eukaryotic organisms each have different three-dimensional structures. A fourth type of ribosomal structure has been reported for the sulfur-dependent archebacteria. On the basis of ribosomal topography, it has been proposed that the sulfur-dependent bacteria constitute a group of equal importance to the eubacteria, archaebacteria, and eukaryotes.

Comparative analysis of the nucleotide sequences of rRNA has provided a means of direct measurement of genealogical relationships. Ribosomal RNAs, particularly 16S rRNA, serve as phylogenetic markers for prokaryotic organisms. Detailed technical and statistical considerations revealed that an oligonucleotide of six bases is unlikely to recur more than once in 16S rRNA. When the 16S rRNAs from different organisms inlude the same six bases, this reflects true homology. By comparing the occurrence of these six-base sequences in various organisms, it is possible to compile a "dictionary" of these six-base sequences characteristic of a given organism. The data can then be analyzed to compare the degree of relatedness among organisms. This type of analysis from a large

---

Fig. 2-24. (*a*) **Summary map of the protein sites on the ribosome.** Lightly shaded areas are located on the far side of the subunits. **P, M,** and **E** are peptidyl transferase, membrane binding, and nascent protein exit sites, respectively; **S** and **L** are small (30S) and large (50S) subunit proteins, respectively; 23S 3', 16S 5', and so on are 3' and 5' ends of the rRNAs. (*b*) **Diagrammatic representation of the ribosome exit and translational domains.** The ribosome shown is bound to the cytoplasmic membrane and is engaged in coupled translation secretion (see also Fig. 2-31). From Oakes, M. I., A. Scheinman, T. Atha, G. Shankweiler, and J. A. Lake. 1990. Ribosome structure: Three-dimensional locations of rRNA and proteins. In *The Ribosome.* W. Hill, A. Dahlberg, R. Garrett, P. Moore, D. Schlessinger, and J. Warner (Eds.). American Society of Microbiology, Washington, DC, pp 180.

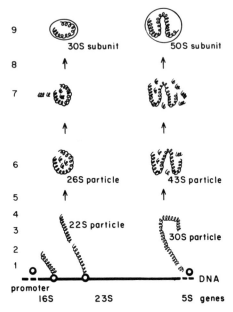

Fig. 2-25. Model for maturation of ribosomal RNAs in *E. coli*. The events occur in the follow-ing general sequence: 1. 16S, 23S, 5S genes are transcribed sequentially. 2. Proteins associate with 16S and 23S precursors before release from DNA. 3. Transcript is cleaved into separate RNAs before transcription is complete. 4. The 16S precursor is 10–20% methylated; 23S precursor is 60% methylated. 5. First precursor particles sediment at 22S and 30S and are probably little folded. 6. Addition of further proteins gives intermediate precursor particles, sedimenting at 26S and 43S; these are folded into more compact structures. 7. Additional sequences in 16S precursor RNA are cleaved; similar breakage may happen to 23S RNA. 8. Remaining methylation takes place. 9. Final set of proteins added to give mature subunits. From Tischendorf, M. 1980. In *Ribosomes: Structure, Function and Genetics*. G. Chambliss, G. R. Craven, J. Davies, L. Kahan, and N. Nomura (Eds.). University Park Press, Baltimore.

number of organisms has shown that certain classical criteria (e.g., Gram stain) used in bacterial classification distinguish valid phylogenetic relationships, whereas other criteria (e.g., morphology) do not. Interestingly, this type of analysis has added some weight to the theory that mitochondria from eukaryotic organisms evolved from prokaryotes. The nucleotide sequence of 16S rRNA from mitochondria relates more closely to bacteria than to the cytoplasmic ribosomes of eukaryotes.

Initiation of Polypeptide Synthesis

The process of initiation and continuation of the elongation of the polypeptide chain involves several specific and somewhat complex mechanisms (Figs. 2-27 and 28). The essential components for peptide chain elongation are ribosomes, several proteins that function in initiation, elongation, and termination, Mg^{2+} and K^+ or NH_4^+ ions, guanosine triphosphate (GTP), and sulfhydryl groups.

The first step in initiation is the formation of a complex between the 30S ribosomal subunit and the three initiation factors (IF-1, 9,000 MW; IF-2, 115,000 MW; and IF-3,

TABLE 2-7. Operons Encoding Ribosomal Proteins of E. coli

Operon Gene	Map Position	Protein Components (S, small; L, large)
α	62	S13, S11, S4 (*rpsD*), α (α subunit of RNA polymerase, L17
spc	72	L14, L24, L5, S14, S8 (*rpsH*), L6, L18, S5, L30, L15, Y, X
S10	78	S10, L3, L4, L23, L2, S19, L22, S3, L16, L29, S17
str	85	S12, S7, G, Tu
	56	S16, *trmD*, L19
	66	dnaG, S21
	70	S15, L21, L27
	72	S9, L13
	82	L28, L33
	83	L34
	87	L31
	88	L11, L1, L10, L7/L12, ββ′
	94	S6, S18, L9
	1	S20
	4	Ts, S2
	21	S1
	31	L32
	38	L20
	47	L25

22,000 MW). In the absence of mRNA, IF-1 and IF-3 function in preventing the association of the 30S and 50S ribosomal subunits. The next events involve the association of the mRNA and initiator tRNA to the 30S subunit (Fig. 2-27.2). The actual order of these steps is not known and may, in fact, vary. IF-3 has been shown to bind to both the 30S subunit and to mRNA. The 30S-[IF-3]-[IF-2]-[IF-1] complex binds to mRNA at the site that includes the initiation codon (in order of preference, AUG, GUG, UUG, CUG, AUA, or AUU). Every mRNA molecule also includes within the untranslated region a ribosome-binding site for each polypeptide in a polycistronic message. The consensus sequence (**5′**)**AGGAGGU**(**3′**), called the **Shine–Dalgarno** (S–D) sequence, is important in the binding of mRNA to the 30S complex. The S–D sequence has been shown to base pair to a region located at the 3′ end of 16S rRNA. This pairing will position the initiating AUG codon so that it may bind to an initiator tRNA anticodon and presumably allows unambiguous distinction between initiation versus internal met codons.

Initiation factor 2 (F-2), complexed with GTP, directs the binding of the initiator tRNA, N-formylmethionyl–tRNA (fMet–tRNA) to the 30S subunit (Fig. 2-27.3). This then permits the association of the 30S and 50S ribosomal subunits (Fig. 2-27.4). Removal of IF-3, which occurs at the same time, is essential, since its presence prevents the association of the two ribosome subunits. The initiation complex, at this point, consists of an association of mRNA with the 30S ribosomal subunit, IF-1, IF-2-GTP, and fMet–tRNA.

Union of the 30S initiation complex with the 50S ribosomal subunit causes the immediate hydrolysis of the bound GTP to GDP. The union itself does not require this hydrolysis since it can be accomplished in the presence of an analog of GTP, 5′-guanylyl (methylene diphosphonate), GDPCP. The hydrolysis of GTP and the subsequent release

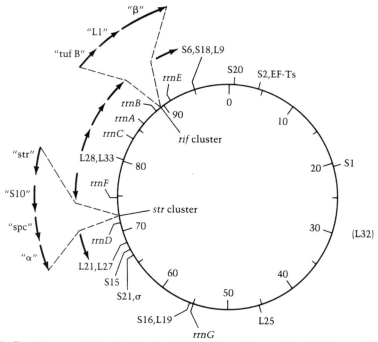

Fig. 2-26. Genetic map of *E. coli* showing the location of genes for rRNA, r-proteins, RNA polymerase subunits, and peptide elongation factors. Where known, the direction of transcription is indicated by arrows. Genes coding for specific proteins are indicated by their product names. Transcription units within the *str* and *rif* clusters are each identified by the operon names. For example, the "B" operon includes genes for L10, L7/L12, B, and B' (see also Fig. 4-17). The rRNA operons are indicated by *rrn*. From Ingraham et al. 1983. *Growth of the Bacterial Cell.* Sinauer Associates, Sunderland, MA.

of GDP is apparently necessary for the IF-1-dependent release of IF-2 from the ribosome (Fig. 2-27.5).

The ribosome has two tRNA binding sites (a third will be discussed later): the aminoacyl–tRNA binding site (A) and the peptidyl (P) binding site. The A site accepts all incoming charged tRNAs while the P site contains the previous tRNA with the nascent polypeptide (peptidyl–tRNA) attached. The initiation tRNA (fMet–tRNA) is unique since it binds directly to the P site without encountering the A site.

Elongation

The addition of amino acids to the growing polypeptide chain occurs in the order specified by the code in the mRNA (~16 residues per second at 37°). At the end of the initiation sequence the 70S ribosome contains the fMet–tRNA in the P site while the A site is free to accept the next aminoacyl–tRNA directed by the triplet base code in the mRNA. It appears that the 5S rRNA molecule recognizes sequences in the T ψ C loop of tRNA thereby aiding in tRNA binding to the A site. The codon directs the binding of a specific aminoacyl–tRNA. This activity is stimulated by the addition of elongation factor T (EF-T) and GTP. The EF–T factor has been shown to consist of two subunits

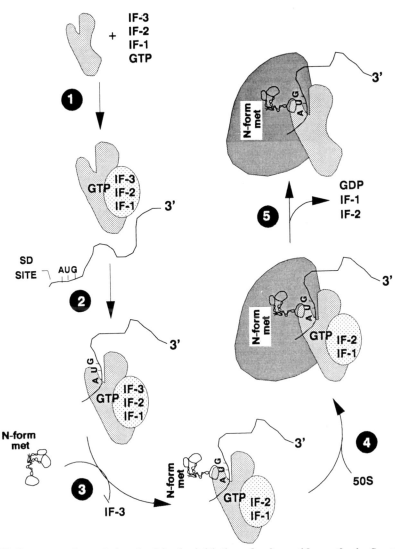

Fig. 2-27. Sequence of events involved in the initiation of polypeptide synthesis. See text for details and abbreviations.

by separation on DEAE-cellulose columns. One subunit (EF-T$_U$, 44,000 MW) is unstable while the other (EF-T$_S$, 30,000 MW) is stable. The EF-T$_U$ subunit is the most abundant protein in *E. coli* comprising 5–10% of the total cellular protein. The GTP binds to EF-T and causes its dissociation into EF-T$_U$–GTP and EF-T$_S$. Now EF-T$_U$–GTP can complex with all aminoacyl–tRNAs except for the initiator tRNA. The resulting complex (GTP–EF-T$_U$–aminoacyl–tRNA) is an intermediate in aminoacyl–tRNA binding to the ribosome. The EF-T$_S$ complex does not appear to play a role in this step (See Fig. 2-28.1).

Upon binding of the aminoacyl–tRNA to the A site, GTP is hydrolyzed and a complex of EF-T$_U$–GDP is released from the ribosome. One GTP is hydrolyzed per aminoacyl–tRNA bound, as shown in Figure 2-28.2. This GTP hydrolysis is not essential for the

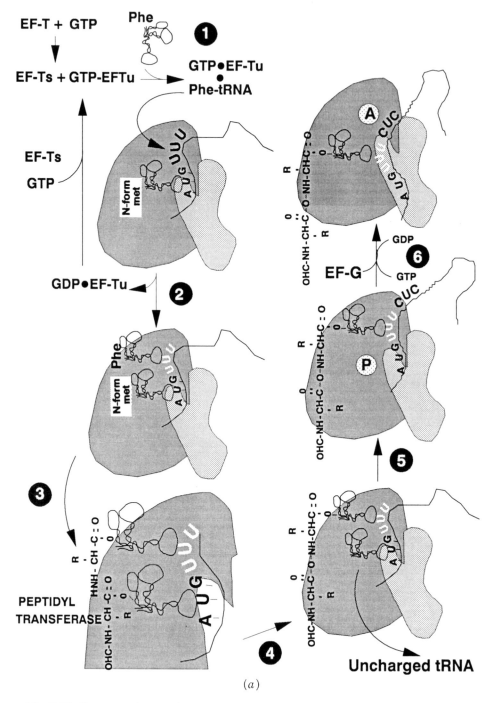

Fig. 2-28. Sequence of events involved in the continuation of peptide chain formation.

aminoacyl–tRNA binding since GDPCP can substitute. The aminoacyl–tRNA can bind to the A site but EF–T$_U$ is not released from the ribosome. Therefore, the result, if not the function, of GTP hydrolysis is the release of EF–T$_U$ from the ribosome.

Subsequently, peptidyl transfer (i.e., peptide-bond formation catalyzed by peptidyl-transferase) occurs (Fig. 2-28.3). Peptide bond formation is not dependent on hydrolysis

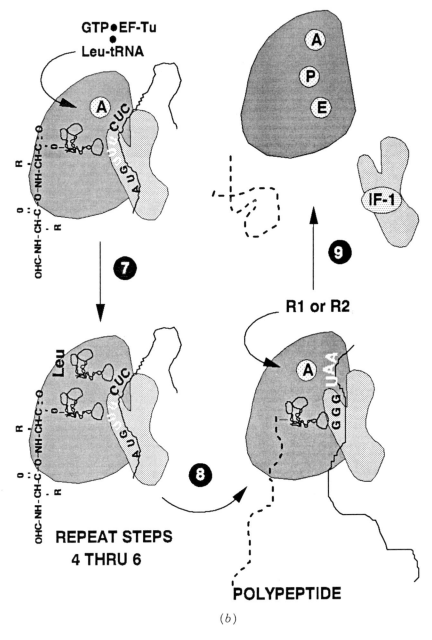

(*b*)

Fig. 2-28. *Continued*

TABLE 2-8. Genetic Nomenclature for Components of the Translation Apparatus[a]

Gene Symbol	Gene Product
Ribosome Components	
*rpl*A to *rpl*Y	50S ribosomal subunit protein L1–L25
*rpm*A to *rpm*G	50S ribosomal subunit protein L27–L33
*rps*A to *rps*U	30S ribosomal subunit protein S1–S21
*rrn*A to *rrn*G	rRNA polycistronic operon
rrf	5S rRNA (encoded by *rrn*)
rrl	23S rRNA (encoded by *rrn*)
rrs	16S rRNA (encoded by *rrn*)
Accessory Factors	
cca	tRNA nucleotidyl transferase
*fus*A	translation factor EF-G
*prm*A, *prm*B	methylation of ribosomal proteins
*rim*A to *rim*J	maturation of ribosome
rna to *rnp*	ribonuclease (I to P)
*trm*A to *trm*D	tRNA methyltransferase
*tuf*A, *tuf*B	translation factor EF-Tu
Transfer RNAs and Their Charging[b]	
*ala*T, *ala*U	alanine tRNA$_{1B}$
*val*T	valine tRNA$_1$
*ala*S	alanyl-tRNA synthetase
*val*S	valyl-tRNA synthetase
etc.	

[a]From Bachman, 1990. *Microbiol Rev.* **54:** 130–197.
[b]Generally, capitals of T and further in the alphabet indicate an *RNA* gene, whereas S and earlier letters are used for synthetases.

of GTP, but it is dependent on the release of $EF-T_U$ from the ribosome. The $EF-T_S$ complex appears to participate in the recycling of $EF-T_U-GDP$ to $EF-T_U-GTP$. However, $EF-T_S$ is not required for release of $EF-T_U$ from the ribosome as is the case with IF-1-dependent release of IF-2.

Peptide-Bond Formation

Once the incoming aminoacyl–tRNA binds to the A site, peptide-bond formation occurs (see Fig. 2-28.3 and 2-28.4). The peptide bond is formed between the amino group of the incoming amino acid and the C-terminal of the elongating polypeptide bound to the tRNA at the P site. The enzyme responsible for peptidyl transfer is peptidyl transferase and is located in the 50S ribosomal subunit. No soluble factors appear to be necessary.

Translocation

After the formation of the peptide bond, the growing peptide chain becomes bound to the tRNA that was carrying the incoming amino acid (see Fig. 2-28.4) and occupies the A site. The uncharged tRNA to which the peptide chain was previously bound is now released from the P site (Fig. 2-28.5). (However, see the E site below.)

The incoming aminoacyl–tRNA must enter the specific ribosomal site in which codon–anticodon recognition takes place (**decoding**). This is the aminoacyl tRNA accepting (A) site. A complex change must occur in the ribosome to allow entry of this incoming aminoacyl–tRNA. The ribosome must advance along the mRNA so that the next codon to be translated is aligned with the decoding (A) site. The peptidyl–tRNA occupying the A site must be transferred to the P site (**translocation**) as shown in Figure 2-28.6.

Up to this point the following steps have been accomplished:

1. Removal of the uncharged (discharged) tRNA from the P site.
2. Movement of the peptidyl tRNA from the A to the P site.
3. Movement of the message by one codon.

Elongation factor G (EF-G, 80,000 MW), in which G = GTPase, and GTP hydrolysis are required for these processes.

Evidence indicates that EF-G binds to the same site as the EF-T_U, hence the requirement for T_U recycling. Once bound, EF-G catalyzes the hydrolysis of another GTP. Therefore, during elongation, two molecules of GTP are hydrolyzed per peptide bond. One hydrolytic step is EF-T dependent and the other is EF-G dependent. After each step of the elongation process EF-G is released from the ribosome. Unless EF-G is released, elongation cannot continue since both EF-G and EF-T utilize the same binding site. It is interesting to note that the three extraribosomal translation factors possessing GTPase activity when bound to ribosomes all interact with L7/L12 (see Fig. 2-24a).

While ribosomes have classically been described as possessing two aminoacyl tRNA binding sites, A and P, recent evidence points to the existence of three binding sites; A, P, and E. The E site is specific for deacylated tRNA and is apparently codon–anticodon specific. The present model is as follows: incoming charged tRNA binds at the A site, the growing polypeptide attached to tRNA bound to the P site is transferred to the tRNA in the A site, the newly deacylated tRNA is not released immediately from the ribosome but, upon translocation, becomes bound to the E site (at this point both the E and P sites are occupied), the next incoming charged tRNA binds to the unoccupied A site causing a reduced affinity of the E site for the deacylated tRNA allowing its release from the ribosome.

Termination

When translocation brings one of the termination codons (UAA, UGA, or UAG) into the A site, the ribosome does not bind an aminoacyl–tRNA–EF–T_U–GTP complex. Instead it binds a peptide release factor protein, R_1 = 44,000 MW or R_2 = 47,000 MW (Fig. 2-28.8 and 2-28.9), which activates peptidyl transferase, thereby hydrolyzing the bond joining the polypeptide to the tRNA at the P site. The result is release of the polypeptide (Fig. 2-28.9). The free polypeptide (*in vivo*) undergoes one of two posttranslational processing steps: 1. The formyl group of the N-terminal fMet may be removed by the enzyme methionine deformylase, or 2. The entire formylmethionine residue may be hydrolyzed by a formylmethionine-specific peptidase (methionyl amino peptidase or MAP). There is apparently some discrimination involved in channeling different polypeptides through these two alternative steps, since not all N-terminal methionines are removed. The discriminating factor appears to be the side-chain length of the penultimate amino acid. The longer the side chain the less likely MAP will remove the methionine. Other processing can occur, such as acetylation (acetylation of L12 to give L7) or adenylylation.

When Nonsense Makes Sense.   There are several examples of proteins, such as formate dehydrogenase (FdhF), that contain a modified form of cysteine called selenocysteine. The gene for formate dehydrogenase contains an in frame TGA (UGA in message) codon that, instead of terminating translation as a nonsense codon, encodes the incorporation of selenocysteine. The requirements for this process include a specific tRNA^{sec} and a unique translation factor, encoded by *selB*, with a function analogous to elongation factor Tu (see **Translocation** above). SelB is a GTPase and binds specifically to selenocysteyl–tRNA^{sec} and not its precursor seryl–tRNA^{sec}. Thus SelB directs selenocysteine to the ribosome. A challenging question is how the in-frame TGA codon is differentiated from TGA codons signaling termination. Apparently there is a specific sequence in *fdhF* present downstream of the TGA codon that is required to recognize this codon as sense. The region may form a loop that could shield the UGA from release factor 2.

Ribosome Editing

During normal translation the number of inappropriate aminoacylated tRNAs that can enter the A site will, of course, outnumber the appropriate aminoacylated tRNAs. Even though the mean lifetime of bound inappropriate aa-tRNA is much shorter than that of appropriate aa-tRNA, the inappropriate aa-tRNA may remain bound to the A site long enough to accept a peptide from the P site. There are two subsequent processes that can reduce errors in surviving completed proteins: ribosome editing and preferential degradation of completed proteins containing erroneous amino acids (see Protein Degradation below).

The ribosome editing hypothesis proposes that an inappropriate peptidyl–tRNA (i.e., one whose structure does not correctly complement the mRNA) dissociates preferentially during protein synthesis. It apears that the alarmone guanosine tetraphosphate (ppGpp), produced by the *relA* gene product, affects the editing process (*relA* is discussed in more detail in Chapter 4 in relation to stringent control). Studies have shown *relA*$^+$ cells to be more efficient at ribosome editing than *relA* cells. One hypothesis devised to account for this is that ppGpp interacts with EFG and leads to a longer lifetime of peptidyl–tRNA (p–tRNA) in the A site and, therefore, enhanced editing. Other mechanisms are also possible.

Once released, the p–tRNA is acted upon by the product of the *pth* gene (peptidyl–tRNA hydrolase). This enzyme catalyzes the hydrolysis of ribosome-free p–tRNA to yield an intact peptide and an intact tRNA. Any defective peptides can be degraded as indicated below. Since p–tRNA hydrolase activity appears to be ubiquitous, it seems likely that dissociation of inappropriate p–tRNA from cellular ribosomes is a universal characteristic of protein synthesis.

Coupled Transcription and Translation

Prokaryotes differ significantly from eukaryotes in that prokaryotic ribosomes can engage mRNA before it has been completely transcribed (see Figs. 1-1 and 1-8). This coupled transcription–translation affords bacteria novel mechanisms for regulation (see Chapter 4, Attenuation). Because the transcription process is faster than translation, the cell must protect itself from situations in which a long region of unprotected mRNA might stretch between the ribosome and RNA polymerase. To avoid that situation most mRNAs contain RNA polymerase pause sites that allow the ribosome to keep pace with polymerase thereby protecting the message from endonucleolytic attack.

Protein Folding and Chaperones

A continuing challenge for biochemists is to unravel the processes required to convert the one-dimensional information encoded by the genetic material into three-dimensional structures that impart biological properties to proteins. Historically, protein biogenesis was once thought to involve only spontaneous folding of polypeptide domains. We realize now that the process is more complex, more intriguing than previously envisioned. Many proteins require assistance to fold properly. This assistance comes from proteins that are not final components of the assembled product. The term given to these ''foldases'' are chaperones (or chaperonins). The proposed function of chaperone proteins is to assist polypeptides to self-assemble by inhibiting alternative assembling pathways that produce nonfunctional structures. During protein synthesis, for example, the amino terminal region of each polypeptide is made before the carboxy terminal region. The chance of incorrect folding of a nascent polypeptide may be reduced through interaction with chaperones. Another process in which chaperones can be invaluable is protein secretion or translocation. Proteins that traverse membranes do so in an unfolded or partially folded state. Often they are synthesized by cytosolic ribosomes and must be prevented from folding into a translocation incompetent state.

Cells also take advantage of chaperonins when faced with an environmental stress that denatures proteins and causes the formation of aggregates. To protect against such stresses, cells accumulate proteins that prevent the production of such aggregates or unscramble the aggregates so that they can correctly reassemble. (see heat shock response)

Examples of chaperones are GroEL(60 kDa), GroES(10 kDa), and DnaK(70 kda) of *E. coli*. They are all abundant constitutive proteins that increase in amount after stresses, such as heat shock. In *E. coli*, DnaK appears to facilitate protein export but also appears capable of renaturing some proteins turned into insoluble protein aggregates (e.g., RNA polymerase). The renaturing function appears dependent on an ATP kinase activity of DnaK. The GroEL chaperone is an oligomer of up to 14 subunits that may form a pore through which improperly folded proteins may be threaded and refolded. This chaperone is essential for cellular viability at any termperature due to its role in facilitating proper protein folding. Future research on molecular chaperones will focus on how chaperones work and the structural basis by which a given chaperone recognizes and binds to some features present in a wide variety of unrelated proteins but which is accessible in the early stages of assembly.

PROTEIN TRAFFICKING

A fundamental question in biology is, ''How do proteins synthesized in the cytoplasm make their way to the inner membrane, outer membrane, periplasmic space, or the external milieux?'' The process, referred to as protein export, involves the insertion of proteins into membranes and the passage of hydrophilic proteins through hydrophobic membrane barriers.

Most of these proteins are first synthesized in a precursor form containing 15–30 amino acids at the N-terminus called a **signal sequence**. Characteristically, the signal sequence includes a hydrophobic region of at least 11 amino acid residues preceded by a short stretch of hydrophilic residues at the N-terminus. Often, the signal sequence is removed during the export process. The role of the signal sequence is either to mediate binding of nascent polypeptides to the membrane or confer a conformation on the precursor that renders it soluble in the membrane.

The main route for export and secretion of signal sequence-bearing proteins in gram-negative organisms is called the general secretory pathway (GSP). The first step involves the translocation of the exported protein across the cytoplasmic membrane. This depends on the *sec* gene products, as outlined in Figure 2-29 and Table 2-9. The protein **SecB** is a ''pilot'' chaperonin that will associate with proteins destined for export (e.g., pro-OmpA) as they are synthesized from the ribosome. Without SecB, proOmpA will aggre-

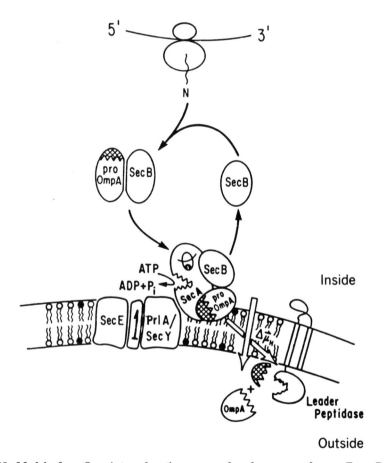

Fig. 2-29. Model of proOmpA translocation across the plasma membrane. From Brondage, L., et al. 1990. *Cell* **62**:649–657.

TABLE 2-9. Components of the *E. coli* Cytoplasmic Membrane Secretory Protein Translocation System and Their Presence in Other Bacteria

Group and Characteristics	Name	Size (kDa)	Location	Presence in Other Bacteria
Secretory chaperonins (pilot protein[a])	SecB	18	Cytoplasm	Widespread in members of *Enterobacteriaceae*
General chaperonins	DnaK	69	Cytoplasm	Universal
	GroEL	62	Cytoplasm	Universal
	GroES	11	Cytoplasm	Universal
Secretory ATPase (pilot protein)	SecA	102	Cytoplasm, ribosome, peripheral cytoplasmic membrane	*Bacilus subtilis*, widespread in members of *Enterobacteriaceae*
Translocase	SecD	67	Cytoplasmic membrane	NI[b]
	SecE	14	Cytoplasmic membrane	*B. subtilis*
	SecF	39	Cytoplasmic membrane	NI
	SecY	48	Cytoplasmic membrane	*B. subtilis*
	Band 1[d]	15	Cytoplasmic membrane	NI
	Ydr[a]	19	Cytoplasmic membrane	NI
Signal peptidases	LepB[e]	36	Cytoplasmic membrane	*S. typhimurium, Pseudomonas fluorescens, B. subtilis* (probably widespread)
	LspA[f]	18	Cytoplasmic membrane	*P. flurescens, Enterobacter aerogenes* (probably widespread)
	Ppp[g]	~25	Cytoplasmic membrane	*P. aeruginosa* (PilD/XcpA), *Vibrio cholerae* (TcpJ), *Klebsiella oxytoca* (PulO), *B. subtilis* (ComC)

TABLE 2-9. Continued

Group and Characteristics	Name	Size (kDa)	Location	Presence in Other Bacteria
Others[h]	4.5 SRNA (ffs)		Cytoplasm	*B. subtilis* (6S), *Thermus thermophilus, P. aeruginosa, Halobacterium halobium* (7S), *Mycoplasma pneumoniae*
	Ffh	48	Cytoplasm	NI
	FtsY	?	Cytoplasmic membrane	*Sulfolobus solfataricus*

[a]See text.

[b]NI, not identified

[c]Genes coding for proteins with greater than 20% overall sequence identity to SecY have been cloned and sequenced from a variety of different sources, ranging from a plastid genome to *S. cerevisiae*. Only in *S. cerevisiae* and *B. subtilis* has any direct participation in protein traffic been demonstrated.

[d]Participation in translocation has not been fully established.

[e]Also called leader peptidase or signal peptidase I.

[f]Also called proliproprotein signal peptidase, lipoprotein signal peptidase, or signal peptidase II.

[g]Prepilin; the gene-based designation vaires according to the organism from which it is derived and its function.

[h]The role of these macromolecules in protein translocation has not been firmly established. They are included here because their homologs in eukaryotes are involved in secretory protein translocation. Adapted from Pugsley, A. P. 1993. *Microbiol. Rev.* **57**:50–108.

gate thereby preventing its membrane insertion. The **SecA** protein is a peripheral membrane ATPase and forms part of a preprotein translocase in collaboration with the integral membrane proteins **SecY/E**. The SecA hydrolysis of ATP is coupled to SecA interactions with other components of the translocation apparatus. The SecA protein serves as the receptor for the proOmpA–SecB complex. Subsequent ATP hydrolysis releases proOmpA into the membrane and drives the overall chaperone and membrane-associated reactions. Once the process has been initiated at the expense of ATP, further translocation by SecY/E appears to proceed through a series of transmembrane intermediates, each with distinct energy requirements. The steps subsequent to SecA are driven by proton motive force rather than ATP hydrolysis (see Chapters 1 and 6). During or immediately following translocation, a signal peptidase (**LepB**) will cleave the signal sequence of the exported protein if it is destined for the periplasm, outer membrane, or external environment. Integral cytoplasmic membrane proteins are usually not cleaved by LepB. The various branches of the GSP that are employed to direct different proteins to their ultimate extracytoplasmic destinations are outlined in Figure 2-30.

As noted above, the export process may involve directing ribosomes that are translating exported proteins to the cell membrane thereby establishing a coordinated translation–secretion process. As soon as the signal sequence emerges from the ribosome it interacts with a cytoplasmic signal recognition particle (SRP). Binding of this particle blocks further elongation until the complex interacts with a membrane-associated docking protein. Translation inhibition is then relieved and synthesis of the protein is completed on

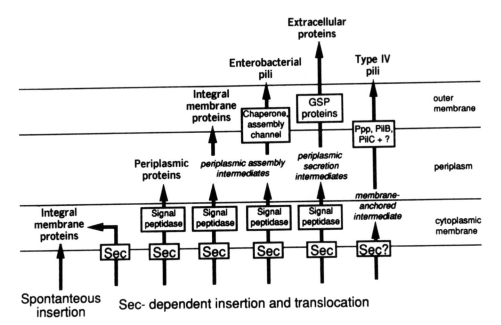

Fig. 2-30. Main branches of the general secretory pathways of gram-negative bacteria. The final destinations for various proteins are shown in relief characters. Intermediate stages are in italics and proteins involved in export and sorting are boxed. Sec proteins are required for the early stages of the GSP. A specific chaperone and a translocator/assembly outer-membrane protein are required for the assembly of most common types of pili found in the *Enterobacteriaceae*. From Pugsley, A. P. 1993. *Microbiol. Rev.* **57**:550–108.

membrane-bound ribosomes. A change in conformation in this precursor protein or interaction with the SRP results in its association with the cytoplasmic side of the membrane. The signal recognition particle is a multifunctional protein–RNA complex including a 48-kDa protein (*ffh* product) and a 4.5S RNA (*ffs* product). The *ftsY* product shows substantial sequence similarity to the α subunit of the SRP receptor in eukaryotic endoplasmic recticulum and may perform a similar function in *E. coli*. This system may provide an alternate route from SecBA for directing presecretory proteins to the translocase.

Figure 2-31 illustrates the various ways presecretion proteins can engage the GSP. They include cotranslational translocation where the ribosome is anchored to the membrane. This translocation may be triggered by spontaneous insertion of the signal sequence into the membrane or by directed insertion via SRP. Chaperone-dependent translation-linked translocation and chaperone-dependent posttranslational translocation are two alternate routes used to engage the GSP.

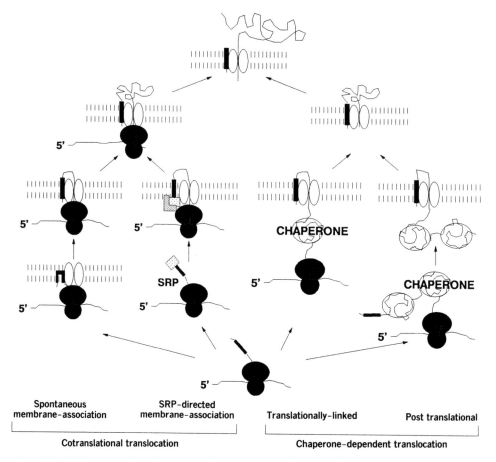

Fig. 2-31. Mechanisms for the presentation of presecretion proteins to the general secretory system. The filled oval figures represent translating ribosomes: the filled rectangle represents a signal sequence present at the N-terminal end of a presecretion protein; SRP = signal recognition particle.

Degradation of Abnormal Proteins

The level of a particular protein present in a cell is determined not only by its rate of synthesis but also by its rate of degradation. During starvation for a carbon or nitrogen source, the overall rate of protein degradation increases severalfold. This degradation of preexisting cell proteins can provide a source of amino acids for continued protein syntheis or for a source of energy. The enhanced proteolysis occurs coordinately with the cessation of stable RNA synthesis and the accumulation of the signal molecule guanosine tetraphosphate (see Stringent Control, Chapter 4).

Interest in bacterial protein degradation has increased due to the observation that abnormal proteins produced as the result of mutations do not accumulate *in vivo* to the levels of their normal counterparts. The breakdown of these proteins occurs even during rapid growth and is independent of nutrient supply. These abnormal proteins include complete proteins resulting from nonsense mutations, complete proteins with amino acid substitutions, and some subunits of large multimeric complexes that are synthesized in excessive amounts. For example, if a ribosomal polypeptide or the σ subunit of RNA polymerase is overproduced in *E. coli*, then it is rapidly degraded.

It appears that a single degradation system is primarily responsible for the breakdown of most of these abnormal proteins. The outline below illustrates a possible pathway for this degradation.

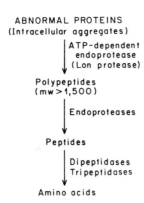

Abnormal proteins, which may exist as intracellular aggregates, are rapidly degraded by the Lon protease. The Lon protease (also called protease LA) is an ATP-dependent proteolytic enzyme with the following properties:

1. Catalyzes the rate-limiting endoproteolytic step in the degradation of most abnormal proteins.
2. Requires Mg^{2+} and ATP for proteolytic activity.
3. It is a serine protease as well as an ATPase.
4. Active form is a tetramer comprised of identical 94,000-Da protein subunits.
5. DNA stimulates proteolysis and protein-activated ATPase.
6. Activity is stimulated by binding to potential substrates.
7. It is a heat shock protein regulated by the *rpoH(htpR)* gene product (see Chapter 4).

In the presence of ATP, lon protease will degrade casein or albumin (and presumably normal intracellular proteins) to acid-soluble peptides usually greater than 1500 Da. *In vivo*, these products are apparently degraded further by the sequential actions of other proteases and peptidases (see Table 2-10). The fact that the Lon protease only recognizes and acts upon unfolded proteins could help explain how intracellular proteases can exist free in the cytosol without destroying essential cell proteins. In addition, ATP hydrolysis results in repeated autoinactivation after each proteolytic event. The ADP remains bound to inactive Lon protease until a new protein substrate induces its release. This phenomenon may also help insure against indiscriminate proteolysis *in vivo*. The Lon protease also plays a role in the SOS response to DNA damage (see Chapter 3) and other regulatory systems (see Chapter 4).

Degradation of Normal Proteins. Another role of proteolytic pathways is to confer short half-lives on proteins whose concentrations must vary with time or alterations in cellular state. For example, many regulatory proteins are unstable due to proteolysis. This contributes to the overall metabolic economy of the cell since a regulatory protein that persists may not allow target genes to react quickly to changing conditions. How is it that some proteins resist degradation while others are degraded rapidly? Certain amino acid sequences, conformational determinants, or chemically modified protein structures confer degradation signals, called **degrons**, on proteins. One example of a degron is the **N-degron**, which is the N-terminal residue of a protein. Manifested as the **N-end rule**, the *in vivo* half-life of a protein was found to be related to the identity of its N-terminal residue. Several gene products of *E. coli* have been associated with enforcing the N-end rule. For example, the *aat* product is required to degrade proteins with N-terminal Arg or Lys. Mutations in *clpA* will stabilize proteins with N-terminal Phe, Leu, Trp, Thr, Arg, or Lys. The *clpA* product is protease Ti, an ATP-dependent serine protease (see Table 2-10). There are many proteases and peptidases in the prokaryotic cell. It is an ongoing challenge to decipher the degradation signals, pathways, and regulatory mechanisms involved in protein turnover.

ANTIBIOTICS THAT AFFECT NUCLEIC ACID AND PROTEIN SYNTHESIS

Much of our knowledge concerning the synthesis of macromolecules was gleaned from studies designed to elucidate the modes of action of several antimicrobial agents. The following sections describe some of these agents and their mechanisms of action.

Agents Affecting DNA Metabolism

Intercalating Agents. Many rigidly planar, polycyclic molecules interact with dsDNA by inserting, or intercalating, between adjacent stacked base pairs of the double helix (see Fig. 2-32). To permit this intercalation, there must be a preliminary unwinding of the double helix providing space between the bases into which the drug can move. Hydrogen-bonding remains undisturbed, however, some distortion of the phosphodiester backbone occurs.

Examples of intercalating agents are proflavine, ethidium, and actinomycin D (Fig. 2-32). Treatment of cells harboring extrachromosomal elements called plasmids (Chapter 8) by proflavine or ethidium bromide will often lead to the disappearance of the

TABLE 2-10. Proteases and Peptidases of *E. coli* and *S. typhimurium*[a]

	Location	Gene (Map Position)	Nature	MWT × Subunit	Inhibitors
PROTEASES					
Do	cytosol		Serine Pr.	52,000 × 10	D
Re	cyt + peri		Serine Pr.	82,000	D,E,O, TPCK
Mi			Serine Pr.	110,000	D,E,O
Fa	cytosol		Serine Pr.	110,000	D,E,O, TPCK
So	cytosol		Serine Pr.	70,000 × 2	D, TPCK
La	cytosol	*lon*(10)	ATP-dependent Serine Pr.	87,000 × 4	D,E, NEM
Ti(Clp)	cytosol	*clpA, clpP*	ATP-dependent Serine Pr.	83,000 × 6 + 21,500 × 12	
I	periplasm		ChTryp-like Pr.	43,000	D
II	cytoplasm		Tryp-like Pr.	58,000	D, TLCK
III (P$_i$)	periplasm	*ptr*(60m)	Metallo-Pr.	110,000	E,O
IV	Inner membrane	*sspA*(38.5m)	Signal Peptide Peptidase	67,000	D
V	Membrane				D
VI	Membrane		Serine Protease	43,000	
OmpT(VII)	Outer membrane	*ompT*(12.5m)	Tryp-like Pr.	36,000	
Ci	cytoplasm		Metyallo Pr.	125,000	E,O
Deg P	Periplasm				
PEPTIDASES					
Dipeptidases					
PepD	cytoplasm	*pepD* (7m)	Broad Specif.	52,000	
PepQ	cytoplasm	*pepQ* (84m)	X-Pro		
PepE	cytoplasm	*pepE* (90m)	Asp-X	35,000	
PepG		*pepG*	Gly-Gly		
Aminotripeptidase					
PepT	cytoplasm		Tripeptides		

TABLE 2-10. Continued

	Location	Gene (Map Position)	Nature	MWT × Subunit	Inhibitors
Signal Peptidases					
I	inner membrane	*lep*(55m)	Precursors of secreted Proteins	36,000	
II	inner membrane	*lsp*(0.5m)	Prolipoproteins	18,000	
Amino Peptidases					
PepN	cytoplasm	*pepN*(20)	Broad	99,000	Amino acids
PepA (PepI)	cytoplasm	*pepA*(97m)	Broad	52,000 × 6	E,Zn^{2+}
PepB	cytoplasm	*pepB*(53m)			
PepP	cytoplasm	*pepP*(63m)	X-Pro-Y	50,000 × 4	E
PepM		*pdpM*(3m)	Met-X-Y	34,000	E
C-terminal Peptidases					
dipeptidylcarboxypeptidase	periplasm	*dep*(28m)	Ac-Ala3	97,000	C
oligopeptidase A	cytoplasm	*optA*(77m)	Ac-Ala4		

[a]Abbreviations: C = Captopril; D = Diisopropyl fluorophosphate; E = EDTA (ethylenediaminetetraacetic acid); O = *o*-phenanthroline; TPCK = *N*-tosyl-phenylalanine chloromethyl ketone; NEM = *N*-ethylmaleimide; ChTryp = chymotrypsin; Tryp = trypsin.

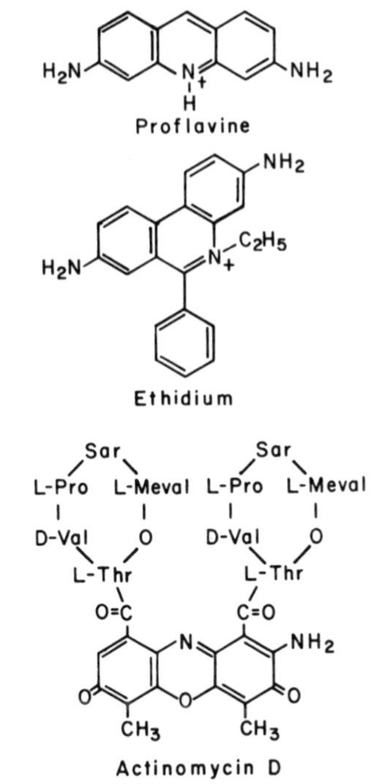

Fig. 2-32. Three agents that intercalate with DNA. Actinomycin D is sometimes called dactinomycin. Thr = threonine; val = valine; pro = proline; sar = sarcosine; meval = N-methylvaline.

plasmids from the cells apparently due to the preferential inhibition of plasmid replication.

Actinomycin D will inhibit DNA replication but inhibits transcription at lower drug concentrations. Certain features of DNA are necessary to interact with actinomycin D: (a) the DNA must contain guanine, (b) the DNA must be a double-stranded helical structure, (c) the sugar moiety must be deoxyribose, dsRNA does not interact with this drug.

Mitomycins. The mitomycins are a group of antitumor and antibacterial agents. Mitomycin C (Fig. 2-33) is one of the more active and more widely studied members of this group. Although mitomycin C produces a number of toxic side effects in mammalian host tissues and is not considered of practical significance as a chemotherapeutic agent against bacterial infections, its action in causing cross-linking between complementary strands of DNA is of considerable interest in molecular biology. The cross-linking action caused by mitomycin C is similar to that produced by nitrous acid, a number of alkylating agents, and irradiation. Several lines of evidence suggest that mitomycin C and related derivatives (e.g., porfiromycin, the ziridine N-methyl analog of mitomycin) form cross-links by functional alkylation of guanine residues in opposing DNA strands. By binding to sites on each of the two complementary DNA strands, replication of a segment of the DNA in the immediate vicinity of the cross-link fails to occur and a loop of DNA is formed. The parent compound is apparently not active in this form but must be converted to the active form by reduction of the quinone structure and elimination of the $-OCH_3$

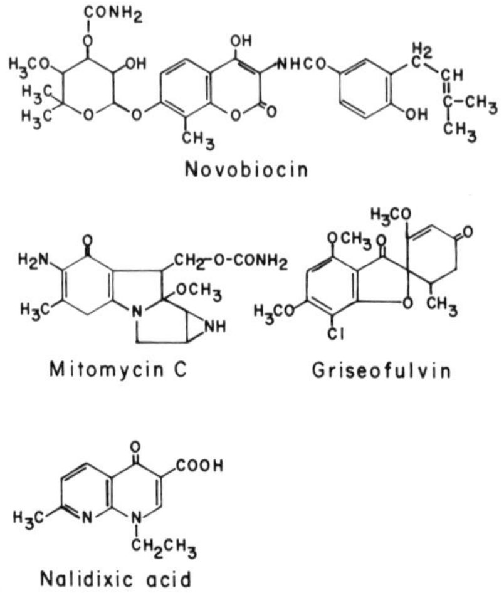

Novobiocin

Mitomycin C   Griseofulvin

Nalidixic acid

Fig. 2-33. Structures of antimicrobial agents affecting the metabolism, structure, or function of DNA.

group. These alterations result in the formation of an aromatic indole ring system with a positive charge that apparently is a major reactive site for linkage with the guanine residues in DNA. Although the exact mode of action of mitomycin C on bacteria may require additional study, prevention of replication of the segment of DNA in which cross-linking has occurred represents a plausible explanation.

Griseofulvin. An antifungal agent with the chemical structure shown in Figure 2-33, griseofulvin (produced by *Penicillium griseofulvum*), exerts its action on DNA replication in fungi. Mitosis is inhibited at metaphase and many hyphal distortions are observed. Griseofulvin usefulness as a chemotherapeutic agent against superficial fungal infections is attributed to its absorption when administered orally and to accumulation in the cornified layers of the skin, hair, and nails as these tissues are formed. The newly synthesized tissues thus acquire resistance to invasion by fungal pathogens, and this prevents spread of the infection into these areas. Because the cornified tissues develop slowly, complete eradication of a mycotic infection with griseofulvin requires a prolonged period of administration.

DNA Gyrase Inhibitors. Nalidixic acid has been widely used in the treatment of urinary tract infections caused by *E. coli, K. pneumoniae, E. aerogenes, Proteus,* and other gram-negative bacilli. Oxolinic acid is a chemically related derivative of nalidixic acid (Fig. 2-33), which has been found to be more potent than nalidixic acid in its antimicrobial action. Both compounds appear to have an identical mode of action, as shown by the fact that mutants isolated as resistant to nalidixic acid (*gyrA*) are also resistant to oxolinic acid. The primary effect of nalidixic acid in intact, living bacteria is an immediate, selective, and reversible cessation of DNA synthesis. Nalidixic acid does not bind to the DNA template, does not cause the production of toxic breakdown products, nor does it inactivate a number of enzymes specifically involved in DNA synthesis. DNA gyrase is the target protein for the action of nalidixic and oxolinic acids. The *gyrA* or ''swivelase'' component of DNA gyrase is sensitive to the action of nalidixic and oxolinic acids. The

DNA gyrase isolated from nalidixic acid-resistant mutants is resistant to the action of nalidixic and oxolinic acids but not to novobiocin or the related compound, coumermycin. Conversely, novobiocin-resistant mutants yield a DNA gyrase that is sensitive to nalidixic acid. Novobiocin binds to the B subunit of DNA gyrase (*gyrB*) and inhibits its ATPase activity.

The quinolone antibiotics (e.g., ciprofloxacin) are structurally related to naladixic acid based upon the presence of a quinolone ring. The structural modifications in the quinolones increase their potency and utility against a broad spectrum of microorganisms, both gram-positive and gram negative. They are also useful clinically because mutants resistant to naladixic acid remain quite sensitive to the quinolones.

Agents Affecting Transcription

Rifamycin. The rifampicin or rifamycin group of agents and the streptovaricins are chemically similar (Fig. 2-34). Rifampin and streptovaricin are semisynthetic compounds, while the rifamycins are naturally occurring antibiotics. All of these agents affect the initiation of transcription by specific inhibition of bacterial RNA polymerase (see Fig. 2-16). The RNA polymerase of eukaryotic cells is not so affected. Since the mechanism of action of all of these agents is similar, it is not surprising to find cross-resistance as a common observation. Rifamycin forms a tight 1:1 complex with the β subunit of RNA polymerase (*rpoB*). Rifamycin resistant variants of *E. coli* contain mutations in *rpoB*. The drug acts to **inhibit initiation of transcription**. While it has little effect upon the formation of the first phosphodiester bond, inhibition of the second bond is almost total. Once transcription is allowed to proceed past the third nucleotide (i.e., to the elongation stage), rifamycin has no effect upon transcription.

Streptolydigin. Streptolydigin is similar to rifamycin in that it inhibits transcription by binding to the β subunit of RNA polymerase. However, this antibiotic will inhibit chain elongation as well as the initiation process *in vitro*. *In vivo*, streptolydigin appears to accelerate the termination of RNA chains. The binding of drug to RNA polymerase may destabilize the transcription complex permitting premature termination.

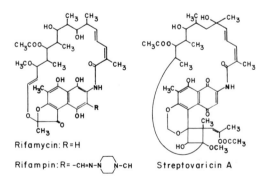

Fig. 2-34. Comparison of the structures of rifamycins and streptovaricins.

Agents Affecting Translation

Chloramphenicol. One of the simplest antibiotic structures (Fig. 2-35), chloramphenicol was originally isolated as a product of *Streptomyces venezuelae*. Chloramphenicol is now produced commercially by organic synthesis. Inhibition of peptidyl transferase activity by chloramphenicol is mediated by binding to the 50S subunit of bacterial ribosomes. Its effect is bacteriostatic and readily reversed with removal of the drug. Experiments with radiolabeled derivatives of chloramphenicol indicate the drug binds to several ribosomal proteins including S6, L3, L6, L14, L16, L25, L26, and L27. Protein L16, which is concerned with peptidyl transferase activity and forms part of the acceptor site, is preferentially labeled. Mammalian cells contain 60S ribosomal subunits in the cytoplasm, hence, the difference in the specificity of the action of chloramphenicol on bacterial versus mammalian cells. Chloroplasts and the mitochondria of mammalian cells and lower eukaryotes contain 70S ribosomes that are comparable to prokaryotic ribosomes. Chloramphenicol affects protein synthesis in these eukaryotic organelles in the same manner as it affects protein synthesis in bacterial cells, a finding that may serve to explain the toxicity toward mammalian cells. Bone marrow toxicity, resulting in aplastic anemia, provides a strong contraindication for the routine use of chloramphenicol, particularly in situations where alternatives in the choice of a therapeutic agent are available.

Cycloheximide (actidione). This antibiotic is composed of two ring structures, one of which contains an imide nitrogen. The two rings are connected by a two-carbon bridge (Fig. 2-35). Cycloheximide inhibits protein synthesis in eukaryotic cells (yeasts, fungi, higher plants, and virtually all mammalian cells), but not in prokaryotic organisms (bacteria), by interfering with the activity of cytoplasmic ribosomes but not the ribosomes in

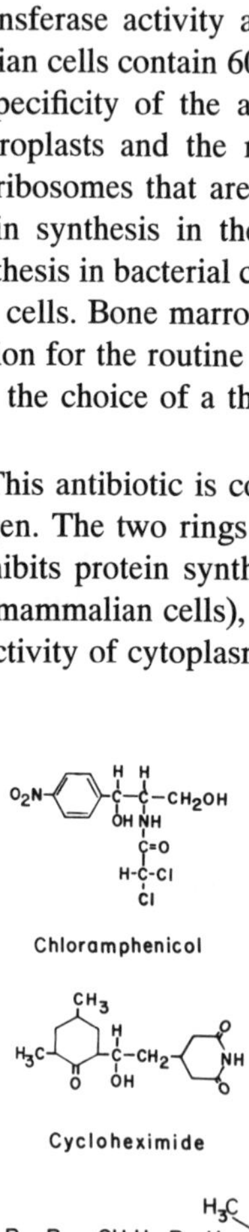

Chloramphenicol

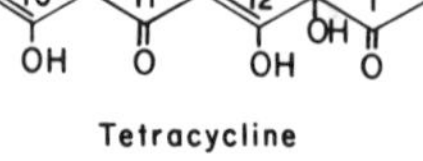

Cycloheximide

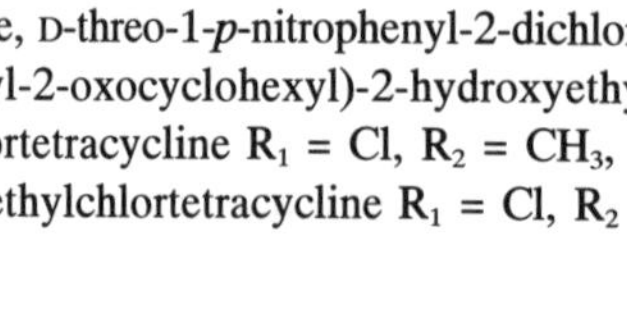

Tetracycline

Fig. 2-35. Structures of chloramphenicol, cycloheximide, and tetracycline derivatives. Chloramphenicol bears the chemical name, D-threo-1-*p*-nitrophenyl-2-dichloroacetamido-1,3-propanediol. Cycloheximide is β-2-(3,5-dimethyl-2-oxocyclohexyl)-2-hydroxyethyl-glutarimide. In tetracycline R_1 = H, R_2 = CH$_3$, R_3 = H. In chlortetracycline R_1 = Cl, R_2 = CH$_3$, R_3 = H. In oxytetracycline R_1 = H, R_2 = CH$_3$, R_3 = OH. In dimethylchlortetracycline R_1 = Cl, R_2 = H, R_3 = H.

mitochondria. The mode of action of cycloheximide resembles that of chloramphenicol. Cycloheximide binds to the 80S ribosomes of eukaryotic cells and prevents the movement of ribosomes along the mRNA. Long-term administration of either cycloheximide or chloramphenicol results in the accumulation of incomplete ribonucleoprotein particles and free RNA. The end result appears to be a lack of structural proteins in the ribosomal components. There appear to be few secondary effects of cycloheximide as shown by studies with cultured mammalian cells. Its action is readily reversible. The general effectiveness of cycloheximide on eukaryotic cells of all types accounts for the lack of applicability to the treatment of fungal infections, despite its high *in vitro* activity against a wide variety of yeasts and molds.

Tetracyclines. Several chemically distinct derivatives of tetracycline are in common clinical usage because this group of antibiotics exhibits broad-spectrum activity (bacteriostatic) against both gram-positive and gram-negative bacteria, as well as rickettsiae, chlamydiae, and mycoplasmas. Chemically, all the tetracyclines contain substituent groups on four linearly fused six-membered rings as shown in Figure 2-35. The mechanism of action of all the tetracyclines is based upon their ability to prevent the binding of aminoacyl–tRNA to the A site on the 30S ribosome. Although the antibiotic can bind to many sites on both ribosomal subunits, the strongest binding occurs to S7, which helps define an area involved in contact between the two subunits (see Fig. 2-24A).

Resistance to tetracycline is plasmid (R factor) mediated in many bacteria (e.g., *E. coli* and *S. aureus*). Plasmid-mediated tetracycline resistance is an inducible property that prevents the active accumulation of the drug in the bacterial cell. Tetracycline-sensitive strains of *E. coli* and *S. aureus* accumulate high intracellular concentrations of tetracycline by what appears to be an active transport mechanism. In the enteric bacteria, expression of tetracycline resistance following induction appears to result from the insertion of one or more proteins into the cell envelope. Similar proteins have been reported to be present in *S. aureus*.

Macrolides. This large group of antimicrobial agents includes angolamycin, carbomycin (magnamycin), chalcomycin, erythromycin, kujimycin, leucomycin, macrocin, megalomycin, and several others including oleandomycin and spiramycin. All contain a large lactone ring (termed an "aglycone") containing from 12 to 22 carbon atoms with few or no double bonds and no nitrogen atoms. One or more amino or nonnitrogenous sugars or both are linked to the aglycone. Erythromycin A, more widely recognized because of its extensive clinical application, is shown in Figure 2-36. The macrolide antibiotics are active mainly against gram-positive bacteria and have relatively limited activity against

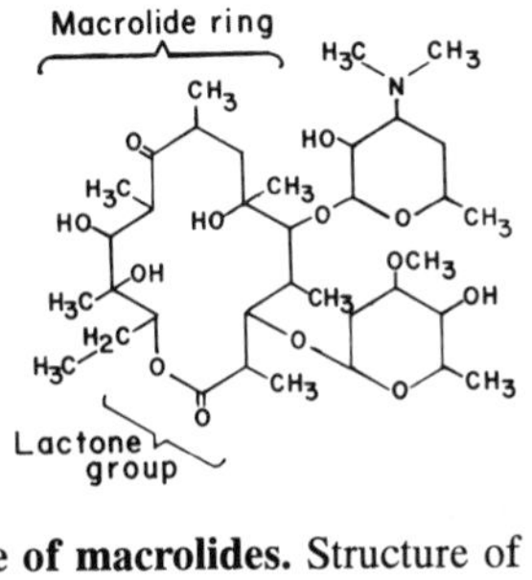

Fig. 2-36. Basic structure of macrolides. Structure of erythromycin A is shown.

gram-negative bacteria. All the macrolides are inhibitors of protein synthesis in bacterial, but not in eukaryotic systems, as a consequence of their interaction with the 50S ribosomal subunit. There is no evidence for effects on aminoacyl–tRNA formation of codon–anticodon interaction at the level of the 30S ribosome subunit. However, there is controversy concerning the specific reactions affected *in vivo*. Members of the erythromycin group do not inhibit peptide-bond formation in most of the reconstituted protein-synthesizing systems. The macrolides act at or near the peptidyl transferase center by blocking peptide bond formation when the donor site is occupied by a peptidyl moiety of a certain length. Erythromycin has been shown to bind to protein L15 and to specific residues in 23S rRNA all of which form part of the peptidyltransferase center. It is proposed that macrolide antibiotics stimulate the dissociation of peptidyl–tRNA from the ribosome by triggering an abortive translocation step. It is known that erythromycin is bacteriostatic in its action, and microbial resistance develops rather readily.

Resistance to erythromycin in *E. coli* depends on an alteration in L4 or L12 of the 50S ribosomal subunit causing reduced affinity for the drug. In *S. aureus*, however, the 23S rRNA of the 50S subunit is modified by an inducible, plasmid-mediated ribosomal RNA methylase.

Aminoglycosides. This is a large group of antibiotics that include the streptomycins, gentamicins, sisomycin, amikacin, kanamycins A and B, neomycins, paromomycins, tobramycins, and spectinomycins. These agents are all chemically similar, containing an inositol ring with two hydroxyl groups replaced by either amino or guanidino substituents and a hydroxyl group to which other sugars are attached. The structure of streptomycin is shown in Figure 2-37. In Figure 2-38, the structures of other aminoglycosides are compared. Although the mode of action of all of these compounds is essentially the same, they differ with regard to their antimicrobial spectrum, potency, and clinical usefulness. Streptomycin has been studied extensively and found to produce a variety of effects on growing bacterial cultures and cell suspensions. The ultimate mechanism of action is considered to be through binding to ribosomes and causing misreading of the amino acid code. In fact, at low concentrations, streptomycin can suppress some missense mutations by allowing misreading of the mutant codon. Streptomycin partially inhibits the binding of aminoacyl–tRNA and peptide synthesis, but it does not affect the reaction

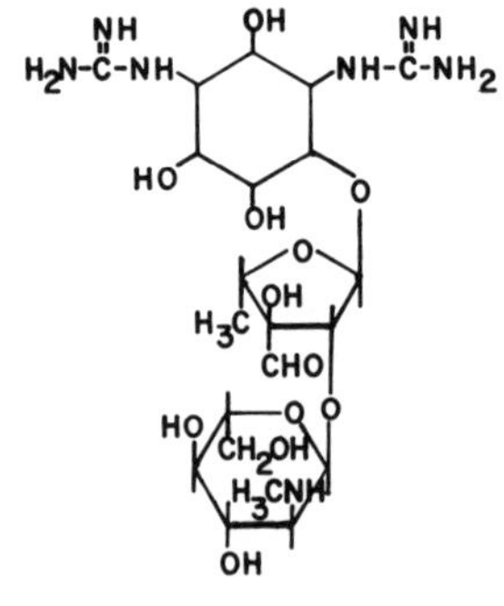

Fig. 2-37. Streptomycin. The basic molecule consists of the streptidine ring (inositol with two hydroxyl groups replaced by two guanidino residues) and the streptobiosamine group containing carbohydrates.

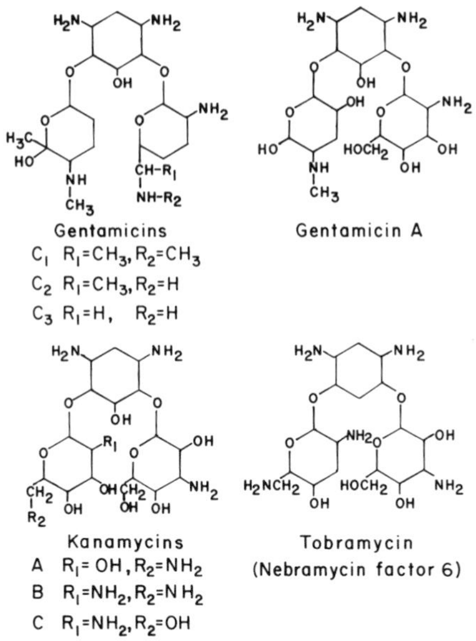

Gentamicins
C_1 R_1=CH_3, R_2=CH_3
C_2 R_1=CH_3, R_2=H
C_3 R_1=H, R_2=H

Gentamicin A

Kanamycins
A R_1= OH, R_2=NH_2
B R_1=NH_2, R_2=NH_2
C R_1=NH_2, R_2=OH

Tobramycin
(Nebramycin factor 6)

Fig. 2-38. Comparison of structures of aminoglycosides.

of pp–tRNA with puromycin. The effect of streptomycin on ribosome activity appears due to interactions with 16S rRNA. Resistance (*rpsL* or *strA*) appears to develop through alteration of the S12 protein of the 30S ribosomal subunit in such a manner that the drug can no longer bind to the 16S rRNA. The drug clearly does *not* bind to S12. Streptomycin dependence has been explained on the basis of a mutationally altered site in S12 that functions normally only with the addition of streptomycin. Other aminoglycosides apparently are sufficiently dissimilar in their action to have little effect on these dependent mutants. Mutations of another type may render microorganisms resistant on the basis of altered permeability to streptomycin or other aminoglycosides. Resistance to the aminoglycosides may also develop as a result of the ability to produce enzymes that inactivate various derivatives by the addition of phosphate groups (aminoglycoside phosphotransferases), the addition of nucleotide residues (aminoglycoside nucleotidyltransferases), or acetylation of various positions (aminoglycoside acetyltransferases). These inactivating enzymes do not react equally with all aminoglycoside derivatives. For this reason, it may be possible to substitute one compound for another. The aminoglycosides exhibit a broad spectrum of activity, being only slightly more effective against gram-negative bacteria than against gram-positive bacteria.

Kasugamycin is an aminoglycoside that acts on the 30S subunit of 70S ribosomes. It inhibits protein synthesis but it does not induce misreading nor can it cause phenotypic suppression as does streptomycin. Kasugamycin-resistant mutants (*ksgA*) are unusual in that they have an altered rRNA instead of protein. The mutation actually affects the activity of an enzyme that specifically methylates two adenine residues to dimethyl adenine in the sequence AACCUG near the 3′ end of 16S rRNA, presumably the region that interacts with the Shine–Delgarno region of mRNAs.

Lincomycin. Lincomycin and a chemically related derivative, clindamycin (Fig. 2-39), inhibit peptidyl transfer by binding to 23S rRNA in the 50S ribosomal subunit. These

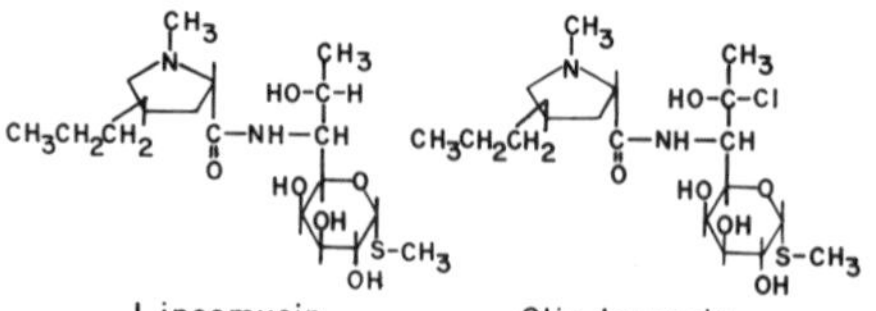

Fig. 2-39. Lincomycin and clindamycin structures.

agents are effective against both gram-positive and gram-negative organisms. Clinda-mycin is especially effective against *Bacterioides* and other anaerobes because of its ability to penetrate into deeper tissues.

Fusidic Acid. Fusidic acid is a member of the steroidal antibiotics (Fig. 2-40). It inhibits the growth of gram-positive but not gram-negative bacteria. The failure to act upon gram-negative bacteria may be due to the inability of fusidic acid to penetrate these cells since it will inhibit protein synthesis from gram-negative ribosomes *in vitro*. Addition of fusidic acid to 70S ribosomes *in vitro* prevents the translocation of peptidyl–tRNA from the acceptor to the donor site. This drug also inhibits the EFG-dependent cleavage of GTP to GDP. The data point to EFG as the target for fusidic acid. The drug forms a stable complex with EFG, GDP, and the ribosome that is unable to release EFG for subsequent rounds of translocation and GTP hydrolysis.

Puromycin. Puromycin is a unique inhibitor of protein biosynthesis since it reacts to form a peptide with the C-terminus of the growing polypeptide chain on the ribosome, thereby prematurely terminating the chain. There is a structural similarity between pu-romycin to the terminal 3' aminoacyl adenosine moiety of tRNA (Fig. 2-41). Since aminoacyl adenosine is the terminal residue of tRNA in both prokaryotic and eukaryotic organisms, it is not surprising that puromycin works equally well on 70S and 80S ri-bosomes. Puromycin only interacts with peptidyl tRNA attached to the P site of the ribosome and not the A site. The drug itself binds to the A site, where it can form a peptide bond through normal functioning of peptidyl transferase linking the amino group of puromycin to the carbonyl group of peptidyl–tRNA.

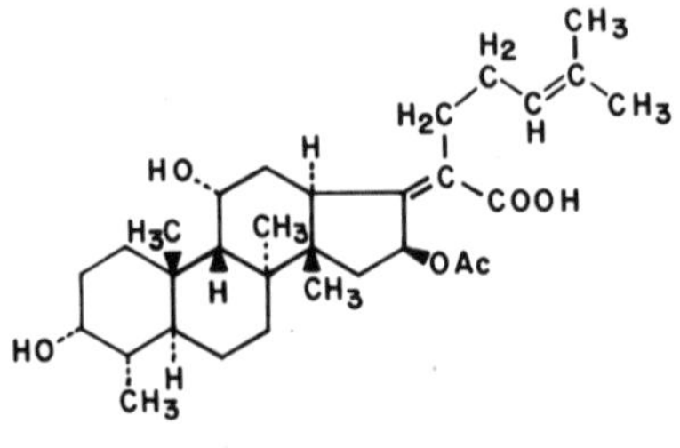

Fig. 2-40. Fusidic acid, an antibiotic with a steroid-like structure.

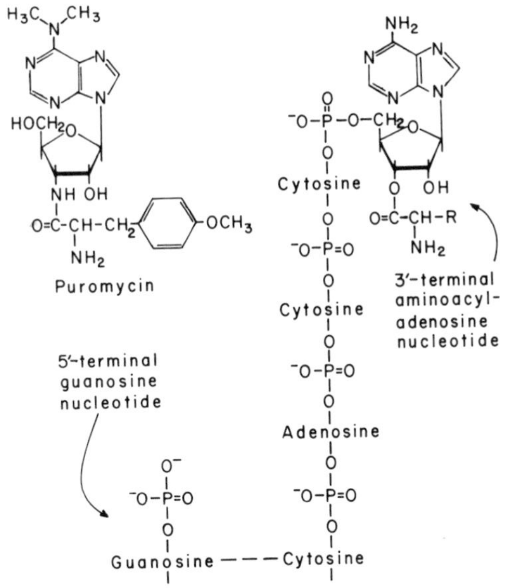

Fig. 2-41. Structure of the terminal end of the amino acid arm of tRNA. The -cytosine-cytosine-adenosine (-CCA) at the 3'-terminus and the 5'-terminal guanosine nucleotide are features that are common to all tRNAs. The structure of puromycin, which binds to the 50S ribosomal subunit and replaces aminoacyl-tRNA as an acceptor of the growing polypeptide, is shown for comparison.

SELECTED REFERENCES

DNA Replication

Alberts, B. 1980. *Mechanistic studies of DNA replication and genetic recombination.* Academic, New York.

Apirion, D. and A. Miczak. 1993. RNA processing in procaryotic cells. *Bio Essays.* **15:**113–120.

Arai, K., R. Low, J. Kobori, S. Schlomai, and A. Kornberg. 1981. Mechanism of *dnaB* protein action. *J. Biol. Chem.* **256:**573–5280.

Baker, T. A. and S. H. Wickner. 1992. Genetics and enzymology of DNA replication in *Escherichia coli. Annu. Rev. Genet.* **26:**447–477.

Cairns, J. 1963. The chromosome of *Escherichia coli. Cold Spring Harbor Symp. Quant. Biol.* **28:**43.

Drlica, K. 1992. Control of bacterial DNA supercoiling. *Mol. Microbiol.* **6:**425–433.

Drlica, K. and M. Riley (Eds.). 1990. *The Bacterial Chromosome.* American Society for Microbiology, Washington DC.

Firshein, W. 1989. Role of the DNA/membrane complex in procaryotic DNA replication. *Annu. Rev. Microbiol.* **43:**89–120.

Hill, T. M. 1992. Arrest of bacterial DNA replication. *Annu. Rev. Microbiol.* **46:**603–633.

Hubbard, T. J. P. and C. Sander. 1991. The role of heat-shock and chaperone proteins in protein folding: possible molecular mechanisms. *Protein Eng.* **4:**711–717.

Hwang, D. S. and A. Kornberg. 1992. Opposed actions of regulatory proteins, DnaA and IciA, in opening in the replication origin of *Escherichia coli. J. Biol. Chem.* **267:**23087–23091.

Jaworski, A., W.-T. Hsieh, J. A. Blaho, J. E. Larson, and R. D. Wells. 1987. Left-handed DNA in vivo. *Science.* **238:**773–777.

Kornberg, A. 1980. *DNA Replication.* W. H. Freeman and Co., San Francisco.

Kornberg, A. 1988. DNA replication. *J. Biol. Chem.* **263:**1–4.

Lehman, I. R. 1974. DNA ligase: Structure, mechanism and function. *Science* **186:**790.

Lehman, I. R., M. J. Bessman, E. S. Simms, and A. Kornberg. 1958. Enzymatic synthesis of deoxyribonucleic acid. 1. Preparation of substrates and partial purification of an enzyme from *Escherichia coli. J. Biol. Chem.* **233:**163.

Marians, K. J. 1992. Prokaryotic DNA replication. *Annu. Rev. Biochem.* **61:**673–719.

McHenry, C. S. 1991. DNA polymerase III holoenzyme. Components, structure and mechanism of a true replicative complex. *J. Biol. Chem.* **266:**1927–1930.

Meselson, M. and F. W. Stahl. 1958. The replication of DNA in *Escherichia coli. Proc. Natl. Acad. Sci. USA* **44:**671.

Weijland, A., K. Harmark, R. H. Cool, P. H. Anborgh, and A. Parmeggiani. 1992. Elongation factor Tu: a molecular switch in protein biosynthesis. *Mol. Microbiol.* **6:**683–688.

Transcription and Translation

Bachman, B. J. 1990. Linkage map of *Escherichia coli* K-12, Edition 8. *Microbiol. Rev.* **54:**130–197.

Das, A. 1993. Control of transcription termination by RNA binding proteins. *Annu. Rev. Biochem.* **62:**893–930.

Crick, F. H. C. 1966. Codon–anticodon pairing: The wobble hypothesis. *J. Mol. Biol.* **19:**548.

Garen, A. 1968. Sense and nonsense in the genetic code. *Science* **160:**149.

Gegenheimer, P. and D. Apirion. 1981. Processing of procaryotic ribonucleic acid. *Microbiol. Rev.* **45:**502–541.

Grossman, A. D., J. W. Erickson, and C. A. Gross. 1984. The *htpR* gene product of *E. coli* is a sigma factor for heat-shock promoters. *Cell* **38:**383–390.

Hill, W. E., A. Dahlberg, R. A. Garrett, P. B. Moore, D. Schlessinger, and J. R. Warner. 1990. *The Ribsome.* American Society for Microbiology, Washington, DC.

Kim, S. H., F. L. Suddath, G. J. Quigley, A. McPherson, J. L. Sussman, A. Wang, N. C. Seeman, and A. Rich. 1974. Three-dimensional structure of yeast phenylalanine transfer RNA. *Science* **185:**435.

Li, J., S. W. Mason, and J. Greenblatt. 1993. Elongation factor NusG interacts with termination factor ρ to regulate termination and anti-termination of transcription. *Genes Dev.* **7:**161–172.

McClure, W. R. 1985. Mechanism and control of transcription initiation in prokaryotes. *Annu. Rev. Biochem.* **54:**171–204.

Meninger, J. R., A. B. Caplan, P. K. E. Gingrich, and A. G. Atherly. 1983. Test of the ribosome editor hypothesis. II. Relaxed (*relA*) and stringent (*relA*⁺) *Escherichia coli* differ in rates of dissociation of peptidyl–tRNA from ribosomes. *Mol. Gen. Genet.* **190:**215–221.

Nierhaus, K. H. and H. G. Wittman. 1980. Ribosomal function and its inhibition by antibiotics in prokaryotes. *Naturwissenschaften* **67:**234–250.

Nirenberg, M. and P. Leder, 1964. RNA codewords and protein synthesis: The effect of trinucleotides upon the binding of sRNA to ribosomes. *Science* **145:**1399.

Pabo, C. O. and R. T. Saucer. 1992. Transcription factors: structural families and principles of DNA replication. *Annu. Rev. Biochem.* **61:**1053–1095.

Pace, N. R. and A. B. Burgin. 1990. Processing and evolution of the rRNAs. In *The Ribosome.*

Structure, Function and Evolution. W. Hill, A. Dahlberg, R. Garrett, P. Moore, D. Schlessinger, and J. Warner (Eds.). American Society for Microbiology, Washington, DC, pp. 417–425.

Persson, B. C. 1993. Modification of tRNA as a regulatory devise. *Mol. Microbiol.* **8:**1011–1016.

Reynolds, R., R. M. Bermudez-Cruz, and M. J. Chamberlin. 1992. Parameters affecting transcription termination by *Escherichia coli* RNA polymerase. I. Analysis of 13 Rho-independent terminators. *J. Mol. Biol.* **224:**31–51.

Rheinberger, H. and K. H. Nierhaus. 1986. Allosteric interactions between the ribosomal transfer RNA binding sites A and E. *J. Biol. Chem.* **261:**9133–9139.

Richardson, J. P. 1983. Involvement of a multistep interaction between Rho protein and RNA in transcription termination. In *Microbiology—1983.* D. Schlessinger (Ed.). American Society for Microbiology, Washington, DC.

Rodriguez, R. L. and M. J. Chamberlin (Eds.). 1982. *Promoters: Structure and Function.* Praeger Publishers, New York.

Russo, F. D. and T. J. Silhavy. 1992. Alpha: The Cinderella Subunit of RNA polymerase. *J. Biol. Chem.* **267:**14515–14518.

von Hippel, P. H., D. G. Bear, W. D. Morgan, and J. A. McSwiggen. 1984. Protein–nucleic acid interactions in transcription: A molecular analysis. *Annu. Rev. Biochem.* **53:**389–446.

Weijland, A., R. Harmark, R. H. Cool, P. H. Anborgh, and A. Parmeggiani. 1992. Elongation factor Tu: a molecular switch in protein synthesis. *Mol. Microbiol.* **6:**683–688.

Wittman, H. G. 1982. Components of bacterial ribosomes. *Annu. Rev. Biochem.* **51:**155–183.

Protein Export and Degradation

Driessen, A. J. M. 1992. Bacterial protein translocation: Kinetic and thermodynamic role of ATP and the proton motive force. *Trend. Biochem. Sci.* **17:**219–223.

Ellis, R. J. and S. M. van der Vies. 1991. Molecular Chaperones. *Annu. Rev. Biochem.* **60:**321–347.

Goldberg, A. L. and S. A. Goff. 1986. The selective degradation of abnormal proteins in bacteria. In *Maximizing Gene Expression,* W. Reznikoff and L. Gold (Eds.). Butterworth, Stoneham, MA.

Gottesman, S. and M. R. Maurizi. 1992. Regulation by proteolysis: Energy-dependent proteases and their targets. *Microbiol. Rev.* **56:**592–621.

Lazdunski, A. M. 1989. Peptidases and proteases of *Escherichia coli* and *Salmonella typhimurium.* *FEMS Microb. Rev.* **63:**265–276.

Pugsley, A. P. 1993. The complete general secretory pathway in gram-negative bacteria. *Microbiol. Rev.* **57:**50–108.

Silhavy, T. J., S. A. Benson, and S. D. Emr. 1983. Mechanisms of protein localization. *Microbiol. Rev.* **47:**313–344.

Antibiotics

Franklin, T. J. and G. A. Snow. 1991. *Biochemistry of Antimicrobial Action.* Chapman & Hall, New York.

Maxwell, A. 1993. The interaction between coumarin drugs and DNA gyrase. *Mol. Microbiol.* **9:** 681–686.

CHAPTER 3

BACTERIAL GENETICS

TRANSFER OF GENETIC INFORMATION IN PROKARYOTES

Genetic recombination is the production of new combinations of genes derived from two different parental cells. Several processes have been described by which prokaryotic cells can exchange genetic information to yield recombinants. They are **conjugation, transformation,** and **transduction.**

Conjugation between two cells of opposite mating type requires physical contact between the cells and results in the transfer of DNA from a donor (male) to a recipient (female).

Transformation involves the transfer of a relatively small fragment of "naked" DNA from one cell (**donor**) to another (**recipient**).

Transduction involves the transfer of genetic information from the genome of a donor cell to a recipient cell mediated by a bacteriophage.

In the following section, these genetic exchange and recombinational processes are considered in greater detail. In each of these processes only a small portion of the donor cell genome is transferred to the recipient cell. The recipient cell temporarily becomes a partial diploid (**merodiploid**) for the short region of homology between the transferred DNA and the DNA of the recepient chromosome. It is the ultimate fate of the donated DNA [whether it becomes incorporated into the recipient genome by recombination processes, degraded by nucleases (host restriction), or maintained as a stable extrachromosomal fragment] that occupies a considerable portion of our attention in the ensuing discussion.

We will first describe the various processes of genetic exchange and explain how the transferred DNA may be recombined into the bacterial chromosome. Next, a discussion of genetic elements called transposons will be used to illustrate how some genes can be made to "jump" from one place to another on a genome or between two genomes in a single cell. Finally, an explanation of DNA mutagenesis and repair will complete our coverage of bacterial genetics.

PLASMIDS

As alluded to above, bacteria can be sexually active organisms! This ability to transfer DNA through cell–cell contact was first discovered in the 1950s using *Escherichia coli*. Transfer of genetic information between different *E. coli* strains was found to depend on the presence in some cells (called donor cells) of a small "extra" chromosome called **F** (fertility) factor. F factor encodes the proteins necessary for the sexual process and will be discussed in more detail below. Subsequent to the discovery of F, many other extra-chromosomal DNA elements, called *plasmids*, have been discovered. All plasmids share some common features. They are generally double-stranded, closed circular DNA molecules capable of autonomous replication (i. e., independent of chromosomal replication). One exception to circularity is the linear plasmid found in the Lyme disease bacterium, *Borrelia burgdorferi*. Some plasmids, called **episomes**, can integrate into the bacterial chromosome. F factor is one example of an episome. The size of plasmids can range from 1 to 2000 kb as compared to the *E. coli* bacterial chromosome that is approximately 5000 kb. A plasmid that can mediate its own transfer to a new strain is called a **conjugative** plasmid, whereas one that cannot is referred to as **nonconjugative**. Plasmids that have no known identifiable function other than self-replication are often referred to as **cryptic** plasmids. However, they may be cryptic only because we have not been clever enough to elucidate their true function.

Interest in plasmids has increased progressively since the 1970s with the advent of recombinant DNA technology. Plasmids are the "workhorses" of this technology serving as the means or vehicle by which individual genes from diverse organisms can be maintained separate from their genomic origin. In addition, many plasmids harbor antibiotic resistance genes. Through conjugation or other means, the transfer of these plasmids among bacterial species led to the rapid proliferation of antibiotic-resistant disease-causing microorganisms.

Why have plasmids been maintained throughout evolution? One reason is that they often provide a selective advantage to organisms that harbor them, such as toxins to kill off competing organisms (see Colicins) or resistance genes to fend off medical antibiotics. Another reason for their persistence is their ability to partition. **Partitioning** assures that after replication each daughter cell gets a copy of the plasmid. The mechanism used for partitioning differs depending on the plasmid. Some may attach to the cell membrane near the equatorial plane where the cell divides. Growth of a septum along the plane would allow for proper distribution. As an alternative, plasmids present in high-copy numbers (e.g., pBR322) may be distributed to daughter cells by random diffusion.

Incompatibility is a property of plasmids that explains why two very similar yet distinct plasmids might not be maintained in the same cell. Two plasmids that share some regulatory aspect of their replication are said to be incompatible. For example, if two different plasmids produce similar repressors for replication initiation, then the repressor of one could regulate the replication of the other and vice versa. The choice as to which plasmid will actually replicate is random. So, in any given cell one plasmid type could outnumber another prior to cell division such that after division one daughter cell will only contain one of the two plasmids. This is called **segregational incompatibility.**

Nonconjugative, Mobilizable Plasmids

Many plasmids smaller than F that are nonconjugative by themselves possess a system that will allow them to be conjugally transferred when they are present in the same cell as a conjugative plasmid. These nonconjugative, mobilizable plasmids include the colicin El (ColEl) plasmid. It has a *dnaA*-independent origin of replication but contains a site called *bom* (basis of mobility) that functions much like *oriT* of the F factor (see below). While some *ColEl* genes are required for mobilization (*mob* genes), transfer will not occur unless some functions are provided by F-factor. Not all conjugative plasmids will fill this role, however. Where members of the incompatibility groups IncF, IncI, or IncP conjugative plasmids will efficiently mobilize ColEl, members of IncW do so inefficiently.

Resistance (R) Plasmids

Many extrachromosomal elements have been recognized because of their ability to impart new genetic traits to their host cells. One important factor, **R-factor** was first recognized by the fact that organisms in which it was present were resistant to a number of chemotherapeutic agents. A single R factor may carry traits for resistance to as many as seven or more chemotherapeutic or chemical agents. The R factors harbored by organisms in the normal flora of human beings or animals may be transferred to pathogenic organisms, giving rise to the sudden appearance of multiply resistant strains.

Different R factors display characteristic groups of resistance markers. Because of the differences in the designation of R factors by various workers, there may be some overlap in terminology, and such factors isolated in widely different geographic localities may be identical. The R factors are composed of two genetically distinguishable units: a **transfer factor** (RTF) and a unit that contains drug **resistance genes (r-determinants)**. The RTF mediates autonomous replication and promotes transfer. The r determinants specify resistance to individual antimicrobial agents. Transfer factor may exist in the absence of the resistance markers, and nontransferable r determinants may be present in the absence of RTF. The stability of the relationship appears to vary for different R factors and different host cells. For example, in *E. coli*, the association between RTF and the r determinants may be quite stable. Segregants have been shown to occur at low frequency, either spontaneously or during transfer. Behavior of the same R factor may be quite different in *Salmonella*, the frequency of segregation being quite high. The R factors may also acquire new resistance markers that may be replicated and transferred. In addition, other genetic determinants may be present along with those for resistance. Evidence has been obtained that R factors did not arise as a direct result of the widespread use of antibiotics. Individuals living in highly sequestered geographic areas who have not been subjected to antimicrobial agents possess normal bacteria flora harboring R factors. Also, organisms maintained in the lyophilized state from time periods prior to the widespread use of antimicrobial agents have been shown to carry R factors.

As a result of the demonstration that R factors may be transferred from normal flora to pathogenic microorganisms under natural conditions, there has been considerable concern with regard to the potential public health hazards that may result from the widespread and indiscriminate use of antimicrobial agents. Because of the close molecular relatedness of the R factors in the enterobacteria of humans and animals, a common pool of R factors exists in these organisms, which requires very careful surveillance.

Plasmids in Other Bacterial Genera

The association of resistance to chemotherapeutic agents and other genetic traits with extrachromosomal elements has been well established in the *Enterobacteriaceae*. It is also possible to demonstrate intergeneric transmission of plasmids among members of this closely related group of organisms. Plasmid-associated resistance factors have been well characterized in the staphylococci and in *Streptococcus mutans* and *Enterococcus faecalis* (now *E. hirae*). Resistance of *E. faecalis* to erythromycin and lincomycin has been shown to be plasmid associated. Self-transferable plasmids determining hemolysins and bacteriocins, as well as multiple antibiotic resistance, have been found in *E. faecalis*. Transfer of plasmids in the streptococci appears to take place through conjugal mechanisms.

Extrachromosomal, covalently closed circular DNA molecules comparable to those observed in *S. aureus* and the enteric organisms are present in *Bacillus pumilis and Bacillus subtilis*. These DNA molecules are present in one or two copies per chromosome. Covalently closed circular DNA derived from *S. aureus* plasmids can be introduced into *B. subtilis*. Once transferred, these plasmids can replicate autonomously in recipient strains of *B. subtilis* and may be transferred between *B. subtilis* strains by transduction or by transformation.

Plasmids have been isolated from strains of *Neisseria gonorrhoeae* obtained from clinical sources. Two different sizes of circular, covalently closed DNA plasmids have been characterized. Small plasmids exhibit a molecular weight of 2.8×10^6, while a large plasmid has a molecular weight of 24.5×10^6. No specific association between the presence of a particular plasmid and antibiotic resistance, piliation, or nutritional requirements has been demonstrated in the strains studied. However, there has been widespread concern over the finding that β-lactamase production by penicillin-resistant strains of *N. gonorrhoeae* is associated with transmissible plasmids and may have originated in the *Enterobacteriaceae*.

A number of studies with the crown gall tumor agent, *Agrobacterium tumefaciens*, provided evidence that a plasmid plays a major role in tumor induction. Irreversible loss of crown gall inducing ability of *A. tumefaciens* resulted from the loss of a large plasmid of 1.2×10^8 Da (Ti plasmid). An avirulent strain that arose spontaneously was also shown to lack the virulence plasmid present in its sibling strain. Transfer of virulence from a donor to an avirulent recipient resulted from the transfer of a plasmid that is related to the donor plasmid, as shown by DNA reassociation measurements. A biochemical marker and sensitivity to a bacteriocin are also located on the virulence plasmid, providing a useful tool for the selection of avirulent plasmid free derivatives from the parent strain.

Plasmid Replication Control

Plasmids can be characterized into separate groups based upon the number of copies normally present in a cell. F factor is a *low-copy number* plasmid whose replication is stringently controlled. On the other hand, ColEl is a *high-copy number* plasmid that can be present at 50–100 copies per cell. Even though there is a wide difference in their copy number, both types of plasmids have mechanisms designed to limit copy number.

The ColEl plasmid undergoes unidirectional replication (one replication fork) and can replicate in the absence of *de novo* protein synthesis. It can do this by utilizing a variety of stable host replication proteins (e.g., Pol I, DNA-dependent RNA polymerase, Pol III, DnaB, DnaC, DnaG, and DnaZ) rather than make its own. Because replication initiation of the bacterial chromosome requires *de novo* protein synthesis, the addition of chloramphenicol to ColEl-containing cells results in the amplification of the plasmid copies relative to chromosome. But even though the ColEl replication system allows for a fairly high copy number per cell, the system is regulated so as not to exceed that number (Fig. 3-1). Replication of ColEl actually begins with a transcriptional event that occurs about 500 bp upstream of the replication origin (*ori*). As the transcript is formed through *ori*, an RNA–DNA hybrid molecule forms that is a suitable substrate for the bacterial host enzyme RNase H. RNase H cleaves the RNA primer leaving a 3'-OH end suitable for DNA Pol III and replication. The processed RNA primer is called RNA II. The copy number is actually negatively controlled by an antisense RNA, a partial countertranscript of RNA II (100 nucleotides). The countertranscript (RNA I) is transcribed near the start of the prepriming transcript (RNA II precursor) but in the opposite direction (Fig. 3-1B). The RNA I and RNA II precursor molecules each contain a significant amount of secondary stem–loop structure and because the two RNA molecules are complementary, the loop areas can base pair with the help of Rom protein. In order for RNA I to prevent replication, it must achieve a concentration that will favor its interaction with the RNA II precursor. It is this aspect of the interaction that allows plasmid copy number to achieve 20–50 per cell before RNA I concentrations are high enough to prevent further replication. Following the initial molecular ''kiss'' of the loop areas, the molecules melt into each other extending the base pairing. The RNA–RNA complex formation disrupts the RNA–DNA hybrid at *ori* and interferes with processing by RNase H thereby aborting

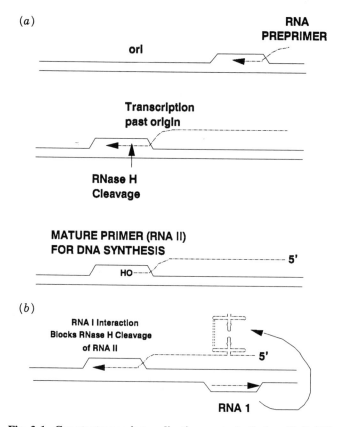

Fig. 3-1. Countertranscript replication control of plasmid ColE1.

replication initiation. The RNA I–preprimer complex is now also susceptible to RNase E (*rnaE*). Another factor affecting the copy number of ColEl is the host protein PcnB, which is involved in the turnover (degradation) of RNA I. Mutations in *pcnB* increase the stability of RNA I, which decreases the copy number of the ColEl plasmid by reducing the number of possible replication events. Other plasmid systems (F factor) utilize a different type of countertranscript control of copy number where the target is the mRNA for a rate-limiting initiator protein (see **FinOP**).

Another type of plasmid copy number control is exemplified by P1, which is actually a bacteriophage DNA that can exist either as a lytic phage or as a low copy number plasmid in its prophage state (Fig. 3-2). How does P1 maintain a low copy number as a plasmid? To understand this process you must first know that the P1-encoded RepA protein serves to initiate P1 plasmid replication. The RepA protein must bind to sites within a region of the p1 genome called *oriR* in order to trigger P1 replication. Within *oriR* there are five copies of a 19 bp sequence, called an iteron, to which RepA can bind. Another nine iterons are found in a separate region called *incA*. Replication will only initiate if RepA binds to the iterons in *oriR*. However, the nine iterons present at *incA* serve to titrate the level of available RepA to a point insufficient for binding to *oriR*, especially when multiple copies of P1 are present in the cell. So with one plasmid in the cell there is a certain probability that RepA will bind the *oriR* sites and trigger replication. However, with two plasmids in the cell the number of competing iterons increases, thereby decreasing the probability that RepA will bind to the *oriR* iterons. Thus, copy number is maintained. Why do cells with two plasmids not synthesize more RepA? After all, with two copies of the *repA* gene, one per plasmid, there should be twice as much RepA and the plasmid should just continue replicating. However, evidence using *repA– lacZ* fusions (see discussion of operon fusions, Global Control Networks in Chapter 4) has revealed that RepA autoregulates its own synthesis. In other words, when too much RepA is present in the cell it shuts off its own synthesis. As the cell grows in size the RepA concentration (molecules per volume) will transiently decrease and thus trigger induction of *repA* such that the RepA concentration is quickly restored. However, the concentration of RepA does not increase above a certain limit. That concentration is not sufficient to allow plasmid copy number to exceed a certain number. The F factor also utilizes an iteron system for copy number control.

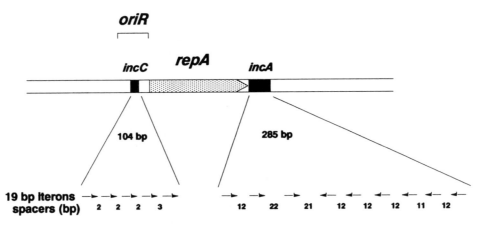

Fig. 3-2. Iteron control of P1 copy number.

***Plasmid Maintenance by Host Killing: The ccd* Genes.** Low copy number plasmids like F factor, as well as R1, R100, and P1, all have the remarkable ability to kill cells that have lost the plasmid. In F, two genes are responsible, *ccdA* and *ccdB* (coupled cell *d*ivision). The *ccdB* product (11.7 kDa) is an inhibitor of cell growth and division. Its activity is countered by the unstable *ccdA* product (8.7 kDa). If plasmid copy number falls below a certain threshold, the loss of CcdA through dilution will unleash the killing activity of CcdB. Affected cells form filaments, stop replicating DNA, and die. How CcdB functions in this regard is not clear.

CONJUGATION

As described above, conjugation involves direct cell-to-cell contact to achieve DNA transfer (Fig. 3-3). For this process, certain types of extrachromosomal elements called plasmids are usually required. The prototype conjugative plasmid is the F, or fertility, factor of *E. coli.*

F Factor. The F factor is an *E. coli* plasmid (100 kb) with genes coding for autonomous replication, sex pili formation, and conjugal transfer functions. In addition, there are several insertion sequences situated at various sites (Fig. 3-4) and F is considered an **episome** since it replicates either independently of the host chromosome or as part of the host genome. Cells containing an autonomous F are referred to as F^+ cells. Replication of the F factor in this situation requires host proteins but is independent of the *dnaA* gene product (see Initiation of DNA Replication section). There are only 1–3 copies of this plasmid per cell, thus, F is an example of a plasmid whose replication is stringently controlled.

The Conjugal Transfer Process. A large portion of F-factor DNA is dedicated to the transfer process. Figure 3-5 presents a portion of the genetic map for F encompassing

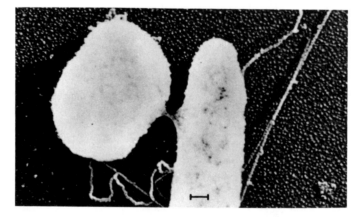

Fig. 3-3. Conjugation: cell-to-cell contact between bacterial cells. The long cell on the right is an Hfr donor cell of *E. coli.* It is attached by the specific F-pilus to a recipient (F^-) on the left. Bar = 100 nm. From Anderson, T. F., et al. 1957. *Ann. Inst. Pasteur Paris* **93**:450.

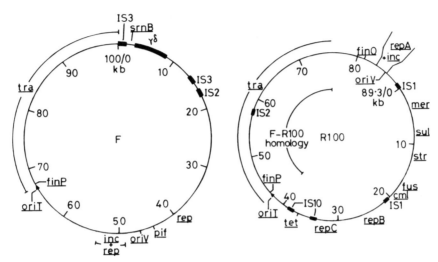

Fig. 3-4. Maps of F and of R100. Kilobase coordinates are shown inside the circles. For F, nontransfer-related markers are the insertion sequences [IS3 (= αβ), γδ, IS3, and IS2], stable RNA degradation (*srnB*), inhibition of replication by T7 and II phages (*pif*), and a region including replication and incompatibility genes and the origins of vegetative replication (*rep, inc,* and *oriV*). The R100 resistance determinants are to mercuric ions (*mer*), sulfonamides (*sul*), streptomycin (*str*), chloramphenicol (*cml*), fusidic acid (*fus*), and tetracycline (*tet*). From Willitts, N. and R. Skurray, 1980. *Annu. Rev. Genet.* **14**:41.

the *tra* region. It will become useful to refer to this diagram during the following discussion of F conjugation.

In the *Enterobacteriaceae*, the presence of specific structural appendages (pili) on the cell surface is correlated with the ability of the cell to serve as a donor of genetic material. Of the 25 known transfer genes, 12 are involved with F pilus formation (*traA,-L,-E,-K, -B,-V,-W,-C,-U,-F,-H,-G*). Evidence for the existence of a pool of preformed subunits that are incorporated into mature sex pili has been provided by experiments showing that F pili can be regenerated in the presence of growth-inhibiting concentrations of chloramphenicol. The site of assembly of the sex pili appears to be the membranous layers of the cell envelope since inhibition of formation of the mucopeptide layer by penicillin does not affect the production of pili. The F-donor cells possess 1–3 sex pili. The tip of the pilus is involved in stable mating pair formation (*traN* and *traG* involvement) interacting with the *ompA* gene product on the outer membrane of the recipient. Once initial contact between the donor pilus and the recipient is established, the pilus is thought to contract, bringing the cell surfaces of the male and female cells into close proximity. This wall-to-wall contact forms a conjugation bridge involving the fusion of the cell envelopes. The DNA strand that is transferred (perhaps with a "pilot" protein bound at the 5′ terminus) travels through this membrane bridge (probably a pore involving the *traD* gene product), not through the pilus itself, as originally believed. Mating mixtures of *E. coli* actually form mating aggregates of from 2 to 20 cells each rather than only mating pairs. Following mating-aggregate formation, the transfer of F⁺ DNA initiates from *oriT* (origin of transfer as opposed to *oriV*, the vegetative replication origin). A plasmid-encoded endonuclease (*traI* gene product, domain Z) nicks the F plasmid at *oriT*

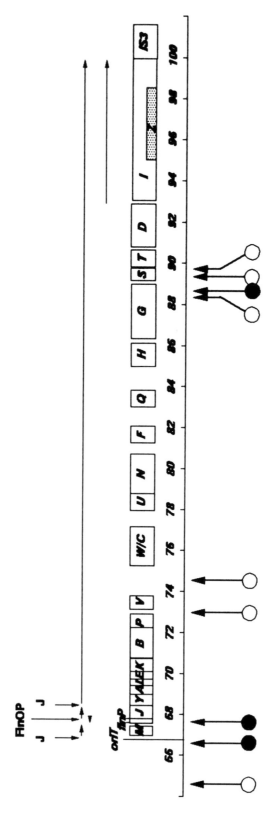

Fig. 3-5 Physical and genetic map of the transfer region of plasmid F. The numbers show kilobase coordinates and the horizontal lines above the genes represent transcripts. The direction of DNA transfer from *oriT* is such that the transfer region is transferred last. The *traM* and *traJ* promoter regions have been sequenced and the *traY-Z* operon has been shown to have its own separate promoter. The *finP* locus may be transcribed from the DNA strand opposite the long leader sequence of the *traJ* mRNA. The vertical arrows with open and closed circles indicate the positions of *Eco*RI and *Bgl*II cleavage sites, respectively. Transcription from the promoters for *traM* and for the *traY-Z* operon is dependent on the product of *traJ*, which is in turn negatively regulated by the FinOP repressor. The *traI* asnd *traZ* genes are transcribed constitutively from a second promoter at about 18% of the level from the *traJ*-induced traY-Z operon promoter. Roles attributed to the genes are regulation, finP and *traJ*; pilus formation, *traA*, -L, -E, -K, -B, -V, -W/C, -U, -F, -Q, -11, and -G; stabilization of mating pairs, *traN*, and *traG*; conjugative DNA metabolism, *traM*, -Y, -D, -I, and Z; surface exclusion, *traS* and *traT*. From Willitts N. and B. Wilkins, 1984. *Microbiol. Rev.* **48**:24.

(Figs. 3-5 and 3-6). The 5′-ended strand is then transferred to the recipient via a rolling circle type of replication with the intact strand serving as a template. TraM appears to trigger conjugal DNA synthesis at the nicked *oriT* site possibly by exposing enough ssDNA to allow binding of the TraI helicase. The *traI* gene product (DNA helicase I) migrates on the strand undergoing transfer to unwind the plasmid duplex (1000 bp/s) with DNA pol III synthesizing the replacement strand of the donor DNA. The 5′ end of the strand, upon entering the recipient, becomes anchored to the membrane. As the donor strand is transferred into the recipient cell it will undergo replication and is then circularized. This process, while *recA* independent, may require a plasmid-encoded recombi-

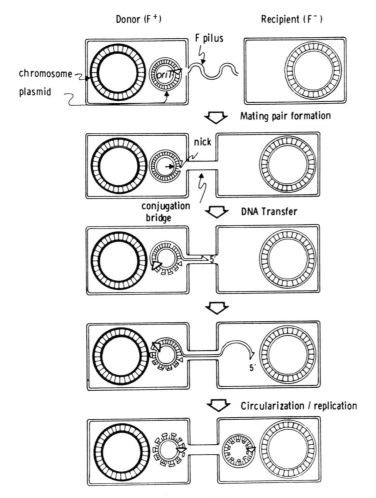

Fig. 3-6. Steps in F-plasmid transfer. The process is initiated by cell-to-cell contact, mediated by the F-plasmid coded pilus on the donor cell. A nick at the origin of transfer site, *oriT*, supplies the 5′ terminus that invades the recipient cell (F DNA synthesis can but need not occur simultaneously on the intact covalently closed circular molecule of single-stranded DNA.) Transfer of a genetic element capable of autonomous replication requires production of the complementary molecule on the transferred strand (continuous DNA synthesis is shown by *dashed lines*, and circularization).

nation system. At this point, the recipient cell is now capable of transferring F to another cell. Figure 3-6 illustrates the overall process of conjugation, while Figure 3-7 presents a more detailed analysis of the proteins involved.

Barriers to Conjugation. It has been known for some time that cells carrying an F factor are poor recipients in conjugational crosses (termed **surface exclusion** or **entry exclusion**). A related phenomenon, **incompatibility** (see above), operates after an F′ element enters into a recipient cell already carrying an F factor and is expressed as the inability of the superinfecting F′ element and the resident F factor to coexist stably in the same cell (see below for an explanation of F′).

Two genes (*traS* and *traT*) are required for surface exclusion with TraT being an outer-membrane protein. It is believed that the *traT* protein might block mating pair stabilization sites or, alternatively, might affect the synthesis of structural proteins (other than *ompA*) necessary for stabilization. Growing F$^+$ cells into late stationary phase im-

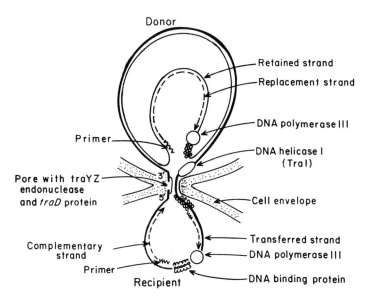

Fig. 3-7. Model for the conjugative transfer of F. A specific strand of the plasmid (thick line) is nicked at *oriT* by the *traYZ* endonuclease and transferred in the 5′ ⟶ 3′ direction through a pore, perhaps involving the *traD* protein, formed between the juxtaposed donor and recipient cell envelopes. The plasmid strand retained in the donor cell is shown by a thin line. The termini of the transferred stand are attached to the cell membrane by a complex that includes the endonuclease. The DNA helicase I (*traI* gene product) migrates on the strand undergoing transfer to unwind the plasmid duplex DNA; if the helicase is in turn bound to the membrane complex during conjugation, the concomitant ATP hydrolysis might provide the motive force to displace the transferred strand into the recipient cell. The transfer of DNA is associated with synthesis of a replacement strand in the donor and of a complementary strand in the recipient cell (broken lines); both processes require de novo primer synthesis and the activity of DNA polymerase III holoenzyme. The model assumes that a single-strand binding protein coats DNA, to aid conjugal DNA synthesis; depending on the nature of the pore, this protein might even be transferred from donor to recipient cell, bound to the DNA. From Willetts and Wilkens, 1984. *Microbiol. Rev.* **48:**24.

parts a recipient ability almost equivalent to an F⁻ cell. These cells are called F⁻ **phenocopies** and are nonpiliated.

Fertility Inhibition. Most conjugative plasmids transfer their DNA at a markedly reduced rate as compared with F. The reason for this is that these plasmids (e.g., R100) possess a regulatory system that normally represses their *tra* genes. It is the FinOP fertility inhibition system that is responsible for their control. The *finO* and *finP* products interact to form a FinOP inhibitor of *tra* gene expression. In contrast to these other plasmids, the F factor is *finO finP⁺* (due to an IS3 insertion, see **Transposons**) and so is naturally derepressed. However, if another plasmid that is *finO⁺* resides in the same cell as F, then the F *finP* gene product and the coresident plasmid's *finO* gene product can interact producing the FinOP$_F$ inhibitor. This will act as a negative regulator of the F-factor's *traJ*, which itself is the positive regulator for the other *tra* genes. Consequently, the fertility of F will be inhibited. Recent evidence suggests that *finP* RNA and *finO* protein form a complex that interacts with the leader portion of the *traJ* transcript and inhibits its transcription and/or translation.

Hfr Formation. The F factor is an example of an episome. An episome is a plasmid that can exist autonomously in a cell or can integrate into the bacterial chromosome. A cell that contains an integrated F factor is referred to as an Hfr cell. The frequency of insertion occurs at about $10^{-5}-10^{-7}$ per generation. In other words, among a population of 10^7 F⁺ cells, 1–100 cells will have an integrated F. The mechanism of integration is illustrated in Figure 3-8. Integration involves homologous recombination between two covalently closed circular DNA molecules forming one circular molecule containing both of the original DNA structures. It is thought that the insertion sequences present in the F genome (F carries two IS3 and one IS2 sequence) and those in the host chromosome (*E. coli* contains 5 each of IS2 and IS3) serve as regions of homology for the insertional event. The F integration is predominantly *recA*-dependent but rare *recA*-independent Hfrs can be formed based upon the transposition functions of the insertion elements. Consequently, Hfr formation is mainly a nonrandom event, primarily occurring in regions of the chromosome containing an insertion sequence (IS) element.

Once integrated, the F DNA is replicated along with the host chromosome. However, in situations where the host *dnaA* gene is inactive, replication of the entire chromosome can initiate from an integrated F (integrative suppression).

An integrated F-factor still has active transfer functions and in an Hfr to F⁻ cross can transfer host DNA to the recipient where the donor DNA can recombine with the recipient DNA. Note, in Figure 3-8, the direction of transfer from *oriT* is such that the *tra* genes are always transferred last. Thus, there is a directionality to conjugal DNA transfer. In this illustration, the proximally transferred host gene is *B* while the distally transferred gene is A (the tip of the arrowhead represents the 5′ leading end of the transferred strand). If the orientation of the chromosomal IS element was in the opposite direction, then the resulting Hfr would transfer gene A as a proximal marker and gene *B* would be one of the last genes transferred.

Since conjugal transfer of the host chromosome in an Hfr cell is time dependent (it takes ~ 100 min to transfer the entire *E. coli* chromosome), a gene can be mapped relative to the position of the integrated F factor simply by determining how long it takes for the gene to be transferred to a recipient. The example provided in Figure 3-9 illustrates how three genes can be mapped by Hfr interrupted matings. The genetic markers are *leu*,

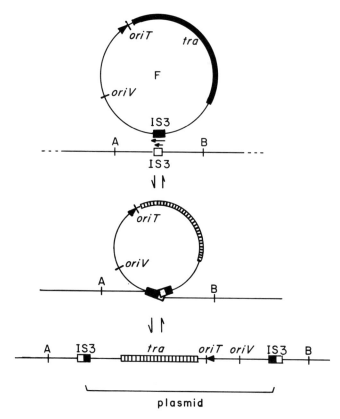

Fig. 3-8. Mechanism of integration of the F factor into the bacterial chromosome.

lac, and *gal.* The Hfr donor is wild type for each, whereas the recipient is leu, *lac,* and *gal* meaning it requires leucine and cannot grow on lactose or galactose as a carbon source (when giving a genotype for a strain *leu*⁺ indicates wild type, *leu* indicates a mutant locus). The Hfr donor contains an integrated F at 97 min on the *E. coli* 100 min map and transfers host markers in a clockwise direction (note position of the arrowhead). In addition, the Hfr is streptomycin sensitive (str^s), while the recipient is resistant to this antibiotic (str^R). After mixing donor and recipient cells (time = 0), aliquots of the mating mixture are removed at different time intervals and mating pairs disrupted in a Waring blender. The mixture is then plated on three types of media: (1) minimal glucose to select for *leu*⁺ recombinants, (2) minimal lactose + leucine to select for *lac*⁺ recombinants, and (3) minimal galactose + leucine to select for *gal*⁺ recombinants. All three media should also contain streptomycin to prevent growth of the donor cells. Colonies that arise on any of these media must be recipient cells into which the donor wild-type gene was transferred and used to replace the mutant gene (see **Recombination** below). A graphic display can be made of the number of recombinants versus conjugation time for each marker. The point where the curve intersects the *X* axis is the map position of that locus relative to the integrated F.

F-Prime Formation. As indicated in Figure 3-8, integration of F factor is a reversible process. Normally, excision of the F factor restores the host chromosome to its original

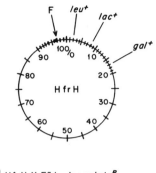

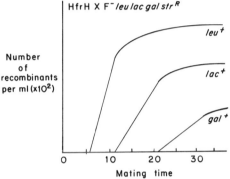

Fig. 3-9. Gene mapping by Hfr interrupted matings

state. However, improper or aberrant excision can occur at a low frequency forming a plasmid containing both F and bacterial DNA. This type of plasmid is called an F prime (F′). There are two types of faulty genetic exchanges that can result in F′ formation. The first involves recombination between a region on the bacterial chromosome and one within the integrated F factor. The resulting F′ has lost some F sequences but now carries some host DNA originally located at one or the other side of the integrated F. This is a *type I* F′. As an alternative, host sequences located on both sides of the integrated F can undergo genetic exchange. The F′ formed in this situation (Type II) contains all of F plus some host DNA from both sides of the point where F was integrated. In both situations the host DNA contained in the F′ is deleted from the host chromosome. However, if the F′ is transferred to a new host, a partially diploid (merodiploid) situation occurs for the host genes contained on the F′. By constructing merodiploids, information can be derived regarding dominance of certain mutations over wild-type alleles of specific genes (see **cis/trans Complementation** below).

cis/trans Complementation Test. When conducting a genetic analysis of mutations that affect a given phenotype, the investigator wants to know whether closely linked mutations reside in a single gene or in different genes of an operon and whether a given mutation resides within a control region (promoter/operator) or within a structural gene. Often one can answer these questions by performing complementation tests using merodiploids. For example, two different mutations are considered to lie in the same gene if their presence on duplicate genes in a merodiploid (one mutation on the F′, the other on the chromosome) fails to restore the wild phenotype, that is, the two mutations fail to complement each other in trans. If the mutations lie in two different genes, then they should complement each other in trans, that is, an F' with the genotype *trpA⁺trpB⁻* will produce a functional TrpA protein that can complement a host with the genotype *trpA⁻ trpB⁺*.* This premise depends, of course, on whether the gene produces a diffusible gene product, that is, protein or RNA. Regulatory genes, such as operators or promotors, do not produce diffusible products, and so mutations in these genes cannot be complemented in trans.

Conjugation and Pheromones in Enterococci

For a long time it was believed that conjugation only occurred in the *Enterobacteriaceae*. Evidence accumulating since 1964, however, has shown that the streptococci also possess conjugative plasmids and conjugative transposons. A number of conjugative plasmids have been characterized, some of which can mobilize nonconjugative plasmids and even some chromosomal markers.

In *Enterococcus faecalis* (*E. hirae*) conjugative plasmids can be placed into two general categories. Members of the first group (e.g., pAD1, pOB1, pPD1, pJH2, pAM1, pAM2, and pAM3) transfer at a relatively high frequency in broth ($10^{-3}-10^{-1}$ per donor cell). Members of the second group transfer poorly in broth but are fairly efficient when matings are carried out on filter membranes (e.g., pAC1, pIIP501, and p5M15346). The reason for this difference involves the production of sex pheromones by streptococci. Plasmids that transfer efficiently in broth use the pheromones to generate cell-to-cell contact. Recipient cells apparently excrete soluble, small peptides (7–8 amino acids) that induce certain donor cells to become adherent. This adherence property facilitates formation of donor-recipient mating aggregates that arise from random collisions.

A model that attempts to explain this phenomenon has been proposed (Fig. 3-10). In this model, a plasmid-free recipient produces two different chromosomally encoded pheromones: cA and cB. Each pheromone is specific for a cell containing the corresponding plasmid type (i.e., pheromone cA will stimulate conjugation with a donor cell containing plasmid pA). In addition to the recipient strain, two isogeneic donor strains harboring conjugative plasmids pA and pB are also present. All three strains have a chromosomally determined **binding substance (BS)**. The pheromones (cA and cB) produced by the plasmidless strain induce synthesis of **aggregation substance (AS)** by the plasmid-containing strains. Once inside the donor cell, pheromone cA interacts with the product of the *RcA* gene (**responding substance**), which in turn activates aggregation substance synthesis. Aggregation substance is a cell-surface proteinaceous microfibrillar substance that will recognize BS on potential recipient cells. The interaction between AS and BS causes aggregation between donor and recipient and stimulates conjugal transfer. Plasmid pA responds exclusively to pheromone cA and possesses an inhibitor of the chromosomal cA pheromone gene (*IcA*) that prevents endogenous production of cA. In other words, it cannot stimulate itself!

It also appears that the role of pheromones goes beyond simple aggregate formation. This can be illustrated through donor–donor matings using two distinguishable plasmids.

---

*An exception to this occurs with intragenic complementation. Two different mutations in a gene whose products form the subunits of a multimeric enzyme can sometimes complement each other in trans. The resulting active enzyme is composed of two different mutant polypeptides where the defect in one polypeptide sequence is compensated by the presence of the correct sequence on the other polypeptide.

(a)

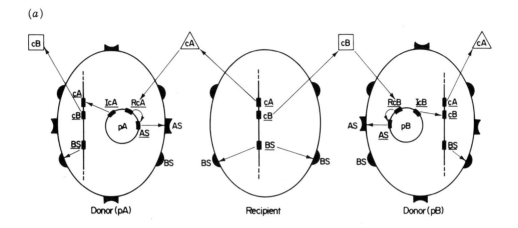

(b)

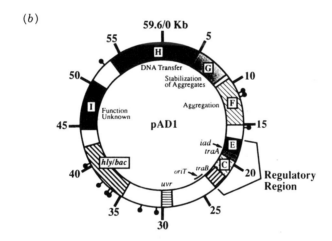

Fig. 3-10. (A) Model proposing various donor and recipient relationships with respect to synthesis of and response to sex pheromones in streptococci. The determinants of the hypothetical sex pheromones, cA and cB, are shown as being located on the bacterial chromosome, along with a determinant for BS. The latter is expressed in both donors and recipients and is located on the cell surface. Both IcA and IcB are determinants (on plasmids pA and pB, respectively) for substances that repress (or inactivate) endogenous cA and cB. Both RcA and RcB are determinants of regulatory proteins that respond to cA and cB, respectively, resulting in a turning on of the determinant AS, which encodes the aggregation substances; the latter is located on the cell surface. Once a donor has responded to the appropriate pheromone, the aggregation substance can bind to the binding substance, initiating conjugal contact. From Clewell, 1981. **(B) Map of pAD1.** Regions associated with different functions are indicated as shaded areas. A hemolysin/bacteriocin determinant is shown as *hly/bac* and *uvr* is a region that confers increased resistance to UV. Here *traA, traB*, and regions C and E are related to regulation of the mating process. This region also contains the determinant *iad* for the pheromone inhibitor peptide iAD1. Regions F, G, H, and I include various structural genes that are induced as a result of exposure to pheromone. Markers on outer circle indicate *Eco*RI restriction sites. From Clewell, 1993. *Bacterial Conjugation*. Plenum Press, New York.

If aggregation is the only role for pheromones, then once cells mate, contact transfer should occur equally in both directions regardless of which donor was induced. This is not the case. When only one of the donors is induced, transfer occurs only in the direction from the induced strain to the uninduced strain. Thus, the pheromone also helps trigger transfer of the plasmid. An illustration of pAD1 showing a variety of associated functions is provided in Figure 3-10B.

Transformation

Not all bacteria are capable of conjugation but that does not stop them from exchanging genetic information. Historically, the first demonstrable system of gene transfer actually demonstrated did not require cell-to-cell contact. The phenomenon, called transformation, was first discovered in 1928 by Griffith in the course of his investigations of *Streptococcus pneumoniae* (pneumococcus). Capsule-producing pneumococci were shown to be virulent for mice, while nonencapsulated strains were avirulent. Griffith discovered that if mice were injected with mixtures of *heat-killed* encapsulated (smooth = S) and *live* uncapsulated (rough = R) cells, a curious phenomenon occurred:

$$\text{Living} \atop \text{R cells} \quad + \quad {\text{heat-killed} \atop \text{S cells}} \quad \xrightarrow[\text{into mice}]{\text{injected}} \quad {\text{dead} \atop \text{mice}} \quad {\text{(recovered living,} \atop \text{virulent S cells)}}$$

The R cells recovered the ability to produce capsules and regained the capacity of virulence! There are different antigenic types of capsular material produced by different strains of pneumococci. Consequently, Griffith showed that if the avirulent R cells injected into the mice were derived from capsular type II and the heat-killed cells were of capsular type III, the viable, encapsulated cells recovered from the dead mice were of capsular type III. This indicated a transformation took place so that type II R cells now produced capsules of antigenic type III. Later, in 1944, Avery, et al. demonstrated that it was DNA from the heat-killed encapsulated strain that was responsible for the transformation. They found that if living, rough type II cells were exposed to DNA isolated from type III encapsulated pneumococci, viable type III encapsulated organisms that were virulent for mice could be recovered. These findings were of exceptional importance because they showed that DNA had the ability to carry hereditary information. Conventional wisdom prior to this time held that hereditary traits were more likely to be borne by protein molecules.

Transformation has subsequently been shown to occur in a number of bacterial genera including *Haemophilus, Neisseria, Xanthomonas, Rhizobium, Bacillus,* and *Stapylococcus.* Considerable effort has been exerted to elucidate the nature of **competence** for transformation on the part of recipient cells. Competence is defined as a physiological state that permits a cell to take up transforming DNA and be genetically changed by it. Organisms that undergo natural transformation can be divided into two groups based upon development of the competent state. Some organisms become transiently competent in late exponential phase, for example, *S. pneumoniae.* Others, such as *Neisseria,* are always competent. These different patterns of competence development belie a complex series of regulatory processes required to control this process. The specific physiological and genetic factors involved will be discussed in relation to specific groups of organisms.

Transformation in streptococci has been studied primarily in *S. pneumoniae* and in members of the streptococci belonging to serological Group H. The competent state is transient and persists for only a short period during the growth cycle of a culture of recipient bacteria. The competent state in pneumococci is induced by a specific protein, the **competence activator protein** (10,000 Da). Binding of this activator protein to receptors on the plasma membrane triggers the synthesis of 10 new proteins within 10 min. After induction by CF, cells develop the capacity to bind DNA molecules. The CF activator protein apparently accelerates a normal process of transport or leakage of autolysin molecules into the periplasmic space. The activity of these autolysin molecules from within serve to unmask DNA-binding sites on the plasma membrane. After binding to recipient cell membranes, donor DNA molecules are acted upon by nucleolytic enzymes located at the cell surface of competent recipient cells. Mutant strains of *S. pneumoniae* totally deficient in the major DNase activity of the cell (endonuclease-1 or *end* mutants) are not transformable. These transformation-defective mutants appear to be blocked in the entry of nucleic acid since they still bind DNA to the cell surface. That portion of the donor DNA that gains entrance into the recipient cell is largely in the form of single strands, supporting the concept that the major endonuclease of the cell may serve as a DNA translocase by attacking and degrading one strand of DNA while facilitating the entry of the complementary strand into the cell (Fig. 3-11). Endonuclease I has been shown to be associated with the cell membrane by examination of membrane fractions of spheroplasts of competent pneumococci. Ethylenediaminetetraacetic acid (EDTA), which inhibits DNase activity, blocks the entry of DNA while still permitting surface binding.

The efficiency of integration of genetic markers into the genome of the recipient cell varies drastically with different genes. This variability is due to a genetic trait possessed by the recipient cell termed *hex* (high efficiency of integration). The *hex* system serves to eliminate a large fraction of **low efficiency (LE)** markers while permitting markers with **high efficiency** of integration (**HE** markers) to be incorporated. The *hex* function, responsible for discrimination among markers, is essentially a mismatched-base correction system. Obviously, donor genes that differ from recipient genes by a single base pair will create a mismatch when initially integrated (remember, only a single strand is incorporated). With low efficiency markers, the *hex* mismatch-repair system can correct either donor strand so there is a 50–50 chance a given marker will be retained. However, only the recipient strand is correctable with high-efficiency markers. The precise mechanism for *hex* repair has not been determined. It appears to be a general property of pneumococci, since all strains are capable of discriminating between LE and HE markers

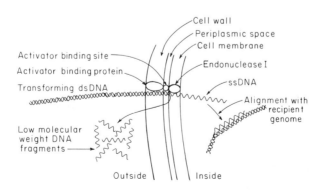

Fig. 3-11. Schematic view of transformation in streptococci.

when transformation is performed with homologous DNA. Mutants that have lost this function (Hex⁻) no longer discriminate between markers and integrate all markers with high efficiency. The hex function is not evident in transformation between heterologous pneumococcal strains, most likely as a result of the sensitivity of the function to saturation by heterologous DNA.

Transformation in Neisseria. Although many reports have appeared in the literature providing evidence for transformation of *Neisseria* using homologous DNA, some of the most interesting studies involve interspecific transformation between species. Auxotrophic mutants of *N. gonorrhoeae* and *N. perflava* can be transformed to prototrophy using either homologous or heterologous DNA. The number of transformants observed in interspecific crosses is invariably much lower than the number of transformants observed in intraspecific crosses. However, certain species appear to be more selective than others with respect to heterologous DNA. Transformation between *N. gonorrhoeae* and *N. meningitidis* has also been described. The efficiency of transformation is usually 10- to 100-fold lower with heterologous DNA than with homologous DNA. Although the ability to transform auxotrophic traits between these two species provides an opportunity for further elucidation of the genetics of these two organisms, it also indicates the existence of a dangerous potential for genetic interaction between two major human pathogens.

Transformation in Haemophilus Influenzae. If exponentially growing *H. influenzae* cells are shifted to nongrowth conditions, 100% of the cells become competent. Competence development appears to be internally regulated. No competence factors have been reported. In *H. influenzae*, changes in the cell envelope accompany the development of the competent state. Envelopes from competent cells exhibit elevated levels of lipopolysaccharide with a composition different from that of log-phase cell envelopes. Six apparently new polypeptides are found in envelopes from competent cells. Most of the polypeptide changes are confined to the outer membrane, although one new polypeptide is associated with the inner cytoplasmic membrane. Structural changes in the envelope also occur in competent cells. Numerous vesicles called **transformasomes** bud from the surface and contain proteins that react specifically with conserved sequences (5′AAGTGCGGTCA3′) present at 4-kb intervals on *Haemophilus* DNA. These vesicles appear to mediate the uptake of transforming DNA. The DNA uptake site is made up of two proteins, 28 and 52 kDa. Several DNA receptor proteins are also produced that recognize the conserved DNA sequence. After binding, the receptor proteins present the donor DNA to the membrane-associated uptake sites. Consequently, DNA binding and uptake is very specific for *Haemophilus* DNA.

In *H. influenzae*, there has been no demonstration of the formation of single-stranded donor DNA during transformation. Donor DNA is taken into competent cells as intact duplex molecules. However, since the final product of recombination is a heteroduplex molecule composed of donor and recipient DNA, the transforming DNA must be incorporated as a single strand. During the competence period, the DNA of the cells is not actively duplicated. Competent cells form a special type of DNA containing single-stranded regions. Competent cells of a mutant of *H. influenzae* that do not permit association of donor and recipient DNA do not contain DNA with comparable single-stranded regions. Shortly after the uptake of donor DNA, single-stranded regions are formed at the ends of donor DNA. Comparable single-stranded regions are formed in recipient cell

DNA. The formation of single-stranded regions in both donor and recipient DNA permits stable pairing between the single-stranded regions.

Transformation in B. subtilis. Unlike *Haemophilus*, *Bacillus* can indiscriminately bind any DNA to its surface. Competence in *B. subtilis* is subject to three types of regulation; nutritional, growth stage specific, and cell-type specific. Competence is developed best in glucose minimal medium. Substitution of glucose with other carbohydrates diminishes the level of competence achieved. Development of the competent state in *Bacillus* occurs postexponentially and may involve nitrogen starvation as a key signal. A variety of studies revealed that competent cells are in an altered metabolic state. This fact is elegantly demonstrated by the ability to separate competent cells from those that are normal based upon buoyant density in Renografin gradients. Competent cells prove to be more buoyant than noncompetent cells reflecting a dramatic physiological change. For example, competent cells are relatively dormant with respect to most forms of macromolecular synthesis. Their chromosomes have completed a round of replication and in essence are all aligned in a terminated configuration. The ability to separate a pure competent population provides a distinct advantage for investigation of the properties of cells in the competent state.

Another fascinating aspect can be observed when using preconditioned competence medium to bring a second culture of *Bacillus* to competence. Preconditioned competence medium is made by growing a culture to competence then removing the cells. The second culture will achieve competence in this medium much faster than did the first culture! This phenomenon is due to some soluble signaling factor since the stimulating activity can be removed by passing the medium through a nitrocellulose column.

Several lines of evidence indicate that donor DNA becomes associated with specific sites on competent cell membranes and remains associated with these membranes until it is integrated. Membrane vesicles isolated from competent cells of *B. subtilis* bind up to 20 μg of double-stranded DNA per milligram of membrane protein in the presence of EDTA (EDTA inhibits the nuclease activities). In addition, the membrane vesicles from competent cells bind up to sixfold more DNA than do membrane vesicles from noncompetent cells indicating DNA binding sites are induced during competence development.

Once bound to the cell a mechanism must exist to transport donor DNA into the recipient cell. As noted earlier with *Streptococcus*, an essential feature of competent cells is a very active cell envelope-associated exonucleolytic activity (products of the *B. subtilis comI* and *comJ* loci). Noncompetent cells are deficient in this activity. Competence-specific exonucleolytic degradation of donor DNA begins 2–3 min after the binding of DNA to the cell. Disruption of *comI* or *J* reduces transformation to 5% of wild type indicating the nuclease complex is an essential function for genetic transformation of *B. subtilis.*

After binding to competent *B. subtilis* cells, donor DNA is converted to double-stranded fragments (20 kb) that can be isolated as early as 30 s after the beginning of the reaction. At this time, these double-stranded fragments are the only recognizable DNA of donor origin. After 1–2 min, the double-stranded fragments are converted to single-stranded DNase-resistant forms indicating membrane association. These single-stranded fragments are intermediates in the transformation process leading to the formation of a complex of donor and recipient DNA. These complexes appear between 2 and 4 min after the beginning of the transformation process. Presumably, the cell-envelope associated exonuclease hydrolyzes one strand of the double-stranded fragment while transferring the sister single-stranded fragment to a membrane-binding site. This degradation may provide some of the energy required to transport DNA across the membrane. However, maintenance of the proton motive force is also required. The identification of several genes that influence transformation in *Bacillus* should lead to a more complete description of this process and its regulation.

Transfection. This phenomenon is defined as the process whereby transformation of cells with purified bacteriophage DNA results in the production of complete virus particles. Transfection has been demonstrated in a number of bacteria, including *B. subtilis, H. influenzae, Streptococcus, Staphylococcus aureus, E. coli,* and *S. typhimurium.*

It should also be noted that organisms not considered naturally transformable (e.g., *E. coli* or *Salmonella typhimurium*) can be transformed under special laboratory conditions. Alterations made in the outer membrane with calcium chloride ($CaCl_2$) or through an electrical shock (electroporation) can be used to transfer DNA, such as plasmids, into cells. This has been an important factor in the success of recombinant DNA research.

Transduction

The transfer of genetic markers from one cell to another mediated by bacteriophage has been termed **transduction**. There are two types of transduction called generalized and specialized transduction.

Generalized transduction is defined as the transfer of any portion of the host or donor cell genome by a bacteriophage. In generalized transduction, the transducing particle contains only bacterial DNA, without phage DNA. During normal nucleic acid packaging (see Chapter 5), the packaging apparatus occasionally packages chromosomal DNA rather than phage DNA. When a transducing bacteriophage binds to a bacterial cell, the donor DNA can be injected into the bacterium by the phage and become integrated into the genome of the new cell through generalized recombination. Thus, in a series of experiments, many different markers from the host cell may be transduced to a recipient cell population. Typical phages that can mediate generalized transduction include P1 (*E. coli*) *and* P22 (*Salmonella*).

A well-studied example of a generalized transduction is that of *Salmonella* phage P22. To understand how generalized transduction occurs one must have some understanding of how this phage replicates and packages DNA into its head. Phage DNA in P22 is linear but circularly permuted and terminally redundant. This means that both ends of a P22 DNA molecule contain duplicate gene sequences but that different P22 molecules within a population contain terminal repeats of different genes. This chromosome structure is due to the headfull packaging method P22 uses to insert DNA into its head. When P22 infects a cell its DNA must circularize through recombination between the terminal redundant ends. The circular molecule replicates forming a concatemer probably via a rolling circle type of DNA synthesis. A concatemer is a long DNA molecule containing multiple copies of the genome. Packaging of P22 DNA into empty heads initiates at a specific region in P22 DNA called the *pac* site, where the DNA is first cut, and then proceeds progressively along the concatemer. Once the P22 head is full, a second cut is made. This second cut also defines the start of packing for the next phage head. This series of events continues until the end of the molecule. The site for the second and subsequent cuts are not sequence specific. To assure that each head receives a complete genome, the system packs a slight bit more than one genome. This is the source of the terminal redundancy of the P22 genome.

When the packaging system encounters a sequence in the bacterial chromosome that is similar to a *pac* site it does not distinguish it from P22 *pac* sites. It will use this homologous site to package chromosomal DNA such that progressive packaging will generate a series of phage particles that carry different parts of the chromosome. The size of DNA packaged is approximately 44 kb, which is about 1/100 of the size of the *Salmonella* chromosome. Therefore, a given P22 will package 1 min worth of chromosomal DNA.

Cotransduction is the simultaneous transfer of two or more traits during the same transduction event. This is an important property of transduction that enables one to map closely linked genes relative to each other. Cotransduction of two or more genes occurs at very low frequencies and requires that the genes be close enough to each other on the host chromosome such that both genes can be packaged into the same phage head. The closer the two genes are to each other the higher the probability they will be cotransduced. This occurs because the DNA available for a potential recombinational (crossover) event that would separate the two genes gets smaller the closer the two genes are to each other.

In **abortive transduction**, the DNA that is transferred to the recipient cell does not become integrated into the genome of the recipient cell. Since this DNA is not replicated, it is transmitted unilaterally from the original cell to only one of the daughter cells and only transiently expresses its genetic information. It is not stably inherited since passage of this DNA eventually is diluted out through subsequent cell divisions.

Specialized transduction is mediated by a bacteriophage that has a high specificity for integrating into a certain site (*att*) on the bacterial chromosome. This specificity limits transfer of genetic material to those markers that are in the immediate vicinity of this site. The most actively studied phage of this type is the lambda (λ) phage. This bacteriophage almost invariably associates with the galactose (*gal*) region of the chromosome and transfers it to recipient cells (see Bacteriophage, Chapter 5).

Both **low-** and **high-frequency** transduction have been described for λ. Figure 3-12 illustrates both classes. First an infecting λ DNA integrates into the bacterial chromosome (Fig. 3-12*a*). In **low-frequency transduction (LFT)**, the integrated phage (**prophage**) may be induced through some forms of stress (e.g., DNA damage) to enter a cycle of lytic infection. As a relatively rare event (10^{-5}–10^{-6}), a portion of the genome of the phage is replaced by a specific segment of the host chromosome (Fig. 3-12*b*). This occurs due to improper excision of integrated prophage DNA in a manner similar to the formation of type I F' factors (see Fig. 3-8). These are defective phages because they lack some portion of the phage genome, but they are capable of transducing and integrating into a new host carrying the original host genes with them (Fig. 3-12*c*). This establishes a merodiploid situation (Fig. 3-12*d*). **High-frequency transduction** usually follows an LFT event. A cell that has been lysogenized by a defective λ (carrying donor DNA) can be induced, following UV irradiation, to yield new phage progeny (see SOS, Chapter 3) if a normal λ has also lysogenized this cell (Fig. 3-12*e*, double lysogen). An HFT lysate is thus produced in which approximately one-half of the particles will be specialized transducing particles (Fig. 3-12*f*).

As with conjugation, transduction usually occurs most readily between closely related species of the same bacterial genus (intrageneric). This preference for related species is due to the need for specific cell surface receptors for the phage. However, intergeneric transduction has been demonstrated between closely related members of the enteric group

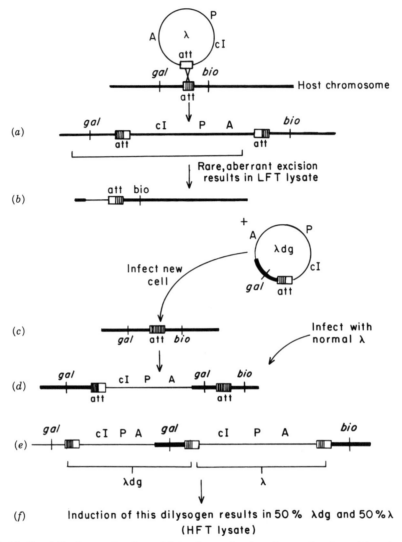

Fig. 3-12. Specialized transduction with λ phage showing the production of low-frequency transducing (LFT) and high-frequency transducing (HFT) lysates.

of organisms, for example, between *E. coli* and *Salmonella* or *Shigella species*. Various genetic traits, such as fermentation capabilities, antigenic structure, and resistance to chemotherapeutic agents, are transducible. Transduction is not limited to the bacterial chromosome. Genetic information residing on plasmids may be transferred by transduction.

Lysogenic conversion is a phenomenon dependent on the establishment of lysogeny between a bacteriophage and the host bacterial cell. Lysogeny occurs when a bacteriophage coexists with its bacterial host without lysing it. In lysogenic conversion, a new phenotypic trait acquired by the host cell is due to a phage gene. Because the gene is part of the normal phage genome, every cell in a population that has been lysogenized acquires the genetic property. This mass conversion of the cell population distinguishes lysogenic conversion from transduction and other genetic-transfer mechanisms.

One of the most interesting and thoroughly investigated examples of lysogenic conversion is the relationship of lysogeny to the production of toxin by the diphtheria bacillus, *Corynebacterium diphtheriae*. Cells that are lysogenized by β phage are designated *tox*⁺. The ability to produce toxin is inherent in the genome of the bacteriophage rather than in any trait that may have been transduced from the original host cell. Thus, cells that are "cured" of the lysogenic state no longer produce toxin. The *tox*⁺ gene is considered to be a part of the prophage genome; however, it is not essential for any known phage function and may be modified or eliminated without any effect on β-phage replication.

The production of erythrogenic toxin by members of the Group A streptococci has also been shown to be the result of lysogenization by bacteriophage. Production of toxin by *Clostridium botulinum*, types C and D, requires the active and continued participation of bacteriophages. Interspecies conversion by bacteriophage has also been described in *Clostridium*. Nonlysogenic strains of *C. botulinum*, type C, can be converted to *C. novyi* by lysogenic phages originally isolated from *C. novyi*. This conversion is apparently the result of alteration of the type toxin produced in response to the presence of the phage genome. Loss of toxigenicity, which is common in some C and D strains of *C. botulinum*, appears to result from reinfection with a nonconverting phage imparting resistance to converting phage.

Lysogenic conversion is also involved in altering the antigenic structure of *Salmonella*. The outer-membrane polysaccharides comprising the O antigens that serve as the basis for serological classification of *Salmonella* can be altered in this manner. Lysogenic conversion of the O antigenic structure can occur among strains of the *S. choleraesuis* Cl group. Conversion of the O antigen of *S. choleraesuis* has been shown to be accompanied by enhanced virulence. A temperate bacteriophage has been shown to alter the O-antigen determinant of *S. newington* imparting a barrier to superinfecting homologous phage. The phage-determining modification results in the addition of glucose units to the repeating mannosyl-rhamnosyl-galactose sequence. Glucose is transferred to the galactosyl unit of the O antigen via a lipid-linked intermediate. Phage mutants that are unable to modify the O antigen are defective either in glucose transfer from uridine diphosphoglucose (UDPG) to a lipid acceptor or in the transfer of glucose from glucosyl-lipid to the O antigen.

RECOMBINATION

The whole key to the success of bacterial genetics is the fortuitous ability of a bacterium to integrate donor DNA into its genome. Without this ability, we as scientists would be considerably more ignorant of the secrets of life. Of course, recombination is not really a philanthropic activity of bacteria toward science but a rather important component of their survival and evolution. There are basically two types of recombination that can occur in *E. coli* and many other bacteria. These are ***recA*-dependent** general recombination (or homologous recombination) and ***recA*-independent** or nonhomologous recombination. General recombination does not produce a net gain of DNA by the recipient genome. Homologous sequences are merely exchanged. The nonhomologous recombinational events require very little sequence homology (as little as 5 or 6 bp) and may be divided into two types: **site specific**, where exchange occurs only at specific sites located on one or both participating DNA molecules, and **illegitimate**, which includes other *recA*-independent events (transposition). In both cases, there is a net gain of DNA by the recipient. These types of recombination have thus been referred to as **additive**.

General Recombination

Most recombination events that occur in *E. coli* following genetic exchange of DNA are mediated by *recA*-dependent pathways and require large regions of homology between donor and recipient DNA. The *recA* product is a *synaptase* that facilitates the alignment of homologous DNA sequences. Lesions in the *recA* locus reduce recombination frequencies by 99.9% Two basic recombinational pathways will be discussed. But before examining the genetics of the systems, an examination of the system itself will be presented along with several models that attempt to explain recombination at the molecular level.

Recombination can be viewed as occurring in five steps: strand breakage, strand pairing, strand invasion/assimilation, chiasma or crossover formation, breakage and reunion and, finally, mismatch repair. A general description of each step is provided below and is followed by a step-by-step outline of events that correspond to the model presented in Figure 3-13.

Initially, a single strand must be produced from the donor molecule that can displace and invade (Meselson–Radding model) one strand of the recipient DNA (see Fig. 3-13). The *recBCD* or *recJ* products may enter at the ends of a molecule and begin to unwind DNA using their helicase activities. The protein RecJ is part of an alternative recombination pathway called the RecF pathway. In either case, the resultant single-strand tail is required to invade the donor DNA and begin the recombination process. The RecA protein (active species is a tetramer of 38,000-Da monomers) is known to promote rapid renaturation of complementary single strands hydrolyzing ATP in the process. It is believed that the *recA* synaptase binds to ssDNA, which spontaneously increases RecA ability to subsequently bind dsDNA. The RecA protein bound to ssDNA then aids in a search for homology between the donor strand and the recipient molecule. Experiments suggest a mechanism in which the RecA-mediated interaction between the searching single strand and the recipient double helix occurs randomly at regions of nonhomology forming nonspecific triple-stranded complexes. The search for homology continues as the RecA protein either reiteratively forms such complexes or translates the two DNA molecules relative to each other until areas of homology are found. Once homologous regions are encountered and the single- and double-stranded DNAs are complexed, a stable D-loop is formed (see Fig. 3-13e). Next, strand assimilation occurs. The donor strand progressively displaces the recipient strand (also called branch migration). Some data suggest that RecA protein will partially denature the recipient DNA duplex and allow the assimilating strand to track in with a $5' \longrightarrow 3'$ polarity. Extensive assimilation seems to involve RecA binding cooperatively to additional areas in the crossover region to form what has been described as ''protein–DNA filament.'' More RecA protein adds to one end of the filament to drive branch migration.

One question that should be addressed is whether the formation of D-loops occur randomly or not. Some data suggest inverted repeats (palindromes) are primary targets for the initial recombination events. Chi sequences, which are naturally present in *E. coli*, clearly enhance recombination and define what are called recombinational hotspots.

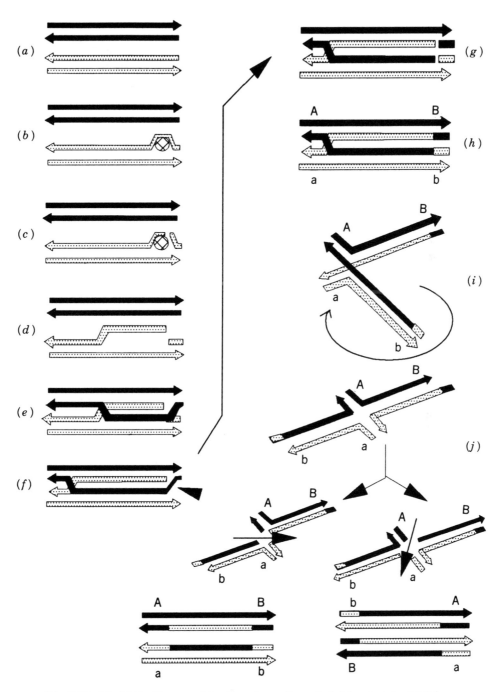

Fig. 3-13. Model for the recombination of genetic material (see text for details).

The Chi consensus sequence is an octomer, 5'-GCTGGTGG-3', and occurs every 5–10 kb. One proposed model suggests that RecBCD enzyme (which possesses unwinding as well as exonuclease and limited, Chi-specific endonuclease activities) travels through and unwinds duplex DNA. Upon encountering Chi, RecBCD enzyme cleaves one strand generating an invasive ss tail. The RecA enzyme and single-strand binding proteins are postulated to synapse this ssDNA tail to a homologous target duplex DNA.

D-loop formation can be coupled with replication of the unpaired donor strand. The fact that certain *dna* mutations cause a decrease in recombination frequencies tends to support the involvement of replication in genetic exchange. Alternatively, a simultaneous assimilation of the sister strand may occur as depicted in Figure 3-13e. While strand assimilation occurs, it is proposed that an endonuclease, such as ExoV (*recBCD* gene product) attacks the D-loop. The endonuclease ExoV possesses both exo- and endonuclease activity. D-loops are known to be susceptible to ExoV. A model that appears to account for most of the experimental evidence (shown in Fig. 3-13) can be described as follows: (a) the RecBCD enzyme moves along a donor DNA duplex until it encounters a Chi sequence (b) at which point a single-stranded nick (c) is introduced creating a single-stranded DNA tail (d). (e) RecA protein, along with ss binding proteins, can bind to the ssDNA region and subsequently bind to recipient duplex DNA searching for regions of homology. (f) Upon encountering a region of homology, D-loop formation occurs. Concomitant to D-loop formation, replication may occur on the unpaired donor strand. Branch migration of the D-loop occurs by cooperative binding of additional RecA tetramers to the donor ssDNA while displacement of the donor strand is driven by replication or by assimilation of the sister strand. At some point a nuclease cleaves and partially degrades the D-loop. The ss tail remaining from the D-loop (f) is trimmed (g) and the exchanged strands ligated (by DNA ligase) to their recipient molecules (h). This structure, as shown in (i), is referred to as the **Holiday intermediate** or **Chiasma** formation and has been observed in electron micrographs (see Fig. 3-14). Mentally rotating the bottom one-half of the molecule from the point of the two-strand crossover to form the structure in (j) helps to visualize the final crossover event. Subsequent breakage and reunion is required to resolve this structure and can occur either in the vertical or horizontal plane. Note the exchange of markers A/a and B/b occur following vertical but not horizontal cleavage. The products of the *ruvC* and *recG* loci are thought to be alternative endonucleases specific for Holiday structures.

Genetics of Recombination.

As alluded to in the previous section, many genes that participate in recombination have been identified in *E. coli*. Table 3-1 provides a list of the known loci associated with recombination in this organism. The identification of all these genes has led to the discovery of multiple recombinational pathways. Mutations in *recA* reduce recombination frequencies to 0.1% of normal levels commensurate with its central role in many homologous recombination systems. One of these pathways was defined by mutations in *recBC*, which reduce the frequency to 1%. Other pathways were discovered following a search for revertants of these *recBC* mutants. Two additional genes were found among the revertants that can suppress the Rec⁻ phenotype. The supressors of *recB* and *recC* (*sbc*) include *sbcA* and *sbcB*. The *sbcA⁺* locus negatively controls the expression of *exoVIII, the recE⁺* gene product. The *sbcA* mutations increase exoVIII activity, which presumably can substitute for exoV (*recBCD*) in the RecBC pathway. It turns out both

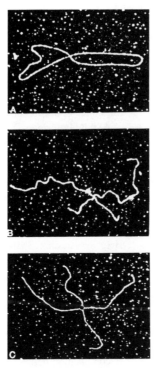

Fig. 3-14. Electron micrographs showing (a) a plasmid figure eight form, (b) a chi form, and (c) a chi form in which the single-stranded connections can be seen in the region of the crossover. Reprinted from Dressler, D. and H. Potter (1982). *Annu. Rev. Biochem.* **51**:727–761, with permission.

sbcA and *recE* are genes present on a lambdoid prophage called *rac*, which maps near the replication terminus.

The fact that *recBC* mutants retain some recombinant ability suggested there were at least two recombinational pathways: one RecBCD dependent, the other RecBCD independent. Support for this theory came from another suppressor of *recBC* mutations, *sbcB*, which encodes exonuclease I. Mutations in *sbcB* decrease exonuclease I activity. How did the discovery of *sbcB* support the existence of multiple recombination pathways? If there are alternate recombinational pathways, one can envision alternative recombinational intermediates. One intermediate being acted upon by one pathway (e.g., the RecBCD-dependent pathway) while a different intermediate is used by the alternative recombinational pathway. Exo I was proposed to shuttle a DNA recombinational intermediate from the alternative pathway to the RecBCD pathway. However, since *recBC* mutants cannot use the RecBC intermediate very little recombination would occur via the alternate, RecBCD-independent, pathway since the important intermediate was siphoned-off by Exo I. Exo I mutants (*sbcB*) would accumulate the alternative intermediate rather than siphon it off, allowing the intermediate to serve as a substrate for the RecBCD-independent system thereby increasing recombination through this pathway. The *sbcB* mutants actually contain a second mutation in *sbcC*, an unlinked gene. Apparently both mutations are required to suppress the *recBC* mutant phenotype although the

TABLE 3-1. Recombination (*rec*) Genes[a]

Gene	Approximate Map Location	Function of Gene, Distinguishing Characteristics of Mutants, or Other Pertinent Information
recA	58	Complete recombination deficiency and many other phenotype defects including suppression of *tif*, DNA-dependent ATPase 38 kDa
recB	61	Structural gene of exonuclease V; 135 kDa, couples ATP hydrolysis to DNA unwinding
recC	61	Structural gene of exonuclease V; 125 kDa
recD	61	α Subunit of exo V; 67 kDa
recE	30	Exonuclease VIII (140 kDa), 5' $\longrightarrow$ 3' dsDNA
recF	83	Recombination deficiency of *recB⁻ recC⁻ sbcB⁻* strains; blocks UV induction of λ prophage
recJ	64.6	Recombination deficiency of *recB⁻ recC⁻ sbcB⁻* strain
recG	82.6	ATPase, disrupts Holiday structures, 76 kDa
recR	11	Help RecA utilize SSB–SSDNA complexes as substrates, 22 kDa
recN	57	Unknown function, 60 kDa
recO	56	promotes renaturation of compl. ssDNA, 31 kDa
ruvA	41.6	Complexes with Holiday junctions
ruvB	41.6	ATPase, dissociates Holiday junctions
ruvC		Endonuclease, Holiday junction, resolvase
recQ	86.5	DNA helicase, 74 kDa
recL (uvrD)	83	Recombination deficiency of *recB⁻ recC⁻ sbcB⁻* strain
sbcA	30	Suppressor of *recB⁻* and *recC⁻* mutations; mutants contain high ATP-independent DNase activity (see *recE*); controlling gene of *recE*
sbcB(Xon)	44	The structural gene for exonuclease I; *recB⁻ recC⁻ sbcB⁻* strains are *rec⁺UV^R mit⁻*

[a]Map positions are derived from the *E. coli* K-12 linkage map published by Bachmann, 1990. Linkage map of *Escherichia coli* K-12 edition 8. *Microbiol. Rev.* **54:**130–197.

reason for this is unclear. The RecBCD-independent pathway is now called the RecF pathway since mutations in *recF* eliminate this alternative recombinational route. The *recF* pathway also requires the participation of the *recJ, N, R, O, Q,* and *ruv* gene products. The role of *recF* in recombination is not yet clear. It may possibly be an endonuclease or, since one *recA* mutation suppresses the recombinational defect of some *recF* mutants, RecF may serve as an accessory protein for the *recA⁺* protein, stimulating RecA binding to single-stranded regions of the chromosome.

Mismatch Repair. Following recombination, the initial duplex product may contain unpaired regions resulting from genetic differences. These regions are unstable and susceptible to mismatch repair. The mechanism involves the excision of one or the other mismatched base along with up to 3000 nucleotides (see p. 142). The RecFJO is also involved in short patch mismatch repair, and so is involved not only in the initial stages of the recombinational process but also at its ultimate resolution, mismatch repair.

Restriction and Modification. The phenomenon of restriction and modification of prokaryotic DNA was discovered when bacteriophage λ was grown on one strain of *E. coli* (*E. coli* K-12) and then used to infect a different strain of *E. coli* (*E. coli* B). It was noticed that the plating efficiency of the virus was less on the B strain relative to the K strain. The plating efficiency is determined by diluting the phage lysate (λ,K) and adding aliquots of diluted phage to tubes containing soft agar and either *E. coli* K-12 or B. The soft agar is then poured onto a normal nutrient agar plate and incubated. After incubation, the plate will be confluent with growth of the *E. coli* except where a phage particle has infected a cell. This area will appear clear and is called a plaque. The plaque appears because progeny phage from the initial infected cell will infect and lyse adjacent cells. An unusual observation was that equivalent amounts of λ,K yielded more plaques on *E. coli* K-12 than on *E. coli* B. However, if λ was isolated from the plaques on *E. coli* B and propagated on *E. coli* B, this lysate (λ,B) had a greater plating efficiency on *E. coli* B than on *E. coli* K. This is the exact opposite of the original observation. The underlying mechanism for this phenomenon is the presence of specific endodeoxyribonucleases known as **restriction endonucleases** in each strain of *E. coli*. Invading bacteriophage DNA originating from the K-12 strain of *E. coli*. will undergo cleavage (**restriction**) by the B endonuclease and then be degraded rapidly to nucleotides by subsequent exonuclease action. Host DNA is not degraded because the nucleotide sequence recognized by the restriction endonuclease has been modified by methylation. The few molecules of bacteriophage DNA that survive the initial infection do so because they have been modified by the host methylase before the restriction enzyme has a chance to cleave the DNA. The methylated phage is thus protected by methylation. Lambda DNA that has been modified in *E. coli* K-12 carries the K modification and is referred to as λ,K, while λ grown on *E. coli* B carries the B modification and is referred to as λ,B. The genetics of this system will be dealt with in more detail below. There are three classes of restriction endonucleases (Table 3-2). The enzymes in the *E. coli* K-12 and B systems belong to Class I while most of the other known restriction enzymes belong to class II.

Class I enzymes (e.g., EcoB, EcoK, and EcoPI) are complex multisubunit enzymes that will cleave unmodified DNA in the presence of *S*-adenosylmethionine, ATP, and Mg^{2+}. These enzymes are also methylases. Each contains two α subunits (135,000 MW), two β subunits (60,000 MW), and one γ subunit (55,000 MW). Prior to binding to DNA, EcoK enzyme binds rapidly to SAM. This initiates an allosteric conformation of the enzyme to an active form that will interact with DNA at random, nonspecific sites. The enzyme then moves to the recognition site with subsequent events depending on the state of the site (Fig. 3-15).

The recognition sequence for both EcoK and EcoB includes a group of three bases and a group of four bases separated by six (EcoK) or eight (EcoB) nonspecific bases. Four of the specific bases are conserved (boxes in Fig. 3-15). The adenines with an asterisk are methylated by the methylase activity. If both strands of the recognition site are methylated when encountered, the enzyme does not recognize it. If one strand is methylated, as would be found immediately on replication, the enzyme binds and methylates the second strand. However, if both strands are unmethylated, the restriction capability is activated. Restriction does not occur at the binding site. Rather, the DNA loops past the enzyme, which does not leave the recognition site, forming supercoils that are cleaved at nonrandom sites approximately 1000 bp from the recognition site.

Elegant genetic experiments have demonstrated the existence of three genes whose products are required for these systems. These include *hsdM* (β subunit, methylase ac-

TABLE 3-2. Characteristics of Types I, II, and III Restriction-Modification Systems

Characteristic	Type I	Type II	Type II
Protein structure	3 different subunits	1 or 2 identical subunits	2 different subunits
Endonuclease and methylase activities performed by	1 multifunctional enzyme	Separate methylase and endonuclease enzymes	1 multifunctional enzyme
Cofactor requirements for endonuclease	ATP, SAM, Mg^{2+}	Mg^{2+}	ATP
Stimulatory cofactors (not required) for endonuclease			SAM, Mg^{2+}
Cofactor requirements for methylase	SAM	SAM	SAM
Stimulatory cofactors (not required) for methylase	ATP, Mg^{2+}		ATP, Mg^{2+}
Cleavage and modification sites	Random, from 1000 bp from recognition site	At or near recognition site	24–26 bp from recognition site
Recognition site	*EcoK*: AACN⁸GTGC *EcoB*: TGAN⁸TGCT	Mostly at sites with dyad symmetry	*EcoP*1: AGACC *EcoP*15: CAGCAG *HinfIII* CGAAT
DNA translocation	Yes	No	No

tivity), *hsdR* (α subunit, restriction endonuclease activity), and *hsdS* (γ subunit, site recognition). Whether EcoK recognizes the K recognition site appears to be dependent on the *hsdS* gene product. Hybrid proteins consisting of HsdM and HsdR from EcoK, and HsdS from EcoB have B site specificity. When characterizing the phenotype of strains carrying mutations in one or more of these genes, r^+m^+ indicates both restriction and modification activities are functioning, r^-m^+ indicates a restrictionless strain that still can modify its DNA, and r^-m^- indicates the loss of both activities. An r^-m^- phenotype can be the result of *hsdR hsdM hsdS$^+$* or *hsdR$^+$ hsdM$^+$hsdS* genotypes since the hsdS product is required for site recognition.

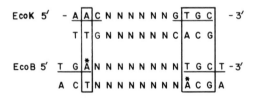

Fig. 3-15. Recognition sequences for the *EcoK* and *EcoB* restriction modification enzymes.

Class II restriction–modification enzymes are less complex and only require Mg^{2+} for activity. These enzymes typically recognize palindromic sequences 4–8 bp in length, depending on the enzyme. Where the Type I enzymes function as a complex including endonuclease and methylase activities the Type II endonucleases and methylases are distinct enzymes. Type II restriction enzymes, some of which are listed in Table 3-3, will only cleave DNA that is unmodified by the cognate Type II methylase. If even one of the strands of the sequence is modified the endonuclease will not recognize it. As opposed to Type I enzymes, these endonucleases do cleave at the recognition site forming either blunt ends (e.g., *Hae*III) by cutting both strands in the center of the site or cohesive ends by introducing staggered, single-strand nicks. An example of cohesive ends is shown below:

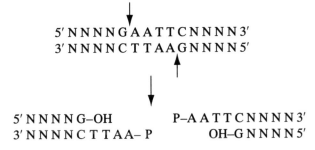

Note that the two cohesive ends can reanneal to each other and that in the presence of DNA ligase will form an intact molecule. These enzymes are heavily responsible for the current explosion in molecular biology in that they are essential in forming recombinant DNA molecules (cloning). The reason is that DNA from two unrelated organisms will both contain sequences that can be recognized by these enzymes *in vitro*. Thus, DNA

TABLE 3-3. Class II DNA Restriction and Modification Enzymes

Enzyme	Restriction and Modification Site[a]	Bacterial Strain
*Eco*RI	G↓A˙ATTC	*E. coli* RY13
	A	
*Eco*RII	↓C˙CTGG	*E. coli* R245
*Ava*I	C↓Py°CGRG	*Anabaena variabilis*
Bam HI	G↓A˙ATTC	*Bacillus amyloliquefaciens II*
*Bgl*II	A↓GATCT	*Bacillus globiggi*
*Hah*I	G°CG↓C	*Haemophilus haemolyticus*
*Hae*III	GG↓CC	*Haemophilus aegyptius*
*Hind*III	˙A↓AGCTT	*Haemophilus influenzae* Rd
*Hinf*I	G↓ANTC	*Haemophilus influenzae* Rf
*Hpa*II	C↓˙CGG	*Haemophilus parainfluenzae*
*Hga*I	GACGCNNNNN↓	*Haemophilus gallinarum*
*Sma*I	CCC↓GGG	*Serratia marcescens Sb*
Dpn	˙AC↓TC	*Diplococcus pneumoniae*

[a]The ↓ shows the site of cleavage and ˙ shows the site of methylation of the corresponding methylase where known. °shows the site of action of a presumed methylase, that is, methylation at this site blocks restriction but the methylase has not yet been isolated.

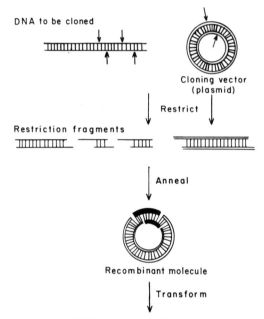

Fig. 3-16. Cloning a fragment of DNA in a plasmid vector. Cloning vector (white) and DNA to be cloned (black) are treated with the same Type II restriction enzyme to generate molecules with complementary, "sticky," end sequences. The cloning vector should only contain a single restriction site for the enzyme being used as is the case in this example. A recombinant molecule is formed by *in vitro* annealing and ligation with a DNA ligase (usually phage T4 DNA ligase). The recombinant molecules are then transformed into suitable *E. coli* hosts (usually r⁻ m⁺ to avoid degrading the foreign DNA insert). An antibiotic resistance gene on the vector will allow selection of cells that received the vector. Clones carrying recombinants may be selected when the DNA inserted complements a mutant defect (e.g., cloning of *arg* genes), insertionally inactivates a readily selectable marker (e.g., tetracyline resistance), or can be detected using labeled DNA probes.

from evolutionarily distinct organisms (e.g., human and *E. coli*) can be cut with *Eco*RI, for example, their DNAs mixed and ligated to each other because of the identical cohesive ends. This forms a recombinant molecule. Figure 3-16 illustrates how a recombinant molecule can be formed *in vitro*. This molecule, because it is a plasmid, can be used to transform *E. coli*.

   Class III restriction–modification systems include those from prophage P1, the *E. coli* plasmid P15, and *Haemophilus influenzae* RF. Unlike the Type I systems, the type III enzymes do not require SAM for restriction and cleave target DNA to fragments of distinct size, typically 25–27 bp away from the recognition sequence.

TRANSPOSABLE ELEMENTS

Transposable elements are discrete sequences of DNA that encode functions to catalyze the movement (**translocation**) of the transposable element from one site to a second, **target**, site (Fig. 3-17A). The target site is duplicated during the transposition event with a copy found to either side of the transposed element. There are essentially two types of

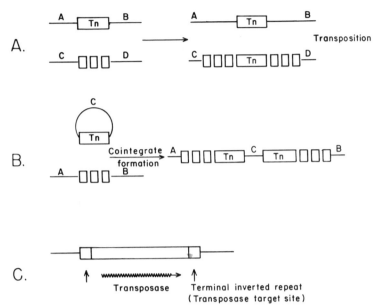

Fig. 3-17. Transposition, cointegrate formation, and a transposable element. (A). Transposition. A schematic presentation of replicative transposition in which a transposable element (Tn) translocates to a new site (▫ ▫ ▫) with concomitant duplication of the site. (B). Cointegrate formation. Replicon fusion mediated by a transposable element. (C). A transposable element. The minimal components of a transposable element include a gene encoding a transposase and terminal target sites for the transposase. These sites are typically the same sequence in inverted orientation. From Reznikoff, W. S. 1983. In *Gene Function in Prokaryotes.* J. Beckwith, J. Davies, and J. A. Gallant (Eds). Cold Spring Harbor Laboratory, Cold Spring Harbor, NY.

transposition. Replicative transposition involves both replication and recombination with a copy of the element remaining at the original site. Conservative transposition does not involve replication. The element is simply moved to a new location. When the target site occurs within a gene, either type of transposition will generate insertion mutations. In addition to causing insertion mutations, transposition of these elements can also cause deletions, inversions, and cointegrate formation in which two distinct replicons (e.g., plasmids) are joined (replicon fusion). Cointegrate formation is illustrated in Figure 3-17B.

There are essentially three classes of transposable elements in *E. coli*:

1. **Insertion sequences,** which encode no function other than transposition, can be simply diagrammed as possessing inverted repeats at either of its ends and a gene encoding the "transposase" responsible for recognizing the terminal repeats and catalyzing the transposition process.

2. **Transposons,** which possess additional genetic information encoding properties, such as drug resistance, unrelated to the transposition process.

3. **Bacteriophage Mu,** a lysogenic bacteriophage that employs transposition as a way of life.

Table 3-4 lists the characteristics of several transposable elements.

TABLE 3-4. Examples of Transposable Elements

Element	Size	Terminal Repeat Element	Target (bp)	Drug Resistance
		Insertion Sequences		
IS1	768	30 bp inverted	9	None
IS2	1,327	32 bp inverted	5	None
IS3	1,400	32 bp inverted	3–4	None
		Transposons		
Tn1	4,957	38 bp inverted	5	Ap^R
Tn5	5,400	1450 bp inverted	9	Km^R
Tn9	2,638	768 (IS1 direct)	9	Cm^R
Tn10	9,300	1400 (IS10) inverted	9	Tet^R
Tn3	4,957	38 bp inverted	5	Ap^R
		Bacteriophages		
Mu	38,000	11 bp inverted	5	None

Before discussing specific examples of transposable elements, it will be useful to discuss a possible model for transposition. The diagram in Figure 3-18 shows the origin of the short, duplicated regions of DNA that flank the insertion element. The target DNA is cleaved with staggered cuts by the transposase, the transposon is attached to the protruding single-stranded ends and the short ends are filled in by repair synthesis leading to the flanking duplications.

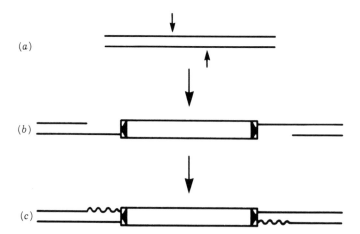

Fig. 3-18. Origin of the short, flanking duplications of target DNA. The target is cleaved with staggered cuts and the extended single strands of the target are then joined to the transposon termini. Repair of the single-stranded gaps completes the duplication.

A model that serves to explain the transposition mechanism is presented in Figure 3-19. The transposase cleaves at either the 5′ or 3′ ends of the transposon (the diagram illustrates the latter) and catalyzes a staggered cut at the target site. Attachment of both ends of the transposon to the target immediately forms two replication forks (Fig. 3-19B and C). Two subsequent paths may be taken. In the first, the transposon is completely replicated with final sealing of the replicated DNA to flanking sequences generating a cointegrate (Fig. 3-19D). Resolution of the cointegrate by genetic exchange between the two transposon copies results in a simple insertion and regeneration of the donor replicon (Fig. 3-19E). While this model appears adequate for Tn3-like transposons, it does not completely hold for IS elements or Mu.

Alternatively, a model has been proposed that would generate simple insertions without requiring cointegrate formation. The model can be seen in Figure 3-19F and G. Repair DNA synthesis could occur at the primer termini in the target DNA and the displaced single strands attaching the transposon to the donor replicon broken. This would result

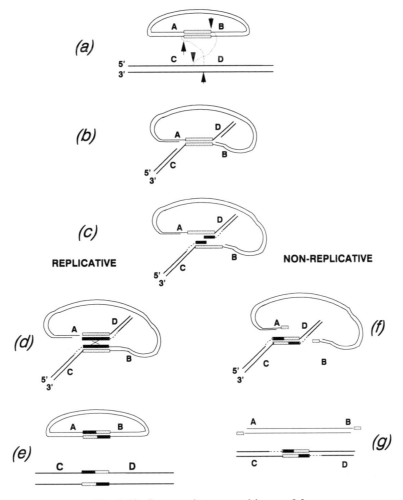

Fig. 3-19. Symmetric transposition model.

in a simple insertion. It would seem that both pathways can be used. However, the ratio of simple insertions to cointegrate formation varies widely from one transposon to another.

Transposon Tn*10*

While there are several examples of transposons that could be discussed in greater detail, the one most clearly understood is Tn*10*. The Tn*10* transposon is actually a composite element in which two IS-like sequences cooperate to mediate the transposition of the entire element. The basic structure of Tn*10* can be seen in Figure 3-20A. There are two IS10 elements that flank a central region containing a tetracycline-resistance locus. The two IS10 elements are not identical, however. It is believed that both started out the same but IS10 left has evolved to the point where its ability to mediate transposition is very low relative to IS10 right.

IS10 right produces a trans-acting function required for transposition (the transposase). There is an inward promoter called p-IN that is required for transcription of this transposition function-locus but there is also an outward promoter (p-OUT) that may be involved in regulating transposition. The p-OUT promoter can also be used to activate genes adjacent to where a Tn*10* has inserted. The IS10-right transposase acts on sites located within the outermost 70 bp of both IS10 elements. This was determined by

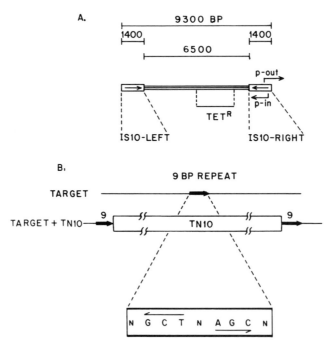

Fig. 3-20. (A). Basic structure of transposon Tn*10*. (B). Position and symmetry of a 6 bp sequence that is responsible for Tn10 specificity. Target DNA is presumed to cleave by staggered single-strand nicks that occur symmetrically at either end of the 9 bp target site sequence that is duplicated during the insertion process. From Kleckner, N. et al. 1982. *Fed. Proc.* **41**:2649–2652.

showing that deletion of this region prevented transposition even when the transposase function was supplied in trans.

The activity of IS10 is regulated by *dam* methylation (see Chapter 2 discussion on replication initiation). The IS10 transposase has two GATC methylation sites, one overlapping pIN . −10 region the other located at the inner end of IS10, where transposase might bind. Methylation of the *dam* site within pIN reduces promoter efficiency of the transposase gene. The reduction in transposase will lower Tn*10* transposition frequency. Methylation of the second *dam* site will lower independent transposition of the constituent IS10 elements thereby increasing the cohesiveness of the composite element. Consequently, the element can only transpose after replication and before the newly synthesized strand is methylated.

Although Tn*10* can insert at many different sites in both *E. coli* and *S. typhimurium* chromosomes, the sites of insertion are not totally random. There is approximately one Tn*10* insertion "hot spot" per thousand base pairs of DNA. The recognition site is a 6 bp consensus sequence (GCTNAGC). This concensus sequence is located within the 9 bp target site that is cleaved by staggered nicks during Tn*10 insertion* (Fig. 3-20B). As originally illustrated in Figure 3-18, the 9 bp target site is duplicated during the insertion process.

As noted earlier, in addition to promoting the movement of a transposon from one site to another, transpositional recombination can result in a variety of genetic rearrangements, such as inversions or deletions. Figure 3-21 illustrates how Tn*10* can generate either deletions (Fig. 3-21A) or inversions (Fig. 3-21B). Note in both cases that the central portion of Tn*10* containing the Tet^R determinant is lost after the rearrangement. In both situations, the inner aspects of the IS10 elements are utilized rather than the outermost sequences, as is the case for transposition. The only difference between whether an inversion or a deletion occurs is the orientation of the target site relative to the transposon.

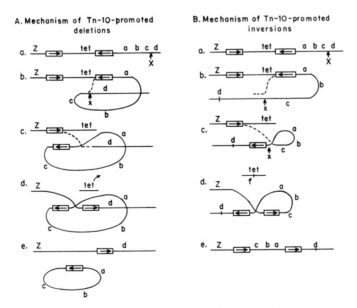

Fig. 3-21. Mechanisms of transposon Tn*10*-promoted rearrangements.

Transposon Tn3

Tn*3* is an example of a complex transposon. It does not have a modular structure like Tn*10*, does not appear to be based on IS elements, and has no obvious evolutionary link to them. But even though Tn*10* and Tn*3* move by different mechanisms, they both go through the same intermediates, as illustrated in Figure 3-19. Figure 3-22 illustrates the basic structure of Tn*3*. The *tnpA* locus codes for the transposase that works at the inverted repeats located at either end of the element. The TnpR protein is the repressor for *tnpA* but also acts as a site specific resolvase. The replicative model for transposition presented in Figure 3-19 involves cointegrate formation in which the transposon has duplicated, joining the donor and recipient molecules. Resolution of this structure requires a recombinational event between the two transposons (Fig. 3-19D). For Tn*3*, the *res* site is the point where recombination, carried out by the TnpR resolvase, occurs.

Conjugative Transposition. Conjugative transposons are characterized by their ability to move between bacterial cells by a process that requires cell-to-cell contact. This phenomenon, seen primarily in gram-positive organisms, was first observed in plasmid-free strains of *Enterococcus faecalis* (*E. hirae*) that contained a tet^R plasmid designated Tn917. This strain could mate and transfer Tn917 to other enterococci. It appears that conjugative transposons can excise from the chromosome, form a "Tn916 circle," and be transferred to a new cell via conjugation. An interesting feature of Tn917 is that the excision process involves dissimilar sequences at either end of the transposon. The Tn916 circle is thus formed by *heteroduplex pairing* of these dissimilar ends. The target sites for transposition usually have significant homologies to these ends of the transposon and so are reminiscent of phage λ integration into the *E. coli* chromosome attachment site by site-specific recombination (see Chapter 5). Little is known about the actual process of cell-to-cell transfer of conjugative transposons and their establishment in the recipient.

Evolutionary Consideration. Insertion sequences and transposons are thought to have been, and probably continue to be, an important component of evolution since they can catalyze major rearrangements of chromosomes as well as serve to introduce new genes into a given organism. For example, there is evidence suggesting that the chromosome of *E. coli*, as we know it, is the result of perhaps two major duplications (which IS-elements can also initiate). One only need note that genes with related biochemical functions often reside in positions 90–180° apart from each other to appreciate this hypothesis. Also, even though *S. typhimurium* and *E. coli* are closely related organisms, a comparison of their genetic maps reveals several inversions and deletions. Conse-

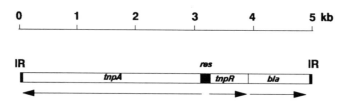

Fig. 3-22. Schematic representation of Tn3. Here IR represents the inverted repeats located at either end; *tnpR*-resolvase locus; *bla*-β-lactamase locus; *res*-site-specific resolution site; transcripts are represented by arrows.

quently, it would appear that insertion elements may be an important key to understanding evolutionary processes.

MUTAGENESIS

During growth of an organism, DNA can become damaged by a variety of conditions. Any heritable change in the nucleotide sequence of a gene is called a mutation regardless of whether there is any observable change in the characteristics (phenotype) of the organism. We will now discuss the various mechanisms by which mutations are introduced and repaired. You may be surprised to learn that it is not usually a chemical agent that causes heritable mutations but rather the bacterial attempt to repair chemically damaged DNA.

But before delving into the molecular details of this process, the terminology of mutations must be defined. For example, a bacterial strain that contains all of the genetic information required to grow on a minimal salts medium is called wild type or **proto-trophic**, whereas a mutant strain requiring one or more additional nutrients is termed **auxotrophic**. Mutations themselves come in a variety of different forms. A change in a single base is called a **point** mutation. The most common type of point mutations are **transition** mutations that involve changing a purine to a different purine (A $\longleftrightarrow$ G) or a pyrimidine to a different pyrimidine. A **transversion** mutation occurs when a purine is replaced by a pyrimidine or vice versa. A point mutation can change a specific codon resulting in an incorrect amino acid being incorporated into a protein. The result is called a **missense** mutation. The codon change could also result in a translational stop codon being inserted into the middle of a gene. Since a mutation of this sort does not code for an amino acid it is called a **nonsense** mutation. A missense mutation allows the formation of a complete polypeptide, whereas a nonsense mutation results in an incomplete (truncated) protein. Nonsense mutations (also referred to as amber, UAG; ochre, UGA; or opal, UAA mutations), if they occur in an operon involving several genes transcribed from a single promoter, may have polar (distal) effects on the expression of genes downstream from the mutation. A mutation of this type is referred to as a **polar mutation**. Figure 3-23 compares a polar nonsense mutation with a nonpolar nonsense mutation.

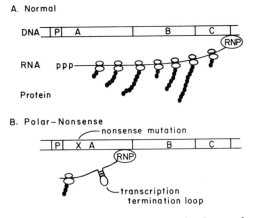

Fig. 3-23. Mechanism of transcription termination by a polar mutation.

Ordinarily, one would not expect a translational stop codon to interfere with the transcription or translation of downstream members of an operon since each gene member of the operon has its own ribosome binding site. The key to understanding polar mutations is that some nonsense mutations will cause premature transcription termination lowering the amount of downstream message produced and available for translation. Premature transcription termination occurs when the nontranslated mRNA downstream from the nonsense codon possesses secondary structure (i.e. stem–loop) that mimics transcription termination signals.

If a process causes the removal of a series of bases in a sequence the result is called a **deletion** mutation. Likewise, the addition of extra bases into a sequence is referred to as an **addition** or **insertion** mutation. Both additions and deletions can result in changing the translational reading frame causing all of the amino acids situated downstream of the mutation to be incorrect. The offending mutation in this event is called a **frameshift** mutation.

One type of mutation that has proven extremely useful is the **conditional** mutation, an example of which is the temperature-sensitive (ts) mutant. These mutants grow normally at a low, permissive temperature (30 °C) but exhibit a mutant phenotype at the higher, nonpermissive temperature (42 °C). An amino acid replacement altering the conformational stability of the gene product is usually thought to be the cause. The mutant protein, being less stable, unfolds at the nonpermissive temperature and becomes inactive.

Spontaneous Mutations

In a population of cells, mutations can arise spontaneously without overt treatment with a mutagen. Spontaneous mutations are rare, ranging from 10^{-6} to 10^{-8} per generation depending on the gene and organism. In addition to the tautomeric considerations discussed below, there are several mispairing schemes that might occur. Recall that the replication apparatus is very accurate with various "proofreading" and repair functions, such as the $3' \longrightarrow 5'$ exonuclease activity of DNA polymerase III. This accounts for the low spontaneous mutation rate. Some mutagens increase the mutation rate by increasing the number of mistakes in a DNA molecule as well as by inducing repair pathways that introduce mutations. Chromosomal rearrangements, such as duplications, inversions, and deletions, can also arise spontaneously.

Mutation rate is calculated as the number formed per cell doubling according to the formula: $a = m$/cell generations = m ln $2/n - n_0$, where a = mutation rate and m is the number of mutations that occur as the number of cells increases from n_0 to n.

THE NATURE OF MUTATIONAL EVENTS

Mutations can arise by a number of molecular events that vary depending in part on the nature of the mutagen. Table 3-5 lists various mutagens, the type of mutation(s) that can arise from each, as well as an indication of the molecular event involved. The following section discusses some of the commonly employed mutagens and their mechanisms for causing mutations. Ultraviolet irradiation at 254 nm causes the production of pyrimidine dimers between thymine-thymine (Fig. 3-24) or thymine-cytosine or cytosine-cytosine pairs. Ultraviolet irradiation may also result in distortion of the backbone of the DNA

TABLE 3-5. Mutagen Action

Mutagen	Specificity	Mechanism[a]
Spontaneous	Substitution	Mispairing
	Frameshift	Slipping
	Multisite	Recombination
	All types	Misrepair
	All types	Misrepair
UV radiation		
Base Analogs		
5-Bromouracil	A-T → G-C[b]	Mispairing
2-Aminopurine	G-T → A-C[c]	Mispairing
Base Modifiers		
Nitrous acid	A-T → G-C	Mispairing
Hydroxylamine	G-T → A-C	Mispairing
Alkylating agents	Mainly transitions[d]	Mispairing
	All types	Misrepair
Intercalators	Frameshift	Slipping
	All types[e]	Misrepair

[a]Mispairing, nonstandard base pairing, may arise spontaneously or through the presence of nucleotide derivatives. Slipping refers to imperfect pairing between complementary strands due to base sequence redundancy. All types of mutations may be induced indirectly by faulty repair mechanisms, misrepair.
[b]G-C → A-T is favored.
[c]A-T → G-C transitions occur 10–20 fold more frequently.
[d]EMS (ethylmethane sulfonate) and MNNG (*N*-methyl-*N'*-nitro-*N*-nitrosoguanidine).
[e]The wide range of lesions induced by ICR compounds may stem from their alkylating side chain.

helix causing replication errors. Ionizing radiations cause instability in the DNA molecule resulting in single-strand breaks.

Chemical mutagens may directly modify the purine or pyrimidine bases causing errors in base pairing. Local distortions in the helix may result in replication or recombination errors. Compounds, such as nitrous acid, cause the deamination of adenine and cytosine and alkylating agents, such as ethylethane sulfonate, produce base analogs that result in **transitions** in base pairing in the nucleic acid structure (Fig. 3-25). Acridine dyes may intercalate between the stacked bases (Fig. 3-26), distorting the structure of the DNA and causing frameshift errors during replication. Abnormal base pairing may also result from tautomeric shifts in the chemical structure of the bases, as shown in Figure 3-27. Normally, the amino and keto forms predominate (~ 85%). Tautomeric shifts occurring during

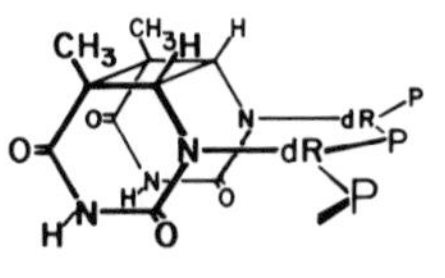

Fig. 3-24. Photodimer of thymine found in UV irradiated DNA. Of four possible stereoisomeric photodimers of thymine, only the *cis* 5,5:6,6 isomer shown is obtained by UV irradiation of DNA. Thymine-cytosine and cytosine-cytosine dimers are also produced. From Camerman, N. and A. Camerman. 1968, *Science* **160**:1451.

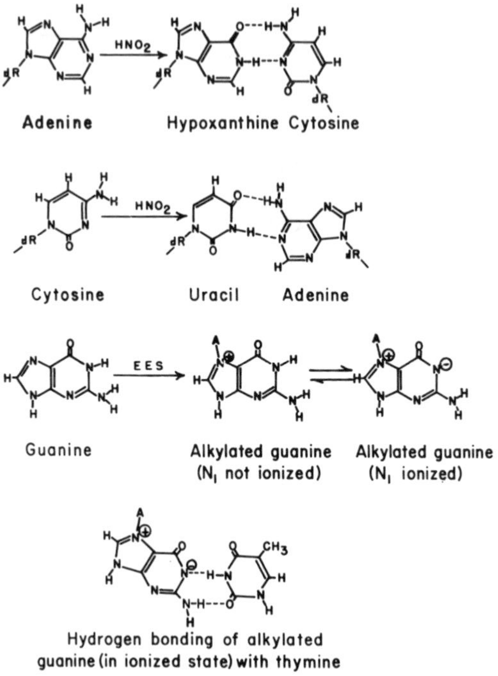

Adenine   Hypoxanthine Cytosine

Cytosine   Uracil   Adenine

Guanine   Alkylated guanine   Alkylated guanine
              (N_1 not ionized)   (N_1 ionized)

Hydrogen bonding of alkylated
guanine (in ionized state) with thymine

Fig. 3-25. Altered base pairing as a result of deamination by nitrous acid or alkylation with ethylethanesulfonate (EES).

the replication of DNA will increase the number of mutational events. Certain analogs of the purine or pyrimidine bases may be incorporated into DNA in place of the normal base. For example, 5-bromouracil, upon incorporation into DNA, may pair with either adenine or guanine, as shown in Figure 3-28, resulting in base transitions in the DNA.

Suppressor Mutations

Suppression is the reversal of a mutant phenotype as a result of another, secondary mutation. The second mutation may occur in the same gene as the original mutation

Fig. 3-26. Diagrammatic view of the manner in which acridine dyes (e.g., proflavin) or chemotherapeutic agents (e.g., ethidium bromide) may intercalate between the stacked base pairs in the dsDNA molecule.

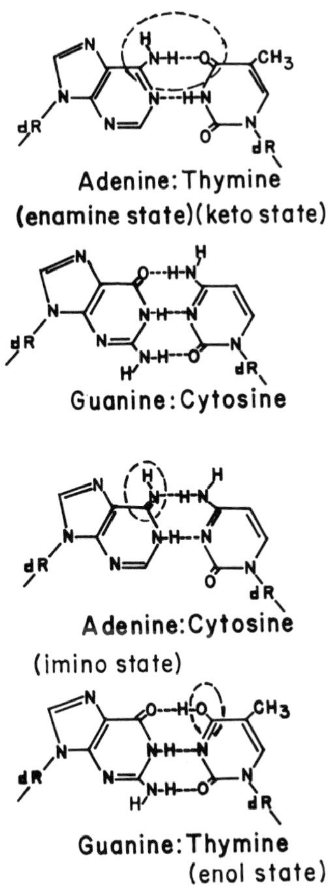

Fig. 3-27. Normal (A=T; G=C) and abnormal (A=C; G=T) base pairing as a result of tautomeric shifting.

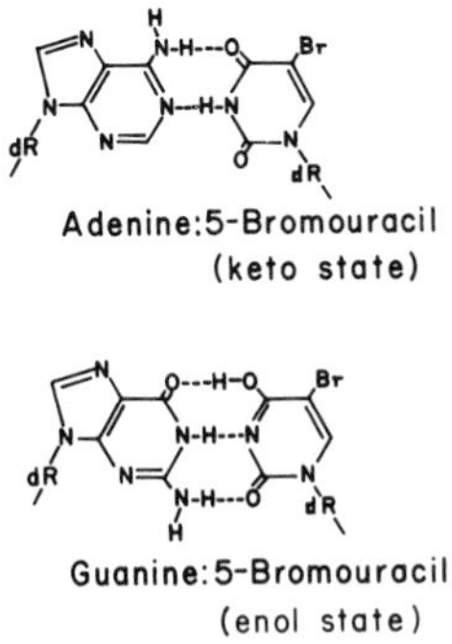

Fig. 3-28. Base pairing of 5-bromouracil (keto state) with adenine and 5-bromouracil (enol state) with guanine.

(**intragenic suppressor**) or in a different gene (**extragenic suppressor**). Intragenic suppressors may cause an amino acid substitution that compensates for the primary missense mutation, thereby partially restoring the lost function or they may be the result of an insertion or deletion that compensates for the original frame-shift mutation. Extragenic nonsense suppressors are usually mutations that alter the genes coding for tRNAs. The revertant phenotypes may be found to contain tRNAs that recognize nonsense codons (e.g., UAA and UAG) and result in the insertion of a particular amino acid at the site of the nonsense mutation. Revertants that are due to suppressor mutations are usually less efficient than the wild type. For example, in the reversion of an auxotroph to wild type as a result of a tRNA suppressor mutation (*sup*), the revertants usually grow more slowly and produce smaller colonies on minimal agar than the wild type. The direct demonstration of a suppressor mutation can be accomplished by genetic crosses since a suppressor mutation does not result in any alteration of the original mutation. The original mutant type can be recovered among the progeny of a cross between wild type and the revertant. Also, by means of appropriate crosses, suppressor mutations can be introduced into other mutants in order to assess their effect on other alleles.

Directed Mutations

Early in this century, scientists began to wonder whether mutations occur spontaneously or were induced or directed when bacteria are exposed to a specific hostile condition. For example, rare phage-resistant variants (mutants) of *E. coli* arise when cells are plated with excess virulent bacteriophage. The question was whether the mutations were caused by the phage or whether the phage simply revealed spontaneous preexisting rare mutants. A classic series of papers published during the 1940s by Luria and Delbruck as well as Joshua and Esther Lederberg demonstrated that the phage resistant mutants were present *before* exposure to virus. The most elegant proof involved replica-plating *E. coli* to identify which colonies were resistant to the phage. The sibling colony, present on the original master plate, had never been exposed to phage yet also proved to be resistant. These experiments seemed to settle the issue, mutations occur spontaneously.

However, in 1988, the issue was reopened in a provocative article by John Cairns. Cairns and colleagues argued the flaw in the earlier work was that a lethal selection was used. Consequently, only preexisting mutants could have been found. The early work clearly demonstrated the occurrence of mutations in the absence of selection, but did not eliminate the possibility that some mutations might be *caused* by selection. To support his argument, Cairns examined the ability of a lactose-negative (Lac⁻) *E. coli* to revert through mutation to Lac⁺. The Lac⁺ revertant colonies will form when Lac⁻ cells are plated on minimal lactose medium. Each day after plating the number of Lac⁺ colonies that appeared increased. The onset of these mutants could be delayed if they were first plated on medium without lactose and subsequently provided with lactose several days later! It seemed many of these late appearing Lac⁺ revertants were being caused (directed) by the presence of lactose!

Additonal work from other researchers confirmed this phenomenon with genes other than *lac*. Several models have emerged that seek to explain these results. One early model required reverse transcription of variable mRNAs produced from a gene with the message producing the best protein being saved as the new gene. A second model proposes that at any given instant during prolonged starvation some fraction of cells in a population might enter into a ''hypermutable'' state, while the remaining cells remain more-or-less immutable. While in a hypermutable state mutations can occur randomly in any gene. If any one of those mutations solves the problem, for example, allowed it to grow on lactose, then that cell would exit the hypermutable state and form a colony. However, if none of the mutations solved the problem then the cell would die from the accumulation of lethal mutations. In the Cairns experiments, cells hypermutable *before* the addition of lactose might still generate *lac*⁺ mutations, but because there is no lactose present they will remain stressed and hypermutable, generating secondary *lethal* mutations. Thus, they will not form colonies once lactose is added. Only cells that have not yet entered the hypermutable state would still be viable when lactose is introduced and, thus have potential for rescuing themselves. This model offers a mechanism with an underlying random basis that does not invoke true directed mutations. An alternative model is that growth of cells on certain carbohydrates (e.g., glucose) catabolite represses error-prone repair pathways. Thus, growth on carbohydrates that do not catabolite repress the error-prone systems will increase mutation rate and the chance that the defect will be corrected. Whatever is the true mechanism, we may no longer be able to view mutations and selections as entirely separate processes.

DNA REPAIR SYSTEMS

Preventing an unacceptably high mutation rate is of extreme importance to the cell. Consequently, a variety of mechanisms have evolved to repair misincorporated residues or bases altered by exposure to radiation or chemical mutagens. Repair pathways can be viewed as either prereplicative or postreplicative and as error-proof or error-prone. The error-prone pathways are often responsible for producing heritable mutations. Table 3-6 lists the genes associated with various aspects of DNA repair in *E. coli.*

Photoreactivation

The cyclobutane ring structure in pyrimidine dimers produced during UV irradiation can be removed enzymatically by the product of the *phr* locus. The *E. coli phr*⁺ gene product (photolyase) will bind to pyrimidine dimers in the dark. However, the reaction that monomerizes the dimer requires activation by visible light (340–400 nm), hence the term photoreactivation. The absorption responsible for photoreactivation is developed only while the enzyme is bound to UV-damaged DNA. The reaction does not require the removal of any bases, just the monomerization of dimers (Fig. 3-24). This activity is, therefore, considered an error-proof repair pathway.

Nucleotide Excision Repair

Bulky lesions, such as UV-induced pyrimidine dimers, can be excised by a complex exonuclease coded for by the *uvrA*, *uvrB*, and *uvrC* genes of *E. coli.* The *uvrABC* nuclease recognizes distortions created by cyclobutane rings (Fig. 3-29). Two single-strand incisions are made, one at the eighth phosphodiester bond 5′ to the dimer and the other at the fourth or fifth phosphodiester bond 3′ to the dimer. The net result is the release of a 12–13 nucleotide DNA fragment that contains the site of damage. Subsequently, DNA polymerase I uses the 3′-OH end of the gapped DNA to synthesize a new stretch of DNA containing the correct nucleotide sequence. Finally, DNA ligase seals the re-

TABLE 3-6. Selected Genetic Loci Associated with Repair

Gene	Map Location	Size	Name and/or Function
uvrA	92	100,000	
uvrB	17	84,000	ATP-dependent endonuclease
uvrC	42	68,000	
uvrD	85	82,000	Helicase II; (*mutU, uvrE, recL*)
ada	48	38,000	Regulates adaptive response, O^8-alkylguanine-DNA alkyl transferase
		18,000	O^6-Alkylguanine-DNA transferase
phr	16	35,800	Photolyase
alkA	43	27,000	3-Methyladenine glycosylase II; sensitive to methylmethanesulfonate
ras	(9)		sensitivity to UV and X-ray
ung	56	24,500	Uracil DNA glycosylase
tag	72	20,000	3-Methyladenine glycosylase I
?		30,000	Hypoxanthine DNA glycohydrolase
xthA	38	28,000	ExoIII, EndoII
nfo	47	33,000	endoIV; AP-specific
?			endoV
xseA	54		ExoVII; $5' \longrightarrow 3'$ exo ss specific
?		30,000	Formamidopyrimidine DNA glycosylase
mutD (*dna*Q)	5	25,000	pol III subunit e
mutT	3		incr. transversion AT $\longrightarrow$ GC
mutH	61	25,000	incr. rate of frameshifts (methyl directed mismatch repair)
mutL	95		incr. rate of AT $\longleftrightarrow$ GC transitions, methyl directed mismatch repair
mutS	59	97,000	incr. rate of AT $\longleftrightarrow$ GC transversions, methyl directed mismatch repair
umuDC	26	16,000/45,000	Error-prone repair
lexA	92	22,700	Repressor of SOS
endA	64		DNA specific endonuclease I
recA	58	38,000	Recombination, effector of SOS regulon.
recBCD	61	Table 3-1	Postreplication repair Exo V
recF	83	37,000	Postreplication repair
dinY	41		DNA-damage inducible, LexA independent
dcm	43		DNA cytosine methylase, 5'-CC(A/T)GG target site
dam	74	32,000	DNA adenine methylase (mismatch correction), 5'-GATC target site
ruvA	41	22,000	See Table 3-1
ruvB	41	37,000	See Table 3-1
ruvC		19,000	See Table 3-1
mutY			Adenine glycosylase excises A from G-A mispair mutants stimulate G-C $\longrightarrow$ T-A transversions
mutA			
			mutants stimulate transversions
mutC			
mutR (*topB*)	38.5		Topoisomerase III, increase in spontaneous deletions

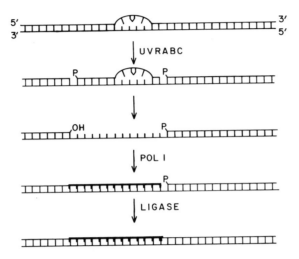

Fig. 3-29. Mechanism of the excision repair system.

maining single-strand nick. The *uvrA* gene product is a DNA-binding protein that also shows ATPase activity. This ATPase activity may reflect a helicase activity for this protein. While the different subunits of the *uvr* endonuclease do not exhibit nuclease activity by themselves, it is proposed that UvrA initially recognizes the lesions and binds at damaged sites; then UvrB and UvrC proteins interact with UvrA catalyzing strand cleavage. The *uvrD* product is required for release of the UvrABC nuclease. Most *uvr*⁺-dependent excision produces a relatively short patch of repair and must be very accurate since UV treatment is far more mutagenic in *uvr* mutants. Thus, excision repair is often referred to as short-patch repair and is error-proof since a proofreading polymerase (pol I) fills in the gap. Short patch repair can also recognize other structural alterations, such as missing bases (AP sites, see below). There is also an error-prone long-patch excision repair pathway that removes several hundred nucleotides that will be discussed below (see **SOS Response**).

An interesting phenomenon occurring both in mammalian cells and in bacteria is that actively transcribing genes are repaired more rapidly than nontranscribed (silent) genes. A current model for this is that the movement of RNA polymerase can be blocked by lesions in the template but not in the coding strand of a gene. The product of the *mfd* (mutation frequency decline) gene, called transcription-repair coupling factor (TRCF), binds to the RNA pol–mRNA–DNA complex causing release of RNA pol and the truncated transcript. The TRCF may replace RNA pol at the lesion site and attract the UvrABC complex by its affinity for UvrA. Then, TRCF and UvrA simultaneously dissociate, leaving the preincision UvrB–DNA complex to bind with UvrC, which makes the dual incisions. This targeting of DNA repair enzymes to transcriptionally active genes is clearly advantageous for cell survival by permitting selective repair of essential genes.

UvrD (Helicase II).

Mutations in the *uvrD* locus result in multiple phenotypes including increased sensitivity to UV, ionizing radiation, and alkylating agents, as well as increased rates of spontaneous and bromouracil-induced mutations. The mutants also exhibit altered rates of recombination and precise transposon excision. Furthermore, double mutants having mutations in both *uvrD* and *polA* are inviable. The *uvrD*⁺ gene product is a polypeptide with a molecular weight of 82,000. It is identical with DNA helicase II and DNA-dependent ATPase I. The UvrD protein is also a component of the DNA damage-indicible SOS system and plays a role in UV repair since *uvrD* mutants present a delayed and incomplete removal of dimers.

Mismatch Repair

Mismatch repair is a postreplicative DNA repair system that recognizes mismatched bases that have eluded the 3′ ⟶ 5′ proofreading function of DNA polymerase. Correction of this sort of mismatch must occur prior to a subsequent round of replication in order to prevent a mutation. The mismatch repair systems require the products of the *mutH*, *mutL*, *mutS*, *mutT*, *dnaQ* (*mutD*) *mutM*, and *uvrD* (*mutU*) genes. Mutations in these genes cause a tremendous increase in the spontaneous mutation rate (10^4 times higher than wild type) referred to as the **mutator** phenotype. To assure that the misincorporated base is removed, some form of discrimination is needed for the repair system to recognize newly synthesized strands from parental strands. One system involves *dnaQ*. The *dnaQ* locus codes for a subunit of DNA pol III involved with 3′ ⟶ 5′ proofreading fidelity. Since proofreading only occurs on the replicating strand, this system naturally discriminates toward the misincorporated base. Another form of discrimination is accomplished through N-6 adenine methylation of DNA by DNA adenine methylase (*dam*) at GATC sequences. Newly synthesized DNA is hemimethylated (methylated on only one strand), while parental DNA is methylated on both strands. Mutants lacking this methylase have a mutator phenotype, that is, since there is no discrimination between unmethylated parent and daughter strands, either the incorrect or correct base may be excised. If the correct base is repaired, then a mutation occurs. The *mutS* gene codes for a 97-kDa protein that will specifically bind to single base pair mismatches and may help direct subsequent repair. The MutH protein is a site-specific endonuclease that cleaves phosphodiester bonds on the 5′ side of *unmethylated* GATC sequences. The MutL protein appears to assist MutH in its activity.

MutY is part of another so-called "short patch" repair system. It is specific for G · A mismatches, which are repaired to G · C pairs. The pathway is independent of MutHLS and DNA helicase II (UvrD). It is not known how MutY discriminates between daughter and parental strands. It is postulated to be part of a two-component system for excluding G · A mismatches from newly synthesized DNA. The other component may be MutT. The *mutT* lesions are also unusual in that they exhibit an increased frequency only of the A-T ⟶ C-G transversion, suggesting that the MutT product is associated with a repair enzyme that specifically recognizes A-G mismatches. Nevertheless, the multitude of gene products involved with mismatch repair implies that a considerable degree of complexity exists.

DNA Glycosylases (Base Excision Repair)

DNA glycosylases catalyze the cleavage of sugar-based bonds in DNA and act only on altered or damaged nucleotide residues. The product of these glycosylase reactions are apurinic or apyrimidinic sites (AP-sites), which can be removed by AP-specific endonucleases that cleave either at the 3′ (Class I) or the 5′ (Class II) side of the AP site

(Fig. 3-30). The AP-specific endonucleases in *E. coli* include the following: (1) exonuclease III (*xth*), although initially identified as an exonuclease, it is also a Class II endonuclease, in which mutations result in an increased sensitivity to hydrogen peroxide; (2) endonuclease IV (*nfo*), a Class II endonuclease with no associated exonuclease activity; and (3) endonuclease V, a Class I endonuclease with an unclear role.

The major DNA glycosylases include **uracil-DNA glycosylase**, the product of the *ung* locus. This enzyme corrects deaminated cytosine residues and is responsible for repairing misincorporated uridine residues during replication. **Hypoxanthine-DNA glycosylase**, in a manner analogous to the uracil-DNA glycosylase, removes spontaneously deaminated adenine residues. No mutant has been found lacking this enzyme. One of the major

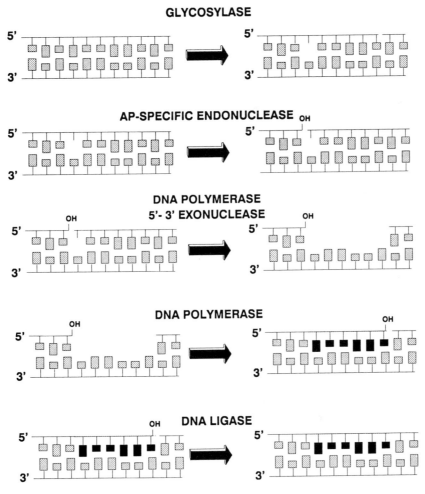

Fig. 3-30. Some enzyme activities implicated in DNA repair. These functions are presented in the sequence in which they are thought to act. Resynthesis (DNA polymerase $5' \longrightarrow 3'$ activity) may or may not take place in conjunction with nucleotide excision (associated $5' \longrightarrow 3'$ exonuclease activity of DNA polymerase I). The enzyme DNA ligase would, of course, be required to complete repair. From Glass, R. E. 1982. *Gene Function. E. coli* and its heritable elements. University of California Press, Berkeley and Los Angeles, CA.

alkylation products in DNA treated with methylating agents is 3-methyladenine. The enzyme **3-methyladenine DNA glycosylase I** (*tag*) removes 3-methyladenine from DNA. There is a second inducible enzyme, **3-methyladenine-DNA glycosylase II**, which is produced during the adaptive response to methylating agents (see below). This enzyme is a product of the *alkA* locus and, in contrast to most DNA glycoslyases, it can release 3-methylguanine, 7-methylguanine, and 7-methyladenine as well as 3-methyladenine from methylated DNA.

The major mutagenic lesion in cells exposed to alkylating agents, such as MNNG, is O^6-methylguanine. Such lesions give rise to mutations by mispairing during replication; repair by SOS functions is not required to produce mutations with this altered base. However, one of the proteins induced by the adaptive response (see below) specifically removes this type of lesion. O^6-**Alkylguanine-DNA alkyltransferase** (*ada*) removes the methyl group from the O^6 position of guanine and transfers it to a cysteine residue of the enzyme itself. Each ''enzyme'' can only act to remove one alkyl group and so is considered a ''suicide'' protein. This raises the semantic question of whether this protein should be referred to as an enzyme since it is ''consumed'' in the process.

Adaptive Response to Methylating and Ethylating Agents

There are three regulatory networks that respond to agents that damage DNA. They are the SOS, heat shock, and adaptive responses. The SOS and heat shock responses will be described later. The adaptive response occurs when *E. coli* is exposed to low concentrations of methylating and ethylating agents (e.g., nitrosoguanidine). Following this exposure, the cells become resistant to the mutagenic and lethal effects of higher doses of these agents. This adaptive response requires protein synthesis and is regulated by the ada^+ gene product, the methyl transferase discussed above. Two of the enzymes induced during the adaptive response are O^6-alkylguanine-DNA alkyltransferase and 3-methyladenine-DNA glycosylase II (*alkA*). The *ada* locus is itself induced during the adaptive response. The ada^+ gene produces a positive regulatory protein that stimulates transcription of the other adaptive response genes including *alkA*, *alkB*, *aid*, as well as *ada* itself. As noted above, Ada is also a methyl transferase, removing methyl and/or ethyl groups from either the O^6 position of guanine or the O^4 position of thymine transfering them to one of the Ada cysteine residues, which inactivates the methyltransferase activity. However, the 39 kDa-methyl Ada continues to activate its own transcription thereby producing more methyl transferase. How is this system ever shut off? the 39 kDa-methyl ada protein is cleaved by a protease into a 19-kDa carboxyl end (possessing the akyltransferase) and a 20 kDa-amino end, which contains the methyl group (Fig. 3-31). The 20 kDa methyl-Ada protein can continue to activate other members of the adaptive response (e.g., *alkA* but cannot activate *ada*. So when the methyl groups are all removed from DNA, no more activator 39 kDa Ada-methyl is made thereby returning *ada* expression back to baseline levels. The 20 kDa methyl-Ada protein will then decrease through proteolytic degradation allowing the whole system to ''stand down.''

Postreplication Daughter Strand Gap Repair

While replication after the introduction of DNA damage often results in a heritable mutation, some DNA damage can be repaired after replication. Figure 3-32 illustrates that repair of DNA damage can be accomplished following replication by two basic routes,

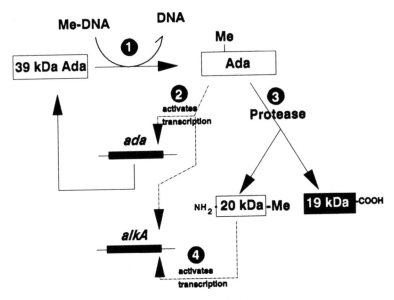

Fig. 3-31. Outline of the adaptive response.

an error-proof recombinational route or an error-prone DNA synthesis route. If a pyrimidine dimer, for example, is not repaired prior to replication, a gap occurs opposite the dimer where normal DNA pol III cannot synthesize DNA because it does not recognize the dimer as template. The gap can be repaired by recombining the appropriate region from the intact sister chain (Fig. 3-32). This pathway is error-proof but requires the $recA^+$,

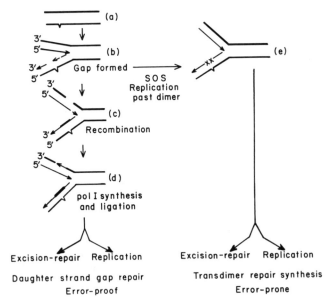

Fig. 3-32. Postreplication repair mechanisms.

recBC[+], *polA*[+], and *uvr* systems. Following recombination, the dimer can either be removed by excision repair or undergo another round of replication. Alternatively, an altered polymerase produced as a result of SOS induction (see below) can be used to synthesize DNA across the dimer. The bases inserted opposite the dimer are randomly chosen, that is, there is a good chance one or another of the inserted bases will represent a mutation.

SOS-Inducible Repair

Exposure of *E. coli* to conditions that either damage DNA or interfere with DNA replication result in the increased expression of genes that are members of the SOS regulatory network. This repair system profoundly affects the mutagenic consequences (i.e., whether a heritable mutation is produced or not) of a number of DNA-damaging treatments. The SOS system is under the coordinate control of *recA* and *lexA*. Aspects of this regulation are outlined in Figure 3-33. In an uninduced cell, the *lexA*[+] gene product acts as a repressor for a number of unlinked genes including *recA* and *lexA*. Several of these genes, such as *recA* and *lexA*, are significantly expressed even in the repressed state relative to the other genes under *lexA* control. The RecA protein, along with its synaptase activity, also possesses a coprotease activity. The RecA protease activity is reversibly activated by some inducing signal generated following DNA damage. The inducing signal probably consists of longer than usual single-stranded regions of DNA stabilized by single-stranded binding proteins. The RecA coprotease can trigger cleavage of the LexA repressor (as well as a few other proteins, such as λ cI repressor) at an Ala-Gly peptide bond yielding two LexA fragments. The RecA coprotease actually activates an autocatalytic cleavage of LexA and CI. In any case, as the quantitative levels of LexA decrease in the cell, the expression of the various SOS genes increases. Genes with operators that bind LexA weakly are the first to become induced. During recovery of the cell from the inducing treatment, RecA molecules return to their proteolytically inactive state allowing LexA repressor to accumulate, thereby restoring repression of the SOS system. Table 3-5 lists some of the genes controlled by the SOS response and the physiological response exhibited by the cell.

One of the physiological responses that occur following UV irradiation is filamentation or a cessation of cell division. A halt in cell division is desirable for DNA damaged cells in order to allow sufficient time for DNA repair systems to fix damage before replication makes the changes permanent. In addition, replication and cell division of heavily UV damaged cells is often lethal. Consequently, one of the facets of the SOS response includes regulation of cell division. The regulatory circuit that controls cell division centers on the *sulA* gene product. The *sulA* gene is normally repressed by the LexA protein but, following mutagenesis, when LexA is cleaved by RecA, SulA protein is synthesized and binds to the *ftsZ* gene product inhibiting its activity. The *ftsZ* gene product normally plays a role in cell division. However, when SulA protein is produced, cell division will not occur. The result is long filamentous cells. As the cell is recovering from irradiation, the *lon* gene product, a protease whose general function is to degrade abnormal cellular proteins (see Chapter 2), will degrade SulA protein allowing cell division to resume.

The major genetic response following SOS induction is the production of mutations. Two genes have been identified that are responsible, at least in part, for mutability. They are *umuD* and *umuC*, which map at 25 min and produce 16,000- and 45,000-Da proteins, respectively. The genes are organized in an operon with *umuD* located upstream of *umuC*

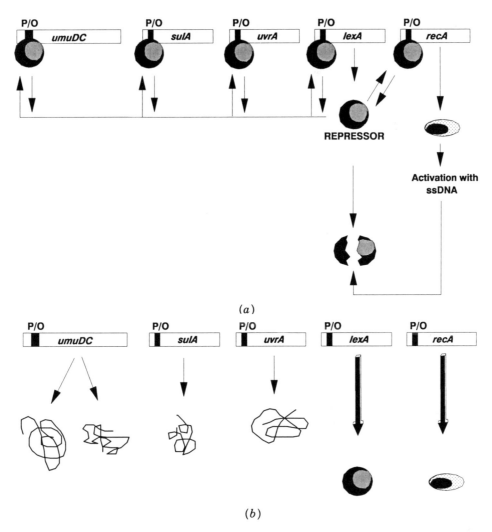

(a)

(b)

Fig. 3-33. Model of the SOS regulatory system. (*a*). The LexA repressor binds to the operator regions of 20 SOS genes (four examples are given including *recA*), inhibiting their expression. The coprotease activity of RecA is activated by DNA damage. Activated RecA will trigger auto-cleavage of LexA. (*b*). The removal of LexA allows the induction of the 20 SOS genes. When DNA damage is repaired the RecA coprotease is deactivated and LexA can reestablish control. As the cell is recovering from irradiation, the *lon* gene product, a protease whose general function is to degrade abnormal cellular proteins (see Chapter 2), will degrade SulA protein allowing cell division to resume.

and are controlled by LexA protein. Apparently the UmuD and UmuC proteins are uniquely required for SOS processing and mutagenesis since mutations in either locus result in a nonmutable phenotype. While other SOS functions are induced in a *umuC* mutant, the mutation rate does not increase. It is interesting to note that elevated levels of UmuCD are not sufficient to promote mutagenesis. The UmuD protein must be posttranslationally modified to an active form. This activation occurs by an autocatalytic cleavage triggered once again by activated RecA.

No one model for UmuCD-induced mutagenesis has received widespread acceptance. However, a plausible model places UmuCD at the site of translesion synthesis. The first step being the insertion of a base opposite a noncoding lesion in the template strand by an altered DNA pol III lacking DnaQ proofreading function. The second step involves UmuCD allowing replication to continue from the mismatched terminus.

Other genes induced by SOS repair, listed in Table 3-7, are involved in long-patch excision repair, excision repair, site-specific recombination, and RecF-dependent recombination.

It should be stressed that many mutations that occur following mutagenic treatments are the result of the error prone $recA^+$ dependent SOS repair systems. It may seem contradictory that a $recA^+ uvrA^-$ strain with its high mutation rate will survive better than a $recA^-$ mutant that does not produce mutations. The apparent paradox can be explained for the following reasons. In $recA^- uvrA^-$ cells, the lack of error-prone repair systems lead to a decreased viability following DNA damage (mutagenic treatments) because of the inability to replicate past gaps in the chromosome thus destroying chromosomal integrity. The mutation rate is lower in $recA^- uvrA^-$ mutants that do survive DNA damage

TABLE 3-7. Some SOS Responses and Genes in *E. coli*

Induced Physiological Response or Gene Function	Induced Genes
Prophage induction	Prophage genes
Weigle reactivation of bacteriophage	*umuDC, uvrA, B,* and *C*
UV mutagenesis of bacterial chromosome	*umuDC, recA*
Filamentation (inhibition of cell division)	*sulA (sfiA)*
$uvrA^+B^+C^+$-dependent repair	*uvrA, uvrB, uvrC*
Long patch repair	*uvrA, uvrB, uvrC*
RecF-dependent recombination	*recN*
Inhibition of DNA degradation by exonuclease V	*recA*
Induced radioresistance	*recA*
Repair of double-strand breaks	*recA*
Induction of RecA protein (roles of homologous recombination; specific protease involved in SOS regulation)	*recA*
Induction of LexA protein (repressor; roles in SOS regulation)	*lexA*
Induction of HimA protein (part of integration host factor; role in site specific recombination)	*himA*
Induction of uvrD protein (helicase II; roles in excision repair and methyl-directed mismatch repair)	*uvrD*
Induction of single-strand DNA-binding protein	*ssb*
Induction of *ruv* locus (unknown role in UV res.)	*ruvAB*
Induction of *dinA* locus (polymerase II)	*dinA-pol II*
Induction of *dinB* locus (function unknown)	*dinB*
Induction of *dinD* locus (function unknown)	*dinD*
Induction of *dinF* locus (in same operon as *lexA*)	*dinF*

because these cells lack the error-prone *recA⁺*-dependent SOS system normally required to synthesize past gaps in DNA. Without the SOS system, incorrect bases are not introduced.

Weigle Reactivation and Weigle Mutagenesis. The experiments that first and most clearly indicated the presence of an inducible repair system involved the infection of UV-irradiated cells with UV-irradiated bacteriophage. Preirradiation of *E. coli* with a low dose of UV was found to greatly increase the survival and mutation rate of UV-irradiated λ phage. This result indicated a repair system that also produced mutations induced in *E. coli* following UV irradiation. These phenomena, discovered by Weigle, have been termed Weigle reactivation and Weigle mutagenesis.

Selected References

Kornberg, A. and T. Baker. 1992. *DNA Replication. Edition two.* W. H. Freeman and Company, New York.

Transformation

Avery, O. T., C. M. MacLeod, and M. McCarty. 1944. Studies on the chemical nature of the substance inducing transformation of pneumococcal types. Induction of transformation by a deoxyribonucleic acid fraction isolated from pneumococcus type III. *J. Exptl. Med.* **79**:137.

Downie, A. W. 1972. Pneumococcal transformation—a backward view. *J. Gen. Microbiol* **73**:1.

Dubnau, D. 1991. The regulation of genetic competence in *Bacillus subtilis. Mol. Microbiol.* **5:** 11–18.

Dubnau, D. 1991. Genetic competence in Bacillus subtilis. *Microbiol. Rev.* **55**:395–424.

Griffith, F. 1928. The significance of pneumococcal types. *J. Hyg.* **27**:113.

Goodgal, S. H. 1982. DNA uptake in *Haemophilus* transformation. *Annu. Rev. Genet.* **16**:169–192.

Kües, U. and U. Stahl. 1989. Replication of plasmids in gram-negative bacteria. *Microbiol. Rev.* **53**:491–516.

Smith, H. O., D. B. Donner, and R. A. Deich. 1981. Genetic transformation. *Annu. Rev. Biochem.* **50**:41–68.

Conjugation

Anderson, T. F., E. L. Wollman, and F. Jacob. 1957. Sur les processus de conjugaison det de recombinaison chez *Escherichia coli. Ann. Inst. Pasteur* **93**:450.

Clewell, D. B. 1981. Plasmids, drug resistance, and gene transfer in the genus Streptococcus. *Microbiol. Rev.* **45**:409–436.

Clewell, D. B. 1992. *Bacterial Conjugation.* Plenum Press, New York.

Clewell, D. B. 1993. Bacterial Sex pheromone-induced plasmid transfer. *Cell* **73**:9–12.

Davies, J. E. and R. Rownd. 1972. Transmissible multiple drug resistance in Enterobacteriaceae. *Science* **176**:758.

Lovett, P. S. and M. G. Bramucci. 1975. Plasmid deoxyribonucleic acid in *Bacillus subtilis* and *Bacillus pumilis. J. Bacteriol.* **124**:484.

Novick, R. P. 1969. Extrachromosomal inheritance in bacteria. *Bacteriol. Rev.* **33**:210.

Willetts, N. and B. Wilkins. 1984. Processing of plasmid DNA during bacterial conjugation. *Microbiol. Rev.* **48**:24–41.

Willetts, N. and R. Skurray. 1980. The conjugation system of F-like plasmids. *Annu. Rev. Genet.* **14**:41–76.

REFERENCES

Recombination

Clark, A. J. 1980. A view of the RecBC and RecF pathways of *E. coli* recombination. pp. 891–899. In B. Alberts and C. F. Fox (Eds.). *Mechanistics of DNA replication and recombination.* Academic, New York.

Dressler, D. and H. Potter. 1982. Molecular mechanisms in genetic recombination. *Annu. Rev. Biochem.* **51**:727–761.

Grindley, N. D. F. and R. R. Reed. 1985. Transpositional recombination in prokaryotes. *Annu. Rev. Biochem.* **54**:863–896.

Umezui, K. N.-W. Chi, and R. D. Kolodner. 1993. Biochemical interaction of the *Escherichia coli* RecF, RecO, and RecR proteins with RecA protein and single-stranded DNA binding protein. *Proc. Natl. Acad. Sci. USA* **90**:3875–3879.

Smith, G. R. 1988. Homologous recombination in prokaryotes. *Microbiol. Rev.* **52**:1–28.

West, S. C. 1992. Enzymes and molecular mechanisms of genetic recombination. *Annu. Rev. Biochem.* **61**:603–640.

West, S. C. and B. Connolly. 1992. Biological roles of the *Escherichia coli* RuvA, RuvB, and RuvC proteins revealed. *Mol. Microbiol.* **6**:2755–2760.

Restriction Modification

Bickle, T. A. 1987. DNA restriction and modification systems. In, *Escherichia coli* and *Salmonella typhimurium:* Cellular and molecular biology. F. Neidhardt, (Ed. in Chief). American Society for Microbiology, Washington, DC.

Transposition

Clewell, D. B. and C. Gawron-Burke. 1986. Conjugative transposition and the dissemination of antibiotic resistance in streptococci. *Annu. Rev. Microbiol.* **40**:635–659.

Grindley, N. D. F. and R. R. Reed. 1985. Transpositional recombination in prokaryotes. *Annu. Rev. Biochem.* **54**:863–896.

Reznikoff, W. S. 1983. Some bacterial transposable elements: Their organization, mechanism of transposition and roles in genetic evolution. In *Gene Function in Prokaryotes.* J. Beckwith, J. Davies, and J. A. Gallant (Eds.). Cold Spring Harbor Laboratory, Cold Spring Harbor, NY.

Scott, J. R. 1992. Sex and the single circle: Conjugative transposition. *J. Bacteriol.* **174**:6005–6010.

Mutagenesis

Balbinder, E., D. Kerry, and C. I. Reich. 1983. Deletion induction in bacteria. I. The role of mutagens and cellular error-prone repair. *Mutat. Res.* **112**:147–168.

Cairns, J., J. Overbauch and S. Miller. 1988. The origin of mutants. *Nature* **335**:142–148.

Drake, J. W. 1991. Spontaneous mutation. *Annu. Rev. Genet.* **25:**125–146.

Foster, P. L. 1992. *Excherichia coli* and *Salmonella typhimurium* mutagenesis. *Encyclopedia of Microbiology.* Vol. 2. Academic, New York, pp. 107–114.

Lederberg, J. and E. M. Lederberg. 1952. Replica plating and indirect selection of bacterial mutants. *J. Bacteriol.* **63:**399.

Luria, S. E. and M. Delbrück. 1943. Mutations of bacteria from virus sensitivity to virus resistance. *Genetics* **28:**491–499.

Repair Mechanisms

Cox, M. M. 1991. The RecA protein as a recombinational repair system. *Mol. Microbiol.* **5:** 1295–1299.

Hannawalt, P. C., P. K. Cooper, A. K. Granesan, and C. A. Smith. 1979. DNA repair in bacteria and mammalian cells. *Annu. Rev. Biochem.* **48:**783–836.

Haseltine, W. A. 1983. Ultraviolet light repair and mutagenesis revisited. *Cell* **33:**13–17.

Lin, J.-J. and A. Sancar. 1992. (A)BC exonuclease: the *Escherichia coli* excision repair enzyme. *Mol. Microbiol.* **6:**2219–2224.

Lindahl, T. 1982. DNA repair enzymes. *Annu. Rev. Biochem.* **51:**61–87.

Little, J. W. 1993. LexA cleavage and other self processing reactions. *J. Bacteriol.* **175:**4943–4950.

Modrich, P. 1991. Mechanisms and biological effects of mismatch repair. *Annu. Rev. Genet.* **25:** 229–253.

Van Houten, B. 1990. Nucleotide excision repair in *Escherichia coli.* *Microbiol. Rev.* **54:**18–51.

Walker, G. C. 1984. Mutagenesis and inducible responses to deoxyribonucleic acid damage in *Escherichia coli.* *Microbiol. Rev.* **48:**60.

Woodgate, R. and S. G. Sedgwick. 1992. Mutagenesis induced by bacterial UmuDC proteins and their plasmid homologues. *Mol. Microbiol.* **6:**2213–2218.

CHAPTER 4

REGULATION OF PROKARYOTIC GENE EXPRESSION

INTRODUCTION

One of the keys to existence for any organism is economy. There is no need for the cell to waste energy synthesizing 20 different carbohydrate utilization systems simultaneously if only one carbohydrate is available. Likewise, it is wasteful to synthesize all the enzymes required to synthesize an amino acid if that amino acid is already present in the growth medium. Wasteful practices like these would jeopardize survival of a bacterium by making it less competitive with the more efficient members of its microbial microcosm. Consequently, regulatory systems designed to maximize the efficiency of gene expression are extremely important and are the focus of this chapter. Prokaryotic gene expression can be controlled basically at two levels, DNA transcription and RNA translation, although mRNA degradation and modification of protein activity also play roles. Most prokaryotic genes that are regulated are controlled at the transcriptional level. This section will deal with the basics of gene expression and describe in detail some of the well-characterized regulatory systems.

TRANSCRIPTIONAL CONTROL

The most obvious place to regulate transcription is at or around the promoter region of a gene. By controlling the ability of RNA polymerase to bind to the promoter or, once bound, transcribe through to the structural gene, the cell can modulate the amount of message being produced, and hence the amount of gene product eventually synthesized. The sequences adjacent to the actual coding region (**structural gene**) involved in this control are called regulatory regions. These regions are composed of the promoter, where transcription initiates, and an operator region, where a diffusible regulatory protein binds. Regulatory proteins may either prevent transcription (**negative control**) or increase transcription (**positive control**). The regulatory proteins may also require bound effector molecules, such as sugars or amino acids, for activity (see *lac* Repressor section in this chapter). Repressor proteins produce negative control while activator proteins are associated with positive control. Transcription initiation requires three steps: RNA polymerase binding, isomerization of a few nucleotides, and finally escape of RNA polymerase from the promoter region allowing elongation of the message. Negative regulators usually block binding while activators interact with RNA polymerase making one or more steps, often isomerization, more likely to occur.

An operon is defined as several distinct genes situated in tandem, all controlled by a common regulatory region (e.g., *lac* operon). The message produced from an operon is **polycistronic** in that the information for all of the structural genes will reside on one mRNA molecule. Regulation of these genes is coordinate, since their transcription depends on a common regulatory region. That is, transcription of all components of the operon either increase or decrease together depending on the situation. Often genes that are components in a specific biochemical pathway do not reside in an operon but are scattered around the bacterial chromosome. Nevertheless, they may be controlled coordinately by virtue of the fact that they all respond to a common regulatory protein. Systems involving coordinately regulated, yet scattered, genes are referred to as regulons (e.g., the arginine regulon).

Figure 4-1 presents a schematic representation of negative versus positive control regulatory circuits. Whether an operon is under negative or positive control, it can be referred to as inducible if the presence of some secondary effector molecule is required to achieve an increased expression of the structural genes. Likewise, an operon can be described as repressible if an effector molecule must bind to the regulatory protein before it will inhibit transcription of the structural genes.

DNA-Binding Proteins

It is evident from Figure 4-1, that regulation at the transcriptional level relies heavily on DNA-binding proteins. Studies conducted on many different regulatory proteins have revealed groups based upon common structural features. The following are brief descriptions of several common families.

Helix–turn–helix (HTH) DNA recognition motifs were the first discovered (see Chapter 5, Lambda). They include the repressors of the *lac*, *trp*, *gal* systems; λCro and CI repressors, and CRP. The proteins consist of an α helix, a turn, and a second α helix. A glycine in the turn is highly conserved, as are the presence of several hydrophobic amino acids. A diagram of how an HTH motif interacts with DNA can be found in Figure 5-14. It should be noted that, unlike some DNA-binding motifs, the HTH motif is not a separate, stable domain but rather is embedded in the rest of the protein. Commonalities among HTH proteins are (1) repressors usually bind as dimers, each monomer recognizing a half-site at a DNA-binding site. (2) The operator sites are B form DNA. (3) Side chains of the HTH units make site-specific contacts with groups in the major groove.

Zinc fingers were first discovered in the *Xenopus* transcription factor IIIA. Proteins in this family usually contain tandem repeats of the 30 residue zinc finger motif (Tyr/phe-X-Cys-$X_{2 \text{ or } 4}$-Cys-X_{12}-His-$X_{3\text{-}5}$-His-X_3-Lys). They contain an anti-parallel β sheet, and an α helix. Two cysteines, near the turn in the β sheet, and two histidines in the α helix coordinate a central zinc ion and hold these secondary structures together forming a compact globular domain. The zinc fingers bind in the major groove of β-DNA and because several fingers are usually connected in tandem they are long enough to wrap part way around the double helix, like a finger.

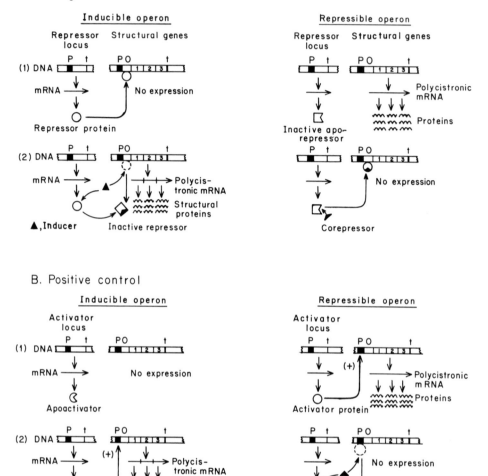

Fig. 4-1. Types of genetic regulatory mechanisms. (A) Negative control mechanisms include inducible and repressible operons. Inducible systems involve a regulatory repressor molecule that prevents transcription of structural genes unless an inducer molecule binds to and inactivates that repressor. Repressible systems also involve a regulatory protein but in this instance the repressor is inactive unless a corepressor molecule binds to and activates the repressor protein. Thus, negatively controlled inducible operons are normally "turned off" while repressible operons are normally "turned on" unless a secondary molecule interacts with the respective regulatory proteins. **(B)** Positive control mechanisms can also be employed with inducible or repressible operons. Inducible systems utilize a regulatory activator protein that requires the presence of a coactivator molecule in order to enhance transcription. Repressible systems produce an activator protein that will enhance the transcription of target structural genes unless an inhibitor molecule is present (P = promoter; O = operator; t = terminator).

Leucine zipper DNA-binding domains generally contain 60–80 residues with two distinct subdomains: The leucine zipper region mediates dimerization of the protein while a basic region contacts the DNA. The leucine zipper sequences are characterized by a heptad repeat of leucines over 30–40 amino acids:

$$\text{L-X}_3\text{-L-X}_2\text{-L-X}_6\text{-L-X}_3\text{-L-X}_2\text{-L-X}_6\text{-L}$$

β-Sheet DNA-binding proteins bind as a dimer with antiparallel β helices filling the major groove (e.g., MetJ). The arrangement is a seven-residue β sheet, a 14–16 residue α helix, and a 15 residue α helix. The β sheet enters the major groove and side chains on the exposed face of the protein contact the base pairs. Both IHF and HU proteins may use β-sheet regions for DNA binding.

The *lac* operon

The operon responsible for the utilization of lactose as a carbon source, the *lac* operon, has been studied extensively and is of classical importance since its examination led Jacob and Monod (1961) to develop the operon model of gene expression. Lactose is a disaccharide composed of glucose and galactose:

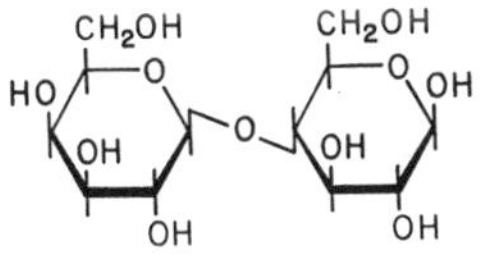

Lactose (4-D-glucose β-D-galactopyranoside)

The product of the *lacZ* gene, β-galactosidase, cleaves the β-1,4 linkage of lactose releasing the free monosaccharides. The enzyme is a tetramer of four identical subunits each with a molecular weight of 116,400. Entrance of lactose into the cell requires the *lac* permease (46,500 MW), the product of the *lacY* gene. The permease is hydrophobic and probably functions as a dimer. Mutations in either the *lacZ* or *lacY* genes are phenotypically Lac⁻, that is, the mutants cannot grow on lactose as a sole carbon source. The *lacA* locus is the structural gene for thiogalactoside transacetylase (30,000 MW) for which no definitive role has been assigned. The promoter and operator for the *lac* operon are *lacP* and *lacO*, respectively. The *lacI* gene codes for the repressor protein and in this system is situated next to the *lac* operon. Usually, regulatory loci that code for diffusible regulator proteins map some distance from the operons they regulate. The *lacI* gene product (38,000 MW) functions as a tetramer.

The *lac* operator is about 28 bp in length and is adjacent to the β-galactosidase structural gene (*lacZ*). Figure 4-2 presents the complete sequence of the *lacP-O* region including the C-terminus of *lacI* (the repressor gene) and the N-terminus of *lacZ*. The operator overlaps the promotor in that the *lac* repressor, when bound to the *lac* operator *in vitro*, will protect part of the promotor region from nuclease digestion. Even so, RNA polymerase can bind to the promotor in the presence of *lac* repressor. The binding of repressor to the operator region situated between the promoter and *lacZ* physically blocks transcription by preventing the release of RNA polymerase to begin transcription.

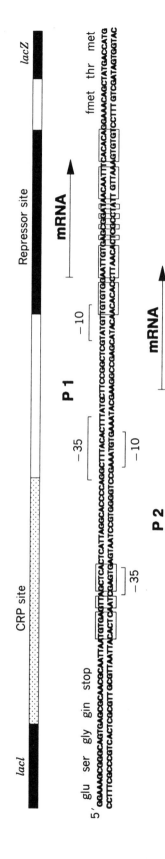

Fig. 4-2. The intercistronic region (122 bp) between the *lacI* repressor and *lacZ* genes. This represents the control site for the entire *lacZYA* operon. It includes a CRP-binding site (or CAP) at positions −84 to −55; an RNA polymerase-binding site (positions −55 to 1), and operator region (1–28). Symmetrical regions in the CRP and operator are boxed. The −10 (Pribnow box) and −35 regions associated with RNA polymerase-binding are indicated by brackets. The symmetrical element of the operator is shown as overlapping the promoter. Modified from Reznikoff, W. S. and J. N. Abelson, 1978. In Miller, T. H. and Reznikoff, W. S. (Eds.). *The Operon*, Cold Spring Harbor Laboratory, Cold Spring Harbor, NY.

Induction of the *lac* operon, outlined in Figure 4-3, occurs when cells are placed in the presence of lactose (however, see Catabolite Repression in this chapter). The low level of β-galactosidase that is constitutively present in the cell will convert lactose to **allolactose** (the galactosyl residue is present on the 6 rather than the 4 position of glucose). Allolactose is the actual inducer molecule. The *lac* repressor is an allosteric molecule with distinct binding sites for DNA and inducer. The binding of inducer to the tetrameric repressor can occur whether repressor is free in the cytoplasm or bound to DNA. Binding of inducer to repressor, however, allosterically alters the repressor lowering its affinity for *lacO* DNA. Once repressor is removed from *lacO*, transcription of *lacZYA* can proceed. Thus, the *lac* operon is a negatively controlled inducible system. It should be noted that experiments designed to examine induction of the *lac* operon usually take advantage of a **gratuitous inducer** (a molecule that will bind repressor and inactivate it but is not itself a substrate for β-galactosidase), such as isopropyl-β-O-thiogalactoside (IPTG). This eliminates any secondary effects the catabolism of lactose may have on *lac* expression (see Catabolite Repression in this chapter).

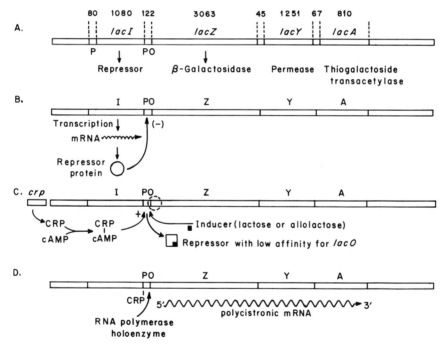

Fig. 4-3. Regulation of the lactose operon. (A) Genetic organization and products of the *lac* operon. Numbers above the horizontal bar indicate the number of base pairs that comprise each gene. (B) Transcription and translation. The *lacI* gene results in the production of the repressor protein which, as a tetramer, binds to the *lac* operator region (see Fig. 9-4) and prevents transcription of the *lac* structural genes. (C and D) In the presence of inducer (i.e., lactose), the inducer binds to and allosterically changes the conformation of the repressor such that it will no longer bind to the operator. This will allow transcription to proceed. Under conditions of high cAMP concentrations, cAMP will bind the cAMP receptor protein (CRP). This complex will bind to a specific CRP site on lacP stimulating polymerase binding and thereby positively affects transcription of the *lac* structural genes.

There are several regulatory mutations that have been identified in the *lac* operon that serve to illustrate general concepts of gene expression. Mutations in the *lacI* locus can give rise to three observable phenotypes. The first and most obvious class of mutations result in an absence of, or a nonfunctional, repressor. This will lead to a constant synthesis of *lacZYA* message regardless of whether inducer is present or not. This is referred to as **constitutive** expression of the *lac* operon. The second class of *lacI* mutations, *lacI*^{s}, produces a **super repressor** with increased operator binding and/or diminished inducer (IPTG)-binding properties. These mutations result in an uninducible *lac* operon. The *lacI* locus contains its own promoter, *lacI*^{p}, distinct from *lacP*. The third class of *lacI* mutations occur in its promoter. Mutations in the *lacI* promoter, called *lacI*^{q}, have been identified as promoter-up mutations that increase the level of *lacI* transcription. The region of the promoter affected by *lacI*^{q} mutations is in the −35 region. Increased transcription due to the *lacI*^{q} mutation occurs by virtue of facilitated RNA polymerase binding. The net result is a 10–50-fold increase in the *lacI* gene product (*lac* repressor). This increased production of LacI repression leads to an almost total absence of basal expression of *lacZYA* yet still allows induction in the presence of inducer.

Mutations in the *lac* operator (*lacO*^{c}) may also result in constitutive expression of the *lac* operon by causing decreased affinity for the repressor. Although Figure 4-3 suggests the presence of a single DNA-binding site for the tetrameric LacI repressor, there are actually three DNA-binding sites, all of which are involved in repression. This introduces the concept of **DNA looping** in the control of transcription. Besides the *lacO* site already introduced, the two other binding sites are O$_1$ at −100 bp (placing it within *lacI*) and O$_z$ at +400 bp (placing it within *lacZ*). Presumably binding of LacI to these sites tethers them together forming competitive loops, meaning either an O–I loop or an O–Z loop can form. The O–I loop represses initiation while the O–Z loop inhibits mRNA synthesis both directly by preventing transcription elongation and indirectly by stabilizing repressor binding at the primary operator, O.

If the *lac* repressor prevents expression on the *lac* operon, then how does the cell allow for production of the small amount of β-galactosidase needed to make allolactose and enough LacP permease to enable full induction when the opportunity arises? A reexamination of Fig. 4-2 reveals the presence of a second *lac* promoter called P$_2$. This promoter binds RNA polymerase tightly but initiates transcription poorly. Once initiation occurs from P$_2$, RNA polymerase can transcribe past the LacI bound operator and into the structural *lacZYA* genes. Note also that the message produced from P2 will contain a large palindromic region corresponding to the operator-binding site. The stem–loop structure that forms will sequester the *lacZ* SD sequence allowing only minimal production of β-galactosidase. This palindromic sequence is not produced from P$_1$.

Catabolite Control

The *lac* operon has an additional, positive regulatory control system. The purpose of this control circuit is to avoid wasting energy synthesizing lactose-utilizing proteins while there is an ample supply of glucose available. The reason for this is that glucose is the most efficient carbon source for *E. coli*. Since the enzymes for glucose utilization in *E. coli* are constitutively synthesized, it would be pointless for the cell to also make the enzymes for lactose catabolism when glucose and lactose are both present in the culture medium. The phenomenon can be visualized in Figure 4-4. This classic experiment shows *E. coli* initially growing on succinate with IPTG as an inducer of β-galactosidase activity

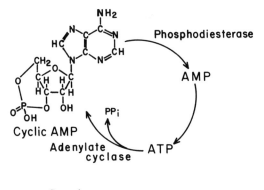

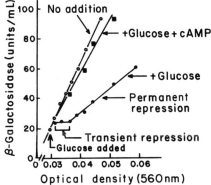

Fig. 4-4. Cyclic AMP (cAMP) formation and reversal of catabolite repression of β-galactosidase synthesis by cAMP. (*top*) Formation of cAMP by adenylate cyclase and its degradation by cAMP phosphodiesterase. (*bottom*) Catabolite repression of **β**-galactosidase synthesis and its reversal by cAMP. Glucose, with or without cAMP, was added to a culture of *E. coli* growing on succinate in the presence of an inducer of the enzyme, IPTG. The cAMP is seen to overcome both the transient (complete) and the permanent (partial) repression by glucose. The brief lag before repression reflects the completion and the translation of already initiated messenger.

(*lacZ* gene product). The early part of the graph shows an increase in β-galactosidase activity due to the induction. Then, at the point indicated, glucose was added to one culture. A dramatic cessation of further β-galactosidase synthesis is observed (**transient repression**) followed by a partial resumption of synthesis (**permanent repression**). It was presumed that a catabolite of glucose was causing this phenomenon, hence the term catabolite repression. However, no catabolite repression was observed if cAMP was added simultaneously with glucose.

The phenomenon is based on the fact that when *E. coli* is grown on glucose, intracellular cAMP levels decrease, but when grown on an alternate carbon source, such as succinate, cAMP levels increase. The enzyme responsible for converting ATP to cAMP is adenylate cyclase, the product of the *cya* locus. How does cAMP increase transcription of the *lac* operon? The evidence indicates that cAMP will bind to the product of the *crp* locus termed the cAMP receptor protein (sometimes referred to as catabolite activator protein, CAP). The CRP–cAMP complex will then bind to the CAP-binding site on the *lac* promoter. Then, CRP will repress transcription from P_2 but activate P_1. Upon binding, it is suggested that the CRP–cAMP bound complex will promote helix destabilization further downstream, facilitating RNA polymerase binding and thus increasing the efficiency of open promoter formation (Fig. 4-2). The result is an increased expression of the *lac* operon. Several studies have shown that CRP causes DNA bending of around 90° or more. The bend enables CRP to directly interact with RNA polymerase. The CRP interaction with RNA polymerase will increase RNA polymerase promoter binding as well as allow RNA polymerase to escape from the promoter and proceed through elongation.

Many operons are affected by cAMP and CRP in a similar manner. These include the *gal*, *ara*, and *pts* operons; all involved with carbohydrate utilization. The genes controlled by CRP are referred to as members of the carbon/energy regulon. Other operons not directly related to carbohydrate catabolism are positively regulated by cAMP as well. These include, among others, *tna*, *ilv*, and *nadA*. The CRP–cAMP complex can also act as a negative effector for several genes including the *cya* and *ompA* loci.

What, then, is the mechanism by which glucose causes catabolite repression? What is the catabolite responsible? A recent model proposes that it is not a catabolite, per se, but rather the inducible component of the phosphotransferase system, III^{glu} (the product of the *pts-crr* gene), which is at the center of this control system. It appears that III^{glu} controls the activity of preexisting adenylate cyclase rather than its synthesis. The model is presented in Figure 4-5. The PTS system employs several proteins, some of which are specific for a given sugar, to transfer a phosphate from phosphoenolpyruvate (PEP) to a carbohydrate during transport of that carbohydrate across the membrane. The phosphate group is transferred from protein to protein until, in the case of glucose, it reaches enzyme

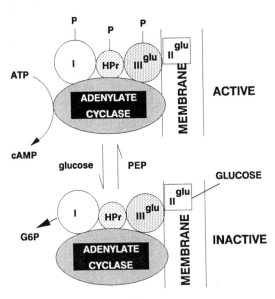

Fig. 4-5. Regulation of adenylate cyclase activity and catabolite repression. Open circles and square represent components of the phosphotransferase system; III, inducible component of phosphotransferase system. (*top*) Disuse of the PTS will lead to the accumulation of phosphorylated-III^{glu} and thus active adenylate cyclase. (*bottom*) Transport of glucose through the PTS will deplete phosphorylated-III^{glu}. The dephosphorylated form of III^{glu} will inhibit adenylate cyclase activity (see text for details).

III^{glu}. In the *absence* of glucose, III^{glu} remains phosphorylated. It can then interact with and activate adenylate cyclase in the presence of phosphate. The result is a dramatic increase in cAMP levels. However, when glucose is available and transported across the membrane, the phosphate on III^{glu} is transferred to the sugar forming glucose-6-phosphate. The dephosphorylated III^{glu} will bind and *inhibit* adenylate cyclase activity causing intracellular cAMP levels to diminish. So when cells are grown on non-PTS sugars (e.g., glycerol), cAMP levels are high and the CRP–cAMP complex forms. This complex then acts at the transcriptional level to alter the expression of several operons either positively or negatively, depending on the operon.

There is another phenomenon called *inducer exclusion* in which nonphosphorylated III^{glu} plays a role. An example of inducer exclusion is the fact that glucose will prevent the uptake of lactose, the inducer for the *lac* operon. It appears that nonphosphorylated III^{glu} inhibits the accumulation of carbohydrates via non-PTS uptake systems (e.g., lactose). It has been shown that a direct interaction occurs between III^{glu} and the lactose carrier (LacP permease); the result being the inhibition of lactose transport.

The *gal* operon

The *gal* operon of *E. coli* consists of three structural genes, *galE*, *galT*, and *galK* transcribed from two overlapping promoters upstream from *galE*, P_{G1} and P_{G2} (Fig. 4-6). Regulation of this operon is complex because aside from being involved with the utilization of galactose as a carbon source, in the absence of galactose the *galE* gene product (UDP-galactose epimerase) is required to convert UDP-glucose to UDP-galactose, a direct precursor for cell wall biosynthesis. While transcription from both promoters is inducible by galactose, it is imperative that a constant basal level of *galE* gene product be maintained even in the absence of galactose. The *gal* operon is also a catabolite repressible operon. When cAMP levels are high the CRP–cAMP complex binds to the -35 region (*cat* site) promoting P_{G1} transcription but inhibiting transcription from P_{G2}. When grown on glucose, however, while cAMP levels are low, transcription can occur from P_{G2} assuring a basal level of *gal* enzymes (Fig. 4-6).

Both of the *gal* promoters are negatively controlled by the *galR* gene product. The *galR* locus is unlinked to the *gal* operon. There are two operator regions to which repressor binds. An extragenic operator (O_E), located at -60 from the P_{G1} transcription start site (S1), and an intragenic operator (O_I), located at position $+55$ from S1 start. It is proposed that a *galR* repressor dimer binds independently to O_E and O_I, and subsequently associate with each other forming a tetramer (Fig. 4-6). This would form a DNA loop that, in effect, sequesters P_{G1} and P_{G2} from RNA polymerase access. Some access must occur, however, since, as mentioned above, basal levels of the *gal* enzymes are synthesized even under repressed conditions.

The Arabinose Operon

Catabolism of the carbohydrate L-arabinose by *E. coli* involves three enzymes encoded by the contiguous *araA*, *araB*, and *araD* genes (Fig. 4-7). The product, xylulose-5-phosphate, can be catabolized further by entering the pentose phosphate cycle, as outlined in Figure 1-11. The transcriptional activity of these genes is coordinately regulated by a fourth locus, *araC*. The *araC* locus and the *araBAD* operon are divergently transcribed from a central promoter region. Both the *araC* promoter (P_c) and the *araBAD* promoter

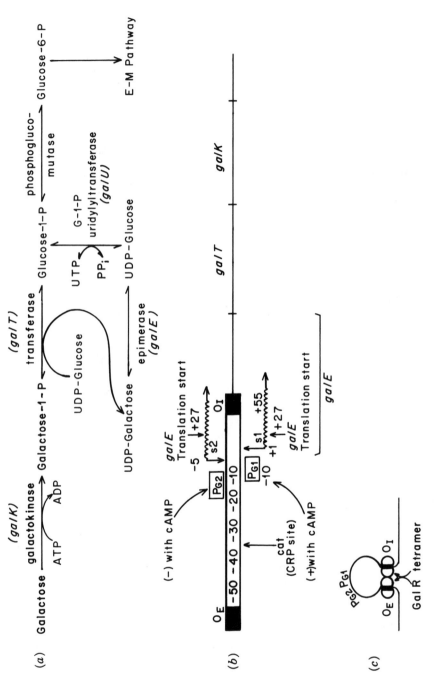

Fig. 4-6. Regulation of the galactose operon. (*a*) Biochemcial pathway. (*b*) Genetic organization of the galETK operon including transcription and translation starts. (*c*) Repression of the *gal* operon by the *galR* product binding to the internal and external operator regions (see text for details).

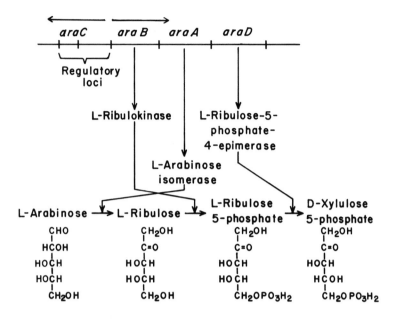

Fig. 4-7. The arabinose operon.

(P_{BAD}) are stimulated by cAMP and CRP. Regulation by *araC* is unusual, however, in that the *araC* product activates transcription of *araP*$_{BAD}$ in the presence of arabinose (inducing condition) yet represses transcription of both promoters in the absence of arabinose (repressive condition). Thus, *araC* protein can assume two conformations: repressor and activator (Fig. 4-8). The *araC* product will also regulate its own expression in the presence or absence of arabinose.

The arabinose operon was one of the first examples of systems that use looped DNA structures to regulate transcription initiation. Figure 4-8 presents the operator and promoter regions. This system uses a series of alternative loops for inducing or repressing the different operons. The AraC protein can bind to four operators; O_1, O_2, I_1, and I_2 (I for induction). Each operator has a different affinity for AraC depending on the condition. In the absence of arabinose, AraC forms a negative repression loop by binding to O_2 and I_1, inhibiting *araBAD* expression (Fig. 4-8B). The binding of sugar to AraC disrupts the O_2–I_1 loop and arabinose-AraC binds to the two I_2 sites, activating transcription of *araBAD*. Occupancy of the I_2 site places AraC near the RNP-binding site perhaps allowing interactions to occur between these two proteins. Repression of *araC* occurs when AraC occupies O_1 and O_2. As the level of AraC increases, its affinity for O_1 must increase. So when enough AraC is made it will repress its own synthesis whether or not it has bound arabinose.

How does cAMP–CRP stimulate transcription from P_{BAD}? *In vitro* evidence indicates that cAMP–CRP complex will disrupt the O_2–I_1 repression loop possibly by bending DNA in a way that is not conducive to forming the repression loop (Fig. 4-8B). The combination of arabinose-binding AraC and binding of cAMP–CRP to its recognition site will cause a dramatic increase in *araBAD* expression.

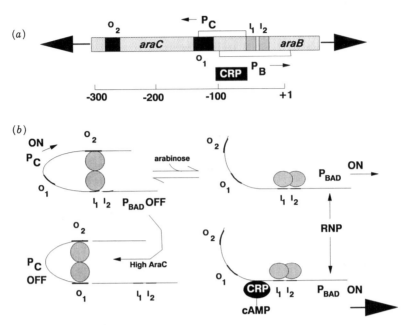

Fig. 4-8. Model for control of the *araBAD* and *araC* promoters. (*a*) The divergent promoter region of the *ara* operon. O_1, O_2, I_1, and I_2 are binding sites for AraC. P_c and P_B represent the promoter regions for *araC* and *araB*, respectively. The CRP box marks the binding region for CRP. (*b*) Alternative AraC-binding configurations that allow differential regulation of this divergently promoted operon (see text for details). RNP, RNA polymerase.

The *trp* Operon: Repression and Attenuation Controls

The genes that code for the enzymes of the tryptophan biosynthetic pathway in *E. coli* are assembled in an operon (Fig. 10-20). There are two regulatory mechanisms that control the expression of this operon at the transcriptional level: transcriptional end-product repression and transcriptional attenuation. Repression involves the product of the *trpR* gene (repressor)-binding tryptophan. This increases the affinity of the repressor for the *trp* operator region, thus blocking transcription. However, there is a second level of regulation placed upon this and many other amino acid biosynthetic operons.

This control mechanism, called attenuation, depends on the coupling that occurs in prokaryotic cells between transcription and translation, as illustrated in Figures 1-8 and 4-11. Between the *trpO* and *trpE* genes is a region called the attenuator region. This region codes for a small, nonfunctional polypeptide called the leader polypeptide. The key to understanding attenuation lies in the secondary structures that can be formed in leader mRNA, depending on the rate of translation relative to that of transcription. Figure 4-9 presents the sequence of the *trp* leader mRNA along with the amino acid sequence of the leader polypeptide. The brackets above the mRNA sequence show regions of mRNA capable of base pairing with each other leading to transcription termination. Region 1 can form a stem structure (Fig. 4-10) with region 2 (1:2 stem) and region 3 can form a stem structure with region 4 (3:4 stem). The 3:4 stem (called the attenuator stem–loop) principally acts as a transcription termination signal. If it is allowed to form, transcription terminates prior to the *trpE* gene. The brackets below the message show

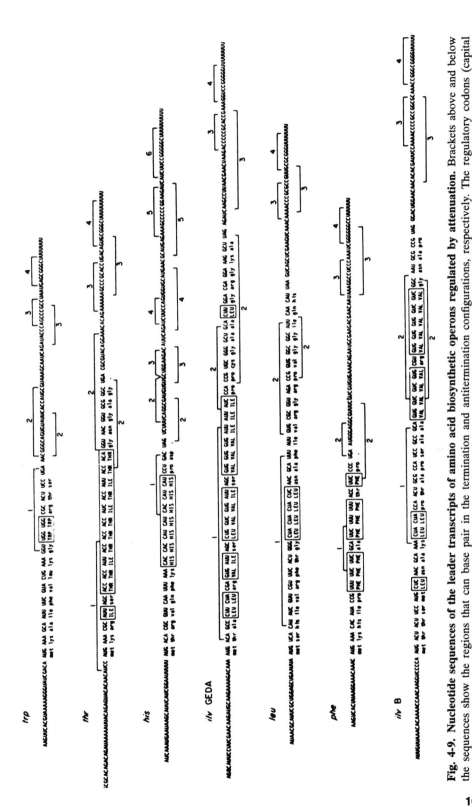

Fig. 4-9. Nucleotide sequences of the leader transcripts of amino acid biosynthetic operons regulated by attenuation. Brackets above and below the sequences show the regions that can base pair in the termination and antitermination configurations, respectively. The regulatory codons (capital letters) are boxed.

165

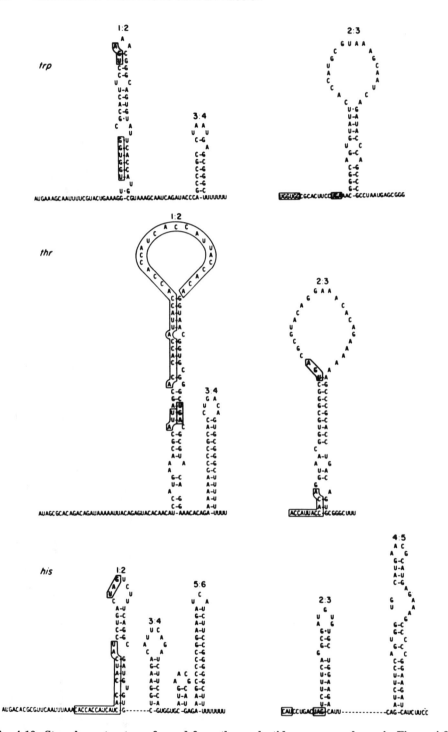

Fig. 4-10. Stem–loop structures formed from the nucleotide sequences shown in Figure 4-9.
From Bauer, C. E., J. Carey, L. M. Kasper, S. O. Lynn, D. A. Waechter and J. F. Gardner. 1983. Attenuation in bacterial operons. In *Gene Function in Prokaryotes*, J. Beckwith, J. Davies, and J. A. Gallant (Eds.). Cold Spring Harbor Laboratory, Cold Spring Harbor, NY. p. 65.

regions that can base pair in an antiterminator configuration (2:3 stem–loop). The formation of a 2:3 loop prevents the formation of the 3:4 terminator stem; thus, transcription is allowed to proceed into the *trp* structural genes.

How, then, does the cell regulate the formation of the 2:3 or 3:4 secondary structures? These mutually exclusive formations occur by varying the rate of translation of the leader polypeptide relative to the rate of transcription (Fig. 4-11). Notice the two adjacent Trp codons in the leader sequence. If tryptophan levels within the cell are low, then the levels of charged tRNA^{trp} will also diminish. As the ribosome translates the leader polypeptide, it will stall over the region containing the trp codons (Fig. 4-11C) overlapping, and thereby preventing region 1 from binding to region 2 and allowing RNA polymerase to synthesize more of the leader transcript. With the 1:2 stem unable to form and once RNA polymerase has transcribed past region 3, the 2:3 stem–loop (called the antiterminator loop) can form, which will prevent formation of the 3:4 terminator. Consequently, RNA polymerase can proceed into the *trp* structural genes, ultimately leading to an increase of tryptophan levels. (There is another Shine-Dalgarno sequence in *trpE* allowing for ribosome binding.)

Once intracellular tryptophan (and, therefore, charged tRNA^{trp}) levels are high, the ribosome will proceed to the translational termination codon present in the leader (Fig. 4-11D). In this situation, the ribosome overlaps region 2 and prevents the formation of a stable 2:3 loop. Consequently, when RNA polymerase transcribes past region 4, the 3:4 transcription terminator stem forms and halts transcription prior to the *trp* structural genes. Attenuation is common for amino acid operon regulation as indicated in Figures 4-9 and 4-10.

Transcriptional Attenuation: The *pyrBI* Strategy

In our discussion of the tryptophan operon, we introduced the concept of attenuation involving ribosome stalling. However, there are other ways a cell can utilize attenuation that do not involve ribosome stalling. In *E. coli* or *Salmonella typhimurium*, the de novo synthesis of uridine monophosphate (UMP), the precursor of all pyrimidine nucleotides, is catalyzed by six enzymes encoded by six unlinked genes and operons (see Chapter 10). One operon within this system is *pyrBI*, which codes for the catalytic (PyrB) and regulatory (PyrI) subunits of aspartate transcarbamylase. The operon is negatively regulated over 100-fold by the intracellular uridine triphosphate (UTP) concentration. Attenuation in this system involves RNA polymerase pausing rather than ribosome stalling. The basic model is presented in Figure 4-12. As with the tryptophan operon, regulation depends on the degree of coupling between transcription and translation. Within the *pyrBI* regulatory region there is an RNA polymerase pause site flanked by U-rich (T-rich in the DNA) areas (Fig. 4-12A). When cellular UTP levels are low (meaning the cell will want to synthesis more UMP), RNA polymerase undergoes a lengthy pause as it tries to find UTPs to extend the message (Fig. 4-12C). This strong pause allows the translating ribosome to translate up to the stalled RNA polymerase. When the polymerase finally escapes the pause region and reaches the transcriptional attenuator region, the rho-independent terminator hairpin cannot form due to the close proximity of the ribosome. This permits RNA polymerase to transcribe into the *pyrBI* structural genes. However, when UTP levels are high, RNA polymerase undergoes only a weak pause, then quickly releases to transcribe the attenuator before the ribosome has caught up (Fig. 4-12B).

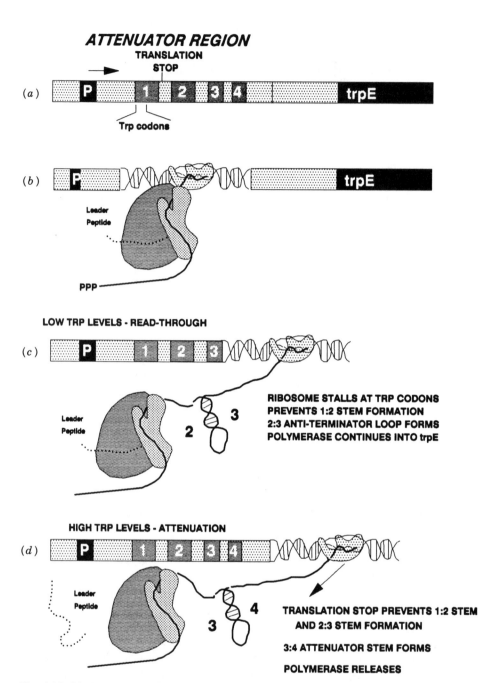

Fig. 4-11. Model for attenuation of the tryptophan operon. (*a*) The attenuator region illustrating positions of promoter(P), tryptophan codons (vertical lines), and regions capable of forming stem–loop structrues (1, 2, 3, 4). (*b*) Coupled transcription and translation of leader polypeptide. (*c*) Under conditions where internal tryptophan levels are low, the ribosome stalls over the *trp* codons in the message due to a depletion of charged tryptophanyl tRNA. This prevents 1:2 stem formation but allows the 2:3 stem to form. Consequently, the 3:4 attenuator loop cannot form and will permit RNA polymerase to read-through into the *trp* structural genes. (*d*) At high levels of internal tryptophan, the ribosome encounters the translational stop codon preventing the formation of the 1:2 and 2:3 stem structures. The RNA polymerase continues, allowing the transcriptional stop 3:4 stem to form. Transcription does not proceed into the *trp* structural genes.

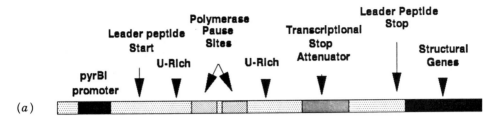

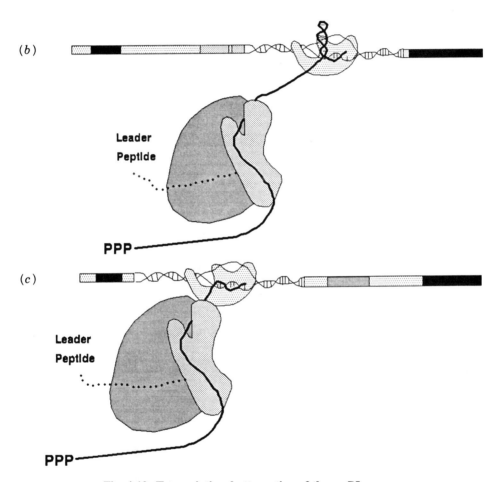

Fig. 4-12. Transcriptional attenuation of the *pyrBI* operon.

Therefore the transcription termination loop forms, terminating transcription before RNA polymerase has entered stuctural genes.

It may seem that attenuation is a very costly and inefficient regulatory system for the cell. Why would nature provide a regulatory mechanism that involves the synthesis of nonfunctional polypeptides or transcripts? Remember, all regulatory systems require information, and so of necessity require the expenditure of energy. The energetic cost of attenuation may be less than that required to synthesize a larger transacting regulatory protein. A clear advantage of attenuation is that it uses information present in the nascent RNA transcript. This is a strategy that quickly and directly translates what is sensed (i.e., low trp levels) into action (i.e., increased transcription).

Translational Control: The *pyrC* Strategy

The *pyrC* locus encodes the pyrimidine biosynthetic enzyme dihydroorotase (see Chapter 10). Expression of this enzyme varies over 10-fold depending on pyrimidine availability. Regulation is not due to the modulation of mRNA levels but is the result of an elegant translational control mechanism. The promoter region for *pyrC* is shown in Figure 4-13. Transcription starts at one of the bases indicated by an asterisk and depends on pyrimidine availability. When grown under pyrimidine excess, transcription starts predominantly at position C_2. Transcripts started at this site can form a hairpin loop due to base pairing between regions indicated by arrows. This hairpin blocks ribosome binding by sequestering a Shine–Dalgarno sequence and lowers production of the *pyrC* product. Alternatively, when pyrimidine availability is low, transcripts preferentially start at G_4. This shorter transcript cannot form the hairpin and so will have an SD sequence available for

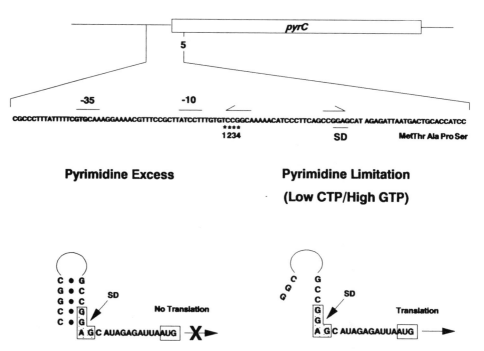

Fig. 4-13. Translational control of *pyrC* expression. The top figure shows the *pyrC* locus with the promoter region expanded. Arrows above the sequence indicate the areas capable of base pairing in the mRNA. The numbered asterisks indicate potential transcriptional start sites. The choice of start site varies with the ratio of CTP to GTP in the cell. The bottom pair of illustrations indicate the mRNA step–loop structure that forms during pyrimidine excess but fails to form during pyrimidine limitation due to the shorter transcript. This stem–loop structure will sequester a ribosome binding site and prevent synthesis of the *pyrC* product. Adapted from Wilson, H. R. et al. 1992, *J. Bacteriol.* **174:**514–524.

ribosome binding. This translational control of *pyrC* is a striking example of a regulatory mechanism employing only the basic components of the transcriptional and translational machinery. The RNA polymerase directly senses nucleotide availability and by differential transcription initiation will control translational initiation of the transcript.

Arginine

The genes involved with the biosynthesis of arginine (see Chapter 10) are scattered around the chromosome of *E. coli* (Fig. 14-14A), yet are all coordinately regulated by the product of the *argR* locus. Hence, this system is referred to as the *arg* regulon. The ArgR protein is unique among known repressors in that it is a hexamer. Most known repressors function as a dimer or as a tetramer (e.g., LacI). Regulation of the *arg* regulon is not strictly coordinated in response to changes in intracellular arginine since repression/derepression ratios for the various enzymes differ dramatically (Table 4-1). The *argR* repressor binds arginine as a corepressor. A possible explanation for the variable repression ratios of the various *arg* genes was discovered following the sequencing of several *arg* regulatory regions. Upstream, yet close to the coding portion of the gene, there are one or two copies of a conserved 18 bp sequence showing dyad symmetry. These sequences are called ARG boxes and are responsible for repressor binding (Fig. 4-14). The base composition and degree of homology between the ARG box sequences, as well as the number of ARG boxes and their positions relative to the promoter, may be the basis of the differential repressor effectiveness in the arginine regulon.

Membrane-Mediated Regulation

The Put System. *Salmonella typhimurium* can degrade proline for use as a carbon and/or nitrogen source. The process requires two enzymes: a specific transport system encoded by *putP* and the *putA*-coded bifunctional membrane-bound proline oxidase, which converts proline to glutamate (Fig. 4-15). The two genes are adjacent but are transcribed in opposite directions from a common controlling region. The *putA* gene product also possesses regulatory properties in that it represses the expression of both *putA* and *putP*. The regulatory and enzymatic properties of *putA* are separable by mutation. That is, some *putA* point mutations result in the constitutive expression of *putAP*, yet do not affect proline oxidase activity.

The PutA protein is a flavoprotein that can exist as a membrane-bound enzyme associated with the electron-transport chain. The PutA protein may require interaction with specific sites in the cell membrane to become enzymatically active. Regulation by PutA appears to involve its reversible sequestration by the cytoplasmic membrane. Proline interaction with PutA allows its interaction with the membrane, in effect removing the protein from the cytoplasm. The depletion of cytoplasmic PutA repressor causes induction of the *put* genes. Once the membrane sites are filled with PutA or the levels of free proline drop, excess PutA may accumulate in the cytoplasm and act as a repressor of the *put* genes to prevent their overexpression.

Recombinational Regulation of Gene Expression (Flagellar Phase Variation)

Salmonella alters the antigenic type of the polypeptide flagellin, the major structural component of bacterial flagella. Two genes, *fliC* (formerly H1) and *fljB* (formerly H2),

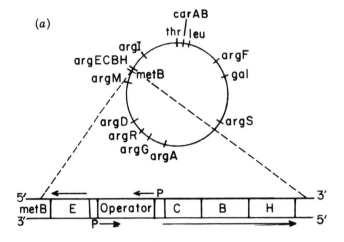

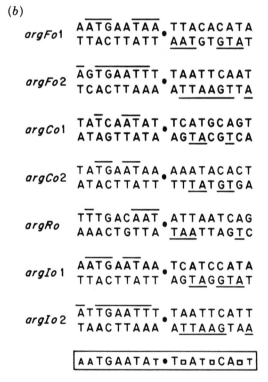

Fig. 4-14. (*a*) **Orientation of the *arg* genes in *E. coli* K-12.** (*b*) **The ARG box; symmetries are underlined.** In the consensus sequence (last line), nucleotides in large type are those showing at least 75% conservation. Nucleotides in small type exhibited 50% conservation. From Cunin, R. 1983. Regulation of Arginine Biosynthesis in Prokaryotes. In *Amino Acids: Biosynthesis and Genetic Regulation,* K. M. Hermann and R. L. Somerville (Eds.). Addison-Wesley, Reading, MA, p. 53.

TABLE 4-1. Repression–Derepression Ratios in the Arginine Regulon of *E. coli*[a]

Transcription Units	Strain	
	K-12	B
argA	250	
argCBH	60 (50–70)	6–10
argD	20	5
argE	17	3
argF[b]	300	
argI	600	18
argG[b]		
carAB	3.8/26[c]	
argR	10	

[a]Specific enzyme activities in extracts of isogeneic *argR/argR*[+] strains grown in excess arginine.
[b]Bacteria grown in excess arginine and uracil.
[c]Two ARG boxes present.

encode different forms of the flagellin protein but only one gene is expressed in the cell at any one time. The organism can switch from the expression of one gene to the other at frequencies from 10^{-3} to 10^{-5} per cell per generation. The two loci do not map near each other with *fliC* at 40 min and *fljB* at 56 min. Figure 4-16 presents the model for the flagellar phase variation switch. The principle of the control involves the protein-mediated inversion of a 993 bp region of DNA called the H region. The H region is flanked by two 26 bp sequences designated *hixL* (repeat left) and *hixR* (repeat right). Each 26 bp sequence is composed of two imperfect 13 bp inverted repeats. Note in Figure 4-16 that the promoter for the *fljB* and *fljA* (the *fliC* repressor) operon resides within the H region. Thus, when this region occurs in one orientation (Fig. 4-16A), the promoter is oriented to transcribe *fljB* and *fljA*. Consequently, the H1 locus is repressed by the *fljA* product (also known as rH1) and H2 flagellin is synthesized. The *hin* gene product (22 kDa) is the recombinase that mediates the inversion of this region. When the region between *hixL* and *hixR* is inverted (Fig. 4-16B), the promoter is reoriented and now reads away from the H2 operon preventing the synthesis of H2 flagellin and rH1. Without rH1, the *fliC* gene can be expressed, producing H1 flagellin. Within the H region are **enhancer** sequences that stimulate the inversion process 100-fold. Enhancers are cis-acting DNA sequences that "magnify" the expression of a gene. They do not require any specific orientation relative to the gene and can function at great distances

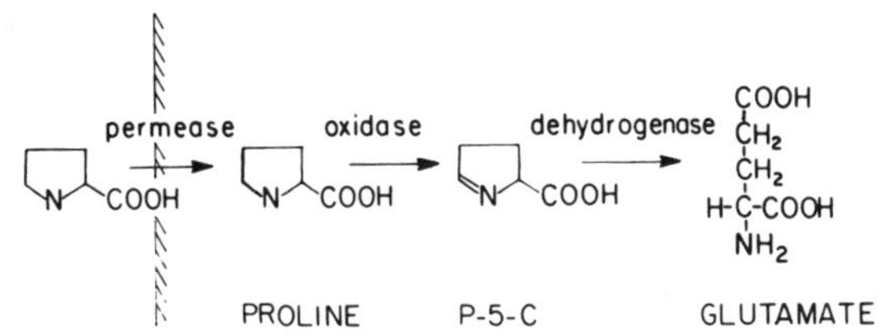

Fig. 4-15. Proline degradation pathway. Pyrroline-5-carboxylic acid (P-5-C). From Maloy and Roth. 1983. *J. Bacteriol.* **154**:561–568.

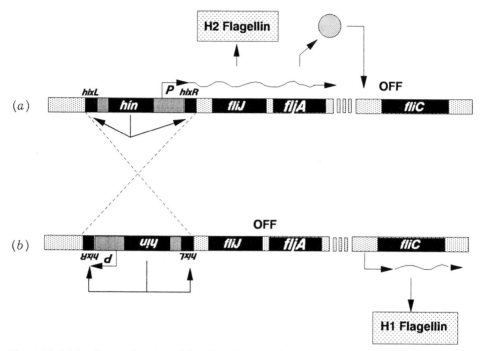

Fig. 4-16. Molecular mechanism of flagellar phase variation in *Salmonella typhimurium* (see text for details). P = promoter regions.

from it. The stages of inversion include Hin binding to the *hix* sites and Fis (a 12-kDa dimeric protein) binding to the enhancer. The *hix* sites are brought together by protein–protein interactions (Hin–Hin and Hin–Fis). The intermediate looped structure is called an **invertasome**. Hin then makes a two-base staggered cut in the center of each *hix* site. Strand rotation probably occurs through subunit exchange since the strands are covalently attached to Hin. After inversion, Hin religates the four strands. Similar mechanisms are used by other organisms to regulate the expressions of alternative genes.

TRANSLATIONAL REPRESSION

The synthesis of ribosomes in *E. coli* is regulated relative to the growth rate of the cell. The synthetic rates of the 52 ribosomal proteins and the three rRNA components of the ribosome are balanced to respond coordinately to environmental changes. The set of ribosomal protein operons comprise approximately 16 transcriptional units having between 1 and 11 genes (Fig. 4-17). The model for the autogenous control of ribosomal protein synthesis states that the translation of a group of ribosomal protein genes encoded on a polycistronic message is inhibited by one of the proteins encoded within that operon. It appears, for example, that L1 of the L11 operon can bind to its own message preventing its translation. Many of the proteins that serve as translational repressors bind specifically to either 16S or 23S rRNA during ribosome assembly. Thus, if growth slows, leading to a decrease in rRNA, the translational repressors will bind instead to their own

	L1	L11	P	L11 OPERON
←	+	+		in vitro
	(+)	+		in vivo

β'	β	L7/12	L10	P	β OPERON
←	−	+	+		in vitro
−	−	+	(+)		in vivo

	EF-Tu	EF-G	S7	S12	P	str OPERON
←	−	+	+	−		in vitro
	−	+	(+)	−		in vivo

S17	L29	L16	S3	(S19, L22)	L2	L23	L4	L3	S10	P	S10 OPERON
−	−	−	−	−	±	+	+	+	+		in vitro
+	+	+	+	+	+	+	(+)	+	+		in vivo

L15	L30	S5	L18	L6	S8	S14	L5	L24	L14	P	spc OPERON
ND	ND	−	−	−	−	+	+	−	−		in vitro
+	+	+	+	+	(+)	+	+	−	−		in vivo

L17	α	S4	S11	S13	P	α OPERON
←	−	+	+	+		in vitro
+	±	(+)	+	+		in vivo

Fig. 4-17. Organization and regulation of genes contained within the *str-spc* and *rif* regions of the bacterial chromosome. Genes are represented by the protein product. (→) For each operon, the direction of transcription from the promoter (P). It has been shown that the L11 and β operons are probably a single operon. The L11 promoter functions as the major promoter for all genes contained within the L11 and β operons in exponentially growing cells. Regulatory ribosome proteins are indicated by boxes. Effects of the boxed proteins on the *in vitro* and *in vivo* synthesis of proteins from the same operon are shown. The L10 protein can function as a repressor in a complex with L12. A + indicates specific inhibition of synthesis; a − indicates no significant effect on synthesis; a ± indicates weak inhibition of synthesis; a (+) is inhibition presumed to occur *in vivo*; ND is not determined. It has not been established how the regulation of the synthesis of ribosomal proteins S12, L14, and L24 is achieved. From Nomura, M., S. Jinks-Robertson, and A. Miuera. 1982. Regulation of Ribosome Synthesis in *Escherichia coli*. In *Interaction of Translational and Transcriptional Control in the Regulation of Gene Expression*. M. Grunberg-Manago and B. Safer (Eds.). Elsevier, New York, p. 92.

message. Homologies have been found between the binding sites on the rRNA and the target site on the mRNA for several operons. From a single target site on a polycistronic message, a repressor can affect the translation of all the sensitive downstream cistrons through "sequential translation." That is, translation of downstream cistrons can be dependent on the translation or termination of the upstream target cistron. Notice that in each operon presented in Figure 4-18, not all of the cistrons in a polycistronic message are affected.

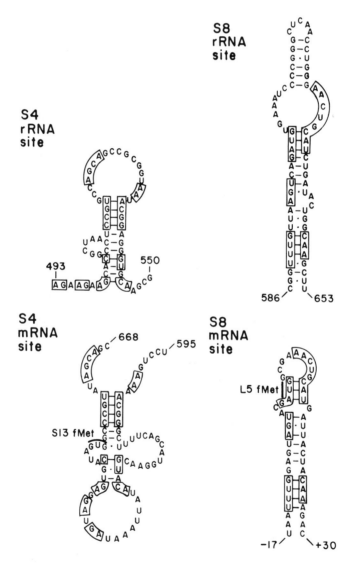

Fig. 4-18. Predicted secondary structures of S4- and S8-binding sites on 16S rRNA and their respective mRNAs. The binding site of S4 on the mRNA overlaps the S13 initiation region; the mRNA site for S8 overlaps the L5 initiation region. Boxed sequences indicate homologies. From Campbell, K. M. et al. 1983. In *Gene Function in Procaryotes.* J. Beckwith, J. Davies, and J. A. Gallant (Eds.). Cold Springs Harbor Laboratory, Cold Springs Harbor, NY.

COMMUNICATION WITH THE ENVIRONMENT: TWO COMPONENT SYSTEMS

Bacteria constantly sense and adapt to their environment in order to optimize their ability to survive and multiply. We have just covered a variety of regulatory systems each of which utilizes a single regulatory protein as a sensor of the internal environment. Over the past few years it has become obvious that distantly related bacteria, such as soil bacteria, commensal or symbiotic bacteria, and pathogenic bacteria often utilize two component signal transduction mechanisms to respond to environmental stimuli. Stimuli that can be sensed by two component systems include pH, osmolarity, temperature, the presence of repellents and attractants, nutrient deprivation, nitrogen and phosphate availability, plant wound exudates, and others. The general theme that has emerged is that a protein spanning the cytoplasmic membrane will sense an environmental signal and then interact with regulatory protein components in the cytoplasm, often through phosphorylation–dephosphorylation reactions, thereby regulating gene expression (Fig. 4-19). This is called a two-component signal transduction system. Although each system only responds to specific stimuli, they are functionally similar. This conclusion is based on the strong amino acid conservation between domains of these proteins (Fig. 4-20). As illustrated in Figure 4-19, the first component is the sensor (or transmitter) protein, usually, but not always, a transmembrane protein. The extra cytoplasmic portion (sensor) senses the environment, then

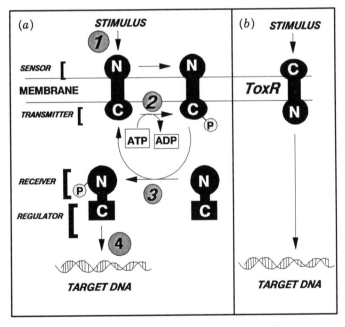

Fig. 4-19. General model for signal transducing systems. (*a*) Two component system for signal transduction. (1) An external stimulus interacts with the periplasmic (usually N-terminus) portion of the sensor/transmitter protein. This activates an autokinase activity in the C-terminal cytoplasmic region (2). The phosphoryl group is then transferred to the receiver regulator protein activating its DNA-binding function (3 and 4). (*b*) One component system of signal transduction. The example, ToxR from *Vibrio cholerae*, is an integral membrane protein that can directly bind to target DNA sequences when activated by an external stimulus.

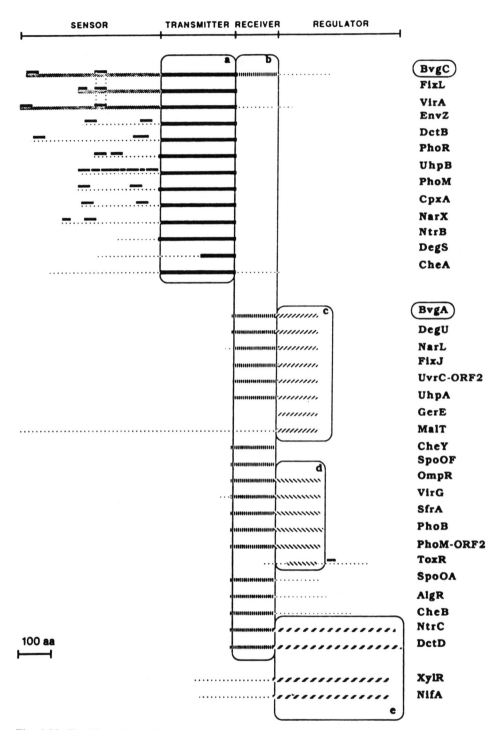

Fig. 4-20. Families of signal transducing proteins based upon sequence homologies. Boxes a–e indicate homologous protein domains with functions indicated on the top bar. Box e indicates transcriptional activators that require the alternate σ factor RpoN as a coactivator. Broken lines represent regions that do not share homology with any of the other members. Black bars above the sequences represent possible transmembrane spanning regions (highly hydrophobic). From Gross, R. et al. 1989. *Mol. Microbiol.* **3**:1661–1667.

transmits a signal via its cytoplasmic domain (transmitter) to the second, regulator component. The second component is located in the cytoplasm. Its amino terminal end (receiver) receives the signal from the sensor/transmitter protein and modifies the carboxyl end (regulator). Examples are listed in Table 4-2.

There are at least two families of sensor proteins. The first shows typical transmembrane topology. These include, among many others, EnvZ, which will be discussed in detail later. The second family of sensor proteins consists of cytoplasmic proteins with no obvious transmembrane domains. Both NtrB and CheA are examples that will be discussed.

Receiver/Regulator components of the sensory transduction systems can be grouped into at least four families. First, there are typical DNA-binding transcriptional activator proteins, such as OmpR from *E. coli* (see Osmotic Control in this chapter). A second family has, in addition to an N-terminal receiver module with conserved regions to the OmpR family, a conserved region in the C-terminus (presumably involved with DNA binding), which is different from OmpR. Examples include BvgA of *Bordetella pertussis* and NarL and UhpA of *E. coli*. The third family of the receiver group exemplified by DNA-binding proteins NtrC and NifA have C-terminal regulator regions similar to each other but distinct from the other families. A fourth family including CheY from *S. typhimurium* and SpoOF of *B. subtilis* only have a receiver module. The *cheY* product is not a DNA-binding protein but interacts with other proteins at the flagellar motor (see Chemotaxis in Chapter 6). The SpoOF system probably collaborates with GerE (a protein only containing a regulator module) to regulate sporulation.

A one-component sensory transduction system has also been characterized that combines the features of the two-component system into one protein. The ToxR system from *Vibrio cholerae* mediates regulation of several virulence components for this intestinal pathogen in response to environmental stimuli. The ToxR protein is a transmembrane sensor that also binds specifically to the promoter regions of virulence-factor genes through its cytoplasmic domain. Because it is a single protein it does not require transmitter or receiver modules. There is homology at the DNA-binding cytoplasmic N-terminus region with the C-terminus of the OmpR-like regulators. Because of this it is classified as a member of the OmpR family.

Because of the high degree of sequence homology between proteins responding to different stimuli, one might wonder about the fidelity of these systems. Can different systems end up "cross-talking" to each other? Under normal *in vivo* circumstances each system is able to specifically transduce its own signal. However, if a system is disturbed by mutations, cross-talk between systems can be observed when a component of one system can, at a low level, substitute for the missing component of another. It seems likely then that the various two-component systems present in the same bacterial cell may communicate with each other and form a complex and highly balanced regulatory network.

OSMOTIC CONTROL OF GENE EXPRESSION

The outer membrane of *E. coli* K-12 contains several major outer-membrane proteins. The relative amounts of two of them, coded by the *ompC* (47 min) and *ompF* (21 min) loci, are mediated by medium osmolarity. Both OmpC (36,500 MW) and OmpF (37,000 MW) are porin proteins that form aqueous pores in the outer membrane allowing polar

TABLE 4-2. Some Properties of Various Two-Component Systems

Organism	Sensor/ Transmitter	Regulator/ Receiver	Response or Regulated Gene	Environmental Stimulus
E. coli/*S. typhimurium*	EnvZ	OmpR	*ompC, ompF*	Changes in osmolarity
E. coli	PhoR	PhoB	*phoA, phoE* etc.	Phosphate limitation
E. coli	PhoM	PhoM–ORF2	*phoA* etc.	Extracellular glucose
E. coli	CpxA	SfrA	*traJ* etc.	Dyes and toxic chemicals
E. coli	UhpB	UphA	*uhpT* etc.	Glucose-6-phosphate
E. coli	?	UrvC–ORF2	Unknown	Unknown
E. coli	NarQ	NarL	*narGHJI*	Nitrate concentration
E. coli/*S. typhimurium*	NtrB(NRII)	NtrC(NRI)	*glnA* etc.	Ammonia limitation
E. coli/*S. typhimurium*	CheA	CheB/CheY	Chemotaxis	Repellents and attractants
S. typhimurium	PhoQ	PhoP	*phoN* etc., virulence	Carbon, sulfur, phosphorus, or nitrogen starvation
K. pneumoniae/*B. parasponaie*	NtrB	NtrC	*nifLA*	Nitrogen limitation
R. meliloti	FixL	FixJ	*nifLA, fixN*	Nitrogen limitation
R. leguminosarum	DctB	DctD	*dctA*	C4-dicarboxylic acids
P. putida	?	XylR	*xylCAB, xylS*	*m*-xylene or *m*-methylbenzyl alcohol
P. aeruginosa	?	AlgR	*algD*	Not known
B. pertussis	BvgC	BvgA/BvgC	*ptx, fha, cya* etc., virulence	Temperature, nicotinic acid, magnesium sulfate (MgSO₄)
A. tumefaciens	VirA	VirG	*virB, virC* etc., virulence	Plant wound exudate
B. subtilis	?	SpoOA/ SpoOF	Control of sporulation	Nutrient deprivation
B. subtilis	DegS	DegU	Regulation of degradative enzymes	Not known

molecules (<600 Da) to cross the outer-membrane barrier. In low osmolarity media, OmpF protein is present in greater quantities than OmpC protein. In media of high osmolarity, the OmpC porin predominates over OmpF. While the quantitative ratios of these proteins vary, their combined levels remain fairly constant.

Regulation of these genes is mediated by the regulatory *ompB* locus (21 min). The *ompB* region actually is comprised of two genes: *ompR* and *envZ*. The products of the *ompR* and *envZ* genes are proteins with molecular weights of 27,400 and 50,300, respectively. Mutations in *ompR* will result in either OmpF⁻OmpC⁻ or OmpF⁺OmpC⁻ phenotypes. Missense mutations in *envZ* are pleiotropic, altering the expression of *ompF* *ompC*, *phoA* (alkaline phosphatase), *mal*, and iron-regulated genes although the only system studied in detail is *ompCF*.

EnvZ is a transmembrane sensor protein that undergoes autophosphorylation under conditions of high osmolarity. Once phosphorylated, EnvZ will then transphosphorylate the receiver module of the true DNA-binding regulator, OmpR, thereby increasing the level of OmpR-P in the cell. The key to understanding how OmpR can differentially regulate *ompC* and *ompF* lies in the knowledge that the *ompF* operator has both high-affinity and low-affinity binding sites (Fig. 4-21). When OmpR-P concentrations are low, this regulator can only bind to the high-affinity binding site that activates *ompF* transcription. The *ompC* gene lacks the high-affinity binding site and so is poorly expressed. However, when osmolarity is high, OmpR-P becomes elevated to levels that can bind to the low-affinity binding site present in both *ompF* and *ompC*. Binding at *ompF* (-60) turns off expression, whereas binding to *ompC* (-105) turns on *ompC* expression. If osmolarity shifts from high to low, an EnvZ phosphatase is activated removing P from both OmpR and EnvZ. How EnvZ actually senses osmolarity is unknown and is an ongoing area of study.

There is an additional control mechanism that appears to regulate the relative amounts of OmpC and OmpF. This control involves a gene (*micF*) adjacent to *ompC*, which is transcribed in the opposite direction from *ompC*. The *micF* transcript (micRNA, 93 nucleotides) is complementary to the 5′ end region of *ompF* mRNA. Under conditions of high osmolarity, the *micF*−*ompC* region is induced and transcribed divergently from a central promoter. The resulting micRNA (also referred to as antisense RNA) inhibits translation of *ompF* mRNA by hybridizing to it. This RNA−RNA interaction is proposed to cause premature termination of *ompF* transcription and/or destabilization of *ompF* mRNA. The end result is that when more OmpC protein is produced, less OmpF protein is synthesized.

Turgor. When bacteria are subjected to changes in medium osmolarity, they experience osmotic stress where the *inward* pressure on the cell membrane can be very high (high-external osmolarity) or the *outward* pressure very high (low-external osmolarity). In order to maintain its shape the bacterium must maintain a minimum outward pressure called turgor (osmotic pressure 0.6 MPa, 250 milliosmol/κg). If external osmolarity increases, the cell possesses homeostatic mechanisms to transport compatible solutes into the cell that will counterbalance the increase in external pressure. This will prevent the collapse of the cell. Compatible solutes are chemicals (K^+) or compounds (proline, glycine-betaine, and glutamate) that are nontoxic to the cell. The *kdpABC* operon of *E. coli* specifies a high-affinity K^+ transport system that is transiently induced by osmotic up-shock. This system will allow an increased uptake of the osmoprotectant K^+. The *proU* operon similarly encodes an uptake system for glycine-betaine and proline. Another operon protectant is trehalose whose uptake and degradation is osmotically regulated. It is

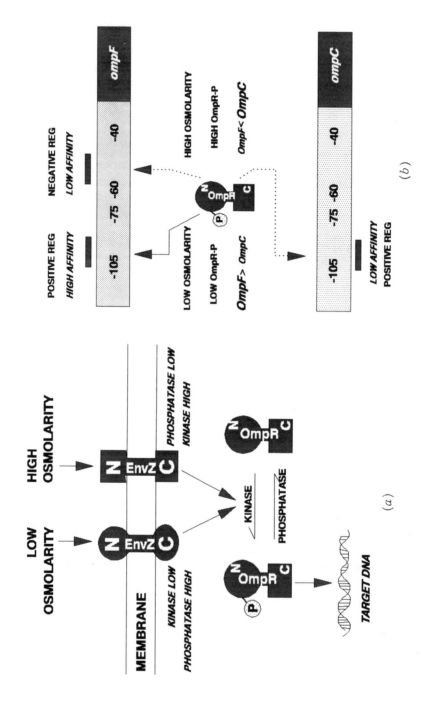

Fig. 4-21. Osmoregulation of major outer-membrane proteins OmpF and OmpC (see text for details).

believed that many of these systems sense pressure changes at the cell wall but the mechanics of the system are unclear.

GLOBAL CONTROL NETWORKS: REGULATION AT THE WHOLE CELL LEVEL

Escherichia coli has the ability to grow in or survive numerous environmental conditions including (1) shifts from rich to minimal media (2) changes in carbon source, (3) amino acid limitations, (4) shifts between aerobic environments, (5) anaerobic environments, (6) heat shock, and (7) numerous starvations, such as phosphate, nitrogen, and carbon source. Through all of these stresses, the cell maintains the required balance between the various physiological components of DNA replication, cell growth, and division. The ability to integrate regulatory circuits involved with single aspects of cellular physiology into a coordinated response to an environmental stress is referred to as **global control.** Global regulatory networks include sets of operons and regulons with seemingly unrelated functions with positions scattered around the chromosome, yet they all are coordinately controlled in response to a particular stress.

Some attention should be given to the terminology used to describe regulatory networks. A **regulon** is a network of operons that are controlled by a single regulatory protein (e.g., the arginine regulon). A **modulon** refers to all the operons and regulons under the control of a common pleiotropic regulatory protein (e.g., the CRP modulon). The simplest way for all of the operons in a modulon to respond to a stress is through the production of a signal molecule (**alarmone**), which accumulates during the stress. Alarmones are often small nucleotides, such as cAMP or guanosine tetraphosphate (ppGpp). The term **stimulon** is used to describe all of the genes, operons, regulons, and modulons that respond to a common environmental stimulus. It is conceivable that some operons that respond to an environmental stress may not share a common regulatory protein. The term is useful when describing all of the coinduced or corepressed proteins that form a cellular response without knowing the mechanism of the response.

Consideration should also be given to the two common methodologies used to identify the components of a global regulatory system. The first method involves the random fusion of the *lacZ* structural gene (β-galactosidase) to host promoters. The usual technique involves a derivative of Mu phage, which contains a β-lactamase locus (ampicillin resistance), and the *lacZ* structural gene lacking its own promoter (*lacP⁻*). Mu is a phage that enters the prophage stage by randomly integrating into the host chromosome (see Chapter 5). Of course, if the phage inserts within a host gene it will inactivate that gene (**insertional inactivation**). The Mu derivative, Mu*d*J, is illustrated in Figure 4-22. The figure shows Mu*d*J inserted within the *aci* structural gene in an orientation such that the *lacZ* structural gene is transcribed from the *aci* promoter. Consequently, whatever conditions regulate transcriptional expression of the *aci* gene will control β-galactosidase mRNA production. The search for *lacZ* operon fusions is usually made on a medium containing an indicator for β-galactosidase production. Components of a global regulon (e.g., *psi*) can be identified as a cell (colony), which produces more β-galactosidase under the stress condition (e.g., phosphate limitation).

The second method used to analyze global control networks involves the two-dimensional separation of cellular proteins labeled with ^{35}S following a stress condition. A polypeptide map of *E. coli* has been compiled and used as an aid to identify protein

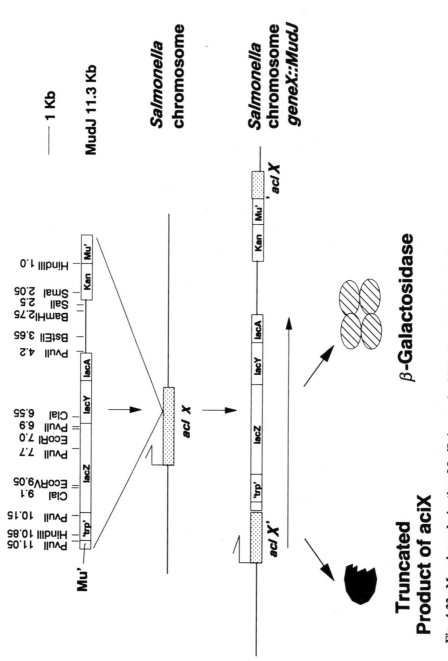

Fig. 4-22. Mu phage derivative, MudJ, inserted within the *aci* (acid inducible) structural gene. The Mu*J* is inserted in an orientation such that the *lacZ* structural gene is transcribed from the *aci* promoter (*bent arrow*). This particular gene is induced by an acid environment. The antibiotic resistance gene has its own promoter. Various restriction enzymes that cleave Mu*J* are shown at the top of the figure along with their position in Kb from the right end of Mu.

components of various stimulons (Fig. 4-23). The following section offers a brief description of several global regulons identified in either *E. coli* or *S. typhimurium.*

The SOS Response

This global control network was described in detail earlier (Chapter 2).

Heat Shock Response

Upon a shift from 32 to 42 °C, *E. coli* transiently increases the rate of synthesis of a set of proteins called heat shock proteins. Many of these heat shock proteins are required

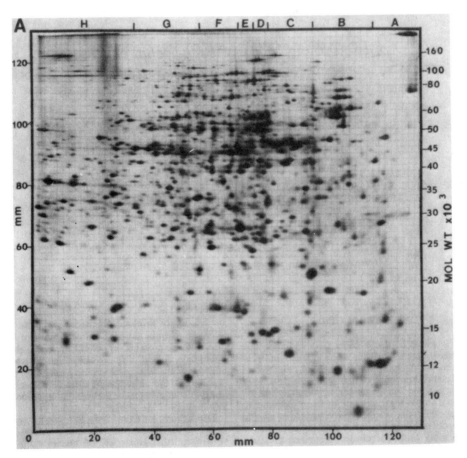

Fig. 4-23. Two-dimensional gel of *E. coli* K-12, strain W3110, cell extract. First-dimension isoelectric focusing to equilibrium (8200 volt-hours), 1.6% pH 5–7 and 0.4% pH 3–10 ampholine mixture; second dimension acrylamide concentration, 11.5%. A grid overlay provides coordinates for individual spots. Letter designations from A to H designate zones of increasing isoelectric points. The approximate molecular weight scale was determined by plotting known molecular weights of identified proteins versus their migration distances. Cells were labeled with $^{35}SO_4$ during exponential growth in glucose minimal medium at 37 °C. From Neidhardt, F. C. et al. 1983. Gene protein index of *Escherichia coli* K-12. *Microbiol. Rev.* **47:**231–284.

for cell growth or survival at more elevated temperature (thermotolerance). Among the induced proteins are *dnaJ* and *dnaK*, the RNA polymerase sigma-70 subunit (*rpoD*), *groES*, *groEL* (see Chaperones in Chapter 2), *lon* (see Proteolytic Control), and *lysU*. There are a total of 17 heat shock proteins identified in *E. coli*. The *rpoH* locus (formerly called *htpR*) codes for a 32-kDa protein, which is a σ subunit for RNA polymerase (sigma-32). The RpoH product regulates the expression of all other *htp* loci. One model to explain the heat shock response proposes that the synthesis of σ^{32} increases during temperature upshifts. The increased levels (or activated form) of σ^{32} can compete more favorably for the appropriate site on RNA polymerase and, upon binding, change the promoter specificity of RNA polymerase toward a preference for *htp* loci. This change will cause the increased production of heat shock proteins. As one would predict, the −10 consensus sequence differs greatly between σ^{70} and σ^{32} promoters. The heat shock response peaks within a few minutes, after which there is a rapid decline in the synthesis of heat shock proteins to a new steady-state level at the higher temperature. Recent reports indicate DnaK itself associates with σ^{32} possibly causing its inactivation, degradation or sequestration (Fig. 4-24). One consequence of heat shock is an increase in the levels of denatured proteins. The DnaK binds to such polypeptides, protecting them from further damage or disaggregation. This use of DnaK would siphon it away from sigma-32 leaving σ^{32} free to associate with RNA polymerase leading to an increase in heat shock gene expression.

The DnaK protein has both ATPase and autophosphorylation activities, which appear important to its role as a chaperone (see Chapter 2). It appears that DnaK may assume a role analogous to a thermometer. This result occurs because the ATPase and autophosphorylation activities are very temperature dependent, relatively inactive at 30 °C

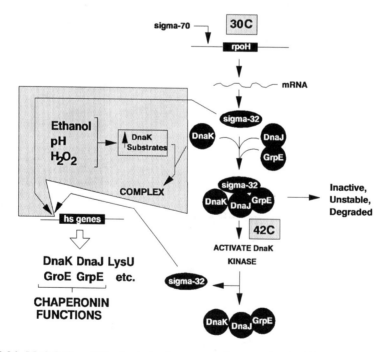

Fig. 4-24. Modulation of the heat shock response (see text for details). hs = heat shock.

but become increasingly active at progressively higher temperatures. It is also proposed that the heat activation of DnaK kinase results in the disassociation of DnaK from σ^{32} leaving σ^{32} more stable.

There are at least three promoters for *rpoH*. Two are recognized by σ^{70} and one by a new sigma (σ^{24}). Upon temperature upshift to 42 °C, an increase in transcription from the σ^{24}-dependent promoter is observed and may be responsible for continued *rpoH* expression at even higher temperatures (e.g., 50 °C). However, it appears the main reason σ^{32} levels increase during temperature upshift is the result of increased translation. Cis-acting mRNA sites occur within *rpoH* message that form temperature-sensitive secondary structures. At high temperatures, these secondary structures melt, thereby enabling more efficient translation of the *rpoH* message. The transient induction of the heat shock system is due to the following: (1) σ^{32} is very unstable with a half-life of 1 min and (2) the heat shock dependent increase in DnaK leads to reassociation with σ^{32} making σ^{32} even more unstable. The result is a decreased level of σ^{32} and thus decreased expression and lower levels of other HSPs.

The heat shock response is also triggered by a variety of environmental agents, such as ethanol, UV irradiation, and agents that inhibit DNA gyrase. Induction by all of these stimuli occur through σ^{32}. How can all of these seemingly diverse stresses activate *rpoH*? The only likely explanation that appears reasonable is the accumulation of denatured or incomplete peptides. There is a potential alarmone that has been implicated in signaling expression of this global network. The molecule is diadenosine $5'$, $5'''$-P^1, P^4-tetraphosphate (AppppA), which is made by some aminoacyl-tRNA synthetases (e.g., *lysU*) at low tRNA concentrations. How this may influence the response is not known.

ELECTRON TRANSPORT: RESPIRATORY PATHWAYS

Facultative microorganisms, such as *E. coli* and *S. typhimurium*, are capable of modifying their metabolism to accommodate growth under either aerobic or anaerobic conditions. The transition between aerobic and anaerobic metabolism is accompanied by alterations in the rate, route, and efficiency of pathways of electron flow. Figure 4-25 illustrates the basic pathways *E. coli* utilizes for aerobic versus anaerobic electron flow. Under anaerobic conditions without alternate electron acceptors pyruvate is converted to formate, acetate or ethanol, carbon dioxide, and dihydrogen gas (mixed acid fermentation, Fig. 4-25*b*). However, the choices and energy yield become more plentiful when alternate electron acceptors are available. *Escherichia coli*, even under aerobic conditions, synthesizes two distinct cytochrome oxidases, cytochrome *o* (*cyo* operon) and cytochrome *d* (*cyd* operon) that are produced under high O_2 and low O_2 conditions, respectively. Under anaerobic conditions at least five more terminal oxidoreductases can be produced (Table 4-3).

Escherichia coli controls the production of the various respiratory pathway enzymes in response, first, to aerobic and anaerobic growth conditions and second, to the availability of alternate electron acceptors. Clearly, there is a hierarchy or preference for substrate use with the following order: oxygen > nitrate > dimethyl sulfoxide (DMSO) > trimethyl-amine-N-oxide (TMAO) > fumarate. When several e^- acceptors are present simultaneously the more energetically favored acceptor will be used first. An interesting question is how does the cell regulate such a complex system?

There are three basic regulators described that sense changes in oxygen level or redox

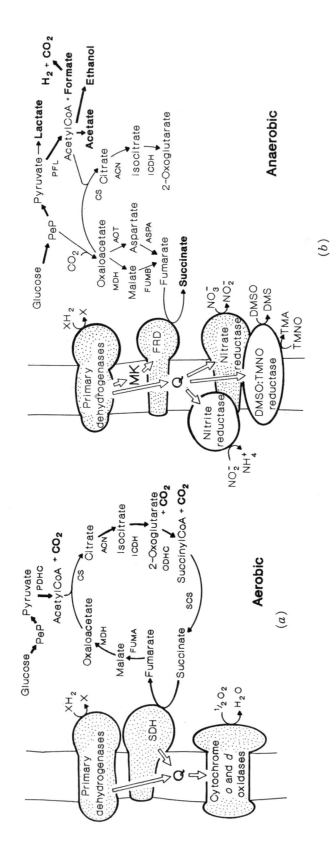

Fig. 4-25. Basic aerobic and anaerobic metabolic pathways of *E. coli*. From S. Spiro and J. Guest. 1991. *Trends Biochem. Sci.* 16:310–314.

TABLE 4-3. Characteristics of Electron-Transport Systems in *E. coli*

Electron Acceptor	(kJ/mol)	Terminal Respiratory Enzyme	Operon	Chromosomal Location (min)
O_2	−233	Cytochrome *o* oxidase	*cyoABCDE*	10
O_2	−233	Cytochrome *d* oxidase	*cydAB*	17
NO_3^-	−144	Nitrate reductase	*narGHIJ*	27
NO_3^-	−144	Nitrate reductase	*narZYWV*	33
DMSO	−92	DMSO/TMAO reductase	*dmsABC*	20
TMAO	−87	TMAO reductase	*torA*	28
Fumarate	−67	Fumarate reductase	*frdABCD*	94

[a] Free energy calculated by using NADH as an electron donor to the indicated electron acceptor. Adapted from Gunsalus, 1992. *J. Bacteriol.* **174:**7069–7074.

conditions. Their overlapping control circuits are illustrated in Figure 4-26. The Fnr protein (*f*ormate-*n*itrate *r*egulation) senses the presence or absence of oxygen. The precise manner in which this is accomplished is unknown but may involve a bound Fe^{+2} molecule that can form an oxygen-sensitive center. The Fnr protein will repress *cyoABCDE* and *cydAD* under anaerobic conditions but transcriptionally activates *frdABCD*, *dmsABC*, and *narGHJI*. Significant sequence homology between Fnr and CRP suggests a similar manner of transcriptional control through direct interactions with RNA polymerase and DNA bending. In contrast to CRP, there is no evidence for cAMP involvement with Fnr.

In addition to Fnr, three additional genes, *narX*, *narL*, and *narQ* are required to regulate respiratory gene expression in response to nitrate availability. Homologies revealed by DNA sequence analysis indicate *narX* and *narL* make up a two-component regulatory system similar to those described for nitrogen (*ntrB/ntrC*) and osmolarity (*envZ/ompR*) NarQ is similar in sequence to *narX*; they represent two membrane bound sensor proteins for nitrate that can transmit a signal to NarL, the cytoplasmic regulator that binds to DNA. This system contributes to the induction of *frd*, *dms*, and *nar* operons but have no effect upon *cyo* or *cyd*.

The system that controls *cyo* and *cyd* is mediated by the *arcA* and *arcB* products (*arc* = *a*erobic *r*espiration *c*ontrol). They, too, comprise a two-component regulatory system. The ArcB protein probably functions as the membrane-bound sensor/transmitter and communicates with ArcA, the receiver/regulator. The Arc system also regulates the coordinate synthesis of tricarboxylic acid (TCA) cycle enzymes. As is the case for most, if not all, two-component systems, ArcB is autophosphorylated and can subsequently transfer its phosphate to ArcA. The nature of the signal sensed, how it is sensed, and how the signal is transduced to activate ArcB kinase are unclear.

Stringent Control

When bacteria experience conditions that limit the availability of one or more amino acids (shift from a rich medium to a minimal medium) or exhaust their primary carbon source, growth stops temporarily and a variety of rapid adjustments in metabolism are made. These include decreasing the rates of RNA accumulation (particularly stable RNA), and DNA replication, as well as reducing the biosynthesis of carbohydrates, lipids, nu-

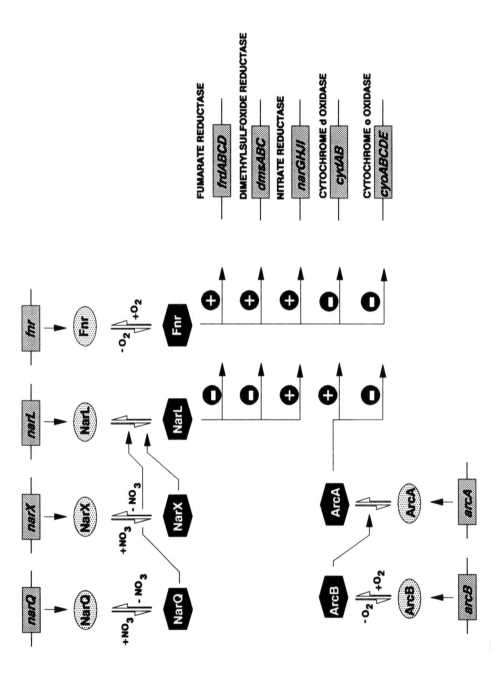

Fig. 4-26. Regulatory scheme for the control of aerobic and anaerobic respiratory pathways in *E. coli*. From Gunsalus, R. B. 1992. *J. Bacteriol.* 174:7069–7074.

cleotides, peptidoglycan, and glycolytic intermediates. The transport of many macromolecular precursors into the cell is also shut down. This set of responses, characterized best as a response to amino acid starvation, is referred to as the **stringent response** or **stringent control.** The stringent response collectively enhances cellular viability during periods of amino acid or energy limitation and allows rapid recovery and reinitiation of growth when conditions improve.

For rapidly growing cells, a major amount of the available energy is used for ribosome synthesis. Therefore, blocking ribosome synthesis under amino acid starvation conditions is a major mode of energy conservation. Conditions causing a stringent response lead to an abrupt change in the rate of ribosome synthesis mediated by specific inhibition of the synthesis of ribosome-associated components, such as rRNA, tRNA, and mRNA, for ribosomal proteins.

When starved for amino acids, bacterial cells rapidly accumulate millimolar concentrations of an unusual nucleotide, guanosine 5'-diphosphate-3'-diphosphate (ppGpp). This nucleotide, first referred to as Magic Spot I (MS-I), has been shown to accumulate in *E. coli* and *S. typhimurium* as well as other bacteria during amino acid limitation. Synthesis of ppGpp is triggered by the binding of an uncharged tRNA molecule (found at increased levels during amino acid limitation) to the ribosome. On the ribosome, the product of the *relA* gene (RelA, stringent factor), a ribosome-bound pyrophosphotransferase (ppGpp synthetase I), catalyzes the formation of ppGpp from GDP or GTP and ATP, as shown in Figure 4-27*a*. A second, ribosome-independent route to ppGpp also exists involving the spoT product (ppGpp synthetase II). The spoT product is also the major ppGpp hydrolase and so appears to have two functions, one in the synthesis and one in the degradation of ppGpp.

Early work on the role of ppGpp in stringent control reported that ppGpp inhibits a step in the initiation of protein synthesis *in vitro*. The AUG (initiator codon)-directed binding of f-Met-tRNA to the ribosome is inhibited by both GDP and ppGpp. The finding that ppGpp accumulates to levels much higher than GDP during the stringent response suggests that, *in vivo*, ppGpp may affect those reactions involving a requirement for GTP in protein synthesis, such as initiation and elongation. It has also been reported that ppGpp blocks the binding of the initiator tRNA complex to the ribosome at the initiation step probably via an interaction with IF-2. The ppGpp might also inhibit the elongation steps of protein synthesis by forming a complex with EF-T_U and EF-G (see Ribosome Editing in Chapter 2). The ratio between the amount of EF-G complexed with ppGpp and GTP, respectively, is 20:1 indicating that ppGpp affects the elongation rate of protein synthesis during stringent control primarily via complex formation with EF-G rather than with EF-T_U. Clearly, without ppGpp to inhibit protein synthesis after exposure to stress, translation errors would increase and precursor molecules would be rapidly depleted making adaptive recovery more difficult.

One major result of the stringent response is a reduction in the rate of stable RNA accumulation, a response strongly correlated with the rise in ppGpp concentration in the cell. Some studies suggest that RNA polymerase is a target of ppGpp action and that the regulated process during stringent control is transcription itself. The probable target is the β subunit of RNA polymerase (see Fig. 2-16C). Both transcription initiation and polymerase pausing (important in certain regulatory mechanisms and transcription termination) are postulated as the major targets of ppGpp action. Thus, ppGpp would reduce the affinity of RNA polymerase for rRNA promoters. This obviously lowers the amount of rRNA available for ribosome synthesis, which can subsequently affect the synthesis

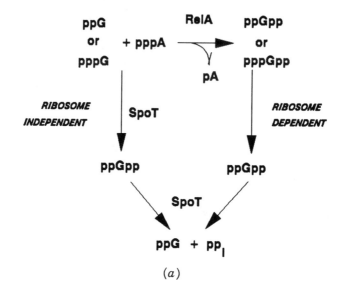

(a)

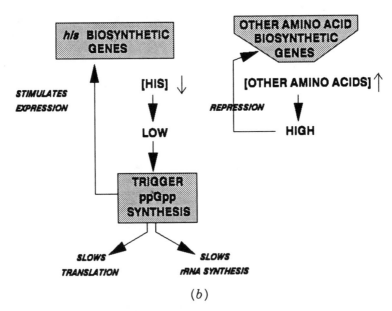

(b)

Fig. 4-27. Stringent control. (a) **Ribosome-dependent and -independent pathways for the bio-synthesis of guanosine 5'-diphosphate-3'-diphosphate.** ppG = guanosine 5'-diphosphate; pppG = guanosine 5'-triphosphate; pppA = adenosine 5'-triphosphate; pA = adenosine 5'-monophosphate; ppGpp = guanosine 5'-diphosphate-3'-diphosphate; pppGpp = guanosine 5'-triphosphate-3'-diphosphate; ppG = guanosine 5'-diphosphate; PP_i = inorganic pyrophosphate. (b) **Role of ppGpp in maintaining an efficiently balanced pool of amino acids.** In this example, histidine levels have fallen below optimum. The ensuing accumulation of ppGpp stimulates expression of the *his* operon while slowing translation and ribosome synthesis. Restoration of histidine levels shut off ppGpp synthesis allowing the resumption of normal translation and ribosome synthesis rates.

of ribosomal proteins (see Translational Control in this chapter). On the other hand, ppGpp has also been found to stimulate transcription of several amino acid biosynthetic operons (e.g., *his*) and possibly other genes. Thus, it appears that ppGpp can act either as a negative effector or a positive effector depending on the target gene.

In vivo, the synthesis of ppGpp on the ribosome is controlled by the charging of total tRNA species as a function of intracellular concentrations of the 20 amino acids. It would appear that ppGpp is a component of a sensing mechanism that functions in adjusting the synthesis, for example, of histidine biosynthetic enzymes with respect to the need for histidine relative to the total amino acid concentration of the cell, as well as in the growth medium (Fig. 4-27*b*). Thus, along with the operon-specific attenuator mechanism that responds to the need for histidine, specifically altering the level of ppGpp enables the organism to sense how the supply of histidine, in this example, compares with the availability of all the amino acids in the cell, a kind of fine-tuning mechanism that maintains the correct relative levels of each amino acid.

REGULATION OF NITROGEN ASSIMILATION AND NITROGEN FIXATION

The preferred nitrogen source for enteric bacteria is ammonia and the principal product of ammonia assimilation is glutamate. Glutamic acid, in turn, is the precursor of several amino acids and ultimately furnishes the amino group of numerous amino acids via transamination (see Chapters 1 and 9). Glutamate is also the precursor of glutamine, which is involved in the synthesis of some amino acids, purines, and pyrimidines. In fact, 85% of the cellular nitrogen is derived from the amino nitrogen of glutamate and 15% from the amide nitrogen of glutamine.

In the absence of ammonia, other nitrogen-containing compounds, such as histidine or proline, can serve as an alternate source of nitrogen. Rapid synthesis of enzymes involved in these alternate nitrogen assimilatory pathways occurs during ammonia deprivation. This section will deal with the regulatory mechanisms controlling various pathways of nitrogen assimilation and nitrogen fixation.

Glutamine synthetase (GlnS) is of central importance to cells growing in medium containing less than 1 mM ammonia (NH_4^+). The pathway is illustrated below:

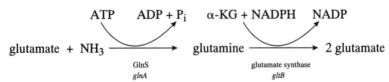

Under conditions where the NH_4^+ concentration is above 1 mM the GlnS/GS system is inactive, but glutamate dehydrogenase (*gdh*) can be used for ammonia assimilation via the following reaction:

$$NADPH \qquad NADP$$
$$NH_3 + \alpha\text{-ketoglutarate} \longleftrightarrow glutamate + H_2O$$

The mechanism for GlnS inactivation under conditions of ammonia excess is outlined in Figure 4-28. This mechanism involves the adenylylation and deadenylylation of glutamine synthetase resulting in an inactive or active form of the enzyme, respectively.

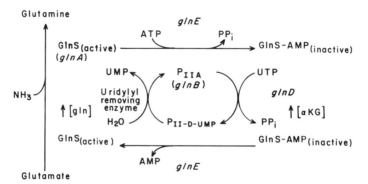

Fig. 4-28. The glutamine synthetase (GlnS) regulatory cascade (see text for details). gln = glutamine; α-KG = α-ketoglutarate; *glnA* = glutamine synthetase; *glnB* = P$_{IIA}$ deuridylylated activating protein; P$_{IID}$ = uridylylated activating protein; *glnD* = uridylyltransferase; *glnE* = adenylyltransferase (ATase).

Adenylylation–deadenylylation of GlnS requires both adenylyltransferase (ATase; *glnE* gene product) and P$_{II}$ (*glnB* gene product). The P$_{II}$ protein can activate either the adenylylation or deadenylylation activities of ATase depending on whether P$_{II}$ is uridylylated or not. The deuridylylated form of P$_{II}$ (P$_{IIA}$) activates the adenylylation of GlnS, while the uridylylated form (P$_{IID}$) stimulates the deadenylylation of GlnS. Thus, P$_{IIA}$ results in an inactive GlnS while P$_{IID}$ results in active GlnS. Under high ammonia conditions cellular glutamine levels will be high while the α-ketoglutarate concentration will be low. If unabated, GlnS will convert all of the cellular α-ketoglutarate to glutamine. Consequently, high glutamine levels will stimulate the production of P$_{IIA}$, which stimulates the adenylylation of GlnS to the inactive form. Low ammonia concentrations, on the other hand, result in elevated α-ketoglutarate levels. This stimulates the formation of P$_{IID}$ which, in turn, promotes the deadenylylation of GlnS yielding active GlnS.

Regulation of this system at the genetic level is elegant and is outlined in Figure 4-29. The *glnA (GlnS), ntrB,* and *ntrC* loci form an operon. The *ntrC* product, also called NR$_{I}$, is a transcriptional activator of this operon. The NtrB protein, also called NR$_{II}$, can either activate or inactivate NtrC depending on glutamine levels and the state of GlnB. When glutamine levels are low (α-KG is high), GlnB is in the uridylylated form (P$_{IID}$) and NtrB is free to activate NtrC. Figure 4-29 illustrates the regulatory cascade dictating the prodution of glutamine synthetase, as well as other genes involved with nitrogen assimilation. Note that the *glnA, ntrB,* and *ntrC* operon has three promoters that can initiate transcription; *glnAp1* (transcriptional start 187 bp upstream of *glnA* translational start), *glnAp2* (73 bp upstream of *glnA* translational start), and *glnLp* (transcriptional start 256 bp downstream of glnA translational termination; 33 bp upstream of *glnL* translational start). The *glnAp1* and *glnLp* promoters both utilize σ^{70} for initiation and have the appropriate −10 and −35 regions. They are not strong promoters but allow the cell to maintain a low level of glutamine synthase and NR$_{I}$ during growth in high nitrogen. Both promoters are negatively regulated by NR$_{I}$. The *glnAp2* promoter utilizes an alternate σ factor, the product of *rpoN*. It contains the consensus sequence TTGGCACAN$_4$TCGCT that is common for other nitrogen-regulated promoters.

The series of events controlling the expression of *glnAp2* are quite elegant (Fig.

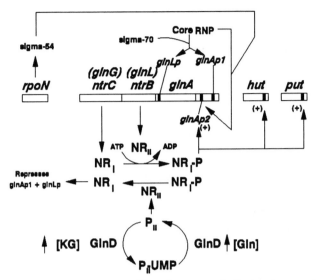

Fig. 4-29. Regulatory circuits controlling nitrogen assimilation (see text for details).

4-29). Under low-nitrogen levels the regulator NR_{II} (*ntrB*) phosphorylates NR_I (*ntrC*). Then, NR_I-P activates transcription at glnAp2 producing a tremendous increase in glutamine synthetase, which enables the cell to better assimilate NH_3. How is the system shut off? As the level of glutamine rises in the cell, the uridylyltransferase activity of GlnD is activated, which converts P_{II}-UMP to P_{II} (GlnB). In addition to inactivating GlnS as described above, P_{II} will also activate the NR_{II} dephosphorylation activity, which removes phosphate from NR_I-P. Since NR_I cannot activate the *glnAp2* promoter but will repress *glnAp1* and *glnLp*, the operon is deactivated and returns to background levels of expression.

It is of considerable interest that the *glnAp2* binding site for NR_I can be moved over 1000 bp away from the original site without diminishing the ability of NR_I to activate transcription from *glnAp2*. Thus, the NR_I sites resemble the enhancer sequences found in eukaryotic cells. It is believed that enhancers work by increasing the local concentration of an important regulator near a regulated gene by tethering it to the region.

The cell has other systems that can capture nitrogen from amino acids that may be available in the environment. These include the *hut* (histidine utilization) and *put* (proline utilization) operons. Both systems require NR_I-P and σ^N RNA polymerase for their expression and so are regulated in a manner similar to the *glnA* operon.

Along with the general nitrogen assimilation regulatory system (*ntr*) just described, *Klebsiella pneumoniae* can fix atmospheric dinitrogen (N_2). Nitrogen fixation in *Klebsiella* involves 17 genes all involved in some way with the synthesis of nitrogenase, the enzyme complex, which converts N_2 to NH_3 (see Fig. 9-3). The *nifHDK* transcriptional unit codes for nitrogenase proper while the remaining *nif* genes are repsonsible for (1) synthesizing the Mo—Fe cofactor, (2) protein maturation, (3) regulation or, (4) unknown functions. The *ntrC* and *ntrB* products will also regulate *nifLA* transcription in a manner analogous to what was described above. The *nifAL* operon is activated by NR_I-P, which results in increased intracellular concentration of NifA. The NifA protein can then activate expression of the other *nif* operons. The activity of NifA is regulated by the NifL product in response to oxygen and probably ammonia, although the mechanism is not known. The *nif* system is active only under anaerobic and low NH_4^+ conditions. For additional discussion of nitrogen fixation and its regulation, see Chapter 9.

Phosphate Starvation-Controlled Stimulon

Another global system studied extensively is the phosphate-controlled stimulon of *E. coli*. Expression of the majority of phosphate-regulated loci is specifically **phosphate starvation inducible** (*psi*). However, other phosphate-controlled loci are phosphate-starvation **repressible.** Many of the *psi* genes encode proteins associated with the outer membrane or the periplasmic space and function either in the transport of P_i across the cell membrane or in scavenging phosphate from organic phosphate esters. Two of these inducible genes, *phoA* (alkaline phosphatase, PhoA) and *phoE* (outer membrane porin protein, PhoE), have been studied and shown to be under the control of a complex regulatory circuit. This circuit involves at least four regulators, the *phoB, phoU, phoR,* and the *pst* system.

The most popular model for phosphate-controlled gene regulation is outlined in Figure 4-30. Both PhoR and PhoB are members of the two-component regulatory systems discussed earlier. The PhoR is an integral membrane sensing protein that can trigger repression of the Pho regulon under excess phosphate conditions. When the cell senses

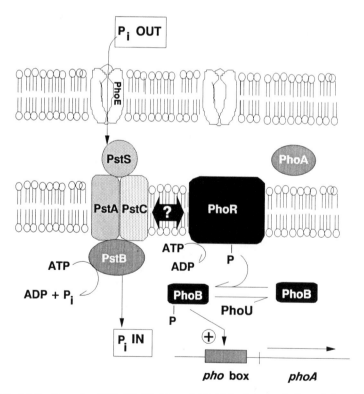

Fig. 4-30. Model for the role of PhoM, PhoR, and PhoB in the regulation of the *pho* regulon. See text for explanation of the model.

low P_i levels, the PhoR protein is autophosphorylated (from ATP). Then, PhoR-P trans-phosphorylates PhoB. The PhoB-P activates transcription by binding to the consensus "Pho-box" sequences (CTTTTCATAAAACTGTCA) upstream of the *pho* regulon pro-moters. As phosphate levels increase, PhoR together with another protein, PhoU, stimulates dephosphorylation of PhoB thereby turning off transcription of the *pho* regulon.

There are two transport systems for P_i, the low-affinity PIT (phosphate inorganic transport) and an inducible high-affinity PST (phosphate-specific transport). In addition to *phoU*, *phoR*, and *phoB*, the *pst* operon must also play a role in regulation. The PST mutants consititutively synthesize alkaline phosphatase even though they can grow well and maintain a high internal phosphate concentration using the PIT system. Clearly, phosphate itself cannot be the corepressor for this system. Presumably some form of communication must occur between the PST system and PhoR. Exactly how that occurs is not clear.

Mu *d*1 (Ap^R *lac*) phage technology was used to identify and characterize 20 separate promoters that show increased transcription initiation under conditions of phosphate star-vation. Such mutants carried the *lac* operon fused to **phosphate starvation inducible** (*psi*) promoters. As mentioned earlier, the term regulon is generally used to describe the control of a group of unlinked genes that display coordinate control under the same physiological condition or stress (e.g., SOS or *pho*). It is further believed that genes within a regulon are subjected to identical controls by a common regulatory protein, such as LexA repressor for the SOS regulon. However, studies of the phosphate-controlled system indicate that the control of some of the unlinked phosphate-regulated promoters occurs in different ways, indicating overlapping global regulons, all of which comprise the phosphate starvation stimulon.

Oxidation Stress

The natural byproducts of aerobic metabolism are the reactive compounds superoxide (O_2^-) and hydrogen peroxide (H_2O_2). These two species can lead to the generation of hydroxyl radicals ($\cdot OH$), which can damage any biological macromolecule. The oxidation stress modulon includes 80 proteins induced during exposure to superoxide. About one-half are also induced by H_2O_2. Investigations have uncovered two regulons within this stress response modulon although there are certainly more than two. The two known systems, OxyR and SoxRS, used by *E. coli* are outlined in Figure 4-31.

The OxyR regulon comprises nine of the proteins induced by H_2O_2. All are controlled by the positive regulator OxyR. Purification of the OxyR protein has shown the protein is activated by oxidation and the formation of a disulfide bond. The reduced form of this regulator can bind to target operator/promoter regions but will *not* activate transcription until OxyR itself is oxidized. Some of the enzymes whose expression is regulated by OxyR include catalase (*katG*), glutathione reductase (*gorA*), and an alkyl hydroperoxide reductase (*ahpC* and *ahpF*). The likely *in vivo* function for alkyl hydroperoxide reductase would be the detoxification of lipid and other hydroperoxides that are produced during oxidative stress.

A second oxidative stress regulon comprises nine proteins induced by superoxide but not H_2O_2. This regulon is under positive transcriptional control by the *soxRS* loci. Genes under *soxR* control include those responsible for Mn^{2+}-containing superoxide dismutase (*sodA*), the DNA repair enzyme endonuclease IV (*nfo*), and glucose-6-phosphate dehydrogenase (*zwf*). The current model for this system has preexisting SoxR protein sensing the stress (superoxide itself?). The activated SoxR then triggers expression of the *soxS* gene continuing the cascade. The *SoxS* product activates the component promoters of the regulon. In contrast to OxyR, there are some clues as to how redox activation of SoxR might occur. There is an iron–sulfur center involving the C-terminal cysteine cluster. It is believed that iron–sulfur centers are especially sensitive to redox reactions with superoxide. The change in redox state of the protein could be sufficient to induce an active conformation of SoxR.

The Lon System: Proteolytic Control

The *lon* locus of *E. coli* codes for a protease that degrades "*abnormal*" proteins (see Chapter 2). Mutations in the *lon* locus have pleiotropic effects including an increased UV sensitivity, filamentous growth, and mucoid colonies due to the over production of capsule. Ultraviolet sensitivitiy and filamentous growth due to *lon*, explained earlier under DNA repair and the SOS system (Chapter 3), results from an inability to degrade the SOS-inducible cell division inhibitor SulA. The mucoid phenotype is the result of decreased degradation of RcsA, a positive regulator of capsular polysaccharide synthesis. In both cases, the proposed role of Lon protease is to assure certain regulatory proteins persist for only a short time. However, if we define a global regulator in terms of **initiating** a coordinated response of a given set of operons to a specific stimulus, then *lon*

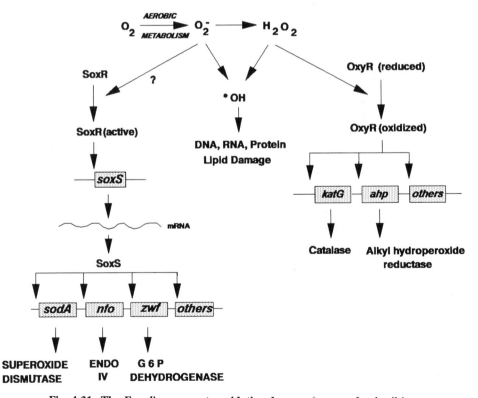

Fig. 4-31. The *E. coli* response to oxidative damage (see text for details).

will not qualify as such. Rather than initiating a response, Lon acts to limit them and may be important as a mechanism for maintaining and returning a cell to equilibrium. For a description of the interaction of *lon* protease with the cell division process, see Chapter 11.

SELECTED REFERENCES

General

Campbell, K. M., G. D. Storms, and L. Gold. 1983. Protein-mediated translational repression. In *Gene Function in Procaryotes*. J. Beckwith, J. Davies, and J. A. Gallant (Eds.). Cold Spring Harbor Laboratory, Cold Spring Harbor, NY.

Dobrogosz, W. J., G. W. Hail, D. K. Sherba, D. O. Silva, J. G. Harman, and T. Melton. 1983. Regulatory interactions among the *cya, crp*, and *pts* gene products in *Salmonella typhimurium*. *Mol. Gen. Genet.* **192**:477–486.

Gottesman, S. and M. R. Maurizi. 1992. Regulation by proteolysis: energy dependent proteases and their targets. *Microbiol. Rev.* **56**:592–621.

Gross, R., B. Arico, and R. Rappuoli. 1989. Families of bacterial signal-transducing proteins. *Mol. Microbiol.* **3**:1661–1667.

Heichman, K. A. and R. C. Johnson. 1990. The Hin invertasome: protein-mediated joining of distant recombination sites at the enhancer. *Science* **249**:511–517.

Jacob, F. and J. Monod. 1961. Genetic regulatory mechanisms in the synthesis of proteins. *J. Mol. Biol.* **3**:318.

Landick, R. and C. L. Turnbough, Jr. 1992. Transcriptional Attenuation. In: *Transcriptional Regulation*, Cold Spring Harbor Press, Cold Spring Harbor, NY.

Liberek, K., T. P. Galitski, M. Zylicz, and C. Georgopolos. 1992. The DnaK chaperone modulates the heat shock response of *Escherichia coli* by binding to the σ-32 transcription factor. *Proc. Natl. Acad. Sci. USA* **89**:3516–3520.

Lim, H. M. and M. Simon. 1992. The role of negative supercoiling in Hin-mediated site-specific recombination. *J. Biol. Chem.* **267**:11176–11182.

Lobell, R. B. and R. Schleif. 1990. DNA looping and unlooping by AraC protein. *Science* **250**:528–532.

Matthews, K. 1992. DNA looping. *Microbiol. Rev.* **56**:123–136.

Majumdar, A. and S. Adhya. 1984. Demonstration of two operator elements in *gal*: In vitro repressor binding studies. *Proc. Natl. Acad. Sci. USA* **81**:6100–6104.

Miller, T. H. and W. S. Reznikoff (Eds.). 1978. *The Operon*. Cold Spring Harbor Laboratory, Cold Spring Harbor, NY.

Nomura, M., S. Jinks-Robertson, and A. Miura. 1982. Regulation of ribosome biosynthesis in *Escherichia coli*. In, *Interaction of Translational and Transcriptional Control in the Regulation of Gene Expression*, M. Grunberg-Manago, and B. Safer (Eds.). Elsevier, New York, p. 92.

Pabo, C. O. and R. T. Sauer. 1992. Transcription factors: Structural families and principles of DNA recognition. *Annu. Rev. Biochem.* **61**:1053–1095.

Peterkofsky, A., I. Svenson, and N. Amin. 1989. Regulation of *Escherichia coli* adenylate cyclase activity by the phosphoenol pyruvate: sugar phosphotransferase system. *FEMS Microbiol. Rev.* **63**:103–108.

Reznikoff, W. S. 1992. Catabolite gene activator protein activation of *lac* transcription. *J. Bacteriol.* **174**:655–658.

Spassky, A., S. Busby, and H. Buc. 1984. On the action of the cyclic-AMP-cyclic AMP receptor protein complex at the *Escherichia coli* lactose and galactose promoter regions. *EMBO J.* **3:** 43–50.

Storz, G., L. A. Tartaglia, and B. N. Ames. 1990. Transcriptional regulator of oxidative stress-inducible genes: direct activation by oxidation. *Science* **248:**189–194.

Wanner, B. L. 1993. Gene regulation by phosphate in enteric bacteria. *J. Cell. Biochem.* **51:** 47–54.

Wilcox, G., S. Al-Zarban, L. G. Cass, D. Clarke, L. Heffern, A. H. Horwitz, and C. G. Miyada. 1982. DNA sequence analysis of mutants in the *araBAD* and *araC* promoters. In *Promoters: Structure and Function.* R. Rodrigues and M. Chamberlin (Eds.). Praeger Publishing Co., New York, pp. 183–194.

Wilson, H. R., C. D. Archer, J. Liu, and C. Turnbough, Jr. 1992. Translational control of *pyrC* expression mediated by nucleotide-sensitive selection of transcriptional start sites in *Escherichia coli. J. Bacteriol.* **74:**514–524.

Zinkel, S. S. and D. M. Crothers. 1991. Catabolite activator protein-induced DNA binding in transcription initiation. *J. Mol. Biol.* **219:**201–215.

Global Regulation

Bukau, B. 1993. Regulation of the *Escherichia coli* heat shock response. *Mol. Microbiol.* **9:** 671–680.

Csonka, L. N. and A. D. Hanson. 1991. Prokaryotic osmoregulation: Genetics and physiology. *Annu. Rev. Microbiol.* **45:**569–606.

Demple, B. 1991. Regulation of bacterial oxidation stress genes. *Annu. Rev. Genet.* **25:**315–337.

Forst, S. A. and D. L. Roberts. 1994. Signal transduction by the EnvZ-OmpR phosphotransfer system in bacteria. *Res. Microbiol.* **145:**363–373.

Gross, R., B. Arico, and R. Rappuoli. 1989. Families of bacterial signal-transducing proteins. *Mol. Microbiol.* **3:**1661–1667.

Parkinson, J. S. and E. C. Kofoid. 1992. Communication modules in bacterial signaling proteins. *Annu. Rev. Genet.* **26:**71–112.

Tokishita, S. -I., A. Kojima, H. Aiba, and T. Mizuno. 1991. Transmembrane signal transduction and osmoregulation in *Escherichia coli.* Functional importance of the periplasmic domain of the membrane-located protein kinase, EnvZ. *J. Biol. Chem.* **266:**6780–6785.

Aerobic–Anaerobic Regulation

Farr, S. B. and T. Kogoma. 1991. Oxidation stress responses in *Escherichia coli* and *Salmonella typhimurium. Microbiol. Rev.* **55:**561–585.

Gunsalus, R. P. 1992. Control of electron flow in *Escherichia coli*: coordinated transcription of respiratory pathway genes. *J. Bacteriol.* **174:**7069–7074.

Gunsalus, R. P. and S. -J. Park. 1994. Aerobic-anaerobic gene regulation in *Escherichia coli* control by the ArcAB and Fnr regulons. *Res. Microbiol.* **145:** 437–450.

Guest, J. R. 1992. Oxygen-regulated gene expression in *Escherichia coli. J. Gen. Microbiol.* **138:** 2253–2263.

Iuchi, S. and E. C. C. Lin. 1993. Adaptation of *Escherichia coli* to redox environments by gene expression. *Mol. Microbiol.* **9:**15.

Neidhardt, F. C., V. Vaughn, T. A. Phillips, and P. A. Bloch. 1983. Gene protein index of *Escherichia coli* K-12. *Microbiol. Rev.* **47:**231–284.

Nitrogen Assimilation

Beynon, J., M. Cannon, V. Buchanan-Wollarston, and F. Cannon. 1983. The nif promoters of *Klebsiella pneumoniae* have a characteristic primary structure. *Cell* **34**:665–671.

Guissin, G. N., C. W. Ronson, and F. M. Ausubel. 1986. Regulation of nitrogen fixation genes. *Annu. Rev. Genet.* **20**:567–591.

Hirschman, J., P.-K. Wong, K. Sei, J. Keener, and S. Justu. 1985. Products of nitrogen regulatory genes *ntrA* and *ntrA* product is a sigma factor. *Proc. Natl. Acad. Sci. USA* **82**:7525–7529.

Magasanik, B. 1993. The regulation of nitrogen utilization in enteric bacteria. *J. Cellular Biochem.* **51**:34–40.

Stewart, V. 1994. Dual interacting two-component regulatory systems mediate nitrate and nitrate-regulated gene expression in *Escherichia coli*. *Res. Microbiol.* **145**:450–454.

Stringent Response

Svitil, A. L., M. Cashel, and J. W. Zyskind. 1993. Guanosine tetraphosphate inhibits protein synthesis in vivo. A possible protective mechanism for starvation stress in *Escherichia coli*. *J. Biol. Chem.* **268**:2307–2311.

CHAPTER 5

BACTERIOPHAGE GENETICS

Bacterial viruses have played an important role in the development of bacterial genetics and remain important today. Their study has revealed many important biological concepts that have proved applicable to higher organisms. For example, T4 and ϕX174 proved crucial to understanding the DNA replication fork, whereas λ was integral to studies dealing with the initiation of DNA replication. T4 phage was one of the first places interchangeable σ factors were discovered, whereas λ proved central to our understanding of transcriptional repressors and activators. Another phage we will discuss, Mu, has taught us much regarding DNA rearrangements. What you should understand as you proceed through this chapter is that what we have learned about each of these viruses has had a dramatic impact on modern biology.

GENERAL CHARACTERISTICS OF BACTERIOPHAGE

The first step in understanding bacteriophage is to learn something about their classification. However, characterization and classification of the bacteriophage viruses poses many problems. From a practical viewpoint, they can be distinguished on the basis of their natural host, host range, and other similar characteristics. Beyond this, they can be characterized on the basis of their RNA or DNA content. Each phage generally contains only one kind of nucleic acid (the filamentous viruses have been shown to be an exception to this general rule). Structural symmetry and susceptibility or resistance to ether and other solvents and to other chemical agents also provide additional criteria for classification (Table 5-1).

Bacteriophages range in size from 20 to 300 nm. Large phage are barely within the realm of resolution of the light microscope. The electron microscope is essential for visualization of any details of viral structure. The nucleic acid may be either double stranded, single stranded, circular, or linear. Morphologically, phage generally display a considerable degree of geometric symmetry. They may be composed of a head that has a protein coat or **capsid** within which the nucleic acid **core** is housed. The capsid is

TABLE 5-1. Characteristics of Some Phages of *E. coli*

Phage	Virion Morphology — Head (nm)	Virion Morphology — Tail (nm)	Nucleic Acid — Type and Amount (Da)	Latent Period (min)	Average Yield per Cell	Growth Cycle at 37 °C — Lysogeny	Growth Cycle at 37 °C — Peculiarities
T1	Icosahedral 50	10 × 150	DNA, 2.5×10^7	13	150		Resistant to drying
T2, T4, T6	Prolated icosahedral 65 × 95	25 × 110	DNA, 1.2×10^8	21–25	150–400		Contain glucosylated HMC
T3, T7	Icosahedral 47	10 × 15	DNA, 2.4×10^7	13	300		Give semitemperate mutants
T5	Icosahedral 65	10 × 17	DNA, 7.5×10^7	40	200		DNA injection in two steps
λ, φ80	Icosahedral 54	10 × 140	DNA, 3.3×10^7	35	100	+	DNA circularizes, *in vitro* or *in vivo*
P1	Icosahedral 65	12 × 150	DNA, 6×10^7	45	80	+	General transduction
P2	Icosahedral 50	10 × 150	DNA, 2.2×10^7	30	120	+	Multiple chromosomal locations
φX174 S13	Icosahedral 30	None	DNA, 1 strand 1.7×10^6	13	180		Circular DNA
f2, MS2	Icosahedral 24	None	RNA, 9×10^5	22	20,000		Male specific attachment to F pili
fl, fd	None	6 × 800	DNA, 1 strand 1.3×10^6	30	100–200 (continuous release)		Male specific, circular DNA
χ[a]	Icosahedral 67.5	12.5 × 230	DNA	60	200		Attaches to motile flagella

[a]Not a coliphage; grows on many strains of *Salmonella*.

composed of subunits called **capsomeres.** Some bacteriophages may also have a prominent tail structure. The capsid structure may assume an icosahedral form (20 sides with 12 vertices). The entire infectious unit is generally referred to as a **virion.** The smallest phage contain nucleic acid (either DNA or RNA) with a molecular weight of the order of 1×10^6. This relatively short-chain nucleic acid can obviously code for only a relatively small number of hereditary units (genes). These genes, of necessity, must code for information that governs the formation of the basic viral subunits. The presence of such a small number of genes reflects the degree to which viruses are dependent on the host cell.

Each species of microorganism has its own set of viruses to which they are susceptible. Each phage also has a specific host range of susceptible microorganisms that it can infect. For example, phage ''A'' can only infect one strain of *E. coli* while phage ''B'' can infect many different strains of *E. coli*. Phage A is viewed as having a narrow host range relative to that of phage B. Host range depends on the presence of specific viral receptors on the host cell surface. These receptors are usually composed of specific carbohydrate groups on lipopolysaccharide (LPS) of cell surface structures. Some viruses, such as M13, attach to sex pili and are referred to as male-specific phage. The salient point here is that receptors are actually normal cellular proteins with a specific function that are coopted by the phage.

Two types of infection cycles may occur following the initial infection. One is a **virulent** or **lytic** infection that ends in lysis and death of the host cell. The other is **lysogenic** infection, which may be quite inapparent since the host cell does not die. Indeed, the general appearance and activity of the lysogenized host cell may not be altered in any overt manner. A lytic type of phage only undergoes virulent infection. Temperate phage, on the other hand, can undergo either lytic or lysogenic infection.

Bacteriophage infection of bacteria follows a characteristic pattern. The growth curve of bacteriophage, sometimes referred to as the ''one-step'' growth curve, is depicted in Figure 5-1. Once a population of bacterial cells has been inoculated with a given number of bacteriophages, the number of detectable infectious particles rapidly decreases. This stage is termed the ''eclipse phase'' and represents that portion of the time following infection during which bacteriophage cannot be detected either in the culture medium or within the cell. The overall period encompassing adsorption and eclipse is referred to as the **latent** period. After the latent period has been completed (~20 min for *E. coli* cells infected with coliphage), infectious viruses begin to be released from the cell. The average number of infectious bacteriophage particles released per cell is referred to as the burst size. The replication cycle may be depicted as follows:

1. Attachment (see Fig. 5-2).
2. Injection of nucleic acid or penetration of the entire virus into the cell.
3. Uncoating (i.e., removal of the capsid surrounding the nucleic acid ''core'').
4. Intracellular synthesis of viral subunits (capsid and nucleic acid).
5. Assembly of the subunits into complete virus particles (maturation). Complete bacteriophages within an *E. coli* cell are shown in Figure 5-3.
6. Release of viruses (burst or lysis).

Steps 1–4 comprise the latent period of which Steps 2–4 represent eclipse.

Lysogenic infection occurs when the nucleic acid of the virus enters the host cell,

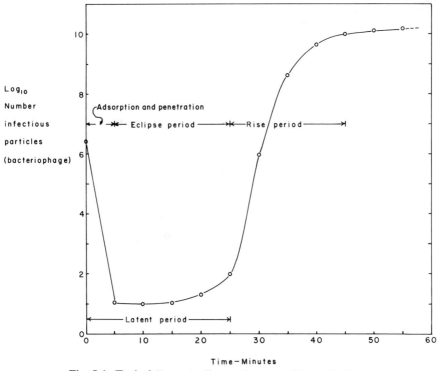

Fig. 5-1. Typical "one-step" growth curve of bacteriophage.

becomes aligned with the genome of the cell, and eventually is integrated into the genome of the cell. In this way, the virus is replicated in unison with replication of the host cell genome. At some later time, information coded within the viral genome may be **induced** to function. This triggers replication of the virus and entry into the lytic stage. Tumor viruses may initiate lysogenic infections of animal cells in a manner comparable to lysogenic infections of bacteria by bacteriophages. Additional discussion regarding the genetics of four specific bacterial viruses, each of which represents a different life style, may be found below.

T4 PHAGE

T4 Phage Structure

The T4 phage is one member of a family of T-even (T2, T4, T6) lytic phages of *E. coli*. Figure 5-4 presents the structure of a typical T-even phage of *E. coli* (T4). The T4 phage possesses an oblong head (80 × 120 nm) and a contractile tail (95 × 20 nm). Within the capsid is a linear, dsDNA molecule (1.3×10^8 Da or 1.69×10^5 bp). Associated with the DNA are a number of polyamines (putrescine, spermidine, and cadaverine). The base composition of T4 DNA is unusual in that all of the cytosine residues have been replaced by a modified base, hydroxymethylcytosine (HMC). The reason for this replacement will become apparent from subsequent discussion. While virions in general are

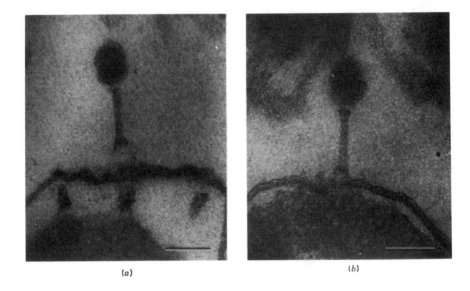

(a)

(b)

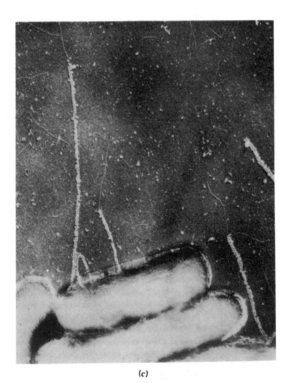

(c)

Fig. 5-2. Bacteriophage attachment to cells and pili (*a* and *b*). Attachments of T-even virions to points of adhesion between cytoplasmic membrane and outer membrane (cell wall) of *E. coli*. Bars = 100 nm. (*c*) *Escherichia coli* cells with F pili showing attached particles of MS2 (icosahedral) and M13 (filamentous) phages. From Luria S. E. et al. 1978. *General Virology*, 3rd ed. Wiley, New York.

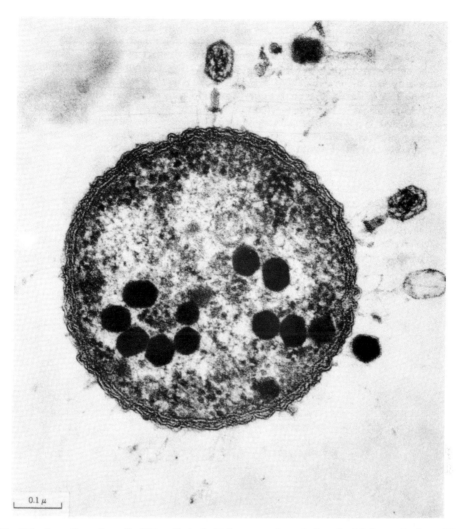

Fig. 5-3. A section of a cell of *E. coli* strain B from a culture infected with bacteriophage T2. Clearly visible are the cell envelopes (wall and cytoplasmic membrane); the clear area of the phage DNA pool containing many condensed phage DNA cores; and attached to the cell surface, three phages, one empty and two still partially filled. The page at the top shows the long tail fibers and the spikes of the tail plate in contact with the cell wall. The tail sheath is contracted and the tail core has apparently reached the cell surface but has not penetrated it. Courtesy Dr. Lee D. Simon.

thought not to possess enzymes, a number of enzymatic activities have been found associated with the T4 virion. Several of these are listed in Table 5-2 along with their possible roles in T4 infection.

The one-step growth experiment described earlier was originally developed with T4. The series of events that occur during T4 infection initiates when the tail fibers of T4 bind to specific *E. coli* cell surface receptors (adsorption). Once the pins of the baseplate contact the outer surface of the cell, the tail sheath contracts and the tail core penetrates the cell wall (Fig. 5-5). The T4 DNA molecule is then injected into the host.

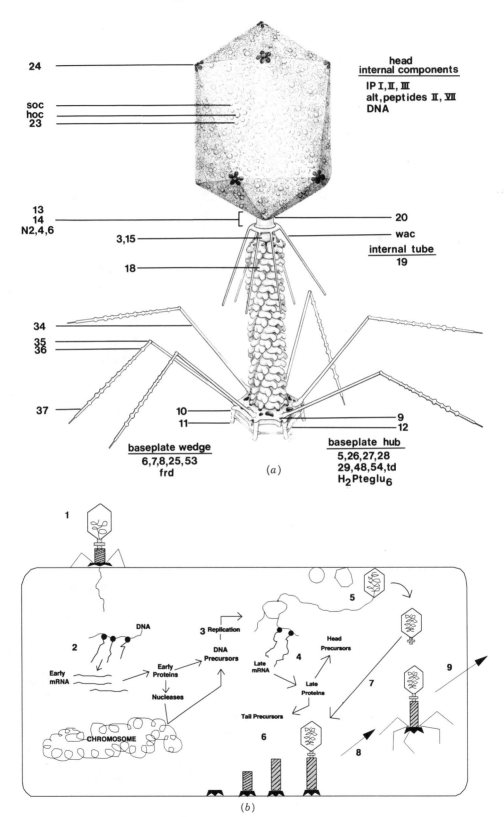

(a)

(b)

TABLE 5-2. Virus-Specific Enzymatic Activities Associated with T4 Virions[a]

Enzyme	Location and/or Function
Dihydrofolate reductase	Located in baseplate, may have role in unfolding tail fibers
Thymidylate synthetase	Found in baseplate; necessary for infectivity
Lysozyme	Possible role in penetration through cell wall
Phospholipase	Possible role in host cell lysis
ATPase	Associated with the tail sheath; presumed to be involved with the contractile process
Endonuclease V	Excision repair (see Section 13.1A) of phage or host DNA
alt function	Alteration of host RNA polymerase

[a]Adapted from Matthews, 1977. Reproduction of large virulent bacteriophages. In *Comprehensive Virology*, Vol. 7. H. Fraenkel-Conrat and R. R. Wagner (Eds.). Plenum, New York, pp. 179–294.

General Pattern of Gene Expression

It is a general strategy of viruses to defer some gene expression to later stages of infection and progeny virus maturation. The T4 phage is no exception. Transcription of certain T4 genes begins almost immediately, the **immediate-early** genes (transcribed at 30 s post-infection) and the **delayed-early** genes (transcribed at 2-min postinfection). The products of these genes are involved with establishing the infection. Since transcription of these genes is directed by σ^{70} promoters, normal host RNA polymerase is used to transcribe these early phage genes. Among the early gene products are nucleases and other T4 proteins that stop host transcription, unfold the host chromosome, and degrade host DNA to nucleotides. Several enzymes are involved with the synthesis of T4 DNA including a T4 DNA polymerase. The T4 DNA is unique in that its DNA contains the unusual base, 5-hydroxymethylcytosine:

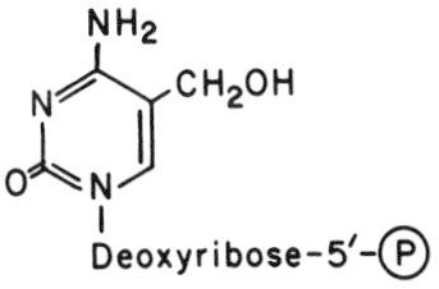

Fig. 5-4. (*a*) **Structure of bacteriophage T4.** Phage structure is based on electron microscopic structure analysis to a resolution of about 2–3 nm. Near the head and tail are shown the locations of the known major and minor proteins. The icosahedral vertices are made of cleaved gp24. The gene 20 protein is located at the connector vertex, bound to the upper collar of the neck structure. The six whiskers and the collar structure appear to be made of a single protein species, gp*wac*. The gp18 sheath subunits fit into holes in the baseplate, and the gp12 short tail fibers are shown in a stored position. The baseplate is assembled from a central plug and six wedges, and although the locations of several proteins are unknown, they are included here with the plug components. From Eiserling, F. A. 1983. Structure of the T4 virion. In Matthews, C. K. et al. (Eds). American Society for Microbiology, Washington, DC, pp. 11–24. (*b*) **Life cycle of T4 Phage.** (1) Attachment and injection of DNA; (2) Transcription of early genes; (3) Replication and concatemer formation; (4) Transcription of late genes; (5) Head assembly; (6) Tail assembly; (7) Attachment of head to tail; (8) Attachment of tail fibers; (9) Cell lysis and release of mature phage.

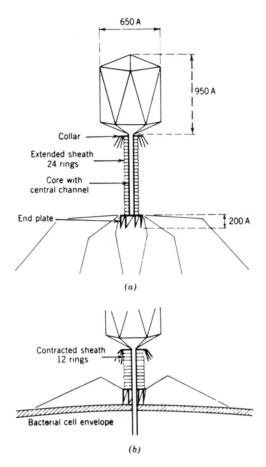

Fig. 5-5. Schematic representation of phage T4 virion and its component parts and of the mechanism of penetration of the phage core through the bacterial envelope. (*a*) Virion with extended tail fibers. (*b*) Virion with the tail sheath contracted and the spikes of the tail plate pointing against the bacterial cell wall. From Luria, S. E., J. E. Darnell, Jr., D. Baltimore, and A. Campbell. 1978. *General Virology*, 3rd ed. Wiley, New York.

This base is synthesized as hm-dCMP (deoxycytidine monophosphate) from dCTP (deoxycytidine triphosphate) by a T4 enzyme prior to incorporation of the base into T4 DNA. Following incorporation, this nucleotide is glucosylated by an HMC glucosyltransferase. These modifications serve two basic functions: (1) phages with HMC modified DNA can use cytosine-specific endonucleases to scavenge deoxyribonucleotides from host DNA without damaging their own DNA and (2) the HMC-modified phages can shuttle back and forth between strains carrying cytosine-specific restriction-modification systems since the phage do not contain cytosine proper (see **Restriction Modification** in Chapter 3).

Synthesis of the early enzymes ceases about 12-min postinfection. However, at 5-min postinfection, DNA replication has begun and transcription of the **late** genes commences. Prior to this transcriptional switch, host RNA polymerase undergoes a series of T4-directed modifications, as shown in Table 5-3. Some of these modifications (e.g., those catalyzed by *alt* and *mod* gene products) appear to be unnecessary for T4 growth while

TABLE 5-3. T4-Evoked Changes of *E. coli* RNA Polymerase (RNAP)

Change	Time First Detectable in RNAP (min after Infection at 30 °C)	Effect or Function
ADP-ribosylation of one of the two subunits (alteration); encoded by *alt* (70K)	< 0.5	Lowers affinity for σ; participation in host shutoff?
Phosphorylation, adenylylation, or ADP-ribosylation of a fraction of σ	< 0.5	Unknown
ADP-ribosylation of both α subunits (modification); encoded by *mod* (27K)	1.5–2.0	Lowers affinity for σ; selective shutoff of some early genes around 4 min; participation in host shutoff?
Binding of 10K protein (AsiA)	< 5	σ Antagonist, anti-σ^{70}
Binding of 15K protein	5	T4 gene 60 codes for a subunit of T4 DNA topoisomerase and may also code for the 15K protein
Binding of gp33 (12K)	5–10	Positive control of late transcription
Binding of gp55 (22K)	5–10	Positive control of late transcription, alternate sigma
Interaction with gp45 (27K)	does not copurify with RNA polymerase)	Gp45 is a component of the core of the T4 replisome and is also directly involved in late transcription

others (e.g., gp33, gp45, and gp55) are required. For example, gp55 is a viral encoded σ factor that is required for late gene transcription. It essentially hijacks the host RNA polymerase by altering its promoter sequence recognition properties. This works because the molecular structure of T4 late promoters is quite different from other known prokaryotic promoters. It uses the consensus sequence TATAAATACTATT at the −10 region but exhibits no consensus sequence at the −35 region. Clearly, the alternate σ and promoter structure will help direct T4 transcription toward the late genes. However, T4 also produces a 10-kDA antiσ^{70} protein (AsiA) that not only prevents transcription of the T4 σ^{70} early gene promoters but will increase the available pool of RNA polymerase for T4 σ^{55} use by decreasing transcription of *E. coli* housekeeping genes. Another curiosity of T4 transcription is that one of the proteins involved with late gene transcription (gp33) serves as a link between RNA polymerase-gp55 and components of the T4 replication machinery. This observation, among others, indicates a close relationship and coordination between replication and late gene expression in T4.

The late gene products are primarily components of viral capsids. Approximately 40 have been identified. In spite of the complexity of the T4 capsid, the assembly of the various components (e.g., heads, tail fibers, and baseplate) is remarkably efficient. It is fascinating that final assembly of the various phage component parts occurs spontaneously, without guidance. This feature is referred to as self-assembly and is due to the remarkable affinity each component has for its mate. The inner portion of Figure 5-6 illustrates the pathways involved with capsid formation and points out where the genes responsible are positioned on the T4 genome. The entire process results in the production of at least 200 particles per cell by 25–30 min postinfection (at 37 °C, in rich medium). At the completion of the T4 multiplication cycle the cell will lyse. Lysis is accomplished by a sudden cessation of respiration and requires the phage *t* gene product plus the phage e product (T4 lysozyme).

An intriguing feature of T4, which is of historical importance, is the still-unexplained phenomenon of **lysis inhibition.** Normal lysis of T4-infected cells can be inhibited by secondary infection of cells at least 3 min after the primary infection. The lysis inhibition by superinfecting phage is effective even if their DNA does not replicate. T4 mutants defective in the rI, rII, and rIII genes are not subject to lysis inhibition and exhibit unusual plaque morphology. In addition to lysis inhibition, rII mutants will not form plaques on an *E. coli* K strain lysogenized by lambda (λ) phage although they can infect these strains. Benzer, in 1955, performed a classic series of experiments in which he constructed a fine-structure genetic map of the rII locus utilizing point mutations and deletions. While an rII mutant will not plaque on an *E. coli* λ lysogen, by infecting these cells with *pairs* of different rII mutants, recombination can occur between the two types of T4 DNAs resulting in a rare r$^+$ plaque (Fig. 5-7). The closer the two independent mutations are to each other the less likely recombination will occur between them (i.e., the more likely recombination will occur to both sides of the pair). In addition, if a deletion in one rII mutant has removed the segment of DNA corresponding to the position of a point mutation on a second rII mutant, then, of course, recombination cannot result in an r$^+$ phage. The first fine structure deletion map of the rII region is shown in Figure 5-8 and has served as the prototype for fine structure genetic maps of numerous genes. Benzer was also the first to use the cis–trans test in prokaryotes (see Chapter 3) to show that the rII region was composed two functional units or cistrons, rIIA and rIIB, that could complement each other in trans.

It should also be noted at this point that while the DNA of T4 phage is linear both in the virus and within the host cell, the genetic map is circular (i.e., genes situated at opposite ends of the linear structure, which should not map close to each other nevertheless, upon genetic mapping, are found to be linked). The resolution of this paradox can be found in the realization that the ends of each T4 DNA molecule are **terminally redundant**, that is, a duplication of genetic information occurs at both ends of a single molecule. This phenomenon is the result of the **headful mechanism** utilized for the packaging of T4 DNA into the T4 capsid. After DNA replication (to be discussed more fully below), replica molecules can recombine with each other at their terminal redundancies creating very long molecules (concatemers) many genomes in length. Each mature virion is then made by packaging a ''headful'' of DNA cut from such a T4 concatemer. One headful, however, is 105% of a single genome. Therefore, the packaging process will generate a population of phage carrying all possible **circular permutations** of the T4 genome (Fig. 5-9). Genetic crosses between members of this population will result in a circular map.

T4 Replication

There are approximately 30 T4 genes known to be involved with some aspect of T4 DNA synthesis. Many of these are not directly involved with the replication fork mech-

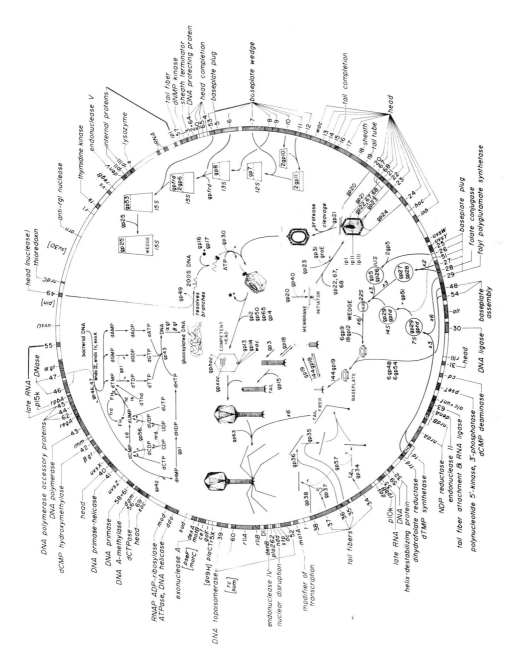

Fig. 5-6. Genomic map of bacteriophage T4. Courtesy of Dr. B. S. Guttman and Dr. E. M. Kutter, Evergreen State College, Olympia, WA.

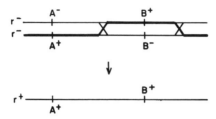

Fig. 5-7. Recombination between two r⁻ mutants yielding an r⁺ T4 phage.

anism but encode enzymes required for nucleotide precursor synthesis. In this section we will only deal with those genes directly involved with the replication process as well as analyze the replication machinery itself.

There is significant evidence that *in vivo* T4 replication can be initiated from multiple origins. One major initiation site occurs between genes *rII* and *42*, and there are up to four additional sites at other locations. As with host DNA replication, there is a leading or continuously synthesized strand and a lagging strand (Fig. 5-10). Okazaki fragments in T4 are about 2000 nucleotides long. The major protein responsible for nucleotide synthesis is gp43, the T4 DNA polymerase (110,000 MW). Accessory proteins include gp44, gp45, and gp62, which collectively form a sliding clamp on the DNA much as DnaB protein does on chromosomal DNA (see Chapter 2). As with any of the known DNA-dependent DNA polymerases, there is a requirement for an RNA primer. In T4, gene *41* (44 kDa) and gene *61* proteins, ATP, and CTP are the minimum requirements for RNA primer synthesis on a ssDNA template. The gene 41 protein is a DNA helicase that will also serve to unwind dsDNA at the replication fork. A complex of gp41 and gp61 is envisioned to move along the lagging strand in the 5′ ⟶ 3′ direction dependent on nucleotide hydrolysis. The gene *32* product, a helix destabilizing protein (ssDNA-binding protein) will bind to the single-stranded regions. At intervals of about 2000 nucleotides, the gp41/61 complex stops at specific sites to make pentaribonucleotide primers (pppApCpNpNpNp). The primers probably remain attached to the priming complex until elongated by T4 DNA polymerase (gene *43* product). Joining the Okazaki fragments to make a continuous DNA chain requires a special DNA repair system including T4 RNase H (44 kDa) to degrade the RNA primer, the T4 DNA polymerase to replace the RNA with DNA and T4 DNA ligase to form a phosphodiester bond between the 3′ end of each fragment and the 5′ end of the preceding one (Fig. 5-10). The gene *44/62* complex and gene *45* proteins appear to keep the polymerase continuously on the growing 3′-OH chain. Replication also requires the gene *39*, *52*, and *60* products that comprise T4 topoisomerase, a Type II topoisomerase. It is not clear exactly what its role is in replication. Recall that *in vivo* the T4 molecule is linear. Thus, the action of T4 helicases at the replication fork would not be expected to introduce tortional stress in DNA ahead of the fork. This is in obvious contrast to circular molecules (e.g., host chromosome), which do not have a free end. However, transient superhelical tension can be placed on a linear molecule if large proteins are bound to it and inhibit rotation. It is also possible that the T4 topoisomerase is directly involved in initiation of replication by introducing a single-stranded nick at specific T4 origins.

The expression of four of the DNA replication genes (*g43*, *g44*, *g45*, and *g62*) appears to be regulated by controlling the initiation and termination of transcription, as well as

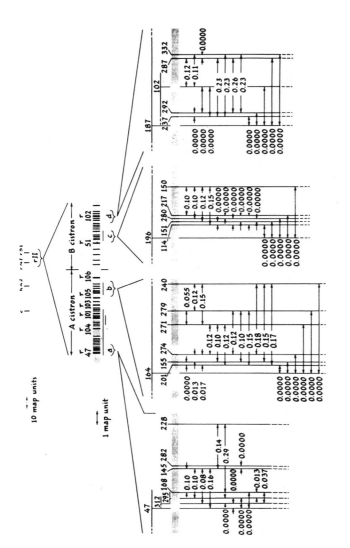

Fig. 5-8. The first fine-structure map of the rII region of T4. The map is based on the frequency of r⁻ recombinants produced in pairwise crosses between a set of 60 rII mutants of independent origin. Successive levels in this figure correspond to progressively greater magnifications of the viral genome. In the lowest level, numbered vertical lines represent individual rII mutations, and the decimals indicate the percentage of r⁺ recombinants found in crosses between two mutants connected by an arrow. The horizontal bars shown in the middle and lowest level represent the genetic extent of long-span mutations, or deletions. From Benzer, S. 1955. *Proc. Natl. Acad. Sci. USA* **41**:344.

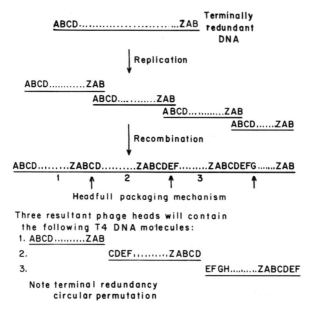

ABCD.........................ZAB Terminally redundant DNA

↓ Replication

ABCD............ZAB
    ABCD.............ZAB
        ABCD............ZAB
            ABCD.......ZAB

↓ Recombination

ABCD.........ZABCD..........ZABCDEF.........ZABCDEFG.......ZAB
   1   ↑   2   ↑   3   ↑

Headful packaging mechanism

Three resultant phage heads will contain
the following T4 DNA molecules:
1. ABCD..........ZAB
2.   CDEF..........ZABCD
3.   EFGH..........ZABCDEF

Note terminal redundancy
circular permutation

Fig. 5-9. Formation of terminally redundant, circularly permuted T4 DNA.

translation (Fig. 5-11). Although these four genes are transcribed by the host RNP, the T4 *mot* product appears to influence promoter selection by RNAP, stimulating transcription of g43 and g45 while inhibiting transcription of g44. The gp43 product (TR-DNA polymerase) also competes with Mot for binding thereby regulating its own transcription. Translational control by *regA*, also indicated in Figure 5-11, is explained below.

Regulation of T4 Gene Expression

Regulation of T4 gene expression has been shown to occur both at the transcriptional and translational levels. At the translational level two systems are recognized; autogenous translational repression by the gene *32* product (helix-destabilizing protein) and transla-

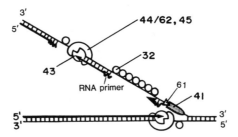

Fig. 5-10. Model of a replication fork showing the roles of bacteriophage T4 proteins in the continuous synthesis on the leading strand and in the discontinuous synthesis on the lagging strand. (see text for details). From Nossal, N. G. and B. M. Alberts. 1983. Mechanism of DNA replication catalyzed by purified T4 replication proteins. In Matthews, C. K. et al. (Eds.) American Society for Microbiology, Washington, DC. pp. 71–81.

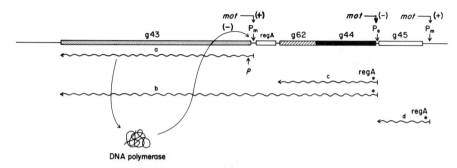

Fig. 5-11. Schematic summary of proposed mechanisms controlling expression of replication genes. Here P_e indicates an immediate early promoter recognized by *E. coli* RNA polymerase, which is independent of *mot*; P_m indicates a middle early promoter that is dependent on *mot* for transcription. The wavy lines indicate length and direction of mRNA synthesis. Transcript *a* accounts for 75% of the cellular level of gene 43 mRNA and the *b* transcript contributes 25%. Translation of transcripts *c* and *d* may be controlled by the product of the *regA* gene. The ρ indicates a potential site for rho-mediated termination, which the *mot* protein may modulate. From Spicer, E. K. and W. H. Konigsberg. 1983. Organization and structure of four T4 genes coding for DNA replication proteins. In Matthews, C. K. et al. (Eds.) 1983. American Society for Microbiology, Washington, DC. pp. 291–301.

tional repression of many T4-induced early mRNAs by the phage *regA* product. In the systems of translational repression already discussed (i.e., Ribosomal Proteins, Chapter 4) each protein repressor was specific to only one species of mRNA. This mechanism appears the same for the T4 gene *32* product. However, translational regulation by RegA (12,000 MW) is unique in that it appears to repress the translation of different mRNA species. The mechanism appears to be the simple recognition of and binding to a consensus mRNA sequence [(AUG)UACAAU-3′] thus blocking ribosome access. Secondary mRNA structures appear to be unnecessary. Although the regulatory significance of *regA* remains a mystery, it would seem to play a role in coordinating the early events in T4 development.

Introns in T4 Genes

Many eukaryotic genes include one or more nontranslated intervening sequences (IVS), or **introns**. These intervening sequences must be excised from the primary transcripts prior to translation. Few examples of introns have been documented in prokaryotes. The first was a 1-kb intron within the thymidylate synthase (*td*) gene of bacteriophage T4. Since then two more T4 genes, *nrdB* and *sunY*, were shown to contain introns. The T4 introns are all self-splicing, the mechanism involving a series of transesterifications, or phosphodiester bond transfers, with the RNA functioning as an enzyme.

A novel form of gene splicing is found in gene *60*, which codes for an 18-kDa subunit of the T4 topoisomerase. Within the message is a 50 nucleotide untranslated region. However, this region is not removed as an intron from the message. It seems a pair of 5 nucleotide direct repeats flanking the IVS can base pair forming a hairpin that brings codons on either side close together. A ribosome can presumably move through or jump across this structure, ignoring the nucleotides in the loop.

LAMBDA PHAGE

The lambda phage (λ, Fig. 5-12) of *E. coli* is a classic example of a temperate phage. That is, it can undergo either a lytic or a lysogenic cycle upon infecting a sensitive *E. coli*. This property manifests itself in the form of a "cloudy plaque" when λ is grown on a lawn of *E. coli*. The cloudy center is due to growing bacterial cells that have been

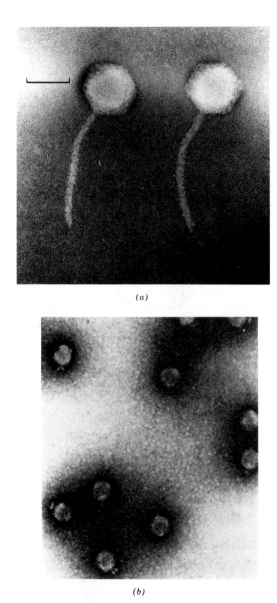

(a)

(b)

Fig. 5-12. Electron micrographs of some temperate bacteriophages. (*a*) Lambda phage negatively stained with potassium phosphotungstate. The length of the bar is 50 nm. Micrograph by E. A. Birge. (*b*) Phage P22, also negatively stained. From King J. and S. Casjeus. 1974. *Nature (London)* **251:**112–119.

lysogenized by the phage DNA and to the fact (below) that these lysogens are resistant to superinfection. Mutations in certain λ genes (*cI, cII, cIII*) will prevent lysogeny so that the mutant phage will produce "clear" plaques in which all infected cells are lysed. The λ genome (48.5 kDa) expresses approximately 50 proteins. The genes of this phage can be grouped into four categories based upon their roles in either lysis or lysogeny. (1) genes involved in lytic development, (2) genes involved with the development of lysogeny, (3) genes participating in both, and (4) genes with unknown functions. The genes in the second and fourth groups are nonessential for phage growth. Figure 5-13A presents a general map of the λ genome.

The DNA within a λ capsid is linear but contains cohesive ends designated *cos* sites. Upon phage attachment and injection, the *cos* ends of λ DNA anneal and are ligated to form a circular molecule. There are two phases of λ replication as the phage undergoes lytic development: an early phase in which the circular molecules replicate bidirectionally and a late phase, approximately 10-min postinfection, consisting of a rolling circle mechanism that generates long concatemeric molecules containing multiple copies of the λ genome. This will be discussed in detail below. The concatemers are then acted upon by the λ packaging system and used to fill preformed capsids with λ DNA. A precise amount of λ DNA is packaged, corresponding to one complete genome, following cleavage (terminase) at the unique *cos* sites mentioned above.

The Lysis–Lysogeny Decision

Upon infection, λ phage must make a molecular decision of whether to multiply in a lytic manner or "*hibernate*" as an integrated prophage in the lysogenic state. As will become evident in the coming sections, it is impossible for λ to go down one path (i.e., toward lysis) without partially going down the other (i.e., toward lysogeny). The multiplicity of infection greatly influences the final "decision." In general, a high multiplicity of infection leads to lysogeny, whereas a low ratio of phage to cells results in lytic multiplication. This correlation should be kept in mind while reading the following sections.

Transcription. Transcription of λ initiates at P_L (leftward promoter) and P_R (rightward promoter) with the **early** transcripts terminating at the rho-dependent terminators at t_{L1} and t_{R1} (Fig. 5-13B). The leftward transcript results in the production of the antiterminator pN (lower case p in this instance refers to a phage gene product or protein), while the rightward transcript codes for Cro protein. Transcription into genes beyond the terminators requires N protein acting as an antiterminator at t_{L1} and t_{R1}. There are six factors produced by *E. coli* that are required for effective N utilization: *nusA* (69 min), *nusB* (11 min), *nusC* (*rpoB*, 88 min), *nusD* (*rho*, 84 min), *nusE* (*rpsJ*, 72 min), and *nusG* (90 min). The *nus* acronym stands for N utilization substance. *Escherichia coli nus* mutants fail to support growth by blocking the *action* of pN rather than its synthesis. The *nusA* and *nusG* gene products and their normal roles in *E. coli* were discussed previously with regard to transcription (Chapter 2). As noted above, the *nusC*, *nusD*, and *nusE* mutations occur in the genes for RNA polymerase β subunit, Rho, and ribosomal protein S10, respectively.

The model for N-action involves a modification of RNA polymerase at the *nut* sites (see Fig. 5-13B). The presence of NusA, NusB, and/or NusG on core polymerase enables pN to modify RNA polymerase when a *nut* site is encountered. The modified RNA

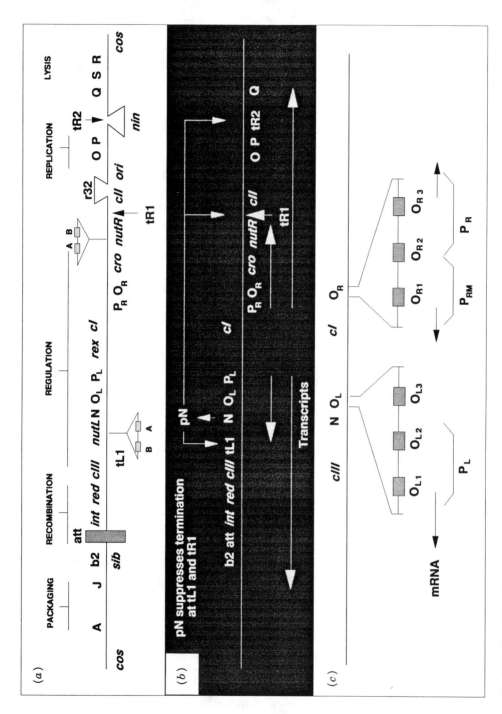

Fig. 5-13. Lambda phage. (A) Genetic organization. (B) Transcriptional organization. (C) Structure of the leftward and rightward promoter/operator regions.

polymerase can then transcribe past t_{R1} and t_{L1} resulting in the synthesis of *cII* and *cIII* message, respectively, and past t_{R2} yielding another antiterminator pQ. The *nut* site (also called boxB) is preceded by another consensus sequence, called boxA, that is involved in NusB association with core polymerase. Both boxA and boxB sequences function as RNA signals. The NusB protein apparently binds to the boxA sequence of nascent message and communicates with RNA polymerase through the S10 collaborator NusE. The N antiterminator protein binds to the boxB hairpin loop in the message, interacts with RNA polymerase through NusA, and that interaction becomes stabilized by NusB–S10 and NusG. This complex interaction will suppress the pausing of RNA polymerase.

Several minutes after λ infection, the synthesis of early (N and cro) and delayed early (*cII, P, O, Q* and *cIII, red, int*) transcripts diminish due to the action of cro repressor on P_L and P_R following its binding to O_L and O_R. However, the pQ antiterminator will allow the P'_R transcript to extend through the late genes involved with capsid production and lysis. This extended transcript eventually terminates at *b* (remember the λ genome is circular at this point). As a general rule, when the multiplicity of infection is low there will not be a sufficient quantity of CII or CIII for the establishment of lysogeny. However, enough pQ is synthesized to transcribe through the late genes producing the lytic cycle of λ phage.

Function of Cro versus CI Repressor and the Structure of O_L and O_R. As opposed to CI, which promotes lysogeny, Cro protein promotes lytic multiplication of λ by preventing the synthesis of CI. Although both Cro and CI bind to the same operators (O_L and O_R), these repressor proteins serve distinct physiological functions due to several unique features of the operators as illustrated in Figure 5-13C. First, within each operator there exist three CI (and Cro) binding sites. Each binding site (designated O_{L1}, O_{L2}, etc.) is similar in sequence and possesses axes of twofold hyphenated symmetry. The two repressors have different relative affinities for each of these binding sites. At *low* concentrations, CI will bind to O_{R1} and O_{R2} blocking transcription from P_K. Actually, CI repressor binds cooperatively to O_{R1} and O_{R2}, binding first at O_{R1}, which increases the affinity of O_{R2} for CI, then at O_{R2}. Binding of CI to O_{R2} will activate P_{rm} (promoter for repressor maintenance) ensuring continued production of CI. In contrast, Cro protein binds more tightly to O_{R3}. This prevents transcription from P_{rm} halting synthesis of CI. Thus, Cro binding to O_R will promote the lytic cycle of development by preventing lysogeny.

The study of both Cro and CI repressors revealed a motif that has proven to be common to many repressor proteins. That is the helix–turn–helix motif. The DNA-binding domain of CI resides at the amino terminal end of this protein. In this region are several alpha helices, two of which participate in DNA binding. As shown in Figure 5-14, the two helices interact forming a helix–turn–helix structure that can bill the major groove of the operator DNA. Both Cro and CI bind the operator as dimers. Figure 5-14 illustrates how two of the HTH regions of a dimer fill the major groove and also shows the importance of the N-terminal arms (last six amino acids) in reaching around the major groove to contact the backside of the operator. The affinity of an armless CI repressor for DNA is reduced about 1000-fold.

Establishment of Repressor Synthesis. If *cI* transcription from P_{rm} requires CI as a positive regulator, how is synthesis of CI established in the first place? After infection, *cI* transcription originates from a different promoter called P_{re} (promotor for repressor estab-

Fig. 5-14. (*a*) **Helix–turn–helix region of the λ CI repressor.** (*b*) View of a pair of λ contacts with operator DNA. The bulk of repressor contacts occur on one side of the DNA, but the N-terminal arms reach around to contact the other side.

lishment). Transcription from P_{re} is positively regulated by the CII protein (Fig. 5-15). The CII protein also activates the integrase gene whose product is required for λ DNA to integrate into the host chromosome (prophage state). Thus, CII coordinately regulates the two transcriptional units required for lysogeny.

One of the main criteria for whether lysis or lysogeny occurs is the amount of CII activator produced upon infection. This amount is dependent on several factors: (1) the level of *cII* mRNA, (2) rate of *cII* mRNA translation, and (3) the rate of CII processing. The level of *cII* mRNA is controlled by CI and Cro repressors as well as by pN, which is required for transcription past t_{R1} into *cII*. Translational control of *cII* expression involves the host HimA protein. Efficient translation of *cII* mRNA requires HimA. Although the mechanism is not well understood, one model states that HimA in some way allows ribosome access to a sequestered Shine–Dalgarno sequence and translation initiation codon in *cII* mRNA. There is a potential stem–loop structure found at the 5′ end of the *cII* message that would protect these sequences.

In addition to the amount of *cII* message produced and the efficiency with which it is translated, the amount of active CII produced in the cell is dependent on its processing. The first two amino acids (*N*-formyl methionine and valine) must be removed in order to form an active CII protein. Also, turnover (or degradation) of CII is very important. The products of the *hflA* locus (93 min) are at least partially responsible for CII degradation. The *hflA* locus consists of three genes, *hflK*, *hflC*, and *hflX*. The Hf1K and C proteins are proteases that can degrade CII *in vitro*. It is the apparent role of CIII protein to inhibit this protease thus allowing sufficient accumulation of CII to promote transcription from *pre* (i.e., CI production). Once again, a high multiplicity of infection will result

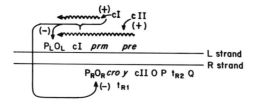

Fig. 5-15. Establishment of lysogeny by λ phage. The CII protein stimulates the transcription of *cI* from *pre*, the promoter for repressor establishment. Once CI protein is produced it will stimulate its own transcription from *prm*, the promoter for repressor maintenance. Note that *prm* overlaps O_R (Fig. 5-13C) such that when CI binds to O_{R1} and O_{R2} transcription is inhibited from P_R but stimulated from *prm*.

in a large amount of CIII produced. This inhibits Hf1A protease from degrading CII. The accumulation of CII promotes the transcription of CI and integrase, which then allows λ to integrate and represses all but CI transcription, establishing the lysogenic state.

It is interesting, too, that the overall energy state of the cell affects the process of lysogenization. Under conditions where cAMP levels are low, lysogeny is the preferred developmental route while the reverse is true where cAMP levels are high. It was discovered that cAMP and CRP appear to inhibit the transcription or activity of Hf1A. Thus, high cAMP levels lead to lowered Hf1A protease activity. This allows for a greater level of CII thereby stimulating CI production. However, there is some evidence that contradicts this model so it is not precisely clear how cAMP affects this system.

Control of Integration and Excision

If molecular events lead toward the lysogenic state, then the circularized genome must integrate into the host chromosome as depicted in Figure 5-16. Integration involves site-specific recombination between the phage *att* and bacterial *att* attachment sites. The integration requires the *int* gene product plus integration host factor (IHF, see below). Prophage attachment normally occurs at the *att* site located between the *gal* and *bio* operons at 17 min. Note the permutation of gene order that occurs during the integration, that is, A and R are distal loci in the linear λ DNA present at infection but end up as proximal loci in the prophage.

The *int* and *xis* loci are tightly linked and their coding regions partially overlap (Fig. 5-17). The *int* locus is controlled by the P_I promoter while transcription of *xis* usually

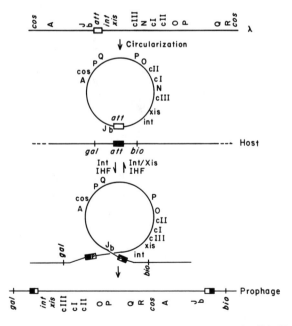

Fig. 5-16 Integration and excision of the λ genome (see text for details). Note the permutation of gene order as a result of integration. (i.e., *b* and *int* are proximal in the vegetative form but distal in the prophage).

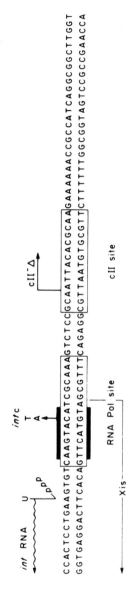

Fig. 5-17. The P$_I$ promoter region and the mechanism of regulation by CII. The CIII protein recognizes the right boxed sequence (CII site) and promotes interaction of RNA polymerase with the left boxed sequence (RNA pol site). The CII-activated RNA chain begins with UTP at adjacent bases with the major start at the point noted above the sequence. Positive regulation of *int* but not *xis* occurs because the coding sequence for Xis begins with the ATG indicated and the CII-stimulated RNA transcript lacks the translation start and other aminoterminal codons for Xis. The base change of an *int-c* mutation to a constitutive promoter is shown above the sequence, as is the beginning point of a deletion that eliminates CII activity. From Echols, H. and G. Guarneros. In *Lambda II*. R. W. Hendrix, J. W. Roberts, F. W. Stahl, and R. A. Weisberg, (Eds.). 1983. Cold Spring Harbor Laboratory, Cold Spring Harbor, NY.

results from P_L. However, as illustrated in Figure 5-17, activation of pI by CII results in polymerase binding at the left box with transcription starting beyond that point. Thus, an incomplete Xis message results from P_I. The end result of CII activation, then, is the preferential synthesis of Int over Xis, which is obviously preferred for integration.

Negative Retroregulation of int by sib. Upon induction of prophage by UV mutagenesis, it is preferred that more Xis be produced than Int in order to achieve lytic replication. The *sib* locus within the *b* region is responsible for this. Upon exposure to UV light, activation of the host *recA* coprotease results in the autocleavage of CI protein (see SOS Response in Chapter 3). Transcription of λ is allowed to occur from P_L and will proceed through *xis* and *int*. Both Int and Xis are needed for prophage excision (host integration factor is also needed). Examination of Figure 5-16 reveals that in the prophage state the sib region is removed from its normal proximal location to *int*. However, upon excision, the sib region is situated near *int* once again (separated by ~200 bp). At this point Int is no longer required and could prove detrimental if it causes reintegration of λ in a cell ravaged by UV irradiation. Regulation by *sib* is posttranscriptional and occurs only in *cis*. The basis for *sib* retroregulation (retro because *sib* is located downstream from *int*) is the formation of a stem–loop structure near the 3′ end of *int* message (Fig. 5-18). This stem–loop structure is a substrate for RNase III, thus RNase cleaves at the stem–loop region and the 3′ end is degraded by a 3′ exonuclease. Retroregulation, however, only regulates the P_L transcript but not the transcript resulting from P_I. The reason for this is that transcription from P_1 is regulated by N protein which, upon binding RNA polymerase, renders it resistant to most termination signals. Thus the P_1 transcript will proceed through the *sib* region bypassing a termination signal. The CII-activated P_I transcript terminates within *sib*. The longer stem–loop structure formed from the P_1 transcript is probably a more suitable substrate for RNase III allowing retroregulation.

Lambda Phage Replication

Upon infection and circularization, early replication is bidirectional with the origin of replication (*ori*) situated within the O gene (θ replication). Evidence suggests that the amino terminal end of the O protein (as four dimers) interacts with *ori* as the first step in a chain of interactions that control replication. Then, pO interacts with pP. The P protein, in turn, interacts with a number of host replication proteins. For example, two DnaB proteins will bind to one pP. The recruitment of DnaB into the λ replication complex sequesters DnaB away from host replication machinery with the consequence of decreasing *E. coli* DNA replication. Activation of DnaB within the λ *ori* complex is different from its activation at the chromosomal origin. Three *E. coli* heat shock proteins, DnaK, DnaJ and GrpE, release pP from the complex that triggers DnaB activation (see Chapter 2, Chaperonins). Following unwinding of λ *ori* by DnaB, DNA replication can proceed upon the addition of primase and DNA polymerase III.

Transcriptional activation of *ori* is also required for replication with transcription through the origin from P_R being a specific requirement. Transcription through *ori* helps melt the region allowing access to replication proteins. Replication is thus directly inhibited by CI repressor binding to the right operator. This is desirable, of course, when choosing lysogeny as a "lifestyle."

As noted previously, early bidirectional replication of λ only occurs for a brief period after which a rolling circle mode of replication, referred to as σ replication, begins to make the long concatemeric DNA required for packaging (Fig. 5-19). Searches for a gene

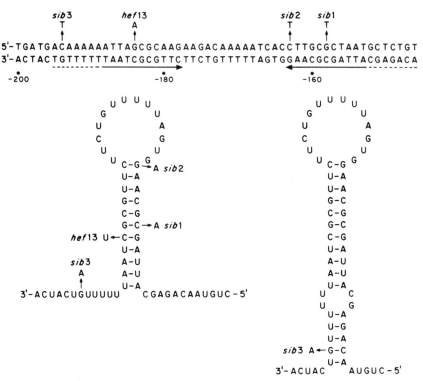

Fig. 5-18. The *sib* region. The locations of four *sib⁻* mutations are shown in the DNA sequence of the *sib* region (*top*). Possible secondary structures of the RNA from this region are shown below. The RNA structure on the right can form from p_L RNA but not p_I RNA because the p_I RNA terminates at the last of the six consecutive U bases (position −193) (see text). From H. Echols and G. Guarneros. In *Lambda II*. R. W. Hendrix, J. W. Roberts, F. W. Stahl, and R. A. Weisberg (Eds.). 1983. Cold Spring Harbor Laboratory, Cold Spring Harbor, NY.

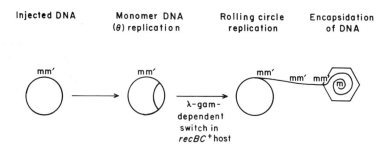

Fig. 5-19. Infective pathway of bacteriophage λ. The phage DNA forms a covalently closed circle through pairing of the cohesive ends, m and m′, followed by ligation. Early DNA replication by the θ mode generates circular monomeric molecules. The switch to the rolling-circle mode of replication is dependent on the product of the *gam* gene, which inhibits the host *recBC* nuclease. The multimeric DNA is the substrate for encapsidation. From Murray, N. In *Lambda II*. R. W. Hendrix, J. W. Roberts, F. W. Stahl, and R. A. Weisberg (Eds.). 1983. Cold Spring Harbor Laboratory, Cold Spring Harbor, NY.

that would mediate the switch from θ to σ replication modes have been fruitless. The *gam* and *red* genes, however, do appear to be important. The host *recBC* product, exonuclease V, will degrade the tails produced during σ-mode replication. The λ *gam* product inhibits exonuclease V activity, thus allowing concatemers to form. A *gam⁻* mutant, therefore, will only undergo θ-mode replication. However, concatemers are required for packaging to occur. The *red* product of λ appears to be involved with recombination between θ molecules forming a large, circular concatemer also suitable for packaging. The packaging process itself is illustrated in Figure 5-20. The linear DNA molecule is inserted into an empty procapsid until the terminase enzyme (gene *A* product) identifies a *cos* site at which point the enzyme introduces a staggered double-strand cut. The filled capsid then has a tail assembly attached and is ready for release from the cell. Release of mature phage particles from an infected cell involves T4 lysozyme (the R gene product), which is a 17K protein with murein transglycosylase activity, and the 11K *S* gene product. The S product is believed to interact with the cell membrane causing pore formation.

Lambda as a Cloning Vector

Lambda phage has proven very useful as a vector for recombinant DNA technologies. It has a very large genome one-third of which is dispensable for phage growth. This means up to 19 kb of foreign DNA can be cloned into λ if the unneeded central "stuffer" region is removed. The maximum genome length that can be packaged is 53 kb because of capsid size limitations. Lambda is not useful for packaging DNA less than 38 kb since the viability of λ decreases when the amount of DNA packaged falls below 78% of wild-type λ. Cosmids are chimeric molecules in which the *cos* site from λ has been incorporated into a 4–6-kb plasmid. Recombinant cosmids can include 25–45 kb of insert DNA and are packaged *in vitro* into λ capsids. Cosmids are not useful for inserts less than 25 kb because, as noted above, the λ packaging system will not efficiently package less than 38 kb.

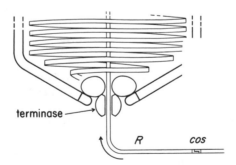

Fig. 5-20. Scanning for the terminal *cos* site, showing part of a nearly filled head with terminase positioned at the connector. The DNA strand being packaged is "scanned" by terminase until the cohesive end symmetry segment of the terminal *cos* site is recognized and cut. Other aspects of packaging, such as the terminase-connector interaction, are purely hypothetical. From Feiss, M. and A. Becker. In *Lambda II*. R. W. Hendrix, J. W. Roberts, F. W. Stahl, and R. A. Weisberg (Eds.). 1983. Cold Spring Harbor Laboratory, Cold Spring Harbor, NY.

MU PHAGE

The genome of phage Mu is a linear molecule of about 38 kb consisting of 36 kb of Mu DNA and 1.5 kb of host DNA at one end and 50–150 bp of host DNA at the other. The attached host sequences reflect the Mu life cycle. When Mu replicates, it undergoes repeated cycles of transposition in the cell (~100 events per cell during the lytic cycle) and is ultimately packaged along with adjacent host sequences into virus particles. While Mu transposition results in the duplication of a 5 bp target sequence, Mu is different from other transposons in that the ends of Mu are not inverted repeats.

The basic map of Mu can be found in Figure 5-21 and Table 5-4. Two gene products are required for the transposition process. The A gene product (70,000 MW) is the transposase and the B gene product (33,00 MW), which is an accessory protein necessary for replication and full transpositional activity. During the lytic cycle where this activity is greatest, Mu transposes by a cointegrate process. This fact indicates that during replication and cointegrate formation inversion-insertions or reciprocal adjacent deletions will occur (i.e., the host chromosome will be completely rearranged).

The Mu phage particle is comprised of an icosahedral head 600 Å long bound to a contractile tail ending in a baseplate containing supporting spikes and fibers (Fig. 5-22). Infection with Mu involves adsorption of the phage tail to a lipopolysaccharide component of the bacerial outer membrane followed by injection of viral DNA into the host cytoplasm. The G segment of Mu is an invertible region that affects host range by changing the type of tail fiber produced. Inversion of the G region occurs via a reciprocal recombination between the short inverted repeats that flank this region. This switching occurs during the prophage state and is controlled by the *gin* product (a recombinase). The switch determines which of two alternative tail fibers are produced in a manner similar to what was described for *S. typhimurium* flagellar phase variation (Chapter 3). The Mu G(+) phages, with S_γ and U at the left end of G, will infect *E. coli* K-12 or *S.*

Fig. 5-21. Map of Mu DNA. The DNA is about 37 kb long. Lengths are shown in kilobases (not drawn to scale). Mature Mu DNA in phage particles is flanked by heterogeneous host sequences, which result from packaging of Mu from different sites. α, G, and β are different segments of the Mu genome. The α segment contains the early genes and the genes for head and tail morphogenesis. The early region contains the repressor gene *c* and a negative regulator (*ner*). In addition to transposase gene *A*, the main replicator gene *B*, and a gene for amplified replication of Mu (*arm*), other genes in this region are involved in DNA metabolism. Gene *C* appears to control the late functions. The G segment contains genes for the Mu host range and is flanked by 34 bp inverted repeats. The *gin* protein, whose gene is located to the right of G, acts on the 34 bp inverted repeats to invert the G segment; hence G is known at the flip–flop segment. The rightmost gene in Mu encodes the DNA modification function (*mom*). From Toussaint, A. and A. Resibois. 1983. Phage Mu: Transposition as a lifestyle. In *Mobile genetic elements*. J. A. Shapiro (Ed.). Academic, New York, pp. 105–158.

TABLE 5-4. Characteristics of the Mu Genes and Functions[a]

Gene	Localization with Respect to the c end (kb)	Function	Molecular Weight of Protein
c	0–0.9	Repressor	26,000
ner (negative regulation)		Shutoff of early transcription	6,000
A	1.3–3.3	Transposase	70,000
B	3.3–4.3	Replication	33.000
cim (control of immunity)		When inactivated in prophage lacking the S end, no repressor synthesis	7,000
kil (killing)		Killing of the host even in the absence of replication	8,000
gam	5–6	Protects DNA from exoV digestion by binding to DNA	14,000
sot (stimulation of transfection)	6–7	When present and expressed in a bacterium, stimulates efficiency of transfection with Mu DNA	
arm (amplified replication of Mu)		When deleted, less replication of Mu DNA	
lig (ligase)		Expression of an activity that can substitute for E. coli and T4 ligase	
C	9–10	Positive regulator of late gene expression	15,550
lys (lysis)		Lysis of host cell	
D		Head protein	
E		Head protein	
H		Head protein	64,000
F			
G			
I			
T		Major head protein	
J		Head protein	
K		Tail protein	
L		Tail protein	55,000
M		Tail protein	
Y		Tail protein	12,500
N		Tail protein	60,000
P		Tail protein	43,000
O		Tail protein	
V		Tail protein	
W		Tail protein	
R		Tail protein	
S	α and G	Tail protein	56,000
S'	α and G	Tail protein	48,000
U	In G	Tail protein	22,000
U'	In G	Tail protein	22,000
gin (G inversion)	Left of β	Inversion of the G region	21,500
mom (modification of Mu)	Middle of β	Modifies Mu DNA protecting against restriction. Is not a methylase	37,000(?)

[a]From Toussaint, A. and A. Resibois, 1983. Phage Mu: Transposition as a lifestyle. In *Mobile Genetic Elements*, J. A. Shapiro (Ed.). Academic, New York, pp. 105–158.

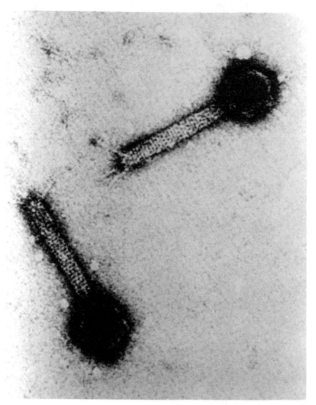

Fig. 5-22. Electron micrographs of the Mu phage particle. From Toussaint, A. and A. Resibois. 1983. Phage Mu: Transposition as a lifestyle. In *Mobile genetic elements*, J. A. Shapiro, (Ed.). Academic, New York, pp. 105–158.

typhimurium. The MuG($-$) phages, with S'_γ and U' at the left end of G, will infect such strains as *E. coli* C, *Shigella sonnei*, *Erwinia cloacae*, or other *Erwinia* species.

Following infection, the linear Mu DNA circularizes but the circular DNA is proteinase-sensitive suggesting that circularization may be the result of proteins covalently attached to the DNA. Much of the infecting Mu DNA is associated with the chromosome 10 min after infection. Transposition to the chromosome from infecting phage involves a nonreplicative simple insertion as opposed to replicative cointegrate formation (see Fig. 3-17, and 5-23). There is a striking difference between Mu and other temperate phages in that Mu integrates into the host chromosome whether or not it enters the lytic or lysogenic cycle. Integration of Mu into the bacterial chromosome occurs at a more-or-less random location, although there are preferred sites for insertion.

Mu DNA synthesis commences 6–8 min after infection or induction of prophage. The original prophage copy is not excised prior to the onset of replication. The newly replicated Mu copies also remain associated with the chromosome (Fig. 5-23). Each new round of replication involves additional convalent attachment of the ends of Mu to a distant target site. In effect, Mu replicates by transposition to numerous sites on the bacterial chromosome. Both the *A* and *B* gene products are required for replication and

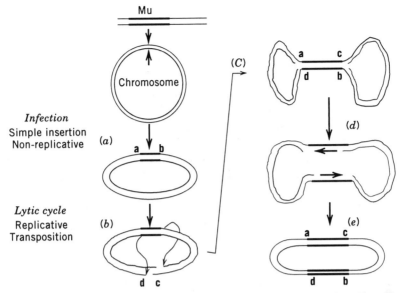

Fig. 5-23. Integration and replication of infecting Mu DNA. (*a*) **Integration.** The paired filled rectangles represent Mu DNA sequences, the lines extending from either side represent variable host DNA that was packaged into the infecting Mu capsid. The circle represents the host chromosome: the arrowheads mark the staggered cut at the target site for integration. Integration occurs as a nonreplicative transposition event. (*b*) **Lytic cycle.** The letters a and b represent the 5 bp repeat formed during integration as a result of the staggered cut in the chromosomal DNA. Both c and d represent host DNA sequences that flank a target for replicative transposition. Curved arrows indicate the attachment points of specific Mu strands at the new target site. (*c*) and (*d*) represent a transposition intermediate and its replication to form two Mu insertions. (*e*) Illustrates the final transposition product. Note the chromosomal rearrangement that has occurred with respect to markers a, b, c, and d.

transposition. In addition, there is a requirement for both the c and S ends of Mu for transposition to occur. Two *c* ends or two *S* ends are not suitable substrates for the transposition enzymes. It appears that the *A* gene product recognizes the ends of the phage genome catalyzing low level of transposition. The *B* product stimulates this process by binding to target DNA and directing Mu transposition.

Transposition by replication is regulated. The Mu *c* repressor protein negatively controls the synthesis of the transposase and other functions involved in transposition. At the end of the lytic cycle, the various Mu DNA copies scattered within the chromosome are cut and packaged into phage particles. These particles are subsequently released into the environment. Recall that Mu DNA in phage particles is linear and contains heterogeneous host DNA at both ends.

Packaging of Mu chromosomes into capsids occurs by a headfull mechanism starting from a *pac* site located at the c end (see Chapter 3, Transduction, P22, for *pac* sites). One cut is made 100 nucleotides outside this end with the second cut made at a fixed length past the *S* end. Thus, cuts are made in the flanking bacterial DNA so that the DNA from each phage particle contains different terminal sequences.

φX174

The φX174 phage is an example of single-stranded DNA phages, which also include M13 and G4. Phage φX174 has been used extensively in studies designed to unravel the mysteries of how the *E. coli* chromosome, 1000 times its size, is replicated.

The viral strand of φX174 contains 5386 bases and is packaged within an icosahedral capsid with a knob at each vertex (Fig. 5-24). The viral strand is designated (+) because it is the sense strand and can be transcribed directly by RNA polymerase. The strand contains genes for 11 gene products. However, the combined molecular weights of these proteins exceed the limit imposed by the available DNA. The ability of some sequences to code for more than one protein (overlapping genes) accounts for this apparent discrepancy.

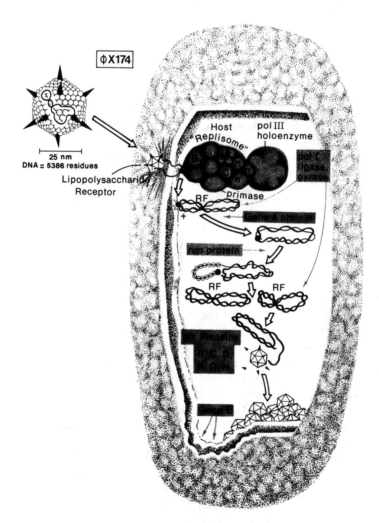

Fig. 5-24. Life cycle of Φ174. From Kornberg, 1980. *DNA replication*, A. H. Freeman & Co., San Francisco.

The genetic map of φX174 is shown in Figure 5-25 and Table 5-5. The region originally assigned to gene *A* is now known to also contain all of gene *B*. This phenomenon of overlapping genes is possible because gene *B* is translated in a different reading frame from gene *A*. In a similar fashion, gene *E* is encoded in its entirety within gene *D*. Another translational control mechanism further expands the use of gene *A*. The 37-kDa gene *A** protein is formed by reinitiating translation at an internal AUG codon within the gene *A* message. The same translational phase is used but the functions of the two proteins are distinct. The manner in which this virus extends the ''one gene-one protein'' hypothesis demonstrates a frugal usage of genetic information in small phage genomes.

During infection (Fig. 5-24), the phage adsorbs irreversibly to the LPS in the outer membrane of rough strains of *E. coli* and *S. typhimurium*. Rough strains lack the O antigen outer chains but retain intact core and lipid A (see Chapter 6). Attachment is probably accomplished via the spikes on the capsid. Following attachment, the virus enters the eclipse stage in which the eclipsed virus has lost infectivity for new cells. Uncoating of the virus is coupled to replication of the viral genome forming a double-stranded replicative form (RFI). The binding of φX174 particles and RFI occurs at areas of adhesion between inner and outer membranes (see Chapter 6) and possibly affords direct access to the host chromosomal replicative apparatus fixed at the inner surface. This SS ⟶ RFI replication is dependent solely on host replication proteins and includes

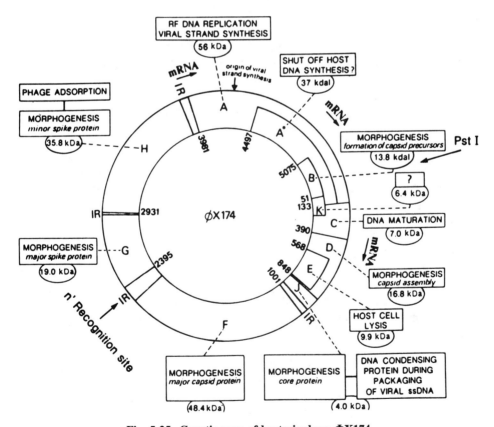

Fig. 5-25. Genetic map of bacteriophage ΦX174.

TABLE 5-5. The φX174 Genes and Functions

Gene	Function	Protein Mass (kDa)	
		SDS Gel[a]	Sequence
A	RF replication; viral strand synthesis	59	56[b]
A*	Shut-off host DNA synthesis	37	
B	Capsid morphogenesis	20	13.8[c]
C	DNA maturation	7	9.4[b]
D	Capsid morphogenesis	14	16.8[c]
E	Cell lysis	10	9.9[b]
F	Major coat protein	48	48.3[b]
G	Major spike protein	19	19.1[c]
H	Major spike protein, adsorption	37	35.8[b]
J	Core protein; DNA condensation	5	4.1[c]
K	Unknown	6	

[a]Sodium dodecylsulfate = SDS.

[b]Calculated from DNA sequence $\left(\dfrac{\text{number of nucleotides}}{3 \times 0.00915}\right)$.

[c]Calculated from amino acid sequence.

prepriming, priming, elongation, gap filling, ligation, and supercoiling. The RFI synthesis commenses at the n' recognition site in the intervening region between genes *G* and *F* (Fig. 5-25).

The first stage of replication (SS ⟶ RF) occurs within 1 min of infection. The second stage involves RF ⟶ RF synthesis and occurs up to 20-min postinfection. Two additional proteins are required for duplex replication. These are the phage-encoded gene *A* protein and *E. coli rep* product.

The gene *A* product nicks supercoiled RFI in the (+) strand between residues 4297 and 4298 forming the open circular form RFII. The gene *A* protein remains attached to the 5' end of the (+) nicked strand (Fig. 5-26). Complexed with *repA* protein (a DNA helicase), the gene *A* protein also remains attached to the (−) strand and proceeds around the circle. When this unwinding is accompanied by replication, the 3' end of the open (+) strand is extended regenerating a duplex, while single-stranded binding proteins (SSB) bind to the (−) strand. After traversing the entire length of the template, the gene *A* protein cleaves the regenerated origin again attaching to the 5' phosphoryl end. The now free 3'-OH end of the (+) strand can attack the original gene *A* protein−5'-phosphoryl bond generating a single-strand circle coated with SSB. This single strand can be reconverted into a new RFI as before. Host DNA synthesis is arrested 20−35 min following infection. This could occur either by the sequestration of a limiting host replication protein to phage DNA replication or by a phage-induced interference protein. The third stage RF ⟶ SS and encapsidation (20−30 min) depends on seven virus-encoded proteins. These proteins complex and encapsidate the (+) DNA making it unavailable for (−) strand synthesis. This strategy only produces new + strands that are used in the new progeny phage. Following morphogenesis of the phage particle (40-min postinfection), cell lysis ensues.

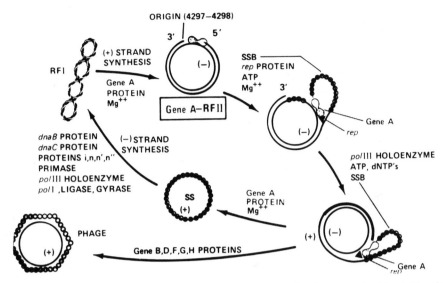

Fig. 5-26. Scheme for ΦX174 RF replication in two stages. Continuous replication initiated by gene *A* protein cleavage generates viral (+) circles, and discontinuous replication of the viral circles by the SS ⟶ RF system produces RF. In the presence of phage-encoded maturation and capsid proteins, viral circles are encapsulated rather than replicated. From Kornberg, A. 1980. *DNA Replication.* Freeman, San Francisco.

CONCLUDING REMARKS

Most, if not all, free-living microorganisms are susceptible to attack by one or more viruses. Learning how viruses usurp their host cells has provided a tremendous insight not only into viruses but into the molecular machinery of the bacterial cell. Take, for example, the N-utilization proteins discovered as a result of investigations with λ phage or the developing model of bacterial DNA replication pieced together from studying the replication of a variety of phage systems. And who could have predicted that a study examining how bacteria attack viral DNA (restriction/modification systems) would lead to the era of recombinant DNA techniques and future gene replacement therapies for human disease. Certainly continued explorations of phage–host interactions promise to yield additional knowledge of the intricacies of life.

SELECTED REFERENCES

General

Birge, E. A. 1981. *Bacterial and Bacteriophage Genetics. An Introduction.* Springer-Verlag, New York.

Luria, S. E., J. E. Darnell, Jr., D. Baltimore, and A. Campbell, 1978. *General Virology*, 3rd ed. Wiley, New York.

Miller, J. H. (Ed.). 1991. Bacterial Genetic Systems. *Methods in Enzymology*, Vol 204. Academic, San Diego, CA.

T4 Bacteriophage

Benzer, S. 1955. Fine structure of a genetic region in bacteriophage. *Proc. Nat. Acad. Sci. USA.* **41:**349–354.

Geiduschek, E. P., 1991. Regulation of expression of the late genes of bacteriophage T4. *Annu Rev. Genet.* **25:**437–460.

Matthews, C. K., E. M. Kutter, G. Mosig, and P. B. Berget (Eds.) (1983). *Bacteriophage T4.* American Society for Microbiology, Washington, DC.

Williams, K. P., G. A. Kassavetis, D. R. Herendeen, and E. P. Geiduschek. 1994. Regulation of late gene expression. In *Molecular Biology of Bacteriophage T4.* J. D. Karam (Ed.). American Society for Microbiology, Washington, DC.

Lambda Phage

Chauthaivwale, V. M., A. Therwath, and V. V. Destrpande. 1992. Bacteriophage Lambda as a cloning vector. *Microbiol. Rev.* **56:**577–591.

Hendrix, R. W., J. W. Roberts, F. W. Stahl, and R. A. Weisberg (Eds.). 1983. *Lambda II.* Cold Spring Harbor Laboratory, Cold Spring Harbor, NY.

ϕX174

Baas, P. D. 1985. DNA replication of single stranded *Escherichia coli* DNA phages. *Biochim. Biophys. Acta* **825:**111–139.

Kornberg, A. and T. Baker, 1992. *DNA Replication.* W. H. Freeman and Co., San Francisco.

Reinberg, D., S. L. Zipursky, P. Weisbeek, D. Brown, and J. Hurwitz. 1983. Studies on the synthesis of the ϕX174 gene A protein mediated termination of leading strand DNA synthesis. *J. Biol. Chem.* **258:**529.

Mu Phage

Bukhari, A. I. 1983. Transposable genetic elements: The Bacteriophage Mu paradigm. *ASM News* **49:**275–280.

Mizuuchi, K. 1992. Transpositional Recombination. Mechanistic insights from studies of Mu and other elements. *Annu. Rev. Biochem.* **61:**1011–1051.

Pato, M. L. 1989. In *Mobile DNA.* D. Berg and M. Howe (Eds.). American Society for Microbiology, Washington, DC, p. 23.

Toussaint, A. and A. Resibois. 1983. Phage Mu: Transposition as a life-style. In *Mobile Genetic Elements*, J. A. Shapiro (Ed.), Academic, New York. pp. 105–158.

CHAPTER 6

CELL STRUCTURE AND FUNCTION

The cytological features of higher eukaryotic cells display greater complexity than lower eukaryotic cells (Table 6-1). Lower eukaryotes (protozoa, algae, and fungi) lack some of the distinguishing features of metazoan cells. Nevertheless, eukaryotic cells do, in essence, appear to be similar versions of the same overall plan. Prokaryotic microorganisms (bacteria or eubacteria and archaebacteria) exhibit a cellular organization that is quite different from that of eukaryotes. In the past, some microbial cytologists maintained that bacterial cells were merely smaller replicas of larger cells. Failure to distinguish subcellular structures comparable to those of higher forms was attributed to the limitations of microscopic and cytological techniques available. Investigations with highly improved techniques and equipment have now made this view untenable. As discussed briefly in Chapter 1, prokaryotic cells differ in certain major ways from eukaryotic cells. It has also become apparent that a group of bacteria referred to as archaebacteria (*Archaea*) differ chemically and metabolically from both eubacteria and eukaryotes.

THE EUKARYOTIC NUCLEUS

Eukaryotes (*Eukarya*) contain a cytologically distinct unit, the nucleus, which is the organizational and regulatory center for virtually all of the biochemical and hereditary processes. With the aid of the electron microscope it has been possible to demonstrate many structural details of the nucleus of eukaryotic cells. Figure 6-1 shows a mammalian cell with a well-defined nuclear membrane composed of at least two distinct layers. The outer surface contains pores with tubules that pass through both membrane layers. Amebae, protozoa, algae, and fungi (yeasts and molds) also contain a discrete nucleus with a well-defined nuclear membrane (Fig. 6-2). The myxomycete *Arcyria cinerea* also contains a double-layered nuclear membrane that exhibits pores (Fig. 6-3).

TABLE 6-1. Major Components of Cells of Various Classes of Organisms

Higher Eukaryotes		Lower Eukaryotes		Prokaryotes
Metazoan	Protozoan	Algae	Fungi	Eubacteria and Archaebacteria[a]
Nucleus	Macronucleus	Nucleus	Nucleus	Nucleoid
Nuclear membrane	Nuclear membrane	Nuclear membrane	Nuclear membrane	No nuclear membrane
Nucleolus	Nucleolar elements			
Mitochondria	Mitochondria	Mitochondria	Mitochondria	
Endoplasmic reticulum	Endoplasmic reticulum	Endoplasmic reticulum	Cytoplasm	Cytoplasm
Golgi apparatus	Dictyosomes	Dictyosomes		
Inclusions	Specialized organelles	Chloroplasts	Inclusions	Inclusions
Lysosomes				
Peroxisomes				Phycobilisomes
Plasma membrane	Plasma membrane	Plasma membrane	Cytoplasmic membrane	Cytoplasmic membrane
No cell wall	No cell wall	Cell Wall chitin, glycans	Cell wall chitin, glycans	Cell wall peptidoglycan

[a]Although comparable to eubacteria in many respects, the archaebacteria (Archaea) do not produce cell wall peptidoglycan comparable to that produced by eubacteria. They also differ from eubacteria in a number of other characteristics not included in this table.

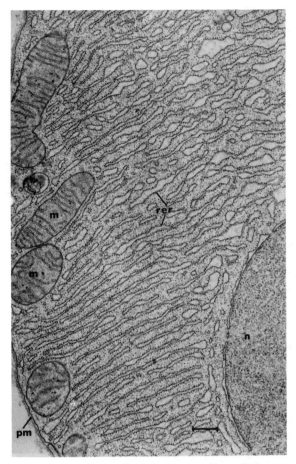

Fig. 6-1. Electron micrograph of a portion of a mammalian cell (pancreatic exocrine cell).
The basal region of the cell between the nucleus (lower right) and the plasmalemma (lower left)
is occupied by numerous cisternae of the rough endoplasmic reticulum and a few mitochondria.
Prominent pores can be seen in the nuclear membrane. The endoplasmic reticulum is lined with
ribosomes. The nucleolus is not visible in this photograph. Bar = 1 μm. From Palade, G. E. *Science*
189, 347, 1975.

BACTERIAL NUCLEOID

Demonstration of a cytologically distinct nuclear region in bacterial cells proved to be a
challenging task. Direct staining with basic dyes resulted in staining of the entire cell.
Removal of RNA by various techniques made it possible to demonstrate structures that
stained with nuclear stains. Improved fixative techniques for electron microscopy permitted visualization of a discrete nuclear region in bacterial cells but no separatory
membrane was observed. The exact shape of the bacterial nucleus has been difficult to
determine because its appearance changes with the conditions under which cells were
grown and the preparative techniques used. As shown in Figure 6-4 (a, b, and c) fixation

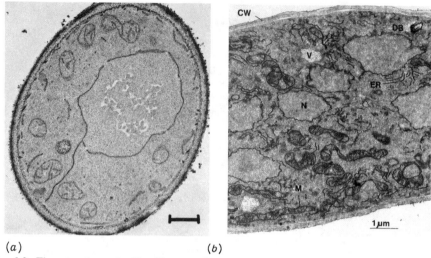

(a)   (b)

Fig. 6-2. Fine structure of cells of lower eukaryotes. (a) General view of a section of a cell of *Saccharomyces cerevisiae*. Prominent features include the cell wall, the nucleus with distinct pores in the nuclear membrane, mitochondria with cristae, and intermembranous structures. Bar = 500 nm. From Avers, C. J., A. Szabo, and C. A. Price, *J. Bacteriol.* **100**, 1044, 1969. (b) Fine structure of a hyphal element of the fungus, *Mucor genevensis*. Nuclei with pores (N), vacuoles (V), mitochondria (M), endoplasmic reticulum (ER), cell wall (CW), and dark bodies (DB) are visible. The cells were fixed in 1% KMnO$_4$ for 1 h. Bar = 1 μm. From Clark-Walker, C. D. *J. Bacteriol.* **109**, 299, 1972.

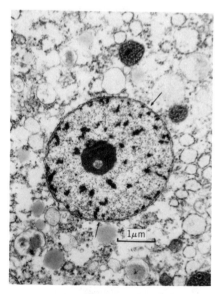

Fig. 6-3. Interphase nucleus of the myxomycete, *Arcyria cinerea*. The typical nuclear envelope is interrupted by prominent pores. The nucleolus near the center is well defined, while the chromatin is randomly dispersed throughout the nucleus. From Mims, C. W. 1972. *J. Gen. Microbiol.* **71:** 53.

with OsO$_4$ and growth in normal or high salt medium results in visualization of a condensed nucleoid, whereas glutaraldehyde fixation (Fig. 6-4d) results in a dispersed appearance of the nucleoid. The development of confocal scanning light microscopy made it possible to observe the shape and substructure of the nucleoid and to compare these images with phase-contrast and electron microscope images. In Figure 6-5, a confocal scanning light microscope image of an OsO$_4$-fixed cell is compared with an electron micrograph and a reconstruction model based on serial sections studied with the electron microscope. The differences observed may reflect the presence of transcription–translation complexes with ribosomes and proteins in the more dense preparations observed with OsO$_4$ fixation.

It has been shown that small, ribosome-free spaces also contain DNA. The nucleoid is observed as a coralline (coral-like) shape as shown in (Fig. 6-6a). By this method the branches of the coralline shaped nucleoid are observed to spread far into the cytoplasm and over the entire area of the cell. By using serial sections it has been possible to reconstruct the ribosome-free area of the nucleoid. Figure 6-6b presents a three-dimensional model of how the coralline nucleoid may appear.

Isolation of the intact bacterial nucleoid has provided some insight into its properties.

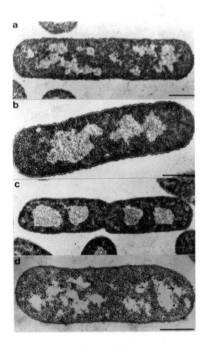

Fig. 6-4. Standard OsO$_4$ fixation of *E. coli* B by the Ryter–Kellenberger technique of prefixation directly in the culture medium. (a) Cell was grown in low-salt medium. Bar marker = 0.5 μm. (b) Same as (a), except that cell was grown in normal salt medium (6 g of KCl/L). The nucleoid is more confined and localized. Bar marker = 0.5 μm. (c) Same as (a), except that the cell was grown in high salt medium (21 g KCl/L). The nucleoid is still more confined. Bar marker = 0.5 μm. (d) Fixation of *E. coli* B with glutaraldehyde embedded in Epon. The cell was grown in high-salt medium. The nucleoid is dispersed into small patches that contain DNA precipitates (arrows). Bar marker = 0.5 μm. From Hobot, J. A. et al. 1985. *J. Bacteriol.* **162:**960–971.

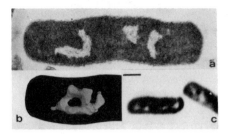

Fig. 6-5. Fast-growing *E. coli* **B/r H266 (generation time, 21 min), showing a comparison of a confocal scanning light micrograph** (c) with a reconstruction of an OsO$_4$ fixed nucleoid (b), based on serial sections studied with the electron microscope. Only one of these sections is shown here (a). Bar = 1 μm. From Valkenberg, J. A. C. et al. 1985. *J. Bacteriol.* **161**:478–483.

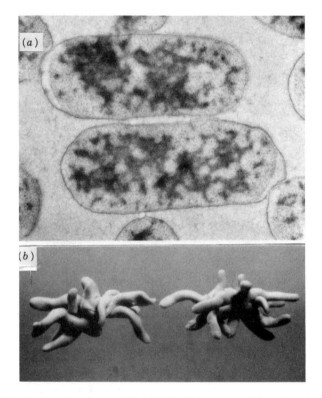

Fig. 6-6. Coralline shape of the bacterial nucleoid. (a) Immunostaining of DNA in exponentially growing *E. coli* B using antidouble-stranded DNA mouse IgM, followed by goat IgG antibodies to mouse IgM stained with a mixture of 1 volume of 2% KMnO$_4$ and 2 volumes of 2% uranyl acetate. Label is observed as gray-stained patches over DNA-rich areas. In this micrograph the upper cell shows two distinct nucleoids. (b) Schematic models of bacterial nucleoids shows a pair of nucleoids of growing cells deduced from thin sections of cryofixed, freeze-substituted *E. coli* B. For practical reasons, the models were flattened on the lower side. From Bohrmann, B. et al. 1991. *J. Bacteriol.* **173**:3149–3158.

Two types of nuclear bodies can be obtained: an envelope-associated nucleoid and an envelope-free nucleoid. Large amounts of RNA, proteins, lipids, and peptidoglycan are found in the envelope-associated nucleoid. The free nucleoid contains lesser amounts of non-DNA components, such as protein and RNA. Examination of the isolated nucleoid under the electron microscope (Fig. 6-7) reveals a number of supercoiled loops of DNA extending from the center. The central core appears to contain RNA that can be diminished by washing the grid with RNase.

Nucleosomes

Eukaryotic chromosomes are organized into well-defined DNA–protein complexes termed **nucleosomes**. The proteins, mainly histones, play an important role in the structure of eukaryotic chromosomes by determining the conformation referred to as **chromatin**. The core particles pack together to form the chromosome. The nucleosome is the repeating unit of DNA organization and is often observed as a "beads on a string" configuration. Extensive digestion of chromatin with micrococcal nuclease releases the nucleosome core, a small, well-defined particle that has been crystallized. The particle mass is equally distributed between 146 bp of DNA and an octomer formed by two each of four major histone proteins (H2A, H2B, H3, and H4). With the aid of X-ray diffraction it has been possible to determine a three-dimensional model of the structure of the nucleosome core (Fig. 6-8).

Nucleosome-like structures, folding and supercoiling of DNA, association of DNA polymerase with the nucleoid, and the presence of histone-like proteins have all been observed in the organization of prokaryotic DNA. However, the details of structural similarity to eukaryotic nucleosomes remain to be completely defined. Several small histone-like DNA-binding proteins have been described in bacteria. These nucleoid-associated proteins include HU protein, IHF, protein H1, FirA, H-NS, and Fis.

The histone-like protein HU, contains two closely related 10-kDa subunits, HU-α and HU-β, encoded by *hupA* and *hupB*. Fluorescein-labeled HU taken up by EDTA-treated cells of *E. coli* concentrates in the nucleoid and is uniformly distributed throughout this structure. The proposed role of HU protein is in stabilizing higher order nucleoprotein structures to confer specificity in DNA interactions. The fact that HU exhibits little sequence specificity suggests that it may be involved in a variety of DNA-protein activities requiring coiling of specific DNA sequences. *In vitro* studies have implicated HU protein in the initiation of DNA replication, binding of repressors, and transposition of bacteriophage Mu. In *E. coli*, *hupAhupB* double mutants that lack HU protein have severe defects in cell division, DNA folding, and DNA partitioning.

The histone-like protein IHF participates in integrative or excisive recombinations of phage DNAs, in transposition by Tn10, and in packaging of DNA in phage heads. The IHF is encoded by *himA* and *hip* genes. Coordinate interactions between HU and IHF are suggested by the fact that *himA* mutants of *Salmonella* are killed when the *hupA* gene is overexpressed by introduction of the cloned *hupA*.

Fis is a small basic DNA-binding protein from *E. coli* that was originally identified by its ability to effect flagellar phase variation by stimulating DNA inversion. Evidence has also been presented to implicate the participation of Fis in rRNA and tRNA transcription. It may also play a role in chromosomal DNA replication as evidenced by its binding to the origin of replication (*oriC*) in *E. coli*. Fis levels vary dramatically during

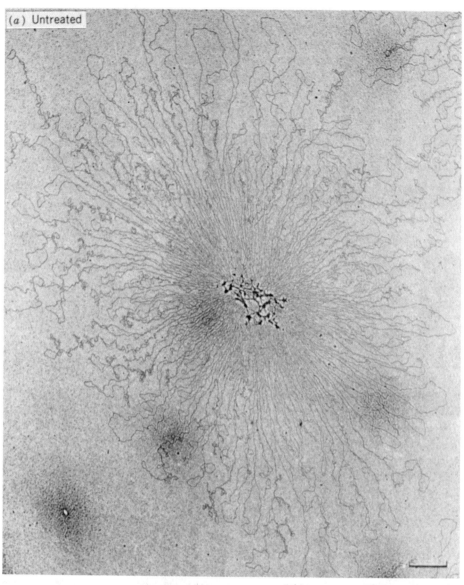

(a) Untreated

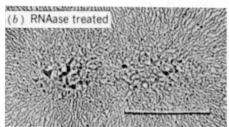

(b) RNAase treated

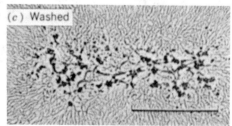

(c) Washed

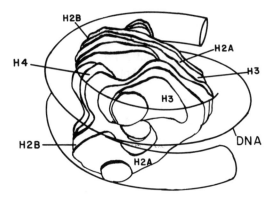

Fig. 6-8. Model of the nucleosome core. The model was made by winding a tube simulating the DNA superhelix on a model of the histone octomer, which was built from a three-dimensional map derived from electron micrographs of the histone octomer. The ridges of the periphery of the octomer form a more or less continuous helical ramp on which a 146 nucleotide pair length of DNA can be wound. The locations of individual histone molecules (whose boundaries are not defined at this resolution) are proposed here on the basis of chemical cross-linking data. From Kornberg, R. D. and A. King. 1981. *Sci. Am.* **250**:52–64.

the course of cell growth and in response to changing environmental conditions. When stationary-phase cells are subcultured into a rich medium, Fis levels increase from less than 100 to over 50,000 copies per cell prior to the first cell division. As cells enter exponential growth, Fis synthesis is reduced, and the intracellular concentration drops as a function of cell division. Concurrent alteration in *fis* mRNA levels suggests that regulation is at the transcriptional level. Fluctuation in Fis levels may serve as an early signal of a nutritional upshift and may be important in the physiological roles played by Fis. Here Fis appears to autoregulate its expression as shown by the fact that binding to its promoter excludes binding of RNA polymerase.

The bacterial histone-like protein H-NS is encoded by *hns*. The H-NS protein inhibits site-specific DNA recombination as evidenced by the dramatic increase in inversion of the *pilA* promoter region in mutants lacking H-NS (see further discussion in the section on pilin later in this chapter).

Fig. 6-7. Cell envelope-free nucleoid from *E. coli* 15 τ-bar. For preparation of the cell-associated nucleoid, cells are first suspended in 20% sucrose, treated briefly with lysozyme–ethylenediaminetetraacetic acid (EDTA), and then lysed with a mixture of the nonionic detergents Brij-58 and deoxycholate in the presence of 1 M NaCl. Centrifugation at 17,000 g for 15 min on a 10–50% sucrose gradient permits isolation of the cell envelope-associated nucleoid. Preparation of the envelope-free nucleoid requires longer lysozyme treatment and use of an ionic detergent, such as Sarcosyl. The sedimentation coefficients are 3200S for the envelope-associated nucleoid and 1600S for the envelope-free nucleoid. The hypophase contained 0.4 M salt. The chromosome had 141 ± 3 loops and possibly a fork on the loop at about 1 o'clock. In (b), the grid was washed with a solution of 40 μg RNAse in 0.15 M ammonium acetate for 2 s. In (c), the grid was washed with the control buffer. Bar markers = 1 μm. From Kavenoff, R. and B. C. Bowen. 1976. *Chromosome* **59**:89–101.

In archaebacteria (*Archaea*) chromosomal DNA has been shown to exist in protein-associated form. Histone-like proteins have been isolated from stable nucleoprotein complexes in *Thermoplasma acidophilum* and *Halobacterium salinarum.* Whole chromosomes of *H. salinarum* consist of regions of both protein-associated DNA and protein-free DNA. In electron micrographs the protein-associated DNA region appears as nucleosome-like fibrous structures. However, the suggestion that these observations represent a major difference between eubacteria and *Archaea* seems untenable in the light of evidence cited above indicating widespread occurrence of protein-associated DNA and nucleosome-like structures in a variety of bacteria.

MITOCHONDRIA

Mitochondria are subcellular organelles that carry out oxidative metabolism. They are highly membranous structures that contain the cytochrome systems and oxidative phosphorylation processes. Lower eukaryotes (yeasts, algae, filamentous fungi, and protozoa) have all been shown to contain mitochondria, although their number and distribution vary widely from one cell type to another. Even within a given species the number of mitochondria per cell and their intracellular arrangement may vary with the cultural conditions and growth phase.

The mitochondria of yeast and other eukaryotes possess two distinct membranes, **outer** and **inner** membranes, which may be tightly associated with each other at certain contact sites. The inner membrane is highly invaginated, forming pleats referred to as **cristae.** Yeasts produce mitochondria actively when grown in an aerobic environment. Under anaerobic conditions, mitochondria are not observed; yet upon shifting back to aerobic conditions, they quickly reappear. Yeast cells grown under strict anaerobiosis contain mitochondria-like particles designated as **promitochondria.** These structures lack most of the characteristic components of the respiratory chain but still contain ATPase, mitochondrial DNA, and various structural proteins. Anaerobic growth apparently induces a dedifferentiation of mitochondria. When yeast cells are grown in the presence of high concentrations of glucose, the respiratory and energy transduction systems are repressed even under highly aerobic conditions. The system responsible for mitochondrial gene replication and expression is completely blocked, and the cell satisifies all its energy requirements by glycolysis.

Mitochondrial DNA consists of a covalently closed double helix of naked DNA comparable to but usually smaller than that found in most bacteria. Despite the presence of its own genetic material, the mitochondrion is not genetically self-sufficient. The mitochondrial genome contributes a limited but essential number of gene products to the biogenesis of the system that carries out oxidative phosphorylation. All mitochondrial genomes encode RNA components of their protein synthetic apparatus (ribosomal RNAs and tRNAs). In yeast and possibly other lower eukaryotes, the structural gene coding for the ribosome-associated protein VarI, is located on mitochondrial DNA. Polypeptides that are part of the electron transport and ATP synthetase complexes are also encoded on mitochondrial DNA. Detailed genetic studies indicate that there has been a transfer of DNA sequences from the mitochondrial to the nuclear genomes. Most other mitochondrial components, including the enzymes of the tricarboxylic acid (TCA) cycle, are coded for by nuclear genes. Transcription of mitochondrial genes appears to be catalyzed by a nucleus-encoded RNA polymerase. One subunit of the yeast RNA polymerase functions as a core enzyme while a second subunit serves as a specificity factor that facilitates recognition of one of the approximately 20 specific mitochondrial transcription initiation signals on mitochondrial DNA. Yeast cells lacking functional mitochondrial DNA (ρ^- mutants) do not synthesize mitochondrial proteins. They exhibit irregular shapes, fewer cristae, and aberrant multilamellar configurations. On the other hand, most nuclear mutations that result in respiratory deficiency (*pet*) do not show a marked effect on the morphology of yeast mitochondria.

A considerable amount of information has been accumulated to substantiate the hypothesis that mitochondria evolved from prokaryotic cells through the establishment of a fortuitous symbiotic relationship between respiratory and fermentative cell types. As discussed in Chapter 2, the primary nucleotide sequences of ribosomal RNA have been highly conserved throughout evolution. A comparison of the nucleotide sequence of various classes of mitochondrial rRNA molecules from humans, other mammals, and lower eukaryotes reveals a very close resemblance to that of bacterial rRNA. This information, together with the revelation that the mitochondrial genome closely resembles that of bacteria, provides convincing evidence of a prokaryotic origin for mitochondria. In a similar manner, chloroplasts are considered to have evolved from prokaryotic origins. In fact, there is reason to believe that chloroplasts are more closely related to cyanobacteria than mitochondria are to bacteria.

CELL SURFACE OF MICROORGANISMS

Eukaryotic Cell Surface

As discussed briefly in Chapter 1, both eukaryotic and prokaryotic cells are often enclosed within a rigid cell wall. This characteristic is not unique to microorganisms nor is it strictly unique to either the plant or animal kingdoms. The cell walls of vascular plants are largely composed of **cellulose**, a β-1, 4-linked polymer of glucose (Fig. 6-9). Many algae, in common with higher green plants, contain cellulose as a major cell wall constituent. The cell walls of some protozoa and many fungi contain **chitin**, a linear polymer

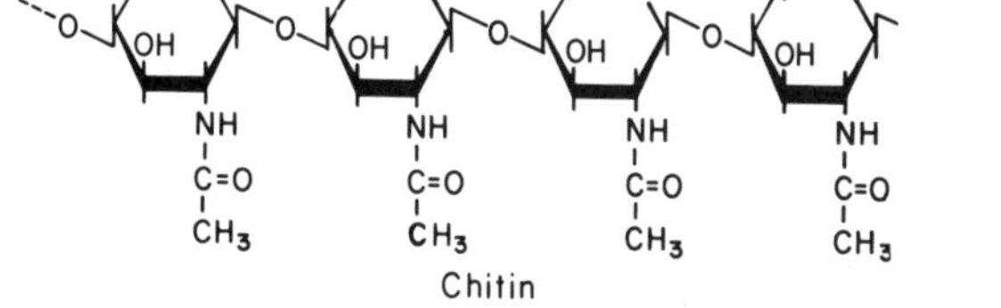

Fig. 6-9. Chemical structure of cellulose, a β-1,4-linked polymer of glucose, and chitin, a linear polymer of *N*-acetyl-D-glucosamine.

of *N*-acetylglucosamine (Fig. 6-9). Most fungi, algae, and higher plants contain micro-fibrils of either cellulose or chitin as a prominent skeletal component of their cell walls. Production of these microfibrils appears to occur at the surface of the cell. Synthesis of chitin microfibrils by an isolated chitin synthetase has been demonstrated. Among pro-karyotes, some species of *Acetobacter* and a few others produce cellulose.

Some yeasts, such as *S. cerevisiae*, produce a wall composed of at least two compo-nents. One contains **glucan**, a glucose polymer with frequent branches (Fig. 6-10). The other contains **mannan**, a branched mannose polymer (Fig. 6-10). Chemical analysis of yeast walls shows that they contain 29% glucan, 31% mannan, and 13% protein. They also contain smaller percentages of lipid and other materials. Many yeasts and filamentous fungi contain sugars, such as galactose, as constituents of their cell walls. However, **chitin** (poly-*N*-acetylglucosamine) is the only common constituent found in the cell walls of all fungi so far examined.

Prokaryotic Cell Surface

In Chapter 1 it was ascertained that a high molecular weight, sugar-containing, rigid structure called **peptidoglycan** formed the major ''backbone'' of the murein sacculus of the cell wall of both gram-positive and gram-negative bacteria. Indeed, the production of peptidoglycan is considered the ''hallmark'' of the true bacteria (**eubacteria**). By contrast, some members of the archaebacteria (***Archaea***) produce a **pseudomurein** and an associated surface (**S**) layer composed of protein or glycoprotein. In many archaebac-

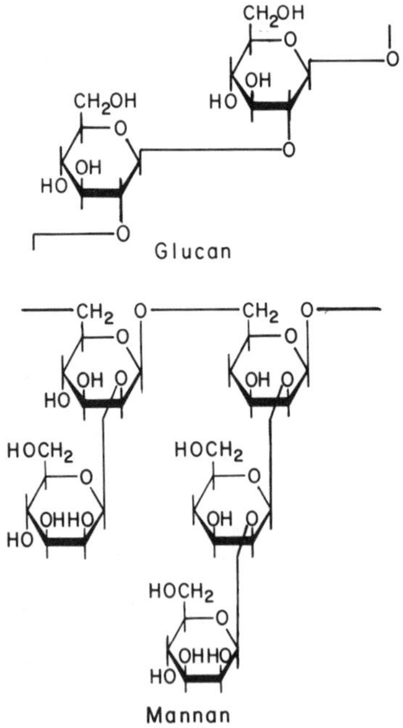

Fig. 6-10. Chemical structures of glucan, a branched polymer of glucose, and mannan, a branched polymer of mannose.

teria, the S layer represents the only wall component outside the plasma membrane, as shown for *Methanococcus voltae* in Figure 6-11.4. In gram-positive eubacteria, such as *B. subtilis*, the S layer is associated with the peptidoglycan-containing sacculus (Fig. 6-11.1). Gram-negative bacteria, such as *E. coli*, produce an outer membrane (Fig. 6-11.2). The S layer, when produced by gram-negative or gram-variable bacteria, is associated with the surface of the outer membrane (Fig. 6-11.4).

Structure and Synthesis of Bacterial Peptidoglycan

The glycan linkages of peptidoglycan are considered to be uniform in all bacteria with every D-lactyl group of the *N*-acetylmuramic acid being peptide substituted. All glycans have short tetrapeptide units terminating with D-alanine or occasionally tripeptide units lacking the terminal D-alanine. The L-alanine at the *N*-terminus can be replaced by L-serine or glycine.

The interpeptide bridges linking peptidoglycans are of four main types as described here and shown in Figure 6-12.

Type I. Direct D-alanyl-R_3 Peptide Bond. This type of bridge is found in *E. coli* and in most other gram-negative bacteria. It is also found in many bacilli.

Type II. Pentaglycine or Other L- or D-Amino Acid Sequences. This type of interpeptide bridge varies from organism to organism, as shown in Table 6-2.

Type III. A bridge composed of one to several peptides, each having the same amino acid sequence as the peptide unit attached to muramic acid. This type of interpeptide bridge is found in *Micrococcus luteus*.

Type IV. A bridge extending between carboxyl groups belonging to either D-alanine or to D-glutamic acid and a diamino acid residue or a diamino acid containing short peptide. This type of bridge is found in *Butyribacterium rettgeri*.

Although the proportion of peptide cross-linking varies considerably, in either case a continuous baglike sacculus is thought to completely surround the cell.

It has been stated that all of the peptide chains are located on the same side of the glycan, whereas the O-6 positions of the *N*-acetylmuramic acid residues are exposed on the other side of the structure and are readily available for substitution with acetyl or phosphodiester groups. However, there is evidence to suggest that the glycans are helically twisted chains from which the peptide bridges radiate in all directions from the axis of the backbone. This latter model would permit interpeptide bridging between many different neighboring chains, forming a supramolecular network (mosaic) of peptidoglycan, such as that observed in gram-positive organisms.

The peptide cross-bridging between peptidoglycan strands is considered to provide structural rigidity to the cell. This certainly appears to be true in gram-positive cells in which a thick network of cross-linked peptidoglycan has been shown to be present. In gram-negative cells, the fact that the peptidoglycan appears to be only one or a few monolayers in thickness may leave some doubt as to whether it is possible for this structure to be solely responsible for withstanding the turgor pressure from within the cell. The presence of the outer membrane with proteins embedded in it may provide some rigidity to the outer surface of the gram-negative cell.

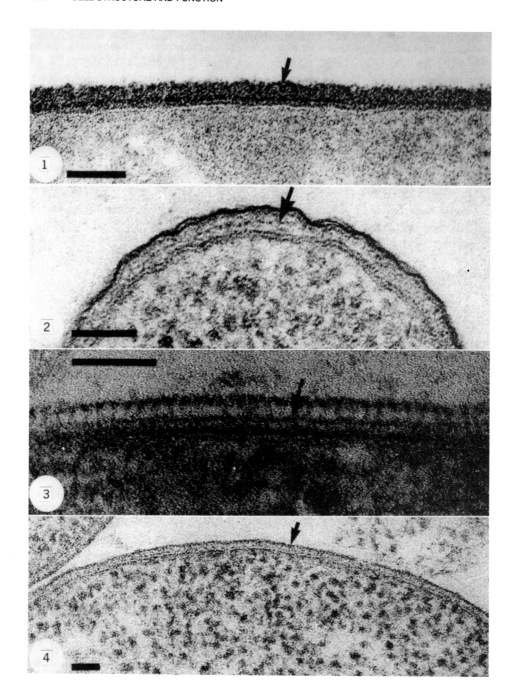

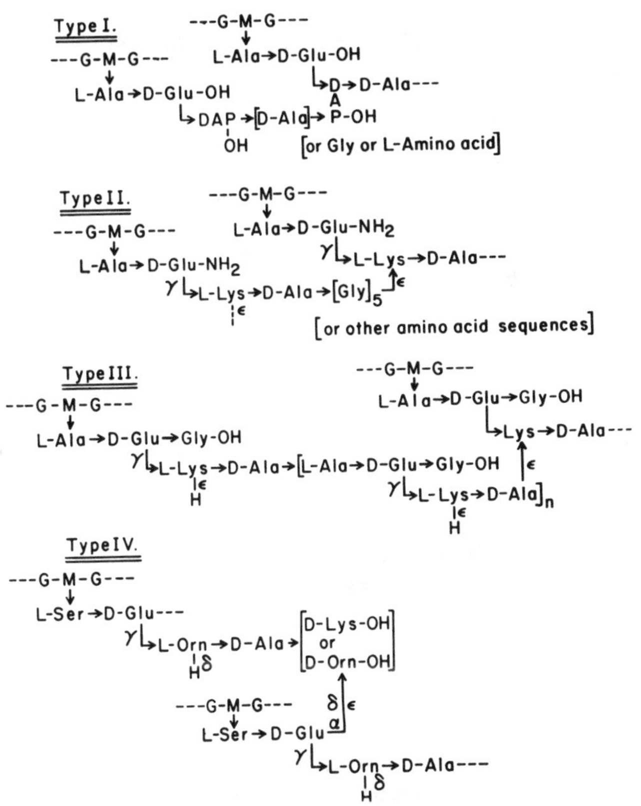

Type I.
```
                    ---G-M-G---
                         ↓
---G-M-G---         L-Ala→D-Glu-OH
     ↓                       ↳D→D-Ala---
L-Ala→D-Glu-OH                A
          ↳DAP→[D-Ala]→P-OH
              |
              OH      [or Gly or L-Amino acid]
```

Type II.
```
                    ---G-M-G---
                         ↓
---G-M-G---         L-Ala→D-Glu-NH₂
     ↓                   γ↳L-Lys→D-Ala---
L-Ala→D-Glu-NH₂              ↑ε
     γ↳L-Lys→D-Ala→[Gly]₅
          |ε
          [or other amino acid sequences]
```

Type III.
```
                         ---G-M-G---
                              ↓
---G-M-G---              L-Ala→D-Glu→Gly-OH
     ↓                            ↳Lys→D-Ala---
L-Ala→D-Glu→Gly-OH
     γ↳L-Lys→D-Ala→[L-Ala→D-Glu→Gly-OH  ↑ε
          |ε                 γ↳L-Lys→D-Ala]ₙ
          H                       |ε
                                  H
```

Type IV.
```
---G-M-G---
     ↓
L-Ser→D-Glu---
     γ↳L-Orn→D-Ala→⎡D-Lys-OH⎤
          |δ        ⎢   or   ⎥
          H         ⎣D-Orn-OH⎦

          ---G-M-G---     δ|ε
               ↓           α
          L-Ser→D-Glu─┘
               γ↳L-Orn→D-Ala---
                    |δ
                    H
```

Fig. 6-12. Major types of interpeptide bridges in peptidoglycans. G = *N*-acetylglucosamine; M = *N*-acetylmuramic acid; DAP = diaminopimelic acid. Other amino acid sequences replacing [Gly]₅ (pentaglycine) in Type II bridges are shown in Table 6-2. From Moat, A. G. 1985. Biology of the lactic, acetic, and propionic acid bacteria. In *Biology of Industrial Microorganisms*, A. L. Demain and N. A. Solomon (Eds.). Benjamin–Cummings, Menlo Park, CA, pp. 143–188.

Fig. 6-11. Thin section envelope profiles of conventionally fixed and embedded bacteria. Bars = 50 nm.

1. Amorphous wall fabric of *B. subtilis* 168 (arrow) lies directly above the plasma membrane.

2. *Escherichia coli* possesses a thin peptidoglycan or murein layer above the plasma membrane (arrow), over which lies a wavy outer membrane. The region between the outer and plasma membranes is called the periplasm or periplasmic space. The waviness of the outer membrane is believed to be an artifact of the conventional fixation-embedding technique, and much of the periplasm has been leached out.

3. *Clostridium thermosaccharolyticum* is an example of a gram-variable bacterium that has a wall profile intermediate between the gram-positive and gram-negative formats in 1 and 2. The peptidoglycan layer (arrow) is thinner than that of *B. subtilis* but thicker than that of *E. coli*. Above this layer is a proteinaceous S layer of periodically arrayed subunits.

4. *Methanococcus voltae* is an archaebacterium and possesses only a thin S layer (arrow) above the plasma membrane as its sole wall layer. From Beveridge, T. J. and L. L. Graham. 1991. *Microbiol. Rev.* **55**:684–705.

TABLE 6-2. Types of Interpeptide Bridges in Addition to Pentaglycine in Various Bacteria

Amino Acid(s)	Organism
-[Gly]₅-	*Staphylococcus aureus*
-[L-Ala]₃-L-Thr-	*Micrococcus roseus*
-[Gly]₅-[L-Ser]₂-	*Staphylococcus epidermidis*
-L-Ser-L-Ala-	*Lactobacillus viridescens*
-L-Ala-L-Ala-	*Streptococcus pyogenes*
-L-Ala-	*Arthrobacter crystallopoietes*
	Enterococcus hirae
-D-Asp-NH₂-	*Enterococcus faecium*
	Enterococcus hirae
	Lactobacillus casei

Autolysins, enzymes that hydrolyze bonds in the peptidoglycan structure, are thought to be produced by all bacteria. These enzymes are of three basic types:

1. **Glycan strand hydrolyzing.**
    a. *Endo-N*-acetylmuramidases.
    b. *Endo-N*-acetylglucosaminidases.
2. **Endopeptidases hydrolyzing.**
    a. Peptide bonds in the interior of the peptide bridges.
    b. Bonds involving the C-terminal D-alanine residue.
3. ***N*-Acetylmuramyl-L-alanineamidase** acting at the junction between the glycan strands and the peptide units.

Cell lysis may occur from within, by autolysins, or from without, by addition of lytic enzymes to cell suspensions. Autolysins may be found in the cytoplasm, associated with the membrane, in the periplasmic region, fixed on the wall, or in the culture medium as excretion products. Certain autolysins are localized in the region of the growing septum. The number of detectable autolysins varies from species to species. In *E. faecalis* (*E. hirae*) and in *Lactobacillus acidophilus* the only autolytic activity detectable is an *endo-N*-acetylmuramidase, but the localization of the enzyme differs in the two organisms. In *S. pneumoniae* and other gram-positive organisms the major, if not the only, autolysin appears to be an *N*-acetylmuramyl:L-alanineamidase. By comparison, *E. coli* has at least nine different hydrolases.

Although most gram-negative organisms are considered to contain autolysins, relatively few specific studies have been conducted with organisms other than the *Entero-bacteriaceae*. Autolysins appear to play an important role in septum and wall extension during cell growth as well as in cell separation, wall turnover, sporulation, competency for transformation, and possibly in the excretion of toxins and exoenzymes.

Peptidoglycan (Murein) Synthesis

Murein biosynthesis involves a number of cytoplasmic, membrane, and periplasmic steps (Fig. 6-13). The cell first activates NAG (N-acetylglucosamine) by coupling it with UDP. A portion of UDP–NAG is converted into UDP–NAM (N-acetylmuramic acid) with its peptide side chains. The bactoprenol (undecaprenol) cycle then couples the two compounds and delivers the disaccharide across the cytoplasmic membrane to the growing cell wall. At the interface between the growing cell wall and the cell membrane, transglycosidation reactions polymerize the growing chain and transpeptidases introduce cross-linking.

In *E. coli* the 2-min region on the chromosome map contains a large cluster of genes that code for proteins involved in various aspects of peptidoglycan synthesis and cell division. Seven genes mapping in this region (*murC, murD, murE, ddl, murF, mraY,* and *murG*) participate in the pathway for peptidoglycan synthesis from UDP-NAG to the formation of the lipid intermediate *N*-acetylglucosaminyl-*N*-acetylmuramyl-(pentapeptide)-pyrophosphateundecaprenol. However, other genes mapping at sites removed from this region also participate in murein sythesis. For example, the *E. coli* enzyme (UDP-*N*-acetylglucosamine enolpyruvate transferase (Reaction 4 in Fig. 6-13) catalyzes the first committed step in peptidoglycan formation. This enzyme, encoded by *murZ*, which maps at 69.3 min, is inhibited by the bactericidal antibiotic phosphomycin (L-*cis*-1,2-epoxypropylphosphonic acid; phosphonomycin), a structural analog of phosphoenolpyruvate:

$$CH_3 - CH - CH - PO_3H_2 \qquad CH_2 = C - CO_2H$$
$$\diagdown \diagup \qquad\qquad\qquad | $$
$$O \qquad\qquad\qquad\qquad OPO_3H_2$$

Phosphomycin   Phosphoenolpyruvate

D-glutamic acid, a specific component of peptidoglycan, is added to the UDP-*N*-acetyl-muramyl-L-alanine by the product of the *murD* gene. Another gene, *murI*, which maps at 90 min on the *E. coli* map, is required for the synthesis of D-glutamate from α-ketoglutarate.

The *mraY* gene at 2 min on the *E. coli* chromosome map encodes the enzyme UDP-*N*-acetylmuramoyl-pentapeptide:undecaprenyl-phosphate   phospho-*N*-acetylmuramoyl-pentapeptide transferase (MraY). The MraY enzyme is involved in the first step of the cycle of reactions leading to the synthesis of UDP-NAM from UDP-MurNAc (Reactions 12-15 in Fig. 6-13). The glycopeptide antibiotics vancomycin and ristocetin appear to inhibit the final polymerization reaction that results in the addition of *N*-acetylglucos-amine-*N*-acetyl-muramoylpentapeptide-pyrophospholipid to the growing point of the peptidoglycan backbone chain. Bacitracin, a member of a group of low-molecular-weight cyclic peptides, inhibits the dephosphorylation of the lipid-P-P carrier involved in the transfer of precursors into the peptidoglycan structure. This prevents the lipid carrier from functioning in the reaction cycle (Reaction 15 in Fig. 6-13).

Penicillin-Binding Proteins

Muramyl pentapeptide is an important structural element of the walls of both gram-positive and gram-negative cells. Peptidases catalyze transpeptidase reactions involving incorporation of the terminal D-alanyl-D-alanine during the final stages of peptidoglycan biosynthesis (Reaction 16 in Fig. 6-13). Bacterial cells contain a variety of penicillin-binding proteins (PBPs), as shown in Table 6-3. These membrane-bound PBPs interact covalently with penicillin and other β-lactam antibiotics. The four high-molecular-weight PBPs (1A, 1B, 2, 3) are involved in peptidoglycan biosynthesis. Both PBP-1A and PBP-1B are associated with cell elongation. The PBP-2 aids in determination of cell shape, and PBP-3 mediates septum formation. The structural similarity between penicillin and the pentapeptide precursor of the bacterial cell wall allows the β-lactam antibiotics to

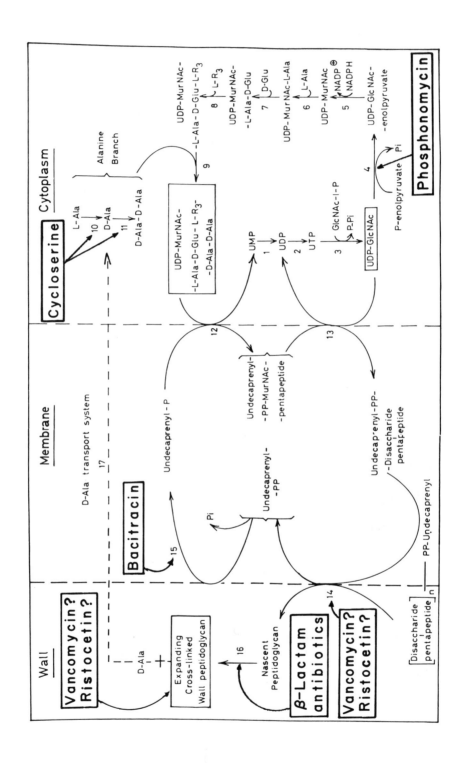

interact with the active site of a transpeptidase, terminating cross-linking:

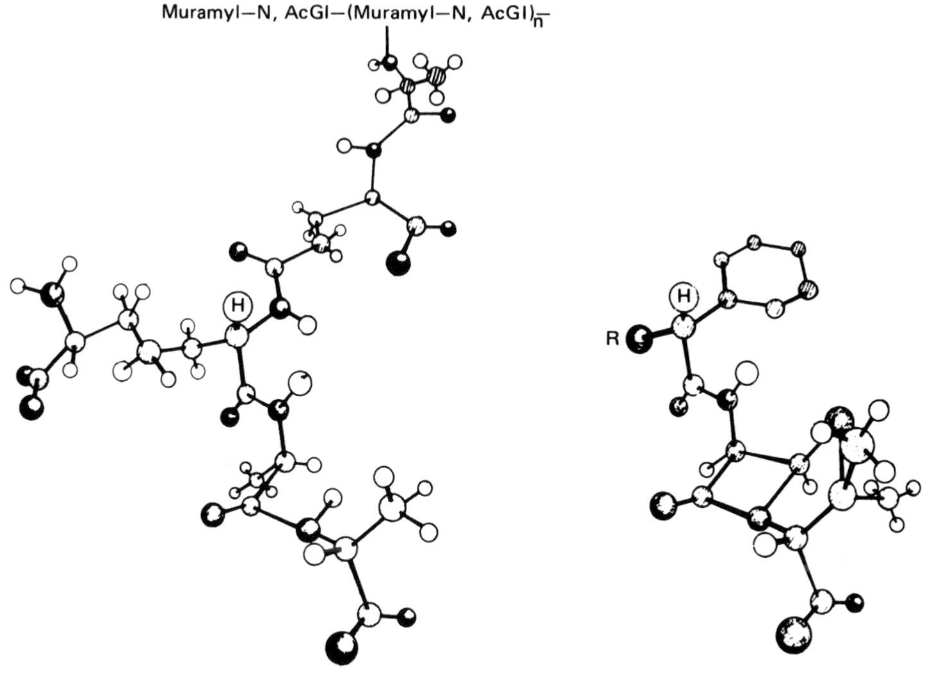

Muramyl—N, AcGI—(Muramyl—N, AcGI)$_{\overline{n}}$

Cell wall pentapeptide   Pencillin G

(Figure from Butler et al., 1970.)

Differences in microbial susceptibility to various derivatives of β-lactam antibiotics suggest that there are multiple sites of action for some of these compounds. Mecillinam,

---

Fig. 6-13. Reactions involved in peptidoglycan biosynthesis and sites of action of antimicrobial agents. The three stages (cytoplasmic, membrane bound, and wall bound) are separated by the dashed vertical lines. GlcNAc = *N*-acetylglucosamine; MurNAc = *N*-acetylmuramic acid; L–R$_3$, for example, *meso*-diaminopimelic acid. The original figure is from Ghuysen, J.-M. 1977. Biosynthesis and assembly of bacterial cell walls. In *Cell Surface Reviews,* Vol. IV, G. Poste and G. L. Nicholson (Eds.). Membrane Assembly and Turnover, ASP Biological and Medical Press, Amsterdam, The Netherlands. Sites of action of the antimicrobial agents have been superimposed for purposes of discussion in this chapter. The structural genes and names of the enzymes are (1,2) *pyrH*, UMP kinase; (3) UDP-*N*-acetylpyrophosphorylase; (4) *murZ*, UDP-*N*-acetylglucosamine enolpyruvate transferase; (5) UDP-*N*-acetylglucosamine enolpyruvate reductase; (6) *murC*, L-alanine adding enzyme; (7) *murD*, D-glutamate adding enzyme; (8) *murE*, *meso*-diaminopimelate adding enzyme; (9) *murF*, D-alanyl:D-alanine adding enzyme; (10) alanine racemase; (11) *ddl*, D-alanine: D-alanine ligase; (12) *mraY*, UDP-*N*-acetylmuramoyl-pentapeptide:undecaprenyl-phosphate phospho-*N*-acetylmuramoyl pentapeptide transferase (first step in lipid-carrier cycle); (13) *murG*, *N*-acetylglucosaminyltransferase (final step in lipid-carrier cycle); (14) transglycosylase and transpetidase; (15) membrane-bound pyrophosphatase; (16) membrane-bound transpeptidase (target of β-lactam antibiotics); (17) D-ala transport system.

TABLE 6-3. Penicillin-Binding Proteins[a]

PBP No.	Gene	Biological Function; Other Properties
1A	***ponB*** *(mrcA)* 2 min	High MW; transpeptidase and transglycosylase in wall elongation; inhibition results in lysis; increased in mecillinam-resistant mutants; Deficiency in both 1A and 1B required for lethality. Inhibition of DNA replication by naladixic acid prevents lytic response to blockade of PBP 1A and 1B by β-lactams, such as cefsulodin.
1B	***ponB*** *(mrcB)* 2 min	High MW (90,000); helps maintain rod shape during wall elongation; lysis triggered by β-lactams (cefsulodin) that bind to 1A and 1B
2	***pbpA*** *(mrdA)* 14.5 min	High MW; with **RodA** conducts transpeptidation and transglycosylation; maintains rod shape; binds **mecillinam** specifically; action essential to prevent lysis in PBP 1B⁻ strains; *rodA* and *pbpA* constitute *rodA* operon.
3	***pbpB*** *(ftsI)* *(sep)* 2 min	High MW (63,850). Mediates **septum** formation; transpeptidation and transglycosylation; interacts with **FtsA, FtsQ, FtsW,** and **RodA**; helps maintain rod shape; overproduced in mecillinam-resistant (*mre*) strains; only transpeptidase is β-lactam sensitive
4	***dacB*** 2 min	Low MW; DD-carboxypeptidase and endopeptidase; postinsertional modification of new murein
5	***dacA*** 2 min	Low MW (41,337); major PB component; D,D-carboxypeptidase I removes terminal D-ala from pentapeptide side chains; regulates peptide cross-linkage
6	***dacC*** 2 min	Low MW (40,804); PBP 5 and PBP 6 account for 85% PB; binds to CM by C-terminal end; bulk of enzyme is in periplasm; regulates peptide cross-linkage, stabilizes PG during stationary phase

[a]Peptidoglycan-synthesizing enzymes in *E. coli.*

a β-lactam with an amidino side chain, as shown in the following structure, appears to have a different mode of action from that of other penicillins:

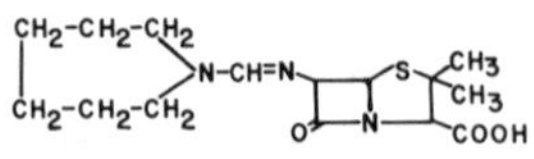

6-β-amidino-penicillanic acid
(Mecillinam or FL-1060)

Gram-negative organisms, such as *E. coli*, are more than 100 times more sensitive to this agent than gram-positive organisms, such as *S. aureus* or *B. subtilis*. Other penicillins may bind with more than one PBP, but mecillinam binds specifically to PBP-2 of *E. coli* causing the formation of protoplasts without effect on the activity of other known penicillin-sensitive sites. The action of different classes of penicillins at different sites suggests that combinations of agents from the two classes of compounds may act synergistically. Synergy has been demonstrated between mecillinam and a variety of other penicillins and cephalosporins both *in vitro* and *in vivo*. Similarly, *N*-formimidoyl thienamycin and cefoxitin exhibit a synergistic killing action on *E. hirae*, suggesting at least two sites of action of these β-lactams. One block in the cell division cycle, induced by *N*-formimidoyl thienamycin or methicillin, occurs before the completion of chromosome replication. A second block, induced by cefoxitin or cephalothin, takes place later in the cell division cycle. Studies with *E. coli* suggest that, on a quantitative basis, *N*-formimidoyl thienamycin and methicillin preferentially inhibit PBP-3, whereas cefoxitin and cephalothin preferentially inhibit PBP-2. The synergistic action of these agents in *E. hirae* suggests a similar mode of action in this organism.

Cycloserine, a compound chemically related to D-alanine, prevents the incorporation of D-alanine into muramyl pentapeptide through its inhibitory action on dipeptide synthetase and alanine racemase (Reactions 10 and 11, Fig. 6-13). These combined activities result in the accumulation of mucopeptides lacking the terminal D-alanine residue.

Insertion of new material into the expanding murein layer requires the activity of enzymes that degrade the glycan strand and endopeptidases that degrade the intact murein polymer. *E. coli* produces two soluble lytic transglycosylases, encoded by *slt* and *mlt*, that catalyze the cleavage of the β-1,4-glycosidic bond between *N*-acetylmuramic acid and *N*-acetylglucosamine residues. At the same time there is also an intramolecular transglycosylation at the *N*-acetylmuramic acid residue, resulting in the formation of a 1,6-anhydro bond. These enzymes apparently play a major role in recycling peptidoglycan components during expansion of the murein layer. For additional discussion regarding the role of genes involved with murein synthesis in the process of growth and cell division, see Chapter 11.

Teichoic Acids and Lipoteichoic Acids

Teichoic acids are found in all gram-positive organisms but are absent from gram-negative bacteria. Teichoic acids are polymers of either ribitol phosphate or glycerol phosphate in which the repeating units are joined together through phosphodiester linkages (Fig. 6-14). Sugars, amino sugars, or amino acids may be condensed to the hydroxyl groups of the ribitol or glycerol to provide wide variations in overall structure (Fig. 6-15). The term teichoic acid has been broadened to include all polymers containing glycerol phosphate or ribitol phosphate associated with the membrane, cell wall, or capsule. Wall teichoic acids are covalently linked to peptidoglycan through muramic acid and the phosphate group of the ribitol or glycerol phosphate. The ribitol teichoic acids of *S. aureus* and some strains of *B. subtilis* are linked to peptidoglycan through a common linkage unit, (glycerol phosphate)$_3$-GlcNAc. The glycerol teichoic acid of *B. cereus*, several strains of *B. subtilis*, and *B. licheniformis* are joined to peptidoglycan through a common linkage disaccharide, *N*-acetylmannosaminyl $(1 \rightarrow 4)N$-acetylglucosamine, irrespective of the structural diversity in the glycosidic branches and backbone chains. Wall teichoic acids are effective antigens and serve as the group- or type-specific substances of many organisms.

The wall ribitol teichoic acids of *S. pneumoniae* that serve as the C-antigen are more complex than those described previously. Choline is present along with glucose, ribitol, phosphorus, galactosamine, and 2,3,6-trideoxy-2,4-diaminohexose. The walls of *Lactobacillus plantarum* contain a mixture of one polymer of glucosylglycerol phosphate and

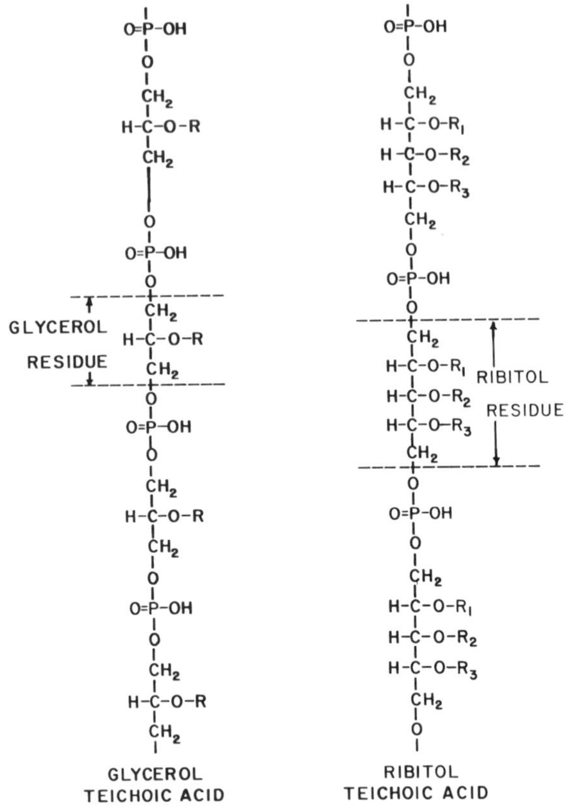

Fig. 6-14. Teichoic acid structures. In glycerol teichoic acid R may be alanine, glucose, or glucosamine. In ribitol teichoic acid R_1 may be glucose or N-acetylglucosamine; R_2 and/or R_3 may be alanine. Some of the variations in the substitutions on teichoic acids are shown in Figure 6-15.

two polymers of isomeric diglucosylglycerolphosphates. In *L. acidophilus* the teichoic acid is a mixture of α- or β-1,6-linked polyglucose polymers with monomeric α-glycerol phosphate side chains attached on the C-2 or C-4 position.

Other anionic polymers that lack polyol phosphate and may be present in cell walls are not, by definition, teichoic acids. Nevertheless, such components are important. For example, the group-specific polysaccharide of *S. pyogenes* group A contains phosphate, glycerol, rhamnose, and N-acetylglucosamine in a molar ratio of 1:1:2:1. Rhamnose and N-acetylglucosamine are incorporated from thymidine-5′-diphosphorhamnose and UDP-N-acetylglucosamine into the group A polysaccharide of *S. pyogenes*. Assembly of the group A polysaccharide occurs at the cell membrane with the participation of a lipoid anchor or acceptor molecule. Immunological specificity of the carbohydrate group antigen of streptococci is based upon the relative amounts of N-acetylglucosamine or N-acetylgalactosamine present.

Lipoteichoic acids (LTAs) are membrane-associated polymers characteristic of gram-positive bacteria. The LTAs are linear polymers of 16–40 phosphodiester-linked glycerophosphate residues covalently linked to a membrane anchor (generally a glycolipid or glycophospholipid). A number of gram-positive bacteria are now known to lack classical

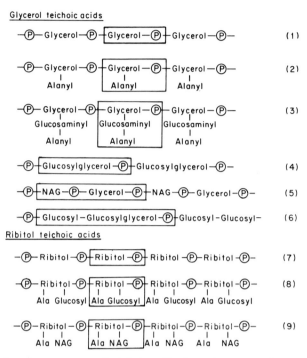

Fig. 6-15. Variations in structures of teichoic acids. In each structure the repeating unit is enclosed in a rectangle. Encircled P = interconnecting phosphate group; NAG = *N*-acetylglucosamine. From Moat, A. G. 1985. Biology of the lactic, acetic, and propionic acid bacteria. In *Biology of Industrial Microorganisms*, A. L. Demain and N. A. Solomon (Eds.). Benjamin-Cummings, Menlo Park, CA, pp. 143–188.

LTAs. In their place polymers with chemical properties similar to those of LTAs are being found with increasing frequency. These polymers are variously referred to as macroamphiphiles, lipoglycans, or cell surface glycolipids.

Membrane lipoteichoic acids of the glycerolphosphate polymer type occur in many gram-positive bacteria. These compounds have long polar glycerolphosphate chains linked to a small hydrophobic glycolipid. The lipoteichoic acid of *S. pyogenes* is a polymer containing glycerophosphate linked to a glycerophosphoryldiglycosyl diglyceride. In *E. faecalis* (*E. hirae*) the glycolipid is phosphatidylkojibiosyldiglyceride. The Forssman antigen of *S. pneumoniae* is similar to lipoteichoic acid. It is found in the cytoplasmic membrane and contains lipids and choline.

Lipoteichoic acids are exposed to the cell surface in many organisms. In *L. plantarum*, specific antiserum to the glycerolphosphate sequence shows the label extending from the outer surface through the wall and even outside the boundary of the cell. Spontaneous release of teichoic acid and lipoteichoic acid has been described in streptococci and lactobacilli. The lipoteichoic acid of *S. pyogenes* binds via its polyanionic backbone to positively charged residues of surface proteins termed M proteins. This protein has been associated with the virulence of group A, β-hemolytic streptococci. This orientation would leave the lipid moiety free to interact with fatty acid binding sites on host cell membranes. In the lactic acid producing lactobacilli, *L. fermenti*, but not *L. casei*, can be agglutinated by the antisera to lipoteichoic acid. The long polar glycerolphosphate chains of lipoteichoic acids probably extend through the network of the wall to evoke an immune response.

Physiological roles for lipoteichoic acid may include: regulation of autolysin activity, scavenging of divalent cations, such as Mg^{2+}, and interaction of bacteria with cells of infected hosts. When staphylococci are grown under the conditions of phosphate limitation, no wall teichoic acid is formed. However, membrane lipoteichoic acids and teichuronic acids are still produced under these conditions.

Teichoic acids also bind divalent cations. Hence, they may serve in concentrating Mg^{2+} or other ions at the cell surface. In *Lactobacillus buchneri*, one Mg^{2+} ion is bound for every two phosphate groups in the wall teichoic acid. Release of teichoic acid and lipoteichoic acid may influence the interaction of bacterial pathogens with cells of the invaded host. In addition to their serological reactivity, teichoic acids may activate the alternative pathway of complement and may play a role in the specific adhesion of bacteria to host epithelial surfaces.

Outer Membrane of Gram-Negative Bacteria

Gram-negative bacteria display a prominent outer membrane that is peripheral to the periplasmic region and the peptidoglycan sacculus (Fig. 6-16). This outer membrane is covalently attached to the peptidoglycan layer through lipoprotein and serves to reinforce the shape of the cell and to provide a protective barrier against the external environment. Under the electron microscope the outer and inner membrane present a similar appearance as trilaminar (double-track) layers. Chemical analysis shows that the inner and outer membranes are similar in lipid content. Some components of the outer membrane are qualitatively similar to those of the cytoplasmic membrane, however, the outer membrane has been determined to be an asymmetric bilayer with the external layer being composed primarily of LPS and the inner layer containing primarily phospholipids. Outer-membrane proteins (OMPs), called porins, form large water-filled pores with diameters of 1–2 nm that traverse the membrane and regulate the access of hydrophilic solutes to the cytoplasmic membrane.

Lipopolysaccharides

Lipopolysaccharide consists of three basic components or regions, as shown in Figures 6-16 and 6-17. **Region I**, the outermost portion contains repeating carbohydrate units that represent the "O" antigen. Alteration in the sugar composition of the O antigen results in a change in the immunological specificity. The sugars found in the O antigen region can occur in a wide variety of combinations, accounting for tremendous antigenic diversity and many hundreds of chemical types or serotypes of *Salmonella*, *Shigella*, and other *Enterobacteriaceae*.

The **core region (Region II)** consists of an outer and an inner core. The outer core shows high-to-moderate structural variability, whereas the inner core shows very low structural variability, particularly within a very closely related group of organisms, for example, the *Salmonella*. The oligosaccharide subunits of the core region of *E. coli* and *Shigella* differ only slightly from those of *Salmonella*. In other gram-negative bacteria less closely related to the enteric bacteria, a greater diversity in the structure of the core region may be encountered. However, the unique octose, 2-keto-3-deoxyoctulosonic acid

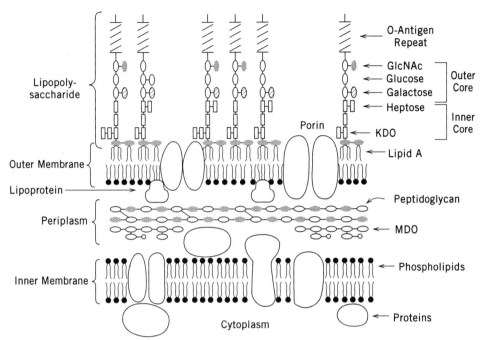

Fig. 6-16. Diagram of the gram-negative bacterial cell surface. Ovals and rectangles represent sugar residues, circles depict the polar head groups of glycerophospholipids (phosphatidyethanolamine and phosphatidylglycerol). MDO = membrane-derived oligosaccharides. The core region of LPS is that of *E. coli* K-12, a strain that does not normally contain an O-antigen repeat unless transformed with an appropriate plasmid. From Raetz, C. R. H. 1993. *J. Bacteriol.* **175:**5745–5753.

(KDO), appears to be a common component of the core region of most gram-negative organisms.

Lipid A (Region III), embedded in the outer membrane, has been extensively studied in *Salmonella*. In this organism, the chemical composition of lipid A has been shown to consist of a chain of D-glucosamine disaccharide units with all of the hydroxyl groups substituted. The substituents are the core polysaccharide units on the one hand and long-chain fatty acids on the other. The most commonly observed fatty acid substituent is β-hydroxymyristic acid (3-hydroxytetradecanoic acid), a C_{14} saturated fatty acid, which is substituted on the amino groups. The hydroxyl groups are also esterified with other long-chain fatty acids, such as lauric, myristic, and palmitic acids. The structure shown in Figure 6-17 is a *monomeric* unit of LPS. An average of three such subunits are linked together through pyrophosphate bridges between the lipid A molecules. The exact manner in which the LPS is linked to other surface structures is uncertain. However, LPS is very tightly associated with OMPs embedded in the outer membrane, particularly OmpA. The OmpA contributes to the stability of the outer membrane since it spans the membrane and is cross-linked to the underlying peptidoglycan layer. OmpA is exposed at the surface where it also serves as a receptor for T-even phage and plays a role in conjugation and the action of colicins K and L.

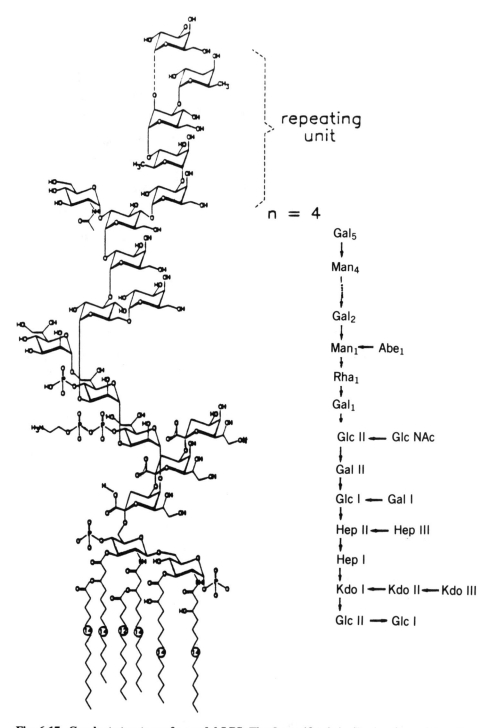

repeating
unit

n = 4

Gal5
↓
Man4
⋮
↓
Gal2
↓
Man1 ◄— Abe1
↓
Rha1
↓
Gal1
↓
Glc II ◄— Glc NAc
↓
Gal II
↓
Glc I ◄— Gal I
↓
Hep II ◄— Hep III
↓
Hep I
↓
Kdo I ◄— Kdo II ◄— Kdo III
↓
Glc II —► Glc I

Fig. 6-17. Covalent structure of a model LPS. The O-specific chain (Region I) consists of four repeating units and a terminal Gal residue. A hexaacyl lipid A component is shown and the two main chain heptoses are substituted in position 4 by phosphate and 2-aminoethylpyrophosphate, respectively. On the right the designations of each of the saccharide residues are given. From Kastowsky, M., T. Gutberlet, and H. Bradaczek. 1992. *J. Bacteriol.* **174**:4798–4806.

Porins

A variety of other OMPs are found in the *Enterobacteriaceae* and other gram-negative bacteria. As shown in Table 6-4, these OMPs play a variety of overlapping roles in the physiology of the cell. Some of these proteins serve to facilitate the entry of specific metabolites, such as vitamin B_{12}, iron, or maltose, while others, such as OmpC and OmpF, constitute components of general porins that allow hydrophilic solutes of less than 700 MW to traverse the outer membrane. Both OmpF and OmpC are similar with respect to functional and structural properties, but expression of their structural genes, *ompF* and *ompC*, is regulated in opposite directions by the osmolarity of the medium.

As shown in Figure 6-18, the channel-forming porin trimers of *E. coli* span the outer membrane of the cell. Electron microscopy and image reconstruction techniques show that the three channels of OmpF on the outer surface merge into a single channel at the periplasmic face. By comparison, the PhoE porin of *E. coli* forms three separate channels that traverse the width of the membrane. The OmpP of *Pseudomonas aeruginosa*, on the other hand, forms a single, small, highly anion-selective channel in which the permeability is related to the presence of a selectivity filter (S) containing three charged lysine molecules (+). A study of 12 different porins from *E. coli*, *P. aeruginosa*, and *Yersinia pestis* revealed that most of them appear to be cation selective. Only 3 of the 12 showed anion selectivity.

TABLE 6-4. Outer-Membrane Proteins of Gram-Negative Bacteria

Protein	Functions
OmpA	Stabilization of outer membrane and mating aggregates in F-dependent conjugation Receptor for phage TuII
Murein lipoprotein (Braun's lipoprotein)	Most abundant surface protein in *E. coli*, *S. typhimurium*; major structural protein, in conjunction with OmpA stabilizes cell surface
OmpB (porin)	Diffusion channel for various metabolites including maltose
LamB (maltoporin)	Specific porin for maltose, maltodextrin Receptor for bacteriophage λ
OmpC (porin)	Diffusion channel for small molecules Receptor for phages TuIb, T4
OmpF (porin)	Diffusion channel for small molecules Receptor for phage TuIa, T2
OmpT	Protease
PhoE (protein E)	Anion-selective diffusion channels induced under phosphate limitation
Protein P	Anion-selective diffusion channel in *Pseudomonas aeruginosa*. Induced under phosphate limitation
TolA	Maintenance of OM integrity; activity of group A colicins
TonA	Ferrichrome siderophore uptake Receptor for phages T1, T5, 80, colicin M
TonB	Siderophore-mediated iron transport; B_{12} transport
Tsx	Nucleoside-specific channel Receptor for T-even phages, colicin K

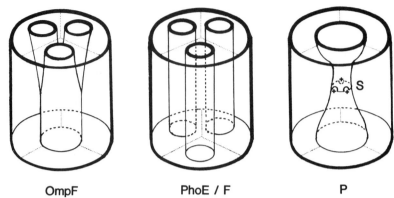

OmpF   PhoE / F   P

Fig. 6-18. Schematic three-dimensional representations of the structures of three different porin proteins. The proteins are oriented such that the top of the figure is the portion exposed to the external environment, whereas the bottom is the portion extending into the periplasmic space. The channels traverse the width of the outer membrane. The OmpF porin of *E. coli* forms coalescing channels. The PhoE porin of *E. coli* and the protein F porin of *Pseudomonas aeruginosa* form three separate and distinct channels. Protein P of *P. aeruginosa* forms a single anion-specific channel containing a selectivity filter (S) consisting of three charged lysine molecules (+). From Hancock, R. E. W. 1987. *J. Bacteriol.* **169**:929–933.

In *E. coli* a major OMP, LamB, serves as a receptor for bacteriophage λ (Chapter 5). Production of this protein is induced by growth on maltose and is involved in the passage of maltose and maltose-containing oligosaccharides through the outer membrane. A comparable OMP with a molecular weight of 44,000 (44K protein) and inducible by maltose is present in the outer membrane of *S. typhimurium* but does not serve as a receptor for λ phage. Immunoelectron microscopy of newly induced LamB at the surface of *E. coli* reveals that LamB is inserted homogeneously over the entire surface.

Murein lipoprotein is a major OMP present in large quantities in *E. coli*. Lipoprotein molecules serve to anchor the outer membrane to the peptidoglycan layer. Mutants lacking *lpp*, the structural gene for lipoprotein, and *ompA*, the structural gene for OmpA, display spherical morphology and abundant blebbing of the outer membrane (Fig. 6-19). In these mutants the murein layer is no longer associated with the outer membrane. These mutants also display an increased growth requirement for Mg^{2+} and Ca^{2+} and are sensitive to hydrophobic antibiotics, such as novobiocin, or to detergents, suggesting a protective function of the outer membrane. For additional discussion of the role of porins in nutrient transport, see Chapter 7.

Lipopolysaccharide Biosynthesis

The three components of LPS (O-antigen, core oligosaccharide, and lipid A) are synthesized independently of each other and later ligated in or on the inner membrane. After assembly, the intact LPS is translocated to the outer membrane. The LPS mutant strains of *S. typhimurium* and of *E. coli* K-12 have been isolated and their LPS structures defined in terms of polysaccharide content (LPS chemotypes), as shown in Figure 6-20. Most of the genetic determinants responsible for oligosaccharide core biosynthesis in *E. coli* and *S. typhimurium* have been found to reside in the *rfa* gene cluster. The five transferases

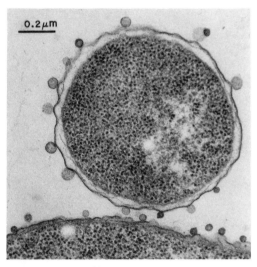

Fig. 6-19. Effect of loss of outer-membrane lipoprotein (Lpp) and OmpA on the cell envelope and shape of *E. coli*. From Sonntag, I. et al. 1978. *J. Bacteriol.* **136**:280.

required for assembly of the outer-core region of *E. coli* LPS have been identified by mutations of the corresponding genes (e.g., *rfaKJIGB*, as shown in Fig. 6-20). Similarly, genes coding for enzymes responsible for inner-core biosynthesis (*rfaC*, *rfaD*) have been identified. The *rfaD* gene codes for ADP-L-glycero-D-mannoheptose-6-epimerase. Mutation in *rfaC* gene (previously designated *rfa-2* in *E. coli*) results in production of a heptoseless LPS structure referred to as chemotype Re. These mutants display increased permeability to both hydrophobic and hydrophilic agents. In *S. typhimurium* the *rfaL* gene encodes a component of the O-antigen ligase and *rfaK* encodes the N-acetylglucosamine transferase. The order of genes within the *rfa* cluster at 79 units on the *S. typhimurium* linkage map has been shown to be *cysE-rfaDFCLKZYJIBG-pyrE*.

The major pathway leading to lipid A biosynthesis takes place in three stages: Stage 1, UDP-GlcNAc acylation; Stage 2, disaccharide formation and 4′ kinase action; and Stage 3, KDO transfer and late acylation. The genes and gene products involved in this pathway are shown along with the intermediates in Figure 6-20. UDP-2,3-diacylglucosamine plays a key role in lipid A biosynthesis, as shown in Figures 6-21 and 6-22.

Enterobacterial Common Antigen

Enterobacterial common antigen (**ECA**) is a cell surface glycolipid synthesized by all members of the *Enterobacteriaceae*. It is an amino sugar heteropolymer containing N-acetyl-D-glucosamine (GlcNAc), N-acetyl-D-mannosaminuronic acid (ManNAcA), and 4-acetamido-4,6-dideoxy-D-galactose (Fuc4NAc) linked together to form chains of trisaccharide repeat units. The biosynthesis of ECA is catalyzed by products of genes in the *rfe* and *rfb* gene clusters as well as the *rff* genes. The *rfe* and *rfb* gene clusters are also involved in LPS biosynthesis. This anomaly is explained by the fact that common intermediates are required for both LPS and ECA.

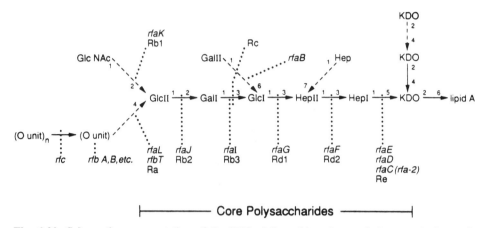

| Core Polysaccharides |

Fig. 6-20. Schematic representation of the LPS of *S. typhimurium* and the genetic determinants (*rfaBCDEFGIJK*) and the LPS structures (chemotypes Re, Rd2, Rd1, Rc, Rb3, Rb2, and Ra) of mutants blocked at various stages of LPS biosynthesis are indicated. The dotted lines indicate the defective LPSs termination points. KDO = 2-keto-3-deoxyoctulosonic acid; Hep = L-glycero-D-mannoheptose; Glc = glucose; Gal = galactose; GluNac = *N*-acetylglucosamine; (O unit)$_n$ = number of O-antigen side chains. The structural genes presumed to be responsible for LPS core biosynthesis are as follows: *rfaE*, specific function unknown; *rfaC* = ADP-heptose:LPS heptosyltransferase 1; *rfaD* = ADP-L-glycero-D-mannoheptose-6-epimerase; *rfaF* = ADP-heptose:LPS heptosyltransferase 1; *rfaG* = UDP-glucose:LPS glycosyltransferase 1: *rfaB* = UDP-galactose:LPS α-1,3-galactosyltransferase; *rfaJ* = UDP-glucose:LPS glucosyltransferase 1; *rfaK* = UDP-N-acetylglucosamine:LPS glucosaminyltransferase. From Chen, L. and W. G. Coleman, Jr. 1993. *J. Bacteriol.* **175**:2534–2540.

CYTOPLASMIC MEMBRANE AND OTHER MEMBRANOUS STRUCTURES

Cytoplasmic Membrane

Removal of the outer membrane and the peptidoglycan layer by the action of various chemical and enzymatic agents does not alter the integrity of the cell with respect to retention of its internal contents. Thus the inner membrane is the **cytoplasmic membrane** and serves the same function as the cytoplasmic membrane of gram-positive cells. The cytoplasmic membrane of bacteria and extensions of it assume many of the functions attributable to specialized organelles in eukaryotic cells. Active transport of metabolites into and out of the cell, oxidative phosphorylation, cell wall biosynthesis, phospholipid biosynthesis, and the secretion of various extracellular enzymes and other proteins have all been associated wth the cell membrane. In addition, the anchoring of DNA to the membrane to aid its distribution to the daughter cells during cell division has also been described. In gram-positive cells, lipoteichoic acids are anchored hydrophobically in the cytoplasmic membrane.

Cell membranes of gram-positive bacteria can be isolated by using any of the enzymes that selectively degrade the peptidoglycan and digest away the wall structure. If cells are treated with lysins while suspended in a solute to which the cell is impermeable (e.g., sucrose) at a concentration that approximately balances the high osmotic pressure of the cell, then an osmotically fragile body (protoplast) can be formed. The

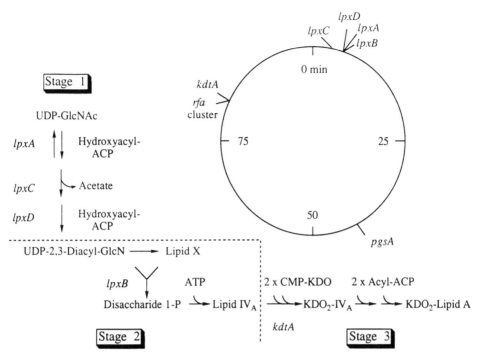

Fig. 6-21. Genes and enzymes involved in lipid A biosynthesis in *E. coli.* Location of the genes on the *E. coli* chromosome are shown along with other relevant genes, such as the *rfa* cluster and *pgsA*. The three stages of the pathway are Stage 1, UDP-GlcNAc acylation; Stage 2, disaccharide formation and 4′ kinase action; and Stage 3, KDO transfer and late acylation. From Raetz, C. R. H. 1993. *J. Bacteriol.* **175:**5745–5753.

protoplast can then be lysed osmotically and the membrane can be isolated. The cytoplasmic membrane contains numerous enzymes that perform various functions. Very few functions involving enzyme activities are localized in the outer membrane of gram-negative bacteria. Terminal electron transport enzymes and metabolite transport functions are found in the cytoplasmic membrane but are entirely lacking in the outer membrane. Only those compounds of less than 700 MW are aided in their passage through the outer membrane by porins.

Initially, it was not an easy task to demonstrate the distinction between the cell wall and the cell membrane with the experimental methods available to early microbiologists. The ability of cells to remain intact as protoplasts or spheroplasts (protoplasts with part of the cell wall still adhering) in an osmotically stable environment after the cell wall had been removed by treatment with enzymes or with antibiotics that inhibit cell wall synthesis (e.g., penicillin) provided the most convincing evidence that the cytoplasmic membrane was distinct and separate from the outer cell wall. Protoplasts are capable of conducting most, if not all, of the various activities attributable to intact cells (e.g., enzyme synthesis, various metabolic activities, bacteriophage synthesis, and spore formation) and are even capable of multiplication in a properly controlled osmotic environment. Some organisms, such as the *Mycoplasma*, exist as stable forms in nature without the capability of forming an outer-cell wall. Sterols serve to stabilize the membrane of

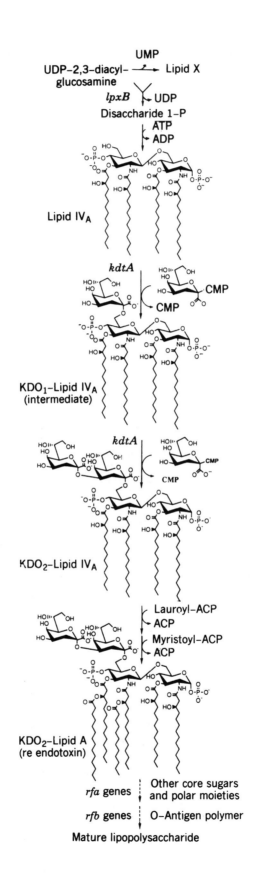

UMP

UDP–2,3-diacyl- → Lipid X
glucosamine

lpxB ⤙ UDP

Disaccharide 1–P

ATP
ADP

Lipid IV$_A$

kdtA

CMP

CMP

KDO$_1$–Lipid IV$_A$
(intermediate)

kdtA

CMP

CMP

KDO$_2$–Lipid IV$_A$

Lauroyl–ACP
ACP
Myristoyl–ACP
ACP

KDO$_2$–Lipid A
(re endotoxin)

rfa genes — Other core sugars
and polar moieties

rfb genes — O–Antigen polymer

Mature lipopolysaccharide

these organisms. All of the available evidence indicates that the cell membrane is the structure responsible for retention of the intracellular contents and is responsible for controlling the entrance and exit of metabolites.

Permeability and Transport

Control mechanisms regulating transport of substances into and out of the cell have been thoroughly investigated and we now understand the permeability properties of the membrane reasonably well. At one time, cell physiologists regarded the cytoplasmic membrane as an osmotic barrier equivalent to a collodion membrane. Metabolites were considered to enter or leave the cell primarily on the basis of the concentration gradient of metabolites from the inside to the outside of the cell. Membranes were considered to have pores of finite size that permitted penetration of molecules through the membrane. The size of the molecule, the relative internal and external concentrations, the charge distribution of the membrane and solute, and the pH and ionic strength of the internal and external environments were considered to be the primary factors governing the entrance of compounds into the cell. While all of these factors have their influences on the passage of materials into and out of the cell, it is now well established that cells have highly specialized transport proteins either inserted in the cytoplasmic membrane, or occasionally, found in the periplasmic region of gram-negative bacteria. The function of these transport systems in the movement of metabolites into and out of the cell will be considered in detail in Chapter 7.

Periplasm

Although membranes can be isolated, photographed, and separated into various fractions and analyzed chemically, we still do not completely understand many of the structural relationships of the transport systems present in the cytoplasmic membrane. As discussed in an earlier section of this chapter, electron micrographs of gram-negative cells reveal two unit membranes: an outer membrane composed of phospholipid, protein, and LPS and an inner membrane composed of phospholipids, glycolipids, and phosphatidylglycolipids. Embedded in the lipid bilayer are various proteins and lipoproteins that function in metabolite transport and the synthesis of macromolecular constituents of the cell wall and the outer membrane. Certain proteins serve to maintain the integrity of the wall–membrane complex, while others provide protection against phagocytosis or act as attachment sites for bacteriophage or chemical agents, such as antibiotics.

The two membranes of gram-negative cells are separated by a periplasmic space and a peptidoglycan layer. Certain types of cytological evidence suggest that there is no truly empty periplasmic "space" between the outer and inner membranes. Peptidoglycan in a hydrated state represents a gel that fills the entire space between the two membranes. The degree of cross-linking of the peptidoglycan probably diminishes toward the cyto-

---

Fig. 6-22. Pathway of lipid A biosynthesis. The *lpxB* and *kdtA* genes code for the lipid A disaccharide synthase and KDO transferase. In wild-type cells, the levels of the intermediates are very low (<1000 molecules per cell). From Raetz, A. 1993. *J. Bacteriol.* **175**:5745–5753.

plasmic membrane resulting in the formation of a **periplasmic gel** in which proteins are freely diffusible. Compression of this gel during osmotic shock would result in the sudden release of periplasmic enzymes.

Other-Membranous Organelles

Examination of a variety of organisms under the electron microscope has shown the presence of other membranous structures. Membranous structures that appear to be sac-like invaginations of the cytoplasmic membrane of bacteria have been observed with remarkable frequency. These structures, termed **mesosomes**, have been studied by a number of workers and several physiological functions attributed to them. However, there is now growing evidence that mesosomes are largely artifacts resulting from the fixation procedures used in the preparation of cells for electron microscopy. By varying the fixation process, it has been possible to increase or decrease the size of mesosomes or to make them disappear altogether. As a result, much of the earlier work on mesosomes and their functions has been discredited. Nevertheless, it is still possible that these invaginated structures may represent **functionally different** membranes.

The marine bacterium, *Nitrocystis oceanus*, possesses a laminar membranous organelle (Fig. 6-23). These membranous organelles consist of approximately 20 vesicles so flattened that the lumen is only 100-Å thick. The outer surfaces are in contact and form a triplet structure with an accentuated center line; these lamellae almost traverse the cell, displace the cytoplasm and the nuclear material, and form the most prominent cytological feature of the cell. *Nitrosomonas europaea* and other nitrifying organisms also produce distinctive membrane systems that differ significantly from one another. The presence of these lamellar organelles in nitrifying organisms (organisms that convert NH_4^+ to NO_2^- and NO_3^-) must somehow be related to the specialized mechanisms for acquiring energy from this process. However, the association of the intricate membrane systems observed in these organisms with the actual oxidation process is largely assumptive. It has been considered that these lamellar membranous organelles may be equivalent to mesosomes, plasmalemmasomes, or chondroids that have been studied in other kinds of bacteria, but the distinctive appearance and high degree of organization suggest that they are unique structures that may ultimately be directly associated with the oxidative processes of the autotrophic nitrifying bacteria.

CAPSULES

Many microorganisms produce an external layer of mucoid polysaccharide or polypeptide material that adheres to the cell with sufficient tenacity to be observed by simple techniques, such as negative staining with India ink (Fig. 6-24). When this outer layer of material reaches sufficient size to be readily visible, it is referred to as a **capsule**. Several examples of well-defined capsules are shown in Figure 6-24. Many microorganisms produce a variety of extracellular materials that may be too sparse or too soluble in the medium to be observed by microscopic techniques. These materials are generally referred to as a **slime layer**. The extracellular polysaccharide colanic acid or M antigen produced by *E. coli* strains and other enteric bacteria (Fig. 6-26) is often referred to as a slime

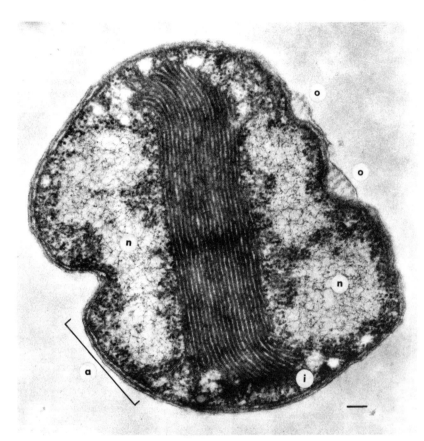

Fig. 6-23. Section of a marine bacterium, *Nitrocystis oceanus*, showing the cell wall and plasma membrane (area a), wall organelles (o), nuclear material (n), and the surrounding ribosome-packed cytoplasm. The lamellae of the membranous organelle traverses the entire cell. Other inclusions (i) may be seen within the cytoplasm. Bar = 0.1 μm. From Murray, R. G. E. and S. W. Watson. 1974. *J. Bacteriol.* **89**:1594.

layer. Most of these surface components can be separated from the underlying cell wall without impairing the function or structural integrity of the cell or its viability. This is not to say that the properties of the cell may not be altered in some manner by removal of the material. For example, strains of *Streptococcus pneumoniae* that fail to produce the polysaccharide capsular material are essentially avirulent. Similarly, the D-polypeptide capsule markedly increases the virulence of the anthrax bacillus (*Bacillus anthracis*).

From time to time, a wide variety of other materials have been shown to be associated with the outer surface of the cell. However, these materials are usually considered to be transient metabolic excretions or chance adherents acquired from the environment and do not represent true capsular material or slime layer. True capsules are generally composed of high-molecular-weight components, are highly viscous materials, and stain poorly if at all, with the usual stains. The best staining procedures employ mordants that cause the precipitation of the capsular material by metal ions, alcohol, acetic acid, and so on. Reaction of the capsule with specific antiserum is frequently employed to enlarge the capsular area so that it may be more readily visualized under the microscope (Fig. 6-24, 6 and 7). Techniques have also been developed for visualization of the capsule under the electron microscope (Fig. 6-24, 8 and 9).

The polysaccharide capsular substances produced by *S. pneumoniae* confer immunological specificity and are also associated with virulence. Type III pneumococcal poly-

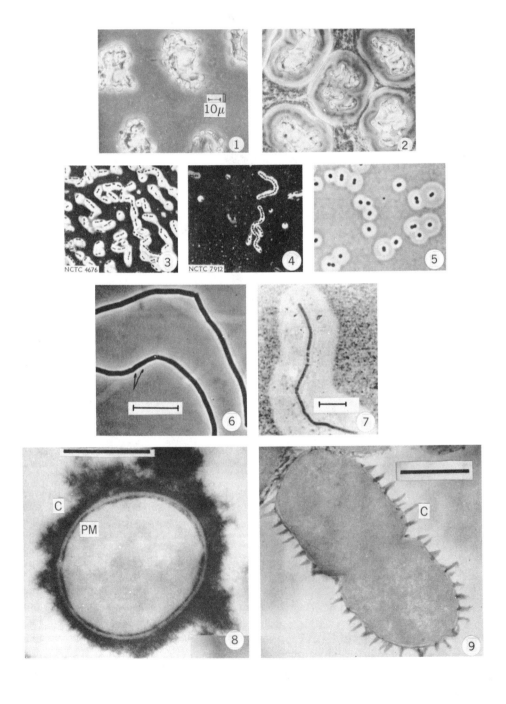

saccharide is composed of glucopyranose and glucuronic acid in alternating β-1,3- and β-1,4- linkages:

Alteration in the immunological specificity results from the incorporation of monosaccharides other than glucose and glucuronic acid in the polysaccharide structure; for example, galacturonic acid, rhamnose, N-acetyl-amino sugars, and uronic acid, in various combinations. Enhancement of virulence by the presence of a capsule results from inhibition of phagocytosis. Although physical size of the capsule correlates with the degree of virulence of the pneumococcus, the precise mechanism whereby capsules interfere with phagocytosis is not known.

Certain members of the streptococci, notably *S. mutans*, produce both soluble and insoluble forms of dextrans. The insoluble form, called a mutan, is considered to be responsible for adherence of this organism to the enamel surfaces of teeth and the formation of bacterial aggregates (plaque). Mutans are branched glucose polymers with linear α-1,3-linked glucose units (Fig. 6-25).

β-Hemolytic streptococci can be classified serologically on the basis of differences in their Lancefield group antigens. The capsular substance of Lancefield Groups A and C streptococci is a polysaccharide containing glucosamine and glucuronic acid units, hyaluronic acid. This substance provides protection from the destructive effects of atmospheric oxygen by virtue of its ability to aid in the formation of cell aggregates. Disruption

---

Fig. 6-24. Representative examples of encapsulated microorganisms. (1 and 2) Dark phase contrast photomicrographs of a pond alga. In (2) skimmed milk run under the coverslip permits visualization of the capsule as a broad, hyaline periphery that is virtually unseen in (1). The μ scale in (1) also applies to (2). From Dondero, N. *J. Bacteriol.* **85:**1171, 1963. (3 and 4) Hyaluronic acid containing capsules of Group C streptococci. The dense capsules are visualized on plates of moist 20% horse serum in 1% glucose agar with India ink background stain. The capsules are absent from smears of hyaluronidase cultures. From MacLennan, A. P. *J. Gen. Microbiol.* **15:**485, 1956. (5) Phase contrast photomicrograph of a gram-negative, capsule-forming coccus. These capsules were formed after the organisms had been stripped of their original capsules. The presence of oxygen and an oxidizable carbon source were the only requisites for resynthesis of the capsular polysaccharide. From Juni E. and G. A. Heym, *J. Bacteriol.* **87:**461, 1964. (6 and 7) Phase contrast photomicrographs of *B. anthracis* capsulated cells. Bars in 6 and 7 = 10 μm. (6) Membranelike outline of capsule in untreated cells. (7) Same cells after treatment with antiserum. Negative staining with India ink. Note the marked swelling of the capsule. From Avakyan, A. A. et al. *J. Bacteriol.* **90:**1082, 1965. (8 and 9) Ultrastructure of the capsules of *Streptococcus pneumoniae* Type II and *Klebsiella pneumoniae* Type I as seen in the electron microscope using ruthenium red in combination with osmium tetroxide. (8) A matlike capsule (C) surrounds each pneumococcal cell. Ruthenium red penetration is evident at the plasma membrane (PM) and into the cytoplasm. Bar = 0.5 μm. (9) *Klebsiella pneumoniae* capsular fibrils (C) are seen at regular intervals along the wall. Bar = 0.5 μm. From Springer, E. L. and I. L. Roth, *J. Gen. Microbiol.* **74:**21, 1973.

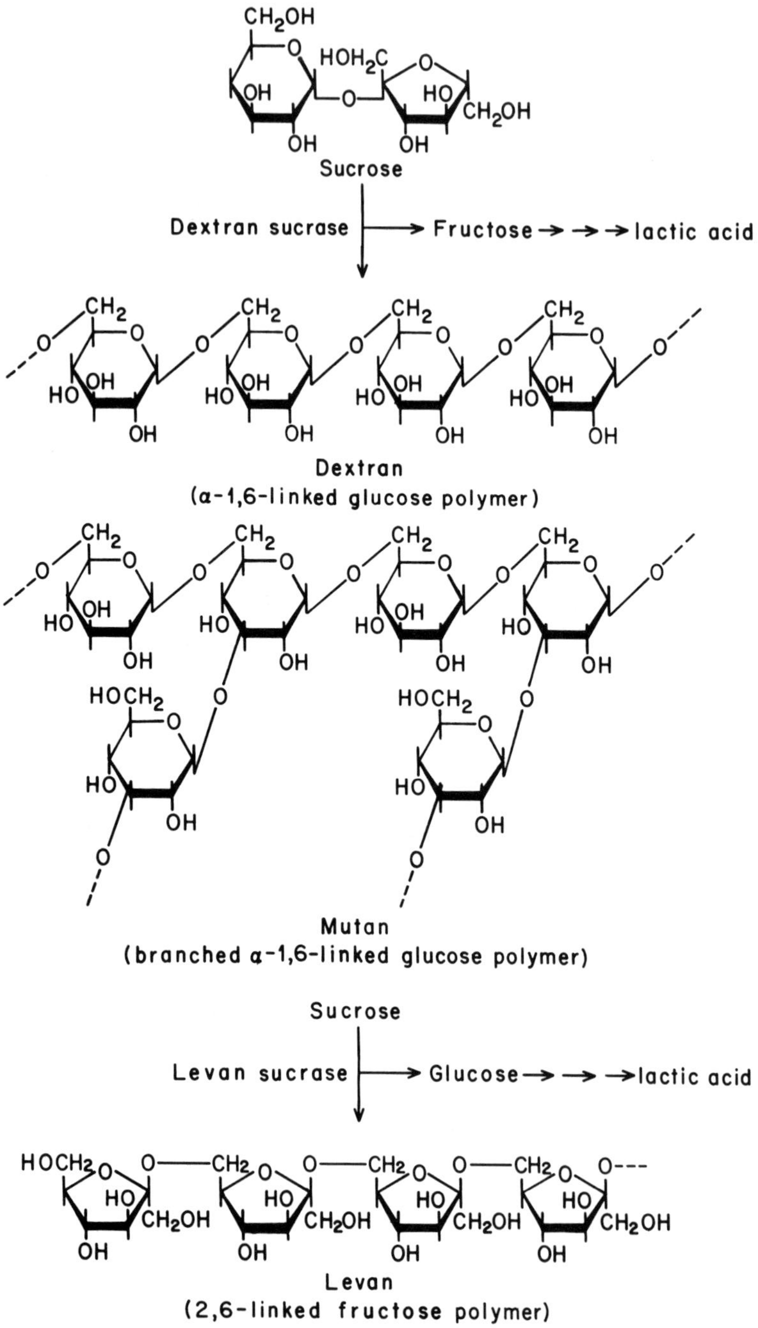

Fig. 6-25. Production of dextran and levan by bacteria. The enzyme detransucrase splits sucrose to form fructose and dextran. The fructose is fermented to lactic acid. Levansucrase splits sucrose to form glucose and levan. The glucose is fermented to lactic acid. The structure of mutan, a branched dextran polymer, is also shown.

of the aggregates with hyaluronidase results in an increased oxygen uptake and the production of toxic levels of hydrogen peroxide. Unencapsulated variants are sensitive to oxygen.

A number of strains of *E. coli* produce capsular polysaccharides commonly referred to as **K antigens**. The K antigens are structurally diverse and give rise to serological specificity. There are over 70 recognized K antigens in *E. coli*. The K antigens are placed in either capsular Group I (heat-stable) or capsular Group II (temperature regulated, expressed at 37°C but not at ≤20 °C). For example, the K30 antigen of *E. coli* is a member of Group I. The K5 antigen is in Group II. The Group II K antigens are also characterized by acidic components, such as 2-keto-3-deoxy-D-mannooctulonic acid, *N*-acetylneuraminic acid (sialic acid), and *N*-acetylmannosaminuronic acid (Fig. 6-26). The K5 repeating unit, shown in Figure 6-26, is identical to that of the first polymeric intermediate in the biosynthesis of heparin. As a consequence, the K5 polysaccharide is practically nonimmunogenic and strains of *E. coli* expressing this capsule are quite virulent since host cells do not mount an active immune response to it. A correlation has been drawn between elevated levels of CMP–2-keto-3-deoxyoctulosonic acid (CMP–KDO) synthetase activity and active production of the K5 polysaccharide.

The capsular polysaccharide 17-kb multigene cluster encodes the proteins required for the synthesis, activation, and assembly of the K antigens of *E. coli*. Three distinct regions have been identified in this gene cluster. The central region (Region 2) is composed of *neu* genes concerned with polysaccharide synthesis and polymerization. The *kps* genes in Regions 1 and 3 are considered to be involved in polymer assembly and transport and govern the same functions in all *E. coli* producing Group II capsules.

A model for polysialic (poly-*N*-acetylneuraminic) acid synthesis and translocation is presented in Figure 6-27. Two genes important in *N*-acetylneuraminic acid (neuAc) synthesis, *neuA* and *neuC*, are located at the 2.7-kb *Eco*R1–*Hin*dIII fragment of the *E. coli* K-1 gene cluster. The *neuA* gene encodes CMP–neuAc synthetase, and *neuC* encodes a 45,000 MW protein (P7) required for the synthesis of neuAc. Region 3 of the *kps* gene cluster contains two genes, *kpsM* and *kpsT*, that encode protein constituents of a system for the transport of polysialic acid.

Under appropriate growth conditions, *E. coli* K-12 strains and other enteric bacteria produce a slime polysaccharide, colanic acid, termed the M antigen (Fig. 6-26). The *E. coli* rcsA$_{K12}$ and rcsA$_{K30}$ genes are transcriptional activators involved in the expression of colanic acid. There is evidence that the RcsA protein may be a relatively widespread regulatory component for the synthesis of enterobacterial extracellular polysaccharides.

Proteus mirabilis produces an acidic capsular polysaccharide that is a high-molecular-weight polymer of branched trisaccharide units composed of 2-acetamido-2-deoxy-D-glucose (*N*-acetyl-D-glucosamine), 2-acetamido-2,6-dideoxy-L-galactose (*N*-acetyl-L-fucosamine), and D-glucuronic acid (Fig. 6-26). This represents the first report of a defined capsule in *Proteus*. *Proteus mirabilis* 2573 also produces an O:6 serotype LPS in which the O-chain component has the same structure as the homologous capsular polysaccharide. The acidic nature of this polysaccharide may play a role in urinary calculi (stone) formation in a similar fashion to the established roles of acid proteins and glycoproteins in mineralization and control of crystal formation in biological tissues. This process is initiated by a minute focus of solid matter that grows continually larger as chemical crystals precipitated from the urine adhere to it.

High-molecular-weight polypeptide capsules are produced by *Yersinia pestis* and *Bacillus anthracis*. The *B. anthracis* capsule is a polypeptide of γ-D-glutamyl subunits,

```
               -4-)-ß-GlcUA-1,4-α-GlcNAc-1,-
```

Repeating unit of K5 capsular polysaccharide of *Escherichia coli*

```
                    OAc
                     ⫶
                    2/3
    -4)α-L-Fucp-(1→3)-ß-D-Glcp-(1→3)-ß-L-Fucp-(1—
                                           4
                                           ↑
                                           1
                                       ß-D-Galp
                                           3
                                           ↑
                                           1
                                       ß-DGlcAp
                                           4
                                           ↑
                                           1
                                       ß-D-Galp=Pyr
```

Slime polysaccharide (colanic acid or M antigen) of
Escherichia coli

```
        -2-α-D-Manp-(1→3)-ß-D-Galp-(1—
                3
                ↑
                1
            α-D-Galp
                4
                ↑
                1
            ß-D-GlcAp
```

K30 (capsular) antigen of *Escherichia coli*

```
        -3)-ß-D-GlcpNAc-(1->4)-α-L-FucpNAc-1(1—
                 3
                 ↑
                 1
            α-D-GlcpA
```

Capsular polysaccharide of *Proteus mirabilis*

```
-[(N-acetyl-D-glucosamine)-(N-acetyl-L-fucosamine)-(D-glucuronic)]-
```

Fig. 6-26. Chemical structures of representative examples of capsules or slime layers of gram-negative bacteria. GlcpNAc = N-acetylglucosamine (italicized *p* indicates pyranose form of the sugar); Fuc = L-fucose; Gal = D-galactose; Glc = D-glucose; Man = D-mannose; FucpNAc = N-acetylfucosamine.

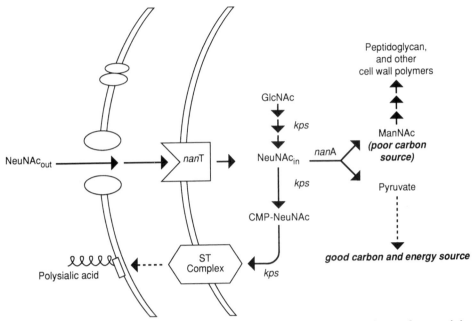

Fig. 6-27. Model for synthesis and translocation of polysialic acid (poly-*N*-acetylneuraminic acid) in *E. coli* K-1. External *N*-acetylneuraminic acid (NeuNAc$_{out}$) diffuses into the periplasm through outer-membrane pores and is transported by the sialic acid permease encoded by *nanT*. Internal sialic acid (NeuNAc$_{in}$), transported or produced endogenously by *kps* gene products, is either degraded by the aldolase encoded by *nanA* or activated to the sugar nucleotide precursor CMP–NeuNAc. Sialic acid is polymerized and translocated to the outer membrane as PSA by the sialyltransferase (ST) complex. The aldolase cleavage products either enter intermediary metabolic pathways or are disseminated into surface polysaccharides. From Vimr, E. R. 1992. *J. Bacteriol.* **174**:6191–6197.

originally thought to be produced only *in vivo* during infection, but later shown to be produced in culture if an excess of carbon dioxide or bicarbonate is present. Other species of bacilli (e.g., *B. subtilis*) produce a polypeptide capsule containing a mixture of D- and L-glutamic acid subunits. The ratio of the two isomers can be controlled by culture conditions. It is not known specifically what governs the ratio of D:L isomers, but aeration tends to favor approximately a 1:1 ratio of D:L units. In addition to carbon dioxide, the presence of DL-isoleucine, DL-phenylalanine, and glutamic acid affect the production of the polypeptide capsule by *B. anthracis*. The polypeptide of *B. anthracis* is of special interest because of its relationship to the virulence of the organism through prevention of phagocytosis.

Acetobacter xylinum synthesizes cellulose, a polymer of β-1,4-linked glucose units:

Cellulose accumulates as an extracellular aggregate of crystalline microfibrils. Intertwined ribbons from large numbers of cells form a tough pellicle on the surface of the culture medium. Although cellulose synthesis has been studied extensively in higher plants as well as in bacteria, the intermediary steps in the pathway have not been entirely elucidated. Resting cells and particulate membrane-bound preparations of A. *xylinum* incorporate [1-^{14}C]-glucose into glucose-6-phosphate, glucose-1-phosphate, uridine glucose-5'-phosphate (UDPG), and cellulose. Labeling studies and demonstration of enzyme activities in cell-free extracts indicate that the sequence of reactions leading to cellulose synthesis is

$$\text{glucose} \longrightarrow \text{G–G–P} \longrightarrow \text{G–1–P} \longrightarrow \text{UDPG} \longrightarrow \text{cellulose}$$

Lipid and protein-linked cellodextrins (partially polymerized glucose units) may function as intermediates between UDPG and cellulose. Cellulose synthetase (UDP–glucose:1,4-β-D-glucan 4-β-D-glucosyltransferase) has been solubilized by treatment of membranes from A. *xylinum* with digitonin (1–10%). The digitonin-solubilized enzyme is specifically activated by GTP in the presence of a protein factor that can be removed from the enzyme by washing the membranes prior to solubilization. Association of the protein factor with the membrane-bound enzyme is promoted by polyethylene glycol or by Ca^{2+}. Cellulose is synthesized intracellularly by a multienzyme complex. Extrusion through pores in the LPS layer results in the aggregation of bundles of cellulose that undergo crystallization by self-assembly at the cell surface.

Many other organisms are known to produce layers of material external to the usual cell surface layers. These external layers may contain carbohydrates, peptides, proteins, lipids, or combinations of these substances (glycoproteins, lipoproteins, lipopolysaccharides, etc.). These external substances contribute to the adherence of the organism to solid surfaces when they are growing in their natural environment. Surface layers may also provide protection to the organism in hostile environments. Studies of organisms grown in artificial media in the laboratory may bear very little relationship to what occurs in the natural environment. In many cases organisms grown in the laboratory lose their ability to produce the protective outer layers observed during growth in their natural environment, apparently because these materials provide no selective advantage in the artificial environment.

ORGANS OF LOCOMOTION

Many microorganisms are motile; that is, they are able to move about in a concerted manner in an aqueous environment. Motile bacteria and protozoa possess flagella or cilia. Some organisms are able to move about by means of pseudopodal or ameboid movement. Spirochetes swim by virtue of a screwlike movement that is effective even in a viscous (semisolid) medium. Motile organisms are placed at an advantage in seeking food, avoiding toxic chemicals or predators, and in colonizing favorable ecological niches.

Motile organisms have a common feature—energy is required for locomotion. Eukaryotes use ATP to activate the contractile movement of complex cilia or flagella. In prokaryotes, proton motive force (PMF) activates a motor that turns the flagellum like the propellor of a boat.

Cilia and Flagella of Eukaryotes

Cilia and flagella of eukaryotic microorganisms consist of paired tubules arranged in a consistent pattern: nine peripheral pairs and one central pair (Fig. 6-28). This 9 + 2 bundle, termed the **axoneme,** develops bending waves as a result of sliding between the doublet tubules produced by ATP-driven cycling of the **dynein** arms that extend between the tubules (Fig. 6-29). In normally beating flagellar axonemes, the ATP-driven cycling of the dynein arms generates longitudinal shear forces between the outer doublet tubules that result in a localized sliding that is opposed by resistive components that convert it into a transverse bending moment. The ATP cyclic cross-bridging of the dynein arms appears to involve the interaction of the projections on the globular heads with tubulin subunits in the tubule of the adjacent doublet (Fig. 6-30). The regulatory mechanism responsible for the control of interdoublet sliding is not known with certainty, but ultrastructural studies suggest that interactions between the radial spokes attached to each doublet tubule and the central complex of the axoneme may be involved in some manner.

Bacterial Flagella

Bacterial flagella consist of three major components: a basal body, a proximal hook, and a long helical filament (Fig. 6-31). The filament and hook are hollow cylindrical structures that extend from the cell surface. The basal body is embedded in the cell surface and contains a motor that turns the helical flagellum to drive the cell through a liquid environment as a propellor drives a boat. This motor is reversible and is under the control of a chemotactic system that enables the cell to move toward favorable or away from unfavorable environmental conditions.

Flagellar assembly and function require 40 or more genes (Table 6-5). The biosynthetic processing in *E. coli* and *S. typhimurium* occurs in a sequence that starts with the assembly of the M and S rings under the influence of regulatory proteins FlhC and FlhD (Fig. 6-32). This step is followed by the addition of switch proteins FliG, FliM, and FliN, assembly of the export apparatus containing FlhA, FliH, and FliI, and formation of the proximal and distal rods containing FlgB, FlgC, FlgF, and FlgG. Mutants defective in any of the genes coding for these proteins result in recovery of the MS ring only. This has been interpreted to mean that intermediary stages of rod assembly are relatively unstable and, as a result, incomplete rod stages have not been found. The next simplest structure isolated is the "rivet" portion of the distal rod that contains FliF, FliE, and four rod proteins (FlgB, FlgC, FlgF, and FlgG). The P and L rings (outer cylinder) contain FlgI and FlgH. These proteins appear to be the only flagellar components exported by the primary cellular export pathway, as shown by the presence of signal peptides. The next stages involve formation of the hook under the influence of *flgD* and addition of the hook proteins FlgE and FliK. Addition of the first hook-filament junction protein (FlgK) and the second hook-filament junction protein (FlgL) followed by the filament capping protein (FliD) complete the formation of components necessary for filament assembly. The addition of FliC (filament protein or **flagellin**) completes the biogenic process of flagellar assembly.

The filament portion of common bacterial flagella is composed of 100% protein (**flagellin**). Refined electron density maps show that the filament is composed of densely packed subunits (Fig. 6-33). The inner part of the filament exhibits strong connections between subunits forming a dense core of 30–50 Å radius. Flagellar growth occurs at

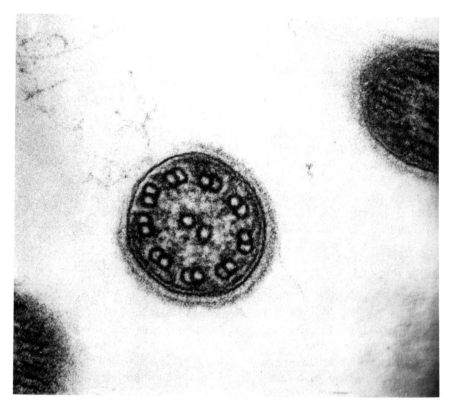

Fig. 6-28. Electron micrograph of a transverse section through a flagellum of the green alga *Chlamydomonas reinhardi* showing the characteristic "9 + 2" arrangement of one axial and nine peripheral pairs of fibrils. Unpublished photograph courtesy of Dr. M. Jacobs and Dr. Anne C. McVittie, East Malling Research Station, Maidstone, Kent, England.

the distal end of the filament. The hook portion of the flagellar structure is composed of at least three types of protein. These **hook-associated proteins (HAPS)** are encoded by three structural genes, *flgK*, *fliD*, and *flgL*, which code for HAP1, HAP2, and HAP3. The HAP2 protein forms a cap structure that presumably serves to prevent arriving flagellin subunits from diffusing away before polymerization occurs. Mutants of *S. typhimurium* lacking the structural genes for HAPs excrete flagellin into the surrounding medium. Presumably the precursor units move through the hollow core of the filament. A refined electron density map shows well-defined subunit packing in the core region and central hole of approximately 60-Å diameter, which is large enough to accommodate the folded flagellin molecule. Most of the flagellar components appear to be exported by a flagellum-specific pathway. Only the P and L proteins of the outer ring seem to be transported by the signal peptide-dependent pathway.

Certain components of the flagellar apparatus, for example, the Mot proteins (MotA and MotB), are not needed for the assembly process. The Mot proteins are associated with the cytoplasmic membrane and surround the MS ring component (Fig. 6-34). Both MotA and MotB are components of torque generators that enable motor rotation. The MotA protein is involved in conducting protons across the cytoplasmic membrane. Al-

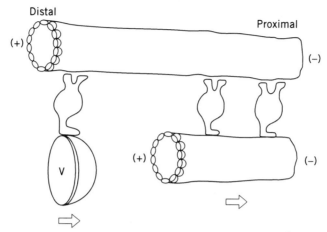

Fig. 6-29. Schematic representation of the dynein-mediated transport along a microtubule of an axoplasmic vesicle, V (left) and of a second microtubule (right). In both cases, the motion directed is toward the minus (proximal) end of the microtube that the dynein is moving along. From Gibbons, I. R. 1988. *J. Biol. Chem.* **263**:15837–15840.

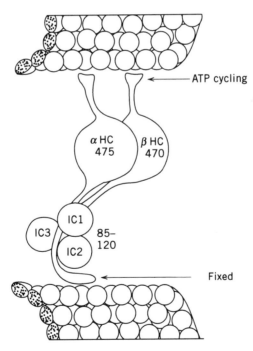

Fig. 6-30. Schematic representation of the subunit organization of a two-headed outer-arm dynein cross-bridging two doublet microtubules of a flagellar axoneme. The ATP-cycling region of the α and β heavy chains is associated with the B tubule of one doublet (top), and the fixed structural attachment of the dynein is associated with the A tubule of the adjacent doublet (bottom). Numbers represent approximate molecular weight in thousands. IC = intermediate chain. From Gibbons, I. R. 1988. *J. Biol. Chem.* **263**:15837–15840.

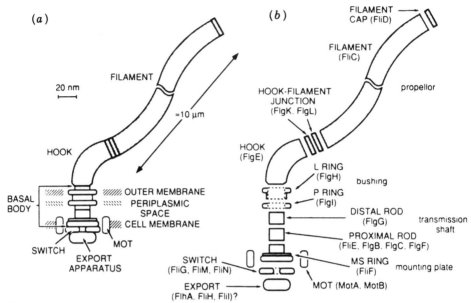

Fig. 6-31. (*a*) **General structure of the flagellum of a gram-negative bacterium, such as *E. coli* or *S. typhimurium*.** The filament–hook–basal-body complex has been isolated and extensively characterized. The location of the export apparatus has not been demonstrated. (*b*) **An exploded diagram of the flagellum showing the substructures and the proteins from which they are constructed.** The FliF protein is responsible for the M-ring feature, S-ring feature, and collar feature of the substructure shown, which is collectively termed the MS ring. The location of FliE with respect to the MS ring and rod, and the order of the FlgB, FlgC, and FlgF proteins within the proximal rod, are not known. From Macnab, R. M. 1992. *Annu. Rev. Genet.* **26:**131–158.

though MotB is associated with the cytoplasmic membrane, most of the protein is located in the periplasmic space. It has been suggested that MotB is a linker that fastens MotA and other components of the torque-generating machinery to the cell wall. This linkage keeps the motor components stationary with respect to the rest of the cell. The MotA and MotB proteins, in conjunction with the flagellar switch proteins FliG, FliM, and FliN, constitute the flagellar motor. The FliG, FliM, and FliN proteins affect motor rotation and the switch from CW to CCW rotation. Additional aspects of the operation of the switch will be discussed later in conjunction with the subject of chemotaxis.

As discussed in some detail in Chapter 4, flagellar phase variation in *S. typhimurium* refers to the alternate expression of the genes for H1 (*fliC*) and H2 (*fljB*) flagellin and results in the production of one of two antigenically distinct proteins. The change in expression from one flagellin gene to the other results from the reversible inversion of a 996 bp chromosomal segment of DNA (Fig. 6-35). Contained within this segment are the coding region for Hin recombinase, which mediates the inversion reaction, and a promoter for the operon encoding one of the two flagellins located immediately adjacent to the invertible segment. Hin mediates recombination between the *hixL* and *hixR* recombination sites, which flank the invertible segment. One orientation of this segment allows expression of the *fljB* gene, encoding H2 flagellin, and the *fljA* gene, encoding rh1, a repressor of the *fliC* gene (H1 flagellin). When inversion of the fragment occurs, neither

TABLE 6-5. Flagellar, Motility, and Chemotaxis Gene Products of *S. typhimurium* and *E. coli* and Their Known or Suspected Functions[a]

Gene Product	Function/Location
Regulatory Proteins	
FlhC, FlhD	Master regulators of the flagellar regulon acting on Class 2 operons
	Transcription initiation (σ) factors?
FliA	Transcription initiation (σ) factor for Classes 3a and 3b operons
FlgM	Anti-FliA (anti-σ) factor. Also known as RflB. Active only when flagellar assembly has not proceeded through completion of the hook
FliS, FliT, FliD?	Repressor of Classes 3a and 3b operons (RflA activity)
FljA	Repressor of *fliC* operon
Hin	Site-specific recombinase, affecting *fljB* promoter
Proteins Involved in the Assembly Process	
FlhA, FliH, FliI	Export of flagellar proteins? FlhA is homolog of various virulence factors
	FliI is homolog of the catalytic subunit of the F_0F_1 ATPase
FlgA	Assembly of basal-body periplasmic P ring
FlgD	Initiation of hook assembly
FliK	Control of hook length
FliB	Methylation of lysine residues on the filament protein, flagellin; function of this modification unknown
Flagellar Structural Components	
FliG, FliM, FliN	Components of flagellar switch, enabling rotation and determining its direction (CCW vs CW)
MotA, MotB	Enable motor rotation. No effect on switching
FliF	Basal-body MS (membrane and supramembrane) ring and collar
FliE	Basal-body component, possibly at (MS-ring)-rod junction
FlgB, FlgC, FlgF	Cell-proximal portion of basal-body rod
FlgG	Cell-distal portion of basal-body rod
FlgI	Basal-body periplasmic P ring
FlgH	Basal-body outer-membrane L (LPS layer) ring
FlgE	Hook
FlgK, FlgL	Hook-filament junction

TABLE 6-5. Continued

Gene Product	Function/Location
FliC, FljB	Filament (flagellin protein). FljB (found in *S. typhimurium* only) is an alternative, serotypically distinct, flagellin
FliD	Filament cap, enabling filament assembly

Flagellar Proteins of Unknown Function

FlgJ, FlhB, FlhE, FliJ, FliL,
  FliO, FliP, FliQ, FliR

Sensory Transduction Components

CheA	CheY and CheB kinase
CheZ	CheY phosphatase. Antagonist of CheY as switch regulator
CheW	Positive regulator of CheA activity
CheY	Switch regulator, placing it in CW state
CheR	Methylation of receptors. Sensory adaptation
CheB	Demethylation of receptors. Sensory adaptation

Chemoreceptors

Tar (aspartate); Tap (peptides, *E. coli* only); Trg (ribose); Tsr (serine); sugar receptors of the phosphotoansferase uptake system.

*a*From Macnab, R. M. 1992. *Annu. Rev. Genet.* **26**:131–158.

the *fljB* gene nor the *fljA* gene is expressed and the gene for H1 flagellin, *fliC*, located 16 min away on the *Salmonella* chromosome, is then expressed from the derepressed *fliC* promoter.

Regulation of the flagellar regulon is under the control of a system referred to as the sigma cascade of flagellar gene expression. As shown in Figure 6-36, the flagellar operons fall into a hierarchy of classes of expression. Expression of an operon in a given class is necessary for expression of operons in a lower class.

Chemotaxis

Many microorganisms, including *E. coli* and *S. typhimurium*, have a wide range of metabolic and physiological versatility enabling to grow and survive in diverse environments. This versatility is employed to greatest advantage when a motile organism is able to direct its movement toward more favorable or away from unfavorable environments. This capability is called **chemotaxis**. The rotation of the flagellar rotor is reversible and can turn in a clockwise (**CW**) or counterclockwise (**CCW**) direction. When all

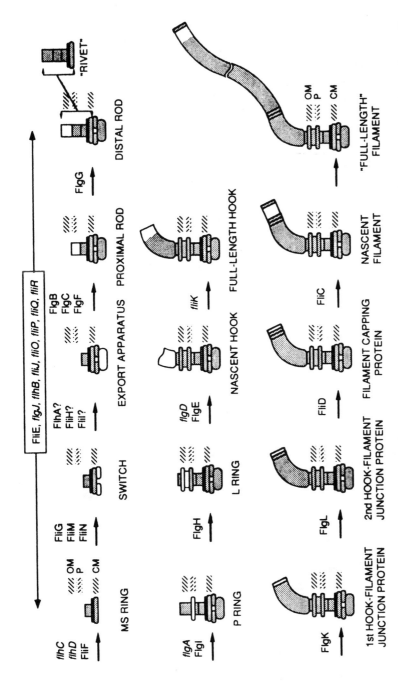

Fig. 6-32. The biogenesis of the bacterial flagellum. Succeeding stages of increasingly complex structure are shown along with the genes needed for each stage. Each incremental feature is shown in white, with all preceding structures shown with stippling. The structure known as the "rivet" is, after the MS ring, the simplest substructure that has been detected by electron microscopy, and has lost the switch and export complex and perhaps other structures during the isolation procedure. The filament does not have a well-defined mature length, "full length" simply implies that the filament is long enough to function in propulsion. Where the gene product is known to incorporate into structure, its symbol is given in roman letters; where this is not known, the gene symbol is given in italics. The gene product and genes indicated in the box are needed at approximately the stages shown, and certainly prior to the assembly of the distal rod. OM = outer membrane; P = periplasmic space and peptidoglycan layer; CM = cell membrane. From Macnab, R. M. 1992, *Annu. Rev. Genet.* **26**:131–158.

285

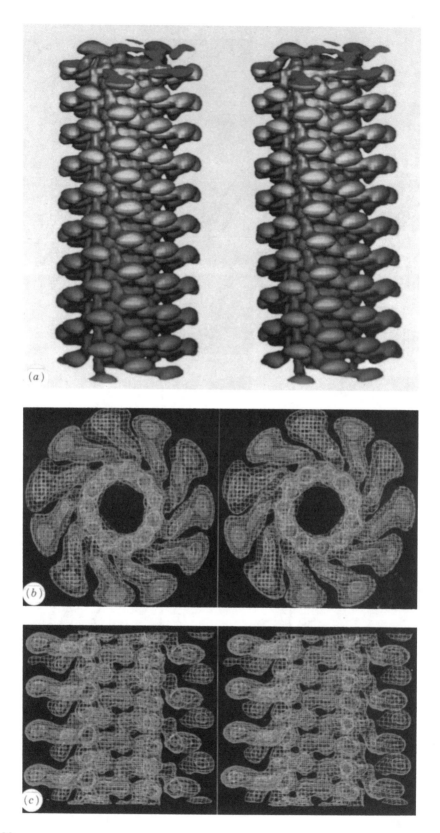

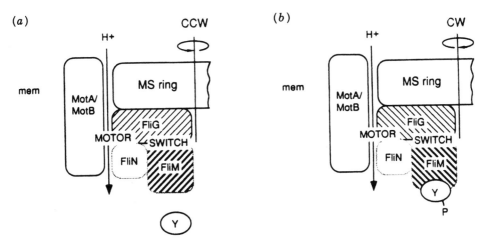

Fig. 6-34. Model for the structure of the flagellar switch and its function. The FliG has been shown to be located in the cytoplasmic face of the MS ring. Both MotA and MotB are integral membrane (mem) proteins that are believed to surround the MS ring. The FliM and FliN are thought on genetic grounds to interact with FliG; the locations shown for them are speculative but are consistent with available evidence. The FliM is postulated to be the target for CheY (Y), based on the large number of positions within its sequence that affect the counterclockwise (CCW) versus clockwise (CW) state of the switch. Both FliM and FliG are postulated to constitute the switching function, while FliG and FliN (perhaps with MotA and MotB) constitute the motor function. (*a*) When CheY is not bound to FliM, both fliM and FliG are in their CCW states and the motor rotates CCW. (*b*) When CheY is phosporylated (Y-P) and binds to FliM, it places FliM in its CW state; this in turn causes FliG to change to the CW state, and so the motor rotates CW. From Irikura, V. M. et al. *J. Bacteriol.* **175:**802–810, 1993.

the motors in a single cell rotate in a CCW direction the flagella sweep around the cell in a common axis forming a concerted bundle (Fig. 6-37A). The result is a smooth swimming motion of the cell. However, if the motors reverse and turn in a CW direction, the flagella disperse (Fig. 6-37B) and the cell undergoes a tumbling motion. This run-and-tumble strategy does not hold for all bacteria. For example, in *Halobacterium* spp., CCW rotation causes movement in one direction and CW movement causes movement in the opposite or reverse direction.

Chemotaxis is the result of regulating the switch between CW and CCW rotation. A cell placed in a medium containing a unidirectional gradient of attractant, such as glucose, will suppress the onset of CW rotation if it is moving toward the attractant but will tumble more often to reorient its direction if it is moving away from the attractant. The ultimate direction of movement will be toward the attractant (see Fig. 6-38).

An important question in biology is how bacteria can interface with their environment

Fig. 6-33. Structure of the core and central channel of bacterial flagella. (*a*) Shaded solid model of the overall surface structure of the flagellar filament. (*b*) Cross-sectional view of the electron density map of the flagellar filament. (*c*) Longitudinal section. From Namba, K., I. Yamashita, and F. Vonderviszt, *Nature (London)* **342:**648–654, 1989.

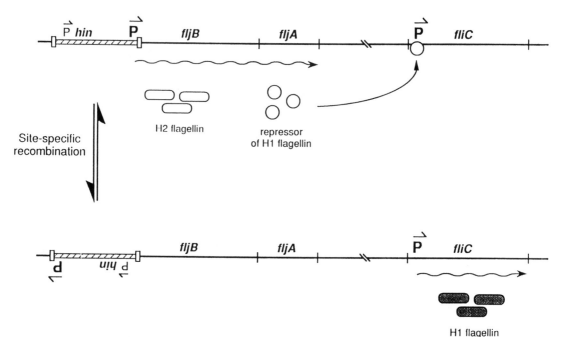

Fig. 6-35. Flagellar phase variation in *S. typhimurium*. This diagram shows the reversible inversion of the promoter of the *fljBA* operon that encodes H2 flagellin (*fljB*) and repressor (*fljA*) of the H1 flagellin structural gene (*fliC*). When this operon is transcribed (top) H2 flagellin is produced and H1 is not. Upon inversion, the *fljBA* operon is not expressed (bottom); H2 flagellin and repressor of *fliC* are not synthesized so that only H1 flagellin is produced. From Gillen, K. L. and K. T. Hughes. *J. Bacteriol.* **173**:2301–2310, 1991.

and sense various attractants and repellents. A first step toward understanding chemotaxis is to view the time course (kinetic) response of a bacterium to the pulse addition of an attractant. As illustrated in Figure 6-39, there is a latency phase (0.2 s) followed by an excitation phase in which CCW flagellar rotation (smooth swimming) is favored. The adaptation phase occurs when the degree of excitation decreases. The phase may last from seconds to minutes. Basically, the cell recognizes the lack of further change in the concentration of attractant and it now requires a higher concentration of attractant to trigger another excitation phase.

In *E. coli*, the mechanism regulating the CW/CCW switch involves a family of membrane receptors and six cytoplasmic proteins called Che proteins that transduce information from the receptors to the motor. The membrane receptors (also called methyl-accepting chemotaxis proteins) are homodimers that span the membrane. The sensory domain lies in the periplasm with the signaling domain residing in the cytoplasm. The signal transducing domain of the membrane receptor proteins interact with CheW and CheA, as depicted in Figure 6-40A. The CheA protein has an autokinase activity that is stimulated 100-fold when it is associated with the receptor and CheW. Binding of attractant to the receptor inhibits formation of this ternary complex, and so diminishes autophosphorylation of CheA (Fig. 6-40B). The CheA-PO$_4$ will transfer its phosphate to CheY (Fig. 6-40A). Then CheY-PO$_4$ interacts with the flagellar motor to stimulate CW

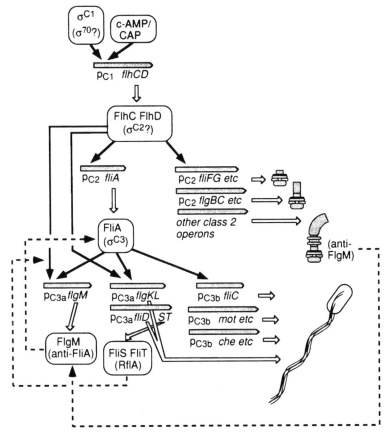

Fig. 6-36. Regulation of the flagellar gene system. The flagellar operons fall into a hierarchy of classes of expression. Operons are indicated by stippled bars, translation to products by open white arrows, positive regulatory controls by solid arrows, and negative regulatory controls by dashed arrows. Promoters of different classes are indicated as p_{C1} (promoter of Class 1 operon), and so on, and σ factors for transcription initiation as $σ^{C1}$, and so on. The negative feedback control called RflA probably involves FliS and FliT but could also involve FliD; the target for the feedback is not known. From Macnab, R. M. 1992. Genetics and biogenesis of bacterial flagella. *Annu. Rev. Genet.* **26**:131–158.

rotation and thus tumbling behavior. Dephosphorylation of CheY-PO$_4$ is stimulated by the CheZ product (Fig. 6-40B). This removes the tumble signal. So, in sum, binding of attractant to the membrane receptor causes a decrease in production of CheA-PO$_4$, which translates into less CheY-PO$_4$ and thus a longer period of smooth swimming.

How does the cell know that it should continue to move toward higher concentrations of attractant? It does this by adapting or densensitizing itself to the most recently encountered concentration of attractant. Chemotactic "memory" involves methylation of the membrane receptor following excitation, hence the designation of these proteins as methyl-accepting chemotaxis proteins (MCPs). Eventually, increased levels of methylation occur after attractant binding. Methylated MCPs favor the formation of an active receptor–kinase complex by increasing MCP affinity for CheW/CheA (Fig. 6-40a). This

(*a*)

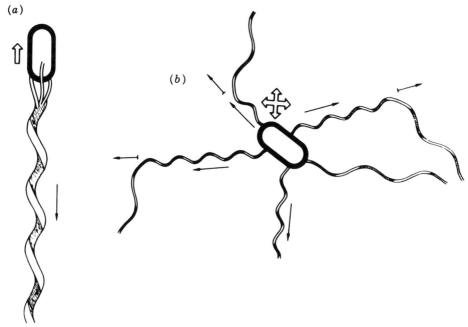

(*b*)

Fig. 6-37. Behavior of flagella of *E. coli* and *S. typhimurium* during swimming and tumbling. (*a*) Swimming: With the motors in CCW rotation, the flagellar filaments form a propulsive bundle, with wave propagation proceeding from proximal to distal. (*b*) Tumbling: CW rotation of the motors causes the normal left-handed form of the filament to undergo a polymorphic transition to the curly right-handed form. While the filaments are undergoing such transitions, the bundle is dispersed and the cell body moves chaotically, end over end; that is, the cell tumbles and reorients randomly, ready for the next swimming interval. Bar = 5 μm. From Kahn, S. 1978. *Proc. Natl. Acad. Sci. USA* **75**:4150–4154; and Macnab, R. M. and M. K. Ornston. 1977. *J. Mol. Biol.* **112**: 1–30.

ultimately leads to increased levels of CheY-PO$_4$ and more tumbling behavior. The cell has adapted to that concentration of attractant. The affinity of MCP for attractant has decreased and higher levels of attractant are then needed in order to bind the membrane receptor and once again disrupt the receptor–kinase complex. Consequently, if the cell has entered an area of higher attractant concentration at the end of tumble, further tumble will be suppressed (not eliminated).

Two proteins are involved with adding and removing methyl groups on the MCP receptors. A methyltransferase (CheR) uses SAM as a methyl donor to methylate specific glutamate residues in the receptor proteins (Fig. 6-40B). Attractant binding to the receptor exposes these glutamyl side chains to CheR, which facilitates their methylation. There is also a methylesterase (CheB) that can remove the methyl groups from the membrane receptors. The CheB methylesterase is part of a feedback loop that ''resets'' the receptor for excitation especially if the organism has moved toward a lower attractant concentration. The amino-terminal end of CheB is homologous with CheY and will compete with CheY for CheA-PO$_4$. The phosphorylation of CheB activates the esterase activity (Fig. 6-40A). Therefore, the increased kinase activity that occurs following reassociation of CheW + CheA with the receptor will also increase esterase activity facilitating removal

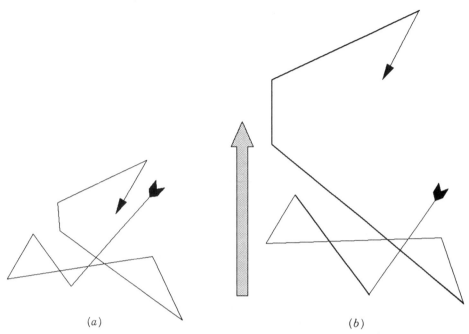

(a) (b)

Fig. 6-38. Idealized trajectory of a swimming cell of *E. coli* or *S. typhimurium* (*a*) in an isotropic medium and (*b*) in a unidirectional gradient of attractant (open arrow). In either case, the cell swims in a straight line, randomizes its direction by tumbling, swims again, tumbles, and so on, yielding a three-dimensional random walk. The effect of an attractant gradient is to extend (by tumble suppression) the mean duration of swimming segments in an up-gradient direction (heavy lines); segments in the down-gradient direction (fine lines) are not appreciably shortened. To illustrate this, the cells (*a*) and (*b*) start at the same point (arrow tail) and execute the same number of tumbles with the same resulting directional changes, but the cell (*b*) in the gradient has up-gradient segments extended. It, therefore, displays migration in the gradient direction by executing a time-biased random walk. From Macnab, R. M. 1979. Chemotaxis in bacteria. In W. Hiaupt and H. C. Feinleib (Eds.). *Encyclopedia of Plant Physiology*, New Series, Vol. 7. Springer-Verlag, Berlin, pp. 310–334.

of methyl groups. This allows a lower concentration of attractant to bind and destabilize the MCP–kinase complex.

Swarming Phenomenon

Some motile gram-negative bacteria that display characteristic swimming and chemotactic behavior when grown in liquid media undergo a differentiation into a swarming type of motility when grown on an agar surface. *Proteus mirabilis, Proteus vulgaris, Vibrio parahaemolyticus, Serratia marcescens*, and some strains of *Bacillus* and *Clostridium* are examples of organisms that display the swarming phenomenon—radially spreading colonies on solid or viscous medium (Fig. 6-41).

When grown in liquid medium, *V. parahaemolyticus* swims by means of a single sheathed polar flagellum that is structurally and functionally similar to the flagella of other gram-negative bacteria. However, on solid surfaces the cell develops into a swarmer

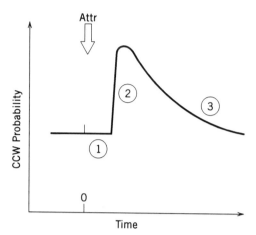

Fig. 6-39. Time course of the response of a bacterium to temporal stimulation by the sudden addition of an attractant at time zero. After an initial latency phase (1) of about 0.2 s, the excitation phase (2) manifests itself as a rapid increase in the probability that the flagellar motors will be in the CCW versus the CW rotational state. The degree of excitation progressively decreases during the adaptation phase (3), which can last from seconds to minutes, depending on the magnitude of the stimulus. From Macnab, R. M. 1985. Transmembrane signalling in bacterial chemotaxis. In Cohen, P. and M. Housley (Eds.). *Molecular Mechanisms of Transmembrane Signalling.* Elsevier, Amsterdam, The Netherlands, pp. 455–487.

cell that is highly elongated and produces extensive lateral flagella (Laf). The polar flagellar system (Fla) is expressed constitutively, whereas the Laf system is expressed only when the organism is grown on a solid surface. Lateral flagella are unsheathed and display a different subunit composition from polar flagella. The gene systems controlling the two distinct motility systems are large, each composed of 40 or more genes without overlap except for shared chemotaxis components. Genes belonging to the Laf system have been identified and ordered in a hierarchical scheme of gene control. These genes encode structural components of the flagellum and the motor and a σ factor of RNA polymerase important for directing swarmer cell development. One gene (*lafA*) codes for lateral flagellin. Other genes required for swarming but not for swimming have been identified by gene replacement mutagenesis. Both types of flagella are driven by reversible motors embedded in the cytoplasmic membrane. However, as compared with the systems of other gram-negative bacteria, the polar flagellar motor of *V. parahaemolyticus* is powered by sodium-motive force, whereas the lateral flagellar motors are driven by proton-motive force. The two independent flagellar organelles appear to be directed by a common chemosensory control system.

Proteus mirabilis is a motile gram-negative bacterium that produces characteristic peritrichous flagella when grown in liquid media. When transferred to solid media the cells begin to elongate, the normal septation mechanism is inhibited, and giant swarmer cells are produced. The swarmer cells undergo extensive production of new flagella (Fig. 6-42). This imparts the ability of the swarmer cell to move over solid media—swarming motility. By definition, individual swarmer cells cannot "swarm." Swarming is the result of the coordinated effort of groups of differentiated swarmer cells that move about in rafts. The swarming process is cyclic or periodic. During swarming, concentric zones of

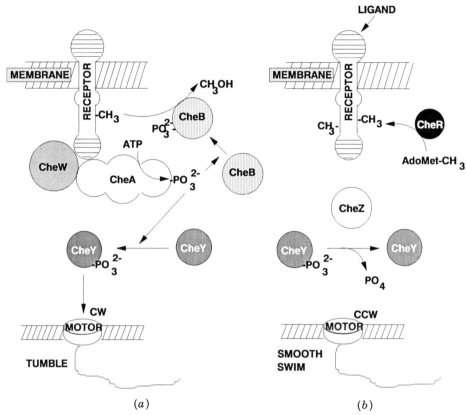

Fig. 6-40. The chemotaxis switch in *E. coli*. (*a*) Tumble mode. Association of CheA and CheW with the membrane receptor stimulates CheA kinase. Transfer of the phosphate to CheY and subsequent interaction of CheY-PO_4 with the flagellar motor causes clockwise rotation and tumble. (*b*) Smooth swimming mode. Binding of attractant to the periplasmic domain of the membrane receptor causes dissociation of the CheW, CheA, MCP complex and a decrease in kinase activity. Dephosphorylated CheY (catalyzed by CheZ) will not interact with the motor and thus allow CCW rotation and smooth swimming. Methylation of the MCP by CheR allows CheA and CheW to reassociate and increase the set point required for ligand binding.

bacterial growth (swarm bands) are formed around a central colony. Swarming occurs for a period of 1–2 h and then stops for several hours before the appearance of a second swarm. At the end of the swarming period the cells slow their activity and divide into short sparsely flagellated cells. Transposon mutagenesis techniques have been used to study the behavior of mutant phenotypes of *P. mirabilis* with altered swarmer cell differentiation. These mutants appear to lack the appropriate intercellular signals required to regulate the cyclic swarming process.

Serratia marcescens also displays two alternate types of motility. As described above for *Proteus* and *Vibrio* species, waves of *S. marcescens* swarmer cells move outward, stop for a few hours, and move out again to form radial bands of growth on solid culture media. A functional chemotaxis system is essential for characteristic swarming motility. At least in this organism, the same flagellar apparatus is utilized for both swimming and

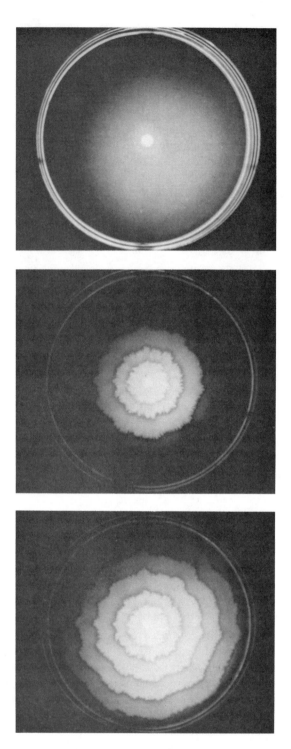

Fig. 6-41. Development of the *P. mirabilis* swarming colony. Multicellular swarming behavior and swarming colony formation were observed by inoculating cells at the center of an agar plate and incubating them at 37 °C. The swarming colony was then photographed at 6 (upper), 24 (middle), and 48 (bottom) hours postinoculation. Consolidation zones are seen as the dark areas separating lighter regions of swarming motility. From Belas, R. 1992. *ASM News* **58**:15–22.

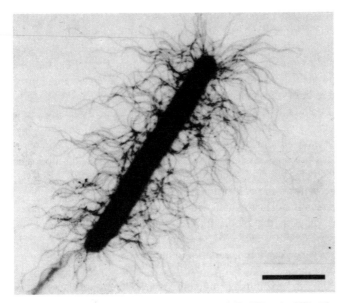

**Fig. 6-42. Electron micrograph of a wild-type swarmer cell of *P. mirabilis.* The cell was taken from the periphery of a swarming colony grown on an agar plate at 37 °C for 6 h and negatively stained with uranyl acetate. Bar = 5 μm. From Belas, R. 1991. *J. Bacteriol.* 173:6279–6288.

swarming motilities. Studies with mutants defective in swarming ability indicate that swarming requires many genes in addition to those required for chemotaxis and flagellar function.

Motility in Spirochetes

Spirochetes are helical in shape. The outermost layer is a membrane sheath that covers a helical cell cylinder. Attached subterminally at each end of the cell cylinder are the **periplasmic flagella** (PFs) (also referred to as axial filaments, axial fibrils, or endoflagella), which are located between the membrane sheath and cell cylinder. Each PF is attached to only one end of the cell. The PFs extend backward toward the middle of the cell and may overlap in the center. The size of the spirochetes, the number of PFs attached at each end, and whether the PFs overlap in the center of the cell are characteristics that vary from species to species.

There are two current models of spirochete motility. A key element to both models, which has not been directly tested, is the assumption that PFs rotate in a manner similar to flagella of rod-shaped bacteria but in the periplasmic region between the cell cylinder and outer-membrane sheath. According to one model, as described for *Spirochaeta aurantia*, rotation of the PFs causes the cell cylinder to rotate in the opposite direction. Both translating and nontranslating cells are observed; the PFs of translating cells are believed to rotate in the same direction, whereas those of nontranslating cells are believed to rotate in opposite directions. The frame of reference used here is a view of the cell from the outside looking from the posterior toward the anterior end.

Members of the family *Leptospiraceae* are believed to swim by a mechanism somewhat different from that proposed for *S. aurantia*. These spirochetes contain one short PF at each end of a right-handed cell cylinder (a right-handed helix spirals CW, and a left-handed helix spirals CCW, moving away from an observer.) The shape of the cell end is determined by the shape and direction of rotation of the PFs; the PFs are assumed to be more rigid than the right-handed cell cylinder. The CCW rotation of a PF causes the anterior end to be spiral shaped and the posterior end to be hook shaped. In translating cells, the PFs at both ends are believed to rotate in the same direction. The CCW rotation of the anterior PF causes the generation of a gyrating, spiral-shaped wave (i.e., propagating a helical wave motion without necessarily rotating, as a bent wire rotating in a rubber tube). This wave is sufficient to propel the cells forward in a pure liquid medium. The right-handed cell cylinder concomitantly rolls around the PFs in the opposite direction (CW), which allows the cell to literally screw through a gellike medium with little slippage. Nontranslating cells are believed to have the PFs rotating in opposite directions. Recent evidence supports this model for the family *Leptospiraceae* and *Treponema phagedenis*. However, *T. phagedenis* differs from the famliy *Leptospiraceae* in that *T. phagedenis* translates very poorly in a pure liquid medium and thus requires a gellike medium for directed movement.

Spirochaeta aurantia exhibits chemotactic behavior and has been used as a model for studies of chemotaxis in spirochetes. Methyl-accepting proteins have been identified and are remarkably similar to the methyl-accepting chemotaxis proteins of *E. coli*. Chemotactic signaling in *E. coli* does not appear to involve membrane potential. In contrast, the *S. aurantia* chemosensory mechanism appears to involve membrane potential in some manner.

Gliding Motility

Members of the Myxobacterales and Cytophagales are commonly referred to as the gliding bacteria because of their unusual manner of gliding on solid surfaces without the aid of flagella or other obvious external organs of locomotion. *Cytophaga*, *Myxococcus*, and *Flexibacter* are among the prominent genera that display this unusual type of motility. Certatin genera of phototrophic cyanobacteria also use gliding as their means of locomotion. Only in a few cases have there been any reports of organized structures extending beyond the cell envelope that might be involved in the locomotion of these organisms. In *Flexibacter polymorphus* short filaments protrude from goblet-shaped structures in the walls of the organism. Some myxobacteria produce fimbriae, but they appear to be associated with cell-to-cell contact rather than with surface locomotion.

Cytophaga johnsonae cells move in groups, known as rafts, away from the point of inoculation to form thin, veil-like colonies with feathery edges. Colony spreading can be visualized on solid media. On solid media colonies of motile cells spread much faster than can be accounted for by growth alone. With the aid of a microscope, raft movement can be observed at the edge of the colony. Motility in this organism has been correlated with the ability to move polystyrene-latex beads over the cell surface. Mutants lacking this trait are all nongliding. Secretion of extracellular slime has been shown to mediate surface adhesion during gliding motility. Slime is often deposited on slime tracks on agar surfaces over which cells move, and these tracks serve as preferred paths of movement for other cells. Extracellular material appears to be secreted ahead and to the sides of the leading organisms in a mass of cells indicating that the cells are moving through the secreted slime. Some gliding motility mutants of *C. johnsonae* have been shown to be defective in the production of sulfonolipids that are found in the outer membranes. Although it is obvious that sulfonolipids are necessary for gliding, the manner in which these compounds function in surface translocation is not known. Membrane proteins associated with gliding motility have been identified by comparative studies with nonmotile mutants.

Vegetative cells of *Myxococcus xanthus* undergo gliding movement, traveling in coordinated rafts or swarms. As in *C. johnsonae*, no obvious organs of locomotion have been identified in this organism. Gliding motility by vegetative cells is controlled by at least two systems of genes. The social (S) system is required for the rafts of peninsulas of cells to move out from the edge of a vegetative swarm. At least 10 genes affecting S motility have been identified. Social motility may be mediated in part by pili, which may touch the surface of adjacent cells and stimulate motility through contact. Cell interactions appear to be required for S motility, because cells guided by S motility are nonmotile when they are separated from each other. The **adventurous** (A) system is required for single cells to move out from the edge of a vegetative swarm. At least 25 genes for A motility, including those for cell surface LPS, have been detected by screening strains for defects in motility. One class of mutants defective in A motility is called the contact-stimulated gliding (*cgl*) class. Mutants in this class become transiently A motile after contact with wild-type cells. There are five complementation groups of *cgl* mutants defective for A motility: *cglB*, $-C$, $-D$, $-E$, and $-F$. Mixing cells from different groups restores A motility. The fact that complementation of mutants is efficient suggests that stimulation for A motility may occur normally between wild-type cells indicating that contacts or signals between neighboring cells may be important in regulating A motility during swarming and possibly during fruiting body formation.

Motility defects caused by **frizzy** (*frz*) mutations affect the frequency of reversal in the direction of gliding. These mutants produce frizzy filaments of aggregating cells during development, which would be characteristic of cells unable to migrate to centers of aggregation. The *frz* genes show striking homology to genes in a pathway for enterobacterial chemotaxis. Mutations in *frz* could block a chemotactic signal transduction pathway that may direct cells to centers of aggregation. Motility may be required not only to bring cells into aggregates, but also to bring cells into contact for cell-to-cell signaling.

PILI OR FIMBRIAE

Bacteria from many different genera produce a multiplicity of nonflagellar, filamentous, proteinaceous appendages called **pili** or **fimbriae**. Conjugative or **F** pili are involved in the mating process. The F pili have been described in conjunction with the discussion of transfer of genetic information in bacteria (Chapter 3). A number of other types of pili, sometimes referred to as common or generalized pili, are also produced by a variety of bacteria. Pili play a role in adherence of microorganisms to surfaces or to specific receptors on eukaryotic cells aiding in the colonization of these ecological niches.

Members of the *Enterobacteriaceae* produce a wide assortment of pili or fimbriae (Table 6-6). In *E. coli* the most prevalent class of fimbriae are of Type 1 encoded by genes *fimA* through *fimH* located at 98 min on the chromosomal map. The major fimbrial subunit is a 17-kDa polypeptide encoded by *fimA*. Type 1 pili are characterized by their ability to mediate **mannose-sensitive** adherence to eukaryotic cells. Expression of Type

TABLE 6-6. Some Pili or Fimbriae Produced by
Enterobacteriaceae

Pilus Type	Properties
1	Encoded by genes (*fimA-fimH*) clustered at 98 min on *E. coli* genetic map; 17-kDa pilin subunits encoded by *fimA*; Agglutination of fowl or guinea pig erythrocytes inhibited by α-D-mannose (mannose sensitive); produced by *Escherichia, Klebsiella, Shigella, Salmonella, Citrobacter, Enterobacter, Edwardsiella, Hafnia, Serratia, Providencia*
2	Nonhemagglutinating; observed in *Escherichia, Salmonella* strains; May be Type 1 antigenic variants
3	Mediate mannose-resistant agglutination of tannic acid treated animal erthyrocytes; Expression in *K. pneumoniae* involves at least six genes (*mrkA-mrkF*). Produced by *Klebsiella, Salmonella, Yersinia, Proteus, Providencia*
P Pap	Produced by uropathogenic or pyelonephritis-associated (Pap) *E. coli*. Mediate binding to digalactoside receptors (galactose sensitive) in cells of urinary tract epithelium. Eleven P pilin (*pap*) genes involved in biosynthesis and expression of P pili; PapE is adhesin located at tips of pili. Expression of Pap pili subject to on–off phase variation involving DNA methylation by deoxyadenosine methylase (Dam)
S Sfa	Mediate mannose-resistant hemagglutination; facilitate binding to α-sialic acid-(2–3)-β-D-Gal residues; encoded by *sfa* gene cluster in *E. coli* K1
F1C	Nonhemagglutinating; encoded by six genes in *foc* gene cluster; *focA* encodes major fimbrial subunit; *focC* gene product required for fimbria formation; *focG, focH* encode minor fimbrial subunits; mediate adherence to collecting ducts and distal tubules of human kidney
K88	Produced by porcine enterotoxigenic *E. coli*. At least six plasmid-borne structural genes (*faeC-faeH*) involved
K99	Produced by bovine enterotoxigenic *E. coli*.
F17	Mediate binding to *N*-acetylglucosamine receptors in calf intestinal mucosal cells; F17 gene cluster contains at least four genes

1 fimbriae in *E. coli* exhibits recombination-mediated on–off phase variation. This phase variation correlates with the orientation of a short, invertible 314 bp DNA element (switch) located immediately upstream of *fimA*. The promoter for *fimA* is believed to reside within this invertible element and to direct the transcription of *fimA* when the element is in one orientation (on) but not when the element is in the alternate orientation (off).

Two genes, *fimB* and *fimE*, mapping immediately adjacent to the *fim* invertible element are believed to encode site-specific recombinases. A model for the roles of *fimB* and *fimE* suggests that *fimB* promotes recombination of the *fim* invertible element in both directions, whereas *fimE* promotes recombination primarily from on to off. However, the on orientation of the *fim* invertible element is necessary but not totally sufficient for fimbrial expression since strains in which the invertible element is locked in the on orientation continue to exhibit phase variable expression of Type 1 fimbriae. Other genes whose products affect the invertible element include *pilG*, which encodes the histone-like protein H1, and *himA* and *himD/hip*, which together encode IHF.

Immediately 3' to *fimA* are two genes (*fimC* and *fimD*) whose protein products are required for polymerization of the *fimA* gene product into pili. The *fimD* product is apparently incorporated into the outer membrane and serves as a scaffold or polymerization channel for pilus assembly. Genes encoding minor components of Type 1 fimbriae, *fimF*, *fimG*, and *fimH*, are located distally to *fimA* and encode products similar to pilin in structure. The *fimH* product appears to be the Type 1 pilus adhesion component that interacts directly with the eukaryotic cell receptor and confers mannose sensitivity. It has been suggested that the *fimF* product aids in starting new pili and that the *fimG* product is an inhibitor of pilus polymerization. An additional gene, *lrp*, is also required for normal

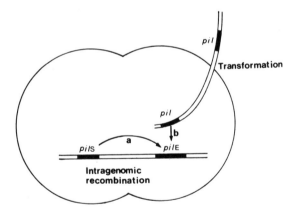

Fig. 6-43. Two routes for generating pilin variation in *N. gonorrhoeae*. (a) Intragenomic recombination and (b) recombination following transformation with exogenous DNA from lysed gonococci: *pilS* = silent pilin sequence; *pilE* = pilin-expression locus complete with promoter; *pil* = any pilin sequence. Note that the incoming donor sequence could be either a second *pilE* locus or any *pilS* locus. Depending on the nature of the *pil* sequence transferred and the outcome of recombination crossovers, the new sequence generated at *pilE* may encode a function and possibly antigenic variable pilin (antigenic variation) or a nonfunctional variant pilin, which is not assembled into pili to effect one form of P$^+$ $\longrightarrow$ P$^-$ phase variation. (Subsequent recombination events with further *pil* sequences can reverse the transition.) From Saunders, J. R. 1989. *Nature (London)* **338:** 622–623.

activity of the *fim* switch. The *lrp* gene product, Lrp, is a site-specific DNA-binding protein that is capable of protecting extended regions of DNA *in vitro* from nuclease digestion. It affects the transcription of many genes, collectively called the leucine-Lrp regulon. It has been suggested that Lrp could influence the *fim* inversion either indirectly by altering the expression of other proteins, such as FimB and FimE or IHF, or directly by participating in the *fim* inversion as an auxiliary factor. The Lrp is known to influence phase variation of Pap fimbriation by blocking the methylation of two *dam* sites in the promotor region of *papAB*. It has been suggested that Lrp is required either for the expression of an additional factor that promotes the *fim* inversion or as a direct participant in the recombination reaction.

The pili of *Neisseria gonorrhoeae* and *N. meningitidis* are important determinants of virulence, as indicated by their ability to mediate adhesion to host cells and to prevent phagocytosis by human polymorphonuclear leukocytes. The major structural subunit of pili in *N. gonorrhoeae*, pilin, is encoded by the chromosomal gene *pilE*. Expression of *pilE* is controlled transcriptionally by the products of two linked chromosomal genes: *pilA* and *pilB*. Production of pili in this organism is subject to on–off phase variation in which the organism can switch from the piliated (P^+) to the nonpiliated (P^-) state. The pilins of this organism are also subject to antigenic variation through the production of altered amino acid sequences in the highly variable C-terminal region. As shown in Figure 6-43, pilin variation can be accomplished by intragenomic recombination or by recombination followed by transformation with exogenous DNA.

REFERENCES

Nucleus

Ball, C. A., R. Osuna, K. C. Ferguson, and R. C. Johnson. 1992. Dramatic changes in Fis levels upon nutrient upshift in *Escherichia coli*. *J. Bacteriol.* **174**:8043–8056.

Barr, G. C., N. N. Bhriain, and C. J. Dorman. 1992. Identification of two new genetically active regions associated with the *osmZ* locus of *Escherichia coli*: Role in regulation of *proU* expression and mutagenic effect at *cya*, the structural gene for adenylate cyclase. *J. Bacteriol.* **174**:998–1006.

Bohrmann, B., W. Villiger, R. Johansen, and E. Kellenberger. 1991. Coralline shape of the bacterial nucleoid after cryofixation. *J. Bacteriol.* **173**:3149–3158.

Brock, T. D. 1988. The bacterial nucleus: A history. *Microbiol. Rev.* **52**:397–411.

Bylund, J. E., M. A. Haines, P. J. Piggot, and M. L. Higgins. 1993. Axial filament formation in *Bacillus subtilis*: Induction of nucleoids of increasing length after addition of chloramphenicol to exponential-phase cultures approaching stationary phase. *J. Bacteriol.* **175**:1886–1890.

Dri, A-M., J. R. Yaniv, and P. L. Moreau. 1991. Inhibition of cell division in *hupAhupB* mutant bacteria lacking HU protein. *J. Bacteriol.* **173**:2852–2863.

Drlica, K. and J. Rouviere-Yaniv. 1987. Histonelike proteins in bacteria. *Microbiol. Rev.* **51**:301–319.

Hobot, J. A., W. Villiger, J. Escaig, M. Maeder, A. Ryter, and E. Kellenberger. 1985. Shape and fine structure of nucleoids observed on sections of ultrarapidly frozen and cryosubstituted bacteria. *J. Bacteriol.* **162**:960–971.

Kawula, T. H. and P. E. Orndorff. 1991. Rapid site-specific DNA inversion in *Escherichia coli* mutants lacking the histonelike protein H—NS. *J. Bacteriol.* **173**:4116–4123.

Kornberg, R. D. and Y. Lorch. 1992. Chromatin structure and transcription. *Annu. Rev. Cell Biol.* **8:**563–587.

Pettijohn, D. E. 1988. Histone-like proteins and bacterial chromosome structure. *J. Biol. Chem.* **263:**12793–12796.

Pettijohn, D. E. and R. R. Sniden. 1985. Structure of the isolated nucleoid. In *Molecular Cytology of E. coli*, Naninga, N. (Ed.). Academic Press, London, pp. 199–226.

Schmid, M. B. 1990. A locus affecting nucleoid segregation in *Salmonella typhimurium. J. Bacteriol.* **172:**5416–5424.

Shellman, V. I. and D. E. Pettijohn. 1991. Introduction of proteins into living bacterial cells: Distribution of labeled HU protein in *Escherichia coli. J. Bacteriol.* **173:**3047–3059.

Takayanagi, S., S. Morimura, H. Kusaoke, Y. Yokoyama, K. Kano, and M. Shioda. 1992. Chromosomal structure of the halophilic archaebacterium *Halobacterium salinarum. J. Bacteriol.* **174:**7207–7216.

Mitochondria

Pon, L. and G. Schatz. 1991. *Biogenesis of yeast mitochondria.* In *The Molecular and Cellular Biology of the Yeast* Saccharomyces: *Vol. I. Genome Dynamics, Protein Synthesis, and Energetics*, J. R. Broach, J. R. Pringle, and E. W. Jones (Eds.). Cold Spring Harbor Press, Cold Spring Harbor, NY, pp. 333–406.

Surface (S) Layers

Beveridge, T. J. and L. L. Graham. 1991. Surface layers of bacteria. *Microbiol. Rev.* **55:**684–705.

Sleytr, E. B. and P. Messner. 1988. Crystalline surface layers in procaryotes. *J. Bacteriol.* **170:** 2891–2897.

Pentapeptide (Murein) Synthesis

Doublet, P., J. van Heijenoort, and D. Mengin-Lecreulx. 1992. Identification of the *Escherichia coli murI* gene, which is required for the biosynthesis of D-glutamic acid, a specific component of bacterial peptidoglycan. *J. Bacteriol.* **174:**5772–5779.

Engel, H., A. J. Smink, L. van Wungaarden, and W. Keck. 1992. Murein-metabolizing enzymes from *Escherichia coli*: Existence of a second lytic transglycosylase. *J. Bacteriol.* **134:**6394–6403.

Ikeda, M., M. Wachi, H. Jung, F. Ishino, and M. Matsuhashi. 1991. The *Escherichia coli mraY* gene encoding UDP-*N*-acetylmuramoylpentapeptide:undecaprenyl-phosphate phospho-*N*-acetyl-muramoylpentapeptide transferase. *J. Bacteriol.* **173:**1021–1026.

Labichinski, H. and H. Maidhof. 1994. Bacterial peptidoglycan: overview and evolving concepts. In *Bacterial Cell Wall.* J.-M. Ghuysen and R. Hackenbeck (Eds.). Elsevier, Amsterdam, The Netherlands, pp. 23–38.

Marquardt, J. L., D. A. Siegele, R. Kolter, and C. T. Walsh. 1992. Cloning and sequencing of *Escherichia coli murZ* and purification of its product, a UDP-*N*-acetylglucosamine enolpyruvate transferase.

van Heijenoort, J. 1994. Biosynthesis of the bacterial peptidoglycan unit. In *Bacterial Cell Wall.* J.-M. Ghuysen and R. Hackenbeck (Eds.). Elsevier, Amsterdam, The Netherlands, pp. 39–54.

Teichoic and Lipoteichoic Acids

Fischer, W. 1994. Lipoteichoic acids and lipoglycan. In *Bacterial Cell Wall.* J.-M. Ghuysen and R. Hackenbeck (Eds.). Elsevier, Amsterdam, The Netherlands, pp. 199–215.

Kaya, S., K. Yokoyama, Y. Araki, and E. Ito. 1984. *N*-acetylmannosaminyl (1 → 4) *N*-acetylglu-cosamine, a linkage unit between glycerol teichoic acid and peptidoglycan in cell walls of several *Bacillus* strains. *J. Bacteriol.* **158**:990–996.

Pooley, H. M. and D. Karamata. 1994. Teichoic acid synthesis in *Bacillus subtilis*: genetic organization and biological roles. In *Bacterial Cell Wall.* J.-M. Ghuysen and R. Hackenbeck (Eds.). Elsevier, Amsterdam, The Netherlands, pp. 187–198.

Sutcliff, I. C. and N. Shaw. 1991. Atypical lipoteichoic acids of gram-positive bacteria. *J. Bacteriol.* **173**:7065–7069.

Outer Membranes

Chen, L. and W. G. Coleman, Jr. 1993. Cloning and characterization of the *Escherichia coli* K-12 *rfa*-2 (*rfaC*) gene, a gene required for lipopolysaccharide inner core synthesis. *J. Bacteriol.* **175**: 2534–2540.

Hancock, R. E. W. 1987. Role of porins in outer membrane permeability. *J. Bacteriol.* **169**:929–933.

Hancock, R. E. W. 1991. Bacterial outer membranes: Evolving concepts. *ASM News* **57**:175–182.

Hancock, R. E. W., D. N. Karunaratne, and C. Bernegger-egli. 1994. In *Bacterial Cell Wall.* J.-M. Ghuysen and R. Hackenbeck (Eds.). Elsevier, Amsterdam, The Netherlands, pp. 263–279.

Kastowsky, M., T. Gutberlet, and H. Bradaczek. 1992. Molecular modelling of the three-dimensional structure and conformational flexibility of bacterial lipopolysaccharide. *J. Bacteriol.* **174**: 4798–4806.

Lew, H. C., P. H. Makela, H.-M. Kuhn, H. Mayer, and H. Nikaido. 1986. Biosynthesis of enterobacterial common antigen requires dTDPglucose pyrophosphorylase determined by a *Salmonella typhimurium rfb* gene and a *Salmonella montivideo rfe* gene. *J. Bacteriol.* **168**:715–721.

MacLachlan, P. R., S. K. Kadam, and K. E. Sanderson. 1991. Cloning, characterization, and DNA sequence of the *rfaLK* region for lipopolysaccharide synthesis in *Salmonella typhimurium* LT2. *J. Bacteriol.* **173**:7151–7163.

Nikaido, H. and M. Vaara. 1985. Molecular basis of bacterial outer membrane permeability. *Microbiol. Rev.* **49**:1–32.

Raetz, C. R. H. 1993. Bacterial endotoxins: Extraordinary lipids that activate eucaryotic signal transduction. *J. Bacteriol.* **175**:5745–5753.

Reeves, P. 1994. Biosynthesis and assembly of lipopolysaccharide. In *Bacterial Cell Wall.* J.-M. Ghuysen and R. Hackenbeck (Eds.). Elsevier, Amsterdam, The Netherlands, pp. 281–317.

Capsules

Beynon, L. M., A. J. Dumanski, R. J. C. McLean, L. L. MacLean, J. C. Richards, and M. B. Perry. 1992. Capsule structure of *Proteus mirabilis* (ATCC 49565). *J. Bacteriol.* **174**:2172–2177.

Finke, A., I. Roberts, G. Boulnois, C. Pzzani, and K. Jann. 1989. Activity of CMP–2-keto-3-deoxyoctulosonic acid synthetase in *Escherichia coli* strains expressing the capsular K5 polysaccharide: implication for K5 polysaccharide biosynthesis. *J. Bacteriol.* **171**:3074–3079.

Jann, K. and B. Jann. 1990. Bacterial capsules. *Curr. Top. Microbiol.* **150**:19–42.

Keenleyside, W. J., P. Jayartne, P. R. MacLachlan, and C. Whitfield. 1992. The *rcsA* gene of *Escherichia coli* 09:K30:H12 is involved in the expression of the serotype specific group I K (capsular antigen. *J. Bacteriol.* **174**:8–16.

Pavelka, M. S. Jr., L. F. Wright, and R. P. Silver. 1992. Identification of two genes, *kpsM* and *kpsT*, in region 3 of the polysialic acid gene cluster of *Escherichia coli* K1. *J. Bacteriol.* **173**: 4603–4610.

Ross, P., R. Mayer, and M. Benziman. 1991. Cellulose biosynthesis and function in bacteria. *Microbiol. Rev.* **55**:35–58.

Vimr, E. R. 1992. Selective synthesis and labeling of the polysialic acid capsule in *Escherichia coli* K1 strains with mutations in *nanA* and *neuB*. *J. Bacteriol.* **174**:6191–6197.

Motility/Organs of Locomotion

Atsumi, T., L. McCarter, and Y. Imae. 1992. Polar and lateral flagellar motors of marine *Vibrio* are driven by different ion-motive forces. *Nature (London)* **355**:182–184.

Belas, R. 1992. The swarming phenomenon of *Proteus mirabilis*. *ASM News* **58**:15–22.

Belas, R., D. Erskine, and D. Flaherty. 1991. *Proteus mirabilis* mutants defective in swarmer cell differentiation and multicellular behavior. *J. Bacteriol.* **173**:6279–6288.

Bourret, R. B., J. F. Hess, K. A. Borkovich, A. H. Pakula, and M. I. Simon. 1989. Protein phosphorylation in chemotaxis and two component regulatory systems in bacteria. *J. Biol. Chem.* **264**:7085–7088.

Charon, N. W., S. F. Goldstein, S. M. Block, K. Curci, J. D. Ruby, J. A. Kreiling, and R. A. Limberger. 1992. Morphology and dynamics of protruding spirochete periplasmic flagella. *J. Bacteriol.* **174**:832–840.

Dreyfus, G., A. W. Williams, I. Kawagishi, and R. M. Macnab. 1993. Genetic and biochemical analysis of *Salmonella typhimurium* FliI, a flagellar protein related to the catalytic subunit of the F_0F_1 ATPase and to virulence proteins of mammalian and plant pathogens. *J. Bacteriol.* **175**:3131–3138.

Fosnaugh, K., and E. P. Greenberg. 1989. Chemotaxis mutants of *Spirochaeta aurantia*. *J. Bacteriol.* **171**:606–611.

Gibbons, I. R. 1988. Dynein ATPases as microtubule motors. *J. Biol. Chem.* **263**:15837–15840.

Gillen, K. L. and K. T. Hughes. 1991. Negative regulatory loci coupling flagellin synthesis to flagellar assembly in *Salmonella typhimurium*. *J. Bacteriol.* **173**:2301–2310.

Godchaux III, W., M. A. Lynes, and E. A. Leadbetter. 1991. Defects in gliding motility in mutants of *Cytophaga johnsonae* lacking a high molecular-weight cell surface polysaccharide. *J. Bacteriol.* **173**:7607–7614.

Heichman, K. A. and R. C. Johnson. 1990. The Hin invertasome: Protein-mediated joining of distant recombination sites at the enhancer. *Science* **249**:511–517.

Iino, T., Y. Komeda, K. Kutsukake, R. M. Macnab, P. Matsumura, J. S. Parkinson, M. I. Simon, and S. Yamaguchi. 1988. New unified nomenclature for the flagellar genes of *Escherichia coli* and *Salmonella typhimurium*. *Microbiol. Rev.* **52**:533–535.

Irikura, V. M., M. Kihara, S. Yamaguchi, H. Sockett, and R. M. Macnab. 1993. *Salmonella typhimurium fliG* and *fliN* mutations causing defects in assembly, rotation, and switching of the flagellar motor. *J. Bacteriol.* **175**:802–820.

Kalos, M. and J. F. Zissler. 1990. Defects in contact-stimulated gliding during aggregation by *Myxococcus xanthus*. *J. Bacteriol.* **172**:6476–6493.

Kihara, M., M. Homma, K. Kutsukaki, and R. M. Macnab. 1989. Flagellar switch of *Salmonella typhimurium*: Gene sequences and defined protein sequences. *J. Bacteriol.* **171**:3247–3257.

Macnab, R. M. 1992. Genetics and biogenesis of bacterial flagella. *Annu. Rev. Genet.* **26**:129–156.

McCarter, L. M. and M. E. Wright. 1993. Identification of genes encoding components of the swarmer cell flagellar motor and propeller and a sigma factor controlling differentiation of *Vibrio parahaemolyticus*. *J. Bacteriol.* **175**:3361–3371.

Namba, K., I. Yamashita, and F. Vonderviszt. 1989. Structure of the core and central channel of bacterial flagella. *Nature (London)* **342**:648–654.

O'Rear, J., L. Alberti, and R. M. Harshey. 1992. Mutations that impair swarming motility in *Serratia marcescens* 274 include but are not limited to those affecting chemotaxis or flagellar function. *J. Bacteriol.* **174**:6125–6137.

Oesterhelt, D. and M. Marwan. 1993. Signal transduction in halobacteria. In *The Biochemistry of Archaea (archaebacteria)*. M. Kates, D. J. Kushner, and A. T. Mathesin (Eds.). Elsevier, Amsterdam, The Netherlands, pp. 173–188.

Pate, J. L. and D. M. DeJong. 1990. Use of nonmotile mutants to identify a set of membrane proteins related to gliding motility in *Cytophaga johnsonae*. *J. Bacteriol.* **172**:3117–3124.

Sar, N., L. McCarter, M. Simon, and M. Silverman. 1990. Chemotactic control of the two flagellar systems of *Vibrio parahaemolyticus*. *J. Bacteriol.* **172**:334–341.

Smith, E. F. and W. S. Sale. 1992. Regulation of dynein-driven microtubule sliding by the radial spokes in flagella. *Science* **257**:1557–1559.

Stock, J. B., A. M. Stock, and J. M. Mottonen. 1990. Signal transduction in bacteria. *Nature (London)* **344**:395–400.

Pili or Fimbriae

Blomfield, I. C., P. J. Calie, K. J. Eberhardt, M. S. McClain, and B. I. Eisenstein. 1993. Lrp stimulates phase variation of type 1 fimbriation in *Escherichia coli* K-12. *J. Bacteriol.* **175**:27–36.

Gibbs, C. P., B.-Y. Reimann, E. Schultz, A. Kaufmann, R. Haas, and T. F. Meyer. 1989. Reassortment of pilin genes in *Neisseria gonorrhoeae* occurs by two distinct mechanisms. *Nature (London)* **338**:655–656.

Kuehn, M. J., H. Heuser, S. Normark, and S. J. Hultgren. 1992. P pili in uropathogenic *E. coli* are composite fibres with distinct fibrillar adhesive tips. *Nature (London)* **356**:252–255.

McClain, M., I. C. Blomfield, and B. I. Eisenstein. 1991. Roles of *fimB* and *fimE* in site-specific DNA inversion associated with phase variation of type 1 fimbriae in *Escherichia coli*. *J. Bacteriol.* **173**:5308–5314.

McClain, M., I. C. Blomfield, K. J. Eberhardt, and B. I. Eisenstein. 1993. Inversion-independent phase variation of type 1 fimbriae in *Escherichia coli*. *J. Bacteriol.* **175**:4335–4344.

Russell, P. W. and P. W. Orndorff. 1991. Lesions in two *Escherichia coli* type 1 pilus genes alter pilus number and length without affecting receptor binding. *J. Bacteriol.* **174**:5923–5935.

Saunders, J. R. 1989. Modulating bacterial virulence. *Nature (London)* **338**:622–623.

Seifert, S. H., R. S. Ajioka, C. Marchal, P. F. Sparling, and M. So. 1988. DNA transformation leads to pilin antigenic variation in *Neisseria gonorrhoeae*. *Nature (London)* **336**:392–395.

CHAPTER 7

CARBOHYDRATE METABOLISM AND ENERGY PRODUCTION

In the study of carbohydrate metabolism, several major factors need to be considered:

1. Pathways used for the degradation of carbohydrates.
2. Mechanisms used for production of biologically useful energy in the form of ATP, that is, whether ATP is generated via
   a. Substrate level phosphorylation
   b. Oxidative phosphorylation
3. Metabolic steps in which reducing activity is generated and used
   a. To reduce pyruvate or other substrates to form fermentative end-products
   b. For biosynthetic reactions requiring reducing action
4. Intermediates of carbohydrate pathways that also serve as biosynthetic precursors for polysaccharides, amino acids, purines, pyrimidines, and other vital cellular constituents
5. Reactions that replenish intermediates of carbohydrate pathways that are also used as biosynthetic intermediates (anapleurotic reactions)
6. Effect of oxygen on metabolic and energy-generating reactions, that is, whether these activities are primarily
   a. Aerobic
   b. Anaerobic
   c. Facultative
7. Metabolic and genetic regulatory systems governing the activity of the various pathways

It should be understood initially that carbohydrates are not the only compounds utilized as sources of energy by microorganisms. A wide variety of other compounds, including fatty acids, lipids, and amino acids can also be utilized as carbon and energy sources by certain microorganisms. Generally, these substances are degraded to compounds that enter the common carbohydrate metabolic scheme at some point.

GLYCOLYTIC PATHWAYS

Fructose Bisphosphate Aldolase (Embden–Meyerhof–Parnas) Pathway

One of the earliest pathways to be investigated was the alcoholic fermentation system of yeast. Yeasts ferment glucose to form carbon dioxide (CO_2) and ethanol (C_2H_5OH). Early chemists considered that the alcoholic fermentation was a chemical process in which the yeast somehow served as a catalyst. In 1860, Pasteur provided the first convincing evidence that it was the biochemical activity of the yeast that resulted in fermentation.

The alcoholic fermentation pathway is the same as shown in Figures 1-11 and 7-1 except that pyruvate is converted to C_2H_5OH and CO_2 rather than lactate. Several important reactions occur in the EMP pathway: (1) **Phosphorylation** of glucose and fructose-6-phosphate by ATP. (2) **Oxidation–reduction** and P_i (inorganic phosphate) **assimilation**. The reduction of pyruvate to lactate or acetaldehyde to ethanol is essential for completion of the fermentation process: That is, it balances the oxidation–reduction status of the overall reaction and serves to regenerate NAD^+ so that the process can continue. (3) Structural rearrangements. (4) Cleavage of fructose-1,6-bisphosphate by a specific aldolase. (5) Energy transfer by phosphoglycerokinase and pyruvate kinase. Thus, the overall equation for the ethanolic fermentation in yeast can be shown as

$$C_6H_{12}O_6 + 2P_i + 2ADP \longrightarrow 2C_2H_5OH + 2CO_2 + 2ATP + 2H_2O$$

The common series of reactions used by many microorganisms for the fermentation of glucose is shown in Figure 7-1. This pathway is often referred to as the EMP pathway in recognition of some of the earliest contributions to its elucidation. The enzyme FBP aldolase is the most critical step in the pathway. In the absence of FBP aldolase activity, glucose or other hexose sugars must be metabolized via one of several alternative pathways, as discussed later.

As noted above, glycolysis in muscle tissue, yeast, and many bacteria, appears to be identical in terms of the intermediates involved. Pyruvate is the last common intermediate. In yeast, pyruvate is cleaved by pyruvate carboxylase to form acetaldehyde and carbon dioxide. The acetaldehyde is then reduced to ethanol by alcohol dehydrogenase. In muscle tissue and in lactic acid bacteria (*Streptococcus, Lactococcus, Lactobacillus*), pyruvate is reduced to lactate by lactate dehydrogenase (see Fig. 7-1). Many other microorganisms that use the EMP pathway convert pyruvate to a wide variety of fermentation end-products. These pathways will be discussed in greater detail following the description of other general pathways of carbohydrate metabolism.

In certain organisms, the glycolytic pathway and tricarboxylic acid (TCA) cycle activities are regulated in either a positive or negative manner by specific metabolic intermediates (**feedback control**). As shown in Table 7-1, a number of enzymes are stimulated

---

Fig. 7-1. The fructose bisphosphate (FBP) aldolase or Embden–Meyerhof–Parnas (EMP) pathway of glycolysis. Symbols for the structural genes are those that have been characterized in *E. coli*. The enzymes required for reversal of carbon flow from pyruvate to hexose during gluconeogenesis (pyruvate carboxylase, phosphoenolpyruvate carboxykinase, fructose-1,6-bisphosphatase, and glucose-6-phosphatase) are also shown. Encircled P or P_i = phosphate group; G-3-P = glyceraldehyde-3-phosphate; OAA = oxaloacetate.

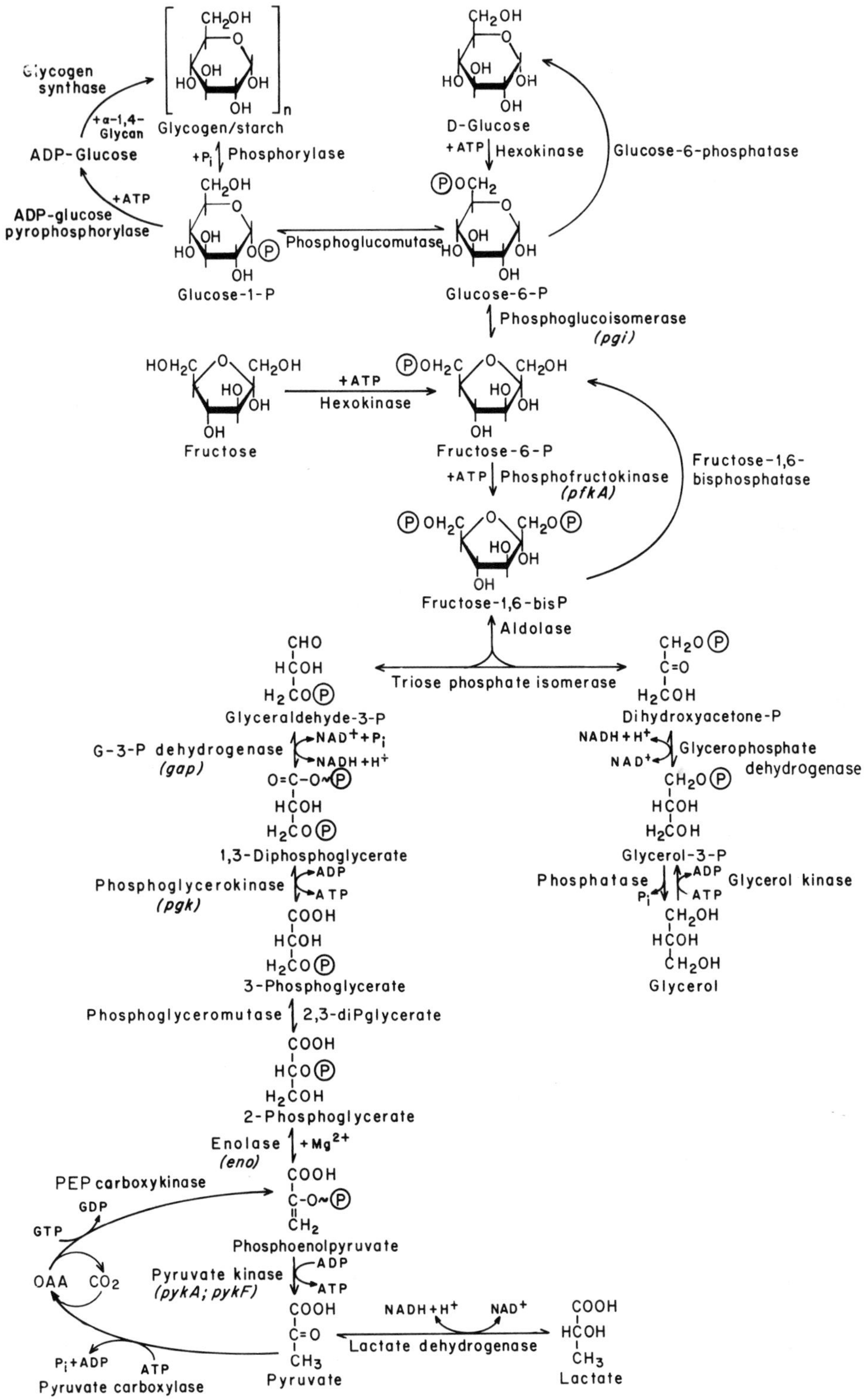

Glycogen
synthase

$+\alpha-1,4-$
Glycan

ADP-Glucose

$+$ATP

ADP-glucose
pyrophosphorylase

Glycogen/starch

$+P_i$ Phosphorylase

Glucose-1-P

Phosphoglucomutase

Glucose-6-P

D-Glucose

$+$ATP Hexokinase

Glucose-6-phosphatase

Phosphoglucoisomerase
(pgi)

Fructose

$+$ATP
Hexokinase

Fructose-6-P

$+$ATP Phosphofructokinase
(pfkA)

Fructose-1,6-bisP

Fructose-1,6-
bisphosphatase

Aldolase

Glyceraldehyde-3-P

Triose phosphate isomerase

Dihydroxyacetone-P

G-3-P dehydrogenase
(gap)

$NAD^+ + P_i$
$NADH + H^+$

$NADH + H^+$
NAD^+

Glycerophosphate
dehydrogenase

Glycerol-3-P

1,3-Diphosphoglycerate

Phosphoglycerokinase
(pgk)

ADP
ATP

3-Phosphoglycerate

Phosphatase

P_i

ADP
ATP

Glycerol kinase

Glycerol

Phosphoglyceromutase $\uparrow$ 2,3-diPglycerate

2-Phosphoglycerate

Enolase $\uparrow + Mg^{2+}$
(eno)

PEP carboxykinase

GDP

GTP

OAA CO_2

$P_i + $ADP   ATP

Pyruvate carboxylase

Phosphoenolpyruvate

Pyruvate kinase
(pykA; pykF)

ADP
ATP

Pyruvate

$NADH + H^+$ NAD^+

Lactate dehydrogenase

Lactate

by AMP or FBP. Some metabolites, such as AMP, phosphoenolpyruvate (PEP), or di-hydroxyacetone phosphate (DHAP), or cofactors, such as reduced nicotinamide adenine dinucleotide (NADH), may have an inhibitory effect on certain catabolic enzymes. An elevation of the AMP concentration may signal a low-energy state. In the case of DHAP or PEP, AMP may regulate the flow of these metabolites into biosynthetic pathways or transport functions. The FBP-activated lactic acid dehydrogenases are characteristic of a number of lactic acid-producing bacteria. In streptococci, 6-phosphogluconate inhibits the activity of phosphohexose isomerase and FBP inhibits the activity of 6-phospho-gluconate dehydrogenase (Table 7-1).

Hexose Monophosphate Pathways

The first evidence for the existence of an alternative pathway for the utilization of hexose sugars was provided by Warburg and Christian who described the oxidation of glucose-6-phosphate to 6-phosphogluconate (6-P-G) via a dehydrogenase reaction (G-6-P dehydro-genase) that they termed *zwischenferment*. They also described the decarboxylation of 6-P-G to form a pentose sugar. For some time afterwards, relatively little attention was paid to the real significance of this pathway as a route to the formation of pentoses, since it was strongly asserted by Meyerhof and others that the FBP aldolase (EMP) pathway was the main route of glucose catabolism. Subsequently, it was shown that ribulose-5-phosphate was the first product formed and that this compound was then converted to ribose-5-phosphate via an isomerase reaction, as shown in Figure 7-2. This series of reactions is common to several distinct hexose monophosphate pathways, as described below.

Phosphoketolase Pathway

One major fermentation pathway follows this core sequence with the subsequent con-version of ribulose-5-phosphate to xylulose-5-phosphate (X-5-P). The X-5-P is then cleaved to form a C_3 compound (glyceraldehyde-3-phosphate) and a C_2 compound (ace-tylphosphate) by the action of a phosphoketolase (Fig. 7-3). Glyceraldehyde-3-phosphate is metabolized via the triose phosphate portion of the EMP pathway to form lactate. Acetylphosphate is converted to acetyl-CoA, which is then reduced to ethanol. The phos-phoketolase pathway serves as a major source of reduced NADP, which provides the reducing activity required in many biosynthetic reactions. However, this pathway is not essential to *E. coli*. Mutants blocked in the pentose phosphate pathway still grow with glucose as the carbohydrate source without other nutritional supplements. Other routes are available for the formation of pentoses and reduced NADP (e.g., the malate enzyme and isocitrate dehydrogenase, as discussed later).

A number of microorganisms utilize the phosphoketolase pathway as the major route of glucose metabolism. *Leuconostoc mesenteroides*, a typical heterofermentative organ-ism, utilizes this pathway yielding lactate, ethanol, and carbon dioxide, as shown in Figure 7-3. Within the genus *Lactobacillus*, it has been possible to clearly differentiate **homofermentative** (*L. casei*, *L. pentosus*) and **heterofermentative** (*L. lycopersici*, *L. pentoaceticus*, *L. brevis*) types. Heterofermentative species are found in the genera *Strep-tococcus*, *Lactococcus*, *Pediococcus*, *Microbacterium*, some *Bacillus* species, and the mold *Rhizopus*. In *Lactobacillus pentoaceticus* and *Leuconostoc mesenteroides*, the basic pathway of glucose conversion leads to the formation of equimolar amounts of lactate, ethanol, and carbon dioxide. A commonly observed variation involves the formation of considerable quantities of glycerol, as discussed in a later section.

TABLE 7-1. Examples of Metabolic Regulation of Enzymes of Glycolysis and Tricarboxylic Acid Cycle[a]

Enzyme	Positive Effector	Negative Effector
6-Phosphogluconate dehydrogenase		FBP
Phosphohexose isomerase		6-PG
Fructose-1, 6-bisphosphatase		AMP
Phosphofructokinase	ADP	PEP
Pyruvate kinase 1	FBP	
Pyruvate kinase 2	AMP	
Citrate synthase	AMP	NADH, α-KG
Malate dehydrogenase		NADH
Malate enzyme		NADH
Phosphoenolpyruvate carboxykinase		NADH
Phosphoenolpyruvate carboxylase	FBP, acetyl-CoA	Aspartate

[a]Abbreviations: FBP = fructose-1,6-bisphosphate; PEP = phosphoenolpyruvate; α-KG = α-ketoglutarate.

The pathway outlined in Figure 7-3 does not reveal some of the details of the reactions involved. In actuality, cleavage of the pentose molecule involves the cofactors thiamine pyrophosphate (TPP) and coenzyme A (CoA) in the following series of reactions:

$$\text{xylulose-5-P} + \text{TPP} \longrightarrow \text{dihydroxyethyl-TPP} + \text{glyceraldehyde-3-P}$$

$$\text{dihydroxyethyl-TPP} + \text{P}_i \longrightarrow \text{acetyl-P} + \text{TPP}$$

$$\text{acetyl-P} + \text{ADP} \longrightarrow \text{acetate} + \text{ATP}$$

$$\text{glyceraldehyde-3-P} + \text{P}_i + \text{2ADP} \longrightarrow \text{lactate} + \text{2ATP}$$

$$\text{Net: xylulose-5-P} + \text{2P}_i + \text{3ADP} \longrightarrow \text{acetate} + \text{lactate} + \text{3ATP}$$

If acetyl phosphate kinase is present, an additional ATP will be generated. In this series of reactions TPP or diphosphothiamine (DPT) functions as a C_2 carrier in a similar manner to its function in the pyruvate dehydrogenase and α-ketoglutarate dehydrogenase complex of reactions in the TCA cycle.

Oxidative Pentose Phosphate Cycle

In some organisms a cyclic mechanism, the **oxidative pentose phosphate cycle**, accounts for the complete oxidation of carbohydrates (Fig. 7-4). In this cycle, G-6-P is converted to ribulose-5-phosphate and CO_2. Ribulose-5-phosphate is maintained in equilibrium with ribose-5-phosphate and xylulose-5-phosphate by the action of ribose phosphate isomerase and ribulose phosphate epimerase. Via the transketolase enzyme, ribose-5-phosphate and xylulose-5-phosphate are converted to sedoheptulose-7-phosphate and glyceraldehyde-3-phosphate. Transaldolase converts sedoheptulose-7-phosphate and glyceraldehyde-3-phosphate to F-6-P and erythrose-4-phosphate. Another reaction, catalyzed by transketolase, converts erythrose-4-phosphate and xylulose-5-phosphate to F-6-P and glyceraldehyde-3-phosphate. Note that by reversal of the FBP aldolase and G-6-P isomerase reactions, glyc-

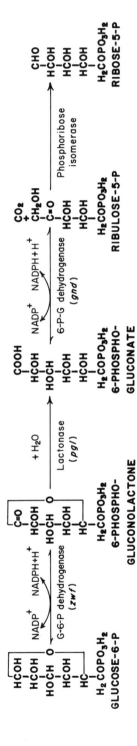

Fig. 7-2. Oxidation of glucose-6-phosphate to ribulose-5-phosphate. Three-letter designations for the structural genes for the enzymes are *zwf*, glucose-6-phosphate (G-6-P) dehydrogenase; *pgl*, lactonase; *gnd*, 6-phosphogluconate (6-P-G) dehydrogenase. Ribulose-5-phosphate is the direct product of the action of 6-P-G dehydrogenase. As shown in Figures 7-3 and 7-4, ribulose-5-phosphate is maintained in equilibrium with ribose-5-phosphate and xylulose-5-phosphate.

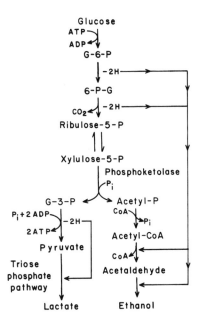

Net: Glucose + P_i + ADP ⟶ Lactate + ethanol + CO_2 + ATP

Fig. 7-3. The phosphoketolase or hexose monophosphate pathway. This pathway has been referred to as the Warburg–Dickens–Horecker heterofermentative pathway in recognition of early contributors to its elucidation. G-6-P = glucose-6-phosphate; 6-P-G = 6-phosphogluconate; G-3-P = glyceraldehyde-3-phosphate.

eraldehyde-3-phosphate and (F-6-P) may be converted to G-6-P. The G-6-P can then reenter the oxidative pentose cycle. During one turn of the cycle the net reaction is

$$\text{G-6-P} + 2\text{NADP}^+ \longrightarrow \text{ribose-5-phosphate} + CO_2 + 2\text{NADPH} + H^+$$

Three turns of the cycle are required to produce one triose phosphate:

$$3\text{G-6-P} + 6\text{NADP}^+ \longrightarrow 3CO_2 + 2\text{F-6-P} + \text{glyceraldehyde-3-P} + 6\text{NADPH} + 6H^+$$

Repetitive action of the cycle could account for the complete oxidation of G-6-P:

$$\text{G-6-P} + 12\text{NADP}^+ \longrightarrow 6CO_2 + 12\text{NADPH} + 12H^+ + P_i$$

However, the cycle does not appear to function in this manner under normal conditions. It is important to note that glycolysis generates NADH, which can be reoxidized by linkage to the electron transport system or, under anaerobic conditions, is used to reduce an oxidized substrate, such as pyruvate, to lactate. By contrast, the pentose phosphate pathway generates NADPH, which is used primarily for reducing power in biosynthetic reactions (e.g., α-ketoglutarate to glutamate; acetate to fatty acids) and is not linked to the terminal respiratory system. Operation of the oxidative pentose phosphate cycle pro-

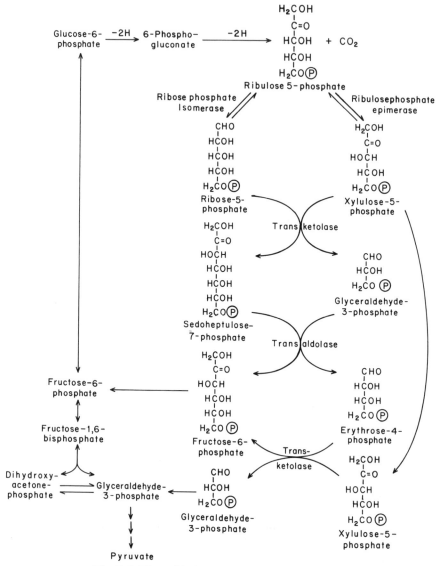

Fig. 7-4. The oxidative pentose phosphate cycle.

vides for the formation of two very important biosynthetic precursors, sedoheptulose-7-phosphate and erythrose-4-phosphate, which function in the aromatic amino acid pathway as discussed in Chapter 10. Under certain conditions in a few unusual organisms, such as *Acetobacter xylinum*, the pentose phosphate cycle may operate to yield a fermentative end-product, such as acetic acid (see later section on acetic acid bacteria).

Entner–Doudoroff or Ketogluconate Pathway

This alternative glycolytic pathway branches from the core HMP pathway shown in Figure 7-2. It consists of two enzymes, 6-phosphogluconate dehydratase, encoded by the *edd* gene, and 2-keto-3-deoxy-6-phosphogluconate (KDPG) aldolase, encoded by the *eda* gene (Fig. 7-5). The Entner–Doudoroff pathway replaces the EMP pathway in many pseudomonads and is described in more detail below.

In *E. coli*, the Entner–Doudoroff pathway is employed for gluconate metabolism. Dehydratase activity is virtually absent in cells grown on glucose and is induced only by gluconate. High basal levels of KDPG aldolase activity are present regardless of the carbon source. This enzyme plays a major role in the metabolism of pectin and aldo-hexuronate by *Erwinia* and other related organisms, as discussed later in this chapter. In *E. coli*, the *edd* and *eda* genes are closely linked to *zwf*, which codes for G-6-P dehydrogenase (Fig. 7-2). However, these genes are regulated under a separate set of regulatory controls. The *zwf* gene is subject to growth rate-dependent regulation at the level of transcription. On the other hand, the *edd*–*eda* operon is regulated by a gluconate-responsive promoter, P_i, located upstream of *edd*, which is responsible for induction of the Entner–Doudoroff pathway. High basal levels of KDPG aldolase are explained by

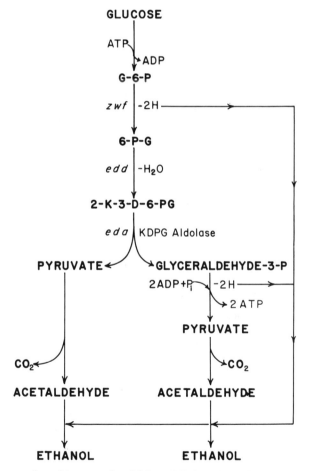

Net reaction: Glucose + P_i + ADP ⟶ 2 Ethanol + 2 CO_2 + ATP + H_2O

Fig. 7-5. The ketogluconate aldolase or Entner–Doudoroff pathway. The structural genes shown are those for *E. coli* and *Z. mobilis*. G-6-P = glucose-6-phosphate; 6-P-G = 6-phosphogluconate; 2-K-3-D-6-PG or KDPG = 2-keto-3-deoxy-6-phosphogluconate.

constitutive transcription of *eda* from additional promoters (P$_2$, P$_3$, and P$_4$) within the *edd*–*eda* region but not from P$_1$.

The Entner–Doudoroff or ketogluconate pathway occurs in several *Pseudomonas* species. It appears to be the sole pathway for the metabolism of glucose in *Zymomonas mobilis* (formerly *Pseudomonas lindneri*) and a major pathway in other members of the pseudomad group of organisms as well as other gram-negative species. However, *Z. mobilis* is unique in that it is the only genus known to utilize the Entner–Doudoroff pathway anaerobically. This organism lacks an oxidative electron-transport system and is, therefore, obligately fermentative. The pathway is inefficient in that it yields only 1 mol of ATP per mol of hexose fermented. It is of special interest for industrial application since the yield of ethanol approaches the theoretical 2 mol/mol of substrate. Rapid production of ethanol by *Z. mobilis* as the sole product of sugar fermentation results from the presence of pyruvate decarboxylase, an enzyme not frequently observed in bacteria. This pathway may also be induced in certain gram-positive organisms (e.g., *Enterococcus hirae*, formerly *E. faecalis*) by growing them in a medium containing gluconate.

GLUCONEOGENESIS

Growth of microorganisms on so-called poor carbon sources, such as L-malate, succinate, acetate, or glycerol, requires the ability to synthesize hexoses needed for the production of cell wall mucopeptide, storage glycogen, and other compounds, such as ribose-5-phosphate, which is needed for nucleic acid biosynthesis. Hexose synthesis involves a reversal of carbon flow from pyruvate (**gluconeogenesis**). This could be achieved by a simple reversal of the enzymes in the EMP glycolytic pathway (Fig. 7-1). However, of the major enzymatic reactions involved in glycolysis, three are insufficiently reversible to allow carbon flow from pyruvate in the direction of hexose synthesis. First, pyruvate kinase is not reversible because the free energy requirement is too large. Instead, the formation of PEP is catalyzed by **PEP carboxykinase**, the first committed step in gluconeogenesis. This Mg^{2+}-dependent enzyme requires GTP as the phosphate donor:

$$\text{oxaloacetate} + \text{GTP} \longleftrightarrow \text{PEP} + CO_2 + \text{GDP}$$

A second irreversible enzyme is phosphofructokinase. To overcome this block in gluconeogenesis, fructose bisphosphatase (fructose-1,6-bisphosphate 1-phosphohydrolase) catalyzes the dephosphorylation of FBP to yield F-6-P and P$_i$:

$$\text{FBP} + H_2O \longrightarrow \text{F-6-P} + p_i$$

The relative rates of phosphofructokinase (glycolytic) and fructose bisphosphate aldolase (gluconeogenic) determines the direction of net carbon flux in the EMP pathway. In enterobacteria, such as *E. coli* and *S. typhimurium*, mutants lacking fructose-1,6-bisphosphatase (*fbp*) cannot grow on L-malate, succinate, glycerol, or acetate (gluconeogenic substrates).

The third bypass reaction required for gluconeogenesis involves the dephosphorylation of G-6-P. The enzyme glucose-6-phosphatase removes P$_i$ from G-6-P to yield free glucose:

$$\text{G-6-P} + H_2O \longrightarrow \text{glucose} + P_i$$

Thus, the overall formation of glucose from pyruvate requires a considerable expenditure of energy:

$$2\text{pyruvate} + 4\text{ATP} + 2\text{GTP} + 2\text{NADH} + 2\text{H}^+ + 4\text{H}_2\text{O} \longrightarrow \text{glucose}$$
$$+ 2\text{NAD}^+ + 4\text{ADP} + 2\text{GDP} + 6\text{P}_i$$

Regulation. A major regulatory step in gluconeogenesis is PEP carboxykinase, encoded by *pckA* in *E. coli*. This enzyme is regulated by catabolite repression, a process in which gluconeogenesis is inhibited when glucose or other carbohydrate carbon sources are available. Maximum levels of PEP carboxykinase are also induced at the onset of the stationary phase of growth, presumably to insure the synthesis of adequate carbohydrate storage reserves or to provide metabolites from the upper part of the EMP as the organism converts proteins to gluconeogenic amino acids. The stationary phase induction of PEP carboxykinase requires cAMP as well as a regulatory signal, the nature of which has not been elucidated. *Escherichia coli* has two pathways available for the synthesis of PEP from C_4 intermediates: PEP carboxykinase and PEP synthetase. The PEP synthetase requires ATP:

$$\text{pyruvate} + \text{ATP} \longrightarrow \text{PEP} + \text{ADP}$$

Double mutants deficient in both of these activities are unable to grow on succinate or other C_4 intermediates, whereas single mutants deficient in only one of these enzymes grow well on these carbon sources.

Glycogen Synthesis. Many organisms store glycogen as an energy reserve. In bacteria, the enzymes ADP-glucose pyrophosphorylase and the branching enzyme, glycogen synthase (1,4-α-D-glucan:1,4-α-D-glucan 6-α D-glucanotransferase yield) glycogen:

$$\text{G-1-P} + \text{ATP} \longleftrightarrow \text{ADP-glucose} + \text{PP}_i \text{ ADP-glucose}$$
$$+ \alpha\text{-1,4-glucan} \longrightarrow \alpha\text{-1,4-glucosyl glucan} + \text{ADP}$$

Escherichia coli mutants defective either in glycogen synthase or in ADP-glucose pyrophosphorylase are unable to accumulate glycogen. Glycogen synthesis in *E. coli* is regulated by both the *relA* gene, which mediates the stringent response to amino acid starvation when the cells are using glucose but not when the cells are using glycerol, and by cyclic AMP (cAMP). These two regulatory controls are independent of each other in that each regulatory process can be expressed in the absence of the other.

Glycogen synthesis in *E. coli* is regulated at the level of ADP-glucose pyrophosphorylase encoded by the structural gene *glgC*. Glucose–ADP synthetase, the first unique step in glycogen synthesis in this organism, is activated by glycolytic intermediates with FBP as the activator and AMP, ADP, and P_i as inhibitors. The ADP-glucose synthetases of *E. coli* and *S. typhimurium* show considerable similarity in that both consist of four identical subunits and have the same spectrum of activators and inhibitors. Genetically, the *glg* genes of both organisms are clustered at the same point (75 min) on their respective genetic maps. Further studies have shown that a number of genes encoding catabolic, biosynthetic, and amphibolic enzymes in enteric bacteria are transcriptionally regulated by a complex catabolite repression–activation mechanism, which may involve enzyme III of the phosphotransferase system as one of the regulatory components. Comparable systems have been described for the regulation of gluconeogenesis in a wide diversity of microorganisms from the yeast *S. cerevisiae*, to the symbiotic nitrogen fixing bacterium *Rhizobium leguminosarum*.

TRICARBOXYLIC ACID CYCLE

Sir Hans Krebs and co-workers demonstrated that C_4 dicarboxylic acids, such as succinate and malate, the C_5 α-ketoglutarate, and the C_6 citrate, were all oxidized by pigeon breast muscle. Citrate was synthesized from added oxaloacetate in these muscle preparations. The fact that the amount of citrate formed was very small seemed odd until it was found that both α-ketoglutarate and citrate were formed. Demonstration of succinate formation from fumarate and oxaloacetate in the presence of malonate (which inhibits succinate dehydrogenase) led to the conclusion that succinate could be formed via oxidative as well as reductive reaction [i.e., reduction of fumarate ($HOOC-CH=CH-COOH + 2H$) or oxidatively from α-ketoglutarate ($HOOC-CH_2-CH_2-CO-COOH - CO_2 - 2H$)]. These observations led Krebs to propose a cyclic mechanism for the oxidation of pyruvate, as shown in Figure 7-6. He theorized that oxaloacetate condensed with a C_3 compound (presumably pyruvate) from the glycolytic sequence to yield a C_7 intermediate that was then converted to citrate via decarboxylation. It was later shown that a C_2 unit (acetyl-CoA) condensed with oxaloacetate to form citrate. The presence of the condensing enzyme has been confirmed in mammalian systems as well as in *S. cerevisiae* and a wide variety of bacteria.

Subsequent studies confirmed that the TCA cycle functions as it is presented in Figure 7-6. The intermediate between citrate and isocitrate apparently exists as a carbonium ion coupled with the enzyme and a metal ion cofactor. As such, *cis*-aconitate is not a true intermediate in the TCA cycle and the enzyme aconitase catalyzes the conversion of citrate to isocitrate. The succeeding two steps, oxidation of isocitrate to oxalosuccinate and decarboxylation of oxalosuccinate to α-ketoglutarate are also catalyzed by one enzyme, isocitrate dehydrogenase.

Malate represents a pivotal point in the cycle. It participates in several alternative reactions. It may be oxidized to oxaloacetate via the classic NAD-linked malate dehydrogenase (Fig. 7-6) as observed in *E. coli* or via a direct cytochrome-linked dehydrogenase, as observed in *Pseudomonas* and *Serratia*. As such, it can contribute to energy production in these organisms, as discussed in a later section. Many organisms can also decarboxylate malate to pyruvate (malic enzyme), and subsequently carboxylate pyruvate to form oxaloacetate. These reactions provide a scavenger system for reclaiming carbon dioxide. A few organisms apparently utilize this system for completion of the TCA cycle in the absence of malate dehydrogenase. However, the malic enzyme, as such, is not a major link in the cyclic operation of the TCA cycle of most organisms. However, since it is NADP-linked, it may play a significant role in providing reducing activity for biosynthetic reactions.

The multienzyme α-ketoglutarate dehydrogenase complex consists of three enzyme components: a thiamine pyrophosphate-dependent α-ketoglutarate dehydrogenase (Enz_1), dihydrolipoyl *trans*-succinylase (Enz_2), and an FAD-dependent dihydrolipoyl dehydrogenase (Enz$_3$), which catalyze the following steps:

$$\alpha\text{-ketoglutarate} + \text{lipoate-S}_2\text{-Enz}_2 \xrightarrow{TPP\text{-}Enz_1} \text{succinyl-S-lipoate-SH-Enz}_2 + CO_2$$

$$\text{succinyl-S-lipoate-SH-Enz}_2 + \text{CoA} \xrightarrow{Enz_2} \text{succinyl-CoA} + \text{lipoate-(SH)}_2\text{-Enz}_2$$

$$\text{lipoate-(SH)}_2\text{-Enz}_2 + NAD^+ \xrightarrow{FAD\text{-}Enz_3} \text{lipoate-S}_2\text{-Enz}_2 + NADH + H^+$$

Net: α-ketoglutarate + CoA + NAD$^+$ $\longrightarrow$ succinyl-CoA + CO$_2$ + NADH + H$^+$

This complex of enzymes is comparable to the pyruvate dehydrogenase complex that catalyzes the oxidative decarboxylation of pyruvate at the initial stage of the TCA cycle (Fig. 7-6). However, the individual components differ from each other in physicochemical properties and specificity except for the third enzyme in which the components are functionally interchangeable and identical. Studies with mutants deficient in various components of the α-ketoglutarate dehydrogenase complex indicate that the dihydrolipoyl dehydrogenase component of both systems is encoded by *lpd*.

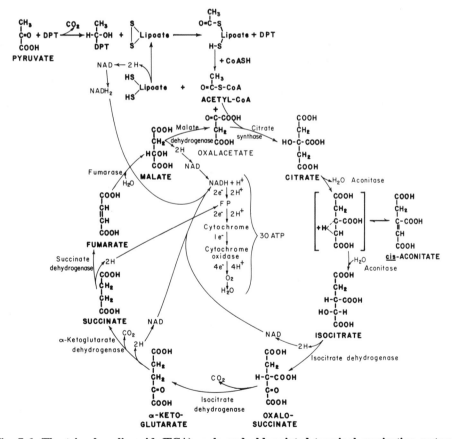

Fig. 7-6. The tricarboxylic acid (TCA) cycle and abbreviated terminal respiration system. DPT = diphosphothiamine; FP = flavoprotein.

Under anaerobic conditions the TCA cycle no longer functions as such because the link to terminal respiration is required to maintain the activities and synthesis of succinate dehydrogenase and the α-ketoglutarate dehydrogenase complex. However, net synthesis of the intermediates is still required for biosynthetic reactions. The activity of fumarate reductase is increased and provides a mechanism for succinate synthesis. Thus, the TCA cycle now functions as a branched biosynthetic pathway, one branch operating as a reductive pathway reversing the usual sequence from succinate to oxaloacetate and the other branch continuing to operate oxidatively to convert oxaloacetate to α-ketoglutarate, as shown in Figure 7-7.

In actuality, the activity levels of a large number of enzymes that serve primarily aerobic functions are markedly reduced when *E. coli* is grown under anaerobic conditions. In this organism, a two-component signal transduction system, consisting of a transmembrane sensor protein ArcB and a cytoplasmic regulatory protein, ArcA, controls the expression of genes encoding enzymes involved in aerobic respiration. Under the influence of this Arc (aerobic respiratory control) system, expression of the structural genes for several flavoprotein-linked dehydrogenases, the cytochrome *o* complex, enzymes of the TCA cycle, glyoxylate shunt, and fatty acid degradation are repressed under anaerobic conditions. Conversely, the oxygen-scavenging cytochrome *d* oxidase (encoded by the *cydAB* operon) is activated under control of the Arc system. The Fnr protein functions as an anaerobic repressor of both the cytochrome *o* oxidase complex (encoded by *cyoABCDE*) and cytochrome *d* oxidase (encoded by *cydAB*). It must, therefore, be assumed that another unidentified regulatory element results in an increase in cytochrome *d* oxidase activity to enable the organism to cope with the conditions of oxygen starvation. Operation of this regulatory system has been discussed in some detail in Chapter 4.

In Chapter 1 it was revealed that all of the components of the cell can arise from only 12 precursor metabolites. A function of the TCA cycle, in addition to its role in the generation of reducing equivalents for the terminal respiratory system, is the formation of precursors for the biosynthesis of amino acids and other compounds. The synthesis

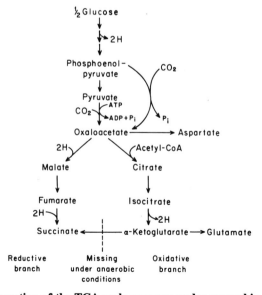

Fig. 7-7. Operation of the TCA cycle enzymes under anaerobic conditions.

of these compounds at the expense of the TCA cycle intermediates tends to diminish the activity of the cycle. However, these metabolites can be replenished through the fixation of carbon dioxide by the following reactions:

Pyruvate carboxylase

$$\text{pyruvate} + CO_2 + ATP + H_2O \xrightarrow{Mg_2^+} \text{oxaloacetate} + ADP + P_i$$

Phosphoenolpyruvate carboxykinase

$$\text{PEP} + CO_2 + H_2O \longrightarrow \text{oxaloacetate} + H_3PO_4$$

Malic enzyme

$$\text{pyruvate} + CO_2 + NADPH + H^+ \longleftrightarrow \text{L-malate} + NADP^+$$

Heterotrophic organisms require at least one enzyme of this nature in order to grow aerobically on hexoses or glycolytic intermediates. The generation of oxaloacetate permits the continued flow of hexose carbon through the TCA cycle under conditions in which intermediates are being removed for biosynthetic reactions. This same purpose is also served when oxaloacetate is used to initiate gluconeogenesis. Enzymes that serve this function are termed **anapleurotic** (from Greek: meaning *to fill up*). Transaminase reactions may also serve to generate oxaloacetate (from aspartate) or α-ketoglutarate (from glutamate). The aspartase reaction can also provide oxaloacetate. The glyoxylate cycle, discussed in the next section, can also serve an anapleurotic role by allowing a bypass of carbon dioxide releasing reactions. However, the glyoxylate cycle can function in this capacity only in organisms capable of utilizing acetate following induction on this substrate. Certain organisms, such as *Arthrobacter pyridinolis*, which are deficient in pyruvate carboxylase activity, exhibit a nutritional requirement for malate in order to grow on glucose.

GLYOXYLATE CYCLE

In the course of investigating the TCA cycle in microbial systems, many investigators obtained evidence that acetate was actively oxidized by many species. In addition, a number of microorganisms were shown to be able to utilize acetate as the sole source of carbon for growth. However, the exact mode of acetate utilization did not become completely apparent until it was realized that direct condensation of 2 mol of acetate did not occur. Direct condensation of 2 mol of acetate to form succinate should result in labeling of both of the methylene carbon atoms of succinate from $^{14}CH_3COOH$:

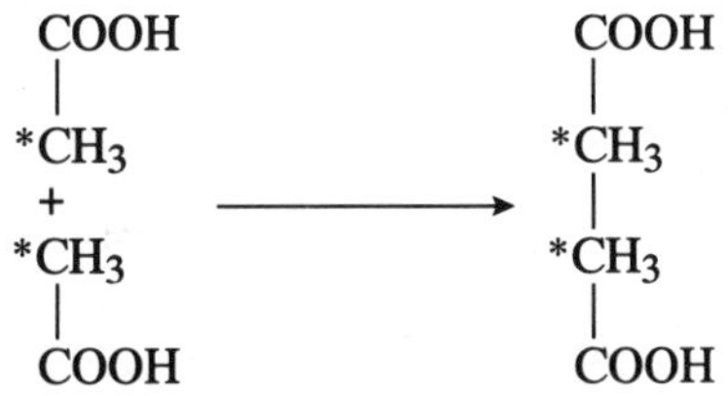

$$
\begin{array}{ccc}
\text{COOH} & & \text{COOH} \\
| & & | \\
*\text{CH}_3 & & *\text{CH}_3 \\
+ & \longrightarrow & | \\
*\text{CH}_3 & & *\text{CH}_3 \\
| & & | \\
\text{COOH} & & \text{COOH}
\end{array}
$$

Examination of the succinate formed showed that a singly labeled ethylene moiety predominated.

The discovery of isocitritase (isocitrate lyase) and of malate synthase provided the foundations for the concept of cyclic utilization of acetate via the glyoxylate cycle, as shown in Figure 7-8. The reactions of these two enzymes yield glyoxylate and malate:

Isocitrate lyase

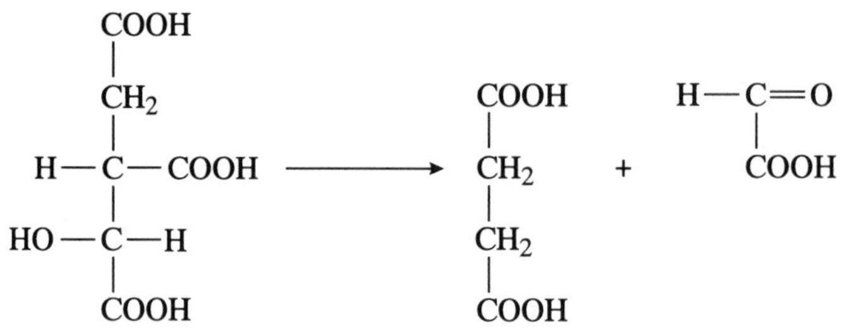

$$
\begin{array}{ccc}
\begin{array}{c}
\text{COOH} \\
| \\
\text{CH}_2 \\
| \\
\text{H} - \text{C} - \text{COOH} \\
| \\
\text{HO} - \text{C} - \text{H} \\
| \\
\text{COOH}
\end{array}
& \longrightarrow &
\begin{array}{ccc}
\begin{array}{c}
\text{COOH} \\
| \\
\text{CH}_2 \\
| \\
\text{CH}_2 \\
| \\
\text{COOH}
\end{array}
& + &
\begin{array}{c}
\text{H} - \text{C} = \text{O} \\
| \\
\text{COOH}
\end{array}
\end{array}
\\
\text{Isocitrate} & & \text{Succinate} \qquad\qquad \text{Glyoxylate}
\end{array}
$$

Malate synthase

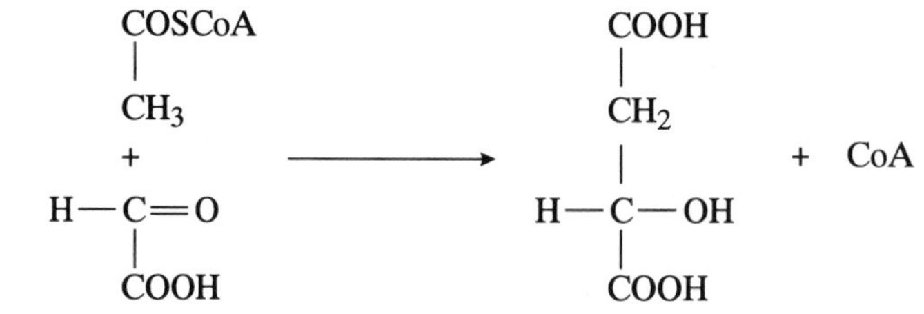

$$
\begin{array}{ccc}
\begin{array}{c}
\text{COSCoA} \\
| \\
\text{CH}_3 \\
+ \\
\text{H} - \text{C} = \text{O} \\
| \\
\text{COOH}
\end{array}
& \longrightarrow &
\begin{array}{c}
\text{COOH} \\
| \\
\text{CH}_2 \\
| \\
\text{H} - \text{C} - \text{OH} \\
| \\
\text{COOH}
\end{array} \quad + \quad \text{CoA}
\\
\text{Acetyl-CoA + glyoxylate} & & \text{Malate}
\end{array}
$$

The net operation of the glyoxylate cycle results in the formation of malate from 2 mol of acetate:

$$
\begin{aligned}
\text{isocitrate} &\rightarrow \text{succinate} + \text{glyoxylate} \\
\text{glyoxylate} + \text{acetate} &\rightarrow \text{malate} \\
\text{succinate} + \text{acetate} - 2(2\text{H}) &\rightarrow \rightarrow \rightarrow \text{isocitrate}
\end{aligned}
$$

$$
\overline{\text{Net: 2 acetate} - 2(2\text{H}) \rightarrow \text{malate}}
$$

Activity of the glyoxylate bypass explains the ability of bacteria, yeast, and other micro-organisms to utilize acetate for the net synthesis of C_4 dicarboxylic acids.

As one might predict, the enzymes of the glyoxylate bypass system are repressed by the presence of glucose or another more rapidly utilized substrate (see Chapter 4). As discussed previously, the Arc (aerobic respiratory control) system and other regulatory factors repress activity of the TCA cycle as well as the glyoxylate cycle under anaerobic conditions. Because of the very low redox potential of ferredoxin (see Fig. 7-13), many

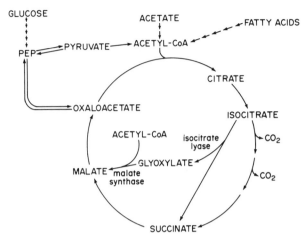

Fig. 7-8. The glyoxylate cycle or glyoxylate shunt. Note that the two CO_2-evolving steps of the TCA cycle are bypassed. Glucose or fatty acids can serve as sources of acetate. From Maloy, S. R., H. Bohlander, and W. D. Nunn. 1980. *J. Bacteriol.* **143**:720.

anaerobes and certain photosynthetic bacteria can form pyruvate, and in some cases, α-ketoglutarate, via reductive decarboxylation reactions. By this means, these organisms can circumvent the irreversibility of the oxidative decarboxylation of pyruvate to acetyl-CoA and CO_2 and of α-ketoglutarate to succinyl-CoA and CO_2 observed in other organisms. Thus, the utilization of acetate can occur in those organisms that produce and use ferredoxin in reductive metabolism.

The oxidative pentose cycle (Fig. 7-4) assumes significance in the energy production of some microorganisms. The cycle has been shown to occur in several organisms but it is sometimes difficult to interpret the relative importance of this pathway as compared to the combined EMP pathway and the TCA cycle. *Escherichia coli*, yeasts, streptomycetes, and fungi have been shown to contain the pentose phosphate cycle. *Gluconobacter suboxydans*, an aerobe, cannot ferment glucose and does not appear to contain the enzymes of the TCA cycle. From the distribution of isotope in various products derived from labeled glucose it has been concluded that a modified pentose cycle is the major pathway of carbohydrate utilization in this organism. It remains a difficult problem to assess the degree to which competing cycles participate in the overall economy of cells that contain both of these pathways and possibly others. It has become increasing apparent that the pentose pathways are of primary importance in providing reducing power in the form of NADPH for biosynthetic reactions and in providing pentoses for the biosynthesis of nucleotides. In *E. coli* it is generally considered that approximately 30% of the carbon is metabolized via the pentose phosphate pathway and 70% via the EMP pathway.

In the discussion of the metabolism of carbohydrates by microorganisms it has been necessary to omit many of the details of the specific metabolic pathways utilized by a number of important organisms. The literature cited at the end of the chapter should provide sources of information for the reader who is interested in exploring the subject in greater detail.

ENERGY PRODUCTION

The second law of thermodynamics states that systems will *spontaneously* change from states of higher order to lesser order. The cell, however, must do the opposite to survive. It must make order out of chaos. To increase order, energy must be introduced into the system. In this section we will discuss how cells accomplish this feat.

The discussion must begin with an appreciation for **entropy**. Entropy (S) is the *measure* of disorder. Chemical reactions that cause a large increase in S (where the difference between the precursor and product entropies [ΔS] is > 0) are favored and will occur spontaneously with the release of energy. Reactions that cause a decrease in S ($\Delta S < 0$) require energy input and so cannot occur spontaneously. The physical properties of a system, including heat, pressure, volume, and energy can be combined into a term called the **Gibbs free energy** (G). The parameter G can be used as an accounting device to deduce the entropy change that occurs in a system following a reaction. The free energy change (ΔG) during a reaction (the G of the products minus the G of the starting materials) is a measure of the *disorder* that is created when the reaction occurs. Energetically favored reactions are those in which a large amount of free energy is released (i.e., have a large *negative* ΔG) and create much disorder. In contrast, reactions with a large *positive* ΔG (such as would occur during peptide-bond formation) create order and so cannot occur spontaneously. The goal of the cell is to harness the ΔG that results from converting glucose (a highly ordered compound) into CO_2 and H_2O. This result is accomplished by a variety of methods primarily designed to synthesize the high-energy compound ATP. The processes used by the cell to generate ATP, substrate level and oxidative phosphorylation, are described below. The ΔG at equilibrium (ΔG_0) for ATP $\leftrightarrow$ ADP + P_i is 7.3 kcal/mol. However, at normal cellular concentrations ([ATP] = 10 [ADP]), ΔG_0 is even higher at -11 to -13 kcal/mol. The cell must then be able to *couple* the release of this energy to enzymatic reactions that create order, that is, biosynthetic reactions. In a hypothetical enzyme reaction that converts substrates A—H and B—OH to A—B and H_2O, the energy from ATP hydrolysis is first used to convert B—OH to a higher energy intermediate, B—O—PO_4. This compound is only transiently formed with the energy released during the decay used by the enzyme to form A—B. Thus, the energy released from the ATP hydrolysis reaction (large $-\Delta G$) is coupled to the synthesis reaction (large $+\Delta G$). In this way the cell can create order.

Substrate Level Phosphorylation

During glycolysis, energy in the form of ATP is generated by substrate level phosphorylation. The oxidative steps in glycolysis or other comparable metabolic systems give rise to this energy. In the EMP pathway (Fig. 7-1), oxidation of glyceraldehyde-3-phosphate and the incorporation of P_i result in the formation of 1,3-diphosphoglyceric acid. One of the phosphate bonds achieves a high-energy state. The energy inherent in this bond is used to transfer the P_i to ADP to form ATP. Subsequent conversion of the 3-phosphoglycerate from this reaction to phosphoenolpyruvate again provides for the transfer of energy by the formation of ATP from ADP (Fig. 7-9).

It is the special physicochemical properties of certain compounds, such as ATP, that enables them to transfer energy into an energy-requiring system. In this case the energy transfer is associated with the transfer of a chemical unit—the phosphoryl group. In other systems the energy transfer is associated with the transfer of electrons and protons

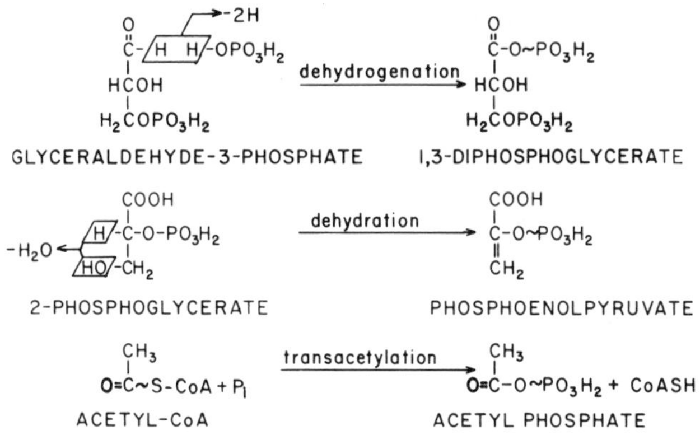

Fig. 7-9. Reactions involved in the formation of high-energy phosphate bonds.

in oxidation–reduction reactions. Although the energy of high-energy compounds is often said to reside in the P—O bond, it is more correct to refer to the **phosphoryl-transfer potential** associated with phosphorylated compounds. The nucleoside triphosphates (ATP, GTP, etc.) are the most universal compounds with energy-transfer potential in biological systems and serve as the immediate energy sources for biosynthesis as well as physicochemical activities, such as flagellar motor rotation or active transport. In the oxidation of triose phosphate, as shown in Figure 7-1, some of the details are not shown. In this reaction, the combined events of thiol addition, hydrogen transfer, and phosphorolysis by glyceradehyde-3-phosphate dehydrogenase lead to formation of a phosphate bond that may be transferred to ADP:

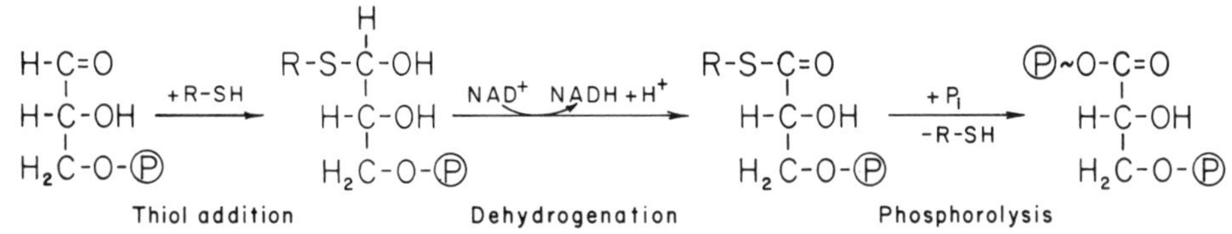

This reaction is initiated by condensation of the thiol (—SH) group at a specific cysteine residue at the catalytic site of the enzyme with the aldehyde carbonyl. Since thiol esters yield considerably more energy on hydrolysis than ordinary oxygen esters, this facilitates the addition of the phosphate group to the carbonyl group. Generation of high-energy phosphate bonds by reactions of this type are referred to as **substrate level phosphorylations** to distinguish them from those generated by electron transport.

The number of moles of ATP generated will depend on the cleavage pathway utilized during the degradation of carbohydrates (Table 7-2). For example, pentose fermentation proceeds via C_2–C_3 cleavage to yield lactate and acetate as products:

$$\text{pentose} + 2P_i + 2ADP \longrightarrow \text{lactate} + \text{acetate} + 2ATP + 2H_2O$$

As shown in the lower part of Figure 7-3, pentose fermentation will yield 2 mol of ATP per mol of pentose used if acetyl phosphate is formed as an intermediate. Phosphoketolase splits the pentose into acetyl phosphate and glyceraldehyde-3-phosphate. Acetyl phos-

TABLE 7-2. Energy Yields from Various Fermentative Pathways

Substrate	Cleavage Type	Products	Moles Triose-P Formed	Net ATP Yield (mol)
Hexose	FBP-aldolase[a]	2 lactate	2	2
Hexose	Phosphoketolase	1 lac, 2 ac, $1CO_2$	1	1
Pentose	Phosphoketolase	2 lac, 1 ac	1	2
Hexose	KDPG-aldolase[b]	2 ethanol, $2CO_2$	1	1
Aldonic	FBP-aldolase + KDPG-aldolase	1.83 lac, $0.5CO_2$	1	1.33^c

[a]FBP aldolase, fructose-1,6-bisphosphate aldolase.
[b]KDPG-aldolase, 2-keto-3-deoxy-6-phosphogluconate aldolase.
[c]Resting cells produce 1.5lactate + $0.5CO_2$ + $0.5C_2$ (actually missing, but calculated as acetate via phosphoketolase. 1.83 lactate formed in the presence of arsenite). From Gunsalus, I. C. and C. W. Shuster. 1961. Energy-yielding metabolism in bacteria. In *The Bacteria*, Vol. II, Metabolism. I. C. Gunsalus and R. Y. Stanier (Eds.). Academic, New York.

phate can be converted to acetate by acetate kinase yielding one molecule of ATP. The glyceraldehyde-3-phosphate is utilized via the triose portion of the EMP pathway to yield two additional molecules of ATP. The phosphoketolase reaction has been shown to occur in *Acetobacter xylinum*, *Lactobacillus* (heterofermentative species), and in *Leuconostoc*. In the genus *Bifidobacterium* **hexose phosphoketolase** cleavage of F-6-P yields acetyl phosphate and erythrose-4-phosphate (E-4-P):

$$\text{F-6-P} + P_i \longrightarrow \text{acetyl-P} + \text{E-4-P}$$

Pentose phosphate is formed by the actions of transaldolase and transketolase. The pentose thus formed is then split by the phosphoketolase to form acetyl phosphate and glyceraldehyde-3-phosphate, as shown previously. The hexose phosphoketolase reaction appears to be unique to the bifidobacteria and has not been found in other heterofermentative organisms. Anaerobic glycolytic organisms do not appear to contain these phosphoketolases.

Other energy-yielding reactions at the substrate level have been described. In *Clostridium cylindrosporum* the fermentation of purines yields ATP via the reaction

$$\text{HN=CH—NH—CH}_2\text{CO}_2\text{H} + \text{ADP} + P_i + 2H_2O \longrightarrow \text{HCO}_2\text{H}$$
$$+ NH_3 + \text{CH}_2\text{NH}_2\text{CO}_2\text{H} + \text{ATP}$$

Reactions that lead to the formation of carbamoyl phosphate via an energy-coupling reaction may eventually yield energy through the transfer of the phosphate group to ADP:

$$\text{H}_2\text{N—COOPO}_3\text{H}_2 + \text{ADP} \longrightarrow NH_3 + CO_2 + \text{ATP}$$

The contribution of inorganic pyrophosphate (PP_i) and polyphosphate to the overall energy yield may be considerable. Although earlier workers had suspected that PP_i could serve as a source of energy, most biochemists considered that hydrolysis of PP_i served to render irreversible reactions in which PP_i was formed. As discussed later in relation to the high-energy yields observed in propionic acid fermentations, PP_i can be used in place of ATP as a source of energy in certain fermentation reactions. Such reactions may be used more widely than is currently appreciated.

OXIDATIVE PHOSPHORYLATION

When a carbohydrate is oxidized via a respiratory mechanism, energy is generated by passage of electrons through a series of electron (e^-) donors and acceptors until ultimately reaching a final e^- acceptor, such as dioxygen (O_2) or nitrate. The energy inherent in the carbohydrate is gradually released during this series of coupled oxidation–reduction reactions and used to pump protons out of the cell. Since membranes are impermeable to protons, this establishes an electrochemical gradient or proton motive force across the cell membrane. Proton motive force ($\Delta\mu H^+$) is composed of electrical (charge) *and* chemical (pH) components according to the following relationship:

$$\frac{\Delta\mu H^+}{\mathcal{F}} = \frac{\Delta\Psi - 2.3RT}{\mathcal{F}} \Delta pH$$

Where $\Delta\Psi$ represents the transmembrane electrical potential and ΔpH is the pH difference across the membrane. The parameters R, T, and $\mathcal{F}$ are the gas constant, the absolute temperature, and the Faraday constant, respectively. Thus, $2.3RT/\mathcal{F}$ at 36 °C equals 60 mV and the total driving force across the membrane generated by proton extrusion in millivolts (mV) is $\Delta pH^+ = \Delta\Psi - 60 (\Delta pH)$. It is this energy the cell uses to synthesize ATP. Proton motive force (PMF) is also used directly to power flagellar movement and some transport systems.

Measurement of PMF

Electrical potential ($\Delta\Psi$) across a cell membrane can be measured using lipophilic cations that will freely pass through the membrane. Since the interior side of the membrane is negatively charged, lipophilic cations, such as tetraphenyl phosphonium ion (TPP^+), will accumulate inside a cell depending on the extent of the negative charge. In a similar manner, ΔpH can be measured experimentally using radiolabeled weak acids, such as benzoic acid. Weak acids dissociate based upon their dissociation constants (pK_a) as follows:

$$HA \longleftrightarrow H^+ + A^-$$

The protonated form (HA) passes freely back and forth across the membrane, whereas the deprotonated form (A^-) does not. At a pH value equal to the pK_a, 50% of the acid will be dissociated. At 1 pH unit *above* the pK_a 90% will be dissociated, whereas at 1 pH unit *below* the pK_a only 10% will be dissociated. So, 10% of a weak acid with a pK_a of 5.5 will be protonated at pH 6.5. If cells are also present, the HA will pass into the cells, equilibrate with external [HA], and dissociate intracellularly according to the intracellular pH. The dissociated form will become trapped and so cannot equilibrate with the outside. This dissociation concomitantly *lowers* internal HA, which will reequilibrate with external HA and bring more HA into the cell. This HA also dissociates according to the pH of the cell, thus continuing the cycle. Consequently, by measuring the accumulation of a radiolabeled weak acid, one can calculate internal pH according to the following formula:

$$pH_i = log \left[\frac{\text{total acid conc inside } (10^{pK} + 10^{pH \ out}) - 10^{pK}}{\text{total acid conc outside}} \right]$$

Under aerobic conditions at pH 6, the $\Delta\mu H^+$ generated is approximately 190 mV with $\Delta\Psi$ and ΔpH contributing equally. However, what happens when $\Delta pH = 0$, that is, ($pH_i = pH_o$)? In this situation, all of PMF is derived from $\Delta\Psi$, the electrical component. In fact, $\Delta\Psi$ increases to compensate for the loss of ΔpH. This illustrates that the two components of PMF can be exchanged according to the needs of the cell. Ion circulations involving the exchange of protons to other cations, such as Na^+ or K^+, are responsible for the interconversion between $\Delta\Psi$ and ΔpH (see later section on transport in this chapter).

Electron-Transport Systems. The cytoplasmic membranes of bacteria contain the electron-transfer system (ETS) that generates PMF by oxidizing NADH and other substrates. These systems are composed of heme-containing components (cytochromes), iron–sulfur cluster enzymes, flavoproteins (containing FMN), and quinones. A typical electron-transport chain is depicted in Figure 7-10. One should keep in mind that different organisms, and even *E. coli* under different conditions, will use an ETS that employs a variety of substrates and terminal electron acceptors. Thus, the respiratory system can be visualized as being branched at the dehydrogenase and terminal oxidase sites, as shown in Figure 7-11. The electron-transfer reactions are initiated by specific modular units. These are NADH, lactate, succinate, formate, or glycerol-3-phosphate dehydrogenases.

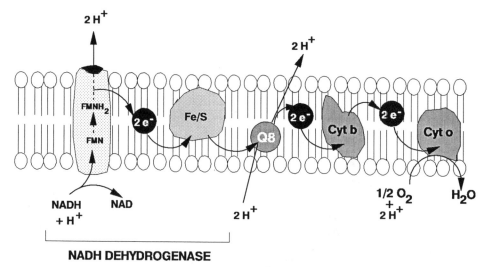

Fig. 7-10. An electron-transport chain of *E. coli* (see text for details). FMN = flavin mononucleotide; Fe–S = iron–sulfur group; Q8 = ubiquinone.

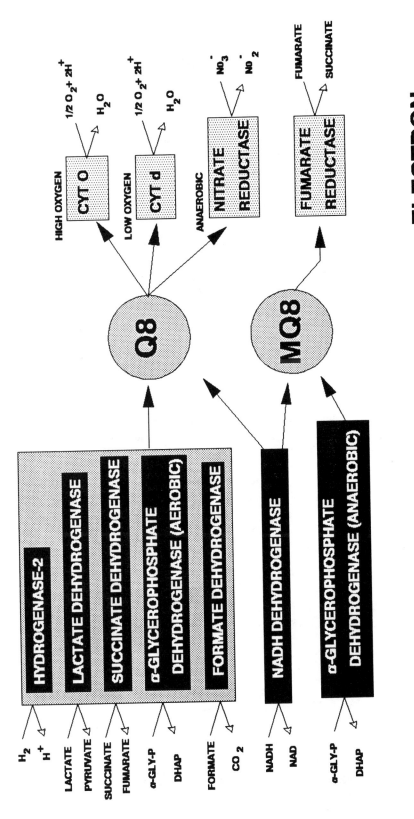

Fig. 7-11. Menu of electron donor and acceptor modules for *E. coli.* Selection (induction) of module combinations depend on the substrate available (left side of figure) and the redox potential of the final electron acceptor (right side of figure). Q8 = ubiquinone; MQ8 = menaquinone.

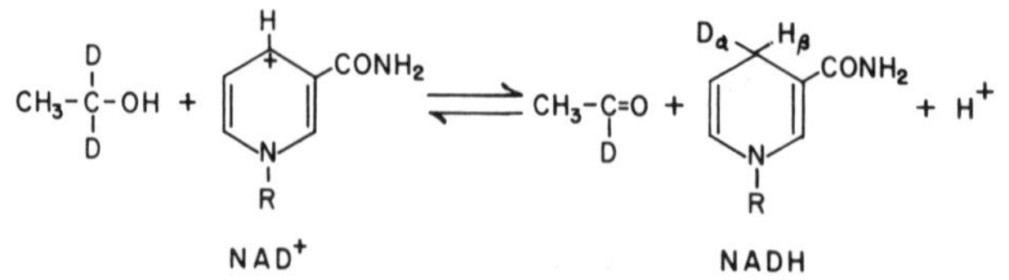

Fig. 7-12. Deuterium labeling of NAD$^+$ during reduction by alcohol dehydrogenase. In the presence of alcohol dehydrogenase, deuterium (D) from deuterium-labeled alcohol is added stereospecifically to NAD$^+$. Here NADP$^+$, which contains an additional phosphate group on the pentose of the adenylic acid moiety, reacts similarly.

Figure 7-12 illustrates how a substrate-bound H (as deuterium) with two e$^-$ is stereospecifically transferred to NAD$^+$ by NAD-dependent dehydrogenases forming NADH. Some enzymes, such as succinate dehydrogenase, contain flavoproteins and are linked directly to the next step in the system, bypassing the need for NAD$^+$. The FAD functions in flavoprotein-linked reactions, as shown in Figure 7-13. Each oxidation–reduction system has a definite oxidation–reduction potential (E_0') at which it functions. By determining the electrode potential for each system, the order in which each functions can be deduced as shown in Figure 7-14.

The electrons are collected from the different donor modules by the quinone pool acting as a universal adapter. The quinone pool passes the electrons on to the appropriate acceptor module depending on what terminal electron acceptor is being used (e.g., dioxygen, nitrate, or fumarate). Quinones are mobile hydrogen carriers that shuttle between the large dehydrogenase and oxidase complexes. The predominant quinone, ubiquinone-8, functions in aerobic respiration while anaerobic respiratory chains may require menaquinone-8. The ETS shown in Figure 7-10 begins with NADH dehydrogenase, a flavoprotein complex containing flavin mononucleoide (FMN) and non-heme iron–sulfur (Fe–S) clusters. The NADH transfers two electrons and protons to FMN, forming FMNH$_2$. Transfer of the electrons to the Fe–S group causes the **release** of the two protons (H$^+$) to the outside generating PMF. The electrons in the Fe–S group are then transferred

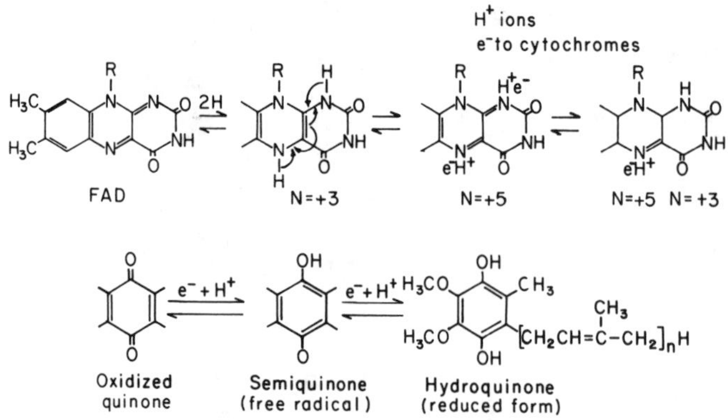

Fig. 7-13. Function of flavin adenine dinucleotide (FAD) and ubiquinone in hydrogen (H$^+$) and electron (e$^+$)-transfer reactions.

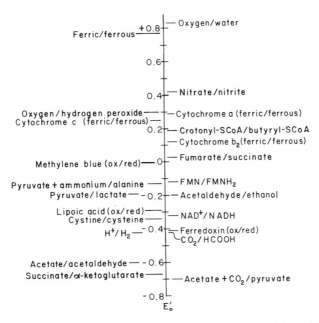

Fig. 7-14. Electrode potentials of some important biological systems. E_0' = oxidation–reduction potential (in V).

to ubiquinone. Ubiquinone picks up the $2e^-$, as well as $2H^+$ from the cytosol, to form hydroquinone.

Under conditions of high oxygenation, hydroquinone will transfer its electrons to the cytochrome o complex and release $2H^+$ to the cell exterior, adding to PMF. The cytochrome o complex is composed of four polypeptides that make up two spectral signals, termed cytochromes b_{555} and b_{562}. The subscript numbers correspond to the maximum wavelength absorbance values. It may be difficult to visualize how proton translocation across the membrane can occur during the transfer of an e^- from hydroquinone to cytochrome o. It seems the ubiquinol oxidase site of the cytochrome o complex faces the **periplasm** while the dioxygen reductase site faces the **cytoplasm**. Thus, the release of two protons to the periplasm is coupled to the consumption of two protons in the cytoplasm to convert $0.5O_2 + 2H^+ + 2e^-$ to H_2O.

When *E. coli* is grown under low-oxygen conditions it requires a terminal oxidase with a higher affinity for oxygen. Thus, the high-affinity cytochrome d complex is induced and the lower affinity cytochrome o complex is repressed. A combination of regulatory proteins including Arc (see Chapter 4) are involved in this control. The cytochrome d complex utilizes cytochrome b_{558} and cytochrome d.

Anaerobic Respiration. When grown under anaerobic conditions, enteric bacteria (such as *E. coli*) cleave pyruvate to acetyl-CoA and formate. If there are no appropriate alternative electron acceptors available, acetyl-CoA and formate will be converted to fermentative end-products, as shown in Figure 7-15. However, as noted above, *E. coli*, and many other organisms, can respire even under anaerobic conditions if alternative electron acceptors are present. Each system for accepting electrons can be viewed as a modular unit. Some of these acceptor molecules are shown in Figure 7-11. They are complex membrane-associated enzymes. For example, fumarate reductase has an FAD-containing subunit, an Fe-S protein, and two membrane anchor proteins. Nitrate reductase is even more complex. A given modular system is induced when cells are grown in medium containing a specific acceptor molecule. There is, in fact, a hierarchy of induction corresponding to the potential energy that can be produced and the electrical potential of each acceptor. The Fnr, NarL, and Arc systems described in Chapter 4 control the expression of the respiratory genes. Anaerobic growth on the specific acceptor molecules also results in the synthesis of cytochromes specific for each acceptor module. Thus, the cytochrome b used for fumarate reductase is different from that used for nitrate reductase. As when oxygen is the terminal electron acceptor, the transmission of electrons to nitrate or fumarate also translocates protons to the cell exterior, generating PMF. Figure 7-15 illustrates the alternative pathways for pyruvate that are possible depending on the type of terminal electron acceptor present.

Conversion of PMF to Usable Energy. How PMF is converted into usable energy by the cell is certainly one of the keys to understanding life. There are several ways the

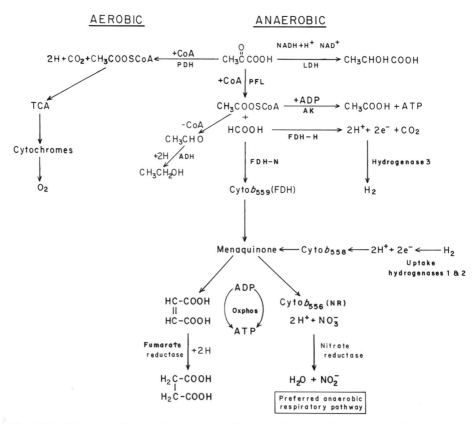

Fig. 7-15. Alternate pathways of pyruvate utilization under aerobic and anaerobic conditions. PDH = pyruvate dehydrogenase; PFL = pyruvate:formate lyase; TCA = tricarboxylic acid cycle; FDH-N = formate dehydrogenase-*N*; FDH-H = formate dehydrogenase-*H*; Oxphos = oxidative phosphorylation; AK = acetate kinase; LDH = lactate dehydrogenase; ADH = alcohol dehydrogenase.

cell can use PMF. The cell can directly generate ATP from PMF by reversing the action of the major H⁺-translocating ATPase. These are the so-called F_1F_0-type ATPases that are widely distributed in the bacterial kingdom. This type of ATPase is composed of two structurally and functionally distinct entities. The F_1 moiety lies extrinsic to the membrane in the cytoplasm and catalyzes the hydrolysis of ATP to ADP. The F_1 is attached to the membrane by a pore complex termed F_0. The F_0 moiety serves as the conductor of H^+ across the membrane. Protons passing into the cell (i.e., down the electrochemical gradient) through F_1F_0 will drive F_1 to synthesize ATP (see Figs. 1-13 and 7-16). Depending on the conditions and the efficiency of a specific system, it generally requires two or three H^+ to generate one ATP.

The PMF can also be used to drive the transport of some metabolites into the cell. Lactose, for instance, is transported into the cell by the integral membrane permease LacY. The energy required to pump lactose into the cell against an increasing lactose gradient comes from the symport of H^+ with lactose and the partial dissipation of the PMF. The flagellar motor is also driven by PMF. Each flagellar rotation requires the influx of $256H^+$!

Another question one might ask is how the cell generates PMF under anaerobic con-

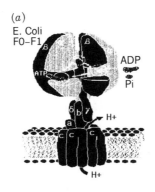

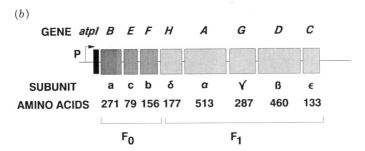

Fig. 7-16. The structure of _E. coli_ F_1F_0 ATPase complex and genetic organization of the _atp_ operon. (_a_) The ATP synthase shown with the cytoplasmic F_1 portion synthesizing ATP as H^+ is transferred through the F_0 pore and into the cell. (_b_) The _atp_ operon is transcribed as a single, polycistronic message from the promoter (P) region. The genetic designations are shown above the operon, the corresponding protein products and the number of amino acids in each are shown below the operon. From Papa, S. et al. 1993. _Ann. N.Y. Acad. Sci._ **671:**345–358.

ditions with no electron-acceptor molecules, that is, under fermentative conditions. Even under these conditions the cell requires a PMF to transport carbohydrates and to activate its flagella. But there would be no active electron-transport chains to pump H^+ out of the cell, so how can an electrochemical gradient be formed? Under these conditions, the F_1F_0 ATPase will use the energy of ATP hydrolysis to pump H^+ out of the cell, thereby generating a PMF. Loss of the F_1F_0 ATPase through mutation will prevent anaerobic growth in instances where the ATPase is required to generate PMF. In addition, these mutants will not grow aerobically on TCA cycle intermediates since substrate level phosphorylation is not possible. The H^+-translocating ATPase is the only way that ATP can be synthesized when TCA cycle intermediates are being used for growth.

Structure of F_1F_0 and the atp Operon.   As noted above, the ATP synthetase consists of two functionally distinct entities, F_1 and F_0. The F_1 moiety retains ATPase function when stripped from the membrane. It consists of five subunits in an unusual stoichiometric ratio of $\alpha_3\beta_3\gamma\delta\epsilon$ (aggregate molecular mass ~ 380 kDa). The F_0 pore consists of three subunits also in an unusual stoichiometry of $a_1b_2c_{10\pm1}$ (aggregate molecular mass ~ 150 kDa). The c subunits form the pore with the a and b subunits being involved with assembly and association between F_1 and F_0. All of the genes encoding F_1F_0 reside in a single operon (Fig. 7-16). The RNA transcription of the entire operon occurs from a promoter region upstream of the first gene such that all of the genes are cotranscribed. With this in mind it seems odd that with the unusual stoichiometries of the ATPase subunits, there is little if any excess a, b, γ, δ, or ϵ subunits in the cytoplasm. The proteins are all synthesized in the ratios as they occur in the assembled complex but there is no differential transcriptional control that would help explain why, for example, more c protein is produced than b or a. The answer appears to lie with the translational efficiencies of the various **translational initiation regions (TIR)** located at the beginning of each gene within the polycistronic message. Genes for subunits required at high stoichiometric ratios contain more efficient TIRs than do those required in smaller amounts. The TIR efficiency can be correlated to the Shine–Dalgarno (SD) consensus sequence as well as to surrounding levels of secondary structure that could block ribosome access to the SD sites.

Energy Yield.   The amount of energy derived from the oxidation of a given carbohydrate during respiration varies depending on several factors. These factors include the type of cytochrome system used to pump protons and generate PMF, the type of terminal electron acceptor, and the efficiency of the ATP synthetase. For example, both *E. coli* and yeast growing aerobically will produce NADH + H^+. Transfer of e^- from NADH to the cytochrome system of *E. coli* will pump $4H^+$ out of the cell while the yeast cytochrome system will pump $6H^+$. Furthermore, the number of protons required to generate ATP via the *E. coli* ATPase is perhaps two or three while the value for yeast mitochondrial ATP synthetase is closer to $2H^+$. So, 1 NADH + H^+ in *E. coli* may yield between 1 and 2ATP molecules, whereas yeast can produce 3ATP molecules. Figure 7-17 illustrates the energy yield from glucose using yeast as the model system.

Energetics of Chemolithotrophs

The chemolithotrophic bacteria derive energy from the oxidation of inorganic substrates. Best known are those oxidizing hydrogen to water, ammonia to nitrite, nitrite to nitrate,

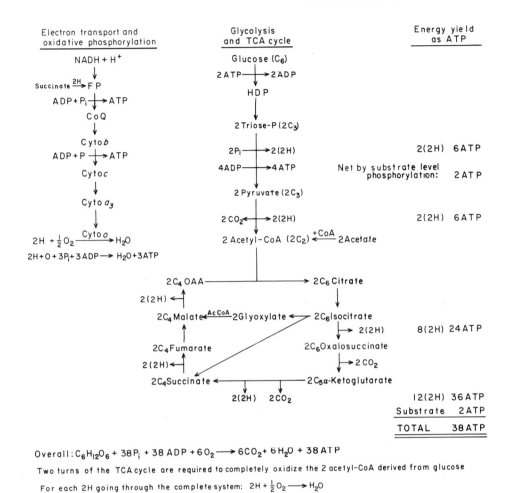

Fig. 7-17. Theoretical energy yield as ATP from glycolysis and the TCA cycle. In the calculations shown here, it is assumed that each pair of hydrogen atoms (2H) released from the substrate yields 3ATP. The reaction shown as ADP + P_i → ATP represents the action of ATP synthetase.

sulfur to sulfate, and iron(II) to iron(III). Most chemolithotrophs contain all the components of a "normal" electron-transport chain and so generate PMF in much the same manner as chemoorganotrophs (e.g., *E. coli*). The inorganic compounds serve as sources of energy in the form of electrons that can be used to pump H^+ (or other cations) out of the cell.

Hydrogen Oxidation. Aerobic hydrogen oxidation is well known in eubacteria, such as *E. coli*. Hydrogen, via hydrogenase 1 or 2 (hydrogenase 3 in *E. coli* is involved with the production of H_2 as part of the formate hydrogen lyase system), is introduced into the electron-transport chain by transfer to NAD^+ forming NADH. Subsequent electron transfers are equivalent to those we have already discussed. **Anaerobic** hydrogen oxidation is known best in the methanogens in which CO_2 serves as the hydrogen acceptor

$$CO_2 + 4H_2 \rightarrow CH_4 + 2H_2O \quad (\Delta G_0' = -130 \text{ kJ/CH}_4)$$

Hydrogen can also be oxidized by some sulfate reducing bacteria as

$$4H_2 + SO_4^{2+} \rightarrow H_2S + 2H_2O + 2OH^- \quad (\Delta G_{20}' = -152 \text{ kJ/mol})$$

Inorganic Nitrogen Compounds. Nitrifying bacteria use either ammonia or nitrite as their energy source

$$NH_3 + 1.5O_2 \rightarrow HNO_2 + H_2O \quad (\Delta G = -272 \text{ kJ/mol})$$
$$NO_2^- + H_2O \rightarrow NO_3^- + 2H^+ + 2e^- \quad (\Delta G = -73 \text{ kJ/mol})$$

Iron Oxidation. Organisms that can derive metabolic energy from Fe^{2+} include *Thiobacillus ferrooxidans* and *Leptospirillum ferrooxidans*. In all cases, oxidation takes place at acidic pH values, physiological optimum pH 2. This allows an increased solubility of the ferric ion product. These organisms, then, are acidophilic, which presents some problems in terms of maintaining internal pH (see pH Homeostasis in the following section). Iron oxidation takes place according to the following equation:

$$4FeSO_4 + O_2 + 2H_2SO_4 \rightarrow 2Fe_2(SO_4)_3 + 2H_2O \quad (\text{at pH 2 } \Delta G = -30 \text{ kJ/mol})$$

These and other forms of chemolithotrophy are discussed in further detail below.

pH Homeostasis

Most enzymes and proteins have rather narrow pH optima for activity or function. Nevertheless, bacteria have an amazing capacity to grow over a wide range of pH values. The bacteria can be divided into three groups depending on their preferred range of growth pH. Organisms that prefer acid, neutral, or alkaline environments are called acidophiles, neutralophiles, or alkalophiles, respectively. *Escherichia coli* is a neutralophile, but even it can grow over a 1000-fold range of pH, from pH 5 to 8, and it can survive over a 100,000-fold range, from pH 4.0 to 9.0!. But, in doing this, the internal pH (pH$_i$) does not vary more than 10-fold, and usually much less than that. At external pH values from 5.5 to 8, pH$_i$ only wavers between 7.2 and 7.7. How this extraordinary degree of homeostasis is accomplished remains one of the fundamental questions of bacterial physiology.

There are a number of factors that can influence pH. Among these are the buffer capacity of the cell, the outward transport of protons associated with respiration or ATP hydrolysis, and electroneutral transport systems that exchange protons for certain cations, particularly Na^+ and K^+. For example, as environmental pH becomes more acidic, any **decrease** in pH$_i$ could trigger a K^+/H^+ antiport system that would bring K^+ into the cell while extruding a proton. This will maintain internal pH near 7.5 and create a larger ΔpH. Likewise, an increase in external pH will cause alkalinization of the cytoplasm that could trigger Na^+/H^+ antiport systems that will extrude Na^+ and import protons to acidify the cytoplasm.

Aside from pH homeostasis, bacteria can induce an acid protection system called the **acid tolerance response** that will increase the resistance to low pH. While the specifics of these systems are not known and may differ among bacteria, it is generally believed that protection or repair of macromolecules exposed to low pH are the principle means of affording acid resistance.

TRANSPORT

The cytoplasmic membrane forms a hydrophobic barrier impermeable to most hydrophilic molecules. Consequently, mechanisms must exist that enable the cell to transport growth solutes into the cell and waste solutes out. The outer membrane of gram-negative cells (e.g., *E. coli*) contains proteins that form narrow channels (porins) that facilitate passive diffusion of hydrophilic compounds, such as sugars, amino acids, and certain ions, to the periplasm. Only compounds smaller than 600–700 Da can pass. Hydrophobic compounds of any size are excluded. The structure of porins, was discussed in Chapter 6 and their regulation was discussed in Chapter 4.

A few compounds essential to bacterial growth, such as vitamin B_{12}, oligosaccharides, and iron chelates, that are too large to enter via the normal pores have specific transport proteins or specific pores in the outer membrane that mediate their passage. For example, maltose traverses the outer membrane via a specific porin termed maltoporin.

Oxygen, carbon dioxide, ammonia, and water all diffuse through both the outer and inner membranes without hindrance. Although most essential nutrients can diffuse through the outer membrane, all require one or more transport systems to cross the inner membrane and gain access to the cytoplasm. The goal of most of these transport systems is to concentrate nutrients inside the cell that may be in short supply externally. Consequently, they work **against** a concentration gradient and require energy in some form to accomplish this. There are several general transport mechanisms involved in membrane transport. These are diagrammed in Figure 7-18 and are divided into four groups: facilitated diffusion, chemiosmotic driven transport, binding protein-dependent transport, and group translocation.

Facilitated Diffusion. This is the only example of a transport system that does not require energy. Facilitated diffusion allows for passage of a substrate down a concentration gradient. Consequently, the substrate will never achieve a concentration inside the cell greater than what exists outside the cell.

Glycerol is one of the few compounds that enters prokaryotic cells by facilitated diffusion. The transport of glucose in *Zymomonas mobilis* and *Streptococcus bovis* appears to be the only other example of facilitated diffusion in prokaryotes. The glycerol facilitator (GlpF) is the only known pore-type protein in the cytoplasmic membrane of *E. coli*. This facilitator is encoded by *glpF*, the first gene in an operon with *glpK* (encoding glycerol kinase). This porin-type channel in the cytoplasmic membrane also allows passage of polyhydric alcohols and other related small molecules, such as urea and glycine, but excludes charged molecules, such as glyceraldehyde-3-phosphate (G-3-P) or DHAP.

Periplasmic-Binding Protein-Dependent Transport System (Shock Sensitive). Periplasmic proteins of gram-negative bacteria can be released into the surround-

(a)

(b)

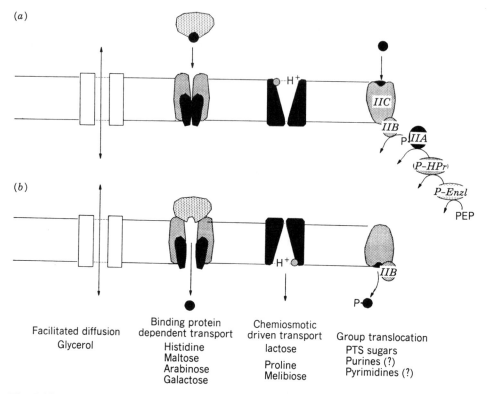

	Binding protein	Chemiosmotic	
Facilitated diffusion	dependent transport	driven transport	Group translocation
Glycerol	Histidine	lactose	PTS sugars
	Maltose		Purines (?)
	Arabinose	Proline	Pyrimidines (?)
	Galactose	Melibiose	

Fig. 7-18. Possible models for cytoplasmic membrane-transport systems. Each of the systems probably has one or more membrane-spanning proteins that form a special channel in the cytoplasmic membrane. Facilitated diffusion and ion-driven transport systems require only one gene product, whereas binding protein-dependent transport and group translocation systems require several gene products. Facilitated diffusion allows entry and exit of substrate through a specific pore. Bonding of substrate to specific sites accessible to the periplasmic side of the active transport systems coupled with an appropriate energy source (a) allows a conformational change in the carrier proteins and release the substrate inside the cell (b). These cartoons are highly speculative; the precise mechanism of transport is not yet understood for any of these systems.

ing medium if cells are subjected to osmotic shock (cells suspended in buffered 20% glucose containing EDTA are centrifuged and rapidly resuspended into cold $MgCl_2$). Among these periplasmic proteins are **binding proteins** for specific nutrients, such as SO_4^{2-}, amino acids, sugars, and other compounds. These proteins function by transferring the bound substrate to a compatible membrane-bound complex of four proteins (Fig. 7-18). The transport process is energized by ATP or other high-energy phosphate compounds, such as acetyl phosphate. Approximately 40% of the substrates transported by *E. coli* involve periplasmic-binding protein-dependent mechanisms. A subset of these transport systems, called **traffic ATPases**, bear striking subsequence homologies to each other. In these systems, it appears that transfer of the substrate to the membrane complex triggers ATP hydrolysis which, in turn, leads to the opening of a pore that allows unidirectional diffusion of the substrate to the cytoplasm. The histidine permease of *E. coli* is probably the best studied example of a traffic ATPase.

Chemiosmotic Driven Transport. These systems accomplish movement of a molecule across the membrane at the expense of a previously established ion gradient, such as proton motive or sodium motive force. About 40% of the substrates that enter *E. coli* involve ion-driven transport. There are three basic types: **symport, antiport, and uniport. Symport** involves the simultaneous transport of two substrates in the **same** direction by a single carrier. For example, a proton gradient can allow symport of an oppositely charged ion or a neutral molecule. Transport of lactose by the LacY permease is such an example. **Antiport** is the simultaneous transport of two like-charged compounds in opposite directions by a common carrier. The Na^+/H^+ antiporters of *E. coli* (NhaA and NhaB) are examples that are believed to be important for generating sodium motive force and in maintaining neutral internal pH under alkaline growth conditions. **Uniport** occurs when movement of a substrate is independent of any coupled ion. Transport of glycerol is an example of uniport and was described above as facilitated diffusion. Table 7-3 lists examples of chemiosmotic transporters and their coupling ions.

Group translocation couples transport of the substrate to its chemical modification (e.g., by attaching a phosphate or CoA group to the substrate). This traps the substrate within the cell in a form different from the exogenous substrate so that the concentration

TABLE 7-3. Representative Chemiosmotic Porters

Uniport	Symport	Antiport
Glycerol	H^+ / **amino acid**	H^+:**cation**
	H^+ / glycine[a]	H^+:Ca^{2+} [a]
	H^+ / histidine[a]	H^+:$CaHPO_4^a$
	H^+ / lysine[a]	H^+:K^+ [a]
	H^+ / phenylalanine[a]	H^+:Na^+ [a]
	H^+ / **organic acid**	K^+:**cation**
	H^+ / DL-lactate[a]	K^+:$CH_3NH_3^{+a}$
	H^+ / pyruvate[a]	$H_2PO^-_4$:organic anion
	H^+ / succinate[a]	$H_2PO_4^-$:hexose 6-P[a]
	H^+ / gluconate[a]	$H_2PO_4^-$:glycerol 3-P[a]
	H^+ / **sugar**	$H_2PO_4^-$:PEP[b]
	H^+ / arabinose[a]	
	H^+ / galactose[a]	
	H^+ / lactose[a]	
	H^+ / **inorganic ion**	
	H^+ / phosphate[a]	
	Na^+ / **amino acid**	
	Na^+, H^+ / glutamate[a]	
	Na^+ (H^+) / proline[a]	
	Na^+ / **sugar**	
	Na^+ (H^+) / melibiose[a]	
	Mg^{2+} / **organic acid**	
	Mg^{2+}, H^+ / citrate[c]	

[a]*Escherichia coli.*
[b]*Salmonella typhimurium.*
[c]*Bacillus subtilis.*

gradient of unmodified substrate never equilibrates. The phosphotransferase system (*pts*) involved in the transport of many carbohydrates functions in this manner.

Establishing Ion Gradients. The establishment of ion gradients is of supreme importance to microorganisms. Proton and sodium gradients are important in various organisms for energy production, as discussed earlier, and in transport. Some consideration should be given at this point to the mechanism by which gradients are established and maintained. The cell membrane is impermeable, for the most part, to charged ions so the cell has an opportunity to control the flow of ions across the membrane through ion-specific transport systems. As described earlier, proton gradients are established principally by the electron-transport systems that pump protons out of the cell as electrons are transferred down the system. There are also specific membrane-bound ATPases that can couple the hydrolysis of ATP with the export of H^+, Na^+, and K^+. In addition, as shown in Table 7-4, there are a variety of antiport systems that can exchange ions such that one gradient can help build another (e.g., Na^+/K^+). So the picture that develops is an interrelated series of ion circulations whose purpose is to provide energy to the cell.

Specific Transport Systems

ATP-Linked Ion-Motive Pumps. The two major classes of these ion pumps are the F-type (F_1F_0) and P-type (E_1E_2) ATPases. As described above, the F-type ATPase is involved with pumping H^+ out of the cell or in coupling H^+ movement into the cell with the generation of ATP. The P-type ATPases are remarkably similar to each other despite the wide range of ions they transport. They include potassium, magnesium, calcium, cadmium (resistance), and arsenate (resistance). All members have the following properties: (1) a phosphorylated intermediate (aspartylphosphate in all ATPases where the phosphorylated amino acid has been determined); (2) two conformational forms of the phosphorylated intermediate, referred to as E_1 and E_2, that differ in reactivity to substrates and proteases; (3) a large (~100 kDa) membrane-bound subunit with six-to-eight membrane-spanning regions and several regions of amino acid sequence homology. Most P-type ATPases have only this single large subunit. An exception is the Kdp ATPase.

The Kdp ATPase of *E. coli* was first identified from the analysis of mutants that could not grow on low concentrations of potassium. It is a three subunit enzyme (Fig. 7-19). The KdpB protein is homologous to other P-type ATPases and is the subunit that is phosphorylated by ATP and couples energy to transport. The KdpA subunit binds K^+ and probably forms most or all of the membrane channel for K^+. The high-energy E_1 state occurs when K^+ is bound at the outer surface. Opening of the membrane channel and movement of K^+ to the inner surface results in the lower energy E_2 form of the transport system. The number of ions transported per ATP hydrolyzed is not known. Neither is the nature of transport, whether electrogenic or electroneutral involving an exchange for another ion. Regulation of the system is unusual in that the *kdpABC* operon is only expressed when cells are unable to maintain the desired pool of K^+. (Another constitutively expressed, non-P-type system, called Trk, is normally involved with K^+ transport). However, it appears that low turgor pressure (see Chapter 4) rather than internal or external concentrations of K^+ is the signal to turn on expression. The membrane sensor, KdpD, somehow measures turgor, possibly by being attached to both the inner and outer membranes. The KdpE sensor is the soluble DNA-binding protein that is presumably transphosphorylated by KdpD and will control expression of *kdpABC*.

TABLE 7-4. Enzymes II of PTSs

PTS	Abbreviation	Organism	Substrates for Growth and/or Transport[a]	Gene(s)[b]	Domain(s)[b]
Glucose class					
Glucose	Glc	*E. coli*	Glc, GlcN, Sor, αMG, 5TG, Atl, Rtl, Man, 2DG	*ptsG, crr*	IICB, IIA
Glucose	GLC	*S. typhimurium*	Glc, αMG, 5TG, Man, 2DG	*ptsG, crr*	?, IIA
Glucose	GLC	*B. subtilis*	Glc, αMG	*ptsG*	IICBA
"Maltose"	Mal	*E. coli*	Mal, Glc	*malX, crr*	IICB; IIA
Trehalose	Tre	*E. coli*	Tre	*treB, crr*	?, IIA
N-acetylglucosamine	Nag	*E. coli*	Nag, Stz, αNag	*nagE*	IICBA
N-acetylglucosamine	Nag	*K. pneumoniae*	Nag, Stz	*nagE*	IICBA
Sucrose	Scr	*Enterobacteriaceae*	Scr, Glc	*scrA, crr*	IIBC, IIA
Sucrose	Scr	*K. pneumoniae*	Scr	*scrA, crr*	IIBC, IIA
Sucrose	Scr	*V. alginolyticus*	Scr	*scrA, ?*	IIBC, ?
Sucrose	Sac	*B. subtilis*	Scr	*scrP, ?*	IIBC, ?
Sucrose	Sac	*B. subtilis*	Scr	*scrX, ?*	IIBC, ?
Sucrose	Scr	*S. mutans*	Scr	*scrA*	IIBCA
β-Glucosides	Bgl	*E. coli*	Bgl, Glc	*blgF*	IIBCA
β-Glucosides	Arb	*E. chrysanthemi*	Bgl	*arbF*	IIBCA
β-Glucosides	Asc	*E. coli*	Arb, Sal, Cel	*ascF, ?*	IIBC, ?
Mannitol class					
Mannitol	Mtl	*E. coli*	Mtl, Gut, Atl, 2DA	*mtlA*	IICBA
Mannitol	Mtl	*S. carnosus*	Mtl	*mtlA, mtlF*	IICB, IIA
Mannitol	Mtl	*S. faecalis*	Mtl	*mtlA, mtlF*	IICB, IIA
Mannitol	Mtl	*S. mutans*	Mtl	*?, mtlF*	?, IIA
Fructose	Fru	*E. coli*	Fru, Xtl, Glc, Sor, Man	*fruF, fruA*	FPr[c], IIBC
Fructose	Fru	*S. typhimurium*	Fru, Xtl	*fruF, fruA*	FPr[c], ?
Fructose	Fru	*R. capsulatus*	Fru	*fruA*	MTP[?], IIBC
Fructose	Fru	*X. campestris*	Fru	*?, fruA*	?, IIBC

TABLE 7-4. Continued

PTS	Abbreviation	Organism	Substrates for Growth and/or Transport[a]	Gene(s)[b]	Domain(s)[b]
Lactose class					
Lactose	Lac	*L. lactis*	Lac, Gal	*lacF, lacE*	IIAB, IICB
Lactose	Lac	*L. casei*	Lac, Gal	*lacE, lacF*	IICB, IIA
Lactose	Lac	*S. aureus*	Lac, Gal	*lacF, lacE*	IIA, IICB
Lactose	Lac	*S. mutans*	Lac	*lacF, lacE*	IIA, ?
Cellobiose	Cel	*E. coli*	Cel	*celA, CelB, celC*	IIB, IIC, IIA
Mannose class					
Mannose	Man	*E. coli*	Man, Nag, GlcN, Fru, 2DG, Glc, Tre, αMG	*manX, manY, manZ*	IIAB, IIC, IID
L-Sorbose	Sor	*K. pneumoniae*	Sor, Fru, Glc	*sorF, sorB, sorA, sorM*	IIA, IIB, IIC, IID
Fructose	Lev	*B. subtilis*	Fru	*levD, levE, levF, levG*	IIA, IIB, IIC, IID
PTS of unknown classification					
Glucitol	Gut	*E. coli*	Gut, 2DA, Mtl, Atl	*gutA, GutB*	II(CB), IIA

[a] Abbreviations: Glc = Glucose; GlcN = glucosamine; Sor = sorbose; αMG = methyl α-glucoside; 5TG = 5-thioglucose; Atl = arabinitol; Rtl = ribitol; Man = mannose; 2DG = 2-deoxyglucose; Mal = maltose; Tre = trehalose; Nag = *N*-acetylglucosamine; Stz = streptozootocin; αNag = methyl-2-acetamido-2-deoxy-α-D-glucoside; Scr = sucrose; Bgl = β-glucoside; Arb = arbutin; Sal = salicin; Cel = cellobiose; Mtl = mannitol; Gut = glucitol; 2DA = 2-deoxyarabinohexitol; Fru = fructose; Xtl = xylitol; Lac = lactose; Gal = galactose.

[b] Not known.

[c] FPr composed of an ELL[Fru] domain (amino terminal) and an HPr-like domain (carboxy terminal); MTP (multiphosphoryl transfer protein), composed of an ELL[Fru] domain, an HPr-like domain, and an EL-like domain.

From Postma, P. W. et al. 1993. *Microbiol. Rev.* **57**:543–594.

Kdp Potassium Transport

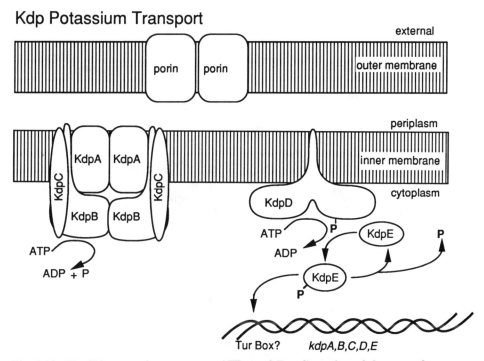

Fig. 7-19. The Kdp potassium-transport ATPase of *E. coli* consists of three-membrane proteins, KdpA, KdpB, and KdpC. KdpB protein is homologous to other E_1 - E_2 ATPases. The outer-membrane porin protein is not specified, and there is no periplasmic K^+-binding protein. The regulatory proteins, KdpD and KdpE, are thought to undergo a cycle of auto- and *trans*-phosphorylation analogous to that for PhoR and PhoB. The Tur box regulatory region for turgor pressure is tentative. Derived from Silver, S. and M. Walderhang. 1992. *Microbiol. Rev.* **56**:195–228.

The Histidine Permease. A member of the traffic ATPases, the histidine permease comprises the histidine-binding protein, HisJ, and three proteins, HisQ, HisM, and HisP, that form the membrane-bound complex. The membrane-bound complex consists of one copy each of HisQ and HisM and two copies of the ATP-binding subunit, HisP. The receptor, HisJ, binds histidine and changes conformation allowing it to interact with the membrane-bound complex. Translocation then occurs through a series of conformational changes concomitant with ATP hydrolysis and is hypothesized to take place through a pore formed by HisQ and HisM.

Iron Transport. Iron uptake by bacteria is unique. It involves the synthesis and secretion of small iron-binding chelates, called **siderophores**, that bind environmental iron and then must be transported back into the cell to release the iron. A major siderophore of *E. coli* is enterochelin. Its chemical structure and biosynthesis is described in Chapter 10. The mechanism by which desferrienterochelin is secreted remains a mystery. Once in the environment, it will bind to Fe^{3+} after which products of the *fep* genes enable movement of ferrienterochelin back into the cell (Fig. 7-20). The outer-membrane FepA protein (79 kDa) functions as a monomer and enables transport of the complex across the outer membrane. It contains as many as 29 β-sheet transmembrane spanning regions

Enterochelin Iron Transport

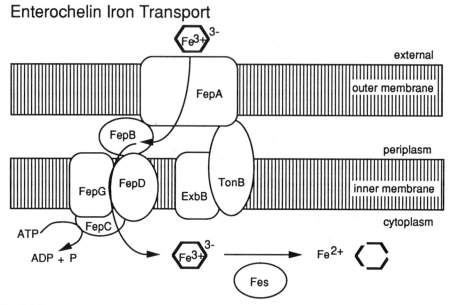

Fig. 7-20. The enterochelin (Ent) iron system. Enterochelin, the trimeric catechol siderophore depicted as a hexagon, passes through the FepA protein of the outer membrane in a process requiring energy coupled through TonB. The periplasmic FepB protein passes the Fe^{3+} enterochelin to the inner-membrane proteins FepG and FepD. Passage across the inner membrane requires ATP energy through FepC. In the cytoplasm, the Fes enterochelin esterase cleaves enterochelin, allowing the release and subsequent reduction of iron. Derived from Silver, S. and M. Walderhang. 1992. *Microbiol. Rev.* **56:**195–228.

and an amino acid sequence called a Ton box. This region is predicted to contact the TonB inner-membrane protein in a way that will "transduce" potential energy of the cytoplasmic membrane to FepA and drive transport of the siderophore across the outer membrane. Another inner-membrane protein, ExbB, functions either to stabilize or activate TonB. Next, a periplasmic ferrienterochelin-binding protein (FebB) delivers the complex to the inner-membrane transport system comprised of FepG, FepD, and FepC. The FepC component is a membrane-associated ATPase that provides the energy required to effect transport of the complex through FepG and FepD. Once inside the cell, the enterochelin backbone is cleaved (by Fes esterase) reducing the affinity for iron from 10^{-52} to 10^{-8}. Regulation of this system is also interesting. The Fe^{3+} released in the cell is reduced by an unknown mechanism to Fe^{2+}. As ferrous iron accumulates in the cell it will bind to the regulatory protein Fur (*ferric up*take regulator). Metalated Fur then binds to a 17 bp consensus sequence (the IRON box) located in front of the seven transcriptional units that contain the *ent*, *fep*, and *fes* genes and represses their expression. Fur, however, has a much more global role in the cell. At least 50 genes experience defective regulation in the absence of Fur. Its impact on the physiology of the cell is enormous.

The Phosphotransferase System (PTS). This system is involved in both the transport and phosphorylation of a large number of carbohydrates, in movement toward these carbon sources (chemotaxis), and in the regulation of several other metabolic pathways (catabolite repression). In this group translocation system, carbohydrate phosphorylation is coupled to carbohydrate transport, the energy required being provided by the EMP intermediate PEP. Figure 7-21 shows that there are two proteins common to all of the PTS carbohydrates. They are enzyme I (*ptsI* in *E. coli*) and the histidine protein, HPr (*ptsH*). They are soluble, cytoplasmic proteins that participate in the phosphorylation of all PTS carbohydrates in a given organism and so are called the **general PTS proteins**. In contrast, the enzymes II (EIIs) are **carbohydrate specific**. The EIIs consist of three domains (A, B, and C) that may be combined in a *single* membrane-bound protein or split into two or more proteins (depending on the system) called IIA, IIB, and IIC. The B and C domains may form one protein called IICB. Likewise, the A and B subunits may form one protein called IIAB. The IIC protein is an integral membrane protein (or domain) and IIA (formerly called enzyme III) is soluble. In any scenario, the phosphate group from PEP is transferred to the incoming carbohydrate via phosphorylated intermediates of EI, HPr, EIIA, and EIIB. The EIIC protein forms the translocation channel and at least part of the specific carbohydrate-binding site. On the basis of sequence homologies, EIIs may be grouped into four classes: **mannitol, glucose, mannose,** and

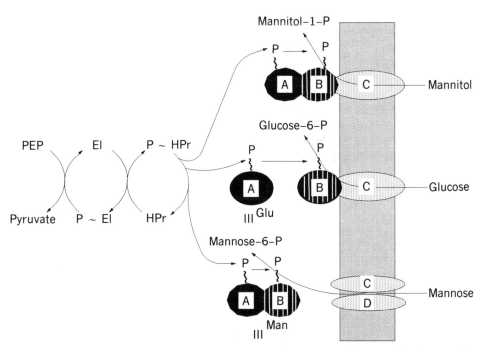

Fig. 7-21. Organization of phosphotransferases (PTSs). The EI and HPr proteins are general proteins for all PTSs. Of the many EIIs, only three are shown: those specific for mannitol (Mtl), glucose (Glc), and mannose (Man). Each contains two hydrophilic domains, IIA (formerly EIII or III) containing the first phosphorylation site (P-His), and IIB containing the second phosphorylation site (either a P-Cys or a P-His residue). The membrane-bound, hydrophobic domain IIC may be split into two domains (IIC and IID). The II^{Mtl}, II^{Glc}, and II^{Man} domains are specific for mannitol, glucose, and mannose, respectively. P, indicates the phosphorylated forms of the various proteins. Derived from Postma P. W. et al., 1993. *Microbiol. Rev.* **57**:543–594.

lactose (see Table 7-5). The lactose class is found in a few gram-positive organisms in which galactose and lactose are PTS carbohydrates. A discussion of the role of the PTS in catabolite repression of gene expression can be found in Chapter 4. The PTS is also involved in chemotaxis, a basic discussion of which is found in Chapter 6. The PTS allows chemotaxis toward PTS carbohydrates. Stimulation corresponds to uptake and phosphorylation of the substrate through EII. No MCP is involved in this process. Neither are CheB nor CheR. While the exact mechanism is not clear, it is thought that dephosphorylated EI or HPr, formed during carbohydrate transport, could directly or indirectly decrease the level of P-CheA or P-CheY with resultant increase in smooth swimming and positive chemotaxis.

PATHWAYS FOR UTILIZATION OF SUGARS OTHER THAN GLUCOSE

Glucose is the preferred carbon source for most of the common heterotrophic bacteria. Nevertheless, microorganisms possess a high degree of versatility in their ability to use a wide range of compounds as sources of carbon and energy. It is important that most (if not all) of these pathways invariably lead to intermediates that can enter one of the major pathways already described (that is, the EMP, pentose shunt, TCA cycle, or glyoxylate cycle). Many of these carbon sources are utilized only after a period of **induction** of the enzymes required for their transport and metabolism. The presence of glucose in the growth medium generally inhibits the expression of the catabolic enzymes for these substrates. This **glucose effect** was first described in 1947 by Monod for the inhibition of β-galactosidase synthesis in *E. coli* and in 1961 the term **catabolite repression** was coined for this phenomenon by Magasanik. In media containing glucose and another carbohydrate, bacteria exhibit two complete growth cycles separated by a lag period. This phenomenon has been termed **diauxic growth**. Glucose prevents entry of the second substrate by a process known as **inducer exclusion** and prevents induction of the genes required for metabolism of the second substrate. In the expression of the lactose (*lac*) operon, cyclic adenosine monophosphate (**cyclic AMP** or **cAMP**) overcomes the different glucose effects to varying extents (see Chapter 4).

Lactose Utilization. In *E. coli*, utilization of lactose requires the induction, under control of the *lac* operon, of a specific permease for its transport into the cell and β-galactosidase, which cleaves it to form D-galactose and D-glucose (Fig. 7-22). The D-galactose is phosphorylated to galactose-1-phosphate and metabolized via the Leloir pathway to yield fructose-6-phosphate (Fig. 7-23). Fructose-6-phosphate is ultimately metabolized via the triose phosphate pathway. Organisms, such as *Lactobacillus casei*, do not contain the system encoded by the *lac* operon of *E. coli* but possess a PEP phosphotransferase system (PTS), which phosphorylates lactose via a specific enzyme II. The intracellular lactose phosphate is split by phospho-β-galactosidase to yield glucose and galactose-6-phosphate. Galactose-6-phosphate is metabolized via the tagatose-6-phosphate pathway to yield glyceraldehyde-3-phosphate and dihydroxyacetone phosphate (Fig. 7-23). In both organisms the D-glucose is metabolized via the EMP pathway. In *Staphylococcus aureus*, the tagatose-6-phosphate pathway is required for the utilization of galactose as well as lactose. The Lac⁻ Gal⁻ strains of *Klebsiella* lacking both its *lac* and *gal* operons give rise to lactose-utilizing mutants that transport lactose via the PEP–PTS and metabolize lactose via a phospho-β-galactosidase. The resultant galactose-6-phosphate is metabolized via the tagatose phosphate pathway known to operate in *Klebsiella*.

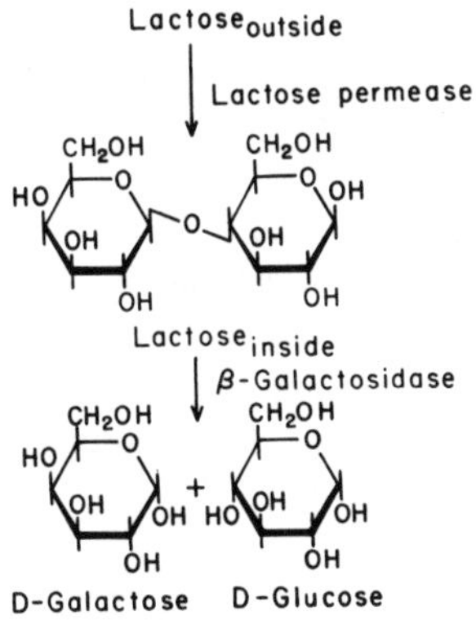

Lactose$_{outside}$

| Lactose permease

Lactose$_{inside}$

| β-Galactosidase

D-Galactose   D-Glucose

Fig. 7-22. Lactose utilization in *E. coli*.

Galactose. The inducible enzyme system required for galactose utilization in *S. cerevisiae* is under the control of a complex system of structural and regulatory genes. The enzymes involved include a galactose permease (encoded by the structural gene *GAL2*) and three enzymes of the Leloir pathway: galactokinase, a transferase (galactose-1-phosphate uridyltransferase), and epimerase (uridine diphosphoglucose-4-epimerase) encoded by the structural genes *GAL1*, *GAL7*, and *GAL10*. Phosphoglucomutase converts glucose-1-phosphate to glucose-6-phosphate, which then enters the glycolytic pathway. The galactose pathway enzymes are coordinately controlled by a positive factor required for the expression of structural genes and a negative factor that interacts with the inducer (galactose) to modulate the function of the positive factor. Expression of the structural genes is controlled by carbon catabolite repression.

Maltose. Utilization of the glucose-containing disaccharide maltose by *E. coli* requires

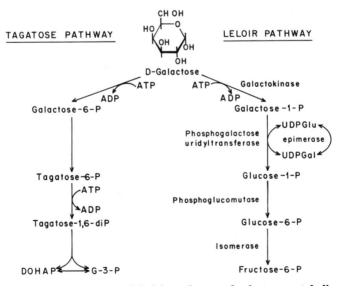

Fig. 7-23. The tagatose and Leloir pathways of galactose metabolism.

the expression of genes concerned with maltose uptake (*malEGF, malK,* and *lamB* located at 91 min) and induction of amylomaltase and maltodextrin phosphorylase (encoded by the *malPG* operon). Expression of these genes is induced by maltose and is mediated indirectly by the *malT* activator. The cAMP receptor protein (CRP) binds to cAMP and positively regulates the *malT* gene and the *malEGF* operon. The action of cAMP–CRP on *malPQ* genes is mediated by the MalT protein. Amylomaltase hydrolyzes maltose with the production of D-glucose and the polysaccharide maltodextrin (Fig. 7-24). Maltodextrin is converted to glucose-1-phosphate by maltodextrin phosphorylase. With the isomerization of glucose-1-phosphate to glucose-6-phosphate by the action of phosphoglucomutase, both products are utilized via the EMP pathway. Similar systems for the utilization of maltose are operative in *Neisseria meningitidis.* In *S. cerevisiae* and *S. carlsbergensis* there are at least five unlinked loci controlling the fermentation of maltose. Each active *MAL* locus is a complex of three genes essential for maltose fermentation: maltose permease (Gene 1), maltase (Gene 2), and transcriptional activator protein (Gene 3). Strains carrying an active allele of any one of these loci are inducible for maltase production. Maltase converts maltose directly to glucose units that are utilized via the glycolytic pathway without the intermediary formation of maltodextrins.

Mannitol. Catabolism of mannitol and other hexitols (D-glucitol or sorbitol, and D-galactitol) by *E. coli* involves transport by specific enzymes II of the PEP–PTS. The phosphorylated derivative is then oxidized to the corresponding sugar phosphate by a specific hexitol phosphate dehydrogenase (Fig. 7-25). The catabolic enzymes for mannitol, glucitol, and galactitol are encoded within three operons (*mtl, gut,* and *gat*) on the *E. coli* chromosome. All three operons have the same gene order: a regulatory gene (C), the hexitol-specific enzyme II of the PEP–PTS (A), and the hexitol phosphate dehydrogenase (D). In the case of mannitol, mannitol-1-phosphate is oxidized by a specific NAD-dependent dehydrogenase to yield fructose-6-phosphate, which enters the general metabolic pathway. *Bacillus subtilis* transports D-glucitol (sorbitol) through an inducible permease. The free intracellular hexitol is oxidized to D-fructose by an inducible D-glucitol dehydrogenase. D-Fructose then enters the glycolytic pathway via fructose-6-

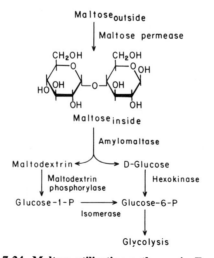

Fig. 7-24. Maltose utilization pathways in *E. coli.*

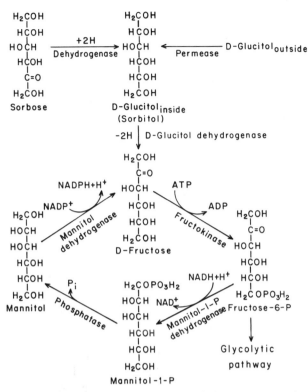

Fig. 7-25. Pathways of metabolism of sorbose and mannitol. In many organisms the mannitol cycle functions as a means of regulating reduced NADP levels.

phosphate. The filamentous fungus *Cephalosporium chrysogenum* produces a hexitol dehydrogenase of broad substrate specificity that can use either NAD or NADP as cofactor. *Gluconobacter oxydans* and *Candida utilis* have also been shown to produce a hexitol dehydrogenase with broad substrate specificity.

Mannitol is also produced as an end-product of glucose metabolism in *E. coli*, *Staphylococcus aureus*, and *Streptococcus mutans*. This hexitol may serve as a storage reserve. In these organisms and in *Lactobacillus brevis*, *Leuconostoc mesenteroides*, *Absidia glauca*, and *Rhodococcus erythropolis*, mannitol dehydrogenase is an NAD-dependent enzyme. However, many organisms utilize a mannitol cycle as a means of regenerating NADP by virtue of the presence of an NADP-dependent mannitol dehydrogenase that oxidizes mannitol to fructose. The cycle operates via the pathway shown in Figure 7-25. The fruiting body of *Agaricus bisporous* accumulates large amounts of mannitol, whereas only low amounts are found in the mycelium. In this ascomycete, a NADPH-specific mannitol dehydrogenase catalyzes the conversion of fructose to mannitol.

Escherichia coli can also convert D-mannitol to D-ribose. Mannitol-1-phosphate is formed when the organism is grown on defined media devoid of mannitol. A substantial portion of the label from D[1-^{14}C]mannitol enters the cell via the PEP–PTS and is phosphorylated equally at C-1 or C-6. The label disappears gradually from the mannitol-1-phosphate pool and approximately 60% can be recovered in the ribose isolated from RNA. The results suggest a pathway of synthesis of ribulose-5-phosphate from mannitol-1-phosphate involving the steps shown in Figure 7-26.

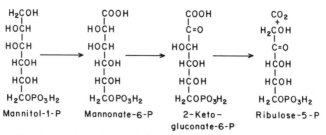

Fig. 7-26. Conversion of mannitol-1-phosphate to ribulose-1-phosphate.

Fucose and Rhamnose Utilization. Interesting parallel pathways are utilized by *E. coli* for the metabolism of L-fucose and L-rhamnose. These pathways (Fig. 7-27) involve the induction of specific permeases, kinases, isomerases, and aldolases. All of the enzymes involved in fucose utilization are encoded by a cluster of genes at 60.2 min on the *E. coli* genetic map. The *fuc* genes are organized into four operons under the influence of *fucR*, a positive regulatory protein. Fuculose-1-phosphate is the true inducer. The genes for the rhamnose system constitute a well-defined operon, whereas the fucose system maintains the gene for aldolase (*ald*) under separate control. Both pathways converge with the formation of identical products: dihydroxyacetone phosphate and L-lactaldehyde. Dihydroxyacetone phosphate is metabolized via the triose phosphate pathway. Under aerobic conditions an NAD-linked lactaldehyde dehydrogenase oxidizes L-lactaldehyde to L-lactate and an FAD-linked dehydrogenase oxidizes L-lactate to pyruvate, which then enters the general metabolic network. Anaerobically, an NAD-linked oxidoreductase is induced with the resultant formation and excretion of L-1,2-propanediol. A single oxidoreductase serves both pathways and is induced by either L-fucose or L-rhamnose. However, each methyl pentose exerts its influence at a different level. L-Fucose is a transcriptional activator of *fucO*, whereas L-rhamnose operates posttranscriptionally. The L-fucose catabolic pathway operative in *E. coli* can be recruited for the metabolism of other carbohydrates including 6-deoxy-L-talitol and D-arabinose. Identical systems for the dissimilation of L-fucose and L-rhamnose are present in *S. typhimurium* and *K. pneumoniae*.

Metabolism of Mellibiose, Raffinose, Stachyose, and Guar Gum. These compounds all contain α-(1→6)-linked galactose residues, as shown in Figure 7-28. Many species of enteric bacteria can ferment mellibiose and raffinose, but relatively few can utilize galactomannans, such as guar gum. *Bacteroides ovatus*, an obligately anaerobic resident of the intestinal tract of humans, develops α-galactosidase activity when grown on mellibiose, raffinose, or galactomannan. However, β-D-mannases, enzymes that degrade the mannose backbone of guar gum (galactomannan), are produced only during growth on this compound. The α-galactosidase I, differs from the α-glactosidase II, which is produced during growth on mellibiose, raffinose, or stachyose.

PECTIN AND ALDOHEXURONATE PATHWAYS

Pectin, a highly methylated form of poly-β-1,4-D-galacturonic acid, is a major constituent of plant cell walls. The ability to degrade pectin and its derivatives contributes to the virulence of bacterial phytopathogens, such as *Erwinia chrysanthemi* and *E. carotovora*,

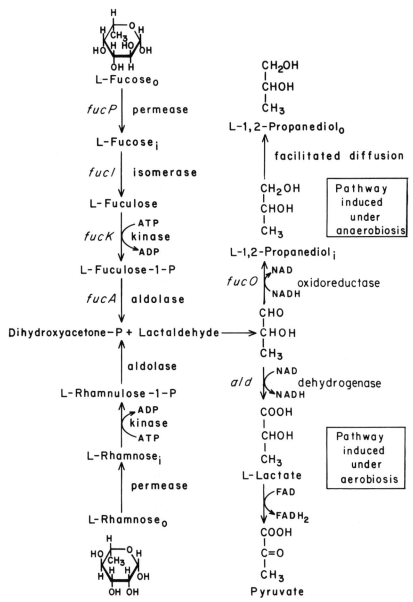

Fig. 7-27. Pathways for the utilization of L-fucose and L-rhamnose. The three-letter structural gene designations are those for *E. coli*.

and fungal phytopathogens, such as *Moloninia fructigena*, *Cladosporium cucumerinum*, and *Botrytis cinerea*. Rumen microorganisms, such as *Butyrivibrio fibrisolvens* and *Lachnospira multiparus*, are able to provide nutrients to the ruminant animal because of their ability to degrade pectin and other plant materials, such as cellulose. *Bacteroides thetaiotaomicron*, a gram-negative anaerobic member of the human colon flora, can also use a wide variety of plant polysaccharides, including pectin, as their source of carbon and energy.

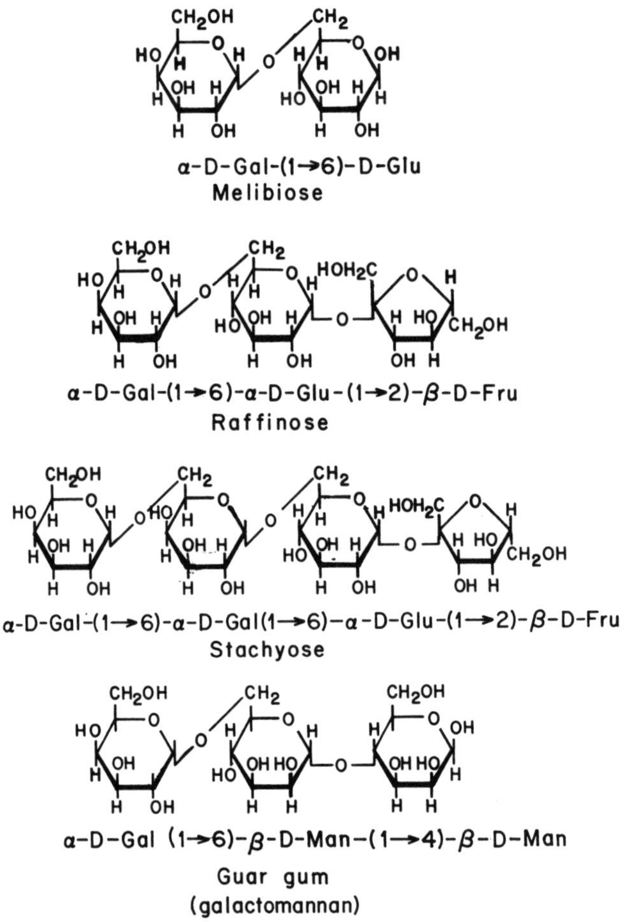

α-D-Gal-(1→6)-D-Glu
Melibiose

α-D-Gal-(1→6)-α-D-Glu-(1→2)-β-D-Fru
Raffinose

α-D-Gal-(1→6)-α-D-Gal(1→6)-α-D-Glu-(1→2)-β-D-Fru
Stachyose

α-D-Gal (1→6)-β-D-Man-(1→4)-β-D-Man
Guar gum
(galactomannan)

Fig. 7-28. Structures of melibiose, raffinose, stachyose, and guar gum (galactomannan).

Erwinia chrysanthemi and *E. carotovora* initiate the degradation of pectin by production of a series of enzymes shown in Figure 7-29. Pectin methylesterase (encoded by the *pem* gene) removes methoxyl groups linked to C-6 to yield polygalacturonate. Polygalacturonate is then degraded by pectate lyases (encoded by *pelABCDE*) to form unsaturated digalacturonate. Alternatively, an exopolygalacturonase produces digalacturonate. The polygalacturonate residues are then degraded via either of the two alternative pathways in Figure 7-30. An oligouronide lyase acts on unsaturated oligouronides to form 3-keto-4-deoxyuronate, which is then converted to 2, 5-diketo-3-deoxygluconate and 2-keto-deoxygluconate. A kinase results in phosphorylation to 2-keto-3-deoxy-6-phosphogluconate that undergoes aldolase cleavage to glyceraldehyde-3-phosphate and pyruvate. Oligouronide lyase also acts on digalacturonate, splitting it to galacturonate. This compound is then metabolized to tagaturonate and altronate, which is then cleaved to 2-keto-3-deoxygluconate. As shown in Figure 7-30, this pathway serves for the utilization of extracellular galacturonate. A parallel pathway, which is under the control of some of the same structural genes, also serves for the utilization of extracellular glucu-

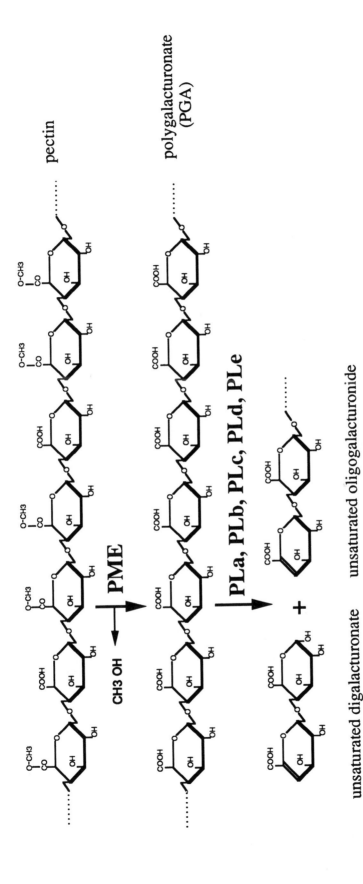

pectin

polygalacturonate (PGA)

unsaturated digalacturonate   unsaturated oligogalacturonide

Fig. 7-29. Initial stages in the degradation of pectin. In *E. chrysanthemi*, pectin methylesterase (PME), encoded by the *pem* gene, demethoxylates pectic polymers to yield polygalacturonate (PGA) and methanol. The PGA is cleaved by the action of polygalacturonate lyases (PL) encoded by the *pel* genes. As many as five *pel* genes (*pelA* to *pelE*) code for lyases secreted by the organism. Further degradation of the oligogalacturonides is conducted by intracellular enzymes shown in Figure 7-30. From Hugouvieux-Cotte-Pattat, N., N. Dominguez, and J. Robert-Baudouy. 1992. *J. Bacteriol.* **174:**7807–7818.

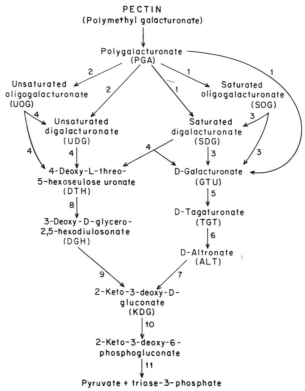

Fig. 7-30. Major pathways of polygalacturonate catabolism in bacteria. Enzymes and genes for the catabolic steps are (1) PG (*pem*), (2) PL (*pel*), (3) α-galacturonidase or oligogalacturonate hydrolase (*ogh*), (4) OGL (*ogl*), (5) uronate isomerase (*uxaC*), (6) altronate oxidoreductase (*uxaB*), (7) altronate hydrolase (*uxaA*), (8) 4-deoxy-L-*threo*-5-hexoseulose uronate isomerase (*kdgK*), (11) 2-keto-3-deoxy-6-phosphogluconate aldolase (*kdgA*). From Chatterjee, A. K., K. K. Thurn, and D. J. Tyrell. 1985. *J. Bacteriol.* **162**:708–714.

ronate, converting it to fructuronate and mannonate. Mannonate is then cleaved to form 2-keto-3-deoxygluconate and proceeds through the Entner–Doudoroff pathway (Fig. 7-5).

Enzymes that degrade pectin or polygalacturonate have been found in a number of other bacteria. These include *Bacillus, Clostridium, Lacnospira, Butyrivibrio,* and *Bacteroides.* Although less thoroughly studied than *Erwinia*, the pathway of pectin degradation in these organisms follows the same general pattern. The hexuronate pathway in *E. coli* (Fig. 7-30) accounts for the conversion of D-galacturonate, D-glucuronate, tagaturonate, and fructuronate to 2-keto-3-deoxy-D-gluconate. These aldohexuronates induce the synthesis of enzymes involved in their transport and metabolism via a pathway that is very similar to that shown in Figure 7-30.

CELLULOSE DEGRADATION

Cellulose is a β-1,4-linked glucose polymer that occurs in many forms. It is found in crystalline as well as amorphous forms and occurs in conjunction with other oligosaccharides in the walls of plants and fungi. Cellulose hydrolysis is important in the physiology of ruminant animals—animals containing a rumen or first stomach. Degradative action of microorganisms in the rumen enables these animals to digest cellulose. A major obstacle to animal protein production has been the inefficient utilization of cellulosic materials by the rumen microflora. The ubiquitous distribution of cellulose in municipal, agricultural, and forestry wastes emphasizes the potential use of this material for conversion to useful products such as single-cell protein or other fermentation products. As a consequence, the degradation of cellulose has been a continuing subject of intense study.

Cellulose degradation requires a complex of three basic types of enzymes (Fig. 7-31). Initially, an *endo*-β-1,4-glucanase cleaves cellulose to produce smaller oligosaccharides with free chain ends. An *exo*-β-1,4-glucanase cleaves dissacharide cellobiose units from the nonreducing ends of the oligosaccharide chains. Cellobiose is then hydrolyzed to glucose by a β-glucosidase. In most cellulolytic organisms, this complex of enzymes is inducible and is subject to catabolite repression. It has been assumed that the cellulolytic enzymes are secreted into the surrounding medium. However, many studies show that these enzymes are not released readily from the cell wall. The concept has developed that the cellulase complex remains in association with the cell and functions in a cell bound, multienzyme complex or **cellulosome**. In *Clostridium thermocellum*, a cellulose-degrading gram-positive anaerobe, the cellulosome is a high-molecular-weight complex that is found both in the culture medium and at the cell surface where it mediates adhesion of the cells to the substrate (Fig. 7-32). Subsequent studies have shown that similar structures occur on the surface of a variety of cellulolytic bacteria indicating that the cellulosome concept may be a general feature of these organisms. The cellulosome of *C. thermocellum* contains endoglucanase, cellobiohydrolase, or hemicellulase activity. A glycoprotein, named CipA (cellulosome integrating protein), promotes binding of the cellulosome to the substrate and also acts as a scaffolding protein around which the catalytic components are organized. A hypothetical model shown in Figure 7-33 indicates the manner in which the cellulosome and its components function at the cell surface. Two major cellulosome components, S_S and S_L, act in a cooperative manner to degrade crystalline cellulose. The S_L component appears to serve as an anchor on the cellulose surface for the S_S component. The gene coding for the S_S component, *celS*, has been cloned and the S_S component is now termed CelS.

A number of cellulolytic bacteria, including *Cellvibrio gilvus*, *Ruminococcus flavefaciens*, and *C. thermocellum*, convert cellobiose into glucose-1-phosphate and glucose by the action of cellobiose phosphorylase (Fig. 7-34). This results in conservation of the β-1,4-bond energy as G-1-P. As a result, organisms that contain cellobiose phosphorylase prefer cellobiose to glucose as an energy source for growth. *Clostridium thermohydrosulfuricum* produces glucose-metabolizing enzymes constitutively and uses glucose as the preferential source of carbon and energy. *Clostridium thermocellum* has a higher degree of metabolic efficiency than *C. thermohydrosulfuricum* on cellobiose because it conserves 1 mol more of ATP per mol of substrate metabolized. Figure 7-34 summarizes the relationship between growth saccharide and enzymatic activities associated with the formation of glucose-6-phosphate in these thermophilic anaerobes. The cellulase component of *C. thermocellum* responsible for the degradation of microcrystalline cellulose (Avicel) is repressed during rapid growth on cellobiose and after adaptation to fructose, sorbitol, and glucose. The missing cellulase component is most likely the exoglucanase that initiates cellulose breakdown.

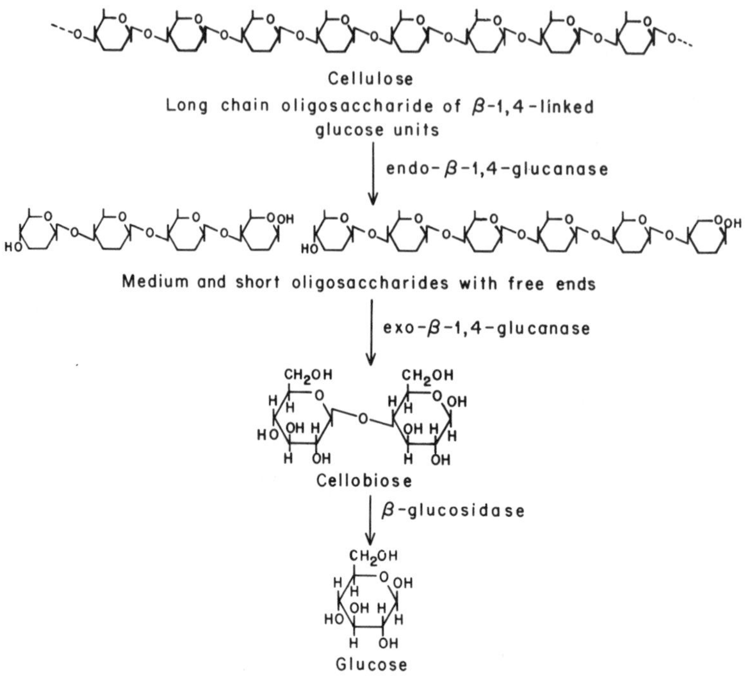

Cellulose
Long chain oligosaccharide of β-1,4-linked
glucose units

endo-β-1,4-glucanase

Medium and short oligosaccharides with free ends

exo-β-1,4-glucanase

Cellobiose

β-glucosidase

Glucose

Fig. 7-31. Enzymatic degradation of cellulose.

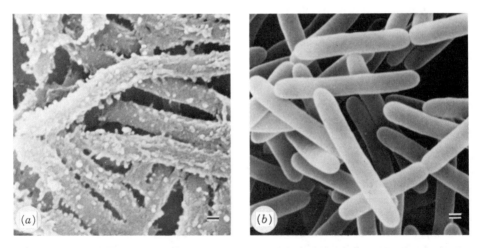

Fig. 7-32. Cellulosomes of *C. thermocellum*. Scanning electron microscopy (SEM) of cationized ferritin-labeled cellobiose-grown cells. Prior to processing for electron microscopy, the cells were treated with cationized ferritin. Wild-type cells (a) are easily distinguishable from mutant cells (b) by the appearance of protuberances that cover the entire cell surface of the wild type. Bars = 200 nm. From Bayer, E. A. and R. Lamed. 1986. *J. Bacteriol.* **167**:828–836.

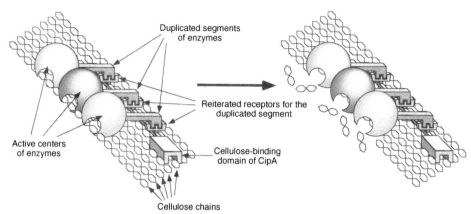

Fig. 7-33. Hypothetical model of the topological organization of subunits within the cellulosome. The diagram shows a portion of the cellulosome with the CBD (cellulose binding domain) of CipA and three of the CipA domains responsible for binding the duplicated segment borne by the catalytic components. Glucose residues are not drawn to the same scale as proteins. From Beguin, P. and J. P. Aubert. 1994. *FEMS Microbiol. Rev.* **13:**25–58.

There is a wide diversity of actively cellulolytic organisms. These organisms are important in industrial applications, in certain ecological niches, such as the rumen, and in the digestive systems of arthropods associated with the utilization of cellulosic materials. Barriers to the degradation of plant cell wall material by rumen bacteria have been a major obstacle to increasing the production of livestock. Treatment of forage materials, such as wheat straw, corncobs, and cornstalks, with dilute sodium hydroxide, potassium permanganate, dilute alkaline solutions of hydrogen peroxide, or chemical equivalents, such as cadmium oxide in aqueous ethylenediamine, renders these materials more susceptible to bacterial colonization and degradation. Arthropods associated with the degradation of wood owe their ability to digest cellulose to the presence of microorganisms in their digestive system. The woodwasp, *Sirex cyaneus*, acquires cellulases and xylanases from a fungal symbiont, *Amylosterum chailletti*. Termites and wood-feeding roaches live on a diet of cellulose as a result of the presence of the cellulose-digesting symbiotic flagellated protozoon, *Trichonympha sphaerica*.

UTILIZATION OF STARCH, GLYCOGEN, AND RELATED COMPOUNDS

Starch, one of the most common storage compounds in plants, is found in two forms. α-Amylose consists of long unbranched chains of D-glucose in α-(1 → 4) linkages. Although not truly soluble in water, amylose forms hydrated micells in which the polysaccharide chain forms a helical coil. By comparison, amylopectin is a highly branched form of starch in which the backbone consists of chains in α-(1 → 4) linkage with α-(1 → 6) linkages at the branch points (Fig. 7-35). Glycogen, the main storage compound in animal cells, is comparable to amylopectin with the main backbone consisting of D-glucose in α-(1 → 4) linkage but with more frequent α-(1 → 6) branches. Pullulan, a starchlike polysaccharide, has the general structure shown in Figure 7-36.

Amylose can be hydrolyzed by α-amylase, which cleaves the α-(1 → 4) linkages

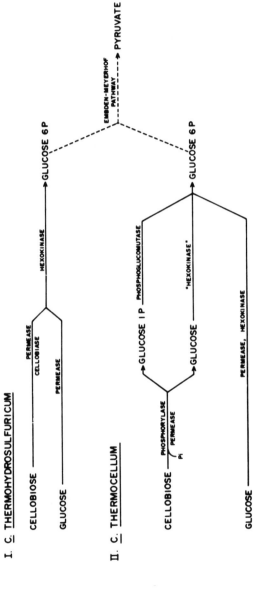

Fig. 7-34. Summary of the relationships between growth saccharide and the enzymatic activities associated with formation of glucose-6-phosphate in thermophilic anaerobes. The cellobiose-metabolizing enzymes are constitutive in *C. thermocellum*, whereas the glucose-metabolizing enzymes are constitutive in *C. thermohydrosulfuricum*. From Ng, T. K and J. G. Zeikus. 1982. *J. Bacteriol.* **150**:1391–1399.

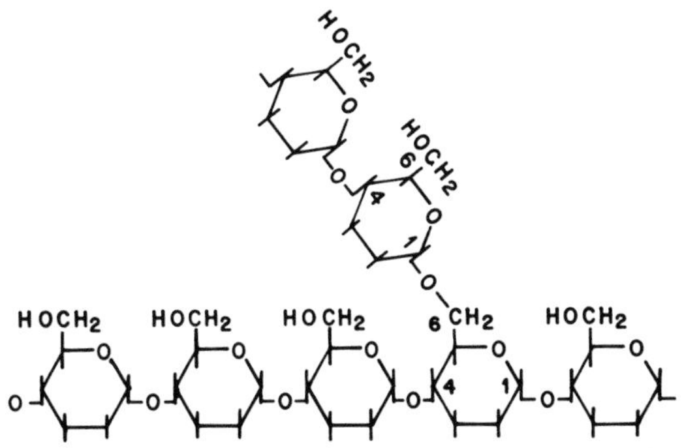

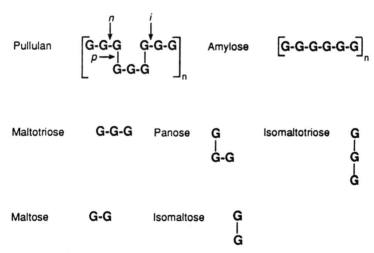

Fig. 7-35. Branched structure of amylopectin. The structure of glycogen is similar except that there are more frequent α-(1 $\rightarrow$ 6) branches.

to yield a mixture of α-glucose and α-maltose. Amylose is also hydrolyzed by β-amylase, liberating β-maltose in stepwise fashion from the nonreducing end of the molecule. Thus β-amylases invert the configuration at the C-1 position during cleavage of the α-(1 $\rightarrow$ 4)-glucosidic bond, whereas α-amylases retain the α-configuration at the C-1 position. These enzymes also hydrolyze amylopectin or glycogen to yield glucose, maltose, and a highly branched core, **limit dextrin**. A debranching enzyme, α-(1 $\rightarrow$ 6)-glucosidase, is capable of hydrolyzing the α-(1 $\rightarrow$ 6) linkages in limit dextrin. The combined action of α-(1 $\rightarrow$ 6)-glucosidase and α-amylase is required to completely degrade amylopectin or glycogen to glucose and maltose. Thus, α-amylases are endo-enzymes that can bypass the α-(1 $\rightarrow$ 6) branch points of amylopectin, whereas β-amylases are exoenzymes that cannot hydrolyze amylopectin internally to the α-(1 $\rightarrow$ 6)

Fig. 7-36. Structure of pullulan, amylose, and related carbohydrates. G = glucose; n = the α-(1 $\rightarrow$ 4) bond cleaved by *Bacillus stearothermophilus* neopullulanase and the *Bacteroides thetaiotaomicron* pullulanase and the *B. thetaiotaomicron* pullulanase I. Horizontal lines indicate α-(1 $\rightarrow$ 4)-D-glucosidic bonds; vertical lines indicate α-(1 $\rightarrow$ 6)-D-glucosidic bonds. From Smith, K. A. and A. A. Salyers. 1991. *J. Bacteriol.* **173**:2962–2968.

branch points. Both α- and β-amylases and debranching enzymes (also called pullulanases) are produced by a number of bacteria and fungi. Microorganisms also produce glucoamylase, an enzyme that degrades starch primarily to glucose. Until recently, it was believed that β-amylase existed only in higher plants. However, extracellular β-amylases have been observed in *B. polymyxa, B. megaterium,* and other *Bacillus* species.

Because of their industrial application, the α-amylases secreted by a variety of *Bacillus* species have been studied intensively. The α-amylases from *B. subtilis* are saccharifying enzymes that produce mostly glucose and maltose from starch. On the other hand, the α-amylases produced by *B. licheniformis* and *B. amyloliquefaciens* are liquefying enzymes that produce mostly maltosaccharides. Under the conditions of nutrient deprivation, *B. subtilis* activates the gene, *amyE,* the structural gene for α-amylase. The synthesis of α-amylase is not inducible in the classical sense in that no compound can be shown to trigger the activation of *amyE.* However, synthesis of the enzyme is repressed by glucose and other readily metabolizable carbon sources. Several genes appear to increase the formation of α-amylase and its secretion. Increasing the number of different genes in a cell results in hyperproductivity of α-amylase.

For ecological as well as industrial reasons, the thermostability of α-amylases has been of considerable interest. Thermophilic species, such as *B. acidocaldarius* and *B. stearothermophilus,* produce amylases that are stable at temperatures ranging from 58 to 80 °C. However, mesophilic species, such as *B. licheniformis* and *B. amyloliquefaciens,* also produce amylases that are active at temperatures in excess of 75 °C. Comparative studies of the enzymes and the structural genes of mesophilic and thermophilic species reveal that there is considerable amino acid homology between the enzymes from the two species and that the proteins are evolutionarily related.

Thermophilic anaerobes have received attention because of their ability to produce highly active thermostable starch-degrading enzymes. *Clostridium thermosulfurogenes* produces an extracellular, thermoactive, and thermostable β-amylase and a cell-bound glucoamylase. It does not produce a debranching enzyme. *Clostridium thermosulfuricum* produces a cell-bound glucoamylase and a debranching enzyme, pullulanase, which are thermoactive and thermostable. Cocultures of these organisms exhibit enhanced production of β-amylase, glucoamylase, and pullulanase, with concomitant increase in ethanol production from starch. The β-amylase of *C. thermosulfurogenes* is expressed at high levels only when the organism is grown on maltose or other carbohydrates containing maltose units. Glucose represses β-amylase synthesis, but cAMP does not eliminate the repressive effect.

Bacteroides thetaiotaomicron, a gram-negative anaerobe found in high numbers in the human colon, can ferment a variety of polysaccharides, including amylose, amylopectin, and pullulan. In this organism the degradative enzymes are not extracellular but are cell associated, indicating that the first step in starch utilization is binding of the substrate to the bacterial surface. This is followed by translocation through the outer membrane into the periplasm where the starch-degrading enzymes are located. *Bacteroides thetaiotaomicron* produces four maltose inducible and two constitutive proteins involved in starch utilization. Three of the maltose-inducible proteins proved to be outer-membrane proteins and one was a cytoplasmic membrane protein. A 115-kDa outer-membrane protein is essential for maltoheptaose utilization, whereas the other outer-membrane proteins are involved in starch utilization. Two constitutively produced outer-membrane proteins are also involved in starch utilization. Four types of pullulan-hydrolyzing enzymes are produced by various bacteria. A glucoamylase hydrolyzes pullulan from the nonreducing ends to produce glucose. A pullulanase (Type I) found in *K. pneumoniae* and *B. thetaio-taomicron* breaks α-(1 → 6)-D-glucosidic linkages to produce maltotriose. An isopullu-lanase found in *Aspergillus niger* hydrolyzes α-(1 → 4)-D-glucosidic linkages in pullulan to produce panose. The structures of these compounds are shown in abbreviated form in Figure 7-36. An α-glucosidase has been described for *B. thetaiotaomicron* in addition to pullulanases I and II.

Cyclodextrins are cyclic oligosaccharides containing from 6 to 12 glycopyranose units bonded through α-(1 → 4) linkages, as shown in Figure 7-37. These compounds are of interest in the food and pharmaceutical industry for several reasons. They can form inclusion complexes with both organic and inorganic molecules that change the physical and chemical properties of the included compounds. Bound to silica they have been found useful as the bonded phase for high-performance liquid chromatography (HPLC) for the separation of a number of organic compounds. Certain organisms produce cyclodextrins in the process of degrading starch. *Bacillus stearothermophilus* degrades starch by means of the enzyme cyclodextrin glucanotransferase. Cyclodextrins are resistant to hydrolysis by many starch-splitting enzymes. *Bacillus macerans, B. coagulans, B. sphaericus,* and *C. thermohydrosulfuricum* produce cyclomaltodextrinase that can hydrolyze cyclodextrins as well as linear maltodextrins.

METABOLISM OF AROMATIC COMPOUNDS

A wide variety of microorganisms can utilize aromatic compounds as their sole source of carbon. They degrade the aromatic ring structure and metabolize the degradation prod-ucts to compounds that can enter the common carbon and energy-yielding pathways. Degradation of aromatic compounds is of major importance because of their widespread occurrence in natural as well as synthetic materials. By degrading these compounds, many different microorganisms facilitate the rentry of the carbon from these aromatic com-

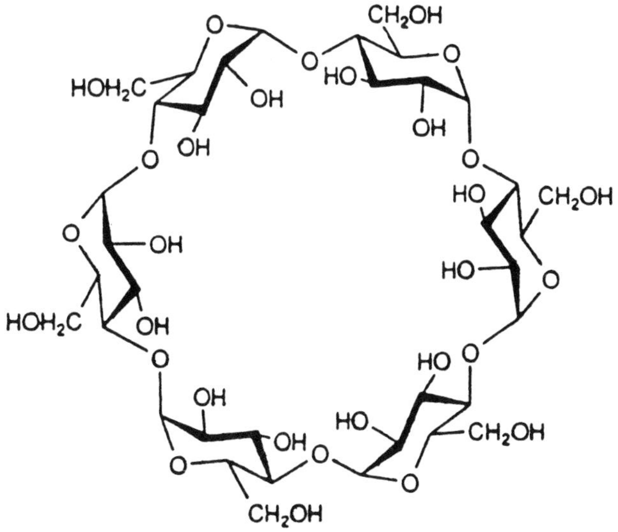

Fig. 7-37. Structure of cyclodextrin.

pounds into the natural biological cycles, preventing their accumulation in the environment. Accumulation of toxic compounds, particularly herbicides, pesticides, and many industrial waste products, can lead to serious contamination of ground water and ultimate ecological problems.

A fundamental aspect of the metabolism of aromatic ring compounds is the manner in which the ring structure is cleaved. Under aerobic conditions, monooxygenases or dioxygenases convert the ring structure to rings containing adjacent hydroxyl groups that facilitate ring cleavage. Several ring oxidation mechanisms are known. *Pseudomonas putida* mt-2 oxidizes the methyl group of toluene to form benzyl alcohol (Fig. 7-38). This is followed by reactions leading to the formation of catechol, the substrate for the meta-cleavage pathway leading to the formation of an α-keto acid (Fig. 7-39). *Pseudomonas mendocina* oxidizes toluene to *p*-cresol, which undergoes ortho cleavage with the formation of β-ketoadipate (Fig. 7-39). *Pseudomonas putida* PpF1 initiates the oxidation of toluene by incorporating oxygen into the aromatic nucleus to form *cis*-toluene dihydrodiol, which is converted to 3-methylcatechol (Fig. 7-38). 3-Methylcatechol is then degraded via the meta-cleavage pathway to yield an α-keto acid (Fig. 7-39). β-Ketoadipate is converted to succinyl-CoA and acetyl-CoA by the action of succinyl-CoA transferase. The α-keto acid intermediates formed via the meta-cleavage route (2-keto-4-hydroxyvalerate) is cleaved by an aldolase to form pyruvate and acetaldehyde.

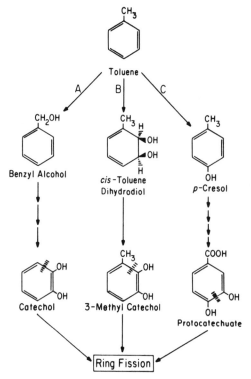

Fig. 7-38. Pathways for the oxidation of toluene to ring fission substrates. (A) *Pseudomonas putida* mt-2, (B) *Pseudomonas putida* PpF1, (C) *Pseudomonas mendocina*. From Finette, B. A., V. Subramanian, and D. T. Gibson. 1984. *J. Bacteriol.* **160**:1003–1009.

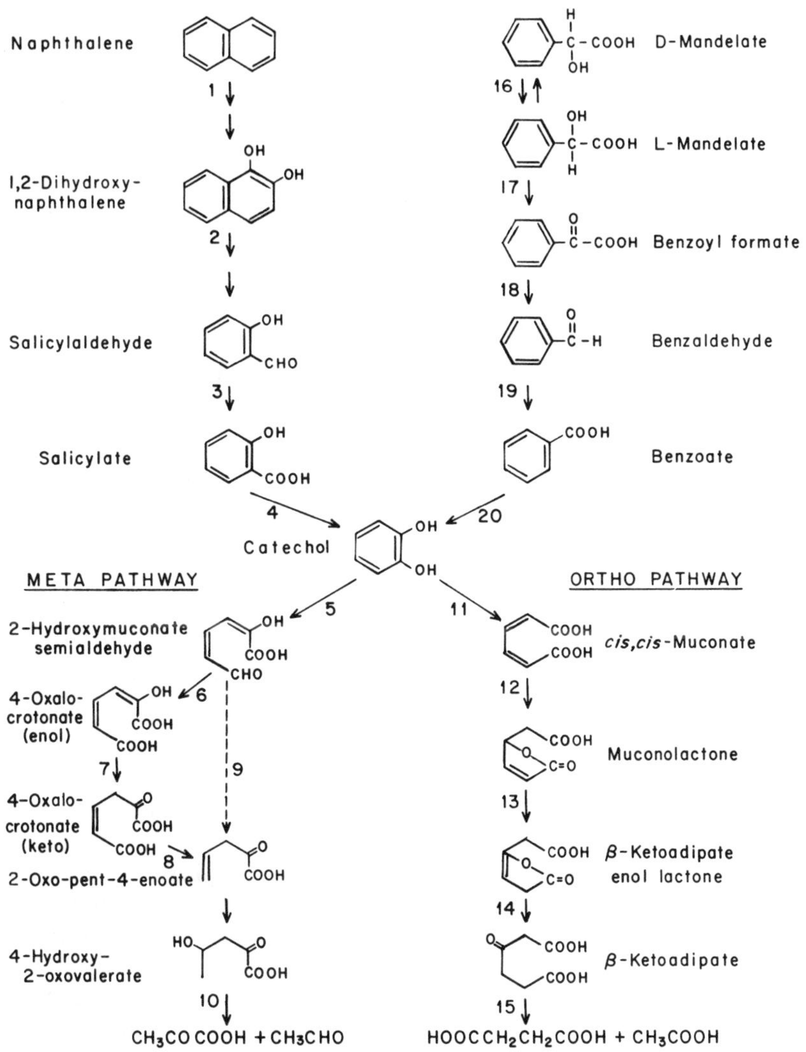

Fig. 7-39. Pathways of metabolism of aromatic compounds by _Pseudomonas_ and other microorganisms. The enzymes are (1) naphtalene oxygenase, (2) 1,2-dihydronapthalene oxygenase, (3) salicylaldehyde dehydrogenase, (4) salicylate hydroxylase, (5) catechol 2,3-oxygenase, (6) 2-hydroxymuconic semialdehyde dehydrogenase, (7) 4-oxalocrotonate tautomerase, (8) 4-oxalocrotonate decarboxylase, (9) 2-hydroxymuconic semialdehyde hydrolase, (10) 4-hydroxy-2-oxovalerate aldolase, (11) catechol 1,2-oxygenase, (12) _cis,cis_-muconate lactonizing enzyme, (13) (+)-muconolactone isomerase, (14) β-keto-adipate enol-lactone hydrolase, (15) succinyl CoA:β-ketoadipate thiolase, (16) mandelate racemase, (17) L-(+)-mandelate dehydrogenase, (18) benzoylformate decarboxylase, (19) NAD- and NADP-linked benzaldehyde dehydrogenase, (20) benzoic acid oxidase.

In *P. putida* mt-2 the enzymes for the complete degradation of toluene are encoded by a set of genes borne on a TOL (for toluene) plasmid pWWO. Toluene, *m*- and *p*-xylene, 3-ethyl toluene, and 1,2,4-trimethylbenzene are oxidized to the corresponding carboxylic acids (benzoate, *m*- and *p*-toluate, 3-ethylbenzoate, and 3,4,-dimethylbenzoate), which are subsequently degraded via the meta-cleavage pathway to form carboxylic acids, pyruvate, and aldehydes. The genes of the upper pathway (*xylABC*) and the meta-cleavage pathway enzymes (*xylDEGF*) are organized into two regulatory units controlled by the products of two regulatory genes (*xylS* and *xylR*).

Degradation of a variety of other aromatic compounds often leads into the pathways just described. For example, mandelic acid and its derivatives are converted to benzoic acid. Benzoic acid is further metabolized by bacteria via the catechol branch of the β-ketoadipate pathway (Fig. 7-39). Most yeasts and filamentous fungi metabolize aromatic compounds through reactions of the β-ketoadipate pathway. An exception is *Trichosporon cutaneum*, which cannot cleave the protocatechuate ring but oxidizes it to 1,3,4,-trihydroxybenzene, which is cleaved to maleylacetate. *Rhodotorula graminis* metabolizes phenylalanine, DL-mandelate, and benzoate through *p*-hydroxybenzoate to protocatechuate in the β-ketoadipate pathway. Salicylate is metabolized through the catechol branch of the β-ketoadipate pathway.

The demonstration by Stanier and colleagues in 1947 of the mechanism of oxidation of mandelate to intermediates of the TCA cycle by *P. putida* represents the first example of sequential induction of enzymes of a catabolic pathway. **Sequential induction** refers to the fact that each of the enzymes in a pathway is inducible by its specific substrate. Compounds with more complex ring structures, such as naphthalene, anthracine, or phenanthrene, undergo sequential ring fission to form ring structures that can then be degraded via the ortho- or meta-cleavage pathways (Fig. 7-39). Chlorinated compounds, such as chlorobenzoate or polychlorinated biphenyls, often require dehalogenation before the aromatic ring can be converted to hydroxylated compounds subject to ring fission.

One aspect of the utilization of aromatic compounds and other complex organic materials that bears mention is the degradation of substances derived from lignin. Lignin is a highly complex heteropolymer found in woody plants. It is generally resistant to attack, but a number of fungi and a few bacteria are capable of degrading lignin or its breakdown products. The wood-rotting fungus, *Phanerochaete chrysosporium*, is able to degrade the pesticide DDT and other complex organic compounds because of their structural similarity to structures found within the lignin complex.

Studies on the degradation of aromatic compounds cited thus far have all involved the incorporation of molecular oxygen into the aromatic nucleus to form dihydroxylated intermediates. Under anaerobic conditions, the aromatic ring structures are attacked reductively. Anaerobic metabolism of benzoate by *Rhodopseudomonas palustris* begins with the reductive saturation of the aromatic nucleus (Fig. 7-40). Subsequent metabolism of aromatic compounds occurs via the CoA derivatives.

FERMENTATION PATHWAYS IN SPECIFIC GROUPS OF MICROORGANISMS

The preceding discussion has delineated the major pathways used by microorganisms for the dissimilation of carbohydrates and the generation of biologically useful energy (as PMF or ATP) and reducing power (as NADH or NADPH). The introductory reactions whereby a variety of alternate sugars and other substrates enter these general metabolic

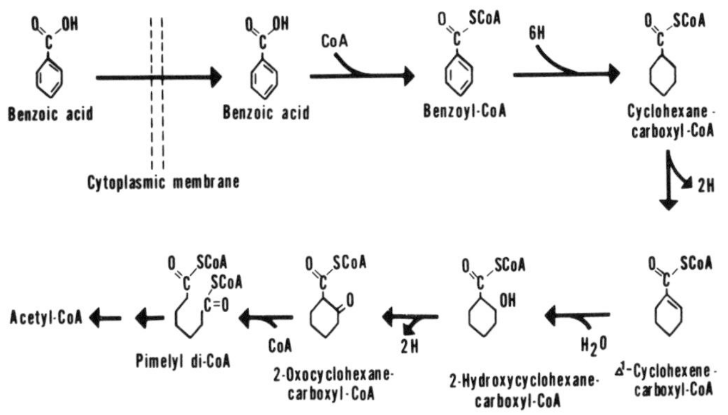

Fig. 7-40. Pathway of anaerobic benzoate metabolism in *Rhodopseudomonas palustris*. From Harwood, G. S. and J. Gibson. 1986. *J. Bacteriol.* **165**:504–509.

routes have also been described. Individual organisms may follow any one or a combination of the fructose bisphosphate aldolase (Embden–Meyerhof), phosphoketolase (pentose shunt), or ketogluconate aldolase (Entner–Doudoroff), pathways or variations of them. Additional fermentative products may be derived through the metabolism of pyruvate or other intermediates. The fermentative pathways occurring in specific groups of microorganisms are outlined in Figure 7-41 and will be considered in detail in the following section.

Fermentation Balances

To fully appreciate the pathways of carbohydrate fermentation, the products should be qualitatively identified and the amount of fermented carbohydrate quantitatively accounted for by the amount of products recovered. To assess the accuracy of the analytical determinations, a **carbon balance** or **carbon recovery** can be calculated, as shown in Table 7-5. Under a number of circumstances, the balance may not be ideal because some portion of the carbohydrate may be assimilated into cellular constituents or utilization of compounds other than the major substrate added to form the recovered products.

Because oxidation–reduction reactions play a major role in the metabolism of carbohydrates and other substrates, it is also advisable to perform an **oxidation–reduction (O–R)** balance on the products formed. The O–R balance provides an indication as to whether the products formed balance with regard to their oxidized or reduced states. It may not be possible to balance the hydrogen and oxygen of the substrate directly because hydrations or dehydrations may occur as intermediary steps in the fermentation pathway. However, a simple device that achieves the same end is to calculate an **oxidation value** for each compound and compare the total amount of oxidized and reduced products, as shown in Table 7-5. If the ratio of oxidized products to reduced products is close to the theoretical value of 1.0, this provides further indication that the products are in balance.

The **oxidation value** of a compound is determined from the number of oxygen atoms less one-half the number of hydrogen atoms. As shown in Table 7-5, for glucose, which has 6 oxygen atoms and 12 hydrogen atoms, the oxidation value is 0. Thus glucose is

TABLE 7-5. Fermentation Balance for *Lactobacillus pentoaceticus*

Compound	mmol	MMol Carbon[a]	Oxidation Value[b]	Oxidized Products	Reduced Products	C₁ Observed	C₁ Calculated[c]
Glucose	100	600	0				
Lactate	96	288	0				
Glycerol	7	21	−1		7		
Ethanol	86	172	−2		172		86
Acetate	7	14	0				7
CO₂	89	89	+2	178		89	
Totals		584		178	179	89	93

Carbon Recovery

$$\frac{\text{mmol product}}{\text{mmol substrate}} \times 100$$

$$\frac{584}{600} \times 100 = 97.4\%$$

Oxidation–Reduction

$$\frac{\text{Oxidized products}}{\text{reduced products}}$$

$$\frac{178}{179} = 1.0$$

C_1 balance

$$\frac{\text{Observed } C_1}{\text{calculated } C_1}$$

$$\frac{89}{93} = 0.96$$

[a] mmol C = mmol compound × number C atoms
[b] Oxidation value = (number O atoms) − (number H atoms/2)
[c] Calculated C_1 = expected amount of products for each C_2 observed. In the fermentation above, for each millimole of pyruvate converted to ethanol, an equal amount of CO_2 is expected:

$$\text{Pyruvate } (C_3) \longrightarrow \text{ethanol } (C_2) + CO_2 (C_1)$$

Similarly, for each mmol of pyruvate converted to acetate, an equal amount of CO_2 is expected:

$$\text{Pyruvate } (C_3) \longrightarrow \text{acetate } (C_2) + CO_2 (C_1)$$

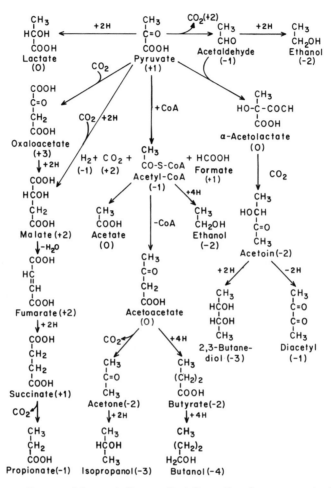

Fig. 7-41. Major pathways of fermentation product formation from pyruvate. Numbers in parentheses are the oxidation values calculated on the basis of the number of oxygen atoms less one-half the number of hydrogen atoms as shown in Table 7-5.

referred to as a **neutral** compound and the fermentation must be equally balanced in that there should be as much oxidized product as reduced product. The carbon in the end-products shown in Table 7-5 balance very closely (97.4%) with the total amount of glucose fermented. In addition, the amount of oxidized product (CO_2) is approximately equal to the amount of reduced products (glycerol and ethanol) so that the O–R balance is very close to the theoretical value of 1.0. Figure 7-41 shows the oxidation values for a number of common fermentation products.

Another concept of fermentation balances is the C_1 **balance**. The amount of C_1 product expected is calculated from the amounts of those products for which CO_2 (or formate) is expected as an accompanying product. For example, if a C_2 compound, such as ethanol or acetate, is among the final products, an equal amount of C_1 product will be expected since ethanol is derived from pyruvate by decarboxylation. Note that in Table 7-5 the C_1 balance (0.96) is very close to the theoretical value of 1.0.

Yeast Fermentation

The yeast *S. cerevisiae* utilizes the EMP pathway of glucose metabolism under the conditions of neutral or slightly acid pH and an anaerobic environment. The major products formed under these conditions are carbon dioxide and ethanol. The sequence of enzymatic reactions involved in this pathway has been presented earlier (Fig. 7-1). However, there are certain facets of alcoholic fermentation in yeast that bear additional consideration.

Alteration in the ethanolic fermentation may occur as a result of a number of changes that may be imposed in the culture medium. In the presence of sodium sulfite, acetaldehyde is trapped as a bisulfite addition complex, and glycerol is formed as a major product:

$$\text{glucose} + HSO_3^- \longrightarrow \text{glycerol} + \text{acetaldehyde-}HSO_3^- + CO_2$$

Under these conditions, acetaldehyde is unable to serve as a hydrogen acceptor. Dihydroxyacetone phosphate becomes the preferred hydrogen acceptor, yielding α-glycerol phosphate, which is then hydrolyzed to glycerol and P_i. The fermentation is not shifted entirely in the direction of glycerol formation as it is not possible to add sufficient bisulfite to bind all of the acetaldehyde without incurring additional toxic effects. Thus some ethanol will be found among the products.

Under alkaline conditions still another type of fermentation pattern is observed.

$$2\text{glucose} \longrightarrow 2\text{glycerol} + \text{acetate} + \text{ethanol} + 2CO_2$$

A dismutation reaction occurs in which 1 mol of acetaldehyde is oxidized to acetate and another is reduced to ethanol. This sequence is catalyzed by two NAD-linked dehydrogenases and requires a balance between the following reactions:

$$CH_3CHO + NAD^+ + H_2O \longrightarrow CH_3COOH + NADH + H^+$$

$$CH_3CHO + NADH + H^+ \longrightarrow CH_3CH_2OH + NAD^+$$

$$\text{G-3-P} + P_i + NAD^+ \longrightarrow \text{1,3-diphosphoglycerate} + NADH + H^+$$

$$\text{dihydroxyacetone-P} + NADH + H^+ \longrightarrow \text{glycerol-3-P} + NAD^+$$

The oxidation of acetaldehyde to acetate yields no ATP since the reaction proceeds directly without the intermediary formation of acetyl-CoA.

In cell-free yeast extracts the addition of P_i results in a marked increase in the fermentation rate. The rate eventually subsides to that of the control without added P_i. However, addition of a second quantity of P_i will again produce an increase in the rate of fermentation. This phenomenon (sometimes referred to as the Harden–Young effect) was found to result from the incorporation of P_i into organic phosphate esters, particularly fructose-1,6-bisphosphate, fructose-6-phosphate, glucose-6-phosphate, and glucose-1-phosphate. Inorganic phosphate is assimilated during the formation of 1,3-diphosphoglyceric acid. In subsequent steps, ADP is phosphorylated to ATP, which in turn permits the phosphorylation of additional glucose. The hexose accumulates as phosphate esters.

This effect can be counteracted by the addition of aresenate ion, which results in the formation of phosphoglyceryl arsenate. The compound hydrolyzes rapidly by a nonenzymatic reaction, preventing the accumulation of phosphorylated intermediates. During the usual procedures of preparing cell-free extracts of yeast, the enzyme ATPase may be inactivated. Preparations in which ATPase is active, or to which exogenous ATPase is added, do not exhaust the P_i and hexose diphosphate does not accumulate. Under these conditions fermentation continues at a constant rate.

It has been asserted that Pasteur reported that the introduction of oxygen to fermenting yeast resulted in the cessation of ethanol formation. This phenomenon, the **Pasteur effect**, has been studied extensively in attempts to elucidate the mechanism involved. *Saccharomyces cerevisiae* does not show a noticeable Pasteur effect when growing in the presence of an excess of sugar and nitrogen source. The Pasteur effect can be demonstrated only in resting cells. Growing cells respire only 3–20% of the catabolized sugar and the remainder is fermented. A shift to anaerobiosis appears to have much greater energetic consequences in resting than in growing *S. cerevisiae*. The main mechanism involved in the loss of fermentation observed in resting cells (i.e., under conditions of nitrogen starvation) is a progressive inactivation of the sugar transport systems that reduces the rate of fermentation to less than 10% of the value observed in growing cells. As a consequence, the contribution of respiration to catabolism, which is small in growing cells, becomes quite significant under the conditions of starvation for ammonia.

Some of the difficulty in studying alcoholic fermentation in yeast resides in the fact that several isozymes of alcohol dehydrogenase (ADH) are present and the activities of these and other enzymes concerned with carbohydrate metabolism are partitioned between the cytoplasm, the mitochondrial membrane, and the mitochondrial matrix. Alcohol dehydrogenase I, the key enzyme of alcoholic fermentation, reduces acetaldehyde to ethanol in the presence of NADH and serves to maintain the redox balance in glycolysis in the cytoplasm of fermenting yeast cells. Alcohol dehydrogenase II catalyzes the formation of acetaldehyde in ethanol oxidation. Alcohol dehydrogenase III is involved in the oxidative utilization of ethanol in the mitochondrion. The inner-mitochondrial membrane is impermeable to NAD^+ or NADH. Yeast mutants lacking ADHI (*adhI* strains) may regenerate cytoplasmic NAD^+ by reducing dihydroxyacetone phosphate (DHAP) to glycerol-3-phosphate. Glycerol-3-phosphate may enter the mitochondrial membrane, where it is reoxidized to DHAP by an FAD-linked glycerol-3-phosphate dehydrogenase. The electrons generated in this oxidation are most likely transferred to the respiratory chain. For this reason, ADHI-deficient cells are unable to grow anaerobically. Glycerol may be an end-product as a result of dephosphorylation of glycerol-3-phosphate. However, yeast deficient in all four of the known ADH isozymes (*adh°*) still produce up to one-third of the theoretical maximum yield of ethanol from glucose. Glycerol is a major fermentation product, but acetaldehyde and acetate are also produced. It has been shown that ethanol production in *adh°* cells is dependent on mitochondrial electron transport associated with the inner-mitochondrial membrane (see Fig. 7-42).

Lactic Acid Producing Fermentations

The discovery of bacterial fermentations in which lactic acid appears as the predominant end-product prompted investigations as to the route of lactic acid formation. Because of the analogy between the production of lactic acid during glycolysis in muscle and the production of lactic acid by various bacteria, it was, at first, assumed that the EMP

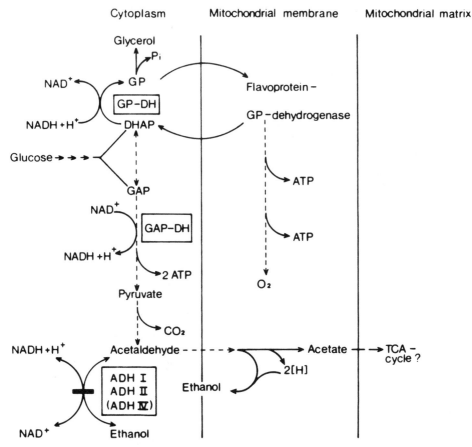

Fig. 7-42. Cytoplasmic and mitochondrial pathways of glucose metabolism in ADH-negative yeast cells. The ADH genes are indicated by a solid line. Cytoplasmic NAD^+, the cofactor for glycolytic glyceraldehyde-3-phosphate (GAP) oxidation, is regenerated by reduction of DHAP to glycerol-3-phosphate (GP), catalyzed by a cytoplasmic glycerol-3-phosphate dehydrogenase (GP-DH), which in part is dephosphorylated to glycerol. Another part of GP presumably is reoxidized in the mitochondrial membrane to DHAP by a flavoprotein GP dehydrogenase. In this oxidation, electrons are transferred directly to the respiratory chain at the level of ubiquinone. Hypothetically, formation of ethanol and acetate in the mitochondrial membrane would require linkage of an acetaldehyde-reducing enzyme to the acetaldehyde-oxidizing enzyme through oxidation and reduction of cofactor. Part of the acetate might enter the TCA cycle after conversion to acetyl-CoA. GAP-DH = glyceraldehyde-3-phosphate dehydrogenase. From Drewke, C., J. Thielen, and M. Ciriacy. 1990. *J. Bacteriol.* **172**:3909–3917.

pathway was the pathway used in both instances. However, it was observed that two major types of lactic acid fermentation occurred. Those species that fermented glucose primarily to lactic acid were termed **homofermentative** and those that produced a mixture of products were termed **heterofermentative**. Some homofermentative species were observed to form a wider variety of products if the conditions of fermentation were altered. At alkaline pH the production of formate, acetate, and ethanol increased at the expense of lactate (see Table 7-6). It became important to determine whether the mixed products of altered homolactic acid fermentation arose via the same pathway(s) as those of normally heterofermentative species. While several investigators provided evidence that the homofermentative species contained intermediates of the EMP pathway, the most convincing evidence was provided by radioisotope labeling studies. *Lactobacillus casei, L. pentosus,* and *S. faecalis* were shown to produce ^{14}C-methyl-labeled lactate from glucose-1-^{14}C with 50% dilution and ^{14}C-carboxyl-labeled lactate from glucose-3,4-^{14}C without dilution of the specific activity. These labeling patterns fit the distribution of carbon atoms expected from utilization of the EMP pathway, as shown in Figure 7-43.

It has been proposed that the term ***homolactic*** be used to refer to those lactic acid bacteria that contain aldolase but not transketolase. True homolactic acid bacteria metabolize glucose via the EMP pathway while those that are truly heterofermentative presumably follow a hexose monophosphate pathway although some data indicate that a combination of both pathways may be operative in certain organisms.

One of the major factors controlling the type of product formed is the amount of reducing equivalents available. In the oxidative step in the triosephosphate pathway, NAD^+ is reduced. The electrode potential ($NAD^+/NADH_2$) of this system is -0.28 V. The electrode potential of the lactic acid dehydrogenase reaction is -0.18 V. Since this is at a higher potential, reduced NAD will donate its hydrogen atoms to pyruvate, reducing it to lactate. However, if some other hydrogen acceptor is available, reduced NAD will donate its hydrogen atoms to this acceptor rather than to pyruvate if the electrode potential of the reaction is higher (see Fig. 7-14). For example, the potential for the reaction is

$$\text{acetaldehyde} + 2H \longrightarrow \text{ethanol} = -0.07 \text{ V}$$

If acetaldehyde is formed, then NADH will donate its hydrogen atoms to this system. Under the usual conditions of lactic acid fermentation, acetaldehyde does not appear to be formed in any appreciable quantity. However, by alteration of the pH, the fermentation may be diverted to a variety of products (Table 7-6). Since the streptococci appear to

TABLE 7-6. Effect of pH on Glucose Fermentation by *S. faecalis*[a]

	pH		
	5.0	7.0	9.0
Products Formed[b]			
Lactic acid	174.0	146.0	122.0
Acetic acid	12.2	18.8	31.2
Formic acid	15.4	33.6	52.8
Ethanol	7.0	14.6	22.4
Carbon rec. %	95.0	90.0	88.0
O/R index[c]	1.02	1.18	1.18

[a]From Wood, W. A. 1961. Fermentation of carbohydrates and related compounds. In *The Bacteria*, Vol. II, Metabolism. I. C. Gunsalus and R. Y. Stanier (Eds.). Academic, New York, p. 59.
[b]Amount of products shown as μmol/100 μmol of glucose fermented.
[c]See Table 7-5 for definition of O/R (oxidation–reduction) index and its calculation.

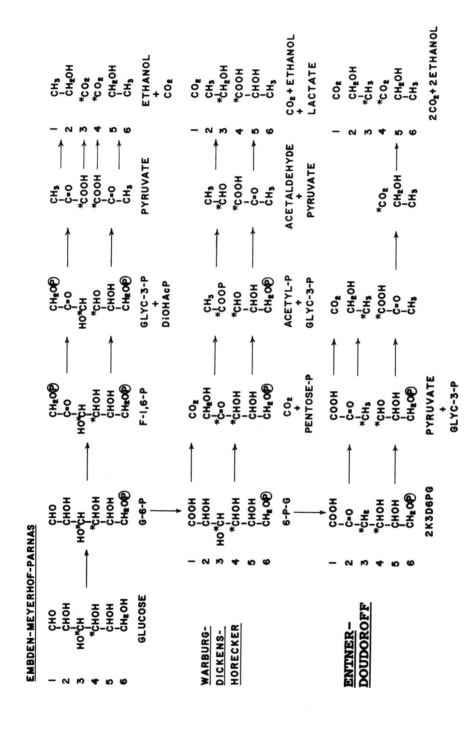

Fig. 7-43. Isotope-labeling patterns in various pathways of carbohydrate metabolism in microorganisms. *C indicates ^{14}C-labeled carbon atoms.

utilize the EMP pathway, the diversion of fermentation must occur at a step after pyruvate formation rather than by metabolizing hexose through an alternative pathway. Under these conditions, a dismutation of pyruvate occurs giving rise to lactate, acetate, and formate through the following sequence:

$$CH_3COCOOH + CoASH \longrightarrow CH_3CO\text{-}SCoA + HCOOH$$

$$CH_3CO\text{-}SCoA + P_i \longrightarrow CH_3COOPO_3H_2 + CoASH$$

$$CH_3COCOOH + 2H \longrightarrow CH_3CHOHCOOH$$

$$CH_3COOPO_3H_2 + ADP \longrightarrow CH_3COOH + ATP$$

Net: 2pyruvate + 2H + ADP

$$+ P_i \longrightarrow lactate + acetate + formate + ATP$$

Ethanol is not formed via pyruvate decarboxylase as observed in other organisms but rather via reduction of acetyl phosphate:

$$CH_3COOPO_3H_2 + 2(2H) \longrightarrow CH_3CH_2OH + P_i$$

Because of a difference in the pyruvate cleaving systems, formate, rather than carbon dioxide, is the major one carbon product. Carbon dioxide is formed in small amounts by some homolactic organisms under conditions in which the fermentation is diverted from the production of a preponderance of lactic acid.

Homofermentative organisms, such as *S. faecalis*, have been shown to possess high levels of glucose-6-phosphate dehydrogenase (G-6-PD) and 6-phosphogluconate dehydrogenase (6-P-GD). However, a homofermentative pattern is maintained through regulatory interrelationships between the EMP and HMP pathways. The lactate dehydrogenase (LDH) of *S. bovis* is specifically activated by FDP. The FDP-activated LDHs are characteristic of many lactic acid producing bacteria. The level of LDH, which is dependent on FDP for activity, decreases as fermentation becomes heterolactic. *Streptococcus faecalis* produces a single NADP-linked 6-P-GD when grown with glucose as the primary energy source. Gluconate-adapted cells produce two 6-P-GDs, one specific for NADP and the other specific for NAD. The NADP-linked enzyme is inhibited by FDP but is insensitive to ATP or other nucleotides. The NAD-linked enzyme is insensitive to FDP inhibition but is inhibited by ATP and other nucleotides (Fig. 7-44). The G-6-P isomerase in inhibited by 6-PG. When there is a cellular demand for ATP, these regulatory activities direct glucose carbon through the energy-producing EMP pathway. Inhibition of G-6-P isomerase by 6-PG prevents the accumulation of F-6-P. Alteration of the fermentative pattern of these organisms by change in pH or other environmental factors may result from the disruption of these regulatory functions. In *S. mutans*, triose phosphate (glyceraldehyde-3-phosphate or dihydroxyacetone phosphate) strongly inhibit pyruvate formatelyase, the first step in the dismutation sequence. Inhibition by triose phosphates in cooperation with a reactivating effect of ferridoxin may regulate pyruvate-formate lyase activity.

The lactic acid producing species of *Streptococcus* (now *Lactococcus*) include *L. lac-*

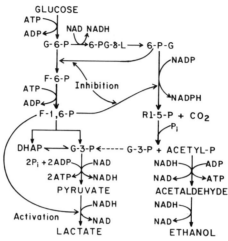

Fig. 7-44. Regulation of metabolic pathways in *S. faecalis*. 6-Phosphogluconate (6-PG) inhibits the activity of hexose phosphate isomerase, which converts glucose-6-phosphate (G-6-P) to fructose-6-phosphate (F-6-P). Fructose-1,6-bisphosphate (F-1,6-P) inhibits the activity of 6-P-GD, which converts 6-P-G to ribulose-5-phosphate (Rl-5-P) and CO_2. The F-1,6-P activates lactic dehydrogenase, which reduces pyruvate to lactate in the presence of reduced NAD. From Moat, A. G. 1985. Biology of lactic, acetic, and propionic acid bacteria. In *Biology of Industrial Microorganisms*, A. L. Demain and N. A. Solomon (Eds.). Benjamin–Cummings, Menlo Park, CA, pp. 143–186.

tis, *L. cremoris*, and *L. diacetylactis*. These species, as well as various species of *Leuconostoc* and *Lactobacillus*, produce acetoin and diacetyl. These compounds are responsible for the characteristic flavor of butter made from sour cream. Several routes have been described for the production of acetoin (Fig. 7-45). With acetate as substrate, acetyl-CoA is formed directly from acetate without the intermediary formation of pyruvate. *Lactococcus diacetylactis*, *Lactobacillus casei*, and other lactate-producing organisms that require lipoic acid for growth activate acetate to acetyl phosphate and then to acetyl-CoA via acetate kinase and phosphotransacetylase. Therefore, in media devoid of lipoic acid, lipoic acid requiring organisms cannot form acetyl-CoA from pyruvate. In a lipoate-free medium containing glucose and acetate, acetoin is formed via both the acetyl-CoA-C_2-TPP and C_2–C_3 routes.

Lactate-producing organisms have generally been considered to metabolize sugars via fermentative pathways. Molecular oxygen is not used as a final hydrogen acceptor and ATP is not generated via oxidative phosphorylation. Organisms that conduct their metabolism in this manner are anaerobes. However, if such an organism grows in the presence of oxygen, it is sometimes referred to as being aerobic or aerotolerant. Toxic products may arise through reaction of oxidative enzymes with molecular oxygen to yield hydrogen peroxide and superoxide anion:

$$FADH_2 + O_2 \longrightarrow FAD + H_2O_2$$

$$O_2 + e^- + \text{oxidative enzymes} \longrightarrow O_2^- \cdot$$

$$O_2^- \cdot + H_2O_2 \longrightarrow OH^- + OH \cdot + O_2$$

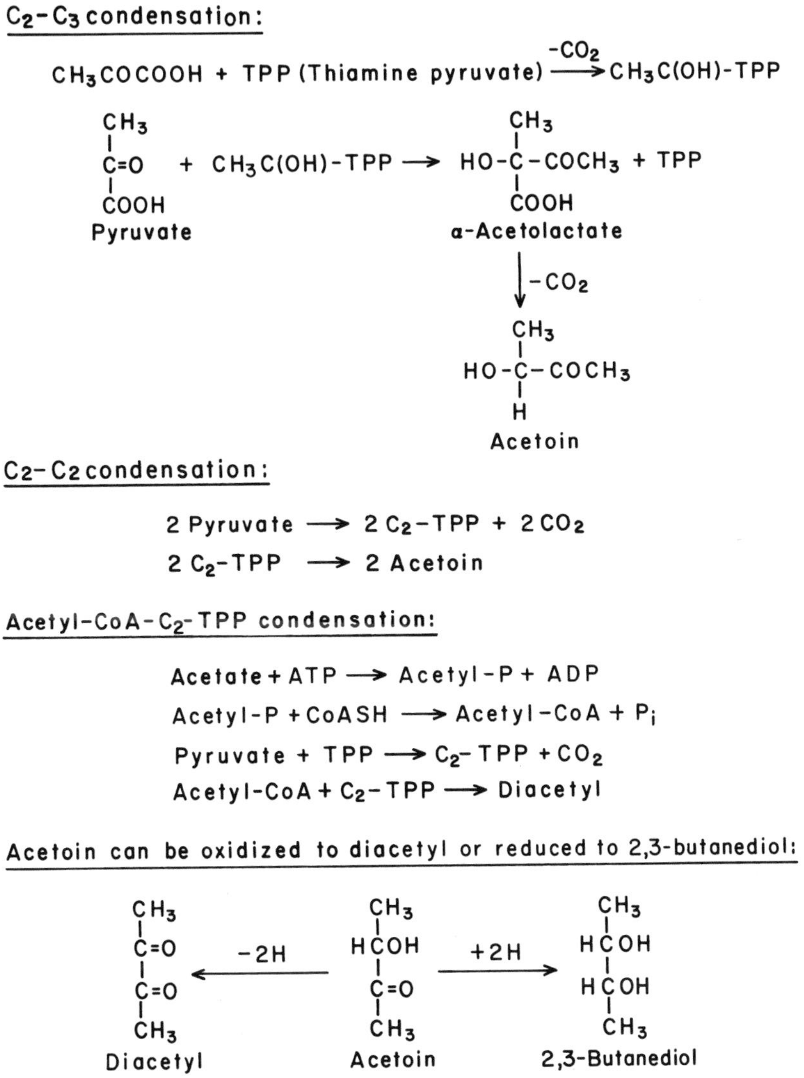

$\underline{C_2-C_3\text{ condensation}}$:

$$CH_3COCOOH + TPP\,(Thiamine\ pyruvate) \xrightarrow{-CO_2} CH_3C(OH)-TPP$$

$$\begin{array}{c} CH_3 \\ | \\ C=O \\ | \\ COOH \\ \text{Pyruvate} \end{array} + CH_3C(OH)-TPP \rightarrow \begin{array}{c} CH_3 \\ | \\ HO-C-COCH_3 \\ | \\ COOH \\ \text{a-Acetolactate} \end{array} + TPP$$

$$\downarrow -CO_2$$

$$\begin{array}{c} CH_3 \\ | \\ HO-C-COCH_3 \\ | \\ H \\ \text{Acetoin} \end{array}$$

$\underline{C_2-C_2\text{ condensation}}$:

$$2\ \text{Pyruvate} \longrightarrow 2\ C_2\text{-TPP} + 2\ CO_2$$

$$2\ C_2\text{-TPP} \longrightarrow 2\ \text{Acetoin}$$

$\underline{\text{Acetyl-CoA-}C_2\text{-TPP condensation}}$:

$$\text{Acetate} + ATP \longrightarrow \text{Acetyl-P} + ADP$$

$$\text{Acetyl-P} + CoASH \longrightarrow \text{Acetyl-CoA} + P_i$$

$$\text{Pyruvate} + TPP \longrightarrow C_2\text{-TPP} + CO_2$$

$$\text{Acetyl-CoA} + C_2\text{-TPP} \longrightarrow \text{Diacetyl}$$

Acetoin can be oxidized to diacetyl or reduced to 2,3-butanediol:

$$\begin{array}{c} CH_3 \\ | \\ C=O \\ | \\ C=O \\ | \\ CH_3 \\ \text{Diacetyl} \end{array} \xleftarrow{-2H} \begin{array}{c} CH_3 \\ | \\ HCOH \\ | \\ C=O \\ | \\ CH_3 \\ \text{Acetoin} \end{array} \xrightarrow{+2H} \begin{array}{c} CH_3 \\ | \\ HCOH \\ | \\ HCOH \\ | \\ CH_3 \\ \text{2,3-Butanediol} \end{array}$$

Fig. 7-45. Alternate routes for the production of acetoin, diacetyl, and 2,3-butanediol by microorganisms. TPP = thiamine pyrophosphate.

Many aerobic and facultative organisms are protected from the toxic action of superoxide anion by the enzyme superoxide dismutase (SOD):

$$O_2^-\!\cdot + O_2^-\!\cdot + 2H^+ \longrightarrow O_2 + H_2O_2$$

Hydrogen peroxide is dissipated by the enzyme catalase:

$$2H_2O_2 \longrightarrow 2H_2O + O_2$$

Some fermentative organisms, such as *L. plantarum,* are aerotolerant but do not produce SOD or catalase. Oxygen is not reduced and therefore superoxide and hydrogen peroxide are not formed. This organism is also able to scavenge superoxide anion due to high intracellular concentrations of Mn^{2+}.

Peptococcus anaerobius, an anaerobic lactate-producing organism, has considerable tolerance to oxygen as a result of its ability to produce high levels of NADH oxidase that reduces oxygen to water:

$$NADH + H^+ + \tfrac{1}{2}O_2 \longrightarrow NAD^+ + H_2O$$

An NADPH oxidase is also produced. This enzyme interacts with molecular oxygen to produce superoxide anion:

$$NADPH + H^+ + 2O_2 \longrightarrow NADP^+ + O_2^- \cdot + H_2O_2$$

The activity of this NADPH oxidase is much lower than that of NADH oxidase so toxic levels of superoxide are not produced.

Some lactate-producing organisms may use molecular oxygen to their advantage (i.e., energy in the form of ATP is produced). Two different mechanisms have been described to account for the utilization of molecular oxygen by these organisms. One is based on the use of oxygen as a hydrogen acceptor for the autooxidation of flavoprotein (FP) coenzymes. *Streptococcus faecalis* exhibits enhanced growth with glycerol as the substrate in aerated cultures. Under these conditions glycerol is oxidized via the triose phosphate pathway to pyruvate. Oxygen is used for the oxidation of reduced FP coenzyme:

glycerol + ATP $\longrightarrow$ glycerol-3-phosphate + ADP

glycerol-3-phosphate + FP $\longrightarrow$ dihydroxyacetone-P + FP-2H

FP-2H + O_2 $\longrightarrow$ H_2O_2 + FP

dihydroxyacetone-P $\longrightarrow$ glyceraldehyde-3-P

glyceraldehyde-3-P + NAD^+ + 2ADP + P_i $\longrightarrow$ pyruvate + NADH + H^+ + 2ATP

pyruvate + NADH + H^+ $\longrightarrow$ lactate + NAD^+

Net: glycerol + ADP + P_i + O_2 $\longrightarrow$ lactate + H_2O_2 + ATP

The energy gained is derived through utilization of the energy-yielding triose phosphate pathway and possibly through the formation of acetyl phosphate, which can yield ATP if acetate kinase is present.

A somewhat comparable situation has been described for *S. mutans* in that this organism generates ATP by substrate level phosphorylation coupled to oxidation of ethanol into acetate in the presence of oxygen. Under anaerobic conditions the conversion of pyruvate to lactate or a mixture of formate, acetate, and ethanol serves to oxidize NADH back to NAD^+. In the presence of oxygen, flavin-containing enzyme(s) catalyze the oxidation of NADH by oxygen, as shown in Figure 7-46. Oxygen is consumed at an appreciable rate and the intracellular ATP level is increased as a result of the oxidation of ethanol to acetate via acetyl-CoA.

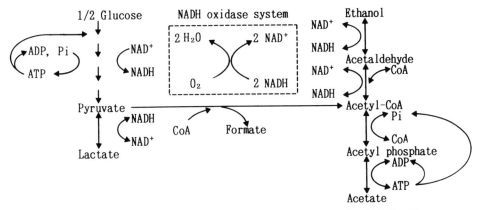

Fig. 7-46. Pathways of glucose utilization in *Streptococcus mutans*. The NADH oxidase system is a flavin enzyme that catalyzes the oxidation of NADH by oxygen. From Fukui, K. et al. 1988. *J. Bacteriol.* **170**:4589–4593.

Another explanation for the utilization of oxygen by some lactic acid bacteria is derived from the production of cytochromes when grown in an aerobic environment in the presence of hematin. Under these conditions, *S. faecalis* produces functional cytochromes that yield ATP through oxidative phosphorylation. In the absence of heme, only a flavin system of electron transport is formed. Distribution of cytochromelike respiration seems to be limited to *S. faecalis* and a few strains of *L. lactis* and its subspecies. *Streptococcus faecium*, *L. cremoris*, and *L. lactis* produce menaquinones, isoprenoid quinones that function in electron transport, and possibly oxidative phosphorylation.

In heterofermentative organisms, the diversity of fermentation products as a result of environmental influences is even greater than in homofermentative organisms. This is the result of the operation of more than one pathway of carbohydrate metabolism. Two main types of fermentation may be superimposed to varying degrees. In certain organisms the basic fermentation appears to take place via a C_2–C_3 cleavage of pentose phosphate:

$$\text{glucose } (C_6) \longrightarrow \text{pentose } (C_5) + CO_2$$

$$\text{pentose } (C_5) \longrightarrow \text{lactate } (C_3) + \text{ethanol } (C_2)$$

In other organisms, glycerol may be a predominant product:

$$\tfrac{1}{2} \longrightarrow 2\text{glycerol} + \text{acetate} + CO_2$$

In this latter group, both pathways operate together to produce the final mixture of products.

It is interesting to compare the diversity of products observed with substrates other than glucose. With fructose as the carbon source, large quantities of mannitol are produced. Fructose serves as the preferred hydrogen acceptor and is reduced to mannitol:

$$3\text{fructose} \longrightarrow 2\text{mannitol} + \text{lactate} + \text{acetate} + CO_2$$

Homolactic species produce only lactate with fructose as substrate, while heterofermentative species convert fructose to acetyl phosphate, acetaldehyde, and dihydroxyacetone phosphate. These intermediates are used as hydrogen acceptors to produce ethanol and glycerol-3-phosphate. Because the pentose phosphate route is used, heterolactic organisms produce CO_2 from the C-1 of glucose. Oxidation of G-6-P and 6-PG are apparently linked to the reduction of acetyl phosphate and acetaldehyde to form ethanol without the formation of additional product.

When the substrate is either more oxidized or more reduced than glucose, there will be a corresponding shift in the products. A reduced substrate, such as mannitol (oxidation value = -1), will elicit the production of more reduced products, such as glycerol, hydrogen, or ethanol. With oxidized substrates, such as citrate (oxidation value = $+3$), there will be more oxidized products (for calculation of oxidation values, see Table 7-5).

Variation in the fermentation products formed is also dependent on pH. Above pH 7, little or no lactate is produced:

$$\text{citrate} \longrightarrow CO_2 + \text{formate} + 2\text{acetate}$$

At acid pH, more lactate and acetoin are formed:

$$2\text{citrate} \longrightarrow \text{lactate} + \text{acetate} + 3CO_2 + \text{acetoin}$$

The initial steps in the fermentation of citrate involve cleavage by the enzyme citritase yielding oxaloacetate and acetate. Citritase is a non-CoA-dependent enzyme that gives rise to acetate directly without the formation of acetyl phosphate. The oxaloacetate formed in this reaction is converted to pyruvate and CO_2, with the pyruvate being reduced to lactate. Acetoin also arises from pyruvate with the production of additional CO_2. The inability of *E. coli* to utilize citrate as a sole source of carbon and energy is useful in distinguishing this organism for *Enterobacter aerogenes* and other gram-negative bacteria. However, care must be taken to avoid contamination with amino acids or other carbohydrates as *E. coli* can utilize citrate in the presence of a cosubstrate.

Butyric Acid and Solvent-Producing Fermentations

Members of the *Clostridium, Butyrivibrio, Bacillus*, and less-well defined flora of anoxic marsh sediments, wetwood of living trees, and anaerobic sewage digestion systems produce butyric acid, butanol, acetone, isopropanol, or 2,3-butanediol. Dihydrogen, carbon dioxide, acetate, ethanol, and other compounds are commonly found among the fermentation products.

Clostridium acetobutylicum utilizes the EMP pathway for glucose catabolism with the formation of C_3 and C_4 products from pyruvate, as shown in Figure 7-41. The fermentation process is biphasic. During growth, the organism first forms acetate and butyrate (acidogenic phase). As the pH drops and the culture enters the stationary phase there is a metabolic shift to solvent production (solvetogenic phase). The pivotal reaction in these fermentations is the formation of acetyl-CoA from either pyruvate or acetate. Acetyl-CoA can then undergo a condensation reaction to form acetoacetate, which may be (a) reduced to butyrate and butanol or (b) cleaved via decarboxylation to acetone. Acetone may be further reduced to isopropanol. Acetyl-CoA may also be reduced to acetaldehyde and ethanol. With other species, such as *Bacillus acetoethylicum* or *B. polymyxa*, lactate or 2,3-butanediol may also be observed. Formation of butyrate from acetyl-CoA involves the steps shown in Figure 7-47. Butanol is formed via the reduction of butyryl-CoA followed by reduction of butyraldehyde via NAD-linked dehydrogenases:

$$CH_3CH_2CO{-}S{-}CoA + NADH + H^+ \longrightarrow CH_3CH_2CH_2CHO + HSCoA + NAD^+$$

$$CH_3CH_2CH_2CHO + NADH + H^+ \longrightarrow CH_3CH_2CH_2CH_2OH + NAD^+$$

In balancing the fermentation products formed in these complex fermentations it is not difficult to account for the carbon (Fig. 7-48). Formulating a theoretical scheme that shows a balance of the oxidized and reduced products may be more difficult. Consider the production of acetate and butyrate and the accompanying CO_2 and H_2 according to the equation in Figure 7-48. The carbon atoms balance but there is a deficit of some reduced product. In most clostridial fermentations there is a compensatory formation of small amounts of several reduced products (see Table 7-7). Production of these compounds appears to be necessary because the condensation of 2 mol of acetyl-CoA to form acetoacetyl-CoA and the subsequent conversion to acetone via acetoacetate decarboxylase results in a sharp decrease in the number of hydrogen acceptors available. Acetyl-CoA can act as a hydrogen acceptor giving rise to ethanol and acetone can be further reduced to isopropanol. In those fermentations in which the balances appear to be the best, ethanol, acetoin, and 2,3-butanediol are found among the final products.

It is important to note that formate is produced in relatively few of the fermentations. Although carbon dioxide and hydrogen gas are produced as in the fermentation of gram-negative organisms, hydrogen gas is apparently formed through the pyruvate:ferredoxin (Fd) oxidoreductase without the intermediary formation of formate:

$$\text{pyruvate (C}_3) \longrightarrow \text{acetyl-CoA} + CO_2 + FdH + H^+$$

Hydrogenase converts reduced ferredoxin (FdH) to H_2:

$$FdH + H^+ \longrightarrow H_2 + Fd$$

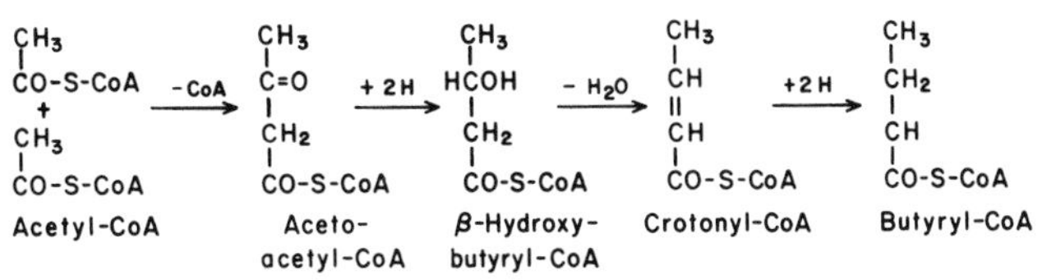

Fig. 7-47. Route of formation of butyryl-CoA from acetyl-CoA. The enzymes involved are acetyl-CoA acetyltransferase (thiolase); β-hydroxybutyryl-CoA dehydrogenase; crotonase; and butyryl-CoA dehydrogenase, respectively.

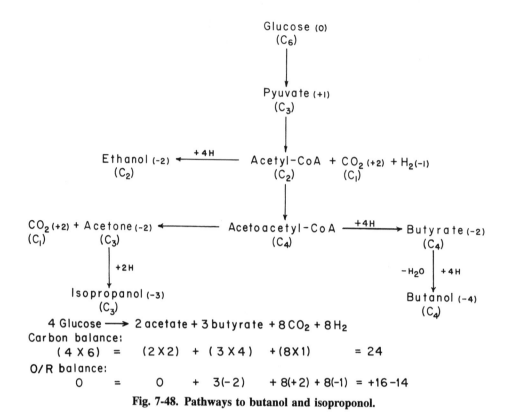

Fig. 7-48. Pathways to butanol and isoproponol.

As shown in Figure 7-49, the flow of electrons from NADH to FdH to H_2 explains why the organism produces large quantities of H_2 and does not consume H_2.

The overall equation for mixed butyrate, acetone, and ethanol fermentation can account for both the carbon and the O–R balance:

$$C_6 \longrightarrow \text{butyrate } (C_4) + 2C_1 + 2H_2$$

$$C_6 \longrightarrow \text{acetone } (C_3) + 3C_1 + 2H_2$$

$$C_6 \longrightarrow 2 \text{ ethanol } (C_2) + 1C_1 + 2H_2$$

Net: $3C_6 \longrightarrow C_4 + C_3 + 2C_2 + 7C_1 + 6H_2$ C balance 18 = 18

O–R bal: $0 \longrightarrow (-1) + (-2) + 2(-2) + 7(+2) + 6(-1) = +14 - 14 = 0$

Generally, small amounts of either oxidized or reduced products will be formed in order to provide for balance of the oxidized and reduced products. The oxidation values are those given in Figure 7-41.

Production of butyrate from butyryl-CoA by *C. acetobutylicum* involves the enzymes phosphotransbutyrylase and butyrate kinase and yields ATP via substrate level phospho-

TABLE 7-7. Examples of Acetone–Butanol and Mixed-Solvent Fermentations

Products[a]	C. saccharobutyricum[b]	C. acetobutylicum[b]	C. butylicum[c]	C. thermosaccharolyticum[d]	B. acetoethylicum[c]	B. polymyxa[c]
CO_2	195.5	220.0	207.0	174.0	215.0	195.0
H_2	233.0	165.9	111.1	229.4	137.0	54.0
Acetate	42.6	24.8	20.3	48.5	16.0	5.0
Butyrate	75.3	7.1	14.5	59.5		
Formate	tr				10.0	
Lactate				25.7		
Enthanol		4.9			122.0	95.0
Butanol		47.4	50.2			4.5
Acetone		22.3			28.0	6.0
Isopropanol			18.0			
Acetoin		5.7				
2,3-Butanediol					12.0	39.0
C. rec. (%)	97.0	98.0	93.5	98.0	105.0	100.0
O/R balance	1.02	1.01	1.05	0.99	0.90	1.13

[a] Amounts are expressed as millimoles per 100 mmol of glucose used.
[b] Donker, H. L., 1926, Ph.D. Thesis, Technische Hoogeschool, Delft, The Netherlands.
[c] Osburn, O. L., R. W. Brown, and C. H. Werkman. 1937. *J. Biol. Chem.* **121:**685.
[d] Sjolander, N. W. 1937. *J. Bacteriol.* **34:**419.

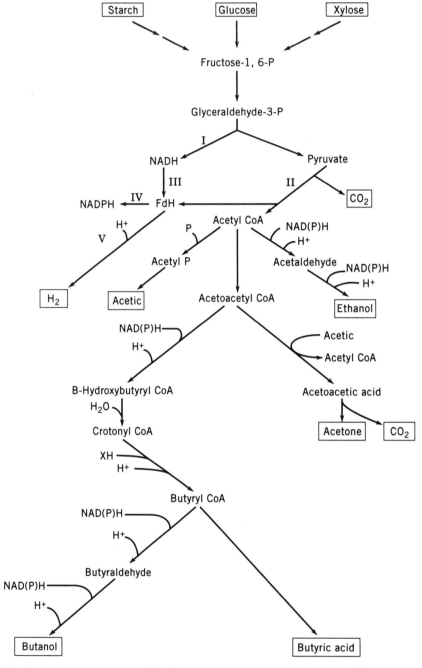

Fig. 7-49. Proposed carbon and electron flow pathway for *Clostridium acetobutylicum* saccharide fermentation. (I) glyceraldehyde-3-phosphate dehydrogenase; (II) pyruvate-ferredoxin oxidoreductase; (III) ferredoxin-NAD oxidoreductase; and (V) hydrogenase. Each arrow represents one or more enzyme-catalyzed reactions. The flow of electrons from NADH to reduced ferredoxin to H_2 explains why the organism produces high partial pressures of H_2 and does not consume H_2. From Zeikus, J. G. 1985. Biology of spore-forming anaerobes. In *Biology of Industrial Microorganisms*, A. L. Demain and N. A. Solomon (Eds.). Benjamin–Cummings, Menlo Park, CA, pp. 79–114.

rylation:

$$\text{butyryl-CoA} + P_i \longrightarrow \text{butyryl-P} + \text{CoA}$$

$$\text{butyryl-P} + \text{ADP} \longrightarrow \text{butyrate} + \text{ATP}$$

Butyrivibrio fibrisolvens is a major butyrate-producing species in the rumen. The pathway to butyrate from acetyl-CoA in this organism is essentially the same as for *C. acetobutylicum* and other saccharolytic clostridia. The Reiter strain of *Treponema phagendenis* produces butyrate, acetate, succinate, and ethanol via pathways similar to those observed in clostridia. Glucose is catabolized to pyruvate via the EMP pathway. Butyrate formation from acetyl-CoA involves an NAD-dependent 3-hydroxybutyryl-CoA dehydrogenase and NAD(P)-independent butyryl-CoA dehydrogenase activities. Butyrate is formed from butyryl-CoA in a CoA transphorase reaction. Adenosine triphosphate is not generated by substrate level phosphorylation during butyrate formation. Phosphate acetyltransferase and acetate kinase are involved in the conversion of acetyl-CoA to acetate. Exogenously supplied fumarate is reduced to succinate. A membrane-associated fumarate reductase uses reduced ferredoxin or flavin nucleotides as electron donors.

Clostridium thermocellum, by comparison with *C. acetobutylicum*, employs the EMP pathway for saccharolytic fermentation. However, lactate, acetate, ethanol, carbon dioxide, and dihydrogen are the major products. Little or no butanol, butyrate, acetone, or isopropanol are formed. As described earlier, this organism utilizes cellulose for the production of high quantities of ethanol and acetate. High H_2 production results from the flow of electrons from NADH to reduced ferredoxin. Other thermophilic species such as *C. thermohydrosulfuricum*, *C. thermosaccharolyticum*, and *Thermoanaerobium brockii* use similar metabolic routes for the production of ethanol, acetate, and lactate from glucose, but quantitative differences in intracellular hydrogen flow and product yields are observed. Some strains of *C. thermohydrosulfuricum* can produce as high as 1.6–1.8 mol of ethanol per mol of glucose fermented. This species has a different flow of hydrogen from reduced ferredoxin to NADH and NADPH. As a result, much less H_2 is produced. High concentrations of H_2 inhibit growth by flowing backwards and overreducing NADP and NAD pools. Studies with mutant strains of *C. thermosaccharolyticum* deficient in acetate production indicate that higher yields of ethanol, the more desirable end-product, are obtainable with these strains.

Production of acetic acid from glucose by *C. thermoaceticum* is particularly efficient in that 3 mol of acetate are produced per mol of glucose. This is achieved by a novel carbon and electron flow involving the synthesis of one acetate molecule by a series of C_1 transformation reactions:

$$C_2H_{12}O_6 \longrightarrow EMP \longrightarrow 2CH_3COCOOH + 4H$$

$$CH_3COCOOH + H_2O \longrightarrow CH_3COOH + CO_2 + 2H$$

$$CO_2 + 2H \longrightarrow HCOOH$$

$$HCOOH + THF + 4H \longrightarrow CH_3THF + 2H_2O$$

$$H_2O + CH_3THF + CH_3COCOOH \longrightarrow 2CH_3COOH + THF$$

$$\text{Net: } C_6H_{12}O_6 \longrightarrow 3CH_3COOH$$

Pyruvate ferredoxin oxidoreductase functions in the formation of reduced ferredoxin. Carbon monoxide (CO) can replace pyruvate as a carboxyl donor for the synthesis of acetyl-CoA with CH_3THF. Methyltransferase, CoA, ATP, carbon monoxide dehydrogenase, and corrinoid enzyme are required. An enzyme-bound form of formate [HCOOH] is a presumed intermediate in the reaction:

$$CO + H_2O \longleftrightarrow [HCOOH] \longleftrightarrow CO_2 + 2H^+ + 2e^-$$

It is of significance that fixation of CO_2 or CO into acetate by this organism was one of the first demonstrations of total synthesis of an organic compound from CO_2 by a heterotrophic organism. It is now well known that practically all, if not all, forms of life use CO_2 for growth. However, most heterotrophs can only add CO_2 to a preexisting compound and form a carboxyl group. The bacteria that grow on CO_2 and H_2 all contain hydrogenase, an enzyme that converts hydrogen gas to two protons and two electrons. These electrons can supply the necessary reductive capacity for the utilization of the CO_2. Extracts of *C. thermoaceticum* contain hydrogenase in addition to the CO dehydrogenase and the corrinoid enzyme. Acetyl-CoA is synthesized from CH_3THF, CO_2, H_2, and CoA. Hydrogenase, with ferredoxin as the electron acceptor, yields reduced ferredoxin, which in turn is used by CO dehydrogenase in the reduction of CO_2 to the enzyme-bound formyl intermediate [HCOOH].

As noted above, **ferredoxins** play an important role in electron-transfer reactions in a wide variety of biological systems. Four classes of ferredoxins have been recognized based upon the number of sulfur-bound (nonheme) iron atoms present and the arrangement of the iron–sulfur cluster at the active center, as shown in Figure 7-50. **Rubredoxins** contain a single iron atom coordinated by four cysteines and have been isolated from *Rhodospirillum rubrum*. Two-iron ferredoxins have an active site containing two irons, two inorganic sulfur atoms, and four cysteines. The two-iron ferredoxins have been isolated primarily from photosynthetic plants and algae but have also been found in some bacteria. Ferredoxins containing four iron atoms have one $Fe_4 S_4$ cluster at the active center. Four-iron ferredoxins have been found in *C. thermoaceticum*, *Desulfovibrio gigas*, *D. desulfuricans*, *B. stearothermophilus*, *B. polymyxa*, and *Spirochaeta stenostrepta*. Eight-iron ferredoxins contain two $Fe_4 S_4$ clusters arranged in approximately twofold symmetry within the ferredoxin protein. *Peptococcus aerogenes*, *Clostridium pasteurianum*, and other clostridia are most commonly found to contain $(Fe_4 S_4)_2$ ferredoxins.

Ferredoxins play an important role in the photophosphorylation process in photosynthetic bacteria (e.g., *Chromatium*, *R. rubrum*), cyanobacteria (e.g., *Nostoc*), and in the eukaryotic algae and green plants. Ferredoxins also function as carriers of reducing or oxidizing equivalents in anaerobic organisms (e.g., *Clostridium*, *Bacteroides*) and in atmospheric nitrogen fixation. Most ferredoxins have an unusually low oxidation–reduction potential (-0.4 V), which is about 100 mV less than that of the $NAD^+/NADH + H^+$ system (Fig. 7-14).

Fermentations of the Mixed-Acid Type

Under anaerobic conditions and in the absence of alternate electron acceptors, members of the *Enterobacteriaceae* (*Escherichia*, *Enterobacter*, *Salmonella*, *Klebsiella*, and *Shigella*) ferment glucose to a mixture of acetic, formic, lactic and succinic acids, and ethanol (Table 7-8). The production of acetate and formate as major products is notable. Indi-

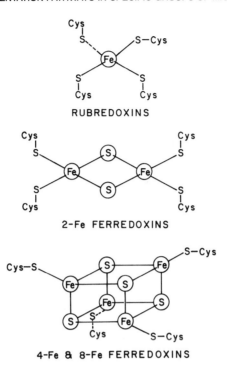

RUBREDOXINS

2-Fe FERREDOXINS

4-Fe & 8-Fe FERREDOXINS

Fig. 7-50. Proposed arrangements of the iron (Fe) and sulfur (S) atoms in ferredoxins. Cys-S represents the linkage with the S in cysteine. X-ray diffraction studies of the molecular structure of the 8Fe/8S ferredoxin of *Peptococcus aerogenes* show that the iron and sulfur atoms exist in two tetrameric clusters, each with 4Fe and 4 inorganic S and 4 cysteine S atoms.

TABLE 7-8. Examples of Mixed-Acid Fermentations

Products[a]	*Escherichia coli*	*Enterobacter aerogenes*	*Salmonella typhi*
Lactate	108.8	53.4	121.7
Ethanol	41.3	59.4	25.4
Acetate	32.0	10.1	25.6
Formate	1.6	5.5	39.3
CO_2	54.0	126.9	0
H_2	45.2	44.2	0
Succinate	18.0	6.0	10.8
Acetoin	0	0.4	0
2,3-Butanediol	0	34.6	0
Carbon recovery (%)	100.0	99.5	93.3
O−R balance	0.99	0.99	0.97

[a]Amounts are expressed as millimoles per 100 mmol glucose used.

cations are that as much as 85% of the glucose fermented by *E. coli* is metabolized via the EMP pathway. Other pathways must contribute to the products to some extent, however. As opposed to clostridial fermentations in which acetyl-CoA, carbon dioxide, and dihydrogen usually arise without formate as an intermediate, formate is consistently found as a product of sugar metabolism by enteric bacteria. A CoA dependent pyruvate formate lyase encoded by the *pfl* gene initiates the sequence:

$$CH_3COCOOH + CoASH \longleftrightarrow CH_3COSCoA + HCOOH$$

Phosphotransacetylase converts acetyl-CoA to acetyl phosphate:

$$CH_3COSCoA + P_i \longleftrightarrow CH_3COOPO_3H_2 + CoASH$$

Acetate kinase (encoded by the *ackA* gene in *E. coli* generates ATP from acetyl phosphate:

$$CH_3COOPO_3H_2 + ADP \longleftrightarrow CH_3COOH + ATP$$

This enzyme results in the production of acetate and generates a major portion of the ATP generated during anaerobic growth. The formate hydrogenlyase system converts formate to H_2 and CO_2. Formate hydrogenlyase consists of two enzymes. Formate dehydrogenase-H (FDH-H) yields CO_2 and a reduced carrier that is acted upon by hydrogenase 3 to yield H_2:

$$HCOOH + carrier \longrightarrow CO_2 + carrier\text{-}2H$$
$$carrier\text{-}2H \longleftrightarrow H_2 + carrier$$
$$\overline{\text{Net: } HCOOH \longrightarrow CO_2 + H_2}$$

As mentioned previously, induction of FDH-H (encoded by *fdhF*) requires the presence of formate, molybdate, the absence of electron acceptors, such as oxygen or nitrate, and acidic pH. Expression of *fdhF* requires an alternate sigma factor (NtrA) and an upstream activating sequence (UAS). Expression of *pfl* is dependent on phosphoglucoisomerase and phosphofructokinase activities.

Lactic acid arises from pyruvate through the activity of a fermentative lactate dehydrogenase (encoded by *ldhA*) induced under the conditions of anaerobiosis and a low environmental pH. Production of ethanol in these fermentations apparently occurs via the reduction of acetyl-CoA to ethanol rather than by formation of acetaldehyde from pyruvate. Acetyl-CoA is converted to acetaldehyde by a CoA-linked acetaldehyde dehydrogenase and the acetaldehyde is then reduced to ethanol by an NAD-linked alcohol dehydrogenase. This enzyme is expressed only under anaerobic conditions.

Acetoin is formed by *Enterobacter aerogenes* but not by *E. coli*. A variety of other organisms (e.g., *Serratia, Erwinia*) also form acetoin or its oxidation (diacetyl) or reduction (2,3-butanediol) products, as shown in Figures 7-41 and 7-45. In gram-negative organisms, thiamine pyrophosphate is involved as a cofactor and α-acetolactate is produced as an intermediate in acetoin formation. Acetate induces the production of the acetolactate-forming enzyme, acetolactate decarboxylase, and diacetyl reductase. Diacetyl reductase also functions as an acetoin dehydrogenase in *E. aerogenes* and serves as a regulator of the balance between acetoin and 2,3-butanediol. In *E. coli*, the direct formation of diacetyl from 2-hydroxyethyl-TPP and acetyl-CoA has been demonstrated. Diacetyl is reduced to acetoin by diacetyl reductase (Fig. 7-45). The reaction is specific for NADP$^+$ and diacetyl and does not express acetoin reductase activity.

Salmonella typhimurium can utilize citrate as the sole source of carbon under anaerobic conditions. *Escherichia coli* cannot. Anaerobically, *E. coli* can degrade citrate if another suitable carbon source, such as glucose, pyruvate, lactate, or fumarate, is present. A cosubstrate is needed because oxaloacetate decarboxylase and malic enzyme are repressed under anaerobiosis. Citrate lyase is induced by citrate but is quickly inactivated by deacetylation of the active center of the enzyme when the cosubstrate is exhausted. *Enterobacter aerogenes* grows well with citrate as the sole carbon source. The gene for citrate lyase cannot be deleted from *S. typhimurium*. The only class of mutants found have a mode of covalent regulation of citrate lyase comparable to that in *E. coli*, that is, the enzyme is activated (acetylated) only in the presence of a cosubstrate.

As shown in Figure 7-51, *E. coli* can transport glycerol via a facilitator encoded by *glpF*. Once inside the cell, glycerol is phosphorylated by glycerol kinase, encoded by *glpK*, and thus is trapped inside the cell as *sn*-glycerol-3-phosphate (G-3-P). The G-3-P is utilized by either an anaerobic dehydrogenase encoded by the *glpACB* operon or the aerobic dehydrogenase encoded by the *glpD* operon. The dihydroxyacetone phosphate formed can then be metabolized via the triose phosphate pathway. The *glpACB* and *glpD* operons belong to the *glp* regulon controlling the utilization of glycerol, G-3-P, and glycerophosphodiesters, as outlined in Figure 7-51. Increased anaerobic expression of *glpA* is dependent on the *fnr* gene product (FNR), a pleiotropic activator of genes involved in anaerobic respiration. Anaerobic repression of the *glpD* operon is relieved by mutations in either *arcA* or *arcB*, genes that encode ArcA, a cytoplasmic regulatory protein, and ArcB, a transmembrane sensor protein. This Arc (aerobic respiratory control) system controls the expression of several genes encoding enzymes involved in aerobic respiration.

A survey of anaerobic fermentation balances with various substrates using a nonintrusive nuclear magnetic resonance (NMR) spectroscopy technique showed that substrates more reduced than glucose yield more of the highly reduced fermentation product ethanol. More oxidized substrates resulted in the production of the less-reduced fermentation product acetate. This survey emphasized the fact that the redox level of the substrate is an important factor governing the ratio of oxidized to reduced products. For example, sugar alcohols, which are more reduced than the corresponding hexoses, must yield a higher proportion of more-reduced fermentation products in order to achieve hydrogen balance.

Propionic Acid Fermentation

Propionate, acetate, and carbon dioxide are the major products of the fermentation of glucose, glycerol, and lactate by *Propionibacterium, Veillonella, Bacteroides*, and some species of clostridia. An early hypothesis for the origin of propionate in bacterial fermentation was by removal of water from lactate to form acrylate with subsequent reduction to propionate. This pathway has been demonstrated in *Clostridium propionicum, Bacteroides ruminicola*, and *Peptostreptococcus*. The reaction sequence is mediated by

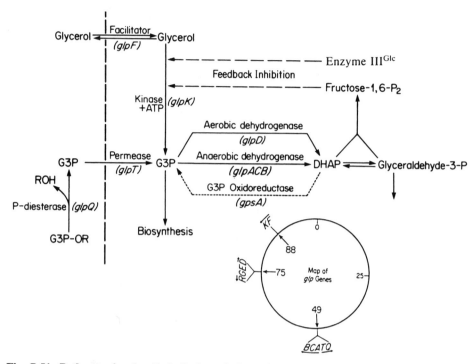

Fig. 7-51. Pathways for the dissimilation of glycerol, G-3-P, and glycerophosphodiesters in E. coli encoded by the glp regulon. The genetic symbols are in parentheses. Dashed arrows indicate feedback inhibitors of glycerol kinase activity: enzyme III^{Glc}, the protein III of the phosphoenolpyruvate phosphotransferase (PEP–PTS) system for the vectorial phosphorylation of glucose, and fructose-1,6-P$_2$, fructose-1,6-bisphosphate. The dotted arrow indicates the pathway for G-3-P biosynthesis. Positions of the glp genes and operons (arrows over genetic symbols show the directions of transcriptions) are indicated on the circular figure. From Iuchi, S., S. T. Cole, and E. C. C. Lin. 1990. *J. Bacteriol.* **172**:179–184.

CoA and the enzyme lactyl CoA dehydrase:

The two hydrogen atoms required for the reduction of acrylyl-CoA to propionyl-CoA arise through the formation of acetate, CO_2, and 2(2H) from a portion of the total lactate utilized:

$$\text{lactate} \longrightarrow \text{acetate} + CO_2 + 2(2H)$$

$$\underline{2\text{lactate} + 2(2H) \longrightarrow 2\text{propionate}}$$

$$\text{Net: } 3\text{lactate} \longrightarrow 2\text{propionate} + \text{acetate} + CO_2$$

In *Propionibacterium* and *Veillonella* the formation of propionate occurs via a more complex series of reactions. The general reaction sequence for the formation of propionate, acetate, and carbon dioxide from glucose is usually given as

$$1.5\text{glucose} + 3\text{ADP} + 3\text{P}_i \longrightarrow 3\text{pyruvate} + 3\text{ATP} + 3(2\text{H})$$

$$\text{pyruvate} + \text{ADP} + \text{P}_i \longrightarrow \text{acetate} + CO_2 + \text{ATP} + 2\text{H}$$

$$2\text{pyruvate} + 2\text{ADP} + 2\text{P}_i + 4(2\text{H}) \longrightarrow 2\text{propionate} + 2\text{ATP} + 2H_2O$$

$$\text{Net: } 1.5\text{glucose} + 6\text{ADP} + 6\text{P}_i \longrightarrow$$
$$\text{acetate} + 2\text{propionate} + CO_2 + 6\text{ATP} + 2H_2O$$

The details of the pathways by which propionate and acetate are formed from glucose by the propionibacteria are shown in Figure 7-52.

A major factor in the elucidation of this route of propionate formation was the discovery of the mechanism of transcarboxylation by methylmalonyl-oxaloacetate transcarboxylase. In this reaction, biotin plays an important catalytic role. Cobalamin (vitamin B_{12}) also serves as a cofactor in the formation of methylmalonyl-CoA from succinyl-CoA. The reaction sequence is

$$\text{Enz-biotin-}CO_2 + \text{pyruvate} \longrightarrow \text{oxaloacetate} \longrightarrow \text{enz-biotin}$$

$$\text{oxaloacetate} + 4\text{H} \longrightarrow \text{succinate} + H_2O$$

$$\text{succinyl-CoA-B}_{12}\text{-enz} \longrightarrow \text{methylmalonyl-CoA} + \text{B}_{12}\text{-enz}$$

$$\text{methylmalonyl-CoA} + \text{enz-biotin} \longrightarrow \text{propionyl-CoA} + \text{enz-biotin-}CO_2$$

$$\text{propionyl-CoA} + \text{succinate} \longrightarrow \text{succinyl-CoA} + \text{propionate}$$

$$\text{Net: pyruvate} + 4\text{H} \longrightarrow \text{propionate} + H_2O$$

As shown in Figure 7-52, the hydrogen atoms required for the reduction of oxaloacetate to succinate are obtained from reactions in the conversion of glucose to acetate and carbon dioxide via pyruvate.

In practice, the ideal yields of propionate, acetate, and carbon dioxide were not always obtained for propionic acid fermentations. The carbon recoveries were often high while the amount of carbon dioxide was usually lower than expected. Another unexpected finding was that the ratio of propionate to acetate was much greater than expected (Table 7-9). The high propionate/acetate ratios were explained by the finding of considerable amounts of residual succinate in the final products, a factor that also improves the carbon recovery. The ATP yields were observed to be higher than expected. This observation has now been explained by the discovery that polyphosphate is used in the glucokinase reaction and inorganic pyrophosphate serves as the energy source for the phosphorylation of F-6-P by phosphofructokinase, as shown in the following sequence:

Glukokinase

$$\text{glucose} + \text{polyP}_n \longrightarrow \text{G-6-P} + \text{polyP}_{n-1}$$

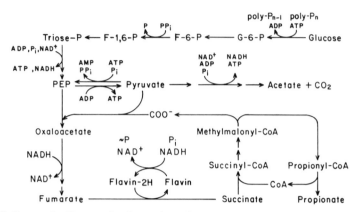

Fig. 7-52. Pathways leading to the formation of propionate, acetate, and CO₂ by *Propionibacterium*. From Wood, H. G. 1986, personal communication.

Phosphofructokinase

$$\text{F-6-P} + \text{PP}_i \longrightarrow \text{F-1,6-BP} + \text{P}_i$$

Pyruvate, orthophosphate dikinase

$$\text{pyruvate} + \text{ATP} + \text{P}_i \longleftrightarrow \text{PEP} + \text{AMP} + \text{PP}_i$$

Carboxytransphosphorylase

$$\text{PEP} + \text{CO}_2 + \text{P}_i \longleftrightarrow \text{oxaloacetate} + \text{PP}_i$$

Net conversion of pyruvate to oxaloacetate

$$\text{pyruvate} + \text{ATP} + \text{P}_i \longleftrightarrow \text{oxaloacetate} + \text{AMP} + 2\text{PP}_i$$

Under appropriate conditions and in the presence of the requisite enzymes, yields of 11 mol of ATP per 3 mol of glucose can be achieved as shown by the following equation:

$$3\text{glucose} + 4\text{PP}_i + 11\text{ADP} \longrightarrow 4\text{propionate} + 2\text{acetate} + 2\text{CO}_2 + 11\text{ATP} + 4\text{P}_i$$

The participation of polyphosphate and PP_i also explains the high cell yields observed in growth of propionibacteria. The demonstration of reactions in which the energy inherent in polyphosphate and PP_i may be utilized and not wasted through hydrolysis may have far reaching implications in other systems since many biological reactions yield PP_i or polyphosphates as products.

Propionispira arboris, an organism causing wetwood disease of cottonwood trees, has been shown to follow essentially the same pathway for the formation of propionate as shown in Figure 7-52 for *Propionibacterium*. This organism uses the EMP pathway for the dissimilation of glucose and forms propionate via the methylmalonyl-CoA route. In

TABLE 7-9. Fermentations by Propionibacteria

Organism	Substrate	Propionate[a]	Acetate	CO_2	Succinate	Propionate Acetate
P. freudenreichii	Glucose	134.0	52.6	49.2	12.6	2.6
P. freudenreichii	Lactate	63.5	35.3	35.8	37.8	1.8
P. freudenreichii	Glycerol	100.0	9.9	6.9	11.7	10.0
P. shermanii	Glucose	140.0	56.8	56.4	12.0	2.46
P. shermanii	Lactate	62.5	36.5	37.0	9.3	1.7
P. shermanii	Glycerol	102.0	10.6	7.3	11.2	10.0
P. peterssonii	Glucose	114.0	54.0	51.0	11.1	2.1
P. arabinosum	Glucose	148.0	10.0	63.6	7.9	14.8

[a]Expressed in micromoles. Data from Van Niel, C. B. (1928) Ph.D. Thesis, Technische Hoogeschool, Delft, The Netherlands; Wood, H. G. and C. H. Werkman, *Biochem. J.* **30**:48, 1936, *Biochem. J.* **30**:618, 1936; and Wood, H. G., R. H. Stone, and C. H. Werkman. 1937. *Biochem. J.*, **31**:349.

this organism, pyruvate oxidation to acetate involves pyruvate-ferredoxin oxidoreductase, a reaction found in many other anaerobic species.

Acetic Acid Fermentation

The acetic acid bacteria are divided into two genera: *Acetobacter* and *Gluconobacter*. Both species are obligate aerobes that oxidize sugars, sugar alcohols, and ethanol with the production of acetic acid as the major end-product. Electrons from these oxidative reactions are transferred directly to the respiratory chain. The respiratory chain of *Gluconobacter suboxydans* consists of cytochrome *c*, ubiquinone, and a terminal cytochrome *o* ubiquinol oxidase. In *Acetobacter aceti* the composition of the respiratory chain varies depending on the cultural conditions, as shown in Figure 7-53.

    Gluconbacter suboxydans lacks a functional TCA cycle although all of the enzymes of this cycle except succinate dehydrogenase are present. This organism cannot ferment glucose or other carbohydrates since neither the EMP nor the Entner–Doudoroff pathway is present. A modified pentose cycle is used for the metabolism of glucose under aerobic conditions (Fig. 7-54). *Acetobacter* species have a functional TCA cycle. The acetyl phosphate resulting from the C_2–C_3 cleavage of pentose is oxidized to CO_2 and energy is generated via oxidative phosphorylation.

    A characteristic activity of *Acetobacter* and *Gluconobacter* is the oxidation of ethanol to acetic acid. Ethanol oxidation occurs via two membrane-associated dehydrogenases: alcohol dehydrogenase and acetaldehyde dehydrogenase:

$$CH_3CH_2OH \longrightarrow CH_3CHO + 2H \longrightarrow CH_3COOH + 2H$$

The electrons generated in ethanol oxidation are thought to be transferred directly to the respiratory chain as described above.

CHARACTERISTICS AND METABOLISM OF AUTOTROPHS

Major Groups of Autotrophs

Organisms that use C_1 compounds (e.g., CO_2 or CH_4) as their major or sole source of carbon and energy are called **autotrophs. Methylotrophs** use methane (CH_4) or methanol

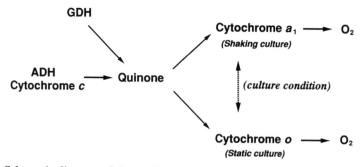

Fig. 7-53. Schematic diagram of the respiratory chain of *Acetobacter aceti*. Alcohol dehydrogenase (ADH) containing cytochrome *c* and glucose dehydrogenase (GDH) are shown to donate electrons directly to ubiquinone, and cytochrome a_1 and cytochrome *o* are shown to oxidize ubiquinone directly. From Matsushita, K. et al. *J. Bacteriol.* **174:**122–129.

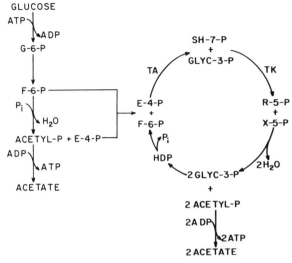

NET: Glucose + 2P$_i$ + 2 ADP $\longrightarrow$ 3 Acetate + 2 ATP + 3H$_2$O

Fig. 7-54. Modified pentose cycle for *Acetobacter* and *Gluconobacter*. TA = transaldolase; TK = transketolase; E-4-P = erythrose-4-phosphate; SH-7-P = sedoheptulose-7-phosphate; R-5-P = ribose-5-phosphate; X-5-P = xylulose-5-phosphate; HDP = hexosediphosphate; glyc-3-P = glyceraldehyde-3-phosphate. From Moat, A. G. 1985. Biology of lactic, acetic, and propionic acid bacteria. In *Biology of Industrial Microorganisms*, A. L. Demain and N. A. Solomon (Eds.). Benjamin–Cummings, Menlo Park, CA, pp. 143–188.

(CH$_3$OH) as their source of carbon. Autotrophic organisms that use light as a source of energy are termed **photoautotrophs**. The source of energy utilized serves as a physiological distinction, as shown in Table 7-10.

Photosynthetic Bacteria and Cyanobacteria

Differentiation between the photosynthetic bacteria and the cyanobacteria (formerly called blue-green algae) is based upon the type of photosensitive pigments produced. True photosynthetic prokaryotes (Cyanobacteria) contain **chlorophyll *a***, which is common to the eukaryotic algae and green plants. Water serves as the electron donor and oxygen is generated by photolysis during the photosynthetic process. The purple bacteria (*Thiorhodaceae*) contain **bacteriochlorophyll *a* or *b***. They utilize H$_2$S and/or organic compounds as electron donors and their metabolism does not involve molecular oxygen (i.e., it is anaerobic). Green bacteria (*Chlorobacteriaceae*) contain **bacteriochlorophyll *c* or *d*** and small amounts of **bacteriochlorophyll *a***. These bacteria utilize H$_2$S and/or organic compounds as electron donors and also metabolize via anaerobic pathways. The structure and biosynthesis of bacteriochlorophyll has been studied in detail and will be discussed in Chapter 10.

Production of light-absorbing carotenoid pigments also represents a differentiating characteristic. All algae and green plants contain β-carotene. The purple sulfur and non-

TABLE 7-10. Principal Groups of Autotrophs

Energy Source	Group	Genera
H_2	Hydrogen bacteria	*Alcaligenes*
		Xanthobacter
		Nocardia
		Pseudomonas
		Derxia
NH_3	Nitrifying bacteria	*Nitrosolobus*
		Nitrosomonas
		Nitrocystis
NO_2	Nitrifying bacteria	*Nitrobacter*
		Nitrospina
		Nitrosococcus
N_2	Nitrogen fixing bacteria	*Azotobacter*
		Rhizobium
		Cyanobacteria
H_2S, S	Sulfur bacteria	*Thiobacillus*
		Sulfolobus
$S_2O_3^{2-}$		*Desulfotomaculum*
		Desulfovibrio
Fe^{2+}	Iron bacteria	*Gallionella*
		Sphaerotilus
		Ferrobacillus
		Leptothrix
CH_4	Methylotrophs	*Methylomonas*
		Methylosinus
CH_3OH		*Methylobacterium*
		Pseudomonas
		Paracoccus
		Hyphomicrobium
$H_2 + CO_2$	Methanogens	*Methanobacterium*
Formate		*Methanobrevibacter*
Methanol		*Methanococcus*
Methylamine		*Methanomicrobium*
Dimethylamine		*Methogenium*
Trimethylamine		*Methanospirillum*
Acetate		*Methanosarcina*
Light	Phototrophs	*Rhodobacter*
		Cyanobacteria
		Algae

sulfur bacteria contain a variety of carotenoid pigments of both aliphatic and aryl types, whereas the green bacteria contain only aryl carotenoids (Fig. 7-55). The carotenoid pigments absorb light energy and transfer it to the chlorophyll molecules of the antenna.

Algae and the cells of higher plants contain chloroplasts. Comparable structures (chromatophores) are observed in the photosynthetic bacteria. The photosynthetic apparatus of *Rhodobacter sphaeroides* consists of a series of **intracytoplasmic membranes** (ICMs), which appear as vesicular invaginations originating from the cytoplasmic membrane. It carries out anoxigenic photosynthesis but is also capable of both aerobic and anaerobic respiration as well as fermentation.

The green bacteria contain vesicles enclosed within a thin nonunit membrane that is not directly associated with the cell membrane. Metabolically, the green bacteria are strict

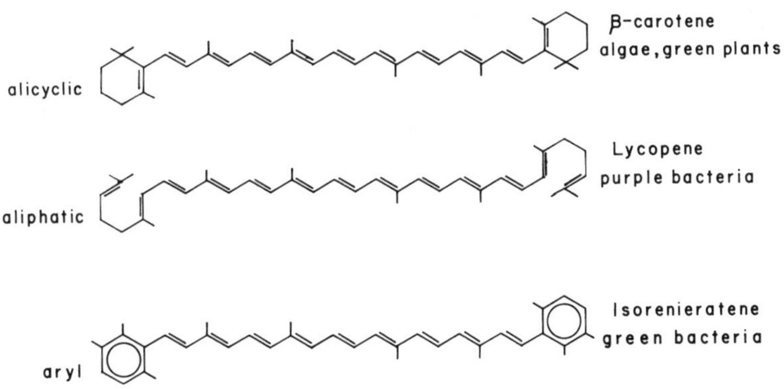

Fig. 7-55. Examples of carotenoid pigments produced by plants, algae, and photosynthetic bacteria. Although there are many variations, all carotenoids are of one of these three basic structural types.

anaerobic organisms that are obligately photosynthetic. They utilize H_2S, thiosulfate, or H_2 as an electron donor and CO_2 as the carbon source:

$$CO_2 + 2H_2S \xrightarrow{light} (CH_2O) + H_2O + 2S$$

$$2CO + Na_2S_2O_3 + 3H_2O \xrightarrow{light} 2NaHSO_4$$

$$CO_2 + 2H_2 \xrightarrow{light} (CH_2O) + H_2$$

The purple bacteria contain two groups: the purple sulfur bacteria (*Thiorhodaceae*), which utilize H_2S as an electron donor and the purple nonsulfur bacteria (*Athiorhodaceae*), which depend on organic compounds for photosynthetic metabolism. Short-chain fatty acids are the best substrates:

$$CO_2 + 2CH_3CHOHCH_3 \xrightarrow{light} (CH_2O) + H_2O + 2CH_3COCH_3$$

$$2CH_3COOH + 2CoASH \longrightarrow 2CH_3COSCoA$$

$$2CH_3COSCoA \longrightarrow H_3CCOCH_2COSCoA + CoASH$$

$$H_3CCOCH_2COSCoA + 2H \longrightarrow \underset{\beta\text{-}Hydroxybutyryl\text{-}CoA}{H_3CCHOHCH_2CoSCoA}$$

$$nH_3CCHOHCH_2COSCoA \longrightarrow CoASH + \underset{Poly\text{-}\beta\text{-}hydroxybutyrate}{(C_4H_6O_2)_n}$$

Poly-β-hydroxybutyrate serves as a major storage reserve material in these organisms. It is also an important reserve energy source in a number of other photosynthetic organisms.

Many photosynthetic bacteria are found in the deeper waters of permanently stratified (**meromictic**) lakes where the conditions are anaerobic but where light is available. The cyanobacteria are considered to be very early evolutionary forms because of their lack of dependence on oxygen and on the basis of molecular evidence derived from 16S rRNA sequencing. Phylogenetic analysis of *c*-type cytochromes and rRNA sequences have established a relationship between cyanobacteria and the chloroplasts of green algae and higher plants. These lines of evidence provide support for the concept of prokaryotic origins of chloroplasts.

Autotrophic CO₂ Fixation and Mechanisms of Photosynthesis

Photoautotrophs and chemoautotrophs, in which CO_2 serves as the sole or principal source of cellular carbohydrate, fix CO_2 via either the reductive pentose phosphate (Calvin) cycle or the reductive C_4-dicarboxylic acid pathway. These systems were first discovered in green plants. Originally, all green plants were thought to assimilate atmospheric CO_2 via the reductive pentose pathway (Fig. 7-56) in which phosphoglyceric acid (PGA) is the first stable product (hence the designation C_3 plants). Description of an alternative pathway of CO_2 fixation in which C_4 dicarboxylic acids (oxaloacetatate and malate) were found as the primary products of photosynthesis led to the division of flowering plants into C_3 plants (PGA as the first stable intermediate) and C_4 plants (C_4 acids as the first stable intermediates). (Within a taxonomic category, plants with C_3 photosynthesis are considered to be ancestral to those with C_4 primary photosynthetic products.)

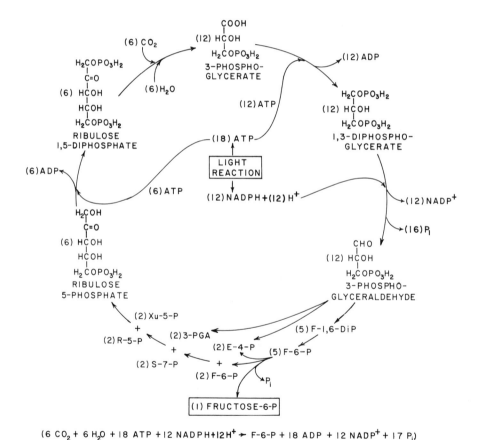

$$(6\ CO_2 + 6\ H_2O + 18\ ATP + 12\ NADPH + 12H^+ \rightarrow F\text{-}6\text{-}P + 18\ ADP + 12\ NADP^+ + 17\ P_i)$$

Fig. 7-56. The reductive pentose phosphate (Calvin) pathway. Since 3-phosphoglycerate is the first stable product of atmospheric CO_2 fixation, this pathway is sometimes referred to as the C_3 pathway. This pathway constitutes the "dark" reaction of photosynthesis because the energy required in the form of ATP has already been generated during photophosphorylation.

In photosynthetic and autotrophic bacteria carbon dioxide fixation has also been considered to occur primarily via the reductive pentose phosphate pathway (Fig. 7-56). In this system reduction of 1 mol of CO_2 to the oxidation level of carbohydrate involves the oxidation of 2 mol of NADPH and the hydrolysis of 3 mol of ATP. Only two of the reactions, phosphoribulokinase and ribulose bisphosphate carboxylase, are specific to photosynthetic or chemoautotrophic organisms. The other reactions are common to the carbohydrate metabolism of nonphotosynthetic organisms. The reductive pentose cycle constitutes the "dark" reaction of photosynthesis. Six turns of the cycle result in the synthesis of 1 mol of hexose (F-6-P). The remainder is recycled through the reductive pathway.

The reductive C_4 dicarboxylic acid pathway (Fig. 7-57) is operative in a number of photosynthetic bacteria. In some organisms, such as the *Chlorobium*, it is the only cyclic pathway for CO_2 assimilation. Organisms that use the C_4 pathway possess the enzyme pyruvate, orthophosphate dikinase, which synthesizes PEP:

$$\text{pyruvate} + \text{ATP} + P_i \overset{Mg^{2+}}{\longleftrightarrow} \text{PEP} + \text{AMP} + PP_i$$

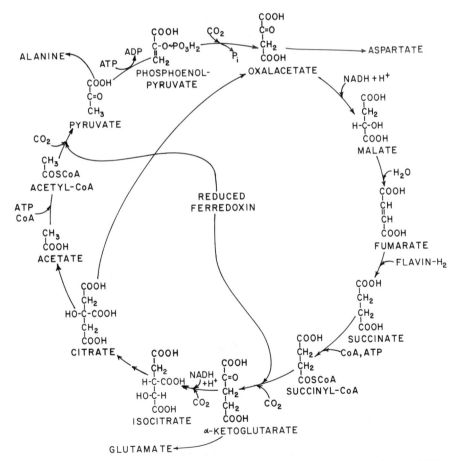

Fig. 7-57. The reductive C_4-carboxylic acid cycle. This is the only cyclic pathway of CO_4 assimilation in certain photosynthetic bacteria, such as *Chlorobium*.

This enzyme differs from the PEP synthase of *E. coli* and other organisms that can utilize C_3 acids in that it produces orthophosphate rather than monophosphate.

$$\text{pyruvate} + \text{ATP} + H_2O \xleftrightarrow{Mg^{2+}} \text{PEP} + \text{AMP} + P_i$$

Enzyme preparations from a representative of each of three major groups of photosynthetic bacteria have been tested for their ability to form PEP from pyruvate and ATP in the presence and absence of P_i. *Chlorobium thiosulfatophilum* (green sulfur bacteria), Chromatium D (purple sulfur bacteria), and *Rhodospirillum rubrum* (purple nonsulfur bacteria) require P_i in addition to Mg^{2+} and ATP for the formation of PEP from pyruvate, supporting the conclusion that photosynthetic bacteria, like C_4 plants, utilize the enzyme pyruvate, orthophosphate dikinase rather than PEP synthase to form PEP from pyruvate in the photosynthetic assimilation of CO_2. The reductive carboxylic acid cycle is essentially a reverse of the TCA cycle in which pyruvate oxidase and α-ketoglutarate oxidase systems are replaced by ferredoxin-dependent pyruvate synthetase and α-ketoglutarate synthetase. This system is also of major importance in the metabolism of anaerobic bacteria.

Photosynthesis, whether in green plants, algae, cynaobacteria, or photosynthetic bacteria, begins with the absorption of light by a pigment molecule and the delivery of the absorbed energy to electron carriers that can transduce the energy into chemical form. The function of the light-harvesting pigments (also called antenna molecules) is to capture photons. The energy contained in the excited pigments is channeled into a complex called the **reaction center**. The components of the bacterial photosynthetic reaction center have been studied in considerable detail (Fig. 7-58). Within the reaction center there are four bacteriochlorophyll molecules. Two of these are referred to as a *special pair* because they absorb light energy and transfer it to an electron. The other two bacteriochlorophyll molecules appear to be inactive. Once the photon has been absorbed by the special chlorophyll molecules and its energy transferred to an electron, the electron moves to a bacteriopheophytin molecule leaving a positive charge on the special pair of chlorophyll molecules. The electron then travels to a quinone. A soluble cytochrome molecule transfers its electron to the special pair. The cytochrome acquires a positive charge and the special pair of bacteriochlorophyll is neutralized. The excited electron is then passed to the second quinone.

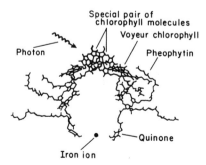

Fig. 7-58. A conceptual view of the bacterial photosynthetic reaction center showing twofold rotational symmetry. From Youvan, D. C. and B. L. Marrs. 1987. Molecular mechanisms of photosynthesis. *Sci. Am.* **256**:42–48.

In cyanobacteria and the eukaryotic red algae, phycobiliproteins are the most prominent light harvesting polypeptides of the cell. These polypeptides are highly pigmented, water-soluble proteins that make up a major portion of the soluble cell protein. The major phycobiliproteins are phycoerythrin, phycocyanin, and allophycocyanin. Phycobilisome complexes appear as rows of closely spaced granules at the outer surface of the photosynthetic (thylakoid) membranes of red algae and cyanobacteria. The composition of the phycobilisome complex of the filamentous cyanobacterium *Fremyella diplosiphon* is altered by growth under red light as compared to green light. As shown in Figure 7-59, the differences in composition of the phycobilisome structure are the result of altered expression of the genes coding for phycobiliproteins.

The terminal steps resulting in the phosphorylation of ADP represent an additional series of electron-transfer reactions involving ferredoxin, $NADP^+$, and cytochromes. It is generally accepted that in photosynthesis ATP is produced by essentially the same mechanism utilized in coupling phosphorylation to electron transport during respiration (i.e., chemiosmotic coupling to an ATPase). In noncyclic photophosphorylation (Fig. 7-60) electrons are transferred from chlorophyll to ferredoxin, flavoprotein, and then to $NADP^+$. An electron donor (water in plants and algae; H_2, H_2S, or various organic compounds in photosynthetic bacteria) transfers electrons to cytochrome, producing the chemical energy needed to phosphorylate ADP. In cyclic photophosphorylation, ATP is generated from ADP and P_i with no other net chemical change (Fig. 7-60). Since cyclic photophosphorylation does not generate $NADH + H^+$, compounds such as H_2, H_2S, or other available reducing compounds provide reducing power.

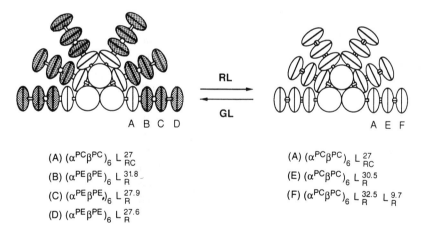

Fig. 7-59. Altered composition of the phycobilisome (PBS) complex in *Fremyella diplosiphon* by growth under red (RL) or green light (GL). The dark speckled double disks represent hexamers of phycoerythrin (PE) $[(\alpha^{PE}\beta^{PE})_6]$, while the light speckled double disks represent hexamers of phycocyanin (PC) $[(\alpha^{PC}\beta^{PC})_6]$. Linker polypeptides are indicated by L. The subscript to L denotes the position of the linker in the PBS substructure (R for rod substructure; RC for rod-core interface). The superscript to L is the molecular mass depicted as white circles that are unspeckled. The core substructure of the PBS is invariant. The composition of the rod substructures varies dramatically between organisms grown under RL and GL. The composition of each of the double disks of the PBS in GL-grown cells (disks A, B, C, and D) and RL-grown cells (disks A, E, and F) is given below the PBS structures. From Grossman, A. R. et al. 1993. *J. Bacteriol.* **175:**575–582.

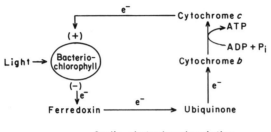

Cyclic photophosphorylation
Photosynthetic bacteria

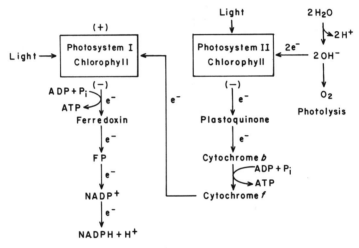

Noncyclic photophosphorylation
Algae, cyanobacteria, green plants

Fig. 7-60. A diagrammatic comparison of cyclic and noncyclic photophosphorylation. In cyclic photophosphorylation ATP is produced but no reducing equivalents are generated. In the noncyclic pathway, two molecules of ATP are produced, reduced NADP is generated and oxygen is produced by photolysis of water.

Hydrogen Bacteria

Members of the hydrogen bacteria utilize H_2 to provide energy and reducing power for growth and CO_2 fixation. They include a number of organisms belonging to the genera shown in Table 7-10. Most of these organisms are facultatively autotrophic and grow readily on organic substrates. Many heterotrophic bacteria are capable of using H_2 to provide reducing power and energy for metabolic purposes, but cannot support CO_2 fixation. Reduction of CO_2 by H_2 can be shown as:

$$2H_2 + CO_2 \longrightarrow (CH_2O) + H_2O$$

Utilizable energy in the form of ADP is generated from the oxidation of hydrogen by hydrogenase:

$$H_2 + 0.5O_2 + NAD^+ \longrightarrow H_2O + NADH + H^+$$

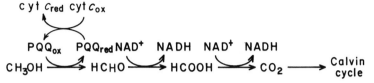

Fig. 7-61. Pathway for the oxidation of methanol by *Xanthobacter.* PQQ = pyroloquinoline quinone.

Carbon dioxide is assimilated autotrophically through the essential reactions of phosphoribulokinase and ribulose-1,5-bisphosphate carboxylase of the Calvin cycle (Fig. 7-56). In *Alcaligenes eutrophus,* phosphoribulokinase is partially inactivated when an autotrophic culture is shifted to heterotrophic growth with pyruvate as the sole source of carbon and energy. Reactivation of phosphoribulokinase occurs after exhaustion of pyruvate from the medium. The hydrogen autotroph, *Xanthobacter,* can grow autotrophically with either hydrogen or methanol as an energy source. Hydrogen is oxidized by a membrane-bound hydrogenase. Methanol is oxidized to formaldehyde, formate, and then to CO_2 by the sequential action of methanol dehydrogenase, formaldehyde dehydrogenase, and formate dehydrogenase, as shown in Figure 7-61.

Nitrifying Bacteria

These organisms are important members of the nitrogen cycle, hence their activities and relationship to the nitrogen cycle are discussed in further detail in Chapter 9. *Nitrosomonas* is the most common organism found in soil that oxidizes ammonia to nitrite:

$$2NH_3 + 3O_2 \longrightarrow 2NO_2^- + 2H^+ + 2H_2O$$

Nitrobacter is the most common organism found in soil that oxidizes nitrite to nitrate:

$$2NO_2^- + O_2 \longrightarrow 2NO_3^-$$

Both *Nitrosomonas* and *Nitrobacter* have been shown to possess specialized mechanisms for production of the ATP and reduced NAD required for the assimilation of carbon dioxide. A portion of the ATP derived from oxidative phosphorylation at the cytochrome level is used to reduce pyridine nucleotides.

Sulfur Bacteria

All members of this group, *Thiorhodaceae,* are capable of growth on elemental sulfur. Many can utilize thiosulfate ($S_2O_3^{2-}$) as well. The bioenergetics of two of the sulfate-reducing bacteria, *Desulfovibrio* and *Desulfotomaculum,* are fundamentally different. In the case of *Desulfovibrio,* the PP_i produced during the formation of adenylyl sulfate (APS) from ATP and sulfate in the first step of sulfate reduction is hydrolyzed to P_i by inorganic pyrophosphatase:

$$ATP + SO_4^{2-} \longrightarrow APS + PP_i$$

$$PP_i + H_2O \longrightarrow 2P_i$$

By this process, the chemical energy in PP_i is not conserved and to obtain a net yield of ATP during growth on lactate plus sulfate, *Desulfovibrio* must carry out electron-transfer coupled phosphorylation. By comparison, *Desulfotomaculum* conserves the bond energy of the pyrophosphate bond produced by ATP-sulfurylase by means of the enzyme acetate: PP_i phosphotransferase and the subsequent formation of ATP by acetate kinase:

$$acetate + PP_i \longrightarrow acetyl\ phosphate + P_i$$

$$acetyl\ phosphate + ADP \longrightarrow acetate + ATP$$

These reactions allow *Desulfotomaculum* to use PP_i as a source of energy for growth with acetate and sulfate. The conversion of APS (adenosine phosphosulfate) to sulfite by APS reductase requires the addition of two electrons:

$$APS + 2e^- \longrightarrow AMP + SO_3^{2-}$$

The further reduction of sulfite to sulfide requires the action of sulfite reductase (a), trithionate reductase (b), and thiosulfate reductase (c), and the recycling of sulfite, as shown in Figure 7-62.

The sulfur-dependent archaebacteria found in the vicinity of hot springs are able to grow chemoautotrophically using CO_2 as the sole carbon source and the oxidation of elemental sulfur with oxygen yielding sulfuric acid:

$$2S^0 + 3O_2 + 2H_2O \longrightarrow 2H_2SO_4$$

However, *Sulfolobus ambivalens* is alternatively able to live by an anaerobic mode of chemoautotrophy using CO_2 as the sole carbon source but using H_2 for the reduction of sulfur to H_2S:

$$S^0 + H_2 \longrightarrow H_2S$$

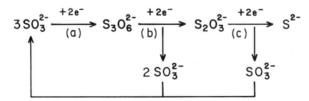

Fig. 7-62. The sulfur cycle as conducted by sulfur bacteria.

Iron Bacteria

Ferrobacillus ferrooxidans and some members of the *Thiobacillus* group (*Gallionella*, *Leptothrix*, and *Sphaerotilus*) are capable of oxidizing ferrous iron to ferric iron as a means of generating biologically useful energy:

$$Fe^{2+} + H^+ + 0.25O_2 \longrightarrow Fe^{3+} + 0.5H_2O + 40 \text{ kcal}$$

Ferrobacillus ferrooxidans is an obligate autotroph. While it can be grown heterotrophically in the absence of an oxidizable iron source, continued cultivation on an organic substrate renders it incapable of growth with ferrous iron as the sole energy source. *Ferrobacillus* differs from other autotrophic organisms that can revert to an autotrophic mode of life after prolonged cultivation on organic substrates. The transition of *F. ferrooxidans* to obligate organotrophy is governed by a number of factors including the pH of the medium, temperature of incubation, availability of oxygen, age of the cells at the time of transition, and the type of energy and carbon source available. Conversion to organotrophy results in a gradual loss of the ability to oxidize Fe^{2+} and cessation of carbon dioxide fixation. *Gallionella*, *Sphaerotilus*, and other iron-oxidizing organisms appear to be facultative and can be readily grown as heterotrophs and then returned to growth on iron.

Methylotrophs

Methylotrophs are able to utilize methane, methanol, methylamine, or formate as the sole source of carbon and energy. The term **methanotroph** designates those methylotrophs that can use methane for carbon and energy. There are also several species of yeasts and molds that can use methane or methanol. Most methylotrophs are obligate methylotrophs in that they can only use C_1 compounds. The general pathway of oxidative reactions in methylotrophs is shown in Figure 7-63. Two types of methylotrophic bacteria have been identified on the basis of the mode of assimilation of formaldehyde. **Type I methylotrophs** use the ribulose monophosphate pathway for formaldehyde assimilation:

$$3HCHO + 3\text{ribulose monophosphate} \longrightarrow 3\text{hexulose-6-phosphate}$$

The hexulose-6-phosphate is metabolized via common pathways. **Type II methylotrophs** use the serine pathway for formaldehyde assimilation:

$$2HCHO + 2\text{glycine} \longrightarrow 2\text{serine}$$

Bacterial methylotrophs include *Paracoccus denitrificans*, several species of *Pseudomonas*, *Hyphomicrobium*, and *Xanthobacter*. However, most of the obligate methylotrophs belong to the genera *Methylophilus*, *Methylobacterium*, *Methylococcus*, *Methylosinus*, or *Methylomonas*.

Methylotrophic yeast include *Hansenula*, *Candida*, *Torulopsis*, and *Pichia*. The met-

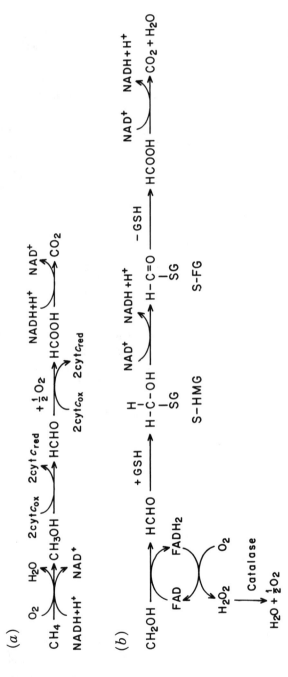

Fig. 7-63. (a) General pathway of oxidative reactions in methylotrophs. (b) Pathway for the conversion of methanol to CO_2 and H_2 in a methylotrophic yeast. GSH = glutathione: S-HMG = S-hydroxymethyl-glutathione; S-FG = S-formylglutathione.

abolic pathway for the conversion of methanol to CO_2 and H_2O appears to be similar for several of these yeasts. The pathway in *Pichia pastoris* involves alcohol oxidase, catalase, formaldehyde dehydrogenase, *S*-formylglutathione, and formate dehydrogenase in the sequence of reactions shown in Figure 7-63.

In yeasts, the alcohol dehydrogenase and catalase reactions take place in peroxisomes, membranous organelles containing flavin-linked oxidases that regenerate oxidized flavin by reaction with O_2. Synthesis of some of these enzymes is tightly regulated and several of the genes involved in methanol utilization appear to be controlled at one level by a glucose catabolite repression-derepression mechanism. The structural genes for alcohol dehydrogenase and two other enzymes in the sequence are regulated by methanol at the level of transcription.

Methane and methanol-oxidizing organisms are useful for the production of single-cell protein, microbial cells used as animal feed supplements. Growing microorganisms on materials that would otherwise be disposed of as waste provides an important means of recycling such materials into useful products. These organisms also display a wide range of biotransformations of potential commercial importance.

Methanogens

Methanogenic organisms gain energy by using H_2 to reduce CO_2 to CH_4. These organisms can also decarboxylate acetate to CH_4 and CO_2. Methane formation represents the terminal portion of a complex series of anaerobic reactions that occur in nature and involve a number of organisms that degrade biopolymers, such as cellulose, starch, or proteins to acetate, H_2, and CO_2. The methanogens belong to the archaebacteria, a kingdom of bacteria distinguished from the eubacteria on the basis of the sequence homology of their 16S ribosomal RNA and other physiological differences. Major genera of methanogens include *Methanobacterium, Methanobrevibacter, Methanococcus, Methanomicrobium, Methanogenicum, Methanospirillum,* and *Methanosarcina.* Methanogens can produce methane from dihydrogen and carbon dioxide, formate, methanol, methylamine, dimethylamine, trimethylamine, or acetate according to the following equations:

$$4H_2 + CO_2 \longrightarrow CH_4 + 2H_2O$$

$$4HCOOH \longrightarrow CH_4 + 3CO_2 + 2H_2O$$

$$4CH_3NH_2Cl + 2H_2O \longrightarrow 3CH_4 + CO_2 + 4NH_4Cl$$

$$2(CH_3)_2NHCl + 2H_2O \longrightarrow 3CH_4 + CO_2 + 2NH_4Cl$$

$$4(CH_3)_3NCl + 6H_2O \longrightarrow 9CH_4 + 3CO_2 + 4NH_4Cl$$

$$CH_3COOH \longrightarrow CH_4 + CO_2$$

Reduction of CO_2 to CH_4 involves attachment of CO_2 to a carrier molecule and the subsequent steps involve the function of several coenzymes unique to methanogens (Fig. 7-64). Carbon dioxide is reduced to CH_4 in a cyclic manner, as shown in Figure 7-65. All methanogens involve the major energy-yielding step associated with the reduction of a methyl group to methane although different species may obtain electrons for the reductive step from the oxidation of a variety of substrates, as mentioned above.

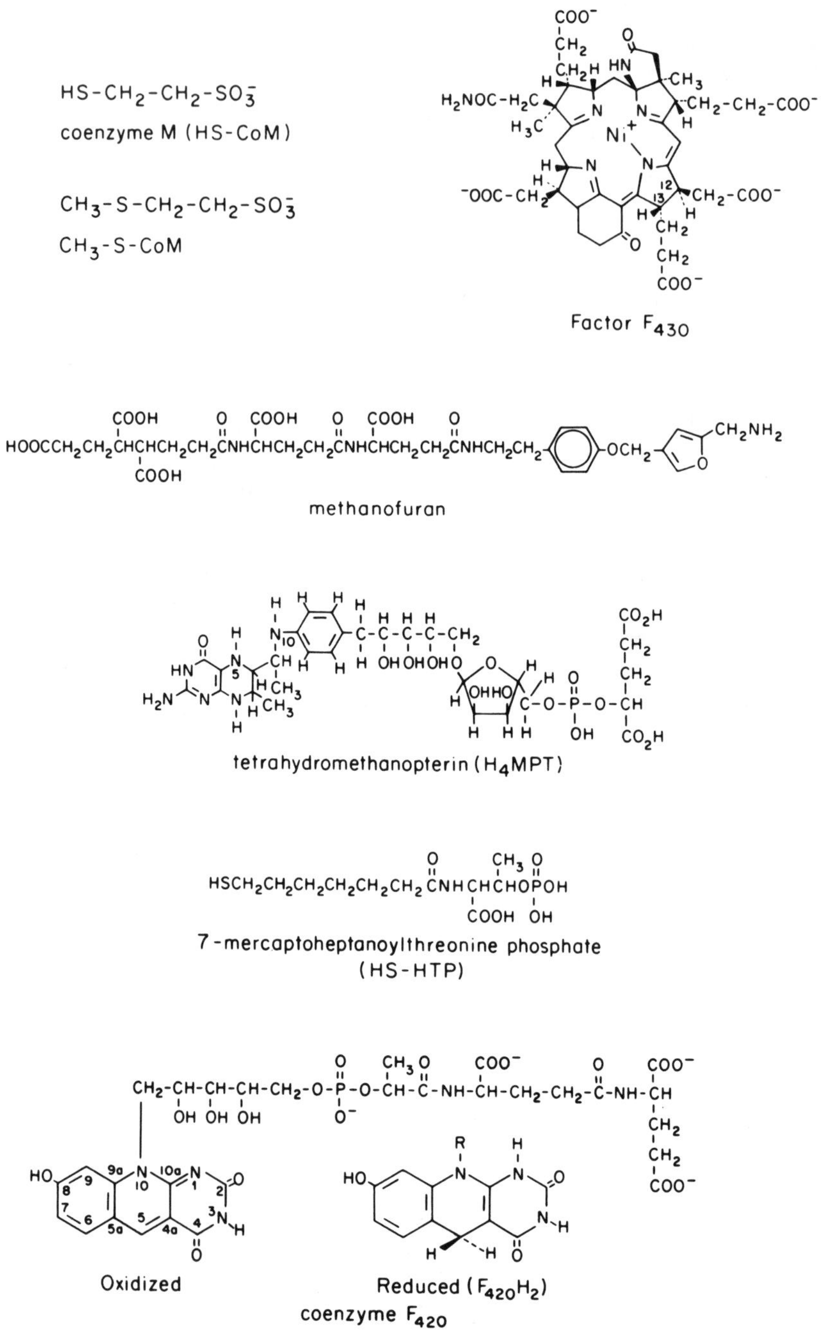

$HS-CH_2-CH_2-SO_3^-$

coenzyme M (HS-CoM)

$CH_3-S-CH_2-CH_2-SO_3^-$

$CH_3-S-CoM$

Factor F_{430}

methanofuran

tetrahydromethanopterin (H_4MPT)

7-mercaptoheptanoylthreonine phosphate
(HS-HTP)

Oxidized

Reduced ($F_{420}H_2$)

coenzyme F_{420}

Fig. 7-64. Structures of the unusual coenzymes that participate in methanogenesis. From Rouvière, P. E. and R. S. Wolfe. 1988. *J. Biol. Chem.* **263:**7913–7916.

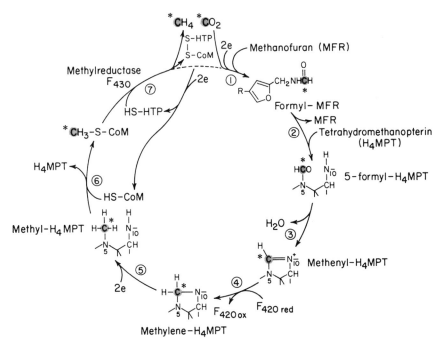

Fig. 7-65. Proposed cycle for the reduction of CO₂ to CH₄ by methoanogenic bacteria. The C₁ unit indicated by the asterisk is sequentially modified, reduced, and transferred in "bucket brigade" fashion bound to coenzymes. Only that portion of the molecule involved in C₁ attachment is shown for methanofuran and tetrahydromethanopterin. The heterodisulfide couples in an unknown manner Reaction 7 to Reaction 1. From Rouvière, P. E. and R. S. Wolfe. 1992. *J. Biol. Chem.* **263**:7913–7916.

REFERENCES

Catabolic Pathways

An, H., R. R. Scopes, M. Rodriguez, K. F. Keshav, and L. O. Ingram. 1991. Gel electrophoretic analysis of *Zymomonas mobilis* glycolytic and fermentative enzymes: Identification of alcohol dehydrogenase II as a stress protein. *J. Bacteriol.* **173**:5975–5982.

Chin, A. M., D. A. Feldheim, and M. H. Sair, Jr. 1989. Altered transcriptional patterns affecting several metabolic pathways in strains of *Salmonella typhimurium* which overexpress the fructose regulon. *J. Bacteriol.* **171**:2424–2434.

Egan, S. E., R. Fliege, S. Tong, A. Shibata, R. E. Wolf, Jr., and T. Conway. 1992. Molecular characterization of the Entner–Doudoroff pathway in *Escherichia coli*: Sequence analysis and localization of promoters for the *edd–eda* operon. *J. Bacteriol.* **174**:4638–4646.

Rowley, D. L., A. J. Pease, and R. E. Wolf, Jr. 1991. Genetic and physical analyses of the growth rate-dependent regulation of *Escherichia coli zwf* expression. *J. Bacteriol.* **173**:4660–4667.

Temple, L., S. M. Cuskey, R. E. Perkins, R. C. Bass, N. M. Morales, G. E. Christie, R. H. Olsen, and P. V. Phibbs, Jr. 1990. Analysis of cloned structural and regulatory genes for carbohydrate utilization in *Pseudomonas aeruginosa* PAO. *J. Bacteriol.* **172**:6396–6402.

TCA Cycle

Cotter, P. A., VC. Cepuri, R. B. Gennis, and R. P. Gunsalus. 1990. Cytochrome *o* (*cyoABCDE*) and *d* (*cydAB*) oxidase gene expression in *Escherichia coli* is regulated by oxygen, pH, and the *fnr* gene product. *J. Bacteriol.* **172**:6333–6338.

Iuchi, S., D. C. Cameron, and E. C. C. Lin. 1989. A second global regulator gene (*arcB*) mediating repression of enzymes in aerobic pathways of *Escherichia coli*. *J. Bacteriol.* **171**:868–873.

Iuchi, S. and E. C. C. Lin. 1992. Mutational analysis of signal transduction by ArcB, a membrane sensor protein responsible for anaerobic repression of operons involved in the central aerobic pathways in *Escherichia coli*. *J. Bacteriol.* **174**:3872–3980.

Iuchi, S. and E. C. C. Lin. 1992. Purification and phosphorylation of the Arc regulatory components of *Escherichia coli*. *J. Bacteriol.* **174**:5617–5623.

Mattevi, A., G. Obmolova, E. Schultze, K. H. Kalk, A. N. Westphal, A. deKok, and W. G. J. Hol. 1992. Atomic structure of the cubic core of the pyruvate dehydrogenase multienzyme complex. *Science* **255**:1544–1550.

Rasmussen, L. J., P. L. Møller, and T. Atlung. 1991. Carbon metabolism regulates expression of the *pfl* (pyruvate formate-lyase) gene in *Escherichia coli*. *J. Bacteriol.* **173**:6390–6397.

Energy Production/Terminal Respiratory Systems

Egan, S. M. and V. Stewart. 1991. Mutational analysis of nitrate regulatory gene *narL* in *Escherichia coli* K-12. *J. Bacteriol.* **173**:4424–4432.

Filingame, R. H. 1990. Molecular mechanics of ATP synthesis by F_0F_1-type H^+-transporting ATP synthases. In *Bacterial Energetics*. T. A. Krulwich (Ed.). Academic, San Diego, CA, pp. 345–391.

Li, J. and V. Stewart. 1992. Localization of upstream sequence elements required for nitrate and anaerobic induction of *fdn* (formate dehydrogenase-N) operon expression in *Escherichia coli* K-12. *J. Bacteriol.* **174**:4935–4942.

Papa, S., F. Guerrieri, F. Zanoti, G. Capozza, M. Fiermonte, T. Cocco, K. Attendorf, and G. Deckers-Hebersteit. 1993. F_0 and F_1 subunits involved in the gate and coupling function of mitochondrial H^+ ATP synthetase. *Ann. N.Y. Acad. Sci.* **671**:345–358.

Stewart, V. 1988. Nitrate respiration in relation to facultative metabolism in enterobacteria. *Microbiol. Rev.* **52**:190–232.

Trumpower, B. L. 1990. Cytochrome bc_1 complexes of microorganisms. *Microbiol. Rev.* **54**:101–125.

pH Stasis

Booth, I. 1985. Regulation of cytoplasmic pH in bacteria. *Microbiol. Rev.* **49**:359–378.

Transport

Doige, C. A. and F.-L. Ames. 1993. ATP-dependent transport systems in bacteria and humans: relevance to cystic fibrosis and multidrug resistance. *Annu. Rev. Microbiol.* **47**:291–319.

Epstein, W. 1990. Bacterial Transport ATPases. In *Bacterial Energetics*, T. A. Krulwich (Ed.). Academic, San Diego, CA, pp. 87–110.

Karpel, R., T. Alon, G. Glasser, S. Schuldiner, and E. Paden. 1991. Expression of a sodium proton antiporter (Nha) pin *Escherichia coli* is induced by Na^+ and Li^+ ions. *J. Biol. Chem.* **266**:21753–21759.

Postma, P. W., J. W. Lengeler, and G. R. Jacobson. 1993. Phosphoenolpyruvate:carbohydrate phosphotransferase systems of bacteria. *Microbiol. Rev.* **57**:543–594.

Senior, A. E. 1990. The proton translocating ATPases of *Escherichia coli*. *Annu. Rev. Biophys. Chem.* **19**:7–41.

Silver, S. and M. Walderhaug. 1992. The regulation of plasmid- and chromosome-determined inorganic ion transport in bacteria. *Microbiol. Rev.* **56**:195–228.

Cellulose Degradation

Bayer, E. A. and R. Lamed. 1986. Ultrastructure of the cell surface cellulosome of *Clostridium thermocellum* and its interaction with cellulose. *J. Bacteriol.* **167**:828–836.

Beguin, P. and J.-P. Aubert. 1994. The biological degradation of cellulose. *FEMS Microbiology Reviews* **13**:25–58.

Demain, A. and J. H. D. Wu. 1989. The *cellulase* complex of *Clostridium thermocellum*. In *Bioprocess Engineering, The First Generation*, T. K. Ghose (Ed.), Ellis Horwood Ltd., Chichester, UK.

Fujino, T., P. Beguin, and J.-P. Aubert. 1993. Organization of a *Clostridium thermocellum* gene cluster encoding the cellulosomal scaffolding protein CipA and a protein possibly involved in attachment of the cellulosome to the cell surface. *J. Bacteriol.* **175**:1891–1899.

Gerngross, U. T., M. P. M. Romaniec, T. Kobayashi, N. S. Huskisson, and A. L. Demain. 1993. Sequencing of a *Clostridium thermocellum* gene (*cipA*) encoding the cellulosomal S_L-protein reveals an unusual degree of internal homology. *Molecular Microbiology* **8**:325–334.

Gilkes, N. R., B. Henrissat, D. G. Kilburn, R. C. Miller, Jr., and R. A. J. Warren. 1991. Domains in microbial β-1,4-glycanases: Sequence, conservation, function, and enzyme families. *Microbiol. Rev.* **55**:303–315.

Wang, W. K., K. Kruus, and J. H. D. Wu. 1993. Cloning and DNA sequence of the gene coding for *Clostridium thermocellum* cellulase S_S (CelS), a major cellulosome component. *J. Bacteriol.* **175**:1293–1302.

Pectin Degradation

Hugouvieux-Cotte-Pattat, N., H. Dominguez, and J. Robert-Baudouy. 1992. Environmental conditions affect transcription of the pectinase gene of *Erwinia chrysanthemi* 3937. *J. Bacteriol.* **174**:7807–7818.

Starch Degradation

Fujiwara, S., H. Kakihara, K. Sakaguchi, and T. Imanaka. 1992. Analysis of mutations in cyclodextrin glucanotransferase from *Bacillus stearothermophilus* which affect cyclization characteristics and thermostability. *J. Bacteriol.* **174**:7478–7481.

Podkovyrov, S. M. and J. G. Zeikus. 1992. Structure of the gene encoding cyclomaltodextrinase from *Clostridium thermohydrosulfuricum* 39E and characterization of the enzyme purified from *Escherichia coli*. **174**:5400–5405.

Smith, K. A. and A. A. Salyers. 1991. Characterization of a neopullulanase and an α-glucosidase from *Bacteroides thetaiotaomicron* 95-1. *J. Bacteriol.* **173**:2962–2968.

Tancula, E., M. J. Feldhaus, L. A. Bedzyk, and A. A. Salyers. 1992. Location and characterization of genes involved in binding of starch to the surface of *Bacteroides thetaiotaomicron*. *J. Bacteriol.* **174**:5609–5616.

Lactic Acid Fermentation

Fukui, K., K. Kato, T. Kodama, H. Ohta, T. Shimamoto, and T. Shimono. 1988. Kinetic study of a change in intracellular ATP level associated with aerobic catabolism of ethanol by *Streptococcus mutans*. *J. Bacteriol.* **170**:4589–4593.

Moat, A. G. 1985. Biology of lactic, acetic, and propionic acid bacteria. In *Biology of Industrial Microorganisms*, A. L. Demain and N. A. Solomon (Eds.), Benjamin–Cummings, Menlo Park, CA, pp. 143–188.

Butyric Acid and Solvent-Producing Fermentations

Cary, J. W., D. J. Petersen, E. T. Papoutsakis, and G. N. Bennett. 1988. Cloning and expression of *Clostridium acetobutylicum* phosphotransbutyrylase and butyrate kinase genes in *Escherichia coli*. *J. Bacteriol.* **170**:4613–4618.

Gerisher, U. and P. Durre. 1990. Cloning, sequencing, and molecular analysis of the acetolactate decarboxylase gene region from *Clostridium acetobutylicum*. *J. Bacteriol.* **172**:6907–6918.

Gerisher, U. and P. Durre. 1992. mRNA analysis of the *adc* gene region of *Clostridium acetobutylicum* during the shift to solventogenesis. *J. Bacteriol.* **174**:426–433.

Jones, D. T. and D. R. Woods. 1986. Acetone–butanol fermentation revisited. *Microbiol. Rev.* **50**: 484–524.

Palosaari, N. R. and P. Rogers. 1988. Purification and properties of the inducible coenzyme A-linked butyraldehyde dehydrogenase from *Clostridium acetobutylicum*. *J. Bacteriol.* **170**:2971 –2976.

Petersen, D. J., R. W. Welch, F. B. Rudolph, and G. N. Bennett. 1991. Molecular cloning of an alcohol (butanol) dehydrogenase gene cluster from *Clostridium acetobutylicum* ATCC 824. *J. Bacteriol.* **173**:1831–1834.

Zeikus, J. G. 1985. Biology of spore-forming anaerobes. In *Biology of Industrial Microorganisms*, A. L. Demain and N. A. Solomon (Eds.). Benjamin–Cummings, Menlo Park, CA, pp. 79–114.

Mixed-Acid Fermentations

Alam, K. Y. and D. P. Clark. 1989. Anaerobic fermentation balance of *Escherchia coli* as observed by in vivo nuclear magnetic resonance spectroscopy. *J. Bacteriol.* **171**:6213–6217.

Fasciano, A. and P. C. Hallenbeck. 1991. Mutations in *trans* that affect formate dehydrogenase (*fdhF*) gene expression in *Salmonella typhimurium*. *J. Bacteriol.* **173**:5893–5900.

Francis, K., P. Patel, J. C. Wendt, and K. T. Shanmugam. 1990. Purification and characterization of two forms of hydrogenase isoenzyme 1 from *Escherichia coli*. *J. Bacteriol.* **172**:5750–5757.

Iuchi, S., S. T. Cole, and E. C. C. Lin. 1990. Multiple regulatory elements for the *glpA* operon encoding anaerobic glycerol-3-phosphate dehydrogenase and the *glpD* operon encoding aerobic glycerol-3-phosphate dehydrogenase in *Escherichia coli*: Further characterization of respiratory control. *J. Bacteriol.* **172**:179–184.

Matsuyama, A., H. Yamamoto, and E. Nakano. 1989. Cloning, expression, and nucleotide sequence of the *Escherichia coli* K-12 *ackA* gene. *J. Bacteriol.* **171**:577–580.

Menon, N. K., J. Robbins, J. C. Wendt, K. T. Shanmugam, and A. E. Przybyla. 1991. Mutational analysis and characterization of the *Escherichia coli hya* operon which encodes [NiFe] hydrogenase 1. *J. Bacteriol.* **173**:4851–4861.

Sawyers, G. and B. Suppmann. 1992. Anaerobic induction of pyruvate formate-lyase gene expression is mediated by the ArcA and FNR proteins. *J. Bacteriol.* **174**:3474–3478.

Truninger, V., W. Boos, and G. Sweet. 1992. Molecular analysis of the *glpFKX* regions of *Escherichia coli* and *Shigella flexneri*. *J. Bacteriol.* **174**:6981–6991.

Propionic Acid Fermentation

Thompson, T. E. and J. G. Zeikus. 1988. Regulation of carbon and electron flow in *Propionispira arboris*: relationship of catabolic enzyme levels to carbon substrates fermented during propionate formation via the methylmalonyl coenzyme A pathway. *J. Bacteriol.* **170**:3996–4000.

Acetic Acid Fermentation

Matsushita, K., H. Ebisuya, M. Ameyama, and O. Adachi. 1992. Change of the terminal oxidase from cytochrome a_1 in shaking cultures to cytochrome o in static cultures of *Acetobacter aceti*. *J. Bacteriol.* **174**:122–129.

Metabolism of Autotrophs

Bassham, J. A. and B. B. Buchanan. 1982. Carbon dioxide fixation pathways in plants and bacteria. In *Photosynthesis*, Vol. II, Govindjee (Ed.). Academic, NY, pp. 141–189.

Giovannoni, S. J., S. Turner, G. J. Olsen, S. Barns, D. J. Lane, and N. R. Pace. 1988. Evolutionary relationships among cyanobacteria and green chloroplasts. *J. Bacteriol.* **170**:3584–3592.

Grossman, A. R., M. R. Schaefer, G. G. Chang, and J. L. Collier. 1993. Environmental effects on the light-harvesting complex of cyanobacteria. *J. Bacteriol.* **175**:575–582.

Kiley, P. J. and S. Kaplan. 1988. Molecular genetics of photosynthetic membrane biosynthesis in *Rhodobacter sphaeroides*. *Microbiol. Rev.* **52**:50–69.

Rouvière, P. E. and R. S. Wolfe. 1988. Novel biochemistry of methanogenesis. *J. Biol. Chem.* **263**: 7913–7916.

Youvan, D. C. and B. L. Marrs. 1987. Molecular mechanisms of photosynthesis. *Sci. Am.* **256**: 42–48.

CHAPTER 8

LIPIDS AND STEROLS

The major lipid-containing components of microbial cells are membranous structures, such as the cytoplasmic membrane, the outer membrane of gram-negative bacteria, the intracytoplasmic membranes of photosynthetic bacteria, and the nuclear and mitochondrial membranes of eukaryotic cells. A small, but significant number of other lipids play an important role as electron carriers, enzyme cofactors, and light-absorbing pigments. A common feature of all lipids is their insolubility in water. Membrane lipids are **amphipathic**; that is, they have **hydrophilic** (water soluble) and **hydrophobic** (water insoluble) regions that cause them to orient into bilayers. This characteristic makes them most suitable as permeability barriers to polar solutes in an aqueous environment. Phospholipids comprise the bimolecular leaflet structure of the cytoplasmic membrane. The cytoplasmic membrane is flexible and self-sealing. Monounsaturated and ring-containing fatty acids contribute to the flexibility and fluidity that enables membranes to undergo changes in shape that accompany cell growth or movement. In the fungi, sterols are also associated with the cytoplasmic membrane and are involved with permeability functions as indicated by the leakage of metabolites following the binding of polyene antibiotics, such as nystatin or amphotericin B, to the sterol groups. Certain lipids, for example, glycosyldiglycerides and lipoteichoic acids, are found in the cell surface of gram-positive but not gram-negative bacteria. Conversely, gram-negative bacteria display an outer membrane whose outer face is composed largely of lipopolysaccharides (LPS) while gram-positive bacteria are devoid of these structures. Glycolipids are present in non-sterol-requiring *Mycoplasma* but not in the sterol-requiring members of this group. This suggests replacement of sterols by glycolipids in the membrane structures of the former group. In fungi there is an association of lipids and sterols with the development of respiratory activity. Although there is still a great deal to be determined with regard to the precise role of lipids in various cellular structures and metabolic activity, our knowledge of the biosynthesis and metabolism of lipids is well developed.

LIPID COMPOSITION OF MICROORGANISMS

Considerable effort has gone into the determination of the nature and quantity of fatty acids found in microbial cells. Representative examples of the major types of naturally occurring fatty acids are shown in Table 8-1. Of the examples shown, not all are found in microbial cells. In contrast to eukaryotic organisms, bacterial cells contain few fatty acids with greater than 19 carbon atoms, the shorter chain-length acids being primarily of importance as metabolic intermediates in fatty acid biosynthesis. The predominant saturated fatty acid is palmitic acid (C_{18}). Lesser quantities of stearic (C_{18}), myristic (C_{14}), and lauric (C_{12}) are observed. The principal unsaturated fatty acids are monoenoic acids (one unsaturated position). Di-, tri-, or polyenoic acids are not found in bacteria. Many unusual fatty acids are observed in smaller quantities. Branched, hydroxylated, or methylated fatty acids and cyclopropane ring-containing fatty acids are present in a number of microorganisms. Corynolic acid (branched, dihydroxy fatty acid with 52 carbon atoms) and mycolic acid (branched, dihydroxy fatty acid with 87–88 carbon atoms) are examples of fatty acids that are unique to the *Corynebacterium* and *Mycobacterium*, suggesting a close taxonomic relationship between these two genera.

Yeasts and molds have a fatty acid composition more closely related to that of plants and their seeds. The common brewer's and baker's yeasts contain about 5–8% lipid (dry weight) with palmitoleic acid being a major constituent. However, when baker's yeast is grown under anaerobic conditions, palmitic acid synthesis increases at the expense of palmitoleic acid. Many yeasts and molds accumulate lipids, usually as triglycerides containing a higher percentage of palmitic acid, particularly near the end of the growth cycle.

The *Archaea* (archaebacteria) produce a number of unusual lipid structures that contain ether linkages. This group of organisms is divided into three major groups: halophiles, methanogens, and thermophiles. As suggested by their names, these organisms thrive in extreme environments that are uninhabitable by any other living forms, because these unusual lipids and other structural and biochemical factors aid in their survival and function under these harsh conditions.

Sterols are commonly present in fungi but are rarely found in bacteria. Exceptions to this general statement are found in the *Mycoplasma* that incorporate sterols from the growth medium into their cell membranes; *Streptococcus pneumoniae*, which incorporate cholesterol into their cytoplasmic membranes; and methanotropic bacteria, which synthesize sterols in significant quantities.

At this point it is important to describe the various classes of fatty acids and lipids that have been observed in microbial cells and to provide some indication as to their occurrence and distribution in microorganisms.

Straight-Chain Fatty Acids

Straight-chain fatty acids are found as the major constituent of membrane phospholipids. Palmitic, stearic, hexadecenoic, octadecenoic, cyclopropanic, 10-methylhexadecanoic, and 2- or 3-hydroxyl fatty acids are those most commonly observed in bacterial cells. Saturated and unsaturated straight-chain fatty acids with less than 10 or 12 carbon atoms are usually intermediates of degradative pathways or biosynthesis of longer–chain fatty acids. The shorter-chain fatty acids are usually found in younger cultures, having been consumed for the formation of longer chain length fatty acids by mid- to late exponential phase.

TABLE 8-1. Representative Examples of Naturally Occurring Fatty Acids

Number C Atoms	Generic Name	Common Name	Composition
	Saturated Fatty Acids (—C—C—C—)		
1	Methanoic	Formic	HCOOH
2	Ethanoic	Acetic	H_3CCOOH
3	Propanoic	Propionic	C_2H_5COOH
4	Butanoic	Butyric	C_3H_7COOH
6	Hexanoic	Caproic	$C_5H_{11}COOH$
10	Decanoic	Capric	$C_9H_{19}COOH$
16	Hexadecanoic	Palmitic	$C_{15}H_{31}COOH$
18	Octadecanoic	Stearic	$C_{17}H_{35}COOH$
	Unsaturated Fatty Acids (Monoenoic) (—C—C=C—C—)		
4	*trans*-2-Butenoic	Crotonic	$C_4H_6O_2$
4	*cis*-2-Butenoic	Isocrotonic	$C_4H_6O_2$
12	5-Dodecenoic	Denticetic	$C_{12}H_{22}O_2$
14	5-Tetradecenoic	Physeteric	$C_{14}H_{26}O_2$
16	*cis*-9-Hexadecenoic	Palmitoleic	$C_{16}H_{30}O_2$
18	*cis*-9-Octadecenoic	Oleic	$C_{18}H_{34}O_2$
18	*trans*-11-Octadecenoic	Vaccenic	$C_{18}H_{34}O_2$
	Unsaturated Fatty Acids (Dienoic) (—C—C=C—C=C—)		
6	2,4-Hexadienoic	Sorbic	$C_6H_8O_2$
18	*cis*-9, *cis*-12-Octadecadienoic	α-Linoleic	$C_{18}H_{32}O_2$
	Unsaturated Fatty Acids (Trienoic) (—C=C—C=C—C=C—)		
18	*cis*-6, *cis*-9, *cis*-12-Octadecatrienoic	γ-Linolenic	$C_{18}H_{30}O_2$
18	*cis*-9, *cis*-12, *cis*-15-Octadecatrienoic	α-Linolenic	$C_{18}H_{30}O_2$
	Unsaturated Fatty Acids (Tetraenoic) (—C=C—C=C—C=C—C=C—)		
20	5,8,11,14-Eicosatetraenoic	Arachidonic	$C_{20}H_{32}O_2$
	Unsaturated Fatty Acids (Penta- and Hexaenoic)		
20	4,8,12,15,18-Eicosapentaenoic	Timnodonic	$C_{20}H_{30}O_2$

Hydroxyalkanoic Acids (—CH₂—CH—COOH)
$$-CH_2-\underset{\underset{OH}{|}}{CH}-COOH$$

12	2-Hydroxydodecanoic	2-Hydroxylauric	$C_{12}H_{24}O_3$
18	2-Hydroxyoctadecanoic	2-Hydroxystearic	$C_{18}H_{36}O_3$

Keto, Epoxy, and Cyclo Fatty Acids

5	4-Ketopentanoic (—C—C—C—) $\overset{\|}{\underset{O}{}}$	Levulinic	$C_5H_8O_3$
19	ω-(2-*n*-Octylcycloprop-1-enyl)-octanoic or (9, 10-methyleneocatadec-9-enoic)	Sterculic	$C_{19}H_{34}O_2$

$$CH_3-(CH_2)_7-\underset{\underset{CH_2}{\diagdown\diagup}}{C}=C-(CH_2)_7-COOH$$

Keto, Epoxy, and Cyclo Fatty Acids

19	ω-(2-*n*-Octylcycloprop)-octanoic (9, 10-methyleneoctadecanoic)	Lactobacillic	$C_{19}H_{36}O_2$

$$CH_3-(CH_2)_7-\underset{\underset{CH_2}{\diagdown\diagup}}{C}=C-(CH_2)_7-COOH$$

Hydroxy Unsaturated Acids

18	D-12-Hydroxy-*cis*-9-octadecenoic	Ricinoleic	$C_{19}H_{34}O_3$

Branched-Chain Fatty Acids

5	3-Methylbutanoic	*iso*-Valeric	$C_5H_{10}O_2$
10	10-Methylhendecanoic	*iso*-Lauric	$C_{10}H_{20}O_2$
16	14-Methylpentadecanoic	*iso*-Palmitic	$C_{18}H_{32}O_2$
19	1-D-10-Methyloctadecanoic	Tuberculostearic	$C_{19}H_{38}O_2$
20	18-Methylnonadecanoic	*iso*-Arachidic	$C_{20}H_{40}O_2$

413

The conditions under which microbial cells are grown can markedly influence their fatty acid constituents. Composition of the growth medium, availability of oxygen, temperature, pH, and the age of the culture can each affect the distribution of fatty acids observed. Young cultures of *E. coli* contain large proportions of C_{16}- and C_{18}-monoenoic acids and only small amounts of C_{17}- and C_{19}-cyclopropane fatty acids. In the late exponential phase of growth the cyclopropane fatty acids are greatly increased at the expense of the unsaturated fatty acids. In *S. aureus* about 5% of the total fatty acids are unsaturated when the organism is grown on a minimal defined medium. When human serum is added to the medium, the proportion of unsaturated fatty acid increases to 27%. Under aeration *S. aureus* displays a decrease in the percentage of saturated fatty acids. In anaerobically grown staphylococci, C_{18} and C_{20} acids comprise 59% of the total fatty acids. Unsaturated fatty acids are absent from the fatty acids of *Bacillus acidocaldarius*, an organism characterized by its tolerance of high temperature and acidity (50–70 °C, pH 2–5). This organism notably produces a high proportion of unusual lipids. The ability of this organism to grow under these extreme conditions is attributed to the changes in fatty acid composition.

Fungi contain the same major classes of lipids and fatty acids as bacteria but the relative amounts of each class may vary. In *Phycomycetes*, yeasts, *Euascomycetes*, and *Basidiomycetes* the C_{16} and C_{18} unsaturated fatty acids predominate. The fatty acids of *S. cerevisiae* are also in the C_{16} and C_{18} class, with only 1–2% of the total fatty acids having chain lengths greater than 18 carbon atoms. These range from C_{19} through C_{34} with C_{26}-unsaturated acids constituting 18% of the fatty acids with greater than 18 carbon atoms at the midexponential phase of growth and 64% at the late exponential phase.

Branched-Chain Fatty Acids

Among bacterial genera, relatively high concentrations of branched-chain fatty acids have been found in *Bacillus*, *Staphylococcus*, *Corynebacterium*, *Mycobacterium*, *Pseudomonas*, and *Spirochaeta*. The most common types of branched-chain fatty acids observed are of the iso and anteiso configuration:

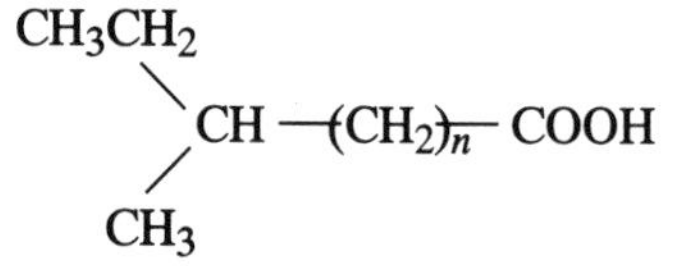

$$CH_3CH_2 \diagdown CH-(CH_2)_n-COOH \qquad CH_3 \diagdown CH-(CH_2)_n-COOH$$
$$CH_3 \diagup \qquad\qquad\qquad CH_3 \diagup$$

Anteiso- 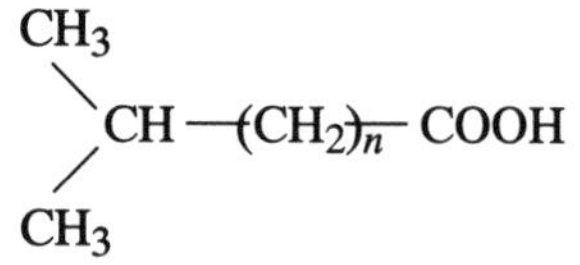   Iso-

Iso- and anteiso-fatty acids have been found as characteristic constituents of lipids in many different organisms. *Bacillus subtilis* was one of the first bacterial species in which branched-chain fatty acids were identified. At least six different branched-chain fatty acids representing 60% of the total fatty acids have been found. Corynemycolenic acid, produced by *Corynebacterium*, has the following structure:

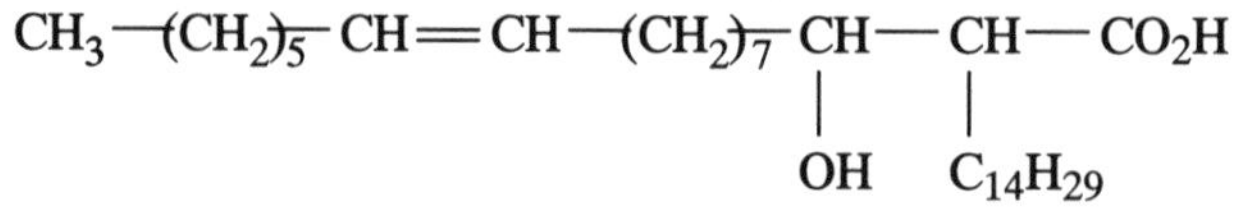

$$CH_3-(CH_2)_5-CH=CH-(CH_2)_7-CH-CH-CO_2H$$
$$| \qquad |$$
$$OH \quad C_{14}H_{29}$$

Three branched-chain hydroxy acids, 2-hydroxy-9-methyl-decanoic acid, 3-hydroxy-9-

methyl-decanoic acid, and 3-hydroxy-11-methyl-dodecanoic acid, have been identified in *Pseudomonas maltophilia*. Several species of *Mycobacterium* have been shown to contain 2-methyl-3-hydroxypentanoic acid:

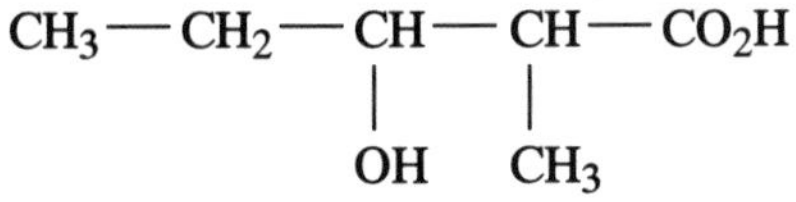

$$CH_3 - CH_2 - \underset{\underset{OH}{|}}{CH} - \underset{\underset{CH_3}{|}}{CH} - CO_2H$$

Many other 3-hydroxy fatty acids have been found in microorganisms. Members of the genus *Bacteroides* have been shown to contain hydroxy fatty acids as major constituents of their total cellular fatty acids. The main fatty acids present in LPS from *B. fragilis* have been identified as 13-methyltetradecanoic, D-3-hydroxypentadecanoic, D-3-hydroxyhexa-decanoic, D-3-hydroxy-15-methylhexadecanoic, and D-3-hydroxyheptadecanoic acids. Analysis of a large number of species of *Bacteroides* reveals that these acids are predom-inantly of the iso-branched D-(−)-3-hydroxy-1-15-methylhexadecanoic acids. Lesser amounts of the iso-branched C_{15}, straight-chain C_{16}, and anteiso-branched C_{17} acids are also found. The relatively high level of hydroxy acid is too great to be accounted for simply as cell-envelope-bound LPS fatty acid. Many of these hydroxy acids may be associated with the sphingolipids present at a high level in *Bacteroides*. In polymeric form, 3-hydroxybutanoate (β-hydroxybutyrate) occurs in prominent granules or spherical inclu-sions, which serve as a repository for reserve energy and carbon. Poly-β-hydroxybutyrate serves as an energy reserve polymer in bacteria, cyanobacteria, and eukaryotic organisms. 3-Hydroxy acids can also serve as precursors to the monoenoic fatty acids.

Ring-Containing Fatty Acids

The first ring-containing fatty acid to be found in bacteria was discovered in 1962 by Hofmann in a *Lactobacillus* and given the name lactobacillic acid. Lactobacillic acid (*cis*-11,12-methyleneoctadecanoic acid) has the following structure:

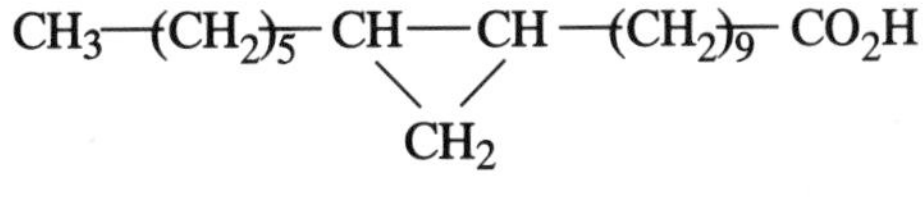

$$CH_3 - (CH_2)_5 - \underset{\underset{\diagdown}{CH}}{} - \underset{\underset{\diagup}{CH}}{} - (CH_2)_9 - CO_2H$$
$$CH_2$$

A considerable variety of ring-containing fatty acids have since been shown to occur in many types of bacteria. Greater than 15% of the fatty acids of most lactobacilli is observed as ring-containing fatty acids, whereas less than 5% of the fatty acids of *Bifidobacterium* is lactobacillic acid. A C_{17} cyclic fatty acid (*cis*-9,10-methylenehexadecanoic acid) is found in *E. coli* and other gram-negative organisms. The marked inhibition of cyclopropane fatty acid synthesis, which accompanies the induction of filamentous forms of *E. coli*, may be an indication that cyclic fatty acids play an important structural role in the cell envelope of gram-negative organisms. The structure of *cis*-9,10-methylenehexadecanoic acid is

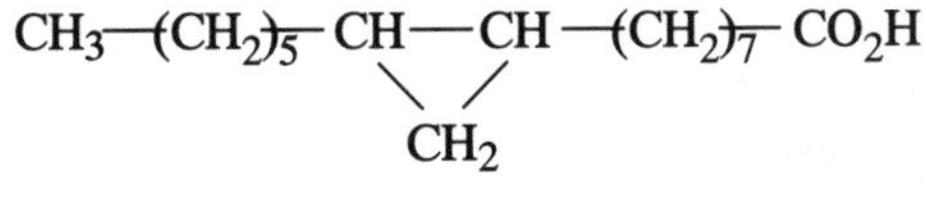

$$CH_3 - (CH_2)_5 - \underset{\underset{\diagdown}{CH}}{} - \underset{\underset{\diagup}{CH}}{} - (CH_2)_7 - CO_2H$$
$$CH_2$$

Cyclopropane fatty acids are formed from unsaturated fatty acids present in phospho-lipids. The unsaturated position is converted to a cyclopropane ring by the transfer of a methyl group from *S*-adenosylmethionine catalyzed by the enzyme cyclopropane fatty acid synthase. This enzyme apparently binds to bilayers of unsaturated fatty acid containing phospholipids and forms cyclopropane rings on both faces of the bilayer. Modification of the unsaturated positions in the fatty acids of the membrane phospholipids occurs at the onset of the stationary phase of growth. The precise reason for the modification of unsaturated fatty acids in phospholipids has not been determined. However, protection of the fatty acid from oxidation or other chemical degradation would seem to provide one logical explanation. Cyclopropane fatty acid synthase can be induced by rapid limitation of oxygen, by initiation of respiration, or by adding nitrate or thiosulfate. Neither removal via mutation nor overproduction via genetic amplification of cyclopropane fatty acid synthase impairs the growth or survival of *E. coli* under a variety of experimental conditions. However, the production of cyclopropane fatty acids suggests that their true importance has yet to be revealed.

Alk-1-enyl Ethers (Plasmalogens)

Phosphoglycerides that yield aldehydes on hydrolysis contain long-chain fatty acids in an ether linkage with the C-1 of glycerol and are called **plasmalogens**. These long-chain aldehydes are otherwise similar in composition to the long-chain fatty acids. For example, the long-chain aldehydes derived from the plasmalogens of *C. butyricum* are comparable to the long-chain fatty acids of this organism. The other carbon atoms of the glycerol backbone are linked to a fatty acid at C-2 and a phosphate ester at C-3 (Fig. 8-1). Although plasmalogens are known to occur in higher organisms, surveys of a wide range of microorganisms have failed to reveal their presence in aerobic, facultative, or microaerophilic bacteria. In anaerobic genera, the plasmalogens occur in addition to, rather than in place of, the usual phospholipids. Strict anaerobes from the genera *Bacteroides*, *Clostridium*, *Desulfovibrio*, *Peptostreptococcus*, *Propionibacterium*, *Ruminococcus*, *Selenomonas*, *Sphaerophorus*, *Treponema*, and *Veillonella* all contain plasmalogens, the molar ratio of aldehyde/phosphorus varying from 0.004 to 1.04.

The major components of the plasmalogens of *C. butyricum* are 16:0 (saturated C_{16} fatty acid), 16:1 (monounsaturated C_{16} fatty acid), 17:cyc (C_{17} cyclopropane fatty acid), 18:0 (saturated C_{18} fatty acid), 18:1 (monounsaturated C_{18} fatty acid), and 19:cyc (C_{19} cyclopropane fatty acid). The fatty aldehydes obtained from the plasmalogens of *Selenomonas ruminantium* consist of normal saturated and monounsaturated fatty aldehydes, with 12–18 carbon chain lengths predominating. A marine bacterium has been shown to contain bisphosphatidic acid mixed with its plasmalogen analogs. Phosphatidylserine and its plasmalogen analog are major phosphoglycerides of the strictly anaerobic rumen bacterium, *Megasphaera elsdenii*. Phosphoglycerides of the anaerobic lactate-fermenting bacteria *Veillonella parvula* and *Anaerovibrio lipolytica* also contain a high proportion of heptadecenoic acyl and alk-1-enyl ether moieties.

Alkyl Ethers

Long-chain alcohols bound to phospholipids in alkyl ether linkage occur in the *Archaea* (archaebacteria), a group of bacteria that exist primarily in extremely harsh environments. Their ability to thrive in these extreme conditions is considered to be due, at least in part, to these unusual lipid structures. Archaebacteria contain predominantly di-*O*-alkyl analogs of phosphatidylglycerophosphate. The lipids of these organisms are formed by

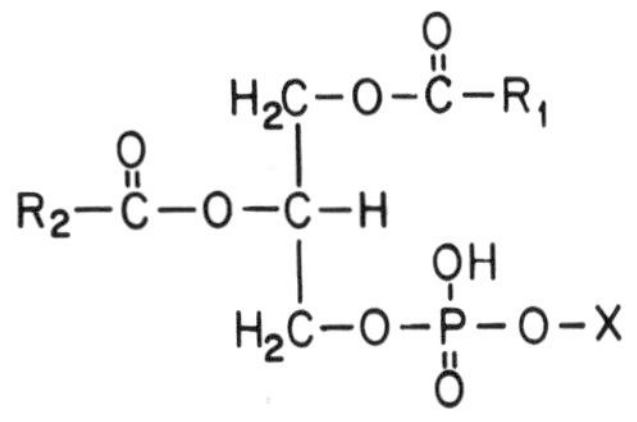

$$
\begin{array}{c}
\qquad\qquad\qquad O \\
\qquad\qquad\qquad \| \\
H_2C-O-C-R_1 \\
\quad O \qquad\qquad | \\
\quad \| \qquad\qquad | \\
R_2-C-O-C-H \\
\qquad\qquad\quad | \quad OH \\
\qquad\qquad H_2C-O-P-O-X \\
\qquad\qquad\qquad\quad \| \\
\qquad\qquad\qquad\quad O
\end{array}
$$

General phosphoglyceride structure

X = H   in phosphatidic acid
   Choline   phosphatidyl choline
   Ethanolamine   "   ethanolamine
   Serine   "   serine
   Glycerol   "   glycerol
   Inositol   "   inositol
R_1, R_2 = Fatty acids

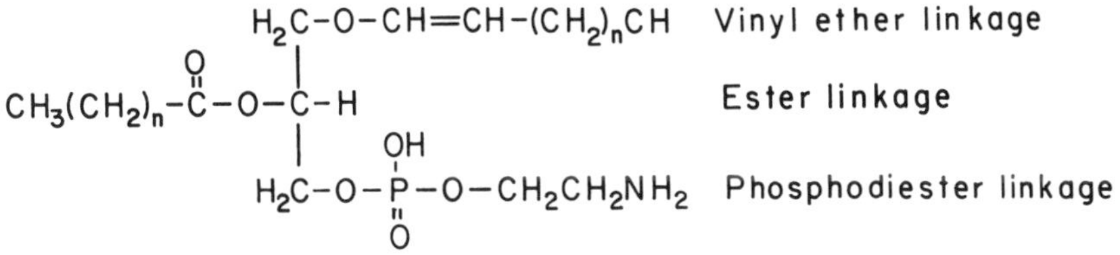

$$
\begin{array}{l}
\qquad\qquad H_2C-O-CH=CH-(CH_2)_nCH \quad \text{Vinyl ether linkage} \\
\qquad O \qquad\quad | \\
\qquad \| \qquad\quad | \\
CH_3(CH_2)_n-C-O-C-H \qquad\qquad\qquad \text{Ester linkage} \\
\qquad\qquad\qquad | \quad OH \\
\qquad\qquad H_2C-O-P-O-CH_2CH_2NH_2 \quad \text{Phosphodiester linkage} \\
\qquad\qquad\qquad\quad \| \\
\qquad\qquad\qquad\quad O
\end{array}
$$

Plasmalogen structure

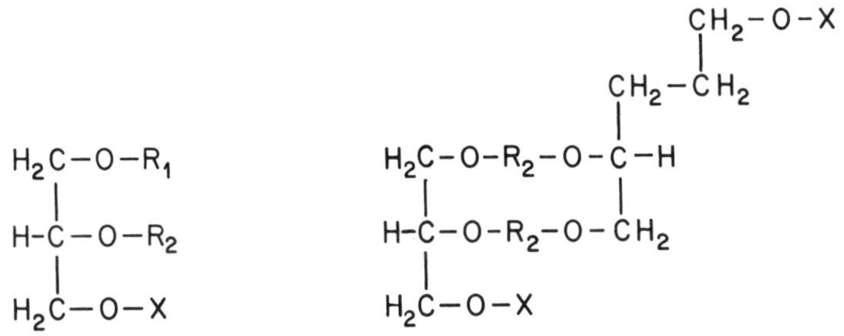

$$
\begin{array}{ll}
& CH_2-O-X \\
& | \\
& CH_2-CH_2 \\
& | \\
H_2C-O-R_1 \qquad\qquad H_2C-O-R_2-O-C-H \\
| \qquad\qquad\qquad\qquad | \qquad\qquad | \\
H-C-O-R_2 \qquad\qquad H-C-O-R_2-O-CH_2 \\
| \qquad\qquad\qquad\qquad | \\
H_2C-O-X \qquad\qquad\quad H_2C-O-X
\end{array}
$$

Glycerol diether   Diglycerol tetraether

Major lipid components of archaebacterial membranes

R_1 = Phytanyl (C_{20} polyisoprenoid alcohol in ether linkage)

R_2 = Biphytanyl (C_{40} polyisoprenoid alcohol in ether linkage)

X = H; saccharide; or phosphate derivative

Fig. 8-1. Structure of phospholipids, plasmalogens, and archaebacterial lipids.

condensation of glycerol or more complex polyols with isoprenoid alcohols containing 20, 25, or 40 carbon atoms. All archaebacterial glycerol ethers contain a 2,3-di-*O-sn*-glycerol as opposed to the *sn*-1,2-glycerol structure of other naturally occurring glycero-phosphatides or diacylglycerols. For additional discussion of some of the unusual characteristics of the archaebacterial lipids, see the next section.

Microbial Lipids

Mono-, di-, and triglycerides are the major lipid components of plants and animals. In contrast, bacteria contain relatively low concentrations of glycerides, suggesting that they play a minor role in microbial structure or function. In mammalian systems, a high percentage of the stored lipids is in the form of glycerides, while in microorganisms, the phospholipids (phosphoglycerides) are the predominant structure observed (Fig. 8-1). Mammalian cells contain phosphatidylcholine as the most common phospholipid. Bacteria are more often found to contain phosphatidylethanolamine, phosphatidylserine, phosphatidylglycerol, and phosphatidic acid. Lysophosphatidyl compounds, which lack one of the three fatty acid chains, are sometimes found in microorganisms. Phosphatidylinositol is low in content in most bacteria, but is found in higher concentration in the mycobacteria and in fungi and protozoa.

Gram-negative bacteria contain large amounts of phospholipids. Together with LPS, they constitute up to 20–40% of the dry weight of the cell envelope fraction (Table 8-2).

TABLE 8-2. Microbial Lipids

Phosphoglycerides (phospholipids)	
Phosphatidylethanolamine	Found in most microorganisms
Phosphatidylserine	Found in most microorganisms
Phosphatidylglycerol	Found in most microorganisms
Phosphatidicacid	Found in most microorganisms
Phosphatidylinositol	Common in fungi, not bacteria
Lysophosphatidyl glycerides (lack one fatty acid chain)	
Lysophosphatidylethanolamine	Marine bacteria
Plasmalogens (ether linkage between fatty acid and C-1 of glycerol)	
	Anaerobic bacteria
Cardiolipin (diphosphatidylglycerol)	
	Absent in certain marine bacteria
Glycolipids	
Lipopolysaccharides (LPS)	Outer membrane of gram-negative bacteria
Mycosyldiglycerides	Gram-positive bacteria only *Acholeplasma* (not in *Mycoplasma*)
Lipoteichoic acid	Gram-positive bacteria only
Cord factor (2 trehalose + mycolic acid)	Mycobacteria

Some of these lipids, especially C_{55}-isoprenols, serve as lipid carriers in the transferase enzymes involved in peptidoglycan synthesis. The phospholipids of the cell membrane are thought to exist in ordered liquid crystalline structures as bimolecular leaflets. A number of marine and estuarine bacteria contain the same phospholipids as nonmarine organisms with the exception that they contain significant quantities of lysophosphatidylethanolamine. A comparison of fermentative and nonfermentative groups of marine bacteria reveals the absence of cardiolipin and cardiolipin synthetase activity in most of the nonfermentative isolates. In those that do possess cardiolipin, it apparently is synthesized by the condensation of two molecules of phosphatidylglycerol, a mechanism similar to that observed in terrestrial bacteria. There appear to be two metabolic pools of phosphatidylglycerol and cardiolipin in *B. stearothermophilus*, *E. coli*, *S. aureus*, and *H. parainfluenzae*. Cardiolipin (diphosphatidylglycerol) is an important constituent of the membrane of *Haemophilus parainfluenzae*.

As mentioned previously, the archaebacteria produce a variety of unusual lipid structures that contain ether linkages. Although quite different in their chemical structure, most of these lipids are comparable to the phospholipids of eubacteria in that they are composed of single or bipolar head groups with hydrophobic alkyl side chains. These lipids still arrange themselves into bilayered membrane structures with the polar head groups towards the aqueous phase to form membranes that are indistinguishable from those of the eubacteria. The alkyl side chains of these compounds in neutrophilic halophiles are C_{20} phytanyl substituents (Fig. 8-1). However, in alkaliphilic halophiles they may be C_{25} sesterterpanylic compounds. In methanogens, the membrane lipids contain diphytanyl-glyceroldiether and di-biphytanyl-diglycerol-tetraether. The content of tetraether lipids in methanogens is correlated with the ease of demonstrating freeze–fracture planes. Above the 45–50% level of tetraether core lipids, the frequency of membrane fracture is markedly reduced. In thermophiles the basic structures are more varied but C_{20} phytanyl and C_{40} biphytanyl chains are found in *Desulfurococcus*, *Thermoproteus*, *Thermofilum*, and *Pyrodictium*. *Sulfolobus* contains a variety of tetraethers of two classes (see Fig. 8-1). The first class is comprised of glycerol-dialkylglycerol-tetraethers that contain two *sn*-2,3-glycerol components bridged by two isoprenoid C_{40} diols through ether linkages. In the second class a branched nonitol replaces one of the glycerols and are referred to as glycerol-dialkylnonitol tetraethers.

Glycolipids

Baddiley and co-workers were among the first to demonstrate glycolipids in microorganisms. Glycosyldiglycerides are widely distributed in gram-positive bacilli, but thus far, have not been found in gram-negative bacilli. They contain glucose, galactose, or mannose. The monoglycosyl derivatives that have been found are believed to represent precursors in the biosynthesis of the diglycosyl derivatives. The structure of the glycosyldiglyceride, α-D-glucopyranosyl-(1,2)-α-D-glucopyranosyl-(2,3)-diglyceride is shown in Figure 8-2. The glycolipids have α,α or β,β configurations and do not seem to appear in α,β form. Glycolipids are found in the nonsterol-requiring *Mycoplasma*. This finding suggests that glycolipids replace sterols in either the structure and/or the function of the membrane.

Most glycolipids have the carbohydrate moiety linked to the glycerol component of the glyceride. However, glycolipid structures have been observed in which the carbohydrate portion is attached directly to the fatty acid. A rhamnolipid found in *P. aeruginosa* contains two molecules of rhamnose and two molecules of decanoic acid, suggesting the structure shown in Figure 8-2.

A glycolipid of major importance in the mycobacteria is the **cord factor**, a compound associated with the virulence of *Mycobacterium tuberculosis* as well as the characteristic serpentine growth pattern. Cord factor has been shown to contain two molecules of mycolic acid attached to trehalose (Fig. 8-2). In human and bovine strains of *M. tuberculosis*, the substituent side chain is $C_{24}H_{49}$, as shown in Figure 8-2. In avian and saprophytic mycobacteria, the chemical composition is $C_{22}H_{45}$. Other variations in the chemical structure of mycolic acids have been observed as a larger number of microorganisms has come to be examined.

A unique phenolic glycolipid isolated from *Mycobacterium leprae* contains 3,6-di-*O*-methylglucose, 2,3-di-*O*-methylrhamnose, 3-*O*-methylrhamnose linked to phenol-dimycoserosyl phthiocerol (Fig. 8-2). This unique antigen of *M. leprae* suppresses the *in*

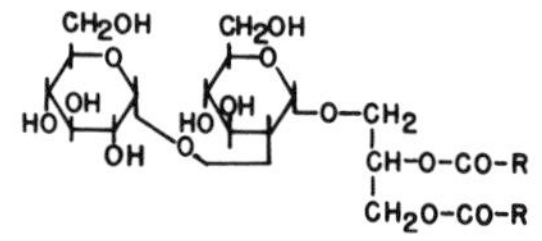

α-D-Glucopyranosyl-(1,2)-α-D-glucopyranosyl-
(2,3)-diglyceride

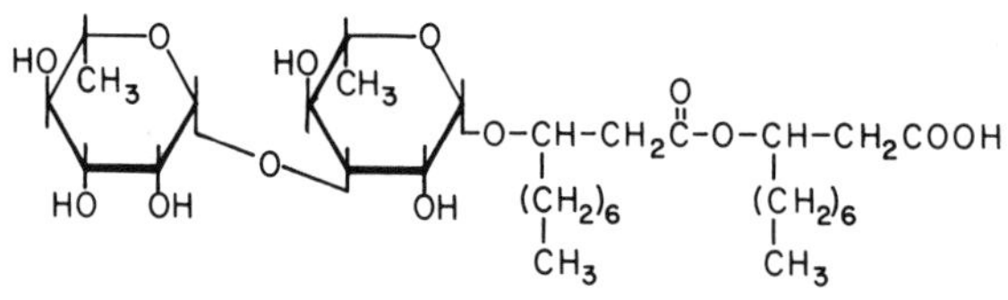

Rhamnolipid containing decanoic acid

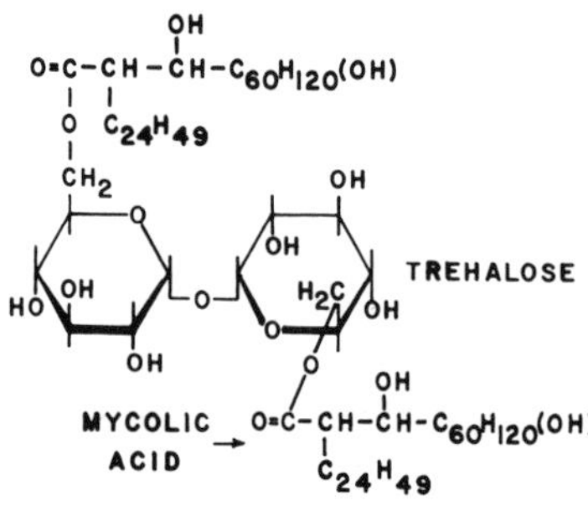

Cord factor from *Mycobacterium tuberculosis*

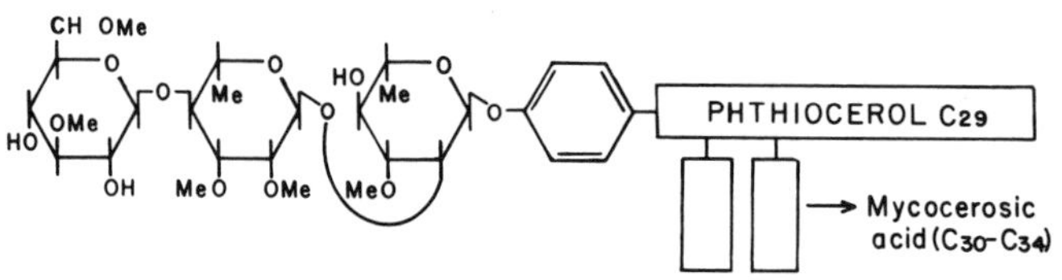

Phenolic glycolipid from *Mycobacterium leprae*

Fig. 8-2. Structure of glycolipids.

vitro mitogenic response of lymphocytes from lepromatous patients. The suppressor T cells of these patients recognize the specific terminal trisaccharide moiety that triggers the suppression. Lymphocytes from patients with tuberculoid leprosy, lepromin-positive contacts, or normal donors showed no comparable suppression of lymphocyte proliferation to concanavalin A. Removal of the mycocerosic acid side chain had no effect on *in vitro* suppression. However, absence of the 3′ terminal methyl group or removal of the terminal sugar abolished or significantly reduced the suppressive effect. Comparable glycolipids from *M. kansasii* and *M. bovis* show no suppressive action.

BIOSYNTHESIS OF FATTY ACIDS

Fatty acid biosynthesis has been studied extensively in bacteria and yeasts. Acetyl-CoA is the ultimate precursor of fatty acids. Acetyl-CoA carboxylase, a biotin-containing enzyme that catalyzes the ATP-dependent fixation of CO_2 into acetyl-, propionyl-, and butyryl-CoA, catalyzes the first committed reaction in the de novo synthesis of fatty acids (Fig. 8-3). The reaction occurs in two stages, the carboxylation of biotin with bicarbonate, catalyzed by biotin carboxylase, and transfer of the CO_2 group from carboxybiotin to acetyl-CoA to form malonyl-CoA, mediated by carboxyltransferase (Fig. 8-4). This initial step is followed by a series of reactions that ultimately lead to the synthesis of long-chain fatty acids at the C_{16} (palmitic) and C_{18} level (Fig. 8-3). Initially, it was puzzling to find that CO_2, an essential reactant in the system, was not incorporated into the fatty acids. This occurs because the CO_2 initially entering the system to form malonyl-CoA is eventually removed in the condensation step. It is actually the liberation of CO_2 that shifts the equilibrium in the direction of synthesis.

Because of its crucial role as the first committed step in fatty acid biosynthesis, the acetyl-CoA carboxylase system has been considered as one of the most likely sites of regulation of fatty acid biosynthesis. In mammalian systems several mechanisms of regulation of acetyl-CoA carboxylase have been proposed. Allosteric regulation by citrate and isocitrate, feedback inhibition by the end-product, long-chain acyl-CoA, covalent modification by phosphorylation and dephosphorylation, and regulation at the level of gene expression have all been implicated as possible mechanisms. In *E. coli* the four components of the acetyl-CoA carboxylase system function as an enzyme complex. The genes that encode the biotin carboxyl carrier protein (BCCP), *accB*, and biotin carboxylase, *accC*, are cotranscribed (*accBC*) and map at 72 min. The genes encoding the carboxyltransferase α (*accA*) and β (*accD*) subunits map at different regions of the chromosome. The rates of transcription of the genes encoding all four subunits of acetyl-CoA carboxylase are directly related to the rate of cell growth. The promoter sequences of these genes and certain features of their respective promoter regions indicate a role in regulation of the complex.

The fatty acid synthetase of *E. coli* exists as a soluble complex (designated **Type II**) from which at least eight individual enzyme components are readily separated and purified. A readily dissociable fatty acid synthetase has been characterized in *Propionibacterium shermanii*. This system requires acetyl-CoA, malonyl-CoA, NADH, NADPH, and a small acyl carrier protein (ACP) similar to that from *E. coli*. **Type I** fatty acid synthase, which occurs in mammals and yeast, is a dimer of two giant polypeptides that cannot be dissociated into the individual components (see below).

The ACP that functions in fatty acid synthesis is very stable to heat and acid treatment.

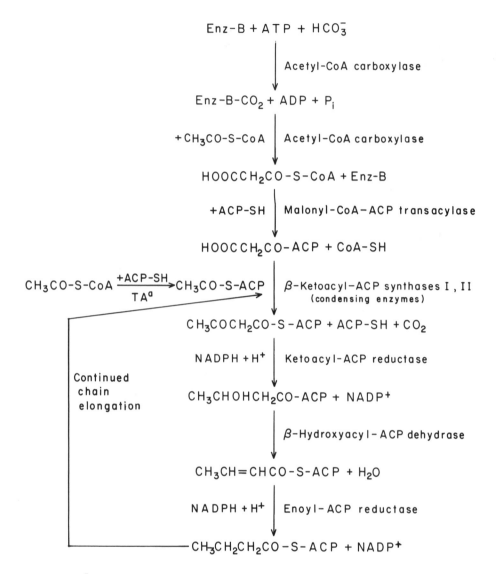

Fig. 8-3. **Reactions involved in the biosynthesis of fatty acids.** Enz-B = enzyme biotin; CoA = coenzyme A; ACP = acyl carrier protein. The *E. coli* structural genes and enzymes in this pathway are *fabE* = acetyl-CoA carboxylase (see Fig. 8-4 for more detail); acetyl-CoA-ACP transacylase; *fabD* = malonyl-CoA-ACP transacylase; *fabB* = β-ketoacyl-ACP synthase I (active with C_2–C_{14} ACP, but inactive with C_{16} ACP, inactive with CoA derivatives); *fabF* = β-ketoacyl-ACP synthase II (active with C_2–C_{14} ACP and 16:1^{Δ9}-ACP); β-ketoacyl ACP reductase; *fabA* = β-hydroxyacyl-ACP dehydrase; enoyl-ACP reductase.

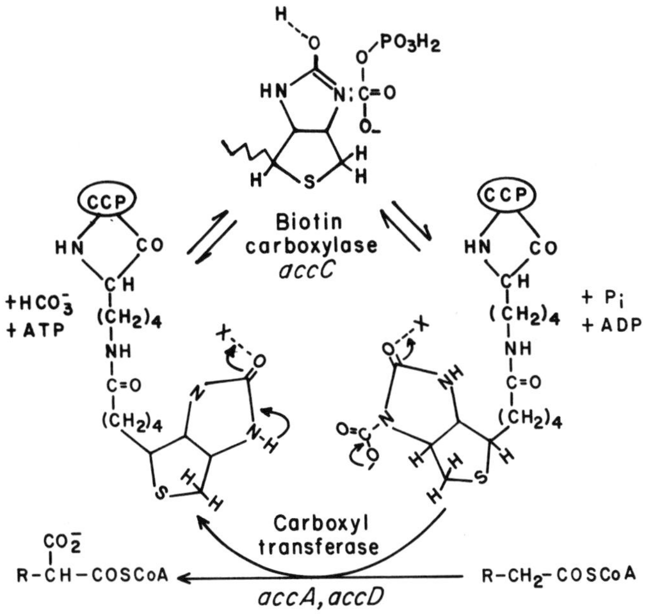

Fig. 8-4. Role of biotin in the acetyl-CoA carboxylase system, the first committed step in fatty acid biosynthesis. The *E. coli* enzyme consists of four subunits. The structural genes for these components are *accB*, biotin-CCP (carboxyl carrier protein); *accC*, biotin carboxylase; *accA*, *accD*, carboxyl transferase.

The composition of the prosthetic group is very similar to that of CoA (Table 8-3). The structure of the prosthetic group of the acyl carrier protein is shown in Figure 8-5. Some of the reactions have been found to be specific for ACP while others can utilize either ACP or CoA. However, the ACP is specific to the fatty acid synthetase system and cannot replace CoA in other systems.

In *E. coli*, at least six genes (*fabA* to *fabF*) encode the enzymes involved in fatty acid biosynthesis shown in Figure 8-3. The enzyme β-ketoacyl-ACP synthase I (en-

TABLE 8-3. Composition of Acyl Carrier Protein

Component	mol/mol Protein	mol/mol CoA
β-Alanine	1	1
Mercaptoethylamine	1	1
Pantoic acid	1	1
Organic P	1	3
Adenylate residue (Ribose + adenine)	0	1

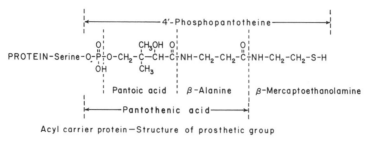

Fig. 8-5. Structure of the prosthetic group of ACP.

coded by *fabB*) is required for the elongation of unsaturated acyl-ACP. Mutants lacking synthase I activity are unable to synthesize either palmitoleic or *cis*-vaccenic acid and require unsaturated fatty acids for growth. β-Keto-ACP synthase II (encoded by *fabF*) is responsible for the temperature-dependent regulation of fatty acid composition. Mutants lacking synthase II activity are deficient in *cis*-vaccenic acid but grow normally. β-Keto-ACP synthase III catalyzes the formation of acetoacetyl-ACP. The *fabA* gene encodes 3-hydroxydecanoyl-ACP dehydratase. The enzyme malonyl-CoA-ACP transacylase (MCT) is encoded by *fabD* and catalyzes the transacylation of ACP with malonate. The *fabD* gene has been shown to be part of an operon consisting of at least three genes involved in fatty acid biosynthesis. The genes encoding a β-ketoacyl-ACP synthase (KAS) and β-ketoacyl-ACP reductase are located immediately upstream and downstream, respectively, of *fabD* and suggests that *fabD* is part of an operon encoding several components of the fatty acid synthetase system.

In eukaryotic cells (animals or yeasts) the series of reactions from acetyltransferase to the completion of the synthesis of a long-chain fatty acid takes place in a multienzyme complex referred to as **fatty acid synthetase**. This multifunctional complex form of fatty acid synthetase is termed **Type I**.

The Type I fatty acid synthetases of yeasts, fungi, and mycobacteria exist as multiple copies of apparently identical subunits with molecular weights of approximately 290,000. The complex from *Neurospora crassa* consists of 11–12 equally sized subunits, 6 of which possess the acyl carrier protein function. The lack of a low-molecular-weight acyl carrier protein component is comparable to the system found in yeast.

During the investigations of the biosynthesis of the saturated fatty acids in *E. coli* it was discovered that unsaturated fatty acids were being synthesized simultaneously. The amount of unsaturated fatty acids formed was actually greater than that of the saturated fatty acids. The divergence point proved to be at the C_{10} stage (β-hydroxydecanoate). At this step there is apparently a competition between β-hydroxyacyl-ACP dehydrase, which forms an α,β-trans double bond, and β-hydroxydecanolylthioester dehydrogenase, which forms α,β,γ-cis double bonds, as shown in Figure 8-6. Continued elongation leads to the formation of unsaturated fatty acids. As a result of this mode of unsaturated fatty acid synthesis, only monounsaturated fatty acids are formed, and *cis*-vaccenic acid ($18:1^{\Delta 11}$) is the major unsaturated fatty acid formed by organisms that use the anaerobic pathway (see Fig. 8-6 and Table 8-4).

In eukaryotes (mammals, yeast, algae, and protozoa) and some highly aerobic bacteria (*Alcaligenes, Mycobacterium, Corynebacterium,* and *Bacillus*) the synthesis of long-chain unsaturated fatty acids involves an oxidative desaturation that introduces double bonds

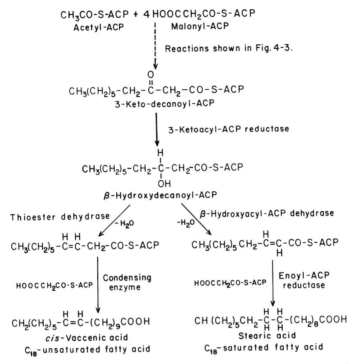

Fig. 8-6. Divergence of the saturated and unsaturated fatty acid biosynthetic pathways under anaerobic conditions. The pathway to saturated fatty acids shown in Figure 8-4 is followed until the C_{10} level (β-hydroxydecanoyl-ACP) at which point there is a competition between the enzymes involved in forming saturated and unsaturated fatty acids.

TABLE 8-4. Distribution of Pathways for Biosynthesis of Long-Chain Monounsaturated Fatty Acids

Anaerobic Pathway	Aerobic Pathway
Escherichia coli	*Alcaligenes faecalis*
Salmonella typhimurium	*Corynebacterium diphtheriae*
Serratia marcescens	*Mycobacterium phlei*
Azotobacter agilis	*Bacillus* (several species)
Agrobacterium tumefaciens	*Micrococcus luteus*
Lactobacillus plantarum	*Beggiatoa*
Staphylococcus haemolyticus	*Myxobacter*
Clostridium pasteurianum	*Leptospira canicola*
Clostridium butyricum	*Saccharomyces cerevisiae*
Caulobacter crescentus	*Neurospora crassa*
Propionibacterium	*Candida lipolytica*
Cloroflexus auranticus	*Stigmatella aurantiaca*
Chlorobium limicola	

after the saturated fatty acid has been synthesized:

$$CH_3(CH_2)_7 \overset{\displaystyle H}{\underset{\displaystyle H}{\overset{|}{\underset{|}{C}}}} - \overset{\displaystyle H}{\underset{\displaystyle H}{\overset{|}{\underset{|}{C}}}} - (CH_2)_7 \, CO - SCoA + 0.5O_2 \longrightarrow$$

Stearoyl-CoA (18:0)

$$CH_2(CH_2)_7 \overset{\displaystyle H}{\overset{|}{C}} = \overset{\displaystyle H}{\overset{|}{C}} - (CH_2)_7 \, CO - SCoA + H_2O$$

Oleoyl-CoA (18:1$^{\Delta 11}$)

Approximately 70% of the fatty acids in membrane lipids of the yeast *Saccharomyces cerevisiae* consist of the unsaturated fatty acids palmitoleic acid (16:1) and oleic acid (18:1). The remaining fatty acids are saturated, consisting primarily of palmitic acid (16:0) and lesser amounts of stearic acid (18:0) and myristic acid (14:0). Most other fungi contain the di- and trienoic acids linoleic acid (18:2) and α-linolenic acid (18:3). Yeast utilize fatty acid desaturases for the formation of unsaturated fatty acids. The Δ-9 desaturase catalyzes the formation of the initial double bond between the ninth and tenth carbon atoms of both palmitoyl (16:0) and stearoyl (18:0) CoA substrates to make 16:1 and 18:1 unsaturated fatty acids. The activity of the Δ-9 desaturase is markedly reduced by the addition of unsaturated fatty acids presumably because the level of desaturase mRNA is repressed under these conditions.

Under anaerobic conditions yeast exhibit a requirement for unsaturated fatty acids. The additional requirement for unsaturated fatty acids for anaerobic growth stems from their requirement for coupling oxidative phosphorylation to ATP synthesis in the yeast mitochondrion. This effect is obvious since yeast use the aerobic pathway (desaturase enzyme) for the formation of unsaturated fatty acids. However, studies with mutant strains show that *S. cerevisiae* also requires unsaturated fatty acids for aerobic growth. The amount of unsaturated fatty acid required for aerobic growth is fourfold higher than that required for anaerobic growth.

BIOSYNTHESIS OF PHOSPHOLIPIDS

The biosynthesis of phospholipids, more correctly referred to as phosphoglycerides, begins with the formation of *sn*-glycerol-3-phosphate (G-3-P). The G-3-P may be formed from dihydroxyacetone phosphate (DOHAP) derived from hexose cleavage by the action of the biosynthetic G-3-P dehydrogenase encoded by *gpsA*:

$$DOHAP + NAD(P)H + H^+ \longleftrightarrow G\text{-}3\text{-}P + NAD(P)^+$$

sn-Glycerol-3-phosphate may also be produced from glycerol via the action of glycerol kinase encoded by *glpK*:

$$glycerol + ATP \longrightarrow G\text{-}3\text{-}P + ADP$$

The first committed steps in phosphoglyceride biosynthesis involve coupling two molecules of fatty acid to G-3-P (Fig. 8-7). This is catalyzed by two acyl transferases. Glycerol-3-phosphate acyltransferase, encoded by the *plsB* gene, adds one saturated fatty acid to form 1-acylglycerol-3-phosphate. An unsaturated fatty acid is added to the C-2 position by 1-acylglycerol-3-phosphate acyltransferase, encoded by *plsK*, to form phosphatidic acid. In bacteria a fatty acid-ACP is the donor rather than a fatty acid-acyl-CoA. In mammalian cells, the seeds of plants, and in lower eukaryotes, triacylglycerides are formed in considerable quantities and serve as storage compounds. In microorganisms, particularly bacteria, phospholipids are more commonly observed. *Escherechia coli* normally contains approximately 70% phosphatidylethanolamine, 25% phosphatidylglycerol, and 5% cardiolipin. The next step involves formation of cytidine diphosphoglyceride (CDP-diglyceride) by phosphatidate cytidyltransferase, as shown in Figure 8-7.

At this point the pathway branches. Addition of L-serine by phosphatidylserine synthase (encoded by *pssA*) yields phosphatidylserine. Decarboxylation of phosphatidylserine by phosphatidylserine decarboxylase (encoded by *psd*) yields phosphatidylethanolamine, one of the major phospholipids found in eubacteria. In the other branch of the pathway, addition of a second molecule of G-3-P to the CDP-diglyceride by phosphatidylglycerophosphate synthase (encoded by *pgsA*) results in the release of CMP and the formation of phosphatidylglycerolphosphate. Removal of P_i by phosphatidylglycerolphosphate phosphatase (encoded by *pgsA*, *pgsB*) yields phosphatidylglycerol. In bacteria, two molecules of phosphatidylglycerol are condensed by cardiolipin synthase (encoded by *cls*) releasing glycerol to form cardiolipin (diphosphatidylglycerol). In mammalian systems, the formation of cardiolipin involves the formation of a CDP-diglyceride as an intermediate.

DEGRADATION OF LIPIDS

Some of the earliest investigations into the metabolism of fatty acids and triglycerides dealt primarily with the degradation of triglycerides by lipases and further oxidation of the fatty acids liberated. Lipases attack triglycerides in the following manner:

$$R_2 - \overset{\overset{\displaystyle O}{\|}}{C} - O - \underset{\underset{\displaystyle CH_2 - O - \overset{\overset{\displaystyle O}{\|}}{C} - R_3}{|}}{\overset{\overset{\displaystyle CH_2 - O - \overset{\overset{\displaystyle O}{\|}}{C} - R_1}{|}}{C}} - H \quad \xrightarrow{\text{lipase}}$$

Triglyceride

$$R_1 - COOH + R_2COOH + R_3COOH + HO - \underset{\underset{\displaystyle CH_2OH}{|}}{\overset{\overset{\displaystyle CH_3}{|}}{C}} - H$$

Fatty acids   Glycerol

The glycerol liberated is oxidized to glycerol-3-phosphate, which is then metabolized via the triose phosphate pathway of the EMP system. The free fatty acid is converted to the

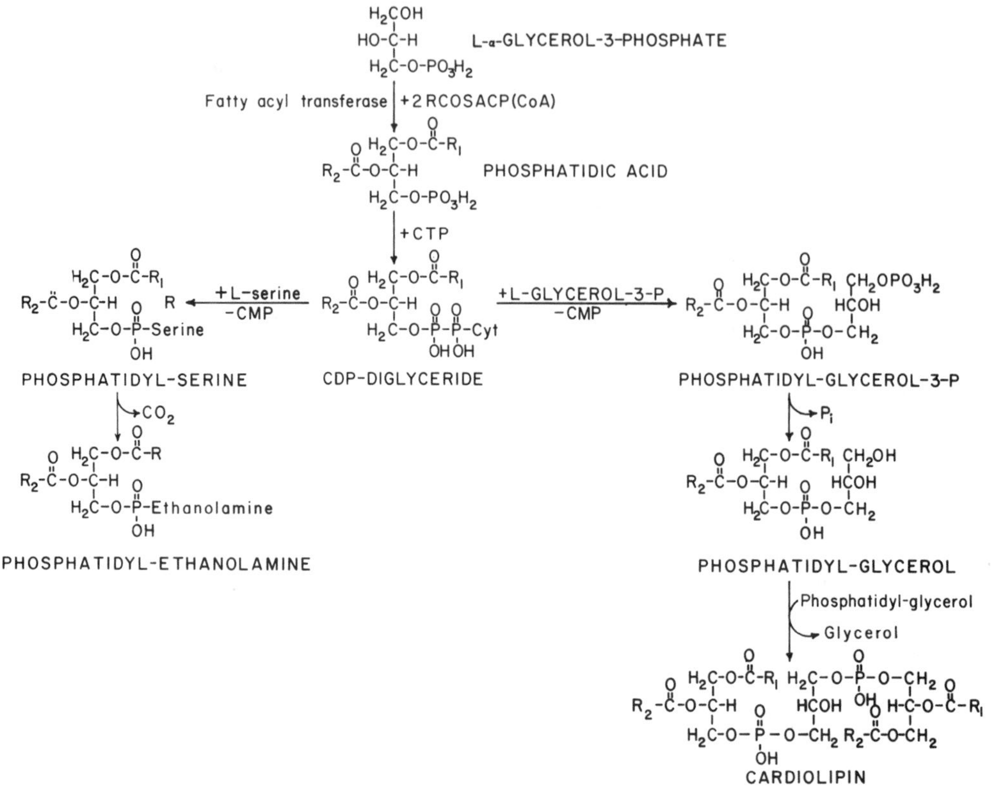

Fig. 8-7. Biosynthesis of phosphoglycerides (phospholipids). The *E. coli* enzymes and structural genes are *plsB* = sn-glycerol-3-3-phosphate acyltransferase; *plsK* = 1-acylglycerol-3-phosphate acyltransferase; phosphatidate cytidyltransferase; *pssA* = phosphatidylserine synthase; *psd* = phosphatidylserine decarboxylase; *pgsA* = phosphatidylglycerophosphate synthase; *pgpA* = phosphatidylglycerophosphate phosphatase; *cls* = cardiolipin (diphosphatidylglycerol) synthase.

acyl-CoA by acyl-CoA synthetase, as shown in Figure 8-8. The free fatty acid, in its acyl-CoA form, is then oxidized via the sequence of reactions shown in Figure 8-8. This sequence of reactions is repeated until the fatty acid has been reduced to acetyl-S-CoA fragments. The two-carbon fragments are then metabolized by the glyoxylate cycle; however, in organisms that do not have an intact glyoxylate cycle or in which some abnormality occurs, acetoacetate, β-hydroxybutyrate, and acetone may be produced. These pathways were discussed previously in relation to anaerobic fermentations.

Most bacteria do not readily degrade fatty acids. However, once induced, they may become very active fatty acid oxidizers. For example, *E. coli* will grow on long-chain fatty acids only after a prolonged lag phase. Cells that are preadapted to growth on a fatty acid, such as palmitate, will grow immediately on fatty acids. These observations suggest that fatty acid oxidation occurs via an inducible system. After culture of *E. coli* with fatty acids of carbon chain lengths of C_{14} or longer, fatty acid oxidation is much more rapid than it is in cells harvested from cultures grown with amino acids or a combination of glucose and amino acids. The activities of the enzymes of fatty acid oxidation shown in Figure 8-8 increase or decrease coordinately, depending on the con-

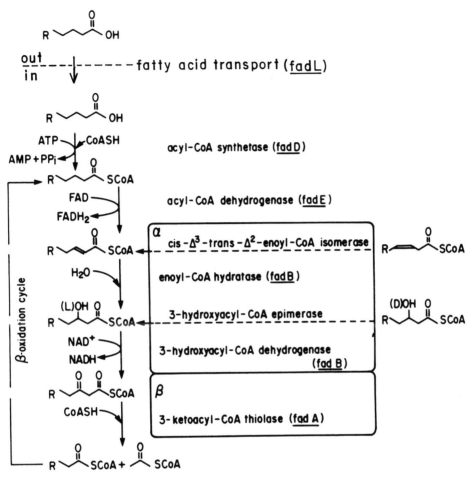

Fig. 8-8. Degradation of fatty acids. The *E. coli* structural genes and enzymes are *fadA*, 3-ketoacyl-CoA thiolase (thiolase I); *fadB*, 3-hydroxyacyl-CoA dehydrogenase; 3-hydroxyacyl-CoA epimerase; *fadB*, enoyl-CoA hydratase; *cis*-Δ^3-C$_{15}$-Δ^2-*trans*-enoyl-CoA isomerase; *fadE*, FAD-linked acyl-CoA dehydrogenase; *fadD*, acyl-CoA synthetase; *fadL*, encodes an outer-membrane protein (FadL) required for the transport of exogenous long-chain fatty acids. The acetyl-CoA is further metabolized via the TCA or glyoxylate cycles. Adapted from Spratt, S. K. et al. 1984. *J. Bacteriol.* **158**:535–542, and Nunn, W. D. 1986, *Microbiol. Rev.* **50**:179–192.

ditions under which the cells are grown. Inhibition of the induction by glucose indicates that the fatty acid oxidation enzymes are subject to catabolite repression. Further evidence for catabolite repression of the degradative enzymes in *E. coli* is provided by the finding that cAMP can partially relieve the strong repression of the fatty acid oxidation system by glucose.

The structural genes (*fad* genes) encoding the enzymes of fatty acid degradation map at four distinct loci on the *E. coli* gene map and encode at least five enzyme activities involved in the transport, acylation, and β-oxidation of medium-chain (C$_6$–C$_{10}$) and long-chain (C$_{12}$–C$_{18}$) fatty acids. Long-chain fatty acids can induce the *fad* genes, whereas medium-chain fatty acids cannot. Therefore, the wild-type *E. coli* can utilize long-chain fatty acids, such as oleate (C_{18}:1), but not medium-chain fatty acids, such as decanoate (C_{10} phenotype), as a sole carbon and energy source. The *fadR* gene is a multifunctional regulatory gene mapping at 25.5 min that exerts negative control over the *fad* regulon. The product of the *fadR* gene (FadR) is a diffusible protein that exerts control over fatty acid degradation by decreasing the transcription of the *fad* structural genes. In addition to the *fad* enzymes, expression of the glyoxylate shunt enzymes is also required for the growth of *E. coli* on acetate or fatty acids as a sole carbon source. In wild-type *E. coli* repression of the acetate (*ace*) operon is under the control of *fadR* and *iclR* (regulatory gene for *aceBA* operon). Both the *fadR* and *iclR* genes regulate the glyoxyate shunt at the level of transcription. It has also been suggested that *fadR* plays a role in the regulation of unsaturated fatty acid biosynthesis as evidenced by the fact that *fadR* mutants synthesize significantly less unsaturated fatty acids than wild-type cells. A functional *fadR* gene is required for optimal synthesis of unsaturated fatty acids.

The *fadAB* and *fadR* genes have been cloned and characterized with regard to their specific functions. In *E. coli* a multifunctional enzyme complex is encoded by the *fadBA* operon. Two protein subunits (42,000 and 78,000 Da) in this enzyme complex exhibit 3-ketoacyl-CoA thiolase, 3-hydroxyacyl-CoA dehydrogenase, Δ^3-C_{15}-Δ^2-*trans*-enoyl-CoA isomerase, and 3-hydroxyacyl-CoA epimerase, and enoyl-CoA hydratase. Two identical small subunits are encoded within the *fabA* gene that contains thiolase activity. The other four activities are associated with the large subunit encoded by *fadB*. The intact complex is required for function. Both *fadA* and *fadB* are transcribed as a single transcriptional unit with the direction of transcription from *fadA* to *fadB*. Plasmids carrying the *fadR*$^\pm$ gene suppress the expression of the *fadAB* operon. The *fadR* gene codes for a 19,000-Da protein, which appears to exert control over the *fad*, *ace*, and *fab* structural genes.

Neurospora crassa also possesses an inducible β-oxidation system. Activities of the β-oxidation enzymes are enhanced after a shift from a sucrose to an acetate growth medium. The induction is even more pronounced after growth in the presence of oleate as the sole carbon and energy source. The enzymes of this pathway are localized in glyoxysome-like microbodies that are distinct from the peroxisomes (subcellular organelles whose main function is to process substances for elimination). The β-oxidation system of *N. crassa* relies on acyl-CoA dehydrogenase rather than acyl-CoA oxidase to catalyze the first step. The *N. crassa* system does not appear to be under the same type of coordinate regulation that has been observed in *E. coli*. By comparison, *Candida tropicalis* contains the enzymes for β-oxidation in peroxisomes. These organelles have a novel long-chain acyl-CoA synthetase whose product is exclusively used for β-oxidation and not for lipid synthesis.

BIOSYNTHESIS OF MEVALONATE, SQUALENE, AND STEROLS

Yeasts and molds contain sterols in their cytoplasmic membranes. A few prokaryotic microorganisms, such as *Mycoplasma*, have sterols in their membranes. Sterols function in membrane permeability. In fungi, an offshoot of the pathway of fatty acid synthesis leads to the formation of mevalonic acid, the first intermediate in the biosynthesis of sterols. Through a complex series of reactions, mevalonic acid is converted to farnesyl pyrophosphate and then to squalene. Squalene then undergoes several cyclization reactions to ultimately give rise to lanosterol, the precursor for the formation of many sterol derivatives. The pathways leading to the formation of the intermediates and to the ultimate formation of sterols are shown in Figures 8-9 and 8-10.

The pathway to mevalonic acid (Fig. 8-9) has been demonstrated in yeast as well as in mammalian liver. The condensing enzyme responsible for the coupling of acetyl-CoA and acetoacetyl-CoA has been isolated and purified from yeast. Hydroxymethylglutaryl-CoA can also arise from malonyl-CoA via reactions that appear to be identical to those responsible for the initiation of the biosynthesis of fatty acids. Free hydroxymethylglutarate is not readily utilized as a precursor of steroids by yeast or mammalian liver, presumably because of the absence of an activating enzyme. *Phycomyces blakesleeanus* can readily incorporate hydroxymethylglutarate, indicating that this organism has the necessary activating enzyme.

The *Mycoplasmataceae* are separated into two groups on the basis of their nutritional requirements for lipid components:

1. *Mycoplasma*—require sterols for growth.
2. *Acholeplasma*—synthesize carotenoid compounds that assume the functional capacity of sterols.

Sterol and carotenol apparently serve in the same functional capacity, since cholesterol added to the culture medium is utilized in place of carotenoids whose synthesis is repressed under these growth conditions (for examples of carotenoid structures, see Fig. 7-52). *Acholeplasma laidlawii* synthesizes carotenoids from acetate. The organism contains a specific acetokinase and phosphotransacetylase, both of which are necessary for the synthesis of acetyl-CoA and a β-ketothiolase and CoA transferase required for the synthesis of acetoacetyl-CoA. The presence of β-hydroxy-β-methylglutaryl-CoA condensing enzyme and reductase activities in *Acholeplasma*, together with the ability to incorporate ^{14}C-labeled acetate into mevalonic acid, indicates that these organisms utilize the same pathway as yeast for the synthesis of mevalonic acid. The absence of β-hydroxy-β-methylglutaryl-CoA condensing enzyme and reductase activities in *Mycoplasma hominis* explains its growth requirement for sterol. In other strains of *Mycoplasma* (e.g., *M. gallisepticum*), the metabolic block occurs subsequent to the mevalonic acid step in the biosynthetic pathway to terpenoids.

Mevalonic acid is incorporated by *Lactobacillus casei* predominantly into bactoprenol, a C_{55}-isoprenol concerned with cell wall biosynthesis:

$$CH_3-\overset{\overset{\displaystyle CH_3}{|}}{C}=CH-CH_2-(CH_2-\overset{\overset{\displaystyle CH_3}{|}}{C}=CH-CH_2)_9\ CH_2-\overset{\overset{\displaystyle CH_3}{|}}{C}=CH-CH_2-O-\overset{\overset{\displaystyle O}{\|}}{\underset{\underset{\displaystyle OH}{|}}{P}}-OH$$

The distribution of tritiated bactoprenol in *L. casei* is relatively uniform throughout the cell, except in organisms that are in the process of undergoing cell division. In dividing cells there is a concentration of labeled bactoprenol in the septal region, indicating that the septal membrane is synthesized at the site of cell division.

In order to function in the biosynthesis of sterols, mevalonic acid must first be converted into the isoprenoid structural form, as shown in Figure 8-9. This sequence of reactions has been demonstrated in yeast. There is some evidence that 5-diphospho-3-phosphomevalonic acid, a very unstable intermediate, is formed, accounting for the re-

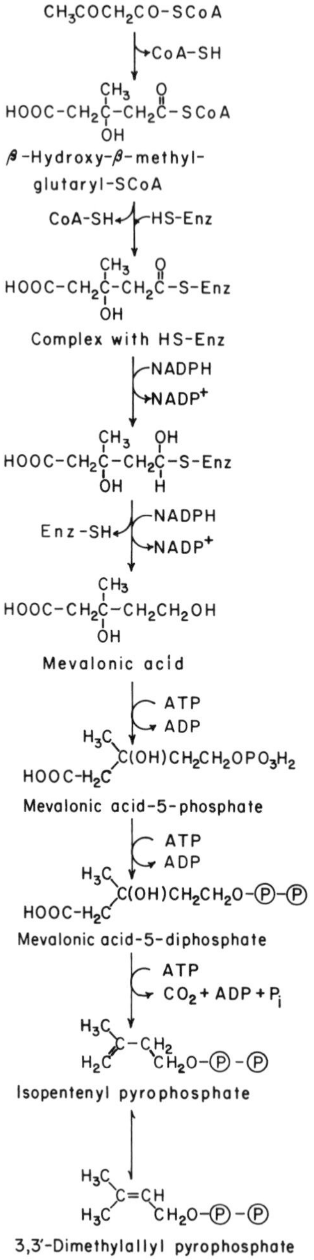

$$CH_3COCH_2CO-SCoA$$

$$\downarrow CoA-SH$$

$$\underset{OH}{\overset{CH_3}{HOOC-CH_2\overset{|}{C}-CH_2\overset{O}{\overset{||}{C}}-SCoA}}$$

β-Hydroxy-β-methyl-
glutaryl-SCoA

$$CoA-SH \nwarrow \!\!\!\! \nearrow HS-Enz$$

$$\underset{OH}{\overset{CH_3}{HOOC-CH_2\overset{|}{C}-CH_2\overset{O}{\overset{||}{C}}-S-Enz}}$$

Complex with HS-Enz

$$\overset{\displaystyle \curvearrowright NADPH}{\underset{\displaystyle \searrow NADP^+}{}}$$

$$\underset{\overset{|}{O}H \;\; \overset{|}{H}}{\overset{CH_3 \;\; OH}{HOOC-CH_2\overset{|}{C}-CH_2\overset{|}{C}-S-Enz}}$$

$$Enz-SH \nwarrow \!\!\!\! \overset{\displaystyle \curvearrowright NADPH}{\underset{\displaystyle \searrow NADP^+}{}}$$

$$\underset{\overset{|}{O}H}{\overset{CH_3}{HOOC-CH_2\overset{|}{C}-CH_2CH_2OH}}$$

Mevalonic acid

$$\overset{\displaystyle \curvearrowright ATP}{\underset{\displaystyle \searrow ADP}{}}$$

$$\underset{HOOC-H_2C}{\overset{H_3C}{}}\!\!\!\!\diagdown C(OH)CH_2CH_2OPO_3H_2$$

Mevalonic acid-5-phosphate

$$\overset{\displaystyle \curvearrowright ATP}{\underset{\displaystyle \searrow ADP}{}}$$

$$\underset{HOOC-H_2C}{\overset{H_3C}{}}\!\!\!\!\diagdown C(OH)CH_2CH_2O-\textcircled{P}-\textcircled{P}$$

Mevalonic acid-5-diphosphate

$$\overset{\displaystyle \curvearrowright ATP}{\underset{\displaystyle \searrow CO_2+ADP+P_i}{}}$$

$$\underset{H_2C}{\overset{H_3C}{}}\!\!\!\!\diagdown \underset{CH_2O-\textcircled{P}-\textcircled{P}}{C-CH_2}$$

Isopentenyl pyrophosphate

$$\uparrow$$

$$\underset{H_3C}{\overset{H_3C}{}}\!\!\!\!\diagdown \underset{CH_2O-\textcircled{P}-\textcircled{P}}{C=CH}$$

3,3'-Dimethylallyl pyrophosphate

Fig. 8-9. Biosynthesis of mevalonic acid and conversion of mevalonic acid to the isoprenoid structural form.

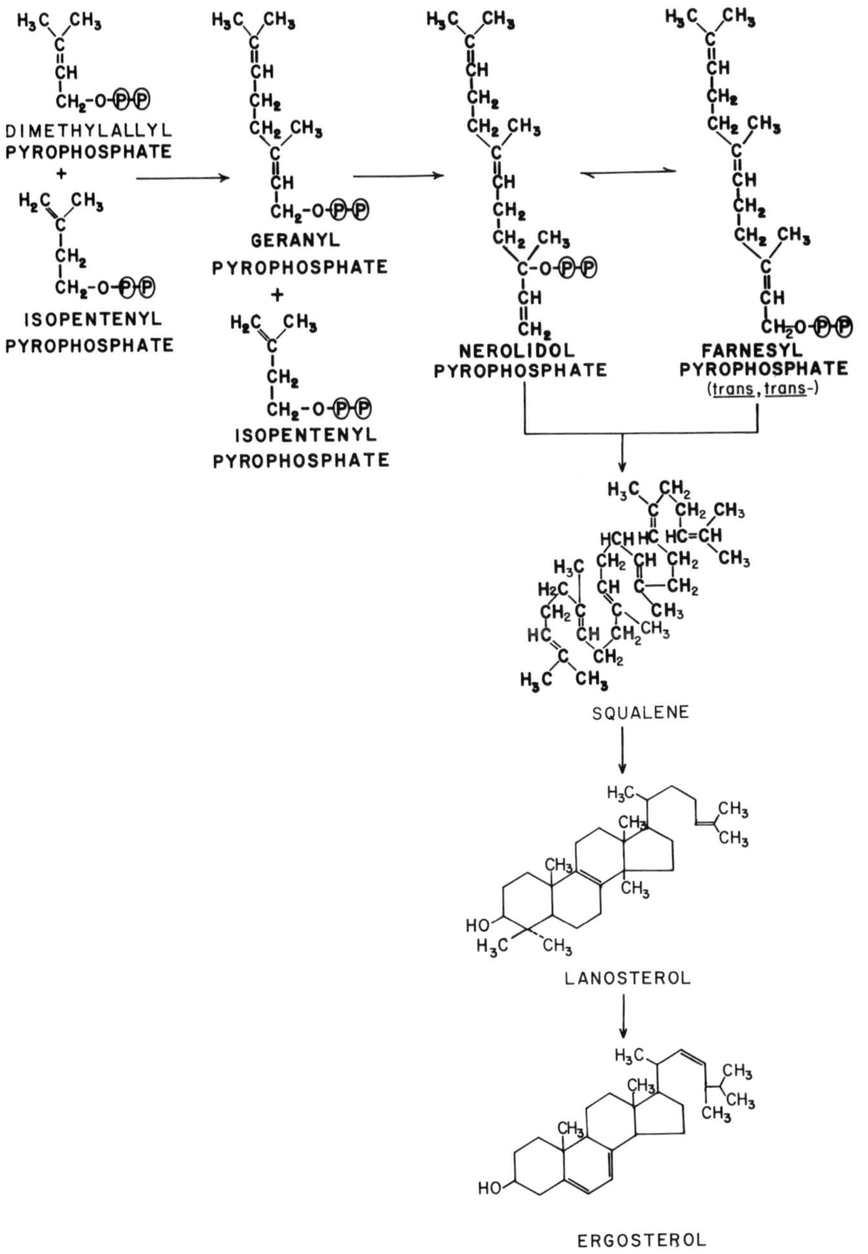

Fig. 8-10. Biosynthesis of nerolidol and farnesyl pyrophosphates and conversion of these intermediates to squalene, lanosterol, and ergosterol.

arrangement of the double bonds in the formation of the isopentenyl structure. Isopentenyl pyrophosphate and dimethylallyl pyrophosphate are condensed with the elimination of pyrophosphate to form geranyl pyrophosphate. This compound condenses with another molecule of isopentenyl pyrophosphate yielding nerolidol pyrophosphate and, ultimately, *trans, trans*-farnesyl pyrophosphate (Fig. 8-10). The reaction sequence leading to the formation of nerolidol and farnesyl pyrophosphate has been demonstrated in yeast.

Squalene is formed by the condensation of either two molecules of farnesyl pyrophosphate or, more likely, one molecule of farnesyl pyrophosphate and one molecule of nerolidol pyrophosphate, as shown in Figure 8-10. The conversion of squalene to ergosterol in yeast or to cholesterol in mammals involves a number of intermediary steps. In yeast, an epoxide intermediate preceeds the formation of lanosterol. At least eight additional steps are required to convert lanosterol to ergosterol. The additional carbon (C_{28}) is donated from *S*-adenosylmethionine at the lanosterol level. In mammalian systems, the conversion of squalene to cholesterol has been shown to require at least two sterol carrier proteins (SCP_1 and SCP_2) that bind the substrate and make it reactive to the sterol synthesizing enzymes present in the microsomes. These sterol carrier proteins differ from acyl carrier protein, which functions in the biosynthesis of fatty acids, in several respects. However, both of these compounds appear to serve the analogous function of maintaining the solubility and reactivity of the substrates.

Saccharomyces cerevisiae and other yeasts require sterols for growth. Under aerobic conditions, most yeasts are able to synthesize the required level of ergosterol. Under anaerobic conditions, yeast cannot synthesize sterols or unsaturated fatty acids and require their addition to the growth medium. Yeast mutants auxotrophic for sterols and unsaturated fatty acids have been found to be altered in the fatty acid composition of their mitochondrial phospholipids. The incorporation of fatty acids into phospholipids varies with the sterol and unsaturated fatty acids supplied. Ergosterol, in the presence of linoleic or linolenic acids or a mixture of palmitoleic and oleic acids, permits excellent growth. Substitution of other sterols, such as cholesterol, or addition of oleic acid as the sole fatty acid, results in poor growth. The genetic basis for limiting sterol synthesis in *S. cerevisiae* resides in the regulation of hydroxymethylglutaryl-CoA reductase activity. A decrease in the specific activity of this enzyme correlates with accumulation of squalene during supplementation with ergosterol or mevalonolactone. The addition of ergosterol results in feedback inhibition of hydroxymethylglutaryl-CoA reductase.

REFERENCES

Beveridge, T. J., C. J. Choquet, G. B. Patel, and G. D. Sprott. 1993. Freeze-fracture planes of methanogen membranes correlate with the content of tetraether lipids. *J. Bacteriol.* **175**:1191–1197.

Bhat, U. R., H. Mayer, A. Yokota, R. I. Hollingsworth, and R. W. Carlson. 1991. Occurrence of lipid A variants with 27-hydroxyoctacosanoic acid in lipopolysaccharides from members of the family *Rhizobiaceae. J. Bacteriol.* **173**:2155–2159.

Black, P. N. 1991. Primary sequence of the *Escherichia coli fadL* gene encoding an outer membrane protein required for long-chain fatty acid transport. *J. Bacteriol.* **173**:435–442.

Black, P. N. and C. C. DiRusso. 1994. Molecular and biochemical analyses of fatty acid transport, metabolism, and gene regulation in *Escherichia coli. Biochimica et Biophysica Acta*, **1210**:123–145.

Bossie, M. A. and C. R. Martin. 1989. Nutritional regulation of yeast Δ-9 fatty acid desaturase activity. *J. Bacteriol.* **171**:6409–6413.

DeRosa, M., A. Gambacorta, and A. Gliozzi. 1986. Structure, biosynthesis, and physicochemical properties of archaebacterial lipids. *Microbiol. Rev.* **50**:70–80.

DiRusso, C. C. 1990. Primary sequence of the *Escherichia coli fadBA* operon, encoding the fatty acid-oxidizing multienzyme complex, indicates a high degree of homology to eucaryotic enzymes. *J. Bacteriol.* **172**:6459–6468.

Dunkley, E. A., Jr., S. Clejan, and T. A. Krulwich. 1991. Mutants of *Bacillus* are deficient in fatty acid desaturase activity. *J. Bacteriol.* **173**:7750–7755.

Furakawa, H., J.-T. Tsay, T. S. Jackowski, Y. Takamura, and C. O. Rock. 1993. Thiolactomycin resistance in *Escherichia coli* is associated with the multidrug resistance efflux pump encoded by *emrAB*. *J. Bacteriol.* **175**:3723–3729.

Jackowski, S., J. E. Cronan, Jr., and C. O. Rock. 1991. Lipid metabolism in procaryotes. *In* D. E. Vance and J. E. Vance (Eds.), *Biochemistry of lipids, lipoproteins and membranes.* Elsevier Science Publishers, Amsterdam, The Netherlands, pp. 43–85.

Kaheda, T. 1991. Iso- and anteiso-fatty acids in bacteria: Biosynthesis, function, and taxonomic significance. *Microbiol. Rev.* **55**:288–302.

Li, S.-J. and J. E. Cronan, Jr. 1993. Growth rate regulation of *Escherichia coli* acetyl coenzyme A carboxylase, which catalyzes the first committed step of lipid biosynthesis. *J. Bacteriol.* **175**:332–340.

Nunn, W. D. 1986. A molecular view of fatty acid catabolism in *Escherichia coli. Microbiol. Rev.* **50**:179–192.

Paltauf, F., S. D. Kohlwein, and S. D. Henry. 1992. Regulation and compartmentalization of lipid synthesis in yeast. In *The Molecular and Cellular Biology of the Yeast* Saccharomyces: Vol. II. *Gene Expression.* E. W. Jones, J. H. Pringle, and J. R. Broach (Eds.). Cold Spring Harbor Laboratory Press, Plainview, NY.

Spratt, S. K., P. N. Black, M. M. Ragozzino, and W. D. Nunn. 1984. Cloning, mapping, and expression of genes involved in the fatty acid degradative multienzyme complex of *Escherichia coli. J. Bacteriol.* **158**:535–542.

Tsay, J.-T., C. O. Rock, and S. Jackowski. 1992. Overproduction of β-ketoacyl-acyl carrier protein synthase I imparts thiolactomycin resistance to *Escherichia coli* K-12. *J. Bacteriol.* **174**:508–513.

Verwoert, I. I. G. S., E. C. Verbree, K. H. van der Linden, H. J. J. Nijkamp, and A. R. Stuitje. 1992. Cloning, nucleotide sequence, and expression of the *Escherichia coli fabD* gene, encoding malonyl coenzyme A-acyl carrier protein transacylase. *J. Bacteriol.* **174**:2851–2857.

Yang, S.-Y., J. Li, X.-Y. He, S. D. Cosloy, and H. Schulz. 1988. Evidence that the *fadB* gene of the *fadAB* operon of *Escherichia coli* encodes 3-hydroxyacyl-coenzyme A (CoA) epimerase, Δ^3-*cis*-Δ^2-*trans*-enoyl-CoA isomerase, and enoyl-CoA hydratase in addition to 3-hydroxyacyl-CoA dehydrogenase. *J. Bacteriol.* **170**:2543–2548.

CHAPTER 9

NITROGEN METABOLISM

Microorganisms play an important part in the nitrogen cycle. A unique group of bacteria is capable of fixing atmospheric nitrogen into ammonia and then assimilating it into organic compounds. One group of these bacteria are harbored in the roots of certain plants where they fix atmospheric nitrogen into forms that can be assimilated by the plants. Degradation of nitrogenous compounds by microorganisms is also an important aspect of the nitrogen cycle. Without the degradation and subsequent return of nitrogen from a wide variety of complex natural and artificial compounds to within this cycle, higher forms of life could not exist. Although many of the degradative reactions carried out by microbial populations have been studied to a considerable extent, our understanding of this aspect of microbial metabolism at a biochemical and molecular level is still far from complete. Greater knowledge of these processes may aid materially in overcoming the imbalances that have been created in the nitrogen cycle by overloading it with waste and excretory materials and improper use of agricultural lands by overabundant applications of fertilizers, herbicides, and pesticides.

In this chapter, we consider the microbial contributions to the nitrogen cycle and the underlying mechanisms for the processes of nitrogen fixation, metabolism of inorganic nitrogen compounds, and assimilation of inorganic nitrogen into amino acids. Reactions involved in the interconversion of the amino acids, especially amino group transfer, and other important reactions are also discussed. The biosynthesis of amino acids, purines, pyrimidines, and other nitrogen-containing compounds is covered in Chapter 10.

BIOLOGICAL NITROGEN FIXATION

Although they were unaware of the precise nature of the relationship, astute Dutch farmers associated the establishment of rich stands of clover with the development of nodules on the root system. They also noticed the improved productivity of other crops grown in soils following several years of planting with clover. These observations provided the foundations for systematic crop rotation first implemented by Townsend in England during the agricultural revolution of the eighteenth century. The photograph in Figure 9-1 shows extensive nodule development, each nodule harboring millions of nitrogen fixing bacteria on the root system of a leguminous plant (a legume is a plant that bears seed pods, such as pea or soybean, and often forms a symbiotic association with bacteria that fix atmospheric nitrogen). Nitrogen fixation is the process by which atmospheric dinitrogen (N_2) is converted to ammonia (NH_3), which is then assimilated into amino acids, purines, and pyrimidines.

The works of Atwater and of Hellriegel and Wilfarth, published in the 1880s, were among the first to provide scientific data supporting the importance of the root nodules and the bacteria within them in the process of nitrogen fixation by clover and other leguminous plants. Despite early criticisms by scientists of the time, their reports initiated a concerted effort to determine the nature of the symbiotic organisms associated with the root nodules of legumes (and other plants), as well as characterization of the numerous

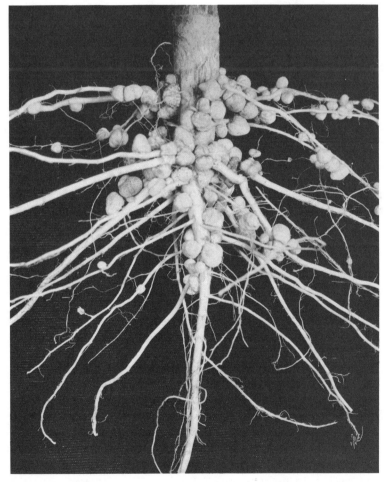

Fig. 9-1. Photograph showing extensive root nodule development in the root system of soybean (*Glycine max*) a leguminous plant. Courtesy of R. S. Smith, Milwaukee, WI.

TABLE 9-1. Examples of Nitrogen Fixing Genera

Symbiotic Association of Rhizobium *and* Bradyrhizobium *spp. with Legumes*

    R. trifolii, clover (*Trifolium, Crotolaria*)
    B. japonicum, soybean (*Glycine max*)
    R. meliloti, alfalfa (*Medicago sativa*)
    R. leguminosarum, cowpea (*Vigna unguiculata*)

Symbiotic Association of Actinomycetes with Angiosperms

    Frankia sp., Alder (*Alnus*)
    Frankia sp., Bog myrtle or sweet gale (*Myrica*)
    Frankia sp., Oleasters (*Shepherdia, Eleagnus, Hippophae*)
    Frankia sp., New Jersey tea (*Ceanothus*)

Symbiotic Association with Leaf Nodulating Plants

    Klebsiella aerogenes

Associative Interaction with Grasses

    Azospirillum brasiliense, tropical grasses
    A. lipoferum, tropical grasses
    A. paspali, tropical grass (*Paspalum notatum*)

Symbiotic Association of Marine Bacteria with Bivalves

    Aerobic chemoheterotrophic sp., bivalves (*Teredinidae*)

Symbiotic Association with Marine Diatoms
    Richelia intracellularis (cyanobacterium), *Rhizoselenia*

Free-Living Bacteria and Cyanobacteria

    Aerobic, heterotrophic
        Azotobacteriaceae (Azotobacter, Derxia, Azomonas, Beijerinkia), Nocardia, Pseudomonas
    Aerobic, phototrophic
        Anabaena, Calothrix, Nostoc, Gleotheca, Cylindrospermum, Aphanocapsa
    Facultative, heterotrophic
        Enterobacter cloacae, Klebsiella pneumoniae, Bacillus polymyxa, Desulfobibrio desulfuricans, D. gigas, Achromobacter
    Anaerobic, heterotrophic
        Clostridium pasteurianum, C. butyricum, Propionispira arboris
    Anaerobic, photoautotrophic
        Chromatium vinosum, Rhodospirillum rubrum, Rhodopseudomonas sphaeroides, R. capsulata, Rhodomicrobium vernielli, Rhodocyclus, Chlorobium limocola
    Nonphotosynthetic, autotrophic
        Methanobacterium, Methylococcus, Methylosinus, Methanococcus, Methanococcus, Methanosarcina

free-living microorganisms capable of nitrogen fixation. A number of examples of nitrogen fixing systems are listed in Table 9-1.

Symbiotic nitrogen fixing organisms and the photosynthetic nitrogen fixers appear to account for most of the atmospheric nitrogen assimilated into organic forms in nature. Under conditions in which fixed nitrogen is low, root-nodulated angiosperms and gymnosperms and the cyanobacteria may be especially valuable in improving soil fertility. Indeed, the estimated amount of nitrogen fixed by symbiotic bacterial-inoculated legumes in the United States approximately equals the amount of nitrogen supplied by farmers as nitrogen fertilizer. In the past, the contributions of free-living forms was probably underestimated. With the introduction of the acetylene reduction technique* for assessing the nitrogen fixing potential of organisms in natural environments, it has been possible to show the presence of nitrogen fixing organisms in a number of settings, including the intestinal tract of nonruminant mammals, and to add significantly to the list of nitrogen fixing organisms. Although free-living organisms, in general, appear less efficient in their ability to fix nitrogen, their number, variety, and ubiquitous distribution suggest that they are of major ecological importance. The cyanobacteria have a distinct advantage in that the energy (as ATP) and the reducing power required for nitrogen fixation can be supplied by photosynthesis, a process that makes it possible for them to become established in environments unfavorable for the development of other nitrogen fixing organisms.

Rapidly rising energy and labor costs have made it less economical to increase plant growth by the use of ammonia fertilizer. Commercial production of ammonia by the Haber–Bosch process (catalyst-mediated reduction of hydrogen with nitrogen under high pressure and temperature) is expensive. Furthermore, continued application of chemical fertilizers at a high rate has threatened our water supply and the ecological balance of rivers and streams. Thus, attention has been focused on the improvement of plant yields through the development of new associations between nitrogen fixing bacteria and plants. In the pursuit of this goal, intensive effort has been placed on the study of the genetics, biochemistry, and ecology of both free-living and symbiotic nitrogen fixing organisms. It has already become apparent that the photosynthetic capacity of plants is one limiting factor in nitrogen fixation. Work on the selection and breeding of plant strains with greater photosynthetic efficiency has also been intensified. Another long-range goal has been to attempt development of a symbiotic association between nitrogen fixing bacteria and the root system of such highly efficient photosynthesizers, such as corn.

THE NITROGEN FIXATION PROCESS

Fixation of atmospheric dinitrogen (N_2 or $N{\equiv}N$) is accomplished by a variety of bacteria and cyanobacteria utilizing a multicomponent **nitrogenase system**. Despite the variety of organisms capable of fixing nitrogen, the nitrogenase complex has been found to be remarkably similar in most of them (Fig. 9-2). Nitrogenase consists of two oxygen-sensitive proteins. **Component I (dinitrogenase)** is a molybdenum–iron protein containing two subunits. **Component II** or **dinitrogenase reductase** is an iron–sulfur protein

---

*Originally, mass spectrometry was used to measure the amount of ^{15}N-labeled nitrogen gas reduced to ammonia by nitrogenase. The discovery that nitrogenase can reduce acetylene to ethylene proved very useful since this reaction can be measured by the somewhat simpler technique of gas chromatography.

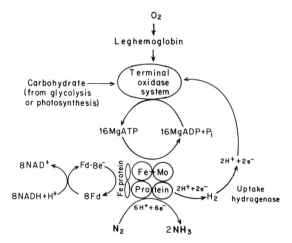

Fig. 9-2. The nitrogenase complex and the associated activities required for nitrogen fixation.
The FeMo protein, dinitrogenase, is also referred to as Component I. The Fe protein, dinitrogenase
reductase (Component II), contains a 4Fe–4S cluster that is not shown in the diagram (Fd =
ferredoxin). The overall reaction requires 8NADH + H$^+$. Six of these are used to reduce N$_2$ to
2NH$_3$ and two are used to form H$_2$. The uptake hydrogenase returns H$_2$ to the system, thus con-
serving energy.

that transfers electrons to dinitrogenase. These proteins, together with ATP, Mg^{2+}, and a
source of electrons are essential for nitrogen fixing activity.

The overall process of nitrogen fixation is accomplished at considerable expense of
energy, requiring between 12 and 16 molecules of ATP and 6 to 8 electrons, depending
on the manner in which the equation is viewed. In one form, the equation may be written:

$$N_2 + 6H^+ + 6e^- + 12ATP + 12H_2O \longrightarrow 2NH_3 + 12ADP + 12P_i$$

This form of the equation does not take into account the fact that dihydrogen (H$_2$) is an
obligate product of the nitrogenase reaction. If H$_2$ is considered, then the equation can
be shown as

$$N_2 + 8H^+ + 8e^- + 16ATP \longrightarrow 2NH_3 + H_2 + 16ADP + 16P_i$$

Thus, the theoretical stoichiometric relationship between nitrogen fixation and dihydrogen
production is related by the equation:

$$N_2 + 8H^+ + 8e^- \longrightarrow 2NH_3 + H_2$$

The production of hydrogen during nitrogen fixation is an energy-expensive process. In
actuality, most aerobic nitrogen fixers rarely evolve H$_2$ because a membrane-bound, uptake hydrogenase system recycles the H$_2$ produced by nitrogenase and produces ATP
through respiration to help support the ATP needs of the system. An anaerobic environment is essential for nitrogenase activity as a result of the oxygen lability of both proteins
in the complex, and hydrogenase coupled to a dioxygen-consuming pathway helps maintain an anaerobic environment. In addition, oxygen represses the formation of the hydrogen uptake system.

Components of the Nitrogenase System

In *K. pneumoniae* the nitrogenase complex contains two separatable proteins. Component I (dinitrogenase) is an $\alpha_2\beta_2$ tetramer of 240 kDa encoded by *nifK* and *nifD* genes. The iron–molybdenum cofactor (FeMo-co) of the dinitrogenase is synthesized under the direction of six *nif* genes (*nifQ, −B, −V, −N, −E,* and *−H*). Mutants of *nifB* and *nifNE* accumulate an inactive apo-Component I (Apo I). Apo I is an oligomer that contains an additional protein, the product of the *nifY* gene, which dissociates from the complex upon activation by the addition of FeMo-co with restoration of the ability to fix N_2. Component II (dinitrogenase reductase, encoded by *nifH*) is an α_2 protein ($\approx$ 60 kDa) containing a single four-iron four-sulfur (Fe_4S_4) center. This protein binds and hydrolyzes MgATP when an electron is transferred from reduced ferredoxin to dinitrogenase. Either ferredoxin and/or flavodoxin can serve as electron donors.

Regulation of the 17 *nif* genes in *K. pneumoniae* is under the direction of the *nifLA* operon. The NtrC protein activates the transcription of the *nifLA* operon under the conditions of nitrogen limitation. The NifA protein is a positive regulatory factor required for *nif gene* transcription (Fig. 9-3). The *nifL* gene product interacts with the NifA protein to prevent NifA activation in the presence of fixed nitrogen (ammonia or amino acids) or oxygen. The functions of *nifZ, −W, −U, −S, −X,* and *−T* are still less well defined. However, there is evidence that the products of the *nifW* and *nifZ* genes may be involved in processing one of the structural components of nitrogenase and that the product of the *nifX* gene is a positive regulator of the *nif* regulon in response to ammonia and oxygen.

The nitrogen fixing system of *Azotobacter* species has also been well characterized. *Azotobacter vinelandii* and *A. chroococcum* produce three different nitrogenases regulated by the Mo (molybdenum) or V (vanadium) content of the culture medium. Nitrogenase 1 is produced by both organisms in the presence of Mo. Dinitrogenase reductase is composed of two identical subunits encoded by *nifH*. Dinitrogenase 1 is a tetramer of two pairs of nonidentical subunits encoded by *nifD* and *nifK*. The three structural genes appear in the order *nifHDK* in an operon. Nitrogenase 2 is produced by both *A. vinelandii* and *A. chroococcum* grown in a nitrogen-free medium in the presence of vanadium. Dinitrogenase reductase 2 is a dimer encoded by *vnfH*. Dinitrogenase 2, encoded by *vnfD* and *vnfK*, is a tetramer composed of two pairs of subunits containing Fe and V. A third nitrogenase, nitrogenase 3, is encoded by alternate nitrogen fixation (*anfHDK*) genes, which are expressed in *A. vinelandii* only in the absence of Mo and V. Dinitrogenase reductase 3 contains two identical subunits and dinitrogenase 3 is present in two active configurations: $\alpha_2\beta_2$ and $\alpha_1\beta_2$. The regulatory genes *nifA*, *vnfA*, and *anfA* are required for the expression of nitrogenases 1, 2, and 3, respectively. An additional regulatory gene, *nfrX*, is required for growth on N_2 (diazotrophic growth) in the presence or absence of Mo.

Purification of the nitrogenase system from *Clostridium pasteurianum* reveals that the enzyme complex from this organism is similar to that found in *K. pneumoniae* and *A. vinelandii*. Two proteins are required for nitrogen fixation and ATP-dependent H_2 evolution. One protein, Component I, is an FeMo protein having dinitrogenase activity. A second protein, the Fe protein, or component II, has dinitrogenase reductase activity. Although there are structural similarities between the nitrogenase complexes of various organisms, the primary structure of those from *Clostridium pasteuurianum* is significantly less related to that of *K. pneumoniae* and *A. vinelandii*. The nitrogenase components from a wide variety of bacteria can interact with one another; complementary functioning of

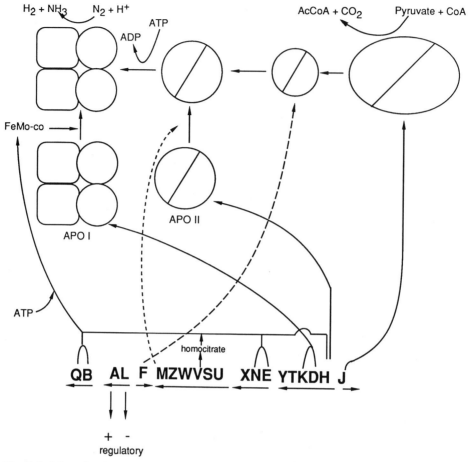

Fig. 9-3. Map of *nif* genes from *K. pneumoniae*. Genes *nifQ,-B,-F,-M,-V-,S,-N,-E,-K,-D,-H*, and *-J* encode proteins essential for effective nitrogen fixation. Genes *nifA* and *nifL* have been shown to be positive (+) and negative (−) regulatory genes, respectively. Of the remaining genes *nifL,-Z,-W,-U,-X,-Y,* and *T, nifX* has been shown to be a positive regulator of the *nif* regulon and *nifW* and *nifZ* have been implicated in the processing of one of the structural components of nitrogenase. The precise function of the rest has not been assessed. From Triplett, E. W., G. P. Roberts, P. W. Ludden, and J. Handelsman. 1989. What's new in nitrogen fixation. *ASM News* **55:**15–21.

the FeMo protein and the Fe protein from different organisms in heterologous complexes shows a remarkably high degree of successful formation of active hybrid nitrogenases. However, the components of *C. pasteurianum* nitrogenase are much less effective in forming heterologous complexes than are mixtures of the components from gram-negative organisms.

SYMBIOTIC NITROGEN FIXATION

The study of nitrogen fixation in symbiotic organisms such as the *Rhizobiaceae* has been difficult because of the complex nature of the interaction between the nitrogen fixing bacteria and the tissues of the host plants. However, *Rhizobium* strain ORS571 is unusual in that it conducts nitrogen fixation during active growth in culture as well as during plant nodule symbiosis. This fortuitous circumstance provided an opportunity to analyze the nitrogen fixation system in a symbiotic organism in the absence of the plant and to compare these activities with those observed under conditions of symbiotic nitrogen fixation. Nitrogenase from *Rhizobium* strain ORS571 contains two protein components that resemble those obtained from free-living nitrogen fixing bacteria, such as *Klebsiella*. In fact, a considerable degree of sequence homology occurs between the nitrogenase components of *K. pneumoniae* and *Rhizobium*. Genes homologous to some of the *K. pneumoniae nif* genes have also been identified in *Rhizobium meliloti* and other symbiotic rhizobia. These include the structural genes coding for nitrogenase components (*nifHDK*), the genes coding for the synthesis of the FeMo cofactor (*nifB*, *nifE*, and *nifN*), and the regulatory gene *nifA*. The *fixABCX* genes were the first genes required for nitrogen fixation in *R. meliloti* that were unrelated to *fix* genes in *K. pneumoniae*. However, these genes have been shown to be required for nitrogen fixation under nonsymbiotic (free-living) conditions in *Bradyrhizobium japonicum* and *Azorhizobium caulinodans* and in a free-living diazotrophic *Azotobacter* species. It has been suggested that the *fixABCX* genes may encode electron-transfer components that may donate electrons to nitrogenase and may be considered analogous to *K. pneumoniae nifF* and *nifLJ* genes, which encode electron carriers.

A *fix* operon containing at least three genes, *fixGHI*, has been identified in *R. meliloti*. A fourth gene, *fixS*, may also be a part of this operon. Evidence has been presented to suggest that FixG is a redox protein and that FixI is a cation pump. The proteins FixG, FixH, FixI, and FixS appear to participate in a membrane-bound complex coupling the FixI cation pump with a redox process catalyzed by FixG.

The developmental steps leading to the initiation of nodule formation include: attachment to root hairs, root hair curling, and the formation of a ''shepherd's crook,'' development of infection threads within root hairs, growth of the threads toward the inner cortex of the root, formation of a nodule meristem in the inner-root cortex. These cytological changes in the root hair provide a means of passage of the organism into the internal root system where it infects a root cell. This infection causes the cell to swell and divide forming a thick mass of cells called the root nodule. Figure 9-4 assigns control of the individual steps in root nodule development to the common and specific *nod* loci of *R. meliloti*. In the root nodule a differentiated form of the bacterium, a bacteriod, is capable of nitrogen fixation. In both symbionts, a portion of the genome is expressed only in the symbiotic state. Considering the complex interaction of the bacterium and the host plant in the development of effective symbiosis, it is not surprising that attempts to develop more efficient strains is hampered by the fact that indigenous strains in the soil are more competitive than laboratory-developed strains for use as inoculants. Nevertheless, continued efforts to engineer more effective strains of *Rhizobium* and other nitrogen fixing bacteria is important because of the continuing increase in the cost of nitrogen fertilizer produced by industrial processes.

Nodule-specific plant genes code for the production of flavonoids that induce the expression of nodulation (*nod*) genes in the bacteria. The early events in root nodule formation by rhizobia in leguminous plants is under the control of genes located on a large plasmid termed **symbiosis or Sym plasmid**. Flavonoids, in concert with the *nodD* gene product, induce the expression of additional *nod* genes. The NodD protein serves as a positive transcription activator. It binds to the *nod* box, located upstream of all

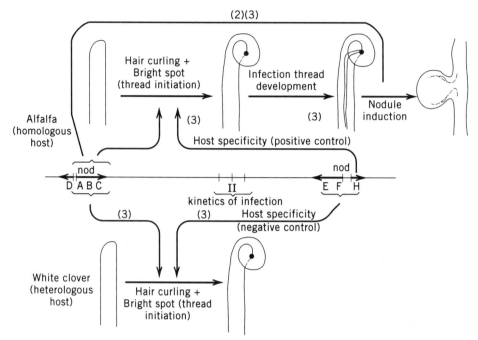

Fig. 9-4. Alfalfa root infection and nodulation. Assignment of development step control to common and specific *nod* loci of *R. meliloti*. Thick arrows represent development steps. Thin arrows represent genetic control. From Debellé, F. et al., *J. Bacteriol.* **168**:1075–1086.

inducible *nod* operons. The *nod* box, a highly conserved DNA sequence, appears to function as a cis-acting regulatory element of *nod* gene expression. The *nodABC* genes common to all rhizobia are required for the synthesis of a lipooligosaccharide signal that triggers root nodule formation. Addition of various substituents to the core compound imparts host specificity to the lipooligosaccharide. The addition of these components occurs under the influence of the *nodH* and *nodQ* genes and the *nodFEL* operon (see Fig. 9-5). The NodI and NodJ proteins are involved in efficiency of nodulation and play a role in the normal development of infection threads. Once bacteroid development is complete, the inducible *nod* genes are no longer transcribed. This transcription switch-off prior to the release of the bacteria from the infection thread is a general phenomenon observed among the rhizobia and is apparently the result of a negative regulatory mechanism that has not yet been fully elucidated.

Under the conditions of symbiosis, the host plant provides the bacteroid with reduced carbon in the form of C_4-dicarboxylic acids, such as succinate, malate, and fumarate. These compounds serve as energy sources for the fixation of nitrogen to ammonia. The ammonia is then released to the plant. The dicarboxylic acids are present in high concentration in the nodule and are actively taken up by the bacteroid and then rapidly oxidized. It is the bacterial symbiont of the nodulated system that is responsible for ATP synthesis used by nitrogenase. Therefore, the bacteroids of the various species of *Rhizobium* and *Bradyrhizobium* display a high activity of TCA cycle enzymes and the dicarboxylic acids are the most effective substrates for respiration and subsequent ATP-utilizing nitrogen fixation. The dicarboxylic acid transport genes (*dct*) are located on a

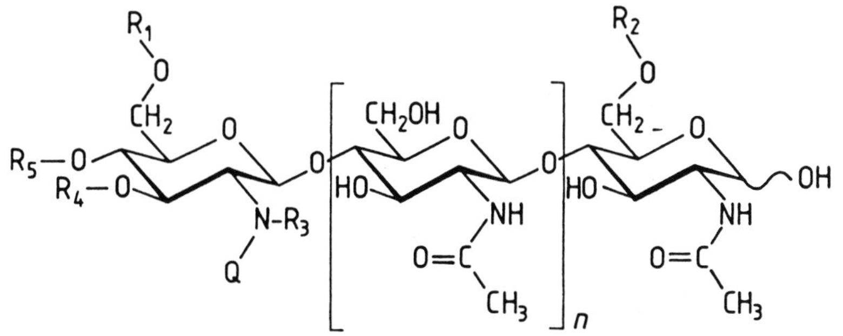

	R. leguminosarum bv. viciae	R. meliloti	R. sp. NGR234
n	2 or 3	1, 2, or 3	3
Q	C18:1 or C18:4	C16:2	C18:1 or C16:0
R_1	CH_3CO or H	CH_3CO or H	H
R_2	H	SO_3H	2-O-methylfucose or substituted with either 3-O-CH_3CO or 4-O-SO_3H
R_3	H	H	CH_3
R_4	H	H	NH_2CO or H
R_5	H	H	NH_2CO or H

Fig. 9-5. Structure of the oligosaccharide Nod factor(s) produced by *Rhizobium* species. The chitin oligomer and the acyl moiety (Q) are present in all Nod factors. The number (n) of *N*-acetyl glucosamine residues can vary. The parameter Q can vary in length and in the number of unsaturated bonds. Several different substitutions to the sugar backbone occur (R_1 to R_5). From Vijn, I., et al. 1993. *Science* **260**:1764–1765.

megaplasmid. Mutants defective in the transport of dicarboxylic acids or aspartate form ineffective nodules. The products of *dctB* and *dctD* regulate the expression of *dctA*, which encodes a transport protein essential for nitrogen fixation by the bacteroid. Here DctB is a sensor protein that activates DctD by phosphorylation. The DctD protein activates transcription at the σ^{54}-dependent *dctA* promoter. In addition to activating *dctA* transcription, DctD can also repress expression of *dctA*. In uninfected cells, inactive DctD binds to the *dctA* promoter and prevents its activation by NtrC.

Nitrogenase synthesis in a variety of organisms is generally subject to the close regulatory controls at the level of *nif* gene transcription. However, many of these organisms also display a posttranslational regulation of nitrogenase activity effected by small extracellular concentrations of ammonia. In some organisms this rapid and reversible inhibition of nitrogenase, termed **ammonia switch-off**, is the result of a covalent modification of the Fe protein (dinitrogenase reductase) in response to the addition of ammonia. Formation of an inactive form of nitrogenase in the photosynthetic bacterium *R. rubrum* results from the NAD-dependent ADP-ribosylation of arginine residues in the Fe protein component of nitrogenase (Fig. 9-6). Activation of the modified Fe protein by removal of the ADP-ribose moiety is catalyzed by an activating enzyme. Studies with purified systems from *Azospirillum brasiliense*, *A. lipoferum*, and *A. amazonense* indicate that the nitrogenase of *A. brasiliense* and *A. lipoferum* is regulated by covalent modification of the Fe protein effected by the addition of ammonia. In *A. amazonense* a different, non-

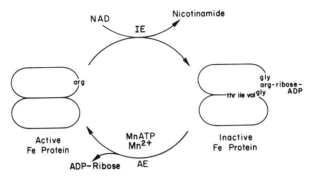

Fig. 9-6. Schematic view of regulation of nitrogenase activity by ADP-ribosylation of the Fe protein in *Rhodospirillum rubrum*. After Lowery, R. G., L. L. Saari, and P. W. Ludden. 1986. *J. Bacteriol. 166*:513–518.

covalent inhibitory mechanism results in only a partial inhibition of nitrogenase activity by ammonia.

The nitrogenase system is repressed when organisms are grown in the presence of excess ammonia. As just described, ammonia has a short-term "switch-off" effect. However, there is little evidence to indicate a direct role for ammonia in *long-term* repression of nitrogenase synthesis. Originally, glutamine synthetase was implicated as a participant in nitrogenase regulation. Mutant strains lacking glutamine synthetase activity showed a lack of ammonia repression of nitrogenase. However, this correlation was also observed with the *nif* genes and *nif* transcription was unaffected by ammonia in strains producing constitutive glutamine synthetase. Further study has shown that expression of *glnA*, the structural gene for glutamine synthetase, is controlled by three genes, originally termed *glnF*, *glnL*, and *glnG* because of their location in the same operon as *glnA*, mediates general nitrogen control rather than having a specific effect on glutamine synthetase. As a consequence, these genes have been redesignated as *ntrA*, *ntrB*, and *ntrC*. The *ntrB* and *ntrC* genes are linked to *glnA* in *K. pneumoniae* and other enteric bacteria (see Chapter 4). The *ntrC* gene product is required for activation of transcription of nitrogen-regulated operons and mutations at this locus fail to synthesize nitrogenase. Thus, there is no evidence that the glutamine synthetase protein has a direct role in positive control of nitrogen-regulated operons. The current model for regulation of the *nif* genes in *K. pneumoniae* is shown in Figure 9-7. The third nitrogen regulatory gene, *ntrA*, has a vital role in nitrogen control since mutations at this locus are unable to activate *nif* genes. It is now apparent that positive control of *nif* transcription is regulated by a complex regulatory cascade mechanism described in detail in Chapter 4.

INORGANIC NITROGEN METABOLISM

The assimilation of inorganic nitrogen ends with the incorporation of ammonia into organic compounds. Since ammonia is the only form of inorganic nitrogen that can be directly assimilated into amino acids, the ability of an organism to utilize other forms of inorganic nitrogen depends on the presence of enzymes or enzyme systems that are able

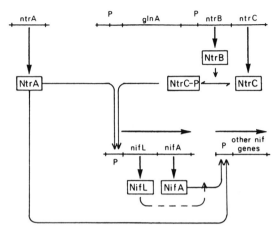

Fig. 9-7. Model for *nif* regulation in *K. pneumoniae.* The thin arrows indicate regulatory functions and the thick horizontal arrows represent transcripts. From Dixon, R. A. 1984. *J. Gen. Microbiol.* **130:**2745–2755.

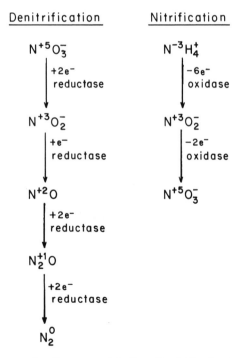

Fig. 9-8. Pathways of inorganic nitrogen metabolism. Denitrification and nitrification. The superscript numbers on the nitrogen indicate the valence state or oxidation level of the nitrogen in the compounds.

to convert these compounds to ammonia. This process is referred to as **denitrification**. The reverse process, converting ammonia to nitrate and nitrite, is termed **nitrification** (Fig. 9-8). Table 9-2 provides examples of organisms known to carry out the reactions indicated.

Many of the organisms that carry out one or more of these reactions do not actually assimilate the nitrogen. In some organisms, nitrate may be used as a terminal electron acceptor in place of oxygen. The end-product is nitrite or N_2. Nitrate respiration yields biologically useful energy under anaerobic conditions. Nitrate assimilation occurs by sequential reduction to nitrite, hydroxylamine, and finally, to ammonia. The ammonia formed is then assimilated by pathways to be discussed in the following section. Organisms that conduct both nitrate respiration and nitrate assimilation contain two nitrate reductases. In *E. coli* and *Neurospora*, the assimilatory enzyme is a soluble cytoplasmic enzyme, whereas the respiratory enzyme is membrane-bound and sensitive to oxygen. Both enzymes contain flavin as the prosthetic group. In most organisms the assimilatory enzyme is repressed by ammonia and induced in the presence of either nitrate or nitrite. Algae readily utilize nitrate as a source of inorganic nitrogen, but the route of assimilation is less well characterized than in bacteria and fungi. Ammonia is considered to be the final product of nitrate reduction by algae.

Denitrification by members of the obligately chemolithotrophic *Nitrobacteriaceae* is considered to be the major source of assimilable nitrogen (ammonia) in soils. The ammonia oxidizer, *Nitrosomonas europaea*, and the nitrite oxidizer, *Nitrobacter winogradskyi*, have been among the most actively studied members of this group. *Nitrosomonas europaea* obtains all of its energy for growth from the oxidation of ammonia to nitrite. The oxidation of ammonia to hydroxylamine is an O_2-dependent reaction catalyzed by ammonia monooxygenase:

$$NH_3 + O_2 + 2e^- + 2H^+ \longrightarrow NH_2OH + H_2O$$

Hydroxylamine is further oxidized to nitrite by hydroxylamine oxidoreductase:

$$NH_2OH + H_2O \longrightarrow NO_2^- + 5H^+ + 4e^-$$

Two of the four electrons generated from hydroxylamine oxidation are used to support the oxidation of additional ammonia molecules, while the other two electrons enter the electron-transfer chain and are used to support ATP synthesis via oxidative phosphorylation and CO_2 reduction. The oxidative chain from NO_2^- to O_2 in *N. winogradskyi* consists of membrane-bound cytochrome *c* oxidoreductase, soluble cytochrome *c*-550, and membrane-bound cytochrome *c* oxidase. *Nitrobacter* is obligately dependent on nitrite as the sole source of nitrogen and energy and requires carbon dioxide for growth. *Nitrobacter* is, therefore, dependent on *Nitrosomonas* or other organisms in the natural environment for its nitrogen supply.

Nitrite formation by heterotrophic soil organisms appears to be rather limited. Nitrate formation, on the other hand, is accomplished by a significant number and variety of heterotrophic species present in the soil microflora. *Aspergillus flavus* is particularly active and must be considered as a major source of nitrate in soils.

Denitrifying bacteria, such as *Pseudomonas stutzeri*, use the reduction of nitrous oxide

TABLE 9-2. Known Biological Reactions of Inorganic Nitrogen[a]

Reaction	Electrons Required	Representative Organisms Performing Reaction	Requirements (comments)
$NO_3^- \rightarrow NO_2^-$	2	*Escherichia coli, Micrococcus denitrificans, Bacillus subtilis, Haemophilus influenzae, Pseudomonas aeruginosa, Neurospora crassa, Achromobacter fischeri*	Varies among organisms—NAD, FAD (FMN), Mo, cytochromes *c*, *b*, NADP
$NH_2^- \rightarrow NH_4^+$ Nitrite assimilation	6	*Bacillus pyocyaneus, Desulfovibrio desulfuricans, Bacillus pumilus, Neurospora crassa, Clostridium pasteurianum*	NADP (NAPH), FAD, Cu, Fe
NO_3^- or $NO_2^- \rightarrow NO$, N_2O, and N_2; Denitrification	Varies from 1 to 10	*Micrococcus denitrificans, Denitrobacillus, Pseudomonas* spp., *Spirillum itersonii, Bacillus licheniformis, Achromobacter* spp., *Thiobacillus denitrificans* and *thioparus*	See requirements for individual reactions of this process
$NO_2^- \rightarrow NO$ Nitrite reduction	1	*Pseudomonas stutzeri, Bacillus subtilis, Pseudomonas aeruginosa, Escherichia coli*	NAD (NADP), FMN (FAD), Fe^{2+}
$NO \rightarrow N_2$ or N_2O Nitric oxide reductase	2 or 4	*Pseudomonas stutzeri, Bacillus subtilis, Clostridium pasteurianum, Pseudomonas aeruginosa*	NAD (NADP), FAD (FMN), (Fe^{2+})
$N_2O \rightarrow N_2$ Nitrous oxide reductase	2	*Pseudomonas stutzeri*	Cu
$H_2N_2O_2 \rightarrow 2NH_4^+$ Hyponitrite assimilation	8	*Neurospora crassa, Escherichia coli (Bn)*	NADPH, Fe^{2+}
$N_2 \rightarrow 2NH_4^+$ Nitrogen fixation	6	See Table 9-1 for representative organisms	Molybdenum, Iron, Vanadium
$NH_2OH \rightarrow NH_4^+$ Hydroxylamine reductase	2	*Neurospora crassa, Azotobacter vinelandii O, Desulfovibrio desulfuricans, Bacillus pumilis, halotolerant bacterium, Clostridium pasteurianum, Pseudomonas aeruginosa*	Varies—NADH, FAD, Mn^{2+}

TABLE 9-2. Continued

Reaction	Electrons Required	Representative Organisms Performing Reaction	Requirements (comments)
$NH_2NH_2 \xrightarrow{?} 2NH_4^+$ Hydrazine reductase	2	*Micrococcus lactilyticus*	Measured hydrogen uptake by whole cells with hydrazine added
$NH_4^+ \rightarrow$ organic compounds	In some cases 2	Large group	Varies
$NH_4^+ \rightarrow NO_2^-$ or $NH_2OH \rightarrow .NO_2$ Nitrification	−6 or −4	*Nitrosomonas* spp.	Acceptor— cytochrome *c* or phenazine methosulfate
$NO_2^- \rightarrow NO_3^-$ Nitrite oxidase	−2	*Nitrobacter* spp.	Cytochrome, Fe^{3+}

[a]Revised from Mortenson, L. E., 1962. Inorganic nitrogen assimilation and ammonia incorporation. In *The Bacteria*, Vol. 3, I. C. Gunsalus and R. Stanier (Eds.). Academic, New York, pp. 115–166.

(N_2O) to dinitrogen (N_2) for the generation of energy. Reduction of N_2O is usually the terminal step of bacterial denitrification proceeding from nitrate to nitrite to nitric oxide to nitrous oxide and the final reduction of nitrous oxide to N_2 by N_2O reductase. A gene cluster containing the genes necessary for the reduction of nitrite (*nir*), nitric oxide (*nor*), and nitrous oxide (*nos*) has been identified in *P. stutzeri*. There are at least 15 genes in this cluster arranged in the order *nos–nir–nor*. The structural gene *nosZ* codes for the copper containing enzyme N_2O reductase. Three other genes, *nosDFY*, are involved in the formation of the copper chromophore of the enzyme. A regulatory gene, *nosR*, has been identified within the *nos* region of the denitrification gene cluster of *P. stutzeri*. It is essential for the expression of the *nosZ* gene and there is indirect evidence that the transcriptional regulator Fnr may also be involved in *nosZ* expression (see Chapter 4 for a discussion of regulatory mechanisms).

ASSIMILATION OF INORGANIC NITROGEN

Many microorganisms synthesize all of their amino acids and other nitrogenous compounds using ammonia and carbon chains derived from carbohydrate metabolism. The available pathways for ammonia assimilation are quite limited, however. Of all the amino acids found in proteins and other cellular constituents, ammonia can be directly assimilated into only a few. These amino acids then serve as donors of their amino nitrogen via transamination to keto acid precursors to form all of the other amino acids. The potential routes of ammonia assimilation are through the synthesis of glutamate, alanine, or aspartate. The major enzymes involved in ammonia assimilation are the glutamate dehydrogenases (GDH) and two enzymes that operate in tandem: glutamine synthetase (GS) and glutamate synthase (the abbreviation GOGAT used for this enzyme is derived from its previous trivial name, glutamine amide-2-oxoglutarate amino transferase) as shown below:

Glutamate dehydrogenases (GDH)

$$\alpha\text{-ketoglutarate} + NH_4^+ + NADH + H^+ \longleftrightarrow \text{L-glutamate} + NAD^+$$

$$\alpha\text{-ketoglutarate} + NH_4^+ + NADPH + H^+ \longleftrightarrow \text{L-glutamate} + NADP^+$$

Glutamine synthetase (GS)—glutamate synthase (GOGAT)

$$\text{L-glutamate} + NH_4^+ + ATP \xrightarrow{\text{GS}} \text{L-glutamine} + ADP + P_i$$

$$\alpha\text{-ketoglutarate} + \text{L-glutamine} + NADPH + H^+ \xrightarrow{\text{GOGAT}} 2\ \text{L-glutamate} + NADP^+$$

Net: $\alpha\text{-ketoglutarate} + NH_4^+ + ATP + NADPH + H^+ \longrightarrow$
$$\text{L-glutamate} + ADP + P_i + NADP^+$$

Enzymes that may play a role in ammonia assimilation in some organisms are the alanine dehydrogenases and aspartase:

Alanine dehydrogenases

$$\text{pyruvate} + NH_4^+ + NADH + H^+ \longleftrightarrow \text{L-alanine} + NADH^+$$

$$\text{pyruvate} + NH_4^+ + NADPH + H^+ \longleftrightarrow \text{L-alanine} + NADP^+$$

Aspartase

$$\text{fumarate} + NH_4^+ \longleftrightarrow \text{L-aspartate}$$

The route(s) of ammonia assimilation vary from one organism to another depending on the ammonia assimilation enzymes present. In the majority of organisms that have been studied, glutamate is the most widely used route for ammonia assimilation. In organisms capable of synthesizing both the NADP-linked glutamate dehydrogenase and the GS–GOGAT system, the NADP–GDH is functional at high concentrations of ammonia while the GS–GOGAT pathway is most active at concentrations of ammonia below 1mM. These pathways are highly regulated as the concentration of ammonia, glutamate, and glutamine are key sensors that relate the nitrogen nutritional status of the organism.

A few organisms appear to be incapable of forming the GS–GOGAT system, and, therefore utilize one or more of the alternative routes for ammonia assimilation. *Rhodospirillum purpureus*, which does not fix N_2, can use only exogenously supplied ammonia via the NADP-linked GDH. The only ammonia assimilation pathway in *Streptococcus sanguis, S. bovis, S. mutans,* and *S. salivarius* is NADP–GDH regardless of the external ammonia concentration. In some *Bacillus* species, the NADP–ADH and NADP–GDH enzymes are highest in activity when ammonia is in high concentration.

Aspartase is sufficiently active in *Klebsiella aerogenes* as to be an important assimilatory enzyme, particularly after growth on C_4 dicarboxylic acids. Under most conditions of growth, however, aspartase appears to serve in a dissimilatory (ammonia-releasing) role. The yeast, *Saccharomyces cerevisiae*, assimilates at least two-thirds of its amino nitrogen requirements via glutamate and, when the ammonia concentration in the culture medium is high, utilizes the NADP–GDH for glutamate synthesis. At low concentrations of ammonia, the GS–GOGAT system is induced. In nitrogen fixing species of *Bacillus* (*B. polymyxa, B. macerans*) NADP–GDH activity is several-fold higher than that of GS and is the predominant pathway for ammonia assimilation.

As described in Chapter 4, the GS–GOGAT system and related enzymes are highly regulated by the Ntr (nitrogen regulation) system, particularly in *E. coli* and *K. aerogenes*. On the other hand, in *Bacillus* spp. there is no evidence for a global regulatory system analogous to the Ntr system. Activity of the GS–GOGAT system in *B. subtilis* and *B. licheniformis* is regulated by the available nitrogen source (feedback inhibition). Transcription of the *B. subtilis* GS structural gene is negatively regulated by the GS regulatory protein GlnR, while the expression of GS is stimulated by GltC. In *B. licheniformis*, the GOGAT enzyme (GltS) consists of two unequal subunits. The larger subunit catalyzes a glutaminase reaction:

$$\text{glutamine} + H_2O \longrightarrow \text{glutamate} + NH_3$$

An ammonia-transfer reaction is catalyzed by the small subunit:

$$NH_3 + \alpha\text{-ketoglutarate} + NADPH + H^+ \longrightarrow \text{glutamate} + NADP^+$$

In *B. subtilis*, GS (GlnA) and a regulatory protein GlnR are encoded in an operon. Regulation of the *glnRA* operon involves the action of both GlnR and GS. Here GlnR is a repressor that interferes with transcription under conditions of nitrogen excess.

In the nitrogen fixing anaerobe *Clostridium kluyveri*, the NADP–GDH pathway plays an important role in ammonia assimilation in ammonia-grown cells but plays only a minor role to that of the GS–GOGAT pathway in nitrogen fixing cells, conditions in which the intracellular ammonia concentration is low. In *C. butyricum* the GS–GOGAT system is the predominant pathway for ammonia assimilation with either ammonia of N_2 as the source of nitrogen. In *C. acetobutylicum* there is no evidence for a global Ntr system and the GS enzyme is not regulated by adenylylation. Instead, the DNA region of *glnA* is characterized by a downstream promoter, P_3, which controls the transcription of an antisense RNA.

In phototrophic bacteria, such as the nonsulfur purple bacterium, *Rhodobacter capsulatus*, the NADP–ADH aminating activity can function as an alternative route for ammonia assimilation when GS is inactivated. The ADH is induced in cells grown on pyruvate plus nitrate, pyruvate plus ammonia, or L-alanine under both light-anaerobic and dark-heterotrophic conditions. Aminating activity is strictly NADPH dependent, whereas deaminating activity is strictly NAD dependent.

GENERAL REACTIONS OF AMINO ACIDS

Amino Acid Decarboxylases

Microorganisms exhibit decarboxylase activity for many amino acids including aspartate, glutamate, ornithine, lysine, arginine, tyrosine, phenylalanine, cysteic acid, diaminopimelic acid, hydroxyphenyl serine, histidine, tryptophan, 5-hydroxytryptophan, and possibly others. The general reaction for all of these enzymes is

$$R-CHNH_2COOH \longrightarrow R-CH_2NH_2 + CO_2$$

As far as is known, pyridoxal phosphate is the coenzyme of all the amino acid decarboxylases with the exception of histidine decarboxylase for which the cofactor is pyruvate. All of the amino acid decarboxylases are essentially irreversible and, therfore, are not of importance in the biosynthesis of most amino acids. However, in the case of diaminopimelic acid (DAP) decarboxylase, lysine is the final product. Amino acid decarboxylases are produced at low pH, and their range of optimal activity is pH 3–5. Thus, excess acidity resulting from the production of acid end products may be regulated by amino acid decarboxylase activity, particularly in anaerobic, proteolytic organisms, such as the clostridia.

The diamine putrescine, the decarboxylation product of ornithine, is an essential growth factor for several organisms and is a biosynthetic precursor of both spermidine and spermine, polyamines found in a wide variety of microorganisms (see Chapter 10). Spermine is present in eukaryotic organisms but is found in only a few bacteria (e.g., *Pseudomonas aeruginosa* and *Bacillus stearothermophilus*). Spermidine is more widely distributed, being found in bacteria and fungi as well as in higher organisms.

Germinating conidia of *N. crassa* produce an active glutamate decarboxylase. This is the first step in a pathway that leads to a rapid increase in aspartate in the amino acid pool.

Amino Acid Deaminases

Deamination of amino acids occurs via several quite different reactions;

1. Oxidative deamination.
   a. NAD^+ or $NADP^+$ linked deamination.
   b. FAD- or FMN-linked deamination.
2. Nonoxidative deamination.

The oxidative deamination of glutamic acid by the reversible glutamate dehydrogenases has been discussed earlier in relation to ammonia assimilation. In most organisms the NAD-specific enzyme operates primarily in the direction of catabolism, whereas the NADP-linked enzyme functions in glutamate synthesis.

Alanine dehydrogenase occurs in members of the genus *Bacillus*, the phototrophic non-purple sulfur bacteria, *Rhodobacter*, and others. The reaction proceeds via an α-imino intermediate:

$$
\begin{array}{ccc}
\text{COOH} & & \text{COOH} \\
| & & | \\
\text{H}-\text{C}-\text{NH}_2 + \text{NAD+/NADP+} & \longleftrightarrow & \text{C}=\text{O} + \text{NADH/NADPH} + \text{H+} + \text{NH}_4\text{+} \\
| & & | \\
\text{CH}_3 & & \text{CH}_3
\end{array}
$$

The ubiquitous distribution and high activity of ADH in the aerobic bacilli has been considered evidence that the reverse reaction catalyzed by this enzyme may serve as a major route of ammonia assimilation in these organisms.

Amino acid oxidases, sometimes termed aerobic or oxidative deaminases, involve reactions catalyzed by enzymes containing FAD (flavin adenine dinucleotide) or FMN (flavin mononucleotide). The reaction proceeds in two stages:

$$R\text{—}CHNH_2COOH + Enz\text{—}FAD \longrightarrow \alpha\text{-keto acid} + NH_3 + Enz\text{—}FADH_2$$

$$Enz\text{—}FADH_2 + O_2 \xrightarrow{\text{nonenzymatic}} Enz\text{—}FAD + H_2O_2$$

Reduced FAD (or reduced FMN) may react nonenzymatically with molecular oxygen or it may transfer the hydrogen to other hydrogen acceptors, as it does in anaerobic organisms. These amino acid oxidases are nonspecific in that a single enzyme may catalyze the oxidation of a variety of amino acids, for example, methionine, phenylalanine, tyrosine, leucine, isoleucine, valine, norvaline, alanine, tryptophan, and cysteine in decreasing order of activity. Another may oxidize proline, hydroxyproline, citrulline, histidine, and arginine. The rate of oxidation may differ for each amino acid, and the order of activity may differ with the source of the enzyme. Both D- and L-amino acid oxidases are known, but a single enzyme is usually specific for one configuration. These reactions yield no useful energy if the reduced FAD is oxidized nonenzymatically by molecular oxygen. It is highly unlikely that these enzymes play any significant role in the assimilation of ammonia.

Nonoxidative deaminases, such as aspartic acid deaminase (aspartase), serine and threone deaminases (dehydratases), and cysteine desulfhydrase, are specific in their substrate requirements.

Aspartase is present in a number of organisms. It converts aspartate to fumarate and ammonia:

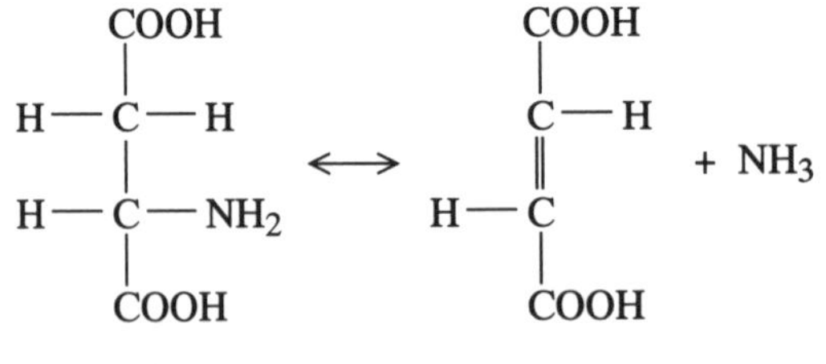

There is no hydrogen exchange in this reaction. Instead, an intramolecular transfer of hydrogen occurs, and pyridoxal-5′-phosphate is not involved as a cofactor in the reaction. The reaction is reversible, providing a potential mechanism for ammonia assimilation, especially if intracellular ammonia needs are high. In gram-negative bacteria, optimal production of the enzyme occurs when the organism is grown in a complex medium containing amino acids and a low concentration of carbohydrate. This finding has been interpreted as an indication that aspartase is involved primarily with catabolic activity. Nevertheless, the equilibrium of the reaction is such that it could serve as an ammonia assimilation pathway under conditions where other enzymes are inactive.

Serine and theronine deaminases (dehydratases) catalzye the following type of reaction:

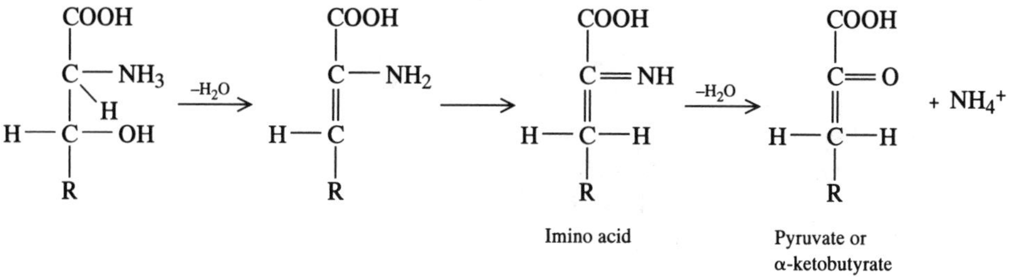

Imino acid   Pyruvate or
α-ketobutyrate

R = H in serine; CH_3 in threonine

An intramolecular transfer of hydrogen atoms occurs and water is removed (dehydration) to produce an imino acid via a β-elimination reaction. In the second stage of the reaction, water reacts nonenzymatically with the imino acid to release ammonia. Pyridoxal-5′-phosphate is the cofactor in this reaction. The two deaminating enzymes have been shown to be distinct. The level of L-serine deaminase varies as a function of nitrogen nutrition, carbon source, and the supply of glycine and leucine. Glycine and leucine induce the formation of serine deaminase. The enzyme seems to play a role in a number of pathways in which serine is generated and further metabolized as part of the main carbon pathway. *Escherichia coli* and *S. typhimurium* produce both a biodegradative threonine deaminase and a biosynthetic enzyme that provides α-ketobutyrate as an intermediate in the biosynthetic pathway to isoleucine. The biodegradative enzyme is induced under anaerobic conditions in amino acid-rich medium, requires cAMP for its synthesis, and is sensitive to catabolite repression by glucose.

Cysteine desulfhydrase has a reaction mechanism similar to that of serine and threonine deaminases except that hydrogen sulfide, rather than water, is removed via a β-elimination reaction:

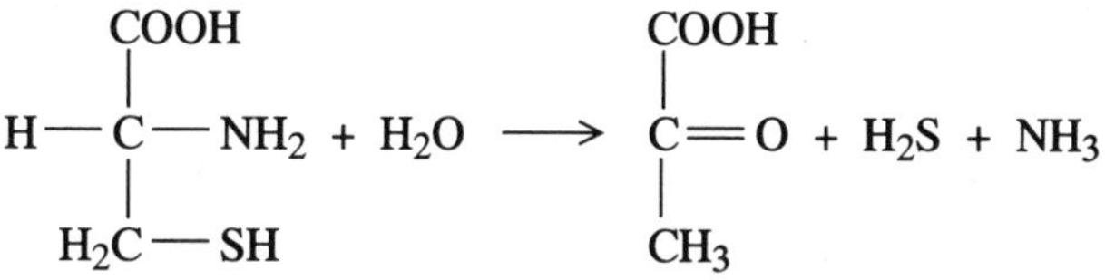

Dehydratases and desulfhydrases are essentially irreversible reactions and do not constitute reactions that could readily participate in ammonia assimilation.

Phenylalanine deaminase (ammonia lyase) occurs in yeasts, molds, and bacteria and catalyzes the nonoxidative deamination of L-phenylalanine to *trans*-cinnamic acid:

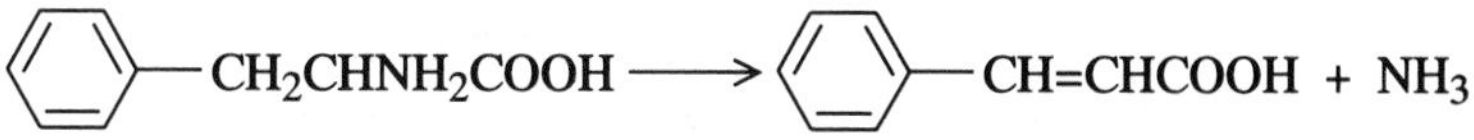

A dehydroalanine group serves as a cofactor in the reaction. The enzyme is of interest because of its potential for use in the treatment and diagnosis of phenylketonuria and has industrial applications in the synthesis of L-phenylalanine from *trans*-cinnamic acid. In *Rhodotorula glutinis* this enzyme serves as the initial step in a metabolic pathway leading to the formation of benzoate and 4-hydroxybenzoate.

Amino Acid Transaminases (Aminotransferases)

In 1945 Lichstein and Cohen first demonstrated transaminase activity in bacteria. Transfer of α-amino nitrogen between glutamate, aspartate, alanine, and their corresponding α-keto acids was shown to be similar to reactions observed in mammalian tissues. The two major transaminases demonstrated were

$$\text{aspartate} + \alpha\text{-ketoglutarate} \longleftrightarrow \text{oxaloacetate} + \text{glutamate}$$

$$\text{alanine} + \alpha\text{-ketoglutarate} \longleftrightarrow \text{pyruvate} + \text{glutamate}$$

Pyridoxal phosphate is the coenzyme for all known transaminase reactions.

Transamination was first thought to be a relatively limited activity confined to these three amino acids and their keto analogs. However, in the early 1950s, Feldman and Gunsalus, and Rudman and Meister showed that a number of amino acids would undergo transamination with α-ketoglutarate to form the corresponding amino acids. Note that they are fully reversible reactions, so that glutamate is a key donor for the synthesis of the other amino acids. This result led to a more generalized view of the transamination in which any amino acid and any keto acid could exchange the amino group:

$$\text{amino acid}_1 + \text{keto acid}_2 \longleftrightarrow \text{keto acid}_1 + \text{amino acid}_2$$

Transaminase activity is quite ubiquitous as evidenced by the fact that keto analogs can replace many of the amino aicds for the growth of amino acid requiring mutants. The reaction is not completely universal, however. Four major transaminases have been iden-tified in *E. coli*, as shown in Table 9-3. By these criteria, strains lacking transaminase C (*avtA* mutants) should have no nutritional requirement, whereas strains lacking trans-aminase B (*ilvE*) should require isoleucine. In practice, most *ilvE* mutants require only isoleucine, but some show additional requirements for valine or leucine as a result of reduced expression of either *avtA* or other genes distal to *ilvE*. These transaminases have not, in general, been well identified in other organisms. However, most of these activities appear to be present in organisms that display a general ability to synthesize the common amino acids.

The asparatate aminotransferase (L-aspartate:2-oxoglutarate aminotransferase) is ubiq-uitous. However, the enzyme has been studied in detail only in *E. coli*, *P. putida*, *S. cerevisiae*, and the archaebacterium, *Solfolobus solfataricus*. A thermophilic *Bacillus* spe-cies produces an aspartate aminotransferase that shows some sequence similarity in the N-terminal region between the eubacterial and archaebacterial enzymes.

Other transaminases are known. For example, *S. typhimurium* displays a glutamine amidotransferase encoded within the *trpD* gene. It serves a dual role in transferring the amino group of glutamine to chorismic acid in the synthesis of anthranilic acid. Gluta-mine amidotransferase activity is also present in a tryptophan gene (*trpE*) in *Bacillus pumilis*.

TABLE 9-3. Major Transaminases in *E. coli*

Transaminase	Interacts with	Gene
Aromatic amino acids	Tyrosine	*tyrB*
	Phenylalanine	
	Glutamate	
	Leucine	
	Aspartate	
	Methionine	
Aspartate (transaminase A)	Aspartate	*aspC*
	Glutamate	
	Tyrosine	
	Phenylalanine	
Transaminase B (branched-chain aminotransferase)	Glutamate	*ilvE*
	Leucine	
	Isoleucine	
	Valine	
	Phenylalanine	
	Methionine	
Transaminase C (alanine–valine aminotransferase)	Alanine	*avtA*
	Valine	
	α-Aminobutyrate	

Amino Acid Racemases

A number of microorganisms contain enzymes that catalyze the conversion of D-amino acids to L-amino acids via the general reaction:

$$\text{H} - \underset{\underset{\text{R}}{|}}{\overset{\overset{\text{COOH}}{|}}{\text{C}}} - \text{NH}_2 \longleftrightarrow \text{H}_2\text{N} - \underset{\underset{\text{R}}{|}}{\overset{\overset{\text{COOH}}{|}}{\text{C}}} - \text{H}$$

Most biochemical compounds are asymmetric and there is a tendency for one form to predominate over another. The amino acids found in naturally occurring proteins are usually in L configuration. However, the cell walls and polypeptide capsules of many organisms contain D-amino acids. Since most biosynthetic reactions lead to the synthesis of the L-amino acids, racemases are necessary for the conversion of certain amino acids to the D configuration for the formation of these specialized cell structures. As an example, D-alanine is a structural component of the cell wall of *Enterococcus faecalis* as well as several other gram-positive organisms. When grown in a medium lacking pyridoxal, D-alanine becomes a specific growth requirement for *E. faecalis* because alanine racemase is inactive under the conditions of pyridoxal deficiency. D-Amino acids are found in the polypeptide capsules of members of the genus *Bacillus* and in peptide-containing antibiotics, providing other indications of the importance of racemases in microbial metabolism.

Role of Pyridoxal-5'-Phosphate in Enzymatic Reactions with Amino Acids

In the foregoing discussion of the various reactions in which amino acids participate, it was mentioned that pyridoxal-5'-phosphate (PLP), the coenzyme form of vitamin B_6, functions in many of these reactions. PLP is the most versatile of all enzyme cofactors in that it can be utilized by various enzymes to catalyze a dozen of so distinct chemical reactions. Such versatility appears to be due, in part, to the PLP cofactor acting as an electron sink. For example, a proton may be removed from the α-carbon atom of the amino acid substrate, with the resultant stabilization of the carbanion at the C_α or C'-4 position; or the electrons may flow into the ring neutralizing a positively charged pyridine nitrogen (i.e., a quinoid structure may be one of the intermediates). It is assumed that appropriate groups on the apoenzyme hold the PLP in the precise alignment necessary for rupture of the α-C—X bond (Fig. 9-9).

Depending on the type of electron shifts that take place, nine main types of reactions are recognized: transamination, β decarboxylation, α decarboxylation, aldol cleavage, cleavage, γ elimination, β elimination; γ displacement, or β displacement may occur. In addition, various PLP enzymes carry out unique reactions that do not fall into this classification scheme; for example, dialkyl amino acid transaminase, tryptophan synthase, threonine synthase, and δ-aminolevulinic acid synthase. Three enzymes are necessary for the reduction of cytidine diphosphate (CDP) sugars to 1,3-dideoxy sugars. One of these enzymes contains pyridoxamine. Pyridoxamine, not pyridoxal, is a required growth factor for certain organisms. These are the only known cases where the cofactor requirement is for the amine rather than the pyridoxal form. The 1,3-dideoxy sugars are important components of certain bacterial cell walls. Interestingly, in mammals, about 40–50% of the body's PLP is bound to glycogen phosphorylase; here, the role of PLP is not catalytic but probably structural.

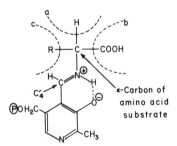

Fig. 9-9. Functions of pyridoxal-5'-phosphate (vitamin B_6) in reactions involving amino acids. The product of the enzyme-catalyzed reaction will depend on which of the four bonds projecting from the α-carbon atom is split. Reaction (a) occurs in racemization, transamination, β elimination, γ elimination, β replacement, γ replacement, and β decarboxylation. Reaction (b) occurs in α decarboxylation. Reaction (c) occurs in aldol cleavage. Although all three types of reactions are well known, almost invariably each enzyme is quite specific as to which bond it will break. This remarkable specificity occurs because the enzyme can "hold" the substrate in such a way that only the required bonds can be broken. This alignment is probably due to the strategic positioning of certain groups within the active site.

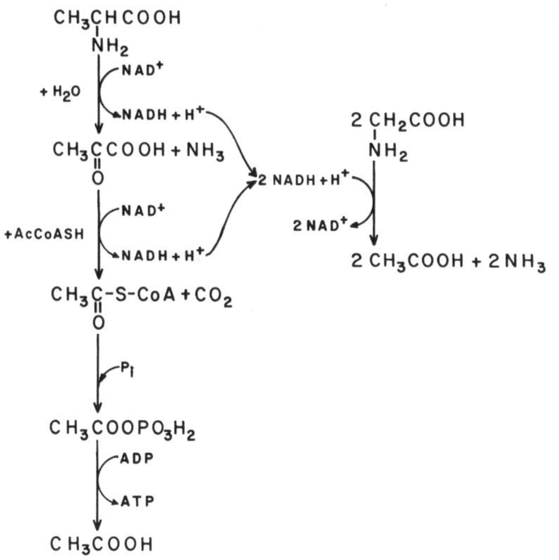

Fig. 9-10. Mechanism of the Stickland reaction.

THE STICKLAND REACTION

Some members of the genus *Clostridium* can utilize amino acids as a source of energy by means of coupled oxidation–reduction reactions involving certain amino acids as hydrogen donors and others as receptors:

$$R_1 \text{---} CH \text{---} CO_2H + R_2 \text{---} CH \text{---} CO_2H + H_2O \longrightarrow R_1 \text{---} CH_2CO_2H + R_2COCO_2H + 2NH_3$$
$$\quad\quad\quad | \quad\quad\quad\quad\quad\quad | $$
$$\quad\quad NH_2 \quad\quad\quad\quad NH_2$$

A specific example of such a reaction involving glycine and alanine would take place according to the following reaction scheme:

$$CH_2NH_2CO_2H + CH_3\text{--}CHNH_2\text{--}CO_2H + H_2O \longrightarrow CH_3CO_2H + CH_3COCO_2H + 2NH_3$$

This type of reaction leads to the formation of short-chain fatty acids and keto acids. The mechanism of the Stickland reaction is not completely understood. It involves several steps in which NAD first accepts a hydrogen atom from the amino acid donor and transfers it to the acceptor amino acid. The latter reaction is catalzyed by an amino acid reductase (Fig. 9-10). Certain amino acids serve preferentially as hydrogen donors and others as hydrogen acceptors. The rate of the reaction is quite rapid and the mechanism is used by proteolytic clostridia to generate energy via substrate level phosphorylation.

REFERENCES

Nitrogen Fixation

Appelbaum, E. 1990. The *Rhizobiium/Bradyrhizobium*-legume symbiosis. *In* P. M. Gresshoff (Ed.). *The Molecular Biology of Symbiotic Nitrogen Fixation*. CRC Press, Boca Raton, FL. pp. 131–158.

Bishop, P. E. and R. Premakumar. 1992. Alternative nitrogen fixation systems. *In* G. Stacey, R. H. Burris, and H. J. Evans (Eds.), *Biological Nitrogen Fixation*. Chapman & Hall, Ltd., London pp. 736–762.

Chan, M. K., J. Kim, and D. C. Rees. 1993. The nitrogenase, FeMo-cofactor and P-cluster pair: 2.2 Å resolution structures. *Science* **260**:792–794.

Debellé, F., C. Rosenberg, J. Vasse, F. Maillet, E. Martinez, J. Dénarié, and G. Truchet. 1986. Assignment of symbiotic developmental phenotypes to common and specific nodulation (*nod*) genetic loci of *Rhizobium meliloti*. *J. Bacteriol.* **168**:1075–1086.

Ehrhardt, D. W., E. M. Atkinson, and S. R. Long. 1992. Depolarization of alfalfa root hair membrane potential by *Rhizobium meliloti* Nod factors. *Science* **256**:998–1000.

Fallik, E., Y.-K. Chan, and R. L. Robson. 1991. Detection of alternative nitrogenases in aereobic gram-negative nitrogen-fixing bacteria. *J. Bacteriol.* **173**:365–371.

Fisher, R. F. and S. R. Long. 1992. *Rhizobium*-plant signal exchange. *Science* **357**:655–660.

Georgiadis, M. M., H. Komiya, P. Chakrabarti, D. Woo, J. J. Komuc, and C. D. Rees. 1992. Crystallographic structure of the nitrogenase iron protein from *Azotobacter vinielandii*. *Science* **257**:1653–1659.

Gosink, M. M., N. M. Franklin, and G. P. Roberts. 1990. The product of the *Klebsiella pneumoniae nifX* gene is a negative regulator of the nitrogen fixation (*nif*) regulon. *J. Bacteriol.* **172**:1441–1447.

Homer, M. I., T. D. Paustian, V. K. Shah, and G. P. Roberts. 1993. The *nifY* product of *Klebsiella pneumoniae* is associated with apodinitrogenase and dissociates upon activation with the iron–molybdenum cofactor. *J. Bacteriol.* **175**:4907–4910.

Jacobitz, S. and P. E. Bishop. 1992. Regulation of nitrogenase-2 in *Azotobacter vinelandii* by ammonium, molybdenum, and vanadium. *J. Bacteriol.* **174**:3884–3888.

Joerger, R. D., M. R. Jacobson, R. Premakumar, E. D. Wolfinger, and P. E. Bishop. 1989. Nucleotide sequence and mutational analysis of the structural genes (*anfHDGK*) for the second alternative nitrogenase from *Azotobacter vinelandii*. *J. Bacteriol.* **171**:1075–1086.

Kahn, D., M. David, O. Domergue, M.-L. Daveran, J. Ghai, P. R. Hirsch, and J. Batut. 1989. *Rhizobium meliloti fixGHI* sequence predicts involvement of a specific cation pump in symbiotic nitrogen fixation. *J. Bacteriol.* **171**:929–939.

Kim, J. and D. C. Rees. 1992. Structural models for the metal centers in the nitrogenase molybdenum–iron protein. *Science* **257**:1677–1682.

Labes, M. and T. M. Finan. 1993. Negative regulation of σ^{54}-dependent *dctA* expression by the transcriptional activator DctD. *J. Bacteriol.* **175**:2674–2681.

Margulis, L. and R. Fester (Eds.). 1991. *Symbiosis as a source of evolutionary innovation: Speciation and Morphogenesis*. MIT Press, p. 4408.

Menon, A. L., L. E. Mortenson, and R. L. Robson. 1992. Nucleotide sequences and genetic analysis of hydrogen oxidation (*hox*) genes in *Azotobacter vinelandii*. *J. Bacteriol.* **174**:4549–4557.

Moshiri, F., A. Chawla, and R. J. Maier. 1991. Cloning, characterization, and expression in *Escherichia coli* of the genes encoding the cytochrome *d* oxidase complex from *Azotobacter vinelandii*. *J. Bacteriol.* **173**:6230–6241.

Nap, J.-P. and T. Bisseling. 1990. Developmental biology of a plant-prokaryote symbiosis: The legume root nodule. *Science* **250**:948–954.

Orme-Johnson, W. H. 1992. Nitrogenase structure: Where to now? *Science* **257**:1639–1640.

Schlaman, H. R. M., B. Horvath, E. Vijgenboom, R. J. K. Okker, and B. J. Lugtenberg. 1991. Suppression of nodulation gene expression in bacteroids of *Rhizobium leguminosarum* biovar viciae. *J. Bacteriol.* **173**:4277–4287.

Spaink, H. P., D. M. Sheeley, A. A. N. van Brussel, J. Glushka, W. S. York, T. Tak, O. Geiger, E. P. Kennedy, V. N. Reinhold, and B. J. Lugtenberg. 1991. A novel highly unsaturated fatty acid moiety of lipo-oligosaccharide signals determines host specificity of *Rhizobium. Nature (London)* **354:**125–130.

Triplett, E. W., G. P. Roberts, P. W. Ludden, and J. Handelsman. 1989. What's new in nitrogen fixation. *ASM News* **55:**15–21.

Vijn, I., L. das Neves, A. van Kammen, H Franssen, and T. Bisseling. 1993. Nod factors and nodulation in plants. *Science* **260:**1764–1765.

Inorganic Nitrogen

Braun, C. and W. G. Zumft. 1992. The structural genes for the nitric oxide reductase complex from *Pseudomonas stutzeri* are part of a 30-kilobase gene cluster for denitrification. *J. Bacteriol.* **174:**2394–2397.

Cuypers, H., A. Viebrock-Sambale, and W. G. Zumft. 1992. NosR, a membrane-bound regulatory component necessary for expression of nitrous oxide reductase in denitrifying *Pseudomonas stutzeri. J. Bacteriol.* **174:**5332–5339.

Ensign, S. A., M. R. Hyman, and D. J. Arp. 1993. In vitro activation of ammonia monooxygenase from *Nitrosomonas europaea* by copper. *J. Bacteriol.* **175:**1971–1980.

Nomoto, T., Y. Fukumori, and T. Yamanaka. 1993. Membrane-bound cytochrome c is an alternative electron donor for cytochrome aa_3 in *Nitrobacter winogradskyi.* **175:**4400–4404.

Solomonson, L. P. and J. Barber. 1990. Assimilatory nitrate reductase: functional properties and regulation. *Annu. Rev. Plant Physiol. Plant Mol. Biol* **41:**225–253.

Ye, R. W., B. A. Averill, J. M. Tiedje. 1992. Characterization of Tn5 mutants deficient in dissimilatory nitrite reduction in *Pseudomonas* sp. strain G-179, which contains a copper nitrite reductase. *J. Bacteriol.* **174:**6653–6658.

CHAPTER 10

AMINO ACIDS, PURINES, AND PYRIMIDINES

AMINO ACID BIOSYNTHESIS

Amino acid biosynthesis is most conveniently discussed on the basis of families of amino acids originating from a common precursor:

1. **Glutamate or α-Ketoglutarate Family.** Glutamate, glutamine, glutathione, proline, arginine, putrescine, spermine, spermidine, and, in fungi, lysine. A tetrapyrrole (heme) precursor, δ-aminolevulinate, arises from glutamate in some organisms.

2. **Aspartate Family.** Aspartate, asparagine, threonine, methionine, isoleucine, and, in bacteria, lysine.

3. **Pyruvate Family.** Alanine, valine, leucine, and isoleucine.

4. **Serine–Glycine or Triose Family.** Serine, glycine, cysteine, cystine. In fungi, mammals, and some bacteria, δ-aminolevulinate is formed by condensation of glycine and succinate.

5. **Aromatic Amino Acid Family.** Phenylalanine, tyrosine, tryptophan. Enterochelin, p-aminobenzoate, ubiquinone, menaquinone, and NAD can also originate via branches from the common aromatic amino acid pathway.

6. **Histidine.**

Each of these families is considered in some detail in the following section. In some cases, it is also of interest to discuss the catabolism of certain amino acids, particularly where this leads to the formation of a carbon skeleton ultimately used for the synthesis of another amino acid or group of amino acids or for the biosynthesis of purines, pyrimidines, or some other vital cell constituents, such as the B vitamins. Throughout this chapter gene designations are indicated by the italicized three-letter code, whereas the gene product is indicated by the same three-letter code without italics and capitalized. Unless otherwise stated, these gene designations are for *Escherichia coli*.

THE GLUTAMATE OR α-KETOGLUTARATE FAMILY

Glutamine and Glutathione Synthesis

The importance of glutamate as a major route of ammonia assimilation has already been discussed in Chapter 9. Glutamate and glutamine play a central role in amino acid biosynthesis by the ready transfer of amino or amide groups, respectively, in the synthesis of other amino acids via transamination or transamidation reactions. Glutamine is synthesized from glutamate with the participation of ammonia and ATP. Glutathione, a disulfide-containing amino acid whose functions have only recently begun to be formulated in a precise manner, is synthesized in two steps. The coupling of L-glutamate and L-cysteine in the presence of ATP to form γ-glutamylcysteine is catalyzed by a specific synthase. In the presence of glycine and ATP glutathione synthase forms glutathione (Fig. 10-1).

The Proline Pathway

The pathway to proline involves formation of γ-glutamylphosphate from L-glutamate and ATP by γ-glutamyl kinase (Fig. 10-1). In the presence of NADPH, γ-glutamylphosphate is reduced to glutamate γ-semialdehyde. Glutamate γ-semialdehyde can cyclize spontaneously to form 1-pyrroline-5-carboxylate. 1-Pyrroline-5-carboxylate is converted to proline by a specific reductase.

Aminolevulinate Synthesis

In *S. typhimurium* and *E. coli* δ-aminolevulinic acid (ALA), the first committed precursor to tetrapyrroles, arises from glutamate. This pathway, termed the C_5 pathway, which was originally thought to occur primarily in plants and algae, is now firmly established as a major route to ALA in several bacterial species. In earlier studies it had been considered that the condensation of glycine and succinyl-CoA by the enzyme ALA synthase was the only route of ALA synthesis. This is still the major route of ALA formation in mammals, fungi, and certain bacteria, such as *Rhodobacter sphaeroides, R. capsulatus,* and *Bradyrhizobium japonicum.*

The C_5 pathway to ALA involves activation of glutamyl-tRNA^{Glu} by glutamyl-tRNA synthetase, reduction to glutamate γ-semialdehyde by an NADPH-dependent glutamyl-tRNA reductase (encoded by *hemA*), and transamination by glutamate γ-semialdehyde aminomutase (encoded by *hemL*) to form ALA:

| COOH | COOH | COOH | COOH |
| Glutamate | Glutamyl-tRNA | GSA | ALA |

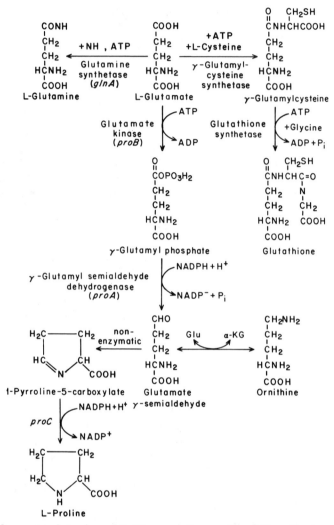

Fig. 10-1. Pathways to glutamine, glutathione, proline, and ornithine. The transaminase that interconverts ornithine and glutamate γ-semialdehyde is reversible. However, it is generally considered that this reaction serves for ornithine degradation. Ornithine is normally synthesized via the pathway shown in Figure 10-2. Gene designations are for *E. coli. proC*=1-pyrroline-5-carboxylate reductase.

The details of the pathway of heme biosynthesis will be considered in a subsequent section.

The Arginine Pathway

Bacteria and fungi synthesize ornithine via a series of *N*-acetyl derivatives (Fig. 10-2). The function of the *N*-acetyl groups is to prevent the premature cyclization of 1-pyrroline-

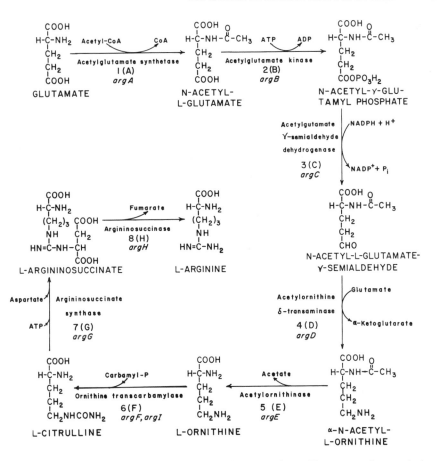

Fig. 10-2. Pathway of arginine biosynthesis in microorganisms. The structural gene designations for the enzymes are for *E. coli*. In yeasts, molds, *E. coli, Proteus mirabilis, Serratia marcescens*, and certain other enterobacteria, Step 5 occurs as shown. In *Pseudomonas fluorescens, Micrococcus glutamicus, Anabaena variabilis*, and several other bacteria, Step 5 involves transacetylation between glutamate and acetylornithine.

5-carboxylate to proline. There is a divergence in the pathway in different organisms depending on the manner in which the acetyl group is removed. In *Enterobacteriaceae* and *Bacillaceae*, *N*-acetylornithine is deacetylated via acetylornithine deacetylase encoded by *argE*. In *Neisseria gonorrhoeae, Pseudomonadaceae*, cyanobacteria, photosynthetic bacteria, and yeasts and molds, the acetyl group of *N*-acetylornithine is recycled by ornithine acetyltransferase encoded by *argJ*.

The eight enzymes involved in the arginine pathway, as shown in Figure 10-2, are found in *N. crassa, A. niger, S. cerevisiae, C. albicans, E. coli, S. typhimurium, B. subtilis, B. sphaericus*, various species of *Proteus, S. bovis, N. gonorrhoeae*, and several species of *Pseudomonas*.

Carbamoyl phosphate is a common precursor in the biosynthesis of arginine and pyrimidine. *Escherichia coli* and *S. typhimurium* produce a single carbamoyl phosphate synthetase that catalyzes the reaction:

$$2ATP + HCO_3^- + \text{L-glutamine} + H_2O \xrightarrow[Mg^{2+}]{K^+} NH_2COOPO_3H_2$$
$$+ 2P_i + 2ADP + \text{glutamate}$$

Ammonia can replace glutamine as a nitrogen donor *in vitro*, but glutamine is the physiologically preferred substrate. The enzyme is composed of two nonidentical subunits. The smaller subunit, encoded by *carA*, acts as a glutamine amidotransferase. The larger subunit, encoded by *carB*, carries out the remaining functions. As might be expected, since carbamoyl phosphate is involved in two major biosynthetic pathways, expression of the *carAB* operon is regulated by both arginine and pyrimidines. The enzyme is also subject to allosteric control by intermediates in both pathways. Whereas ornithine stimulates enzyme activity, UMP is inhibitory. The enzyme is also activated by IMP and PRPP, coordinating its activity with purine biosynthesis as well. In some earlier studies certain organisms appeared to be dependent on the nonenzymatic formation of carboxylamine and its conversion to carbamoyl phosphate by carbamoyl phosphokinase:

$$NH_4^+ + HCO_3^- \xrightarrow{\text{nonenzymatic}} NH_2COO^- + H_3O^+ \; NH_2COO^-$$
$$+ ATP \xrightarrow{Mg^{2+}} NH_2COOPO_3H_2 + ADP$$

The relative importance of this reaction in ammonia assimilation as compared to other reactions, such as the GS–GOGAT or GDH systems, has never been adequately assessed.

In fungi, such as *S. cerevisiae* and *N. crassa*, there are two carbamoyl phosphate synthetases. One enzyme, carbamoyl phosphate synthetase A, is linked to the arginine pathway and is subject to repression by arginine. The other enzyme, carbamoyl phosphate synthetase P, is linked to the pyrimidine biosynthesis pathway and is subject to both repression and feedback inhibition by pyrimidines. Localization of the arginine pathway carbamoyl phosphate synthetase in mitochondria of *N. crassa* seems to play a major role in channeling of this precursor (see further details later). In yeast there seems to be little channeling of carbamoyl phosphate since both carbamoyl phosphate synthetases are in the cytoplasm and contribute to a common pool of carbamoyl phosphate.

Some microorganisms are able to generate usable energy in the form of ATP by the degradation of arginine via the arginine deiminase pathway. In *E. coli*, this complex consists of three enzymes: (1) arginine deiminase, encoded by *argA*, converts arginine to citrulline and ammonia; (2) ornithine carbamoyl transferase, encoded by *argB*, forms ornithine and carbamoyl phosphate from citrulline and P_i; and (3) carbamate kinase, encoded by *argC*, which converts carbamoyl phosphate and ADP to ATP, CO_2, and NH_3:

$$\text{arginine} + H_2O \longrightarrow \text{citrulline} + NH_3$$
$$\text{citrulline} + P_i \longrightarrow \text{ornithine} + \text{carbamoyl phosphate}$$
$$\text{carbamoyl phosphate} + ADP \longrightarrow ATP + CO_2 + NH_3$$

Arginine deiminase activity is widely distributed in the aerobic spore-forming bacilli and a number of other organisms including *Mycoplasma*, *Streptococcus*, *Lactococcus*, *Lactobacillus*, *Clostridium*, and *Pseudomonas*. In *P. aeruginosa*, *argD* encodes a membrane protein that serves as an arginine-ornithine exchanger. The ArgD protein mediates proton motive force-driven uptake of arginine and ornithine and the stoichiometric exchange between arginine and ornithine.

In mammals and fungi, arginine is degraded to urea and ornithine via arginase activity:

$$\text{L-arginine} \longrightarrow \text{urea} + \text{L-ornithine}$$

Arginase activity in mammals is apparently so rapid that insufficient amounts of arginine are available for protein synthesis and arginine is thus required as an essential nutrient. In yeasts and molds, the presence of an alternative route of arginine synthesis and other regulatory factors control arginase activity so that a nutritional requirement for arginine does not arise. In *N. crassa*, arginase activity is inhibited both *in vitro* and *in vivo* by citrulline as well as ornithine, eliminating a potentially wasteful catabolism of endogenous arginine. Compartmentation of arginine metabolism, as shown in Figure 10-3, further regulates the intracellular flux of arginine. In this organism, glutamate is synthesized in the mitochondrion. After formation of ornithine and its condensation with carbamoyl phosphate, the resultant citrulline enters the cytosol where arginine synthesis is completed. Large amounts of arginine and other amino acids are also stored in vacuoles. *Neurospora crassa* catabolizes arginine via enzymes located in the cytoplasm. Regulatory mechanisms of feedback inhibition, induction, and repression in *N. crassa* are comparable to those found in bacteria. However, these regulatory activities must function between the various cellular compartments. During active growth in minimal medium, arginine is synthesized but not degraded. Most of the arginine synthesized is sequestered in the vacuoles and only low concentrations of arginine are present in the cytosol and the mitochondrion. If arginine is added to the growth medium, the cytosolic level of arginine rises rapidly and degradation occurs at a high rate and de novo biosynthesis in the mitochondrion is curtailed. If the exposure to exogenous arginine is short term, then

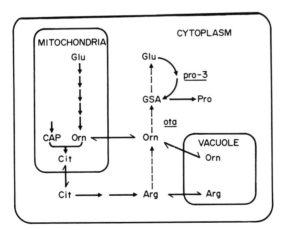

Fig. 10-3. Arginine and proline metabolism and their organization in *N. crassa*. Solid arrows are biosynthetic reactions, dashed arrows are catabolic reactions, and double-headed arrows represent transport processes. Glu = glutamate; CAP = carbamoyl phosphate; Orn = ornithine; Cit = citrulline; Arg = arginine; GSA = glutamate 5-semialdehyde; Pro = proline. The *pro-3* and *ota* (ornithine transaminase) symbols represent structural genes for the indicated steps. From Goodman, I. and R. L. Weiss. 1986. *J. Biol. Chem.* **261:**10264–10270.

arginine degradation stops and de novo biosynthesis starts shortly after arginine is depleted. On longer exposure to exogenous arginine, a lag period occurs before biosynthesis is reinitiated. This lag is due to an extended period of feedback inhibition that controls the mitochondrial level of arginine.

Saccharomyces cerevisiae, which degrades arginine and proline via a common intermediate, 1-pyrroline-5-carboxylate, also has a compartmentalized system for arginine and proline metabolism. The proline biosynthetic enzyme 1-pyrroline-5-carboxylate reductase is located in the cytoplasm, whereas the proline degradading enzyme, 1-pyrroline-5-carboxylate dehydrogenase is in a particulate fraction of the cell, presumably in the mitochondrion. Arginase is encoded by *CAR1*, which is inducible by arginine. Regulated expression of the *CAR1* gene requires three upstream activation sequences and an upstream repression sequence.

Polyamine Biosynthesis

Polyamines are widely distributed in bacteria, yeasts, and molds as well as in higher forms. Although a number of growth processes are affected by polyamines, their precise role in governing cell growth and differentiation has not been established. The major pathway to polyamines is via the decarboxylation of ornithine to putrescine, as shown in Figure 10-4. Ornithine decarboxylase is the rate-limiting step in the pathway to polyamines. The ornithine decarboxylase pathway appears to be common to all cells.

In bacteria and plants an alternate pathway to putrescine proceeds via decarboxylation of arginine to agmatine by a biosynthetic arginine decarboxylase (encoded by *speA* in *E. coli*). Agmatine ureohydrolase removes urea from agmatine to yield putrescine:

$$
\begin{array}{ccc}
\text{COOH} & & \\
| & & \\
\text{H}_2\text{N}-\text{CH} & \xrightarrow{-CO_2} \text{HN}_2-\text{CH}_2 & \xrightarrow{-urea} \text{H}_2\text{N}-\text{CH}_2 \\
| & | & | \\
(\text{CH}_2)_3 & (\text{CH}_2)_3 & (\text{CH}_2)_3 \\
| & | & | \\
\text{NH} & \text{NH} & \text{NH}_2 \\
| & | & \\
\text{H}_2\text{N}-\text{C}=\text{NH}_2 & \text{H}_2\text{N}-\text{C}=\text{NH}_2 & \\
\text{Arginine} & \text{Agmatine} & \text{Putrescine}
\end{array}
$$

Exogenous arginine acts as a signal for the selective utilization of this pathway in *E. coli*. This organism lacks arginase and cannot convert arginine to ornithine. When arginine is added to the growth medium, ornithine levels decline due to inhibition of arginine biosynthesis.

Lysine Biosynthesis in Fungi

In yeasts and molds the biosynthetic pathway to lysine emanates from α-ketoglutarate (Fig. 10-5). The bacterial pathway to lysine, which is initiated by the condensation of pyruvate and aspartate-β-semialdehyde, is part of the aspartate family of amino acid biosynthetic routes, and is shown in Figure 10-6. These completely divergent routes of lysine biosynthesis represent a major phylogenetic difference between bacteria and fungi.

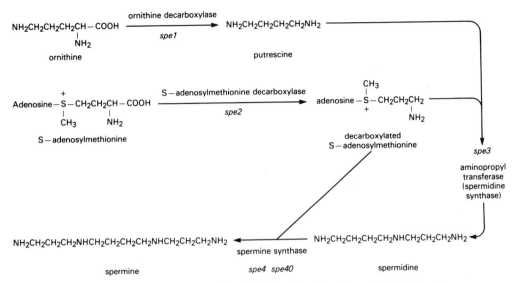

Fig. 10-4. Biosynthetic steps and genes involved in polyamine biosynthesis in *Saccharomyces cerevisiae*. From Tabor, C. W. et al. 1982. *Fed. Proc.* **41**:3084–3088.

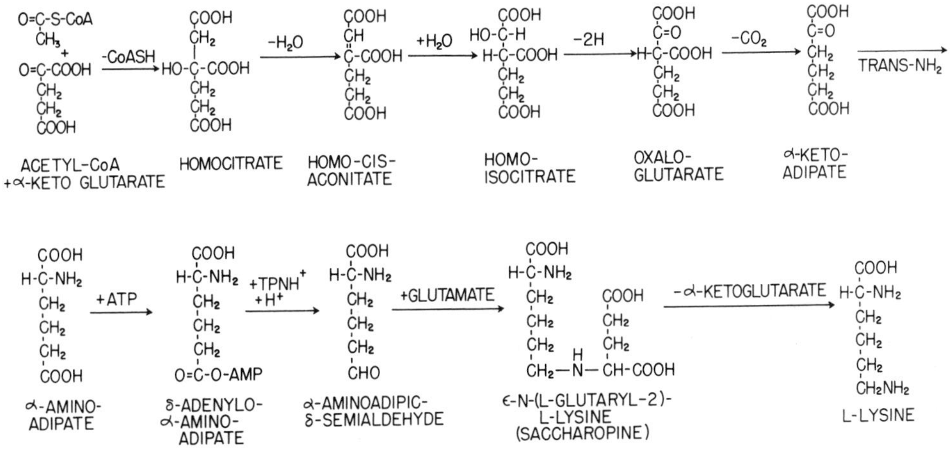

Fig. 10-5. Biosynthesis of lysine in yeasts and molds. This pathway is termed the α-aminoadipate pathway. Genetic studies of this pathway have been conducted mainly in *Saccharomyces cerevisiae*, *S. lipolytica*, and *Neurospora crassa*. PALP = pyridoxal phosphate. The structural genes shown are for *S. cerevisiae*: *LYS7* = homocitrate dehydratase; *LYS4* = homo-cisaconitate hydratase; *LYS12* = homoisocitrate dehydrogenase; *LYS2*, *LYS5* = α-aminoadipate reductase; *LYS9* = saccharopine reductase; *LYS1* = saccharopine dehydrogenase.

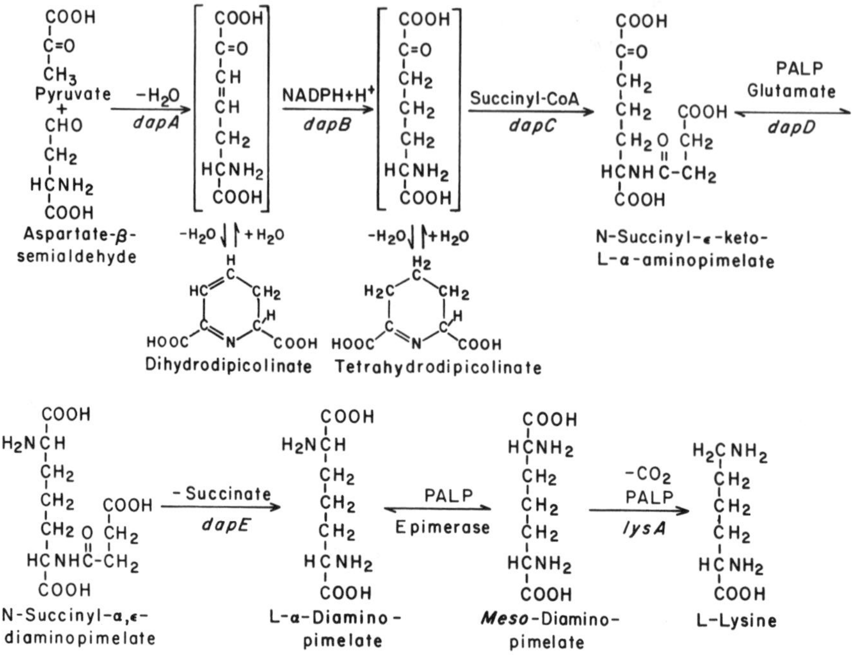

COOH
|
C=O
|
CH₃   Pyruvate   −H₂O
+   ——→
CHO   dapA
|
CH₂
|
HCNH₂
|
COOH
Aspartate-β-
semialdehyde

[COOH
|
C=O
|
CH
‖
CH
|
CH₂
|
HCNH₂
|
COOH]   NADPH+H⁺
            ——→
            dapB

−H₂O ↓↑ +H₂O

[COOH
|
C=O
|
CH₂
|
CH₂
|
CH₂
|
HCNH₂
|
COOH]   Succinyl-CoA
            ——→
            dapC

−H₂O ↓↑ +H₂O

COOH
|
C=O
|
CH₂
|
CH₂   COOH
| |
CH₂ O   CH₂
| ‖
HCNH C-CH₂
|
COOH
N-Succinyl-ε-keto-
L-α-aminopimelate

PALP
Glutamate
←——
dapD

Dihydrodipicolinate

Tetrahydrodipicolinate

COOH
|
H₂N CH
|
CH₂
|
CH₂   COOH
| |
CH₂ O   CH₂
| ‖
HCNHC-CH₂
|
COOH
N-Succinyl-α,ε-
diaminopimelate

− Succinate
——————→
dapE

COOH
|
H₂N CH
|
CH₂
|
CH₂
|
CH₂
|
HC NH₂
|
COOH
L-α-Diamino-
pimelate

PALP
←————
Epimerase

COOH
|
HCNH₂
|
CH₂
|
CH₂
|
CH₂
|
HCNH₂
|
COOH
Meso-Diamino-
pimelate

−CO₂
PALP
————→
lysA

H₂C NH₂
|
CH₂
|
CH₂
|
CH₂
|
HC NH₂
|
COOH
L-Lysine

Fig. 10-6. Biosynthesis of lysine in bacteria. The structural gene designations given are those for *E. coli. dapA* = dihydrodipicolinate synthase; *dapB* = dihydrodipicolinate reductase; *dapC* = tetrahydrodipicolinate succinylase; *dapD* = succinyl diaminopimelate aminotransferase; *dapE* = succinyl diaminopimelate desuccinylase; *lysA* = diaminopimelate decarboxylase. PALP = pyridoxal phosphate.

The series of reactions from homocitrate to α-ketoadipate are analogous to the reactions involved in the conversion of citrate to α-ketoglutarate in the citric acid cycle. The formation of homocitrate from α-ketoglutarate and acetyl-CoA is inhibited by lysine, indicating a feedback control mechanism in the pathway. The initial step in the pathway is also subject to repression by lysine.

The biosynthesis of β-lactam antibiotics (penicillins and cephalosporins) occurs by a branch from the lysine pathway in *Penicillium chrysogenum, Cephalosporium acremonium, Streptomyces clavuligerus,* and related organisms. Diminution of penicillin yields by excess lysine may be related to feedback inhibition of homocitrate synthase, which limits the supply of L-α-aminoadipate, a common precursor required for lysine and penicillin synthesis. In virtually all organisms that synthesize penicillins and cephalosporins the pathway is initiated by the multistep condensation of L-α-aminoadipate, L-cysteine, and L-valine to form the tripeptide, δ-(L-α-aminoadipyl)-L-cysteinyl-D-valine (ACV). This condensation is carried out by an ATP-dependent enzyme, ACV synthetase. This enzyme is quite labile, but has been purified from *C. acremonium* and its characteristics studied in some detail. The enzyme is induced by methionine. The ACV synthetase functions mechanistically in a similar manner to the nonribosomal peptide synthetases that give rise to such metabolites as glutathione and the peptide antibiotics gramicidin, tyrocidin, and bacitracin. Synthesis of benzylpenicillins and cephalosporins take place via one or more branches from this common pathway.

THE ASPARTATE AND PYRUVATE FAMILIES

The aspartate and pyruvate families of amino acids are discussed together since there is a distinct overlap in the enzymes involved in the terminal steps of the biosynthesis of the branched-chain amino acids, as shown in Figure 10-7. Valine and leucine carbon chains are derived from pyruvate. In bacteria, threonine, isoleucine, methionine, and lysine all emanate from aspartate. Fungi utilize similar pathways with the exception of lysine, which is synthesized via a completely different pathway as discussed in the previous section.

Asparagine Synthesis

Asparagine is synthesized from aspartate, ammonia (or glutamine), and ATP by the en-

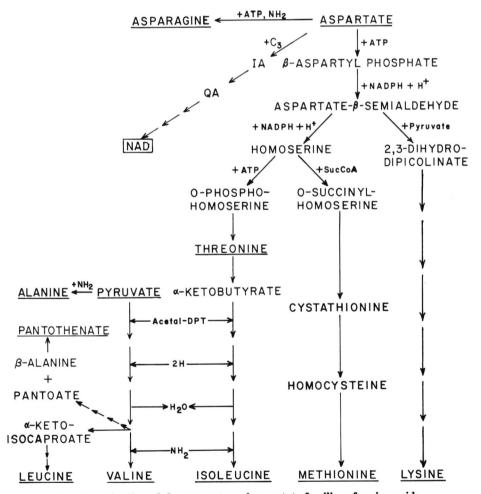

Fig. 10-7. Outline of the pyruvate and aspartate families of amino acids.

zyme asparagine synthetase:

$$
\begin{array}{c}
\text{COOH} \\
|\\
\text{CH}_2 \\
|\\
\text{HCNH}_2 \\
|\\
\text{COOH}
\end{array}
\quad + \text{NH}_3 + \text{ATP} \xrightarrow[\text{AsnB}]{\text{AsnA}}
\begin{array}{c}
\text{CONH}_2 \\
|\\
\text{CH}_2 \\
|\\
\text{HCNH}_2 \\
|\\
\text{COOH}
\end{array}
\quad + \text{ATP}
$$

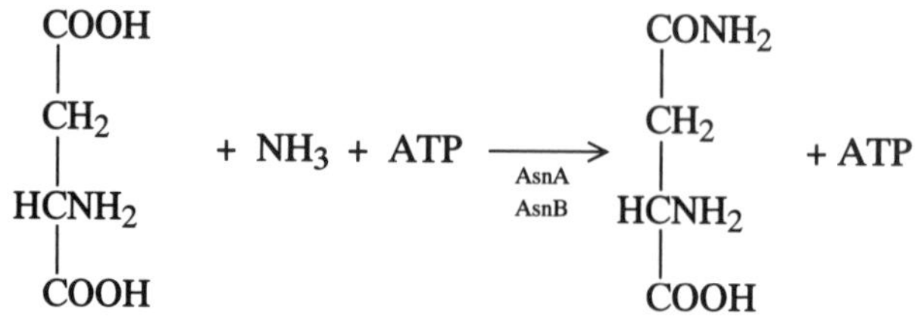

Two asparagine synthetases are present in *E. coli*. They are encoded by two unlinked structural genes designated *asnA* (84 min) and *asnB* (15 min). Asparaginase A (AsnA) is ammonia dependent, whereas asparaginase B (AsnB) can use either ammonia or glutamine as the nitrogen donor. Mutant strains carrying only the *aspB* gene can assimilate ammonia into asparagine in the presence of excess ammonia. The glutamine-dependent enzyme is regulated by the asparagine content of the intracellular asparagine pool. Ammonia derepresses the ammonia-dependent asparagine synthetase B. There are also multiple asparagine synthetases in yeast. Although the yeast enzymes can use ammonia, affinity studies show that glutamine is the preferred substrate.

The Aspartate Pathway

In both bacteria and fungi, the enzymes aspartokinase and aspartate semialdehyde dehydrogenase initiate the aspartate pathway of amino acid biosynthesis (Fig. 10-8). In *E. coli* there are three aspartokinases, each of which is specific to the end-product of the pathway involved in its synthesis. Aspartokinase I (ThrA) is specific for the threonine

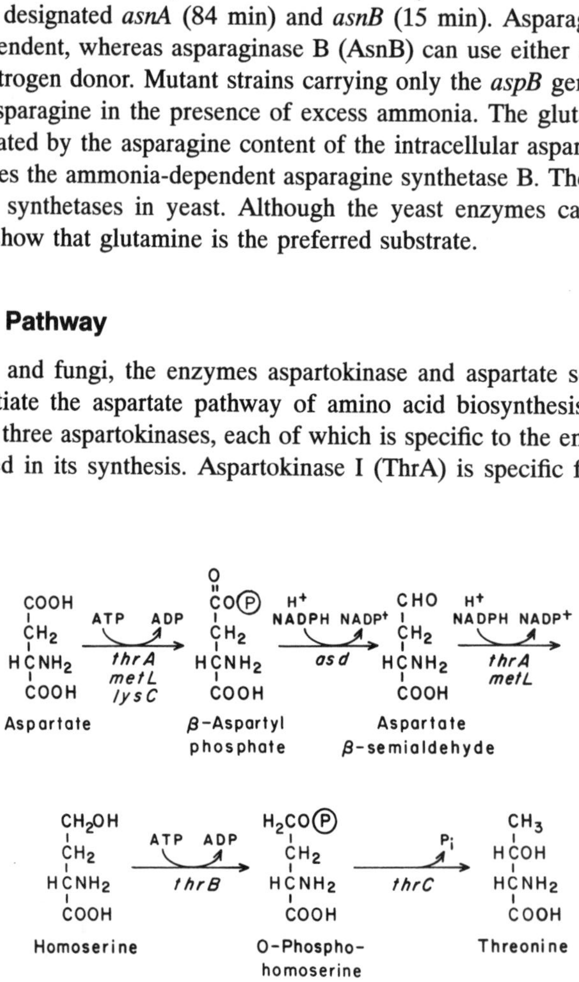

Fig. 10-8. Initial steps in the aspartate pathway of amino acid biosynthesis. In *E. coli* there are three aspartokinase enzymes. Their designated structural genes are *thrA* = aspartokinase I; *metL* = aspartokinase II; *lysC* = aspartokinase III. The structural gene for aspartate semialdehyde dehydrogenase is *asd*. *Escherichia coli* contains two homoserine dehydrogenases: *thrA*, which is specific for the threonine branch of the pathway and *metL*, specific for the methionine branch. Homoserine is converted to *O*-phosphohomoserine by homoserine kinase (ThrB). Threonine synthetase (ThrC) converts *O*-phosphohomoserine to threonine.

branch. Aspartokinase II (MetL) is specific for the methionine branch and aspartokinase III (LysC) is specific for the lysine branch. Aspartate semialdehyde dehydrogenase (Asd) forms aspartate β-semialdehyde by removal of phosphate from aspartyl phosphate and reduction using NADPH. Aspartate β-semialdehyde stands at the first branch point in the multibranched pathway. Reduction of aspartate β-semialdehyde to homoserine is catalyzed by homoserine dehydrogenase I (ThrA), which is specific for the threonine branch of the pathway, or homoserine dehydrogenase II (MetL), which is specific for the methionine branch.

The Bacterial Pathway to Lysine

Condensation of aspartate β-semialdehyde with pyruvate yields dihydrodipicolinate, the first intermediate in the bacterial pathway to lysine (Fig. 10-6). This pathway is of special interest because dipicolinic acid is produced during sporulation in *Bacillus* species and diaminopimelic acid or lysine are present in the peptidoglycan structures of all prokaryotes that produce a rigid cell wall. This pathway to lysine has been shown to occur in *E. faecalis, S. aureus, E. coli, S. typhimurium, B. subtilis*, and the cyanobacteria, and, by inference, considered to occur in most other bacteria.

The biosynthetic pathway depicted in Figure 10-6 involves the formation of succinylated derivatives of α-keto-L-α-aminopimelate and diaminopimelate as intermediates in the pathway. This pathway is utilized by *E. coli, S. typhimurium*, and *S. aureus* and a number of other bacteria. Certain *Bacillus* species appear to form acetylated derivatives exclusively. A few organisms utilize a pathway in which tetrahydrodipicolinate (piperideine-2,6-dicarboxylate) is converted in a single step to D,L-diaminopimelate by diaminopimelate dehydrogenase. This pathway is operative in *B. sphaericus* and *Corynebacterium glutamicum*. However, in *C. glutamicum* the pathway using succinylated intermediates appears to function along with the direct dehydrogenase pathway. This organism is of interest because of its use for the commercial production of lysine. The dehydrogenase pathway is apparently a prerequisite for handling increased flow of metabolites to D,L-diaminopimelate and lysine.

Because of the insolubility of iron at physiological pH, microorganisms have evolved a variety of systems to facilitate the acquisition of iron (see Chapter 7). *Escherichia coli, Shigella*, and *Salmonella* can synthesize an iron-chelating siderophore, aerobactin, from lysine. The iron-regulated aerobactin operon is found on a ColV-K30 plasmid. This operon consists of at least five genes for synthesis (*iuc*, iron uptake chelate) and transport (*iut*, iron uptake transport) of aerobactin. The biosynthetic pathway starts with oxidation of lysine to N^ϵ-hydroxylysine by N^ϵ-lysine monooxygenase (IucD), acetylation of N^ϵ-hydroxylysine to form N^ϵ-acetyl-N^ϵ-hydroxylysine by the action of a N-acetyltransferase enzyme (IucB), as shown in Figure 10-9. The final stage of citrate addition and condensation to form the aerobactin ring is catalyzed by aerobactin synthetase (IucC).

Threonine, Isoleucine, and Methionine Formation

Another branch point in the aspartate pathway occurs at homoserine. One branch leads to the formation of threonine and, ultimately, isoleucine. Homoserine can also be converted to methionine via several alternate routes. In *E. coli, S. typhmurium*, and other enteric organisms, conversion of homoserine to homocysteine involves either of two alternative pathways (Fig. 10-10). The enzyme homoserine succinyltransferase (MetA)

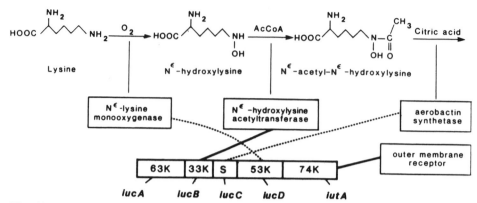

Fig. 10-9. Pathway of aerobactin synthesis from lysine. The enzyme designations are for *E. coli.* IucD = N^6-lysine monooxygenase; IucB = *N*-acetyltransferase; IucC = aerobactin synthase. From deLorenzo, et al., 1986. *J. Bacteriol.* **165**:570–578.

condenses succinyl-CoA and homoserine to yield *O*-succinylhomoserine. Cystathionine γ-synthase is apparently capable of catalyzing the direct reaction of hydrogen desulfide (H$_2$S) or methylsulfide with *O*-succinylhomoserine to form homocysteine.

The methylation of homocysteine to form methionine may occur via either of two enzymes. One enzyme (MetH) is vitamin B$_{12}$ dependent and requires NADH, FAD, SAM, and either 5-methyltetrahydrofolate or its triglutamyl derivative. A vitamin B$_{12}$-independent enzyme (MetE) requires only Mg^{2+} and 5-methyltetrahydropteroyltriglutamate. The 5-methyltetrahydrofolate is formed from serine and tetrahydrofolate (THF) through the action of serine hydroxymethyltransferase (GlyA), which converts THF to 5,10-methylene-THF. The enzyme, 5,10-methylenetetrahydrofolate reductase (MetF) uses reduced FAD (FADH$_2$) to convert 5,10-methylene-THF to 5-methyl-THF.

In *Neurospora*, the conversion of homoserine to homocysteine occurs via an analogous sequence of reactions in which homoserine and acetyl-CoA are condensed to form *O*-acetylhomoserine. The reaction is catalyzed by homoserine acetyltransferase. In this sequence, cystathionine, γ-synthase exchanges cysteine for the acetyl group in *O*-acetylhomoserine to release acetate and form cystathionine. This pathway is also utilized by *Aspergillus nidulans.*

An alternate route of homocysteine formation involving the direct sulfhydrylation of *O*-acetylhomoserine, *O*-succinylhomoserine, or *O*-phosphohomoserine, is mediated by homocysteine synthase:

$$O\text{-acetylhomoserine} + H_2S \longrightarrow \text{L-homocysteine} + \text{acetate}$$

$$O\text{-succinylhomoserine} + H_2S \longrightarrow \text{L-homocysteine} + \text{succinate}$$

$$O\text{-phosphohomoserine} + H_2S \longrightarrow \text{L-homocysteine} + P_i$$

One or both of these activities has been reported in *N. crassa, S. cerevisiae,* and *E. coli.* The first of these reactions appears to be the main pathway in yeast; however, alternative routes for methionine biosynthesis are also available.

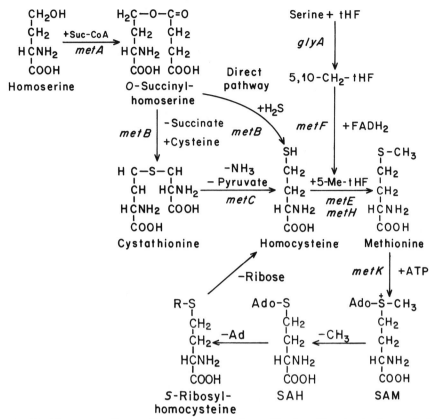

Fig. 10-10. Alternate pathways from homoserine to methionine. The structural gene designations are for *E. coli. metA* = homoserine acyltransferase; *metB* = cystathionine-γ-synthase; *metC* = cystathionase. 5-Methyltetrahydrofolate = 5-Me-THF, is formed from serine and tetrahydrofolate (THF) via serine hydroxymethyltransferase (*glyA*), which yields 5,10-methylene-tetrahydrofolate (5,10-CH$_2$-THF) upon reduction by reduced flavin adenine dinucleotide (FADH$_2$) catalyzed by 5,10-methylenetetrahydrofolate reductase (*metF*). The vitamin B$_{12}$-dependent homocysteine methylase (*metH*) requires catalytic amounts of SAM and FADH$_2$. *S*-adenosylmethionine synthetase (*metK*) catalyzes the synthesis of SAM from methionine and ATP.

Isoleucine, Leucine, and Valine Biosynthesis

The isoleucine carbon skeleton is derived, in part, from aspartate via the deamination of threonine. It is convenient to discuss the biosynthesis of all three branched-chain amino acids (isoleucine, leucine, and valine) because of the close interrelationship of the pathways. In the isoleucine–valine pathway, four of the steps in both sequences are catalyzed by the same enzymes shown in Figure 10-11. The immediate precursor of valine, α-ketoisovalerate, represents another branch point. Condensation with acetyl-CoA initiates the series of reactions leading to leucine synthesis, as shown in Figure 10-11. Via another series of reactions, α-ketoisovalerate is converted to pantoic acid, a precursor of pantothenic acid. The basic series of reactions leading to the formation of the branched-chain amino acids appears to be quite similar in most microorganisms examined to date. For example, the isopropylmalate pathway to leucine is widespread among diverse or-

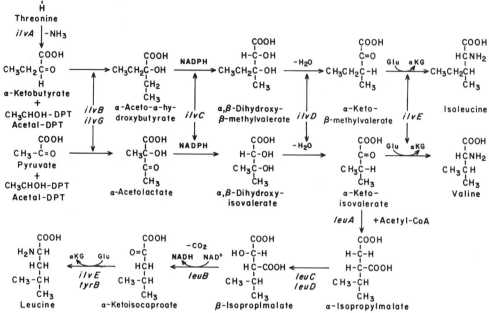

Fig. 10-11. Biosynthesis of the branched-chain amino acids, isoleucine, leucine, and valine.
The structural gene designations and the enzymes for *E. coli* are *ilvA* = threonine deaminase; *ilvB* = valine-sensitive acetohydroxy acid synthase I; *ilvG* = acetohydroxy acid synthase II; *ilvC* = acetohydroxy acid isomeroreductase; *ilvD* = dihydroxy acid dehydratase; *ilvE* = branched-chain amino acid aminotransferase; *leuA* = α-isopropylmalate synthase; *leuC, leuD* = isopropylmalate dehydratase; *leuB* = β-isopropylmalate dehydrogenase; *tyrB* = aromatic amino acid aminotransferase.

ganisms capable of leucine biosynthesis. Occasionally, organisms are found with novel routes of biosynthesis. In *Leptospira*, isoleucine carbon arises via a pathway involving citramalate as an intermediate (Fig. 10-12), whereas the origin of the carbon skeletons of the other amino acids are consistent with pathways common to other prokaryotes.

Regulation of the Aspartate Family

Regulation of the biosynthesis of the amino acids of the aspartate family is complex because of the multiple branches and, to some extent, the interrelationships between the aspartate and pyruvate families. In *E. coli* K-12, primary regulation is exerted at two points. The major one is the aspartokinase reaction, which regulates the flow of carbon to all of the amino acids involved (Fig. 10-13). The second site of primary regulation is the conversion of aspartate-β-semialdehyde to homoserine, a reaction catalyzed by homoserine dehydrogenases I and II. Aspartokinase I and homoserine dehydrogenase I activities are associated with a single multifunctional enzyme, ThrA, which is subject to allosteric inhibition by threonine. Aspartokinase I-homoserine dehydrogenase I synthesis is repressed by a combination of threonine and isoleucine. The structural genes for aspartokinase I-homoserine dehydrogenase I and homoserine kinase (*thrB*), and threonine synthase (*thrC*) lie in close proximity to one another on the *E. coli* map and constitute

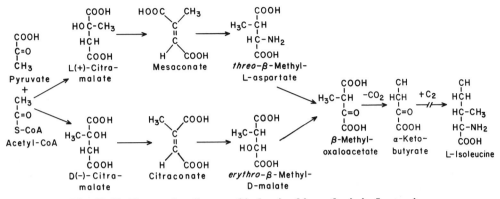

Fig. 10-12. Proposed pathways of isoleucine biosynthesis in *Leptospira*.

an operon. Aspartokinase II and homoserine dehydrogenase II activities are also catalyzed by a bifunctional protein (MetL). The MetL transcription is regulated by methionine. However, neither of the two activities is feedback-inhibited by methionine, threonine, SAM, nor by combinations of these compounds. Aspartate kinase III (*lysC*), is regulated by lysine by both feedback and repression mechanisms. Comparable regulatory systems in the aspartate pathway have been found in other members of the *Enterobacteriaceae*, particularly *Salmonella*, *Enterobacter*, *Edwardsiella*, *Serratia*, and *Proteus*.

In *E. coli*, feedback inhibition plays a prominent role in regulation of the lysine pathway. A constitutive mutant for the synthesis of the lysine-sensitive aspartokinase III (*lysC*) has been isolated. The aspartokinase III protein may serve as an inducer of diaminopimelate decarboxylase synthesis and thus regulate lysine synthesis in this manner.

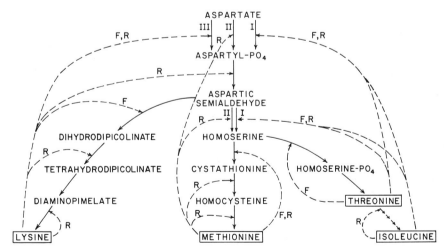

Fig. 10-13. End-product regulation of the aspartate family of amino acids in *E. coli*. F = feedback inhibition; R = repression of enzyme synthesis. Aspartokinase I-homoserine dehydrogenase I is a bifunctional enzyme encoded by *thrA*. Aspartokinase II-homoserine dehydrogenase II is a bifunctional enzyme encoded by *metL*. Aspartokinase III is encoded by *lysC*.

In some organisms **concerted feedback** or **multivalent** inhibition is a more common type of regulation of the enzymes of the aspartate family. In this type of inhibition, all of the products of the pathway act together to effect the inhibition. Single aspartokinase enzymes that are subject to inhibition by lysine and threonine have been reported.

By comparison, *B. subtilis* also produces three isozymes of aspartokinase. Aspartokinase I is selectively inhibited by diaminopimelate. Aspartokinase II is selectively inhibited by lysine. Aspartokinase III is synergistically inhibited by threonine and lysine. Synthesis of aspartokinase III is induced by lysine and repressed by threonine. Aspartokinase I activity is relatively constant under all growth conditions, whereas the levels of aspartokinases II and III vary widely with growth conditions and between different strains of *Bacillus*. The loss through mutation of either aspartokinase II or aspartokinase III does not result in an auxotrophic requirement, but the loss of both enzymes causes a growth requirement for all three major end-products of the aspartate pathway (lysine, threonine, methionine). These findings imply that aspartokinase I alone cannot provide adequate supplies of precursors for the three major amino acids during exponential growth. In the lysine pathway, the synthesis of diaminopimelate decarboxylase is repressed by lysine. Lysine can thus regulate its own synthesis. Although most of the genes in the lysine pathway in *Bacillus* are coordinately regulated by lysine, the synthesis of dihydrodipicolinate synthase is not repressed by lysine.

In the corynebacteria, regulation of the aspartate family does not involve isozymes of aspartokinase or other enzymes in the pathway. Instead, the relative specific activities of the enzymes and patterns of inhibition and repression of enzymes at the major branch points serve to regulate the amount of individual end-products formed. A considerably higher specific activity of homoserine dehydrogenase over that of dihydrodipicolinate synthetase results in a preferential flux toward the threonine-methionine branch (Fig. 10-7). Methionine represses the synthesis of homoserine dehydrogenase while threonine inhibits its activity, thus directing the flow into the lysine branch. Concerted feedback inhibition of aspartokinase activity by lysine and threonine also governs the activity of the overall pathway. In *Corynebacterium flavum* two promoters, *ask*P1 and *ask*P2, are required for expression of the *askasd* genes, which encode aspartokinase and aspartate-β-semialdehyde dehydrogenase.

Genes of the aspartate pathway that are unique to methionine biosynthesis in *E. coli* are repressed by methionine. Although the structural genes coding for these enzymes are not contiguous, they are strongly regulated by a coordinated mechanism. *S*-Adenosylmethionine is also under the control of some of the regulatory elements governing the methionine biosynthetic enzymes. *Pseudomonas denitrificans*, which uses the B_{12}-dependent pathway for methionine synthesis (Fig. 10-10), has been shown to contain the enzyme betaine-homocysteine transmethylase. This serves as an alternative route for the methylation of homocysteine to methionine. The observation that many organisms have more than one mechanism for methionine formation provides evidence that these reactions help to satisfy the demands for methionine as a methyl donor as well as for protein synthesis. Various yeasts can also use the folate-dependent pathway or the SAM-homocysteine transmethylase reaction for methionine biosynthesis.

In the methionine pathway, both the regulatory effector and the site of action may differ widely from one organism to another. *S*-Adenosylmethionine appears to be the regulatory signal in *E. coli* and *S. typhimurium*. Methionyl-tRNA^{met} (met-tRNA^{met}) has been ruled out as a possible corepressor in these organisms because met-tRNA^{met} synthetase-defective mutants are still repressible by methionine. In *Aspergillus nidulans*, either cysteine or homocysteine rather than methionine or SAM, are involved in regulation of methionine biosynthesis. In *S. cerevisiae*, both SAM and met-tRNA^{met} have been implicated as separate regulators of this system. However, studies with a mutant in which modification of met-tRNA synthetase was the only major defect indicate that met-tRNA^{met}, more likley than methionine, is involved in regulating methionine biosynthesis.

Regulation of the enzymes involved in the biosynthesis of the branched-chain amino acids is under the control of two sets of closely linked genes (clusters) in *E. coli, S. typhimurium*, and other related gram-negative bacteria (Fig. 10-11). The structural genes associated with the isoleucine-valine (ilv) pathway are governed by five structural genes (*ilvABCDE*). These genes correspond to threonine deaminase (*ilvA*), which is specific for the isoleucine pathway, and the four enzymes that catalyze the corresponding steps in the isoleucine and valine pathways (Fig. 10-11). The *ilvB* gene codes for a valine-sensitive acetohydroxy acid synthetase I. Additional acetohydroxy acid synthetases are coded for by *ilvG* and *ilvHI*. An NADPH-requiring acetohydroxy acid isomeroreductase is coded for by *ilvC*. Dihydroxy acid dehydratase is coded for by *ilvD* and the branched-chain amino acid aminotransferase is coded for by *ilvE*. An additional aminotransferase (alanine-valine aminotransferase) coded for by *avtA* can substitute for the branched-chain aminotransferase in the valine pathway. Regulation of the isoleucine and valine biosynthetic enzymes occurs at the level of transcription. Full repression of the synthesis of these enzymes requires an excess of all three branched-chain amino acids. Regulation of carbon flow through these parallel pathways to valine and isoleucine depends on the end-product inhibition of threonine deaminase and acetohydroxy acid synthetase.

The structural genes that regulate leucine biosynthesis in *E. coli* and *S. typhimurium* are arranged in a contiguous group at 4 min on the *E. coli* gene map and represent an operon. α-Isopropylmalate synthase (*leuA*) condenses acetyl-CoA and α-ketoisovalerate to form α-isopropylmalate. Isopropylmalate dehydratase (*leuC, leuD*) converts α-isopropylmalate to β-isopropylmalate. β-Isopropylmalate dehydrogenase (*leuB*) removes carbon dioxide and hydrogen atoms to form α-ketoisocaproate. The branched-chain amino acid transferase (*ilvE*) or the aromatic amino acid transferase (*tyrB*) catalyzes the transamination of α-ketoisocaproate to leucine. The activity of α-isopropylmalate synthase, the first enzyme unique to leucine biosynthesis, is inhibited by leucine.

In the *Enterobacteriaceae* a complex system of multivalent control by both repression and feedback inhibition governs the synthesis and activity of the biosynthetic enzymes and the tRNA synthetases in a common and parallel pattern (**transcriptional control**). The details of these genetic regulatory activities have been discussed in Chapter 4. In some organisms, such as the cyanobacteria, regulation of branched-chain amino acid synthesis appears to be regulated primarily at the level of feedback inhibition.

In a eukaryotic cell, such as *S. cerevisiae*, the genes encoding the enzymes of biosynthetic pathways are often spread over the entire genome and are under the control of their own individual promoters. Starvation for a given amino acid results in transcription of many genes in pathways unrelated to the limiting amino acid. This regulatory effect has been termed **general amino acid control**. This system is under the control of a series of *trans*acting regulatory genes with both positive (*GCN1* to *GCN9*) and negative (*GCD1* to *GCD13*) effects. A recessive mutation in any one of the nine *GCN* genes results in impairment of enzyme derepression under the conditions of amino acid starvation (**Gcn⁻ phenotype**, for general control nonderepressible). A recessive mutation in any one of the 12 *GCD* genes results in constitutive derepression of enzymes subject to general control under nonstarvation conditions (**Gcd⁻ phenotype**, general control derepressed).

The promotor regions of the structural genes of *S. cerevisiae* contain elements necessary for the basal level of transcription in the presence of amino acids (*BAS* genes) and elements necessary for an additional stimulation of transcription in the absence of amino acids. For most of the 30 or more amino acid biosynthetic genes subject to general control, this starvation response is mediated by the GCN4 activator protein. The GCN4 protein binds specifically to 5'-TGA(C/G)TCA-3' upstream activation sequences (UASs) in the promoters of these genes. Removal of these UASs or the introduction of a *gcn4* mutation prevents activation of the corresponding genes under the conditions of amino acid starvation. The translational derepression of *GCN4* mRNA is dependent on the functions of the *GCN1*, *GCN2*, and *GCN3* gene products as well as four small open reading frames located in the 5'-untranslated region of the mRNA. Expression of amino acid biosynthetic genes in *S. cerevisiae* is also subject to pathway-specific repression by the end-products of the pathway.

THE SERINE–GLYCINE FAMILY

In bacterial, fungal, and mammalian systems, metabolic relationships between glycine and serine may be described by the general scheme shown in Figure 10-14. Serine is a precursor of L-cysteine (Fig. 10-15). Glycine contributes carbon and nitrogen in the biosynthesis of purines, porphyrins, and other metabolites. As shown in Figure 10-14, serine hydroxymethyltransferase (SHMT), the *glyA* gene product, converts serine to gly-

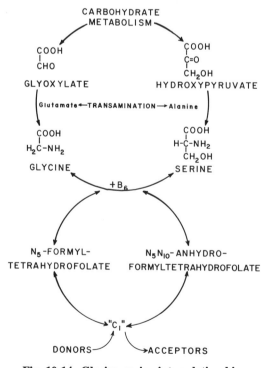

Fig. 10-14. Glycine–serine interrelationships.

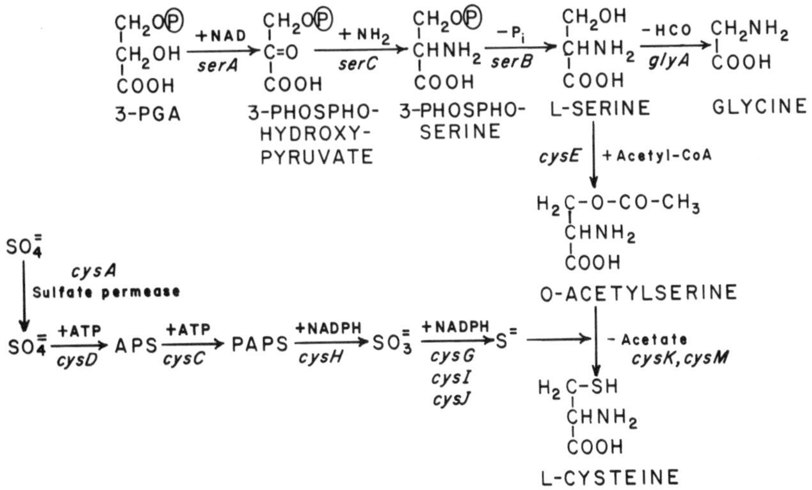

$$
\begin{array}{ccccccccc}
\underset{\text{3-PGA}}{\begin{array}{c}\mathrm{CH_2O\textcircled{P}}\\\mathrm{CH_2OH}\\\mathrm{COOH}\end{array}}
& \xrightarrow[serA]{+\mathrm{NAD}}
& \underset{\begin{array}{c}\text{3-PHOSPHO-}\\\text{HYDROXY-}\\\text{PYRUVATE}\end{array}}{\begin{array}{c}\mathrm{CH_2O\textcircled{P}}\\\mathrm{C{=}O}\\\mathrm{COOH}\end{array}}
& \xrightarrow[serC]{+\mathrm{NH_2}}
& \underset{\begin{array}{c}\text{3-PHOSPHO-}\\\text{SERINE}\end{array}}{\begin{array}{c}\mathrm{CH_2O\textcircled{P}}\\\mathrm{CHNH_2}\\\mathrm{COOH}\end{array}}
& \xrightarrow[serB]{-\mathrm{P_i}}
& \underset{\text{L-SERINE}}{\begin{array}{c}\mathrm{CH_2OH}\\\mathrm{CHNH_2}\\\mathrm{COOH}\end{array}}
& \xrightarrow[glyA]{-\mathrm{HCO}}
& \underset{\text{GLYCINE}}{\begin{array}{c}\mathrm{CH_2NH_2}\\\mathrm{COOH}\end{array}}
\end{array}
$$

cysE | +Acetyl-CoA (from L-SERINE down to:)

$$
\underset{\text{O-ACETYLSERINE}}{\begin{array}{c}\mathrm{H_2C-O-CO-CH_3}\\\mathrm{CHNH_2}\\\mathrm{COOH}\end{array}}
$$

$$
\mathrm{SO_4^=} \qquad \underset{\text{Sulfate permease}}{cysA}\ \downarrow
$$

$$
\mathrm{SO_4^=} \xrightarrow[cysD]{+\mathrm{ATP}} \mathrm{APS} \xrightarrow[cysC]{+\mathrm{ATP}} \mathrm{PAPS} \xrightarrow[cysH]{+\mathrm{NADPH}} \mathrm{SO_3^=} \xrightarrow[\substack{cysG\\cysI\\cysJ}]{+\mathrm{NADPH}} \mathrm{S^=} \xrightarrow[cysK,\,cysM]{-\text{Acetate}}
$$

$$
\underset{\text{L-CYSTEINE}}{\begin{array}{c}\mathrm{H_2C-SH}\\\mathrm{CHNH_2}\\\mathrm{COOH}\end{array}}
$$

Fig. 10-15. Biosynthesis of serine, glycine, and cysteine. The structural gene designations in *E. coli* are *serA* = 3-phosphoglycerate (3-PGA) dehydrogenase; *serC* = phosphoserine aminotransferase; *serB* = phosphoserine phosphatase; *glyA* = serine hydroxymethyltransferase; *cysA* = sulfate permease; *cysD* = sulfate adenylyltransferase; *cysC* = adenylylsulfate (APS) kinase; *cysH* = 3′-phosphoadenylyl sulfate (PAPS) reductase; *cysG, cysI, cysJ* = sulfite reductase; *cysE* = serine acetyltransferase; *cysK* = acetylserine sulfhydrylase A; *cysM* = acetylserine sulfhydrylase B. *cysB* is a regulator of the *cys* operon.

cine and 5,10-methylenetetrahydrofolate, a major contributor of one-carbon units to formation of methionine, purines, and thymine. Oxidative cleavage of glycine by the glycine cleavage (GCV) enzyme system provides a second source of one-carbon units:

$$\mathrm{CH_2NH_2COOH} \longrightarrow \mathrm{NH_3} + \mathrm{CO_2} + 5,10\text{-methylenetetrahydrofolate}$$

The GCV system has been described in mammalian liver and in several bacterial species including *E. coli*, *Peptococcus glycinophilus*, and *Arthrobacter globiformis*. It consists of four proteins: the P protein, a pyridoxal phosphate enzyme; the H protein, a lipoate-containing hydrogen carrier; the T protein that transfers the methylene carbon to THF; and the L-protein, a lipoamide dehydrogenase encoded by *lpd* that is common to the pyruvate and α-ketoglutarate dehydrogenase complexes. The genes encoding these enzymes form an operon that maps at 65.2 min on the *E. coli* chromosome. The regulatory proteins Lrp, GcvA, and PurR are involved in the regulation of the glycine-inducible GCV system.

Serine, glycine, and cysteine are synthesized via a pathway emanating from 3-phosphoglycerate and proceeding through a series of phosphorylated intermediates, as shown in Figure 10-15. Serine may be derived from glycine formed by transamination of glyoxylate. Deamination of threonine to α-ketobutyrate with cleavage to acetyl-CoA and glycine can also give rise to serine.

Bakers' yeast possesses the phosphorylated pathway to serine from 3-phosphoglycerate as well as the glyoxylate pathway. Regulation of the phosphorylated pathway by serine feedback inhibition of 3-phosphoglycerate dehydrogenase and regulation of the glyoxylate transaminase has been demonstrated in *S. cerevisiae*. In bacteria the phosphorylated pathway appears to be the main route of serine formation. The bacterial pathway is also regulated by feedback inhibition of 3-phosphoglycerate dehydrogenase.

The converging pathways of sulfate reduction and the formation of *O*-acetylserine shown in Figure 10-15 appear to be the most common route for the synthesis of cysteine. However, cysteine may also arise from serine and hydrogen sulfide by the action of serine sulfhydrase (cysteine synthase). Cysteine is also formed by transsulfuration between homocysteine and serine:

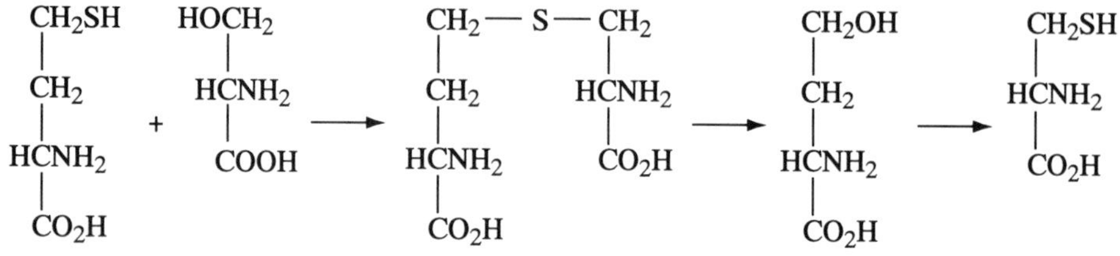

$$
\begin{array}{ccccccccc}
CH_2SH & HOCH_2 & & CH_2\!-\!S\!-\!CH_2 & & CH_2OH & & CH_2SH \\
| & | & & | \qquad\quad | & & | & & | \\
CH_2 & HCNH_2 & & CH_2 \quad\; HCNH_2 & & CH_2 & & HCNH_2 \\
| & | & + & | \qquad\quad | & \longrightarrow & | & \longrightarrow & | \\
HCNH_2 & COOH & & HCNH_2 \quad CO_2H & & HCNH_2 & & CO_2H \\
| & & & | & & | & & \\
CO_2H & & & CO_2H & & CO_2H & & \\
\end{array}
$$

Homocysteine   Serine   Cystathionine   Homoserine   Cysteine

Under conditions where methionine serves as the sole source of sulfur, homocysteine is formed by a pathway involving SAM:

$$\text{Methionine} + \text{ATP} \longrightarrow \text{SAM} \xrightarrow{\;-CH_3\;}$$

$$S\text{-adenosylhomocysteine} \xrightarrow{\;-ado\;} \text{homocysteine}$$

Cysteine biosynthesis is regulated by the genes that code for the enzymes in the pathways of sulfate reduction and the formation of *O*-acetylserine from serine (Fig. 10-15). In *S. typhimurium* each of the enzymes in the sequence is repressed by cysteine. The first enzyme in the sulfate reduction pathway (sulfate permease, CysA) and serine acetyltransferase (CysE) are also inhibited by cysteine via a feedback mechanism. *O*-Acetylserine induces the transport of sulfate by sulfate permease and also induces the rest of the enzymes in the sulfate reduction sequence. Although the biosynthesis of cysteine and methionine are obviously interrelated, it is the sulfate transport system and the pathway involved in the activation of sulfur that regulate the synthesis of cysteine through positive control mechanisms, whereas methionine biosynthesis is under negative control. However, it has been proposed that the *O*-acetyl derivatives play a parallel role in the regulation of the synthesis of methionine and cysteine by inducing the initial steps in the pathway. Biochemical and genetic studies indicate that three of the genes controlling the pathway of sulfate activation (*cysIJH*) are arranged in a cluster forming an operon. Other regulatory interrelationships exist between the cysteine and methionine pathways.

Fig. 10-16. Tetrapyrrole biosynthetic pathway. The enzymatic steps in the pathway are catalyzed by (1) glutamyl-tRNA synthase; (2) NAD(P)H:glutamyl-tRNA reductase; (3) GSA (glutamate γ-semialdehyde) 2,1-aminotransferase; (4) PBG (porphobilinogen) synthase; (5) HMB (hydroxymethylbilane) synthase; (6) UroIII (uroporphyrinogen) synthase; (7) uroporphyrinogen III decarboxylase; (8) coproporphyrinogen III oxidase; (9) protoporphyrinogen IX oxidase; (10) ferrochetolase; (11) ALA synthase.

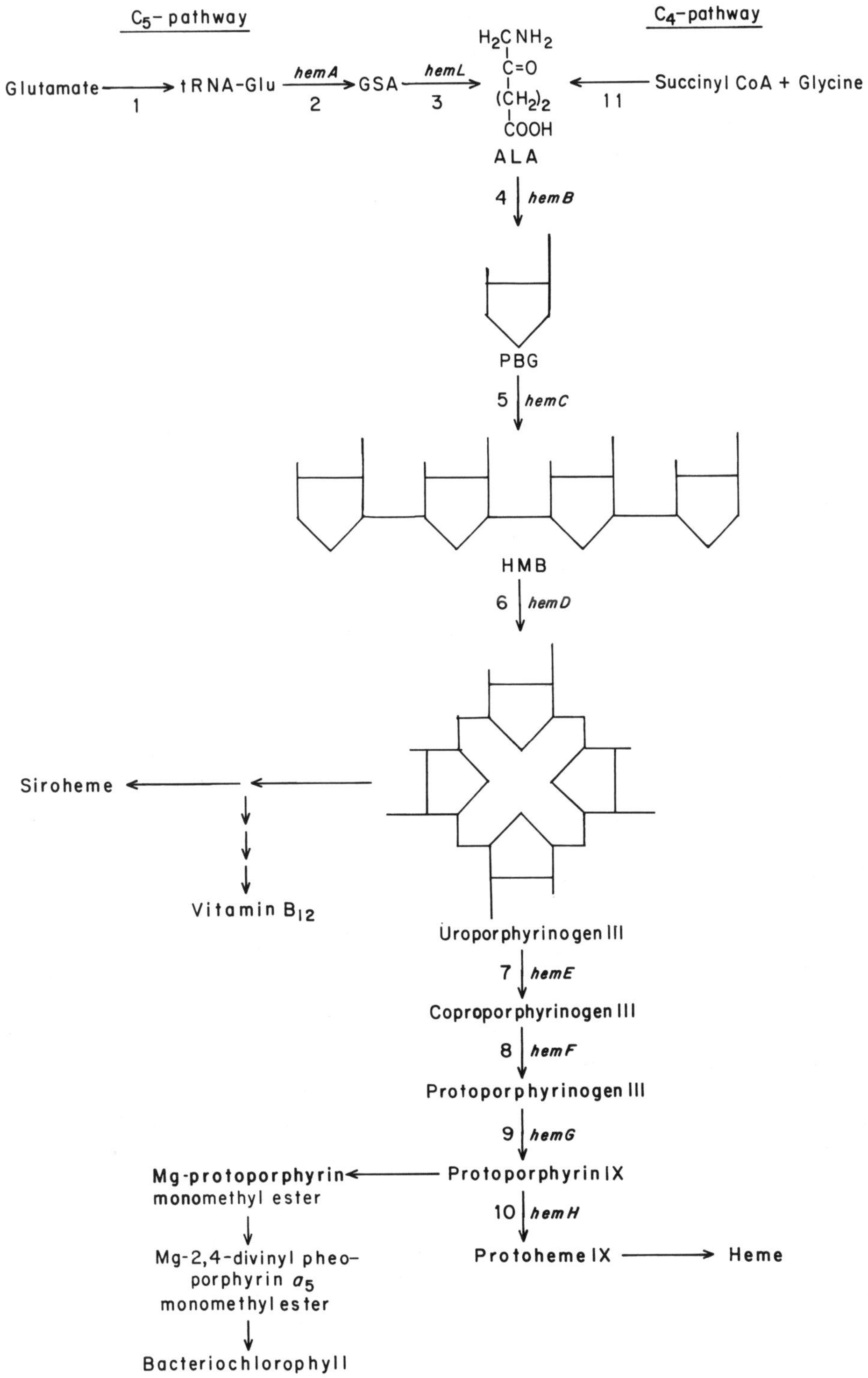

C$_5$-pathway   C$_4$-pathway

Glutamate ——→ tRNA-Glu —$\xrightarrow{hemA}$— GSA —$\xrightarrow{hemL}$—
1   2   3

H$_2$CNH$_2$
C=O
(CH$_2$)$_2$
COOH
ALA

←—— 11 ←— Succinyl CoA + Glycine

4 │ hemB

PBG

5 │ hemC

HMB

6 │ hemD

Siroheme ←———←

Vitamin B$_{12}$

Uroporphyrinogen III

7 │ hemE

Coproporphyrinogen III

8 │ hemF

Protoporphyrinogen III

9 │ hemG

Mg-protoporphyrin ←—— Protoporphyrin IX
monomethyl ester

10 │ hemH

Mg-2,4-divinyl pheo-
porphyrin a_5
monomethyl ester

Protoheme IX ————→ Heme

Bacteriochlorophyll

Aminolevulinate and the Pathway to Tetrapyrroles

Aminolevulinate (ALA) can be synthesized via the C_4 route through the condensation of glycine and succinyl-CoA by ALA synthetase or the C_5 pathway from glutamate. In either case, ALA is the first committed step in the formation of tetrapyrroles, as shown in Figure 10-16. The pathway diverges at uroporphyrinogen III (UroIII), one branch leading to the synthesis of siroheme and vitamin B_{12} and the other leading to the synthesis of protoporphyrin IX. At this point the pathway branches again, one branch leading to the synthesis of heme and the other to bacteriochlorophyll formation. Some of the complexities of the divergent pathways to heme, siroheme, and cobinamide are shown in Figure 10-17.

THE AROMATIC AMINO ACID PATHWAY

Phenylalanine, Tyrosine, and Tryptophan

Demonstration of shikimate, anthranilate, and indole as precursors of tryptophan and other aromatic amino acids in fungi and bacteria provided the earliest clues as to the intermediates in the aromatic amino acid pathway. Inhibition by analogs of the aromatic amino acids, isolation of mutants with multiple requirements for aromatic compounds, and other nutritional and biochemical findings indicated that these compounds shared a common origin. Shikimate substituted for the aromatic amino acid requirements of many auxotrophs, indicating that it was the common precursor of at least five different aromatic compounds. As a result of these findings, shikimate was considered to be the major branchpoint compound in the aromatic pathway. However, in both bacteria and fungi it was ultimately shown that chorismate serves in this capacity (Fig. 10-18).

The Common Aromatic Amino Acid Pathway

The aromatic amino acids phenylalanine, tyrosine, tryptophan, and several other related aromatic compounds are produced via a common pathway that begins with the condensation of erythrose-4-phosphate and phosphoenolpyruvate to form 3-deoxy-D-arabino-heptulosonate 7-phosphate (DAHP). The enzymes and structural genes for the common aromatic pathway and its branches to tyrosine and phenylalanine are shown in Figure 10-18. DAHP is converted to shikimate and then to chorismate. Chorismate is so-named because it represents the branch-point leading to the formation of other aromatic amino acids and several other related metabolites. The formation of many additional essential aromatic compounds via branches of this pathway emphasizes the need for a complex regulatory system. Furthermore, a mutant blocked in the common pathway prior to the branch-point may exhibit multiple nutritional requirements for the end-products of the branches. Although there are many similarities in the aromatic amino acid pathway in all bacteria and fungi, there are several differences in detail regarding the manner in which tryptophan, tyrosine, and phenylalanine are formed. Even more apparent are the number of variations in the ways in which other aromatic compounds are formed through branches or extensions from the common pathway.

In *E. coli* and *S. typhimurium*, regulation of the common aromatic pathway is modulated through three unlinked genes, *aroF*, *aroG*, and *aroH*, that encode isozymes of the

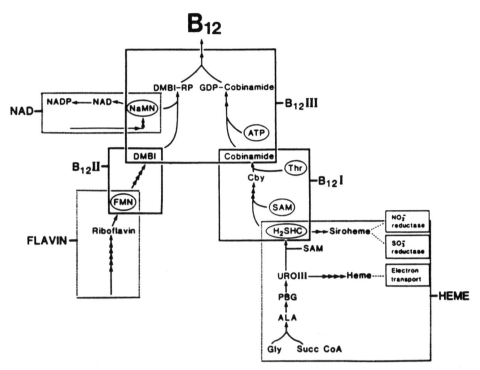

Fig. 10-17. The tetrapyrrole pathway showing the complexities of the branches leading to heme, siroheme, cobinamide, and vitamin B$_{12}$. Branch I represents cobinamide synthesis, Branch II represents DMBI (5,6-dimethylbenzimidazole), and Branch III represents cobalamin (vitamin B$_{12}$) synthesis from these precursors. Interconnections with the pyridine nucleotide, flavin, and heme biosynthetic pathways are depicted. ALA = δ-aminolevulinate; Cby = cobyric acid; DMBI-RP, 1-α-D-ribofuranosido-DMBI; H$_2$SHC = dihydrosirohydrochlorin; NaMN = nicotinic acid mononucleotide; PBG = porphobilinogen; SAM = S-adenosylmethionine; Succ CoA = succinyl coenzyme A; URO III = uroporphyrinogen III. From Jeter, R. M. and J. R. Roth. 1987. *J. Bacteriol.* **169:**3189–3198.

first enzyme, DAHP synthetase, that are sensitive to tyrosine, phenylalanine, and tryptophan, respectively. Although all three DAHP synthetases are regulated transcriptionally, feedback inhibition is quantitatively the major control mechanism *in vivo*. A detailed study of the three isozymes of DAHP synthetase and the genes specifying their synthesis in *S. typhimurium* showed that the structural gene for the phenylalanine-inhibitable enzyme (*aroH*) is linked to *gal*. The structural gene for the tryptophan-inhibitable isozyme (*aroF*) is linked to *aroE*, whereas that for the tyrosine-inhibitable isozyme (*aroG*) is linked to *pheA* and *tyrA*. The *pheA* and *tyrA* gene products are the respective branchpoint enzymes leading to the phenylalanine and tyrosine pathways. Only the tyrosine-related isozyme (*aroG*) was completely repressed, as well as feedback inhibited, by low levels of tyrosine.

Comparable studies in *E. coli* revealed a similar pattern of isozymic control of DAHP synthetase and interrelationships of the genes governing the aromatic pathway. Expression of *aroF* and *aroG* is repressed by the Tyr repressor, the *tyrR* gene product, mediated by tyrosine and phenylalanine, respectively. The TyrR protein also regulates the expres-

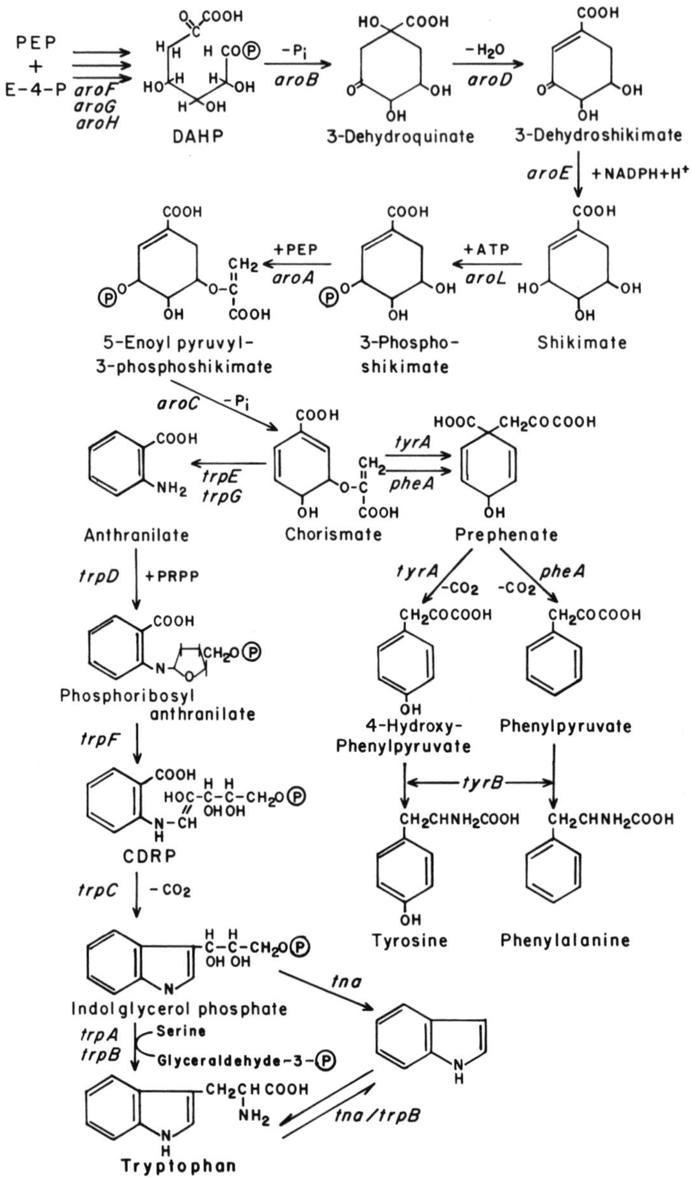

Fig. 10-18. Biosynthesis of the aromatic amino acids. The structural gene designations for the enzymes in *E. coli* are: *aroF*, *aroG*, *aroH* = 3 hydroxy-D-arabino-heptulosonate 7-phosphate (DAHP) synthetase; *aroB* = 3-dehydroquinate synthase; *aroD* = 3-dehydroqinate dehydratase; *aroE* = shikimate dehydrogenase; *aroL* = shikimate kinase; *aroA* = 5-enolpyruvoylshikimate 3-phosphate synthase; *aroC* = chorismate synthase; *pheA* = chorismate mutase P; *tyrA* = chorismate mutase T; *pheA* = prephenate dehydratase; *tyrA* = prephenate dehydrogenase; *tyrB* = tyrosine aminotransferase; *trpE*, *trpG* = anthranilate synthase; *trpD* = anthranilate phosphoribosyl transferase; *trpF* = phosphoribosyl anthranilate isomerase; *trpC* = indolglycerol phosphate synthase; *trpA* + *trpB* = tryptophan synthase; *tna* = tryptophanase. CDRP = 1-(*O*-carboxyphenylamino)-1-deoxyribulose 5-phosphate; PEP = phsophoenolpyruate; E-4-P = erythrose-4-phosphate.

sion of several other genes concerned with aromatic amino acid biosynthesis or transport. These include *aroL, aroP, tyrB, tyrP,* and *mtr* (resistance to the tryptophan analog, 5-methyltryptophan). Expression of *aroH* is controlled by the Trp repressor (encoded by *trpR*), which also regulates the expression of the *trp* operon, the *mtr* gene, as well as expression of *trpR* itself.

In a wide variety of fungi, the genes coding for enzymes 2–6 in the prechorismate pathway (Fig. 10-18) are arranged in a cluster, designated the *arom* gene cluster. In all of the fungal species examined to date, these enzymes sediment as a complex aggregate on centrifugation in sucrose density gradients. The sedimentation coefficients for these enzymes are very similar in representative examples of Basidiomycetes (*Coprinus lagopus, Ustilago maydis*), Ascomycetes (*N. crassa, S. cerevisiae, A. nidulans*), and Phycomycetes (*Rhizopus stolonifer, Phycomyces nitens, Absidia glauca*). By comparison, the five enzymes catalyzing the conversion of DAHP to chorismate in bacteria are physically separable in *E. coli, S. typhimurium, E. aerogenes, Anabaena variabilis,* and *Chlamydomonas reinhardii.* In *Euglena gracilis* an enzyme aggregate with five activities has been isolated. This aggregate can be dissociated into smaller components.

In some bacteria, the genes coding for enzymes of the chorismate pathway appear to be scattered on the chromosome. However, in *B. subtilis,* a cluster of contiguous genes for tryptophan biosynthesis represent a polycistronic operon. The *trpEDCFBA* genes form an operon similar to operons found in enteric bacteria. However, in *B. subtilis* the *trp* genes are part of a "**supraoperon**" containing *trpEDCFBA-hisH-tyrA-aroE,* as shown in Figure 10-19A. It is likely that *aroFBH* may also be present at the 5′ end of this supraoperon. All *aroF* mutants (chorismate synthetase) also lack dehydroquinate synthase (*AroB*) activity. The gene that specifies AroB is closely linked to the gene coding for the AroF enzyme. Both genes are part of the *aro* gene cluster. Mutants lacking chorismate mutase activity also lack DAHP synthetase and shikimate kinase activity, presumably as a result of their aggregation in a multienzyme complex. As an indication of the complexity of the "supraoperon," the *mtrAB* operon of *B. subtilis* encodes GTP cyclohydrolase I (MtrA), an enzyme involved in folate biosynthesis. The *mtrB* gene is a transacting RNA-binding regulatory protein activated by tryptophan. Transcription termination at the attenuator preceding the *trp* gene cluster presumably occurs as a consequence of binding of the activated MtrB protein to the nascent transcript (see later discussion on interpathway regulation).

The genes controlling tryptophan synthesis in coliform organisms are also arranged in an operon. Arrangement of the *E. coli* operon and the enzymes under its control are shown in Figure 10-19B. *Escherichia coli* and *S. typhimurium* each have five structural genes for the enzymes in the operon and they are induced coordinately in response to the availability of tryptophan. In *E. coli* the activities of the first two enzymes are catalyzed by an aggregate formed from the products of the first two genes in the operon. The activities of the next two steps (3 and 4) are catalyzed by a single protein. Tryptophan synthase converts indolglycerol phosphate to indol and then couples indol to serine to form tryptophan. The enzyme is a complex formed from the products of the last two genes (*trpA* and *trpB*). A regulatory site preceding the structural genes is found to regulate the synthesis of mRNA^{Trp} and the enzymes of the tryptophan pathway, thus reducing operon expression. This **attenuator** function apparently occurs at the level of transcription by providing a region in which transcription is terminated (see Chapter 4 for additional details concerning regulation of gene expression).

(a)

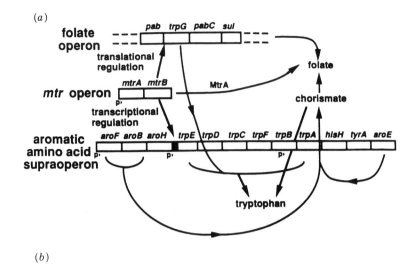

(b)

| Chorismate | → | AA | → | CP–PRA | → | CDRP | → | InGP | → | Tryptophan |

| | | 1 | | 2 | | 3 | 4 | | 5 | |

| p1, 0 | a | trpE | | trpG/D | p2 | trpC/F | | | trpB | trpA |

| | | 1 | | 2 | | 3 | 4 | | 5a | 5b |

p1 = promoter
o = operator
a = attenuator
p2 = internal promoter

Anthranilate	Phosphoribosyl	Phosphoribosyl	Indoleglycerol	Tryptophan	Tryptophan
synthase	transferase	anthranilate	phosphate	synthase	synthase
		isomerase	synthase	β	α

Responds to:

Anthranilate					
Indole	Indole	Indole	Indole	Indole	
Tryptophan	Tryptophan	Tryptophan	Tryptophan	Tryptophan	Tryptophan

Accumulates: None / Anthranilate / Anthranilate / CDR / Indoleglycerol / Indole

Fig. 10-19. Organization of the genes for tryptophan biosynthesis in *Bacillus subtilis* and *Escherichia coli*. (a) The aromatic amino acid supraoperon of *B. subtilis* showing interrelationships with the genes from the folic acid operon and the *mtr* operon. The MtrA protein (GTP cyclohydrolase I) catalyzes formation of the initial pteridine ring (dihydroneopterin triphosphate) of folic acid. The proteins AroF, AroB, and AroE are required for chorismate synthesis. The proteins Pab, TrpG, and PabC are involved in *p*-aminobenzoate synthesis from chorismate. Sul catalyzes condensation of *p*-aminobenzoate and the pteridine ring. MtrB is considered to bind to a target sequence that overlaps the *trpG* ribosome-binding site, resulting in a translational regulation, and to the *trpE* leader transcript, resulting in transcriptional regulation of the *trp* gene cluster. The *trpEDCFBA* and *trpG* gene products are necessary for tryptophan biosynthesis from chorismate. Note that AroH is involved in phenylalanine and tyrosine biosynthesis from chorismate in some strains of *Bacillus*. p> marks the positions of known promoters. ■ denotes the position of the *trp* attenuator, = = = indicates that open reading frames exist upstream of *pab* and downstream of *sul*. From Babitzke, P., P. Gollnick, and C. Yanofsky. 1992. *J. Bacteriol.* **174**:2059–2064. (b) Organization of the genes of the tryptophan operon of *E. coli*. AA = anthranilate; CP-PRA = *N*-(*o*-carboxyphenyl)-phosphoribosylamine; CDRP = 1-(*o*-carboxyphenylamino)-1-deoxyribulose 5′-phosphate; InGP = indoleglycerol phosphate.

Pathways to Tyrosine and Phenylalanine

The biosynthesis of L-tyrosine and L-phenylalanine from chorismate can occur via alternate routes in many organisms (Fig. 10-20). Chorismate is converted to prephenate by chorismate mutase. Prephenate is converted to phenylpyruvate by prephenate dehydratase and then to phenylalanine or to 4-hydroxyphenylpyruvate and then to tyrosine. Alternatively, prephenate may be converted to arogenate by transamination with arogenate dehydratase forming phenylalanine or arogenate dehydrogenase forming tyrosine. Utilization of these alternate pathways varies greatly from one organism to another. In *E. coli, S. typhimurium, B. subtilis,* and other common bacteria, the prephenate pathway to phenylpyruvate and phenylalanine and to 4-hydroxyphenylpyruvate and tyrosine is the predominant route. On the other hand, *Euglena gracilis* uses arogenate as the sole precursor of both tyrosine and phenylalanine by a pathway in which arogenate rather than prephenate is the branch point. The presence of the arogenate pathway is common in the cyanobacteria. However, the degree to which it is used and the manner in which the alternative pathways are regulated differ widely from one species to another. *Pseudomonas aeruginosa* and other members of the pseudomonad group possess coexisting alternative pathways for the formation of tyrosine and phenylalanine. These alternative pathways account for the unusual resistance of these organisms to analogs of the aromatic amino acids. Under certain circumstances, the arogenate pathway appears to function as an unregulated overflow route for tyrosine production. In some members of the coryneform group of bacteria (*Corynebacterium glutamicum, Brevibacterium flavum,* and *B.*

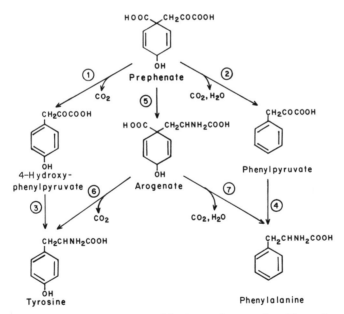

Fig. 10-20. Alternate pathways to L-phenylalanine and L-tyrosine. The pathways shown are observed in *Pseudomonas aeruginosa*. Various organisms may express any combination of these pathways. The dehydrogenase pathways may be either NAD^+ or $NADP^+$ linked. The enzymes are (1) prephenate dehydrogenase, (2) prephenate dehydratase, (3) 4-hydroxyphenylpyruvate aminotransferase, (4) phenylpyruvate aminotransferase, (5) prephenate aminotransferase, (6) arogenate dehydratase, and (7) arogenate dehydrogenase.

ammoniogenes) the arogenate pathway is an obligatory route for the formation of tyrosine. By comparison, arogenate seems to be a dead-end metabolite in *N. crassa*, the prephenate pathway being used as the major route to tyrosine and phenylalanine.

Salmonella typhimurium and *E. coli* synthesize tyrosine and phenylalanine by virtually separate pathways in that there are separate DAHP synthetases and chorismate mutases coded for by genes specific to either the tyrosine or phenylalanine pathways. The chorismate mutase T and prephenate dehydrogenase coded by *tyrA* are specific for tyrosine formation while chorismate mutase P and prephenate dehydratase coded by *pheA* are specific for phenylalanine biosynthesis. In *S. typhimurium* the genes *aroF* and *tyrA*, which specify the tyrosine-repressible DAHP synthase (AroF) and chorismate mutase T-prephenate dehydrogenase, comprise an operon. The genes *aroG* and *pheA*, which specify phenylalanine-repressible DAHP synthase (AroG) and chorismate mutase P-prephenate dehydratase (PheA), regulate the synthesis of phenylalanine, as shown in Figure 10-21. Chorismate mutase T-prephenate dehydrogenase and chorismate mutase P-prephenate dehydratase exist as bifunctional enzymes in *S. typhimurium*, *E. coli*, and *E. aerogenes*. These same enzymes are not found in aggregate forms in *N. crassa*, *S. cerevisiae*, or *B. subtilis*. Within the genera *Pseudomonas*, *Xanthomonas*, and *Alcaligenes*, variation in the enzymes of tyrosine biosynthesis comprises five groups that compare favorably with rRNA–DNA hybridization groups. The rRNA homology groups I, IV, and V all lack activity for arogenate/NADP dehydrogenase. Group II species possess arogenate dehydrogenase (with lack of specificity for NAD or NADP) and are sensitive to feedback inhibition by tyrosine. The arogenate dehydrogenase of species in Group III is insensitive to feedback inhibition. Group IV displays prephenate/NADP dehydrogenase activity, whereas Groups I and V lack this activity.

Mammalian species appear to have the ability to convert phenylalanine to tyrosine through the action of phenylalanine hydroxylase:

$$\text{phenylalanine} + 0.5 O_2 \longrightarrow \text{tyrosine} + H_2O$$

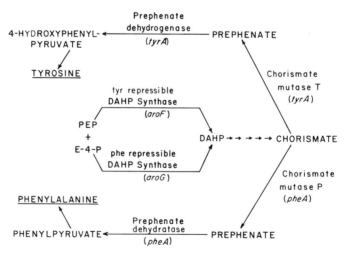

Fig. 10-21. Genes and enzymes of phenylalanine and tyrosine biosynthesis in *E. coli* and *S. typhimurium*. PEP = phosphoenolpyruvate; E-4-P = erythrose-4-phosphate; DAHP = 3-deoxy-D-arabinoheptulosonate 7-phosphate.

In microorganisms this activity is not common, dual requirements for tyrosine and phenylalanine generally accompanying mutational loss of enzyme activity at a point prior to the branch point in the synthesis of these amino acids. *Pseudomonas* forms an inducible phenylalanine hydroxylase. Since hydroxylation reactions often represent the initiating step in the degradation of aromatic compounds, as discussed in Chapter 7, it seems plausible that a highly degradative organism, such as *Pseudomonas*, would possess such activity for degradative purposes and only secondarily to provide tyrosine from phenylalanine.

p-Aminobenzoate and Folate Biosynthesis

Many bacteria and fungi, including *E. coli, B. subtilis, Streptomyces griseus*, and *Pseudomonas acidovorans* convert chorismate to *p*-aminobenzoate (PAB) via PAB synthase (Fig. 10-22). In *E. coli* the enzyme complex termed PAB synthase consists of two nonidentical subunits designated Components I and II. Component I is encoded by *pabB* and catalyzes the synthesis of 4-amino-4-deoxychorismate from chorismate and ammonia. Component II, encoded by *pabA*, is a glutamine amidotransferase that uses the amide nitrogen of glutamine to convert chorismate to 4-amino-4-deoxychorismate. 4-Amino-4-deoxychorismate is converted to PAB by aminodeoxychorismate lyase, encoded by *pabC*. The PAB is then condensed with 6-hydroxymethyl-7,8-pterin pyrophosphate by dihydropteroate synthase (SulA). 7,8-Dihydropteroate is then converted to 7,8-dihydrofolate by the addition of glutamate by dihydrofolate synthetase (FolC). After 7, 8-dihydrofolate is reduced to tetrahydrofolate by dihydrofolate reductase, the synthase enzyme then serves to add additional glutamate units to form folylpolyglutamate.

The pteridine component of folate is derived from guanosine triphosphate (GTP). The first step in this sequence is the removal of C-8 from GTP as formic acid by the enzyme GTP cyclohydrolase I to form dihydroneopterin triphosphate. In *B. subtilis* this enzyme is the product of the *mtrA* gene. The GTP cyclohydrolase I has been identified in a number of other bacterial species including *E. coli* and *S. typhimurium*. Dihydroneopterin triphosphate is converted to 7,8-dihydroneopterin by removal of the phosphate residues. 7,8-Dihydropterin is then converted to 6-hydroxymethyl-7,8-dihydropterin by the action of 6-hydroxymethyl-7,8-dihydropterin kinase, the product of the *folK* gene in *E. coli* (*sulD* in *Streptococcus pneumoniae*). In *S. pneumoniae*, Sul D is a bifunctional enzyme with both 6-hydroxylmethyl-7,8-dihydropterin kinase and 7,8-dihydroneopterin aldolase activities.

Enterobactin Biosynthesis

The catechol siderophore enterobactin (enterochelin) is an iron-chelating compound that facilitates the transport of iron (see Chapter 7). It is a cyclic trimer of *N*-2,3-dihydroxybenzoylserine, which is synthesized from chorismate. The first stage of this pathway is the formation of 2,3-dihydroxybenzoate via the intermediates isochorismate and 2,3-dihydro-2,3-dihydroxybenzoate via the intermediates isochorismate and 2,3-dihydro-2,3-dihydroxybenzoate, as shown in Figure 10-23. The enzymes catalyzing these steps are isochorismate synthetase (EntC), 2,3-dihydro-2,3-dihydroxybenzoate synthetase (EntB), and 2,3-dihydro-2,3-dihydroxybenzoate dehydrogenase (EntA). In *E. coli* eight genes govern the biosynthesis of enterobactin and its function. These genes occur in a cluster at 13 min on the *E. coli* linkage map. The second stage of enterobactin

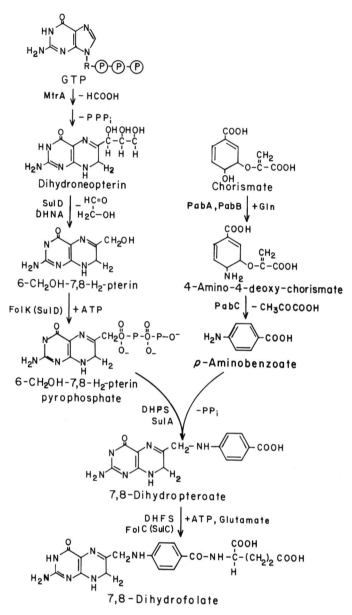

Fig. 10-22. *para***-Aminobenzoate and folate biosynthesis.** PabA/PabB, PAB synthase, consists of Component I, PabA, a glutamine amidotransferase; and Component II, PabB, which uses ammonia to form 4-amino-4-deoxychorismate; PabC = aminodeoxychorismate lyase. FolA (SulD) = 6-hydroxymethyl-2,8-dihydropterin kinase (SulD also serves as a dihydroneopterin aldolase in *Streptococcus pneumoniae*); SulA = dihydropteroate synthase; FolC = dihydrofolate synthase; MtrA = GTP cyclohydrolase. (Pterin is the trivial name for 2-amino-4-hydroxypteridine; neopterin is the trivial name for 6-[D-*erythro*-1′,2′3′-trihydroxypropyl]pterin.)

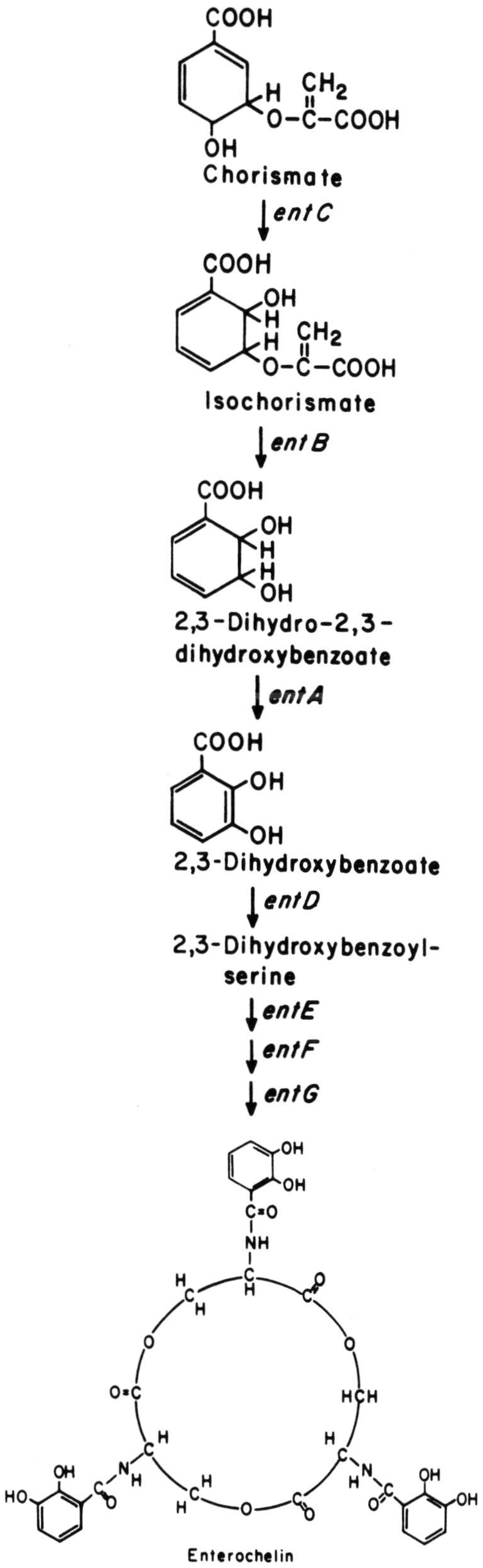

Fig. 10-23. Enzymatic steps in the biosynthesis of enterobactin (enterochelin). EntC = isochorismate synthetase; EntB = 2,3-dihydro-2,3-dihydroxybenzoate synthetase; EntA = 2,3-dihydro-2,3-dihydroxybenzoate dehydrogenase; EntD = 2,3-dihydroxybenzoylserine synthetase; EntE, EntF, and EntG serve to condense three molecules of 2,3-dihydroxybenzoylserine into the cyclic enterobactin structure.

synthesis is catalyzed by a sequence of enzymes (EntD, EntE, EntF, and EntG) present in a multienzyme complex. The EntD couples 2,3-dihydroxybenzoate and serine to form 2,3-dihydroxybenzoylserine. The remaining enzymes catalyze the coupling of three molecules of 2,3-dihydroxybenzoylserine to form the cyclical enterobactin structure.

Biosynthetic Pathway to Ubiquinone

Ubiquinone (coenzyme Q), the only nonprotein component of the electron-transport chain, has the structure:

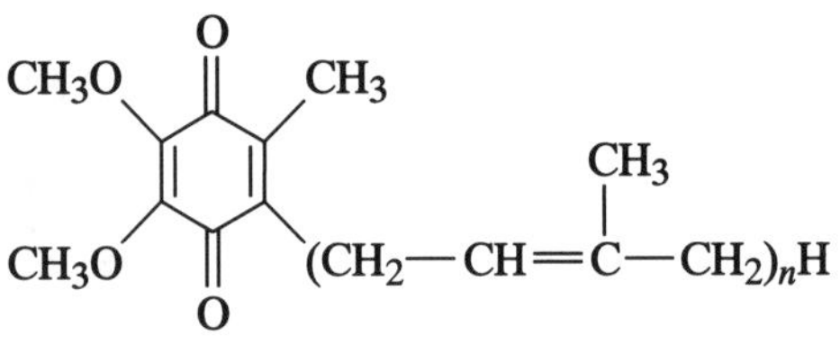

$$CH_3O\text{—}\underset{O}{\overset{O}{\underset{\|}{\overset{\|}{C}}}}\text{—}CH_3 \quad (CH_2\text{—}CH\text{=}\underset{CH_3}{\overset{|}{C}}\text{—}CH_2)_nH$$

This important compound is synthesized from chorismate via the pathway shown in Figure 10-24. The first precursor specific to the ubiquinone pathway, 4-hydroxybenzoate, is formed by the action of chorismate lyase. In some organisms tyrosine can serve as a source

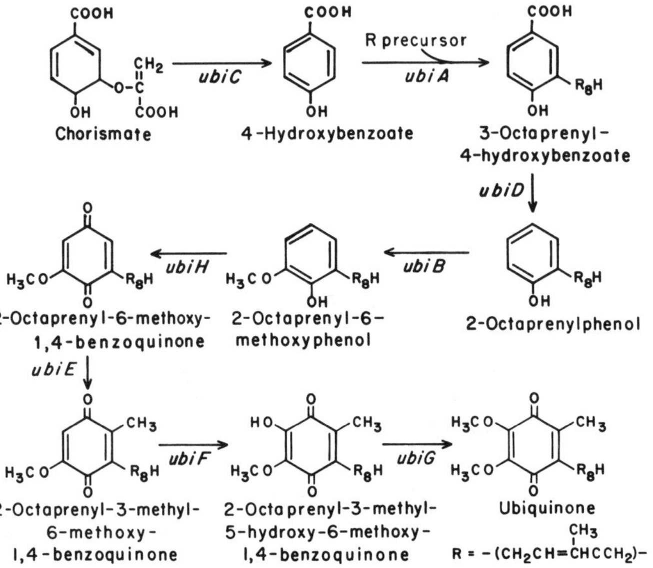

Fig. 10-24. Biosynthetic pathway to ubiquinone (coenzyme Q).

of 4-hydroxybenzoate through the intermediary formation of 4-hydroxyphenylpyruvate. The functions of ubiquinone are discussed in Chapter 7.

Menaquinone (Vitamin K) Biosynthesis

Menaquinone (vitamin K_2) is a methylnaphthoquinone with the structure shown in Figure 10-25. The biosynthesis of menaquinone originates with the addition of α-ketoglutarate to the ring of chorismate and concomitant removal of pyruvate and CO_2. The details of the pathway are shown in Figure 10-25. All naturally occurring menaquinones have *trans* configurations in the double bonds of the prenyl side chain. Menaquinone serves as an

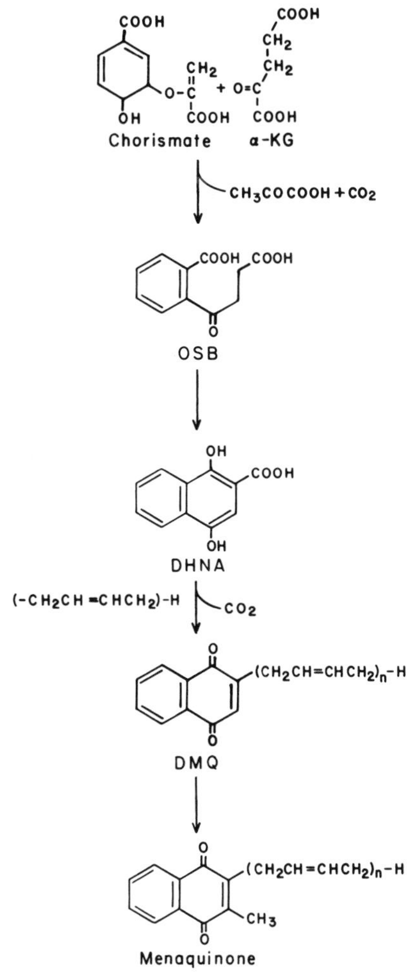

Fig. 10-25. Biosynthetic pathway to menaquinone. α-KG = α-ketoglutarate; OSB = o-succinyl-benzoate (4-[2'-carboxyphenyl]-4-oxobutyrate); DHNA = 1,4-dihydroxy-2-naphthoate; DMQ = demethylmenaquinone.

important electron carrier during anaerobic growth of *E. coli* (see Chapter 7). It is essential for electron transfer to fumarate.

Biosynthesis of Nicotinamide Adenine Dinucleotide (NAD)

Both NAD and NADP, the functional forms of nicotinic acid, are formed from tryptophan in mammals, *N. crassa*, and *S. cerevisiae*, as shown in Figure 10-26. The bacterium *Xanthomonas pruni* also synthesizes NAD from tryptophan. However, most bacteria utilize an entirely different pathway to NAD involving the condensation of aspartate and a three-carbon compound and subsequent conversion of the condensation product into quinolinate, as shown in Figure 10-26. The pathway from quinolinate to NAD appears to be identical in all organisms regardless of the mode of quinolinate formation. *Escherichia*

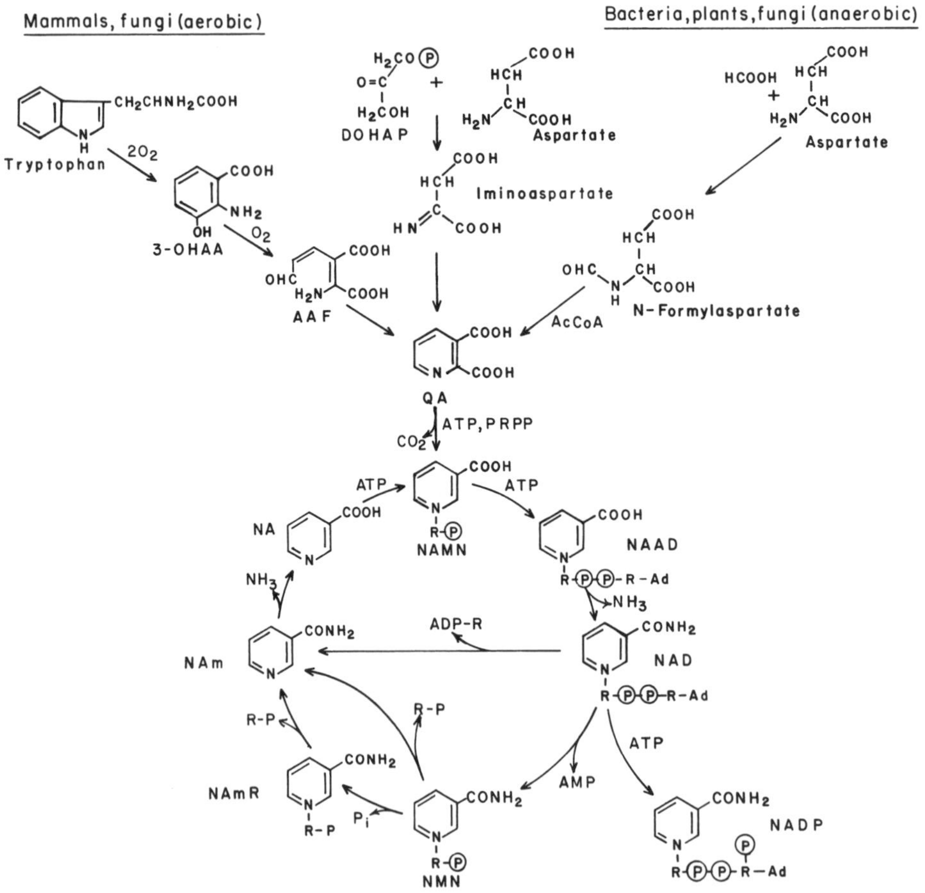

Fig. 10-26. Alternate pathways for NAD biosynthesis and the pyridine nucleotide cycles. 3-OHAA = 3-hydroxyanthranilate; AAF = 2-acroleyl-3-aminofumarate; DOHAP = dihydrodoxy-acetone phosphate; AcCoA = acetyl coenzyme A; QA = quinolinate; NAMN = nicotinic acid mononucleotide; NAAD = nicotinic acid adenine dinucleotide; Ad = adenine; NAD = nicotinamide adenine dinucleotide; NADP = nicotinamide adenine dinucleotide phosphate; NMN = nicotinamide mononucleotide; NAmR = nicotinamide riboside; NA = nicotinic acid; NAm = nicotinamide.

coli, S. typhimurium, M. tuberculosis, and other common bacteria use the aspartate-dihydroxyacetone phosphate pathway. *Clostridium butylicum,* follows a unique pathway via condensation of aspartate and formate into the intermediate *N*-formylaspartate. Further addition of acetate and ring closure yields quinolinate.

Organisms that demonstrate a specific nutritional requirement for nicotinamide utilize a unique route of NAD biosynthesis that involves conversion of nicotinamide to NAD without prior deamidation:

$$\text{nicotinamide} + \text{PRPP} + \text{ATP} \longrightarrow \text{nicotinamide mononucleotide (NMN)}$$

$$\text{NMN} + \text{ATP} \longrightarrow \text{NAD}$$

Lactobacillus fructosus and *Haemophilus influenzae* utilize this route for NAD biosynthesis. This unusual pathway is used for NAD biosynthesis in the nucleus of mammalian cells. The second reaction, catalyzed by NAD pyrophosphorylase (ATP:NMN adenylyltransferase) is considered to be located exclusively in the nucleus of mammalian cells and can be used as a *marker* for testing the purity of isolated nuclei.

The NAD serves as a substrate in several important metabolic reactions. In bacterial cells, NAD serves as a substrate of DNA repair by DNA ligase:

$$\text{nicked DNA} + \text{NAD} \longrightarrow \text{repaired DNA} + \text{NMN} + \text{adenosine}$$

Reactions in which the ADP-ribosyl moiety of NAD is transferred to an acceptor to yield mono-or poly-ADP-ribosylated derivatives are catalyzed by enzymes referred to as ADP-ribosylases:

$$\text{NAD} + \text{acceptor} \longrightarrow \text{acceptor-(ADP-ribose)}_n + \text{nicotinamide}$$

Utilization of NAD as a substrate emphasizes the need for rapid recycling of the degradation products by means of salvage pathways or pyridine nucleotide cycles (PNCs). A number of alternative PNCs have been shown to operate in microorganisms, as shown in Figure 10-26.

Some organisms that degrade NAD to nicotinamide and nicotinic acid cannot recycle these products and accumulate them in the culture medium. The human strains of *Mycobacterium tuberculosis* and the anaerobe *Clostridium butylicum* are examples.

Histidine Biosynthesis

The biosynthesis of histidine occurs via a unique pathway that is more closely linked to the metabolism of pentoses and purines than to any of the other amino acid families. The pathway of histidine biosynthesis, shown in Figure 10-27, is initiated by the coupling of phosphoribosylpyrophosphate (PRPP) with ATP at N-1 of the purine ring followed by opening of the ring. The N-1 and N-2 of the imidazole ring of histidine are thus derived from the N-1 and N-2 of the adenine ring. The aminoimidazolecarboxamide ribonucleotide derivatives are normally bound to the enzymes involved in their formation and rearrangement. Cleavage of the open-ringed structures gives rise to imidazoleglycerolphosphate (IGP) and aminoimidazolecarboxamide ribonucleotide (AICRP), which is recycled via the purine biosynthetic pathway to reform the purine nucleotide. There appear

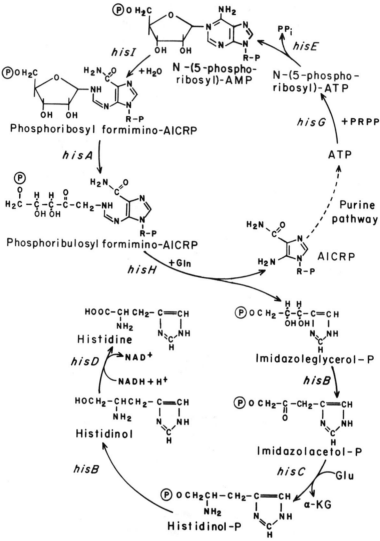

Fig. 10-27. Biosynthesis of histidine. AICRP = 5-aminoimidazole-4-carboxamide ribonucleotide; PRPP = 5-phosphoribosyl 1-pyrophosphate. The structural gene designations for *E. coli* are *hisG*, ATP-PRPP transferase; *hisE* = phosphoribosyl-ATP pyrophosphohydrolase; *hisI* = phosphoribosyl-AMP cyclohydrolase; *hisA* = phosphoribosyl formimino-5-amino-1-phosphoribosyl-4-imidazole carboxamide ribotide isomerase; *hisH* = amidocyclase; *hisC* = histidinol phosphate aminotransferase; *hisB* = imidazoleglycerolphosphate dehydrase and histidinol phosphate phosphatase; *hisD* = histidinol dehydrogenase.

to be no branches from this pathway to other end-products. The chemical intermediates and the enzymes involved are virtually identical in a variety of bacteria and fungi, but the arrangement of the genes controlling the synthesis and operation of the pathway differ markedly from one organism to another.

In *S. typhimurium* the 10 genes controlling the synthesis of the enzymes of the histidine pathway are arranged in a single operon (Fig. 10-28). This is also true for the genes in *Staphylococcus aureus*.

In yeast, the *his4* region specifies three of the enzyme activities of the histidine pathway, namely, the second, third, and tenth steps. The enzymes responsible for these reactions are tightly associated in a multifunctional complex. There is no evidence for aggregation of the remainder of the enzymes, nor are the genes specifying the other enzymes contiguous with these or with each other. This contrasts with the close linkage of the 10 *his* genes in *S. typhimurium* but is comparable to many other gene–enzyme relationships in fungi. The equivalent genes coding for the second, third, and tenth steps in *Neurospora* are also arranged in a cluster. They code for a complex composed of nonidentical isomers. Purified preparations of the complex contain all three enzyme activities.

The arrangement of the genes in the *his* operon of *S. typhimurium* and the *his4* region of *S. cerevisiae* are shown in Figure 10-28. The genes in the *his* operon do not correspond to the order of the reactions in the pathway. Mechanisms for regulation of histidine biosynthesis include (a) operon repression mediated at an attenuated site in response to the availability of histidyl-tRNA^{his} (b) transcriptional control responsible to the general availability of amino acids involving the general signal molecule guanosine

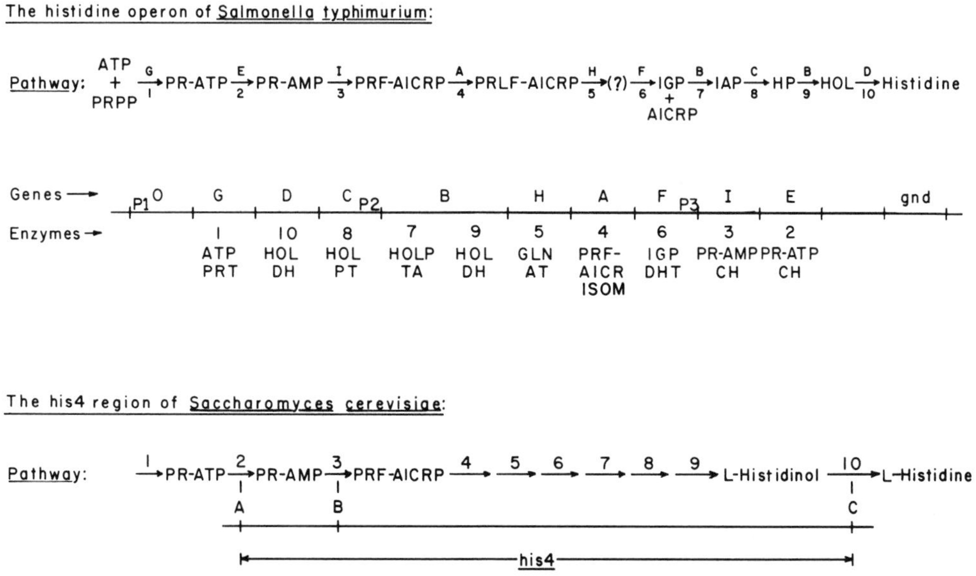

Fig. 10-28. Arrangement of the genes in the histidine operon of *Salmonella typhimurium* and the *his4* region of *Saccharomyces cerevisiae*. The equivalent genes for the second, third, and tenth steps in *Neurospora crassa* are clustered in a manner similar to those in yeast. Abbreviations: (1) ATP-PRT = ATP-phosphoribosyltransferase; (10) HOL-DH = histidinol dehydrogenase, (8) HOL-PT = histidinol phosphatase; (7) HOLP-TA = histidinol phosphate transaminase; (9) histidinol dehydrogenase, (5) GLN-AT = glutamine amidotransferase, (4) PRF-AICR ISOM = phosphoribosyl-formimino-5-aminoimidazolecarboxamide ribotide isomerase; (6) IGP-DHT = imidazoleglycerolphosphate dehydrogenase; (3) PR-AMP CH = phosphoribosyl-AMP cyclohydrolase; (2) PR-ATP CH = phosphoribosyl-ATP-cyclohydrolase. P1, P2, and P3 are promoters. In *S. typhimurium*, the genes are transcribed in a clockwise direction for O through E.

tetraphosphate, and (c) feedback inhibition of the first enzyme of the pathway (ATP-phosphoribosyltransferase, ATP-PRT). Details of the various mechanisms of regulation were discussed in Chapter 4.

PURINES AND PYRIMIDINES

Historically, our knowledge of the nature and structure of nucleic acids and their subunits was gained by the isolation and chemical characterization of these structures from a variety of biological materials. The basic purine and pyrimidine structures and the numbering system recommended by *Chemical Abstracts* are shown in Table 10-1. The most commonly encountered purines are **adenine** (6-aminopurine) and **guanine** (2-amino-6-oxopurine). A number of methylated purines have been found in a variety of sources and their significance will be discussed in conjunction with their role in specialized functions. Several pyrimidine derivatives have been characterized. **Uracil** (2,4-dioxopyrimidine), **cytosine** (4-amino-2-oxopyrimidine), and **thymine** (5-methyl-2,4-dioxopyrimidine or 5-methyluracil) are most common. Thymine is found only in DNA; uracil only in RNA, 5-hydroxymethylcytosine is found in the DNA of T-even (T2, T4, T6, etc.) bacteriophages of *E. coli* and not in the DNA of the host cell (see Chapter 5). This most significant discovery by Wyatt and Cohen in 1953 proved to be of inestimable value in studying the synthesis of bacteriophage DNA, as it provided the distinguishing feature necessary to show that the synthesis of host cell DNA is terminated when bacteriophage DNA synthesis is initiated (see Chapter 4).

Biosynthesis of Purines

Definitive evidence as to the origins of the purine molecule was obtained from studies with isotopically labeled precursors. Birds were used in the original experiments because of the high percentage of nitrogen excreted as uric acid or its derivatives (>80%). From these studies it was possible to determine that glycine was incorporated intact into the 4, 5, and 7 positions of the purine ring. Formate was consistently incorporated into the 2 and 8 positions and carbon dioxide into position 6 (Fig. 10-29). Pigeon liver homogenates were found to synthesize purines de novo from these same precursors, and the basic scheme for the formation of purines was considered to be

$$CO_2 + NH_3 + glycine + HCOOH + R\text{-}1\text{-}P \longrightarrow riboside\ intermediates$$

$$riboside\ intermediates + HCO_2H + P_i \longrightarrow inosine\text{-}5'\text{-}phosphate$$

$$inosine\text{-}5'\text{-}phosphate \longrightarrow inosine + P_i \longrightarrow hypoxanthine + R\text{-}1\text{-}P$$

One of the first compounds suggested as a potential intermediate in the biosynthesis of purines by microorganisms was aminoimidazolecarboxamide (AICA). Somewhat ironically, AICA did not serve as an intermediate in the synthesis of IMP by the avian system. In microorganisms, considerable variation was observed in the ability of AICA to support the growth of purine-requiring auxotrophs. Eventually, it was demonstrated that the true intermediates in the pathway were the ribonucleotides rather than the free bases in many systems and concerted efforts were initiated to show that other imidazole derivatives could serve as intermediates in the pathway. Aminoimidazole and AICA were

TABLE 10-1. Nomenclature of Purines and Pyrimidines[a]

Structure and Common Name	Recommended IUPAC Name	Synonym(s)
Adenine	1 *H*-Purin-6-amine	6-Aminopurine
Guanine	2-Amino-1, 7-dihydro-6 *H*-purin-6-one	2-Amino-6-oxopurine (2-amino-hypoxanthine)
Uracil	2,4(1*H*,3*H*)-Pyrimidinedione	2,4-Dioxopyrimidine
Thymine	5-Methyl-2,4-(1*H*,3*H*)-pyrimidinedione	5-Methyl-2,4-dioxopyrimidine (5-methyl-uracil)
Cytosine	4-Amino-2(1*H*)-pyrimidinone	4-Amino-2-oxopyrimidine

[a]The chemical structures of these compounds are often shown as the tautomeric form with the maximum number of double bonds within the ring(s). However, this is true only for adenine. The most commonly encountered tautomeric form is reflected in the name assigned by the IUPAC commission and given in the table. The 1977 *Chemical Abstracts* listing of chemical names gives the IUPAC recommendations for purine and pyrimidine nomenclature.

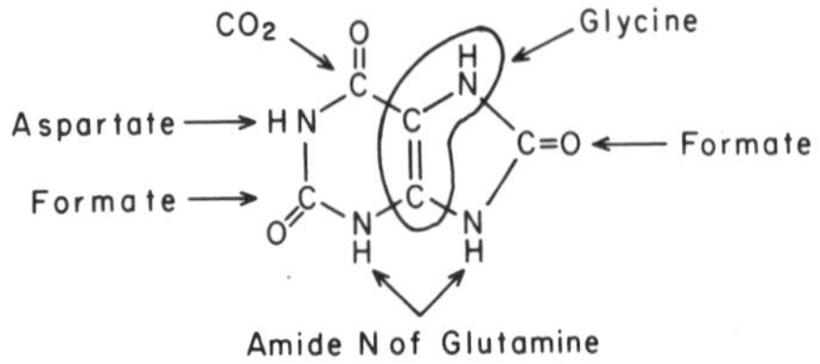

Fig. 10-29. Precursors of the purine molecule. In avian systems, the contribution of the amino nitrogen of aspartate to N-1 and the amide nitrogen of glutamine to N-3 and N-9 has been verified with ^{15}N-labeled compounds. Comparable isotope-labeling studies have not been performed with microorganisms. However, other evidence indicates that these same donors provide the nitrogen for positions 1, 3, and 9 in the purines of microbial systems.

among the first compounds to be identified as degradative metabolites of purines by *Clostridium cylindrosporum.* A purine-requiring mutant of *E. coli* accumulated an arylamine that differed from AICA in several respects and was converted to AICA by an AICA-accumulating mutant. It was found to be aminoimidazole or its ribonucleotide. Adenine auxotrophs of yeast or yeasts grown under the conditions of biotin deficiency accumulated aminoimidazole derivatives. Further work in a variety of systems eventually confirmed the participation of AIR, AICAR, and FAICAR as intermediates in the purine pathway (Fig. 10-30).

In the pigeon liver system, the formation of AIR from FGAM and the intermediates involved in the pathway to IMP and AMP were identified. Several microbial systems were also shown to follow the same pathway shown in Figure 10-30. The enzyme responsible for the conversion of AMP to AMP-S was shown to be the same enzyme that catalyzed the conversion of SAICAR to AICAR. The formation of GMP from IMP was also elucidated in microbial systems. Inosine 5–monophosphate first undergoes a dehydrogenation to form XMP. Then, via an amidotransferase reaction, the amido-nitrogen of glutamine is transferred to the 2 position to form GMP.

The amino acid histidine and the essential B vitamins, folate, riboflavin, and thiamine, may be considered products of the purine biosynthetic pathway in that a portion of each of their structures is derived from purines. The regulatory systems required to coordinate the flow of metabolites through these multiple branches of the purine pathway will be discussed later.

In histidine biosynthesis, as described earlier, ATP is the initial substrate. Coupling of PRPP at N-1 of the purine ring of ATP and ring cleavage results in formation of the histidine precursor, imidazoleglycerol phosphate and AICAR, as shown in Figure 10-27. The AICAR is recycled through the purine biosynthetic pathway.

Synthesis of the pterin moiety of folate starts with GTP (Fig. 10-22). The GTP cyclohydrolase I initiates the pathway leading to the formation of a pterin derivative which, when coupled to PAB, forms 7,8-dihydropteroate, the precursor of 7,8-dihydrofolate. Details of folate synthesis were discussed earlier in conjunction with the derivation of PAB from chorismate as one of several branches of the common aromatic amino acid biosynthetic pathway.

The pyrimidine ring and the ribityl side chain of riboflavin are derived from GTP, as shown in Figure 10-31. The GTP cyclohydrolase II, encoded by *ribA*, cleaves the imidazole ring of GTP to form 2,5-diamino-6-ribosylamino-4(3*H*)-pyrimidinone 5'-

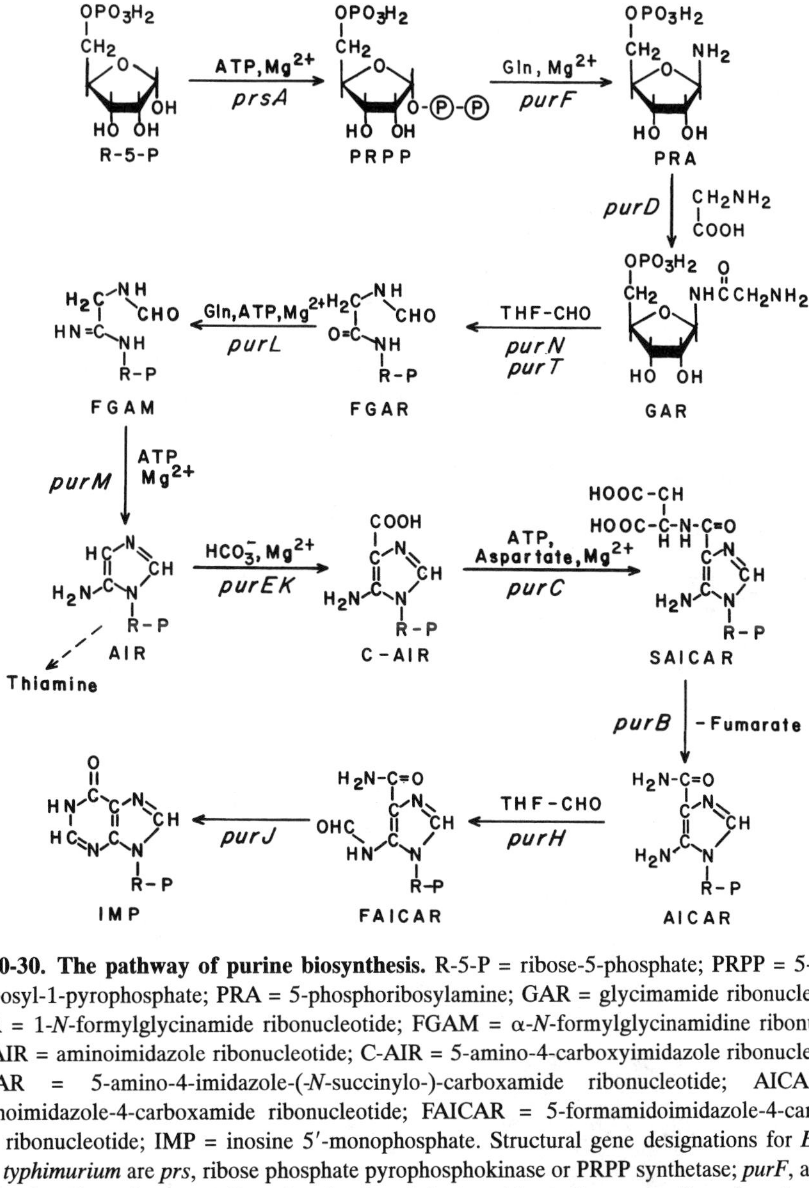

Fig. 10-30. The pathway of purine biosynthesis. R-5-P = ribose-5-phosphate; PRPP = 5-phosphoribosyl-1-pyrophosphate; PRA = 5-phosphoribosylamine; GAR = glycimamide ribonucleotide; FGAR = 1-N-formylglycinamide ribonucleotide; FGAM = α-N-formylglycinamidine ribonucleotide; AIR = aminoimidazole ribonucleotide; C-AIR = 5-amino-4-carboxyimidazole ribonucleotide; SAICAR = 5-amino-4-imidazole-(-N-succinylo-)-carboxamide ribonucleotide; AICAR = 5-aminoimidazole-4-carboxamide ribonucleotide; FAICAR = 5-formamidoimidazole-4-carboxyamide ribonucleotide; IMP = inosine 5'-monophosphate. Structural gene designations for *E. coli* and *S. typhimurium* are *prs*, ribose phosphate pyrophosphokinase or PRPP synthetase; *purF*, aminophosphoribosyltransferase; *purD* = phosphoribosylglycinamide synthetase; *purI*, *purL* = phosphoribosylglycinamide formyltransferase; *purG* = phosphoribosylformylglycinamidine synthetase; *purI*, *purM* = phosphoribosylaminoimidazole synthetase; *purE* = phosphoribosylaminoimidazole carboxylase; *purC* = phosphoribosylaminoimidazole succinocarboxamide synthetase; *purB* = adenylosuccinate lyase; *purH* = phosphoribosylaminoimidazolecarboxamide formyltransferase; *purJ* = IMP cyclohydrolase.

phosphate (phosphoribosylaminopyrimidine or PRP) with the release of formate and PP_i. Ultimately, condensation of 5-amino-6-ribitylamino-2,4-($1H,3H$)-pyrimidinedione with 3,4-dihydroxy-2-butanone 4-phosphate yields 6,7-dimethyl-8-ribityllumazine, the direct precursor of riboflavin (Fig. 10-31). Ribulose 5-phosphate is the precursor of 3,4-dihydroxy-2-butanone 4-phosphate that forms the xylene ring of the riboflavin molecule.

The pyrimidine ring of thiamine (vitamin B_1) is synthesized from AIR as a branch of the purine biosynthetic pathway. Thiamine contains a pyrimidine ring and a thiazole ring. Two precursors, [4-amino-5-hydroxymethyl-2-methylpyrimidine pyrophosphate and 4-methyl-5-(β-hydroxyethyl)thiazole monophosphate] are synthesized separately and coupled to form thiamine monophosphate (THI-P). This compound is then phosphorylated to form thiamine pyrophosphate (THI-PP), as shown in Figure 10-32. *Salmonella typhimurium* mutants blocked prior to the AIR step require both purine and thiamine for growth. Isotope labeling studies confirmed that AIR is the source of the carbon and

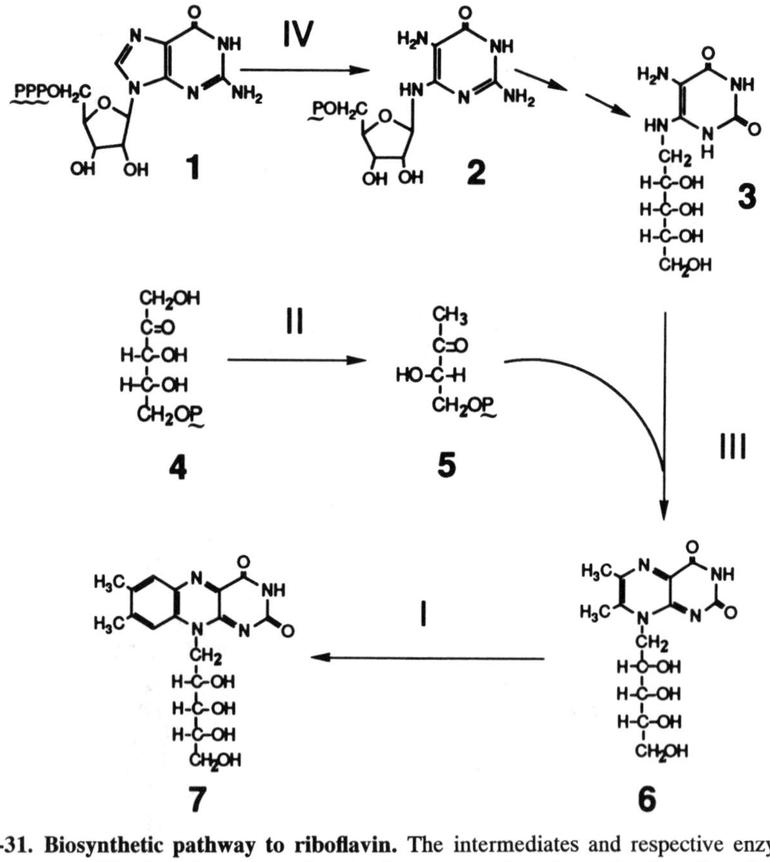

Fig. 10-31. Biosynthetic pathway to riboflavin. The intermediates and respective enzymes involved are 1 = GTP; 2 = 5-amino-6-phosphoribosylamino-8-aminopyrimidine; 3 = 5-amino-6-ribitylamino-2,4-($1H,3H$)-pyrimidinedione; 4 = ribulose 5-phosphate; 5 = 3,4-dihydroxy-2-butanone-4-phosphate (DHBP); 6 = 6,7-dimethyl-8-ribityllumazine; 7 = riboflavin. I = riboflavin synthetase; II = DHBP synthetase; III = lumazine synthetase; IV = GTP cyclohydrolase II, encoded by *ribA*. PO_3 groups are indicated by **P**. From Lee, C. Y., D. J. O'Kane, and E. A. Meighen. 1994. *J. Bacteriol.* **176**:2100–2104.

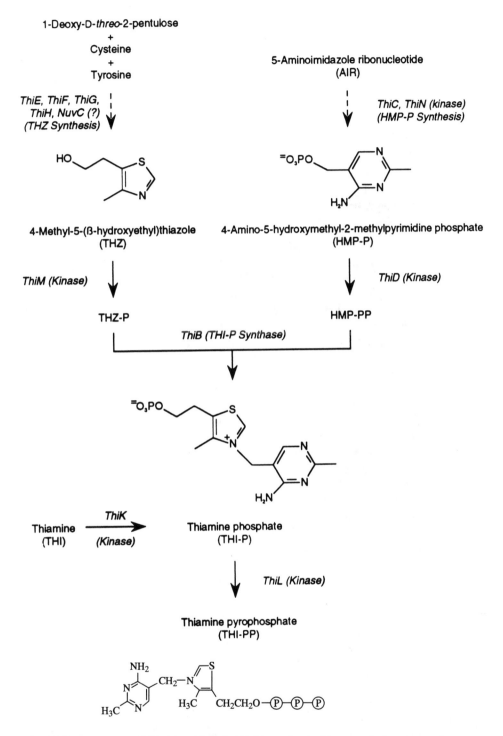

Fig. 10-32. Pathways of thiamine (vitamin B₁) biosynthesis. The gene designations and enzymes are those in *E. coli* and *S. typhimurium*. Modified from Vander Horn, P. B., A. D. Backstrom, V. Stewart, and T. D. Begley. 1993. *J. Bacteriol.* **175**:982–992.

nitrogen of the pyrimidine ring of thiamine. The genes coding for enzymes in the THI-PP pathway comprise a tightly linked cluster, *thiCEFGH*, at 90 min on the *E. coli* genetic map. The *thiC* gene product is required for the synthesis of the hydroxymethyl-pyrimidine precursor of THI-PP. The *thiEFGH* gene products catalyze the synthesis of the thiazole precursor.

Under anaerobic conditions, *S. typhimurium* can form the pyrimidine moiety of thiamine independently of the *purF* locus. In the absence of oxygen, exogenous pantothenate satisfies the thiamine requirement of *purF* mutants. Only the PurF enzyme is bypassed, however, and PurD, -G, and -I are still required for the formation of AIR and its conversion to the pyrimidine ring of thiamine.

Biosynthesis of Pyrimidines

Pyrimidine biosynthesis is initiated by the formation of carbamoyl phosphate (Fig. 10-33). In view of the importance of carbamoyl phosphate in both arginine and pyrimidine synthesis, coordinating the channeling of this intermediate into the two pathways is critical. In *Neurospora*, two carbamoyl phosphate gradients are maintained by two separate carbamoyl phosphate synthetases. Carbamoyl phosphate synthetase A provides a pool of carbamoyl phosphate specifically for the arginine pathway, as discussed in the section on amino acid biosynthesis, while carbamoyl phosphate synthetase P provides a carbamoyl phosphate pool specifically for the pyrimidine pathway. Other organisms, for example, yeast, apparently achieve by regulation what *Neurospora* accomplishes by compartmentation.

Aspartate and carbamoyl phosphate are coupled to form carbamoyl aspartate (ureidosuccinate), which is cyclized by a separate enzyme to form dihydroorotate (Fig. 10-33). After oxidation to orotate, the nucleotide is formed by phosphoribosyl transferase. Note that, in comparison to purine biosynthesis, the nucleotide stage is established **after** completion of ring formation. Uridine-5′-phosphate (UMP) is converted to UTP. Then, UTP is aminated to form cytidine triphosphate (CTP). There is no known enzyme that converts cytosine to CMP. For cytosine to serve as a nutritional source of pyrimidines, it must be deaminated to uracil.

The deoxynucleoside diphosphates are formed from the corresponding ribonucleoside diphosphate (NDP) by the action of thioredoxin, a sulfhydryl-containing protein cofactor, and ribonucleotide diphosphate reductase:

$$\text{thioredoxin } (-SH)_2 + NDP \longrightarrow \text{thioredoxin } (-S\!-\!S-) + dNDP + H_2O$$

Thymidylate (TMP) is formed from dUMP by transfer of a methyl group from N_5,N_{10}-methylene tetrahydrofolate. In the formation of hydroxymethyldeoxycytidylate, apparently no reduction accompanies the C-1 transfer from N_5,N_{10}-methylene tetrahydrofolate so that the complete hydroxymethyl group is added. Both the purine and pyrimidine derivatives must be converted to the trinucleotide stage before they can be incorporated into nucleic acid.

In *E. coli* and *S. typhimurium*, dCTP deaminase forms most of the dUTP, the precursor of thymidylate. The dUTP is degraded by dUTPase, the product of the *dut* gene, yielding PP_i and dUMP, the substrate for thymidylate synthase. The breakdown of dUTP prevents its incorporation into DNA by DNA polymerase in place of dTTP. Accumulation of dUTP or incorporation of uracil into DNA is probably not lethal until at least 10% of DNA

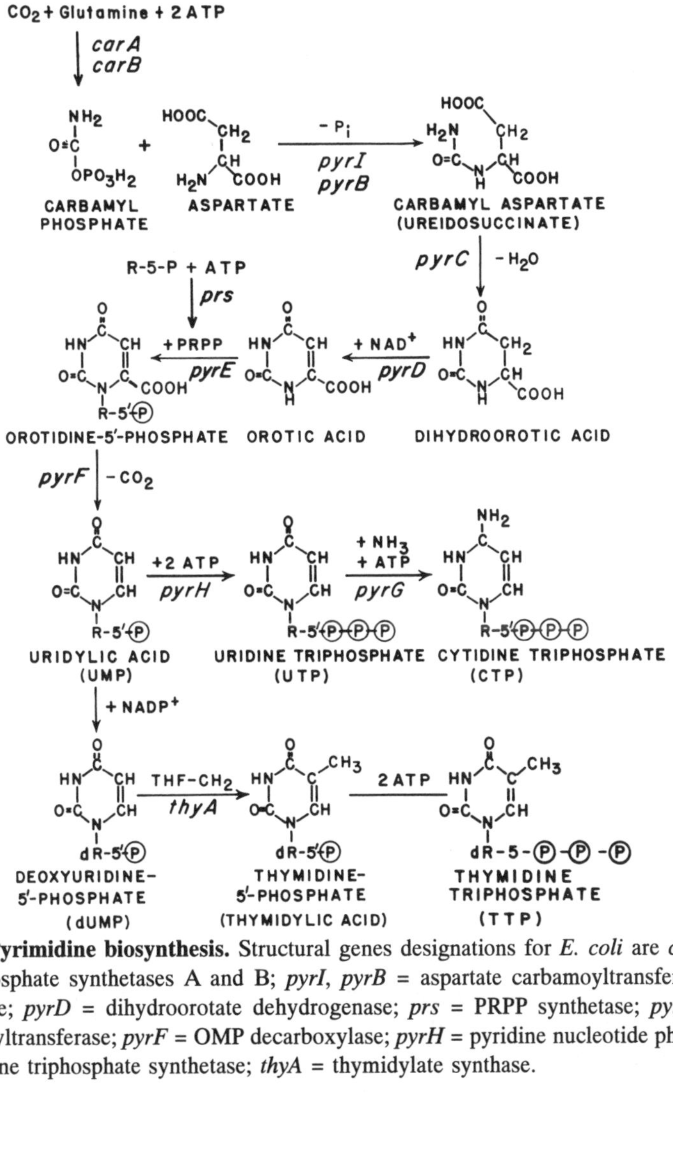

Fig. 10-33. Pyrimidine biosynthesis. Structural genes designations for *E. coli* are *carA*, *carB* = carbamoylphosphate synthetases A and B; *pyrI*, *pyrB* = aspartate carbamoyltransferase; *pyrC* = dihydroorotase; *pyrD* = dihydroorotate dehydrogenase; *prs* = PRPP synthetase; *pyrE* = orotate phosphoribosyltransferase; *pyrF* = OMP decarboxylase; *pyrH* = pyridine nucleotide phosphorylase; *pyrG* = cytidine triphosphate synthetase; *thyA* = thymidylate synthase.

thymine is replaced by uracil. Above this level excess uracil in DNA may result in degradation by repair enzymes causing lethal double-strand breaks. Uracil-containing DNA may not be recognized by DNA-binding enzymes or may turn-off protein synthesis by a regulatory system. Note the existence of a DNA repair system that deals with this problem (see Chapters 2 and 3).

Regulation of Purine and Pyrimidine Biosynthesis

Both feedback inhibition and repression–derepression control systems regulate the activity of the purine and pyrimidine biosynthetic pathways. As shown in Figure 10-34, the steps in the purine pathway subject to regulation are the first step, encoded by *purF*, and

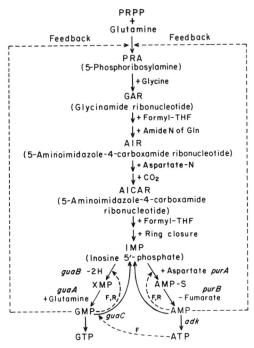

Fig. 10-34. Regulatory mechanisms in purine biosynthesis. PRPP = phosphoribosylpyrophosphate; XMP = xanthosinemonophosphate; GMP = guanosinemonophosphate; GTP = guanosinetriphosphate; AMP-S = succinyloadenosinemonophosphate. The structural genes for *E. coli* are *guaA* = GMP synthase; *guaB* = IMP dehydrogenase; *guaC* = GMP aminotransferase; *purA* = adenylosuccinate synthase; *purB* = adenylosuccinate lyase; *adk* = adenylate kinase. Each of the individual enzymatic steps is repressed by AMP.

the enzymes immediately after the branch point at IMP (encoded by *purA*, *purB*, *guaB*, and *guaA*). The activity of PurF is inhibited by either AMP or GMP. Purine-requiring mutants that accumulate intermediates in the culture medium no longer do so when purines or the nucleotides are added. The enzymes at the IMP branch point are affected by the specific nucleotide being synthesized. The IMP dehydrogenase is inhibited by feedback and its synthesis is repressed by GMP. The two enzymes that convert IMP to GMP are coded for by the *guaBA* operon and are coordinately controlled. The two genes encoding the enzymes responsible for conversion of IMP to AMP, *purA* and *purB*, are subject to end-product repression. The AMP can inhibit PRA synthesis by a feedback mechanism. Regulation of the enzymatic steps leading from IMP to AMP and GMP forms a metabolic "figure 8" insofar as the end-products of the two branches loop back to inhibit formation of the XMP or AMP-S intermediates (Fig. 10-34).

The purine pathway genes are distributed throughout the *E. coli* chromosome both as single genes and small operons. They are negatively regulated at the transcriptional level by the PurR protein and its corepressors. The *purR* gene encodes an aporepressor that combines with the purine corepressors hypoxanthine and guanine, resulting in an increased affinity for a 16 bp palindromic operator in each of the eight operons of the purine regulon. Other genes that supply intermediates for purine nucleotide synthesis or are involved in the synthesis or salvage of pyrimidine nucleotides are also part of the purR-regulated regulon. This cross-pathway regulation is necessary to assure the proper supply of precursors for each of the pathways involved. These genes include *glyA* (encoding serine hydroxymethyltransferase, a major contributor of C-1 units), *codA* (encoding cytosine deaminase, a pyrimidine salvage enzyme), *pyrC* (encoding dihydroorotase), *pyrD* (encoding dihydroorotate dehydrogenase), *prsA*, (encoding PRPP synthase), *speA* (encoding arginine decarboxylase), and *glnB* (encoding PII protein in nitrogen regulation). Each of these enzymes contains a site to which PurR binds *in vitro* and are coregulated *in vivo* by *purR*. The expression of *E. coli purR* is autoregulated. Autoregulation at the level of transcription requires two operator sites, designated $purR_{o1}$ (O_1) and $purR_{o2}$ (O_2). Operator O_1 is in the region of DNA between the transcription start site and the site for translation initiation, and O_2 is in the protein-coding region. Operator site O_2, located within the *purR* coding sequence, binds the repressor with sixfold lower affinity than O_1, but still appears to make an important contribution to *in vivo* autoregulation.

In *B. subtilis* the genes encoding enzymes for the 10 steps to IMP and the four genes encoding the conversion of IMP to AMP and GMP form a 12-gene *pur* operon. The *pur* operon is subject to dual control by adenine and guanine compounds. An adenine compound represses transcription initiation and a guanine compound regulates transcription termination–antitermination in a mRNA leader region preceeding the first structural gene.

In considering the regulation of purine biosynthesis, complexities arise as a result of the branch points leading to the pyrimidine ring of thiamine, the pterin moiety of folate, the ribitylamino pyrimidine of riboflavin, and a portion of the ring structure of histidine (Fig. 10-35). The PRPP synthetase, encoded by *prsA*, participates in several biosynthetic pathways. It is involved in the biosynthesis of purine and pyrimidine nucleotides, tryptophan, histidine, and pyridine nucleotides [NAD(P)]. The PRPP synthetase is subject to repression by pyrimidine nucleotides and is also under the regulatory control of PurR. In view of the increasing number of genes that are subject to cross-pathway regulation, it seems likely that an even greater number of genes containing sequences related to PurR binding sites will be revealed through further research in this area.

Aspartate transcarbamylase (ATCase), the first enzyme specific to the pyrimidine pathway, represents one major site of regulatory control. In *E. coli*, the *pyrBI* operon encodes the catalytic (*pyrB*) and regulatory (*pyrI*) subunits of ATCase. ATCase activity is subject to allosteric regulation by the activator ATP and the inhibitors CTP and UTP, which bind to the regulatory subunit and alter substrate binding at the catalytic site. This dual control by purines and pyrimidines provides the cell with an efficient mechanism for maintaining the proper ratio of these compounds. This system serves as a prototype for allosteric regulatory systems. ATCase activity is also controlled by the level of *pyrBI* expression, which is negatively regulated by pyrimidine availability. This regulation occurs primarily through UTP-sensitive attenuation control, with additional pyrimidine-mediated regulation occurring independently at the level of transcription initiation (see Chapter 4). Of two promoters, P_1 and P_2, located upstream of *pyrB*, greater than 95% of the *pyrBI* transcripts are initiated at promoter P_2, with only a small portion of the level of ATCase influenced by the P_1 promoter.

In *S. typhimurium* and *E. coli*, expression of *pyrC* (encoding dihydroorotase) and *pyrD* (encoding dihydroorotase dehydrogenase) is regulated in response to fluctuations in the intracellular CTP/GTP pool ratio. High CTP/GTP pool ratios repress expression by production of an mRNA initiated with a CTP downstream of the leader region. As shown in Figure 10-36, this transcript is inefficiently translated because of its capacity to form

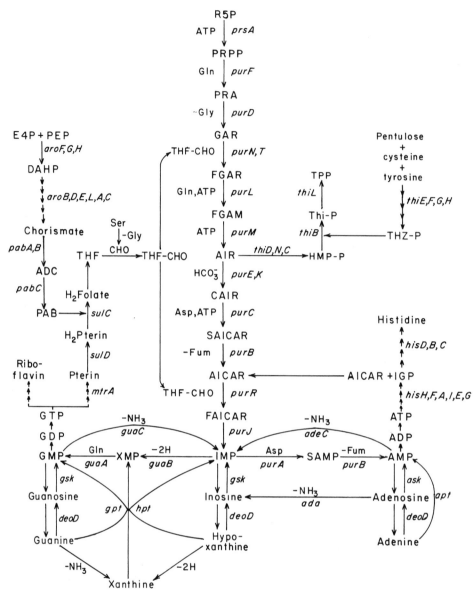

Fig. 10-35. Biosynthetic pathways leading from the de novo purine biosynthetic pathway.
R5P = ribose 5-phosphate; PRPP = 5-phosphoribosylpyrophosphate; PRA = 5-phosphoribosylamine;
GAR = glycinamide ribonucleotide; FGAR = 1-*N*-formylglycinamide ribonucleotide; FGAM = α-*N*-
formylglycinamidine ribonucleotide; AIR = 5-aminoimidazole ribonucleotide; C-AIR = 5-amino-4-
carboxyimidazole ribonucleotide; SAICAR = 5-amino-4-imidazole-(-*N*-succinylo-)-carboxamide
ribonucleotide; AICAR = 5-aminoimidazole-4-carboxamide ribonucleotide; FAICAR-5-formamidoimi-
dazole-4-carboxamide ribonucleotide; IMP = inosine 5'-monophosphate; SAMP = *N*-succinylo-AMP;
AMP = adenosine 5'-monophosphate; XMP = xanthosine 5'-monophosphate; GMP = guanosine 5'-
monophosphate; GTP = guanosine triphosphate; ATP = adenosine triphosphate; RF = riboflavin; H₄Fol-
ate or THF = tetrahydrofolate; THF—CHO = N_5, N_{10}-tetrahydrofolate; E4P = erythrose 4-phosphate;
PEP = phosphoenolpyruvate; DAHP = 3-hydroxy-D-arabino-heptulosonate 7-phosphate; ADC = 4-
amino-4-deoxychorismate; PAB = *p*-aminobenzoate; H₂pterin = 6-CH₂OH-7,8-dihydroxypteroate; H₂fol-
ate = 7,8-dihydrofolate; IGP = imidazoleglycerol phosphate; HMP-P = 4-amino-5-hydroxymethylpyr-
imidine phosphate; THZ-P = 4-methyl-5-(β-hydroxyethyl) thiazole phosphate; Thi-P = thiamine
monophosphate; TPP = thiamine pyrophosphate.

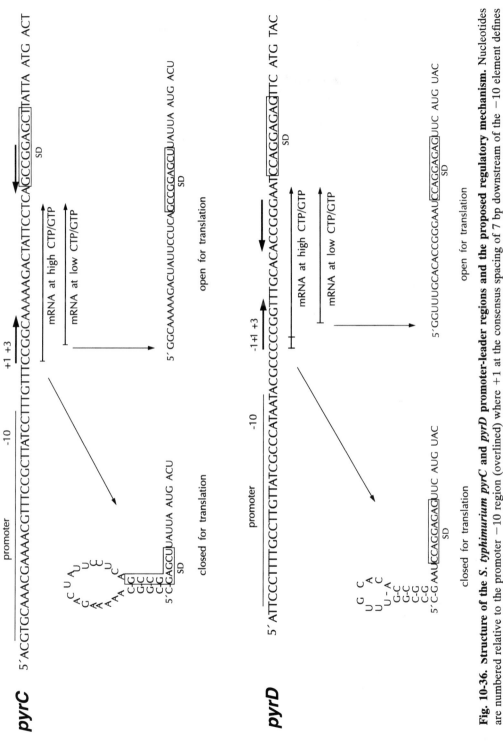

Fig. 10-36. Structure of the S. typhimurium pyrC and pyrD promoter-leader regions and the proposed regulatory mechanism. Nucleotides are numbered relative to the promoter −10 region (overlined) where +1 at the consensus spacing of 7 bp downstream of the −10 element defines the *in vivo* transcriptional initiation point in repressing conditions. The Shine–Dalgarno (SD) regions are boxed, and the regions of hyphenated dyad symmetry are indicated by arrows above the sequence. Arrows below the sequence represent the transcripts arising in conditions of different CTP/GTP pool ratios. The putative secondary structures formed at the 5′ ends of the transcripts are shown. From Sørensen, K. I. et al., 1993. *J. Bacteriol.* **175:**4137–4144.

a stable secondary structure (hairpin) at the 5′ ends of the transcripts, thus sequestering sequences required for ribosomal binding. The potential for hairpin formation is controlled through CTP/GTP-modulated selection of the transcriptional start site. See Chapter 4 for additional discussion of regulatory mechanisms.

Interconversion of Nucleotides, Nucleosides, and Free Bases: Salvage Pathways

Although it is often assumed that adenine and guanine or their ribonucleotides are freely interconverted by microorganisms, this is not actually the case. There is a limited ability to interconvert one base with another and an even more limited ability to interconvert nucleotides and nucleosides (Figs. 10-37 and 10-38). As shown in Figure 10-37, adenine and guanine cannot be directly converted from one to the other by any means. The free purine bases are converted directly to their nucleotides by pyrophosphorylases (phosphoribosyltransferases) by reaction with PRPP. Two distinct purine-utilizing enzymes,

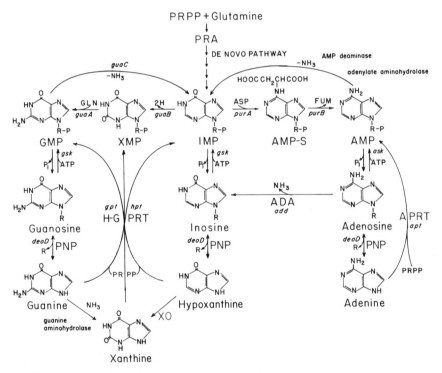

Fig. 10-37. Interconversion of purines and their derivatives. PRPP = 5-phosphoribosylpyrophosphate; PRA = phosphoribosylamine; GMP = guanosine monophosphate; XMP = xanthine monophosphate; IMP = inosine monophosphate; AMP-S = succinyloadenosine monophosphate; AMP = adenosine monophosphate; *guaB* = IMP dehydrogenase; *guaA* = GMP synthase; *guaC* = GMP reductase; *purA* = adenylosuccinate synthetase; *purB* = adenylosuccinate lyase; *hpt* = hypoxanthine phosphoribosyltransferase (H-PRT); *gpt* = guanine/xanthine phosphoribosyltransferase (G-PRT); A-PRT = adenine phosphoribosyltransferase (*apt*); XO = xanthine oxidase; ADA = adenosine deaminase (*add*); GLN = glutamine; ASP = aspartate; FUM = furmarate, PNP = purine nucleoside phosphorylase (*deoG*); ask = adenosine kinase; gsk = guanosine-inosine kinase.

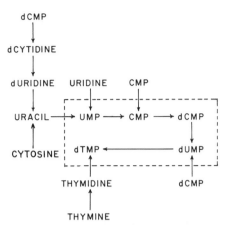

Fig. 10-38. Interconversion of pyrimidines and their derivatives. dCMP = deoxycytosine mono-phosphate; CMP = cytosine monophosphate; UMP = uridine monophosphate; dTMP = thymidine monophosphate.

one specific for the conversion of adenine to AMP and the other specific for the conversion of hypoxanthine and guanine to IMP and GMP, respectively, are present in a variety of bacteria. These enzymes are subject to feedback inhibition and may play a role in the regulation of purine transport and incorporation into nucleic acids. Nucleotidases and nucleosidases are involved in the sequential degradation of the nucleotides to the free bases and are primarily catabolic enzymes. Glycohydrolases may degrade the nucleotides to the free bases and ribose phosphate.

As shown in Figure 10-37, the interconversion of the purine bases may be quite complex and often occurs only through rather devious routes. As a specific example, consider the pathway by which *S. typhimurium* converts exogenous adenine to guanine nucleotide (GMP). Adenine is converted to adenosine, followed by deamination to inosine and subsequent phosphorolysis to hypoxanthine. Hypoxanthine is then converted to inosine monophosphate (IMP) by inosine monophosphate pyrophosphorylase. The IMP is then converted to XMP and GMP.

At the nutritional level, the ability of a given organism to interconvert adenine and guanine is reflected in the ability of the two purines to promote the growth of purine-requiring organisms. Adenine serves quite readily as the sole purine nutrient for many purine-requiring organisms while guanine shows a limited capability in this regard. The capacity of organisms to interconvert purines and their nucleotides also influences the action of a variety of purine antagonists. Regulation of the *gua* operon (which controls the conversion of IMP to XMP to GMP) is effected by adenine and guanine nucleotides rather than the free bases, a finding that may explain the diverse effects noted for various bases.

Some microorganisms, such as the lactobacilli, possess nucleoside-*N*-glycosyl transferases, which can catalyze ribosyl- and deoxyribosyl-transfer reactions from nucleosides to free bases. However, a wide variety of bacteria, including many of the enteric bacteria, appear to interconvert nucleosides and free bases through the coupling of nucleoside phosphorylases and do not possess nucleoside-*N*-glycosyl transferases. Both purine and pyrimidine bases may be converted to the nucleoside and nucleotide stages by the combined action of nucleoside phosphorylases and nucleoside kinases:

$$\text{pyrimidine} + \text{R-5-P} \longleftrightarrow \text{pyrimidine-ribose} + \text{P}_i$$

$$\text{pyrimidine-ribose} + \text{ATP} \longleftrightarrow \text{pyrimidine-ribose-P} + \text{ADP}$$

REFERENCES

Amino Acids

General References

Herrmann, K. M. and R. L. Somerville (Eds.). 1983. *Amino Acids: Biosynthesis and Genetic Regulation*. Addison-Wesley, Reading, MA.

Hinnebusch, A. G. 1992. General and pathway-specific regulatory mechanisms controlling the synthesis of amino acid biosynthetic enzymes in *Saccharomyces cerevisiae*. In *The Molecular and Cellular Biology of the Yeast* Saccharomyces: *Gene Expression*, Vol. II., Jones, E. W., J. R. Pringle, and J. R. Broach (Eds.). Cold Spring Harbor Laboratory Press, pp. 319–414.

Glutamate Family

Cunin, R., N. Glansdorff, A. Piérard, and V. Stalon. 1986. Biosynthesis and metabolism of arginine in bacteria. *Microbiol. Rev.* **50**:314–352.

Davis, R. H. 1986. Compartmental and regulatory mechanisms in the arginine pathways of *Neurospora crassa* and *Saccharomyces cerevisiae*. **50**:280–313.

Davis, R. H., D. R. Morris, and P. Coffino. 1992. Sequestered end products and enzyme regulation: The case of ornithine decarboxylase. *Microbiol. Rev.* **56**:280–290.

Demain, A. L. 1991. Production of beta-lactam antibiotics and its regulation. *Proc. Natl. Sci. Council, ROC. Part B: Life Sci.* **15**:252–265.

Han, B.-D., W. G. Nolan, H. P. Hopkins, R. T. Jones, J. L. Ingraham, and A. T. Ardelal. 1990. Effect of growth temperature on folding of carbamoylphosphate synthetases of *Salmonella typhimurium* and a cold-sensitive derivative. *J. Bacteriol.* **172**:5089–5096.

Ilag, L. L., D. Jahn, G. Eggertsson, and D. Söll. 1991. The *Escherichia coli hemL* gene encodes glutamate 1-semialdehyde amino-transferase. *J. Bacteriol.* **173**:3408–3413.

Kovari, L. Z., I. Kovari, and T. G. Cooper. 1993. Participation of RAP1 protein in expression of the *Saccharomyces cerevisiae* arginase (*CAR1*) gene. *J. Bacteriol.* **175**:941–951.

Martin, P. R. and M. H. Mulks. 1992. Sequence analysis and complementation studies of the *argJ* gene encoding ornithine acetyltransferase from *Neisseria gonorrhoeae*. *J. Bacteriol.* **174**:2694–2701.

Moore, R. C. and S. M. Boyle. 1990. Nucleotide sequence and analysis of the *speA* gene encoding biosynthetic arginine decarboxylase in *Escherichia coli*. **172**:4631–4640.

Mora, J. 1990. Glutamine metabolism and cycling in *Neurospora crassa*. *Microbiol. Rev.* **54**:293–304.

Picard, F. J. and J. R. Dillon. 1989. Cloning and organization of seven arginine biosynthesis genes from *Neisseria gonorrhoeae*. *J. Bacteriol.* **171**:1644–1651.

Tabor, C. W. and H. Tabor. 1985. Polyamines in microorganisms. *Microbiol. Rev.* **49**:81–99.

Verhoogt, H. J. C., H. Smit, T. Abee, M. Gamper, A. J. M. Driessen, D. Haas, and W. N. Konings. 1992. *arcD*, the first gene of the *arc* operon for anaerobic arginine catabolism in *Pseudomonas aeruginosa*, encodes an arginine-ornithine exchanger. *J. Bacteriol.* **174**:1568–1573.

Zhang, J. and A. L. Demain. 1992. ACV synthetase. *Crit. Rev. Biotechnol.* **12**:245–260.

Wong, S. C. and A. T. Abdelal. 1990. Unorthodox expression of an enzyme: Evidence for an untranslated region within *carA* from *Pseudomonas aeruginosa*. *J. Bacteriol.* **172**:630–642.

Aspartate and Pyruvate Families

de Lorenzo, V. A. Bindereif, B. H. Paw, and J. B. Neilands. 1986. Aerobactin biosynthesis and transport genes of plasmid ColV-K30 in *Escherichia coli* K-12. *J. Bacteriol.* **165**:570–578.

Follettie, M. T., O. P. Peoples, C. Agoropoulou, and A. J. Sinskey. 1993. Gene structure and expression of the *Corynebacterium flavum* N13 *ask-asd* operon. *J. Bacteriol.* **175**:4096–4103.

Graves, L. M. and R. L. Switzer. 1990. Aspartokinase III, a new isozyme in *Bacillus subtilis* 168. *J. Bacteriol.* **172**:218–223.

Schrumpf, B., A. Schwarzer, J. Kalinowski, A. Pühler, L. Eggeling, and H. Sahm. 1992. A functionally split pathway for lysine synthesis in *Corynebacterium glutamicum*. **173**:4510–4516.

Zhang, J.-J., F.-M. Hu, N.-Y. Chen, and H. Paulus. 1990. Comparison of the three aspartokinase isozymes in *Bacillus subtilis* Marburg and 168. *J. Bacteriol.* **172**:701–708.

Zhang, J.-J and H. Paulus. 1990. Desensitization of *Bacillus subtilis* aspartokinase I to allosteric inhibition by *meso*-diaminopimelate allows aspartokinase I to function in amino acid biosynthesis during exponential growth. *J. Bacteriol.* **172**:4690–4693.

Serine–Glycine Family

Avissar, Y. J. and S. I. Beale. 1989. Identification of the basis for δ-aminolevulinic acid auxotrophy in a *hemA* mutant of *Escherichia coli*. *J. Bacteriol.* **171**:2919–2924.

Biel, A. J. 1992. Oxygen-regulated steps in the *Rhodobacter capsulatus* tetrapyrrole biosynthetic pathway. *J. Bacteriol.* **174**:5272–5274.

Elliot, T., Y. L. Avissar, G.-E. Rhie, and S. J. Beale. 1990. Cloning and sequence of the *Salmonella typhimurium hemL* gene and identification of the missing enzyme in *hemL* mutants as glutamate-1-semialdehyde aminotransferase. *J. Bacteriol.* **172**:7071–7084.

Escalante-Semerena, B. C., S.-J. Suh, and J. R. Roth. 1990. *cobA* function is required for both de novo cobalamin biosynthesis and assimilation of exogenous corrinoids in *Salmonella typhimurium*. *J. Bacteriol.* **172**:273–280.

Hansson, M. and L. Hederstedt. 1992. Cloning and characterization of the *Bacillus subtilis hemEHY* gene cluster, which encodes protoheme IX biosynthetic enzymes. *J. Bacteriol.* **174**:8081–8093.

Hansson, M., L. Rutberg, I. Schröder, and L. Hederstedt. 1991. The *Bacillus subtilis hemAXCDBL* gene cluster, which encodes enzymes of the biosynthethetic pathway from glutamate to uroporphyrinogen III. *J. Bacteriol.* **173**:2590–2599.

Jeter, R. M. and J. R. Roth. 1987. Cobalamin (Vitamin B_{12}) biosynthetic genes of *Salmonella typhimurium*. *J. Bacteriol.* **169**:3189–3198.

Wilson, R. L., L. T. Stauffer, and G. V. Stauffer. 1993. Roles of the GcvA and PurR proteins in negative regulation of the *Escherichia coli* glycine cleavage enzyme system. *J. Bacteriol.* **175**:5129–5134.

Aromatic Amino Acid Family

Babitzke, P., P. Gollnick, and C. Yanofsky. 1992. The *mtrAB* operon of *Bacillus subtilis* encodes GTP cyclohydrolase I (MtrA), an enzyme involved in folic acid biosynthesis, and MtrB, a regulator of tryptophan biosynthesis. *J. Bacteriol.* **174**:2059–2064.

Blakley, R. L. and S. J. Benkovic (Eds.). 1984. *Folates and Pterins, Vol. 1, Chemistry and biochemistry of folates*. Wiley, New York.

Crawford, I. P. 1989. Evolution of a biosynthetic pathway: the tryptophan paradigm. *Annu. Rev. Microbiol.* **43**:567–600.

Green, J. M., W. K. Merkel, and B. P. Nichols. 1992. Characterization and sequence of *Escherichia coli pabC*, the gene encoding aminodeoxychorismate lyase, a pyridoxal phosphate-containing enzyme. *J. Bacteriol.* **174**:5317–5323.

Hagervall, T. G., Y. H. Jönsson, C. G. Edmonds, J. M. McCloskey, and G. R. Björk. 1990. Chorismic acid, a key metabolite in modification of tRNA. *J. Bacteriol.* **172**:252–259.

Lopez, P., B. Greenberg, and S. A. Lacks. 1990. DNA sequence of folate biosynthesis gene *sulD*, encoding hydroxymethyldihydropterin pyrophosphokinase in *Streptococcus pneumoniae*, and characterization of the enzyme. *J. Bacteriol.* **172**:4766–4774.

Lopez P. and S. A. Lacks. 1993. A bifunctional protein in the folate biosynthetic pathway of *Streptococcus pneumoniae* with dihydroneopterin aldolase and hydroxymethyldihydropterin pyrophosphokinase activities. *J. Bacteriol.* **175**:2214–2220.

Moat, A. G. and J. W. Foster. 1987. Biosynthesis and salvage pathways of pyridine nucleotides. In *Pyridine Nucleotide Coenzymes: Chemical, Biochemical, and Medical Aspects*, Vol. 2B, D. Doplhin, R. Poulson, and O. Avramovic (Eds.). Wiley, New York.

Muday, G. K. and K. M. Herrmann. 1990. Regulation of the *Salmonella typhimurium aroF* gene in *Escherichia coli. J. Bacteriol.* **172**:2259–2266.

Ozenberger, B. A., T. J. Brickman, and M. A. McIntosh. 1989. Nucleotide sequence of *Escherichia coli* isochorismate synthetase gene *entC* and evolutionary relationship of isochorismate synthetase and other chorismate-utilizing enzymes. *J. Bacteriol.* **171**:775–783.

Pyne, C. and A. L. Bognar. 1992. Replacement of the *folC* gene, encoding folylpolyglutamate synthetase-dihydrofolate synthetase in *Escherichia coli*, with genes mutagenized in vitro. *J. Bacteriol.* **174**:1750–1759.

Richter, G., H. Ritz, G. Katzenmeier, R. Volk, A. Kohnle, F. Lottspeich, D. Allendorf, and A. Bacher. 1993. Biosynthesis of riboflavin: Cloning, sequencing, mapping, and expression of the gene coding for GTP cyclohydrolase II in *Escherichia coli. J. Bacteriol.* **175**:4045–4051.

Sharma, V., K. Suvarna, R. Meganathan, and M. E. S. Hudspeth. 1992. Menaquinone (vitamin K$_2$) biosynthesis: nucleotide sequence and expression of the *menB* gene from *Escherichia coli.* **174**:5057–5062.

Talarico, T. L., P. H. Ray, I. K. Dev, B. M. Merrill, and W. S. Dallas. 1992. Cloning, sequence analysis, and overexpression of *Escherichia coli folK*, the gene coding for 7,8-dihydro-6-hydroxy-methylpterin-pyrophosphokinase. *J. Bacteriol.* **174**:5791–5977.

Tran, P. V. and B. P. Nichols. 1991. Expression of *Escherichia coli pabA. J. Bacteriol.* **173**:3680–3687.

Zalkin, H. and D. J. Ebbole. 1988. Organization and regulation of genes encoding biosynthetic enzymes in *Bacillus subtilis. J. Biol. Chem.* **263**:1595–1598.

Histidine

Delorme, C., S. D. Ehrlich, and P. Renault. 1992. Histidine biosynthesis genes in *Lactococcus lactis* subsp. *lactis. J. Bacteriol.* **174**:6571–6579.

Purines and Pyrimidines

Bacher, A. 1990. Biosynthesis of flavins, In *Chemistry and Biochemistry of Flavoenzymes*, Vol. 1. F. Müller (Ed.). CRC Press, Boca Raton, FL, pp. 215–259.

Downs, D. M. 1992. Evidence for a new, oxygen-regulated biosynthetic pathway for the pyrimidine moiety of thiamine in *Salmonella typhimurium.* **174**:1515–1521.

He, B., K. Y. Choi, and H. Zalkin. 1993. Regulation of *Escherichia coli glnB, prsA,* and *speA* by the purine repressor. *J. Bacteriol.* **175:**3598–3606.

Kowasaki, Y. 1993. Copurification of hydroxyethylthiazole kinase and thiamine pyrophosphorylase of *Saccharomyces cerevisiae*: Characterization of hydroxyethylthiazole kinase as a bifunctional enzyme in the thiamine biosynthetic pathway. *J. Bacteriol.* **175:**5153–5158.

Kawasaki, Y., K. Nosaka, Y. Kaneko, H. Nishimura, and A. Iwashima. 1990. Regulation of thiamine biosynthesis in *Saccharomyces cerevisiae.* **172:**6145–6147.

Liu, C., J. P. Donahue, L. S. Heath, and C. L. Turnbough, Jr. 1993. Genetic evidence that promoter P_2 is the physiologically significant promoter for the *pryBI* operon of *Escherichia coli* K-12. *J. Bacteriol.* **175:**2363–2369.

Reuke, B., S. Korn, W. Eishenreich, and A. Bacher. 1992. Biosynthetic precursors of deazaflavins. *J. Bacteriol.* **174:**4042–4049.

Richster, G., R. Volk, C. Krieger, H.-W. Lahm, U. Röthlisberger, and A. Bacher. 1992. Biosynthesis of riboflavin: Cloning, sequencing, and expression of the gene coding for 3,4-dihydroxy-2-butanone 4-phosphate synthase of *Escherichia coli. J. Bacteriol.* **174:**4050–4056.

Sørensen, K. I., K. E. Baker, R. A. Kelln, and J. Neuhard. 1993. Nucleotide pool sensitive selection of the transcriptional start site in vivo at the *Salmonella typhimurium pyrC* and *pyrD* promoters. *J. Bacteriol.* **175:**4137–4144.

Vander Horn, P. B., A. D. Backstrom, V. Stewart, and T. P. Begley. 1993. Structural genes for thiamine biosynthetic enzymes (*thiCEFGH*) in *Escherichia coli* K-12. *J. Bacteriol.* **175:**982–992.

Wang, L. and B. Weiss. 1992. *dcd* (dCTP deaminase) gene of *Escherichia coli*: Mapping, cloning, sequencing, and identification as a locus of suppressors of lethal *dut* (dUTPase) mutations. *J. Bacteriol.* **174:**5647–5653.

Wilson, H. R., C. D. Archer, J. Liu, and C. L. Turnbough, Jr. 1992. Translational control of *pyrC* expression mediated by nucleotide-sensitive selection of transcriptional start sites in *Escherichia coli. J. Bacteriol.* **174:**514–524.

CHAPTER 11

GROWTH AND ITS REGULATION

In previous discussions, it was emphasized that myriad regulatory mechanisms control the flow of metabolites required for macromolecular synthesis and to coordinate the processes of chromosome replication, partitioning, and cell division. These various regulatory mechanisms have been considered primarily as isolated activities controlling the formation of one or, at most, only a few closely related metabolic end-products. The timing of all these events must be exquisitely coordinated so they lead to the orderly process of cell division. Accurate measurement of the events occurring in this cyclical process of growth has proven difficult for a number of technical reasons.

In its simplest aspect, the cell cycle can be envisioned as occurring in phases that include:

B = the interval between cell division and the initiation of DNA synthesis.

C = the period of DNA synthesis (chromosome replication).

D = the period between termination of chromosome replication and the end of cell division.

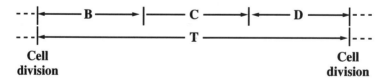

T, the total cell cycle, is equal to B + C + D = 1

Several lines of evidence indicate that the bacterial chromosome is attached to the cell surface and that, as the surface extends during growth, these attachment sites serve as a primitive mitotic apparatus to guide the two new chromosomes into each of the two daughter cells. The final stages of synthesis of the cell wall polymers (peptidoglycan) occur at the surface of the cell membrane. Wall polymers that are covalently linked to peptidoglycan appear to be synthesized concurrently. Other external constituents, such as the LPS of gram-negative organisms, appear to be synthesized independently. One major difficulty encountered in studies of the synthesis of cell wall constituents is the inability to distinguish between thickening of the wall as compared to its extension or elongation as new wall. In gram-positive cocci, the addition of new material to the wall occurs primarily at the septal region. By comparison, there appears to be less agreement regarding the site of extension of the walls of rod-shaped bacteria. Contributing to this difficulty is the considerable turnover of peptidoglycan that occurs in some organisms. Also, in rapidly growing cells, multiple rounds of nucleoid replication of multiple sites of surface extension. Because of its apparently greater simplicity, growth of gram-positive streptococci will be discussed first.

GROWTH OF GRAM-POSITIVE STREPTOCOCCI

In dividing cells of *Enterococcus hirae* (formerly *Streptococcus faecium*) an equatorial band develops at the initiation of cell division (Fig. 11-1). This fortuitous occurrence permits visual description of the cell cycle in this organism on rather precise terms. A cross-wall is assembled at each of these sites. As a constricting division furrow bilaterally splits this cross-wall, the two cleaved cross-wall layers separate and expand to form two new polar caps. As the formation of new poles nears completion, new sites must be initiated for continued cell growth. These nascent septa appear early in the cell cycle. The absence of significant turnover of peptidoglycan in *E. hirae* aids in distinguishing

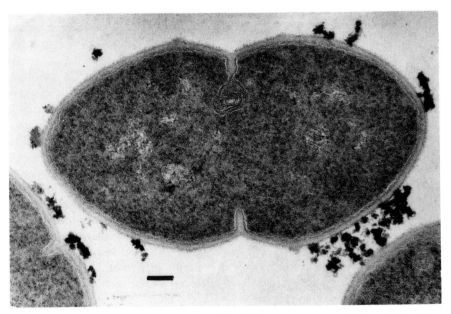

Fig. 11-1. Electron micrograph of dividing cells of *Enterococcus hirae* (formerly *Streptococcus faecium*). Raised bands are observed around the circumference of the cell wall. Nascent septa appear at these sites early in the cell cycle. The wall bands mark the meeting of polar caps, made during a previous cell cycle, with equatorially located wall synthesized during the most recent cell cycle. Bar = 0.1 μm. From Higgins, M. L. and L. Daneo-Moore. 1974. *J. Cell Biol.* **61**:288–300.

preexisting wall surface from newly extended areas. The raised wall bands, at relatively fixed distances from the pole, are observed at all stages of the division cycle. With cells of increased length, an increased area of wall is observed between the bands and an equatorially located nascent septum. The wall bands mark the meeting of polar caps, made during a previous cell cycle, with equatorially located walls synthesized during the most recent cell cycle. The events that occur during cell division are more or less continuous. However, for the convenience of description they can be divided into stages as outlined here and in Figure 11-2:

Stage 1: Centripetal penetration of the wall is initiated under an equatorial band. A notch is formed in the external surface of the wall directly above the nascent cross-wall, and two external *wall bands* appear.

Stage 2: A thin cross-wall penetrates a short distance (70–80 nm) into the cell as the two new wall bands are separated by the insertion of newly synthesized peripheral wall.

Stage 3: The nascent cross-wall thickens somewhat as the new wall bands become separated still further.

Stage 4: When the two new hemispheres have reached adequate size, and the wall bands are near the equators of the new daughter cells, the penetration of the equatorial cross-wall resumes. As it penetrates into the cytoplasm, the cross-wall continues to peel apart into two layers of wall that become the poles of the newly formed cells. In rapidly growing cells, the entire process begins again at the two new subequatorial bands.

Stage 5: Final separation of the two new daughter cells occurs. Although the separation may begin as early as the beginning of Stage 4, the event that can be quantitated as a doubling in cell number is not completed until the end of Stage 5.

In a rich growth medium the mass doubling time of *E. hirae* is about 30 min. Since DNA synthesis takes longer than this (C = 50–52 min), a new round of chromosome replication must be initiated before the old round of synthesis is completed (**dichotomous replication**). It follows, therefore, that wall band splitting and initiation of chromosome replication do not occur simultaneously. The rounds of replication must have been completed early enough so that segregation can take place. Consequently, both cell division and chromosome replication are indirectly controlled by wall band splitting. Thus, it may be concluded that growing streptococci respond to alteration of nutritional conditions mainly by altering the rate at which they fashion the cell wall. Only to a small extent do they respond by shifting the time of wall band splitting to earlier portions of the cycle. By comparing the cell cycle events of band splitting (BS) and cell division (CD), as shown in Figure 11-3, it can be concluded that the cell initiates wall band splitting when the growth zones cannot function rapidly enough to allow the increase of surface area required to accommodate continuing production of cytoplasm. Although initiation of new growth sites seems to be independent of normal chromosome replication, it has been shown that chromosome replication is necessary for the terminal events of growth site development that result in the division of a site into two separate poles. Mitomycin C, which inhibits DNA synthesis rapidly and with a minimal effect on the synthesis of other macromolecules, causes an eventual cessation of cell division. However, the growth sites

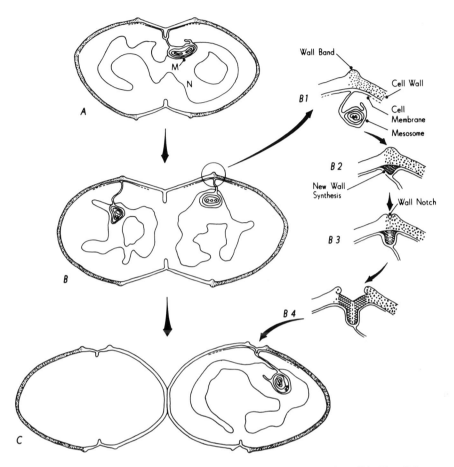

Fig. 11-2. Stages in the cell division cycle of *E. hirae* (formerly *S. faecalis*). The diplococcus in (*A*) is in the process of growing new wall at its cross-wall and segregating its nuclear material (N) to the two nascent daughter cocci. In rapidly growing exponential phase cultures before completion of the central cross-wall, new sites of wall elongation are established at the equators of each of the daughter cells at the junction of old, polar wall (stippled) and new equatorial wall beneath a band of wall material that encircles the equator (*B*). Beneath each band a mesosome (M) is formed while the nucleoids separate and the mesosome at the central site is lost. The mesosome appears to be attached to the plasma membrane by a thin membranous stalk (*B*1). Invagination of the septal membrane appears to be accompanied by centripetal cross-wall penetration (*B*2). A notch is then formed at the base of the nascent cross-wall creating two new wall bands (*B*3). Wall elongation at the base of the cross-wall pushes newly made wall outward. At the base of the cross-wall, the new wall peels apart into peripheral wall, pushing the wall bands apart (*B*4). When sufficient new wall is made so that the wall bands are pushed to a subequatorial position [e.g., from (*C*) to (*A*) to (*B*)] a new cross-wall cycle is initiated. Meanwhile, the initial cross-wall centripetally penetrates into the cell, dividing it into two daughter cocci. At all times the body of the mesosomes appears to be associated with the nucleoid. Doubling of the number of mesosomes seems to precede completion of the cross-wall by a significant interval. Nucleoid shapes and the position of mesosomes are based on projections of reconstructions of serially sectioned cells. From Higgins, M. L. and G. D. Shockman. 1971. *Crit. Rev. Microbiol.* **1**:29–72.

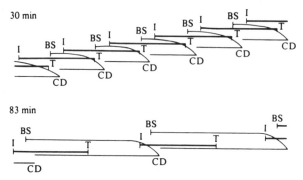

Fig. 11-3. Representation of the cell cycle of *E. hirae* (formerly *S. faecalis*) at two growth rates. 1 = initiation of chromosome replication; T = termination of chromosome replication; BS = band splitting events; CD = cell division events. From Koch, A. L. and M. L. Higgins. 1984. *J. Gen. Microbiol.* **130**:735–745.

formed either before or after inhibition of DNA synthesis enlarge until they reach about 0.25 μm^3 of cell volume and do not increase further in size. When these sites approach this 0.25-μm^3 limit, new sites are initiated. These observations suggest that regardless of whether chromosome replication is inhibited or not, sites have the same finite capacity to enlarge to about 0.25 μm^3, and when this capacity is reached, new sites are initiated.

GROWTH OF GRAM-NEGATIVE RODS

The mode of cell wall extension in gram-negative rods differs considerably from that of the gram-positive cocci. In *S. typhimurium* and *E. coli* the cell walls are extended by a process of diffuse intercalation of new wall material. The newly formed cross-walls are synthesized in a manner that differs from the extension of the peripheral wall. Under a variety of circumstances, cell division can be inhibited without inhibiting the continued linear elongation of the cells. The action of certain β-lactam antibiotics, a variety of chemical and physical agents, as well as alterations in the nutritional environment may result in filament formation regardless of whether they selectively affect the synthesis of the wall or membrane layers of the cell surface. Agents and conditions that directly affect the formation of septa or the growth (elongation or extension) of the wall surface are useful for elucidation of the mechanism of cell division.

Gram-negative rods form a division septum by circumferential invagination of the three layers (cytoplasmic membrane, murein sacculus, and outer membrane) of the cell envelope at the division site. Prior to the onset of septum formation, two rings, termed periseptal annuli, are produced at the eventual site of division. The cytoplasmic membrane and the outer membrane are in close association with the murein layer. Electron microscopy reveals a continuous zone of membrane murein attachment at the innermost edge of this nascent septum.

One very fruitful approach to the study of the processes of growth and cell division has been the selection of mutants that are defective in cell surface elongation or cell division. A very large cluster of genes coding for these activities is located at the 2-min region of the *E. coli* genetic map (Table 11-1). As might be expected, many of these

TABLE 11-1. Genes in the 2-min (*mra*) Region Involved in Cell Wall Peptidoglycan Synthesis and the Cell Division Process[a]

mraA	D-Alanine carboxypeptidase
mraB	Peptidoglycan-synthesizing enzyme
mrbA	UDP-*N*-acetylglucosaminyl-3-enolpyruvate reductase
mrbB	Peptidoglycan-synthesizing enzyme
mrbC	Peptidoglycan-synthesizing enzyme
mrcA	**PBP1A (*ponA*)**; PG-synthesizing enzyme
mrcB	**PBP1B (*ponB*)**; PG-synthesizing enzyme
mraY	First step in lipid cycle of PG-synthesizing reactions: UDP-mur-NAc-pentapeptide + undecaprenyl-P ⟷ undecaprenyl-pyrophosphate-mur-NAc-pentapeptide + UMP
ftsL	Essential cytoplasmic membrane protein involved in cell division. Depletion of FtsL results in Y-shapes, etc.
fts36	(*lts33*), filamentous; maps upstream of *ftsI*
ftsI	**PBP3 (*pbpB; sep*)**, membrane protein, septum PG formation
murE	meso-DAP-adding enzyme
murF	D-alanyl-D-alanine adding enzyme
murD	D-glutamic acid adding enzyme
ftsW	Septum PG formation; may pair with PBP-3 like RodA/PBP-2
murG	*N*-Acetylglucosaminyl transferase; final step in lipid cycle of PG synthesis
murC	L-Alanine adding enzyme
ddlB	D-alanyl-D-alanine ligase
dacA	**PBP-5;** D-alanine carboxypeptidase
dacB	**PBP-4;** D-alanine carboxypeptidase
dacC	**PBP-6;** D-alanine carboxypeptidase
ftsQ	Membrane protein extends into periplasm, required throughout septation process; forms filaments with no sign of septa
ftsA	Forms filaments with constrictions at septal sites
ftsZ	Mg^{2+}-Dependent GTPase; FtsZ forms ring structure at cell division site; facilitates integration of PBP-3 into CM; FtsZ overproduction induces minicell formation; forms filaments with no sign of septa; reduced D-ala carboxypeptidase activity; interacts with SOS response-associated cell division; inhibits **SulB (SfiB)** action
sulB	(suppressor of *lon*, or **sfiB**, suppressor of filamentation) Allelic to *ftsZ*; functions in septation process
envA	(***divC***); forms chains of unseparated cells; reduced levels of NAc muramyl-L-ala amidase
ftsM	(***supU; serU***); encodes $tRNA_2^{Ser}$; regulation of cell div
secA	(*azi, pea*); secretion of envelope proteins; **terminal end of 2-min region**

[a]Genes are presented in their map order within the 2-min region.

genes code for enzymes involved in the synthesis of peptidoglycan (PG) or murein (mur). Six of the seven genes that code for penicillin-binding proteins (PBSs) have been mapped in this 2-min region. Two additional genes (located at 14.5 min on the *E. coli* gene map), *rodA* (*mrdB*) and *pbpA* (*mrdA*), constitute the *rodA* operon and are involved in elongation of the peptidoglycan chain.

The proposed steps in the morphogenesis of *E. coli* cells are shown in Figure 11-4. Three main systems are involved. One system contains enzymes capable of forming the complete peptidoglycan sacculus around the growing cell. In the absence of modifying

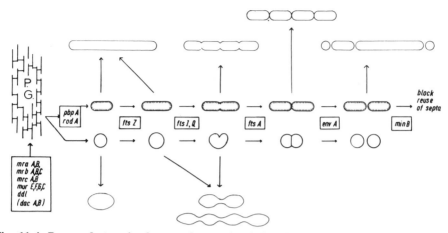

Fig. 11-4. Proposed steps in the morphogenesis of *E. coli* cells. Three main morphogenetic systems are represented by genes shown in the boxes. (1) A set of enzymes involved in the net synthesis of peptidoglycan (PG). Genes specifying these enzymes include *mraA*, *mraB*, *mrbA*, *mrbB*, *mrbC*, *mrcA*, *mrcB*, *mraY*, *murE*, *murF*, *murG*, *murC*, *ddl*, *dacA*, *dacB*, *dacC*, and possibly others. (2) The products of two genes, *rodA* and *pbpA*, are required to shape the cylindrical PG sacculus. In the absence of either of these products, the cells grow as spherical cocci. (3) At least six gene products (those of *ftsZ*, *ftsI*, *ftsQ*, *ftsA*, *envA*, and *minB*) are required for the initiation, formation, completion, separation, and inactivation (as potential division sites) of new cell poles. The action of each of these gene products can be assigned to successive steps. The left-hand side of the diagram shows a generalized PG net produced by system (1) in the absence of the other two systems. System (2) modifies the structure of this net so as to give it a cylindrical shape. Whether or not system (2) is operating, system (3) is activated periodically to form a pair of new cell poles at the center of the old cell. At the top of the diagram are shown the phenotypes of cells that have been blocked at different stages in this system (3) division process (e.g., cells with mutations in the various genes involved). Thus, *ftsZ*, *ftsI*, or *ftsQ* mutants form elongated cylinders without visible signs of septa. The *ftsA* mutants form elongated cylinders with constrictions at the septal sites, *envA* mutants form chains of unseparated cells, and *minB* mutants form the correct number of completed septa but partition them at random between newly formed and preexisting division sites. At the bottom of the diagram are shown the two main phenotypes of mutants that are blocked both in system (2) and, in particular, in steps in system (3). Thus *ftsZ*(Ts) double mutants grow as enlarging prolate elipsoids after a shift to 42 °C, whereas *ftsA*(Ts), *ftsQ*(Ts), or *ftsI*(Ts) double mutants grow as swollen balloons with regular constrictions. From Begg, K. J. and W. D. Donachie, 1985. *J. Bacteriol.* **163**:615–622.

activities, this set of reactions produces the minimum amount of peptidoglycan necessary to cover the cell volume completely: That is, it produces spherical cells that enlarge without division. The genes required for this process are shown in Figure 11-4. A set of modifying reactions, for example, the RodA-PBP2 system, result in the peptidoglycan taking a cylindrical shape. These functions must be active over the entire cell surface (except at the poles) because preexisting cylindrical surfaces are converted to spherical surfaces after inactivation of either of these two gene products. Also, a spherical surface formed in the absence of these products is replaced by or converted to a cylindrical surface after reactivation of this system. The third system, responsible for cell division through the formation of new cell poles, presumably acts periodically and its action is localized to a particular part of the peptidoglycan surface.

The gene *mraY* encodes the first step in the lipid cycle of peptidoglycan synthesis, UDP-*N*-acetylmuramoyl-pentapeptide:undecaprenyl-phosphate phospho-*N*-acetylmuramoyl-pentapeptide transferase (see discussion of peptidoglycan synthesis in Chapter 6). This enzyme catalyzes the following reaction:

$$\text{UDP-MurNAc-pentapeptide} + \text{undecaprenyl-phosphate}$$
$$\longleftrightarrow \text{undecaprenyl-pyrophosphoryl (PP)-MurNAc-pentapeptide} + \text{UMP}$$

The *N*-acetylglucosaminyltransferase, responsible for the final step in the formation of the lipid-linked disaccharide pentapeptide, is the product of the *murG* gene. This enzyme is associated with the inner face of the cytoplasmic membrane (CM) suggesting that the peptidoglycan subunit is completely assembled before it traverses the cytoplasmic membrane. Figure 11-5 provides a schematic illustration of the formation of the lipid-linked subunit catalyzed by MraY and MurG. The mechanism of transport of the peptidoglycan precursor across the cell membrane has not been determined.

The sequential action of the gene products of *murC*, *murD*, *murE*, *murF*, and *ddl* results in the formation of the nascent pentapeptide chain as shown in Figure 11-6. In *E. coli* and *S. typhimurium* four of the five amino acids in the pentapeptide are D-amino acids. This pentapeptide precursor is transferred from UDP to a lipid phosphate carrier for translocation into the membrane and subsequent cross-linking.

Peptidoglycan (murein) synthesis is of two types: (1) cell elongation or extension to maintain the cylindrical or rod configuration, and (2) septum or cross-wall formation that results in cell division. Elongation of the murein layer is specifically inhibited by mecillinam (amidinopenicillanic acid). Prevention of cell surface extension results in the development of spherical forms. All cells divide normally at least once following the addition of mecillinam, indicating that cell surface elongation or cell division that is in progress at the time of addition is not affected. The fact that only new rounds of elongation and division are affected indicates that an early event specific to the initiation of

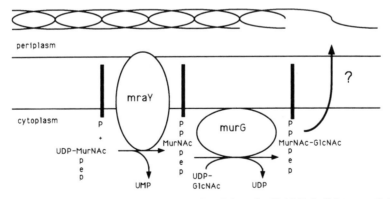

Fig. 11-5. Schematic illustration of the steps involving the lipid-linked intermediates in *E. coli* peptidoglycan biosynthesis. The netlike structure at the top represents the peptidoglycan in the process of being polymerized at the right end of the figure. The thick vertical lines located within the cytoplasmic membrane represent the undecaprenyl phosphate carrier alone or attached to the peptidoglycan subunit precursors via a pyrophosphoryl linkage. pep = the pentapeptide moiety. From Bupp, K. and J. van Heijenoort. 1993. *J. Bacteriol.* **175**:1841–1843.

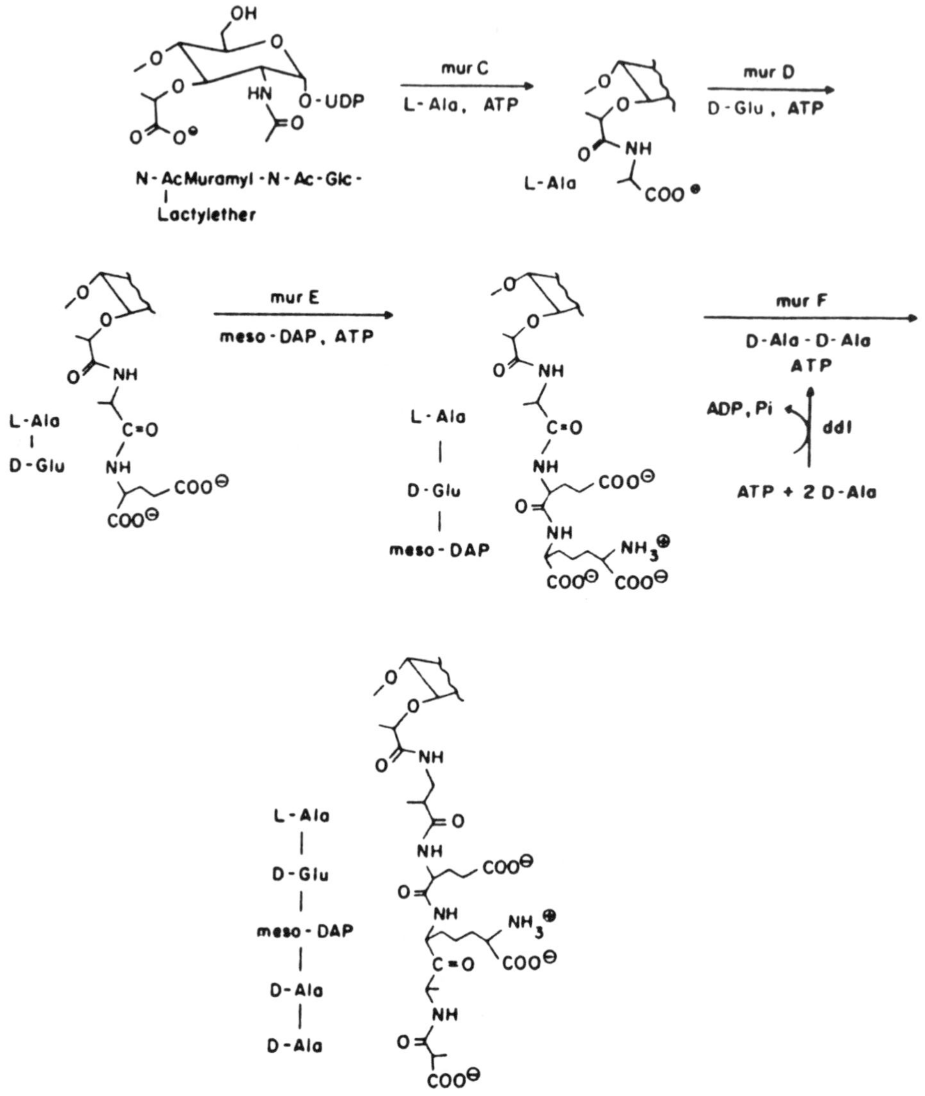

Fig. 11-6. Sequential action of the *murCDEF* and *ddl* gene products to produce nascent pentapeptide chains during peptidoglycan synthesis. From Walsh, C. T. 1989. *J. Biol. Chem.* **264:** 2393–2396.

a round of cell division is the site of mecillinam action. Mecillinam exerts its effect by binding to a cytoplasmic membrane protein, **PBP2**, a high-molecular-weight protein responsible for the initiation of new elongation sites in the murein sacculus. The PBP2 protein is a peptidoglycan synthetase encoded by *pbpA* that adds peptidoglycan at primers formed by the action of **PBPs 1A** and **1B**. The cytoplasmic membrane protein RodA is essential for the action of PBP2. The two genes *rodA* (*mrdB*) and *pbpA* (*mrdA*), map at 14.5 min on the *E. coli* gene map and constitute the *rodA* operon. The **PBP 1A** protein, encoded by *mrcA* (*ponA*), is a high-molecular-weight protein with transpeptidase and transglycosylase activity. The **PBP 1B** protein, encoded by *mrcB* (*ponB*), is a high-molecular-weight bifunctional enzyme involved in polymerization of peptidoglycan strands and their insertion into the murein sacculus through both transglycosylase and DD-transpeptidase activities. A deficiency of both PBP 1A and 1B is required for lethality. Cefsulodin is a β-lactam antibiotic that binds specifically to PBPs 1A and 1B causing cell lysis in *E. coli*. The activities of PBPs 1A and 1B provide the peptidoglycan units to which the PBP2–RodA system initiates new elongation sites.

The type of murein synthesis required for septation occurs late in the *E. coli* cell cycle. Inhibition by β-lactam agents give rise to the formation of long, multinucleate, nonseptate filaments. β-Lactam antibiotics, such as benzylpenicillin and cephalexin, selectively bind to **PBP3**, a high-molecular-weight protein specifically involved in the septation process. The PBP3 protein is encoded by *pbpB* (*ftsI*) and displays both transpeptidation and transglycosylation activity. The *pbpB* gene interacts with other genes in the 2-min region (*ftsA*, *ftsQ*, *ftsW*, and *rodA*) to maintain the rod shape of the organism. The FtsW pairs with PBP3 in the same manner that RodA and PBP2 interact to regulate peptidoglycan synthesis. The PBP2-RodA and PBP3-FtsW pairs of membrane proteins carry out alternating morphogenetic modifications of the growing murein sacculus. It has been suggested that switching from cell elongation to septum formation may result from the change in the relative availability of the different peptide chains required by the two competing systems PBP2-RodA and PBP3-FtsW. In Figure 11-7, an outline of peptidoglycan synthesis and the proposed switch between cell elongation and cell division is presented.

Filament-forming, temperature-sensitive mutants (designated *fts*) mapping within the 2-min gene cluster continue to grow but fail to divide at the restrictive temperature. The contiguous genes *ftsQ*, *ftsA*, and *ftsZ* are transcribed in the same direction (clockwise on the genetic map) and each has at least one associated promoter that allows it to be transcribed independently of neighboring genes. A cell separation gene (*envA*) located within this same region is also transcribed clockwise relative to the *E. coli* genetic map. An *envA* mutant causes *E. coli* to form chains of cells during fast growth in rich medium. The mutation is associated with low levels of *N*-acetylglucosaminyl-L-alanine amidase, an enzyme involved in splitting peptidoglycan molecules between the *N*-acetylmuramic acid residue and the pentapeptide side chain.

The FtsZ protein acts at an early stage of the septation process (Fig. 11-8). It is a Mg^{2+}-dependent GTPase dispersed throughout the cell. At the time of onset of septal invagination, FtsZ is recruited from the cytoplasm to the division site, where it forms a ring that remains associated with the leading edge of the invaginating septum until septation is completed. The FtsZ system is present in a broad spectrum of eubacteria and may represent a conserved cell division mechanism.

The FtsZ protein is the target of the cell-division inhibitor SulA (SfiA) produced in response to DNA damage (the SOS system). In *E. coli*, damage to DNA induces the SOS response. One of many genes derepressed is *sulA* (suppressor of *lon*). The SulA protein interacts directly with the cell division protein, FtsZ (Fig. 11-9). This inhibition is transient because SulA is degraded by the Lon protease.

The *min* operon, mapping at 26 min on the *E. coli* gene map, contains the *minB*, *minC*, *minD*, and *minE* genes. The *minB* mutant is the original minicell-forming mutant isolated by Adler et al. in 1967. In Figure 11-10, a model is presented for the interaction of the MinCDE system with FtsZ in the cell division process. Depicted in the center of the diagram is a cell that is ready to divide. The cell has three potential division sites or

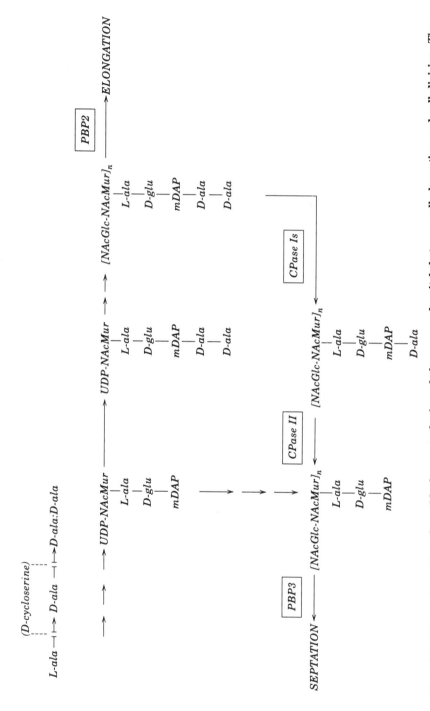

Fig. 11-7. Schematic outline of peptidoglycan synthesis and the proposed switch between cell elongation and cell division. The UDP-N-acetylmuramyl tripeptide is converted to pentapeptide by the addition of D-ala-D-ala dipeptide. Subsequent reactions link these units together, alternating with long N-acetylglucosamine chains, which are then cross-linked by transpeptidation to form the peptidoglycan sacculus. Pentapeptides (or possibly tetrapeptides) are always required as donors in transpeptidation reactions. In this scheme, synthesis of the cylindrical peptidoglycan (causing cells to grow by elongation) requires the action of PBP2, which utilizes tetrapeptides as acceptors in transpeptidation. Septum formation requires transpeptidation carried out by PBP3, which utilizes tripeptides as acceptors. The availability of tripeptides depends on both (1) the production of D-ala-D-ala dipeptide and (2) the activity of D-ala carboxypeptidase I and II, which together remove the terminal D-ala residues from pentapeptide side chains. D-Cycloserine inhibits the synthesis of D-ala-D-ala dipeptide and causes the production of peptidoglycan with increasing numbers of tripeptide side chains. It is proposed that the switch from cell elongation to septation (and vice versa) depends on changes in the availability of tripeptide acceptors brought about by periodic changes in the relative activities of the competing systems **PBP2-RodA** and **PBP3-FtsW**. Modified from Begg, K. J. et al. 1990. *J. Bacteriol.* **172:** 6697–6703.

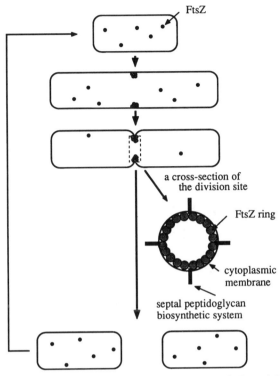

Fig. 11-8. Model for the role of FtsZ in cell division. From Bi, E. and J. Lutkenhaus. 1991. *Nature (London)* **354**:161–164.

periseptal annuli: one in the center of the cell and one at each pole remaining from the previous division. Which and how many of these sites will be used depends on the levels of FtsZ and MinCDE. Under normal physiological conditions, division occurs at the cell center. This division is localized through the action of the minB locus. A cell-division inhibitor MinC, acts in concert with MinD to inhibit cell division topologically regulated by MinE. The membrane ATPase MinD is required for the topographical specificity conferred by the action of MinE. Then MinE causes MinCD inhibitor to act at the cell

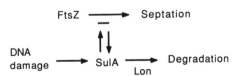

Fig. 11-9. Model for inhibition of FtsZ by the cell division inhibitor SulA. In this model, SulA is synthesized in response to DNA damage and interacts reversibly with FtsZ to inhibit its function. The presence of the Lon protease results in the rapid disappearance of SulA when its synthesis ceases following repair of DNA damage. The FtsZ(Rsa) mutant proteins are resistant to SulA inhibition. In addition, SulA inhibition can be overcome by higher levels of FtsZ. From Bi, E. and J. Lutkenhaus. 1990a. *J. Bacteriol.* **172**:5602–5609.

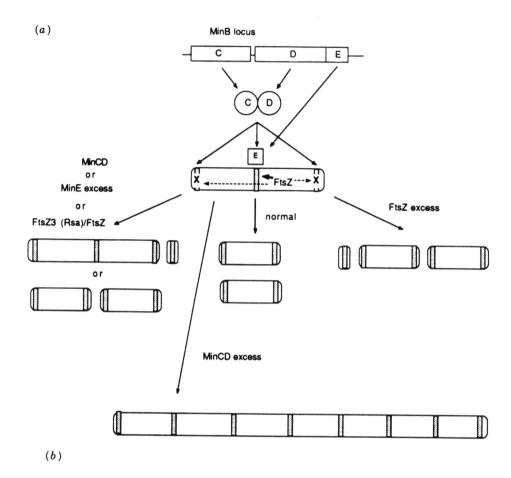

(a)

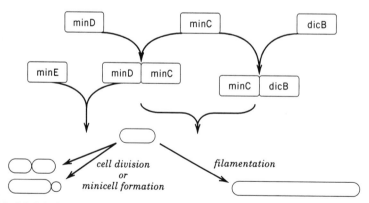

(b)

Fig. 11-10. Models for division inhibition by various regulatory systems. (a) Model for division inhibition by interaction of the MinCDE system with FtsZ (see text for details). From Bi, E. and J. Lutkenhaus. 1990b. *J. Bacteriol.* **172:**5610–5616. **(b) Model for division inhibition by interaction of MinC and MinD or MinC and DicB proteins.** Only MinC–MinD division inhibition can be suppressed by MinE overproduction. From Mulder, E. et al. 1992. *J. Bacteriol.* **174:**35–39.

poles. Division occurs when a critical FtsZ concentration is reached and the nascent site is ready. Overproduction of MinCD results in inhibition of cell division at all sites, resulting in filamentation. Loss of MinCD through mutation or an excess of MinE leads to activation of the poles and the minicell phenotype. Minicell production is also seen with alleles of *ftsZ*, *such as ftsZ3* (Rsa), that appear to be resistant to MinCD. In such cases, division occurs at any one of the three sites. Overproduction of FtsZ can overcome MinCDE inhibition at the poles and this results in a minicell phenotype. This minicell phenotype is distinct in that more than one division event can occur per mass doubling. Overproduction of FtsZ can suppress the filamentation caused by an excess of MinCD.

The DicB protein, encoded by *dicB*, located at 34.9 min, is another cell division inhibitor unrelated to MinC, MinD, or MinE, but interacts with MinC to cause division inhibition by a MinD-independent process. A model for the division inhibition by the interaction of the MinCD and DicB proteins is shown in Figure 11-11.

As discussed in Chapter 2, prokaryotic cell division is ultimately controlled by DNA replication since cell division cannot take place until a round of DNA replication has been completed. Thus, the rate of growth is controlled by the rate of initiation of new rounds of replication at the origin of replication, *oriC*. Initiation of DNA replication is coupled to the cell division cycle by the concentration of DnaA, an initiator protein. The DnaA is the product of the *dnaA* gene mapping at 83 min on the *E. coli* gene map. Initiation of replication requires the formation of a complex of 20–40 molecules of DnaA with *oriC* at specific sites called *dnaA* boxes. Increased synthesis of DnaA stimulates the initiation of replication. Initiation of replication is coupled to the cell cycle by methylation of adenine residues in the GATC sites in or near *oriC* and in the *dnaA* promoter region. This methylation is catalyzed by Dam methyltransferase, the product of the *dam* gene. One or more GATC sites must be fully methylated for the origin to function in initiation. Guanosine tetraphosphate (ppGpp) regulates the *dnaA* operon. It has been suggested that ppGpp provides a signal that couples changes in the rate of protein synthesis to the mechanism governing the rate of initiation of DNA replication. Since cell division cannot take place until a round of replication has been completed, it follows that cell division is triggered by replication termination events. The most important of these events is the decatenation of the replicated chromosomes. The separated chromosomes are then partitioned into the two daughter cells. Division occurs subsequent to or simultaneously with partitioning. The *mukB* gene encodes a protein (MukB) involved in chromosome partitioning in *E. coli*. A suppressor gene *smbA* (suppressor of *mukB*), is essential for cell proliferation in the temperature range of 22–42 °C. Cells lacking the SmbA protein cease macromolecular synthesis.

The most attractive model for positioning of potential division sites remains that proposed by Jacob and co-workers in 1963. As described in Chapter 1, this model proposes that daughter chromosomes are segregated by direct attachment to specific cell envelope sites. Cell envelope growth between the sites would then separate the chromosomes by continued extension. In this regard, it has been shown that hemimethylated DNA from the *E. coli* replication origin (*oriC*) binds with high specificity to a membrane fraction, OCB (*oriC*-binding fraction). This membrane fraction is unique and displays properties that distinguish it from either outer or inner membranes.

As introduced in an earlier discussion, two circumferential rings, referred to as periseptal annuli, are formed at sites where septal ingrowth will eventually occur. Each annulus is a continuous zone of adhesion in which the inner membrane, murein, and outer membrane lie in close apposition to each other. New periseptal annuli are generated

(a)

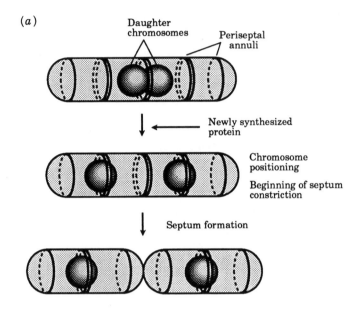

(b)

Inhibition of Protein Synthesis

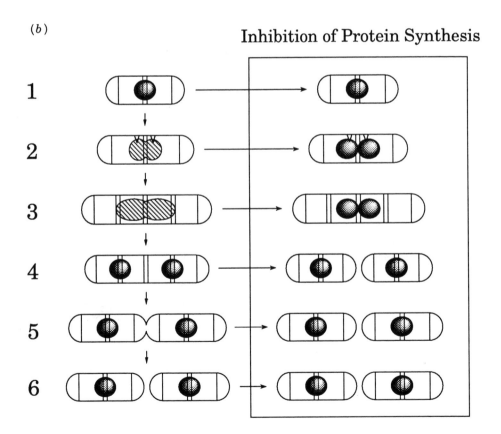

from annuli that are already positioned at the midpoint of the cell. The nascent annuli are then positioned at one-fourth and three-quarters of the cell length during elongation (Fig. 11-11A). This positioning process is inhibited when protein synthesis is inhibited by chloramphenicol or rifampin or by starvation for amino acids. Chromosome replication continues but replicated chromosomes remain close to each other as a single nucleoid mass at midcell. When protein synthesis resumes, the daughter cells move from midcell to the cell quarters before any detectable increase in cell length is observed. Chromosome positioning occurs even when the initiation of chromosome replication is inhibited or when DNA gyrase is inactivated (Fig. 11-11B).

GROWTH OF GRAM-POSITIVE BACILLI

Biosynthesis of the complex polymers that comprise the peptidoglycan layers of gram-positive bacilli involves the nucleoside diphosphate-activated *N*-acetylmuramyl peptide compounds and uridine diphospho-*N*-acetylglucosamine. These precursors are formed in the cytoplasm by soluble enzymes and transferred by membrane-bound enzymes to form a disaccharide-peptide compound joined by a pyrophosphate bond to a C_{55} isoprenoid alcohol. In gram-positive bacilli there is also a rather active system for the turnover of wall components, making it difficult to differentiate between the processes of extension and thickening of the existing wall and localized wall formation in the region of the developing septum. It is tempting to assume that autolysins play a role in preparing sites for the addition of new wall material, but their role in this process has not been clearly established. Lysins could serve to provide acceptor ends of mucopeptide or to break bonds permitting realignment and rearrangement of existing mucopeptide for the insertion of new wall material. In *B. subtilis* and *B. licheniformis*, the most prominent enzyme is an amidase that splits peptides from saccharide chains by hydrolysis of the bonds between L-alanine and muramic acid. Logically, it seems more likely that this enzyme is involved in the separation of cells that have already completed septation. The process of cell division in the gram-positive bacilli undoubtedly requires an overlapping series of interacting activities that control membrane, wall, and chromosome synthesis as well as nucleoid segregation.

The morphogenetic approach to the investigation of the process of cell division in gram-negative organisms has been useful for studies with the gram-positive bacilli. As shown in Figure 11-12, there is considerable similarity between the genes in the 133° region of the genetic map of *B. subtilis* and the genes in the *mra* (2-min) region of the *E. coli* genetic map that are involved in the synthesis of cell wall peptidoglycan and the cell division process. Mutants of *B. subtilis* have been found that form filaments at high

---

Fig. 11-11. (a) Model of positioning of daughter chromosomes from midcell to one-fourth and three-fourths of the cell length. (b) Schematic representation of nucleoids and plasmolysis bays under the inhibition of protein synthesis. Hatched areas represent replicating nucleoids. Meshed areas represent nucleoids that did not undergo DNA replication. Vertical lines in the cells represent periseptal annuli. The two triangles in Stage 2 represent newly generated periseptal annuli. After incubation under the inhibition of protein synthesis, two replicating daughter chromosomes were located close to each other and observed as one nucleoid mass. From Hiraga, S. et al. 1990. *J. Bacteriol.* **172**:31–39.

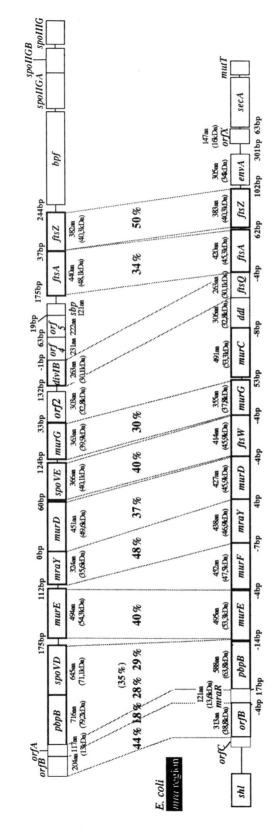

Fig. 11-12. Comparison of the 133° region of the *B. subtilis* genetic map with the *mra* (2-min) region of the *E. coli* genetic map. Similar pairs of genes are joined by dotted lines and similarities in predicting amino acid sequences are indicated as percent (%) identity. The gaps between adjacent coding regions are indicated in base pairs (bp). From Buchanan, C. E., A. O. Henriques, and P. J. Piggot. 1994. Cell wall changes during bacterial endospore formation. In Ghuysen, J.-M. and R. Hackenbeck (Eds). *Bacteriall Cell Wall*, Elsevier, Amsterdam, The Netherlands, pp. 167–186.

534

temperature (40–50 °C) but grow normally at low temperature (20–35 °C). These filamentous mutants can then be screened to eliminate those that affect division because of
defects in DNA replication, nutritional requirements, or steps in the final separation of
cells that have already been compartmentalized.

A number of division mutants of *B. subtilis* have been identified and designated *divA*,
divB, *divC*, and *divD* on the basis of genetic and morphological criteria. Members of this
group of mutants are defective in some stage of septum synthesis. It is often difficult to
distinguish between defects in septum initiation and septal growth. Two mutants of *B.
subtilis* 160, designated *ts1* and *ts12* have been definitively characterized as division
initiation mutants. The *ts1* gene product is required both for normal division and for
division leading to the production of anucleate cells. The action of the gene product is
required up to the time of commencement of septal growth. Studies on recovery of septa
after transfer of filamentous cells from the nonpermissive to the permissive temperature
show that septa are formed at sites that are discrete fractional lengths of the filaments.
The first septum is formed at the most polar of these sites. These observations have been
interpreted as an indication that potential division sites are formed at the nonpermissive
temperature and, upon recovery, the most recently formed site is used preferentially for
septum formation, as shown in Figure 11-13. Recovery of septa at the permissive temperature occurs in the absence of DNA synthesis, but is blocked completely by inhibitors
of RNA and protein synthesis. The only protein synthesis required for recovery of septa
may be to form the *ts1* gene product itself.

A fundamental feature of the replicon hypothesis is that replicating daughter chromosomes attached to the membrane would be segregated by zonal insertion of newly
synthesized membrane between the points of attachment. Zonal growth of the wall has
been clearly demonstrated in gram-positive cocci. In bacilli, evidence both for zonal and
diffuse growth has been reported. However, the cell wall of *B. subtilis* consists of a
limited number of large subunits that segregate during cell growth. Mutants with reduced
cell wall turnover allow long periods of comparison of the numbers and relative positions
of segregation units of cell wall and DNA along chains of cells. Continuous labeling of
cell wall and DNA show that cell wall units are localized according to a symmetrical
pattern, whereas those of DNA are distributed in an asymmetrical but highly regular
manner. This type of segregation pattern indicates symmetrical zonal insertion of new
cell wall material in agreement with the replicon model. For every replication point of
DNA (two new segregation units) four segregation units of cell wall are synthesized and
each DNA strand is attached to only one out of two cell wall units synthesized at the
same time, as shown in the model in Figure 11-14. In this model the DNA cosegregates
with cell wall according to a regular but asymmetrical pattern and serves as a primitive
mitotic apparatus. In another approach, it has been possible to examine the major features
of the nucleoids and cross-walls of cells under the electron microscope. Growth rates for
nucleoid movement are parallel to those of total length extension. The summation of
both types of experiments envisages that nucleoid extension is coupled to length extension. Cell length extension in *B. subtilis* occurs spontaneously and at no stage of the cell
cycle is there any conspicuous change in growth rate.

As in the case of other rod-shaped organisms, gram-positive bacilli usually extend in
length with little, if any, change in diameter. At a certain point in this process, the
elongating cell is partitioned into two nearly equal halves by the formation of a cross-
wall. Temperature-sensitive morphological mutants that change from rod to coccus shape
upon shift from 20 to 42 °C (Rod⁻) display an altered growth pattern from the normal

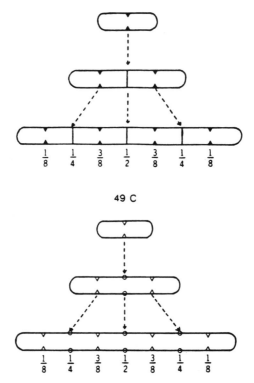

$$\frac{1}{8} \quad \frac{1}{4} \quad \frac{3}{8} \quad \frac{1}{2} \quad \frac{3}{8} \quad \frac{1}{4} \quad \frac{1}{8}$$

49 C

$$\frac{1}{8} \quad \frac{1}{4} \quad \frac{3}{8} \quad \frac{1}{2} \quad \frac{3}{8} \quad \frac{1}{4} \quad \frac{1}{8}$$

Fig. 11-13. Model for the formation of potential division sites in filaments at 49 °C. Division is initiated at 34 °C (▼) when cells reach a specific cell length, surface area, or mass. (The actual growth parameter that determines cell division is as yet unknown.) The first septation divides the bacillus into two cells; the second into four cells; the third, into eight cells, and so on. The sites at $^1/_2$, $^1/_4$, $^1/_8$, and $^3/_8$ of the bacillus length are shown accordingly. At 49 °C, the positional information is still generated, but division initiation (▽) is defective in the absence of an active *ts1* gene product so that septum formation cannot proceed. Symbols: ▲ = most recently formed potential division sites; ● = unused sites formed at an earlier stage of growth. From Callister, H., T. McGinness, and R. G. Wake. 1983. *J. Bacteriol.* **154**:537–546.

cylindrical extension of the wall to one in which the incorporation of newly synthesized wall material leads only to an increase in diameter (Fig. 11-15). In these mutants, the wall forming the septum and nascent poles can be distinguished from the surface distal to the division site by the presence of raised areas on the cell surface that appear to be analogous to the wall bands of streptococci. Mathematical calculations that provide a three-dimensional reconstruction of growing cells indicate that the proportion of septal wall increases during the shape change. In the coccal forms, all surface growth seems to arise from septal growth sites.

EFFECT OF ENVIRONMENTAL CHANGES ON MICROBIAL GROWTH

As we have already described in previous chapters, changes in the cultural environment represent **stresses** imposed upon the organism. Ultimately, microbial physiology is the

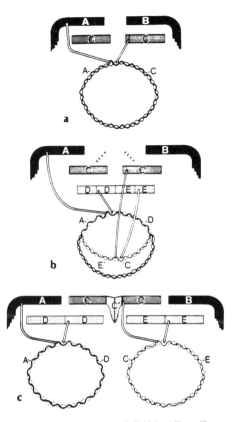

Fig. 11-14. Model of cell wall organization and DNA cell wall association. Upon sufficient chase periods, this model provides segregation patterns of cell wall and DNA that indicate cosegregation of DNA with cell wall according to a regular but asymmetrical pattern, which permits the system to serve as a primitive mitotic apparatus. a, b, and c correspond to three different stages in the cell cycle. A, B, C, D, and E refer to segregation units based on whether radioactive labeling in a precursor labeling experiment is observed at the ends of chains of cells (A and B), in the middle of chains (C), between A and B (D), or between C and B (E). Here C′ designates the new septum. From Schlaeppi, J.-M. and D. Karamata. 1982. *J. Bacteriol.* **152:**1231–1240.

study of the inherent ability of the organism to respond to these various environmental stresses. The molecular response to these stresses at the cellular level has been termed **global control** and is discussed in some detail in Chapter 4. The environment in which a microbial cell is grown influences the physiological manifestations observed, that is, it is unable to cope and dies or it is able to mobilize enzymes and other factors enabling survival or even renewed growth at a near normal rate. Factors in the environment that stimulate or inhibit growth, whether they are specific or general in their mode of action, play a composite role in the final outcome of the amount and the nature of the metabolic activity and subsequent growth observed.

For convenience, environmental factors that influence microbial growth are considered as either **physical** (e.g., temperature) or **chemical** (nutrients). In some cases it is not

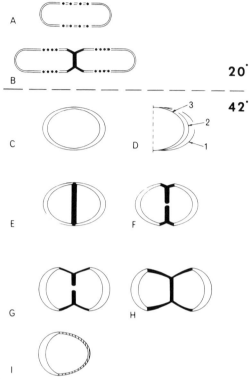

Fig. 11-15. Diagram illustrating the possible nature of shape change in cells of *Bacillus subtilis* *rodB* **switched from 20 to 42 °C.** At 20 °C (A and B) sites of cylindrical extension (●), shown as discrete sites purely for convenience, continue to extend the cell until a cross wall and nascent poles (shown in black) are formed. New cylindrical sites are assumed to exist in daughter cells. On a shift-up to 42 °C, the cylindrical sites are proposed to become modified so that cells enlarge in diameter C. Progressive increases in diameter D may displace wall from the center of the cell toward the distal poles (i.e., the surface labeled 1 is the oldest wall, and surface 3 is the youngest). As the cells become coccal in shape, a progressively greater proportion of wall may be assembled at the septal growth sites (E–H) shown in black. In coccal forms (G and H) pole production may also occur before septal closure. After cell separation the newly formed pole (1, hatched area) will be of different shape and thickness compared with the other, presumably younger pole. During subsequent growth, the shape and thickness of the pole may be modified by localized wall thickening. From Burdett, I. D. 1979. *J. Bacteriol.* **137:**1395–1405.

possible to distinguish between these two categories, and it is the combined **physico-chemical** effect that results in alteration of the activities of the cell and influences the growth of the cell population. A distinction is often made between alterations in the cultural conditions within a reasonable range and the imposition of toxic or extremely harsh conditions, sometimes referred to as the **hostile environment**. On some of these topics we may need only to remind the reader of previous discussions in earlier chapters. In other cases it may be useful to provide additional examples of microbial responses to environmental stress.

Water

Of all the various factors that influence the growth of microorganisms, water may be considered the most important. Indeed, water may really be considered a nutrient since it forms the bulk of the cellular substance ($\sim$ 75%). When organisms are grown on surfaces such as an agar plate, high humidity can provide conditions favorable to the development of the colony. Water acts as a solvent, and most metabolic activities are conducted within an aqueous environment within the cell. Water also serves as a catalyst by aiding or actually entering into many enzymatic reactions.

Turgidity of the cell is dependent on the presence of water. In turn, turgidity is affected by the **osmotic tension** (surface tension) of the medium in which the organism is suspended. As described in the next section, microorganisms can regulate their internal osmotic tension by various means. In so doing, they are able to control their internal water content to some extent.

The effects of desiccation on the viability of microorganisms provides a good example of the importance of water. Slow desiccation in the presence of air is most detrimental, since oxidative stresses are superimposed on the effects of water removal. In contrast, the rapid removal of water at low temperature (-35 °C or lower) under a high vacuum (**lyophilization** or freeze-drying) and subsequent storage under a partial vacuum in sealed ampules is an effective method for the preservation of viability over long periods of time. Successful retention of the viability of microorganisms by this method has led to the false assumption that the lyophilization process is highly efficient. This is not entirely true. Loss of viability is frequently quite high and only by using the best possible conditions of added stabilizers (protein-containing materials, such as milk or serum) and adherence to the optimal conditions of low temperature and high vacuum is it possible to obtain a respectable percentage of viable cells.

Osmotic Pressure and Osmoregulation

Most microorganisms display optimal growth in a culture medium of low osmotic strength. Many of the commonly encountered microorganisms, when placed in an environment containing a high concentration of solute, fail to initiate growth. In hypertonic solution, the cell loses water and shrinks (**plasmolysis**). In a hypotonic solution, there may be swelling (**plasmoptysis**) and the cell may actually burst if its integrity is not maintained by the rigid cell wall. Most living cells respond to changes in the osmotic strength of the culture medium by accumulating **compatible solutes** that regulate the internal osmolarity of the cell. These **osmoregulatory compounds** or **osmoprotectants** may be inorganic cations, such as K^+, amino acids, such as glutamate or proline, amino acid derivatives, or carbohydrates, such as trehalose (see Chapter 4).

Of these various compounds, betaines and, to some extent, proline, are the most effective osmoprotectants for *E. coli* placed in culture media at inhibitory osmolarities. Although choline and betaine aldehyde are also osmoprotective for *E. coli*, the action of these compounds is the result of their conversion to betaine (N,N,N-trimethylglycine):

$$(CH)_3{}^+NCH_2CH_2OH + 0.5O_2 \longrightarrow$$
Choline

$$H_2O + (CH_3)_3{}^+NCH_2CHO \xrightarrow{\;NAD(P)^+ \quad NAD(P)H\;} (CH_3)_3{}^+NCH_2COOH$$
Betaine aldehyde $\qquad\qquad\qquad\qquad\qquad\qquad\qquad$ Betaine

The first step in this sequence is catalyzed by a membrane-associated, oxygen-dependent choline dehydrogenase coded for by *betA* in *E. coli*. The second step is mediated by an NAD(P)-dependent betaine aldehyde dehydrogenase. Both enzymes are osmotically regulated. Only cells grown aerobically in choline-containing medium at an increased osmotic strength display full enzyme activities. Because of the oxygen requirement for choline dehydrogenase, choline functions as an osmoprotectant only under aerobic conditions, whereas betaine aldehyde and betaine function under aerobic or anaerobic conditions.

Exogenously supplied proline or metabolic overproduction of proline enhances osmotolerance in *S. typhimurium*. Three genetically distinct proline transport systems have been characterized in this organism. The major proline permease, PutP, is encoded by the *putP* gene, functions primarily for proline catabolism, and is not involved in osmoregulation. A second proline permease, ProP, has a very poor affinity for proline ($K_m = 300 \mu M$). However, it plays a major role in betaine uptake and mutations in *proP* reduce the ability of betaine to serve as an osmoprotectant. Kinetics of betaine uptake via ProP indicate that betaine, rather than proline, is the primary physiological substrate for this enzyme. The third proline permease, ProU, is osmotically induced. Proline transport by this enzyme is negligible, but it serves as a high-affinity transport system for betaine transport. Regulation of the *proU* gene occurs at the transcriptional level, increasing over 100-fold with increasing osmotic pressure. The osmotic reduction of *proU* expression is independent of the *ompB* locus, which is involved in the osmotic regulation of porin expression, indicating that there are at least two distinct pathways for osmoregulation of gene expression in *S. typhimurium* (see Chapter 4).

Halotolerant bacteria produce some unique osmoprotective compounds. One such compound, ectoine (1,4,5,6-tetrahydro-2-methyl-4-pyrimidine carboxylic acid), a cyclic amino acid, is produced by a phototrophic bacterium, *Ectothiorhodospira halochloris*. It is also produced by *Brevibacterium linens*. Ectoine purified from cultures of *B. linens* was found to stimulate growth of *E. coli* in media of inhibitory osmotic strength. It was equivalent to betaine in osmoprotective ability. Both ProP and ProU systems are involved in uptake and accumulation of ectoine by *E. coli*.

Trehalose, a nonreducing disaccharide of glucose, is a stress metabolite whose synthesis is evoked in yeast as well as bacteria in response to a variety of environmental stimuli (elevated temperature, chemical agents, dessication, changes in osmolarity). In *E. coli* the osmoregulatory trehalose pathway is encoded by the *otsBA* genes. Mutations is *otsA* and *otsB* block production of trehalose-6-phosphate synthase and a phosphatase that converts trehalose-6-phosphate to trehalose (see Chapter 7). The *otsA* and *otsB* genes are transcriptionally activated by osmotic stress. The synthetase is also activated by potassium glutamate. The *otsA* and *otsB* genes constitute the *otsBA* operon whose transcription is dependent on RpoS, a putative σ factor for starvation- and stationary phase-induced genes.

Complex sugars, termed membrane-derived oligosaccharides, containing 6–12 glucose units joined by β(1 → 2) and β(1 → 6) linkages forming a branched structure, are found in the periplasmic space of *E. coli* and other gram-negative bacteria. Membrane-derived oligosaccharides are substituted with *sn*-1-phosphoglycerol and phosphoethanolamine residues derived from membrane phospholipids and also with *O*-succinyl ester residues. Synthesis of these compounds is osmotically regulated being evoked by growth in medium of low osmolarity. Membrane-derived oligosaccharides may play a role in other osmoregulated systems, such as chemotaxis, motility, production of the colanic acid-containing capsules, glycogen accumulation, and the production of outer-membrane porins.

Surface tension, per se, may affect the growth of some organisms. In a given medium, certain organisms grow readily in uniform suspension while others form a surface pellicle. Pellicle formation may not necessarily reflect the highly aerobic nature of the organism since the same organism may be grown in uniform suspension if surface tension depressants are added to the medium. The use of Tweens (e.g., Tween 80, a nonionic surfactant composed of sorbitan monooleate polyoxy-alkylene) added to liquid culture media permits uniformly suspended growth of mycobacteria. Conversely, surfactants may serve as highly effective antimicrobial agents. Cationic agents are particularly effective in this regard if they are coupled with a positively charged hydrophilic group. Agents of this type disrupt the cell membrane by reacting with the negatively charged phospholipids, while the nonpolar portion of the molecule penetrates into the hydrophobic layer of the membrane.

Hydrogen Ion Concentration

As discussed briefly in Chapter 1, microorganisms exhibit a range of pH at which growth is most readily initiated, but many organisms are able to function over a wide range of pH values (Table 1-4). It has been known for a long time that microorganisms respond to a decrease in external pH by the production of enzymes that convert acidic metabolites to neutral compounds. As examples, the expression of genes encoding lysine decarboxylase (*cadA*) and arginine decarboxylase (*adi*) is markedly increased at low external pH. In Chapter 7, the concept of **pH homeostasis**, the ability of organisms to regulate their intracellular pH, was introduced. It has been shown that intracellular pH (pH_i) is maintained by modulation of primary cellular proton pumps as well as K^+/proton and Na^+/proton antiport systems. However, a superimposing genetic response can further protect cells of *S. typhimurium* and *E. coli* from acid stress. This process, referred to as the **acid tolerance response (ATR)**, is triggered by pH values between 6.0 and 5.0, but can protect against much stronger acid (pH 3.0–4.0) when nonadaptive pH homeostasis normally fails. The ATR process includes two distinct stages. The first stage, called preacid shock, is induced at pH 5.8 and involves the production of an inducible pH homeostasis system functional at external pH values below 4.0. The second stage occurs following a shift to pH 4.5 or below and is called the postacid shock stage. During this stage, 43 acid-shock proteins (ASPs) are synthesized. Some of these proteins are believed to be responsible for prevention or repair of acid damage to macromolecules. Although both stages of the ATR response are required for maximum protection against low pH, a single brief period (15 min) of exposure to pH 4.3 enables cells to tolerate a subsequent challenge pH of 3.3. The ATR system is sensitive to external protein synthesis inhibitors, whereas the classical constitutive pH homeostasis is unaffected by protein synthesis inhibitors and is effective only at pH values above 4.0. Physiological studies indicate an important role for the Mg^{2+}-dependent proton-translocating ATPase in the ATR system.

There exist in nature a number of **acidophilic bacteria** whose pH optima for growth are generally between 2 and 4. These organisms maintain their cytoplasmic pH at 6.0 or higher. Consequently, there has been considerable interest in determining the mechanism whereby this large pH differential (ΔpH) is maintained. It is generally agreed that maintenance of this large transmembrane ΔpH is energy dependent and that its maintenance requires a transmembrane electrical potential (Δψ) that is positive inside (the reverse of that found in neutrophilic bacteria). A reverse transmembrane potential has been observed in both *Bacillus acidocaldarius* and *Thiobacillus acidophilis*. If this transmembrane potential is abolished, then the ΔpH collapses. Despite the large ΔpH, the cytoplasmic pH is extremely stable. The cytoplasmic buffering capacity of *T. acidophilis* is responsible for the cytoplasmic pH homeostasis in metabolically comprised cells. When a large influx of H⁺ occurs, the cytoplasmic buffering capacity prevents drastic changes in pH. In addition, the resultant increase in positive membrane potential due to this influx of H⁺ eventually leads to cessation of further H⁺ influx.

Studies with a heterotrophic, mesophilic, obligate acidophilic bacterium that presents a more usual type of physiology than the chemolithotrophic *T. acidophilus* provides additional insights into the nature of changes in membrane potential under active and inactive conditions (Fig. 11-16). Starving cells of this acidophile continue to show a ΔpH of about 1.7, but exhibit changes in membrane potential (Δψ) and proton motive force (Δp) that are just opposite of those seen under conditions of optimal nutrition. The linkage of the transient H⁺ influx with the rise in Δψ and the cytoplasmic buffering capacity play central roles in acidophilism. It is considered that the same impermeant cellular macromolecules can account for both. Thus the Δψ represents a Donnan potential that in active cells is offset by energy-dependent H⁺ extrusion.

Many bacteria display optimal growth in alkaline media. Although a number of these are *Bacillus* species, alkalophilic strains of *Micrococcus*, *Pseudomonas*, *Clostridium*, the photosynthetic bacterium, *Ectothiorhodospiras*, and many others have been reported. These alkalophiles can grow only at pH levels of 8.0–11.5. A true alkylophile maintains a cytoplasmic pH of 9.0 or lower even at an external pH of 11.0, thus growing optimally under conditions in which the cytoplasm is more acidic than the external medium. The alkalophiles, *B. firmus* and *B. alcalophilus*, both depend on the activity of a Na⁺/H⁺ antiporter to catalyze acidification of the cytoplasm relative to the exterior. A nonalkalophilic strain of *B. firmus* isolated after mutagenesis of the alkalophilic strain, lacks the Na⁺/H⁺ antiporter. The inability of *B. firmus* to grow at neutral pH is not due to excessive acidification but is related to a failure of respiratory activity to generate a trans-

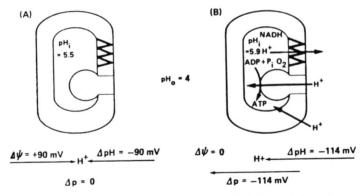

Fig. 11-16. Model for bioenergetic parameters in an obligately acidophilic bacterium at an outside pH (pH$_o$) of 4.0 in inactive (A) and respiring (B) cells. Left and right arrows indicate the forces impelling the H⁺ into and out of the cells and are drawn to scale. There is no H⁺ flux in inactive cells. In active cells H⁺ influx equals H⁺ eflux. From Goulborne, E., Jr. et al. 1986. *J. Bacteriol.* **166**:59–65.

membrane electrical potential that is high enough to maintain certain cellular functions, such as Na^+/solute symport and motility. Nonalkalophic mutant strains of *B. alcalophilus* exhibit loss of Na^+/H^+ antiport activity and Na^+ coupling of solute transport as well as decreased contents of membrane-bound cytochromes and a membrane-bound chromophore.

Temperature

In considering the effect of variations in temperature as well as other environmental influences on microorganisms, it should be kept in mind that there is a fundamental difference between the ability of an organism to initiate and sustain growth and the ability to simply survive exposure to a particular set of environmental conditions. Also, in the process of evaluating the effect of a given environmental parameter on growth, it is important to keep all other environmental factors constant.

In Table 1-3, microorganisms were grouped over certain temperature ranges. When extremes of temperature are considered, the effects on microbial growth may be explained on the basis of inactivation of enzymes or other vital cell structures, such as membranes. In Chapter 4, the heat shock response was described in some detail. *Escherichia coli* is able to survive over a temperature range from 10–49 °C. Increases in temperature throughout this range result in the induction of the heat shock response. As many as 20 or more genes are transiently expressed at increased levels following a shift-up in growth temperature. These genes form a regulon under the coordinate control of the positive transcription regulator σ^{32}. The σ^{32} polypeptide allows the RNA polymerase enzyme to specifically recognize heat shock promoters, which have a consensus sequence distinct from that recognized by the normal σ^{70} factor. Some of the heat shock genes are essential for the growth of *E. coli* at all temperatures indicating an important role in the normal cell physiology of this organism. Mutants of *E. coli* resulting in the inability to grow at high temperatures are designated *htr* for high-temperature requirement. The *htrA* gene is under the control of σ^E (σ^{24}), which is induced only at very high temperature (>42 °C). The HtrA protein is a periplasmic endopeptidase (protease) required only at elevated temperatures.

In an analogous manner, the **cold shock response** in *E. coli* follows an abrupt shift in growth temperature from 37–10 °C. Growth ceases and over the next 4–5 h there is a severe reduction in synthesis of a number of proteins. During this same time period there is an increase in certain proteins designated as **cold shock proteins**. These proteins include CS7.4, NusA, RecA, H-NS, polynucleotide phosphorylase, translation initiation factors 2β and 2α, pyruvate dehydrogenase, and the dihydrolipoamide acetyl transferase of pyruvate dehydrogenase. The A subunit of DNA gyrase has also been identified as a cold shock protein. The duration of the lag period in synthesis and the synthetic rate of the A subunit is dependent on the synthesis of the induced cold shock protein, CS7.4. The promoter of the *gyrA* gene contains specific binding sites for the CS7.4 protein, suggesting that CS7.4 acts as a transcriptional activator of the A subunit of DNA gyrase.

Thermophilic organisms have been studied intensively in an effort to determine the mechanisms whereby these organisms not only survive but prefer higher temperatures for growth. Chemical differences in the lipids found in thermophiles, higher metabolic rates facilitating rapid resynthesis of heat-denatured cellular components, and macromolecules with higher thermostability have all been suggested as mechanism for greater heat tolerance in thermophiles. Thermophilic organisms contain lipids with higher melting points than those found in mesophiles, suggesting that the temperature at which the major lipid components of the cell melt may establish the upper limit for growth. In yeast, a direct correlation has been established between the growth temperature limits and the degree of unsaturation in mitochondrial lipids. The lower the temperature limit of growth, the greater the degree of lipid unsaturation. The membrane lipid composition of thermophilic yeasts are distinguished by the higher percentage (30–40%) of saturated fatty acids as compared with mesophilic and psychrophilic species. Psychrophilic yeasts contain approximately 90% unsaturated fatty acids, 55% of which is linolenic acid ($C_{\alpha-18:3}$).

Certain lines of evidence indicate that thermophilic organisms possess an intrinsic thermostability that is independent of any transferable, stabilizing factors. Membranes with greater heat stability, more rapid metabolic and growth rates, and factors that impart structural stability and greater inherent heat stability to individual protein macromolecules have been implicated as contributing factors. The greater heat stability of protein molecules from thermophilic organisms appears to reside in their ability to bind certain ions more tightly, thereby enhancing the establishment of more stable conformations. Proteins from thermophiles also contain increased levels of hydrophobic amino acids as compared to those from mesophiles. Although macromolecules from thermophiles are more thermostable, they are, for the most part, physicochemically similar to their mesophilic counterparts with regard to molecular weight, subunit composition, allosteric effectors, amino acid composition, and major amino acid sequences. Certain attributes of the thermostability of proteins have been shown to be due to subtle changes in structure and to alterations in hydrogen bonding, hydrophobic interactions, and other noncovalent activities. For example, the thermostability of the tryptophan synthetase of *E. coli* is increased by amino acid substitutions that alter the hydrophobicity of the molecule in the absence of gross changes in conformation. In the thermophile *B. stearothermophilus*, a marked increase in thermostability is engendered by the presence of lysine in place of threonine in the plasmid-encoded enzyme that inactivates kannamycin. The nucleotide sequence of the plasmid encoding this enzyme in *B. stearothermophilus* differs by only one base from the plasmid coding for the identical enzyme in the mesophilic *S. aureus*. The lysine substitution permits increased electrostatic bridging with little significant change in the three-dimensional structure.

Isolation of thermophilic mutants of *B. subtilis* and *B. pumilus* and transformation of the thermophilic trait to mesophilic strains provides direct genetic evidence that the thermophilic trait is a phenotypic consequence of at least two unlinked genes. Additional studies of this type should provide information that could shed light on the specific nature of thermophily.

Psychrophilic organisms, those that exhibit an optimal temperature of growth at 15 °C or lower and a maximum temperature for growth at 20 °C, are defined as **obligate psychrophiles** to differentiate them from organisms that can grow at low temperatures but are actually mesophiles in terms of their optimum growth temperature. Although some confusion still exists with regard to the proper nomenclature and cardinal temperatures that differentiate psychrophiles from organisms that are *psychrotolerant* or *psychrotropic*, those organisms that grow well at temperatures at or below the freezing point of water are considered to be true psychrophiles.

The psychrophilic yeast *Leucosporidium* grows well at −1 °C. In this organism there is a positive correlation between the growth temperature and the unsaturated fatty acid composition of the cell lipids. At subzero temperatures (−1 °C) with ethanol as substrate, 90% of the total fatty acid is unsaturated with linolenic (35–50%) and linoleic (25–

30%) predominating. At temperatures close to the maximum for growth, linolenic acid accounted for less than 20% of the total fatty acid and oleic (20–40%) and linolenic (30–50%) were the major components. Marked changes also occurred in the cytochrome composition as the growth temperature is altered. In *L. frigidum*, the ratios of cytochromes $a + a_3:b:c$ were 1:1:2.9 at 8 °C with glucose as substrate whereas at 19 °C they were 1:2.3:16.7. It can be concluded that changes in membrane structure and composition are fundamental to temperature adaptation in psychrophilic yeasts.

The psychrophilic bacterium, *Micrococcus cryophilus*, also undergoes alteration in its lipid composition with changes in growth temperature. Cultures of this organism continue to grow without a lag following a sudden increase in temperature from 0 to 20 °C (shift-up) or a reciprocal decrease (shift-down). The growth rate changes gradually to that typical of cultures grown isothermally at the final temperature. After a temperature shift-down, the phospholipid acyl chain length begins to change immediately, whereas there is a delay following shift-up. However, the final fatty acid composition is attained within the same number of cell division times after a shift-up or a shift-down. It appears that this psychrophile is more stressed by a sudden increase in growth temperature than by a sudden decrease. Studies on the viscosity and phase transition temperatures of lipids isolated from psychrophilic, psychrotropic, and mesophilic organisms are able to adjust their lipid-phase transition temperature to the growth temperature. By comparison, a psychrophilic *Clostridium* synthesizes lipids that have the same phase transition temperature after growth at different temperatures. This lack of growth temperature-inducible regulation of lipid-phase transition temperature appears to be a molecular determinant for psychrophily in this organism. Comparisons of the properties of triosephosphate isomerase purified from psychrophilic, mesophilic, and thermophilic clostridia indicate that the purified enzymes have the same molecular weight, subunit molecular weight, and susceptibility to the active site-directed inhibitor, glycidol phosphate. However, their temperature and pH optima, as well as stabilities to heat, urea, and sodium dodecyl sulfate (SDS) differ markedly.

Nutrition–Starvation Induced Proteins

Sporulation by some organisms is a developmental response to nutrient limitation that increases the survival of the species. Nonsporulating organisms also exhibit a stress response to nutrient starvation that is designed to increase survival. When *E. coli* is starved for carbon in a minimal medium, over 50 proteins are induced. Several different regulons are involved in this response. The Cst (carbon starvation) proteins require the cyclic AMP (cAMP)–cAMP receptor protein system for synthesis. Strains defective in either adenylate cyclase (encoded by *cya*) or the cAMP receptor protein (encoded by *crp*) fail to make 19 of the prominent starvation-induced proteins. Another set of starvation proteins, designated Pex (postexponential) proteins, is also induced upon entry of the cells into stationary phase or by starvation for nitrogen or phosphate. Many of the Pex proteins are regulated by RpoS, an alternate σ factor whose transcription increases in stationary phase cells. A heat shock protein RpoH, that acts as an alternate σ factor (σ^{32}), can also function in carbon starvation protein synthesis and survival. Cross-resistance against heat, oxidative stress, and osmotic stress challenge develops following carbon starvation, indicating a broad overlap in the stress response of the organism to different types of environmental stimuli. Similar studies in *S. typhimurium* have revealed several key proteins (Sti) important for surviving a variety of nutrient limitations. These genes are regulated by Crp (some in a cAMP-independent fashion) and RpoS.

Hydrostatic Pressure

Marine organisms are subject to wide variations in hydrostatic pressure. The growth of most bacteria is inhibited by increases in hydrostatic pressure, but considerable differences are observed in the degree of tolerance even among bacteria isolated from the marine environment. With the exception of a few species from the deep sea, most marine bacteria fail to grow at pressures ranging from 200 to 600 atm, the limiting pressure being dependent on the species, size of inoculum, chemical composition of the medium, temperature, and a number of other physiological factors. Bacteria isolated from depths of 7000–10,400 m are capable of growth at pressures ranging from 700 to 1000 atm: however, a number of unexplained differences in growth and metabolic activity are observed in the pressure responses of deep sea microflora.

The rate of production of several proteins by some hydrothermal vent archaebacteria and the degree of saturation of membrane lipids in other deep sea bacteria have been found to change as a result of cultivation at high temperature. Studies with a gram-negative eubacterium isolated from a depth of 2.5 km show that the mRNA encoded by the gene *ompH* is expressed when cells are grown at 280 atm but not at 1 atm. This finding suggests that transcription of *ompH* is controlled by hydrostatic pressure.

REFERENCES

Bacterial Growth and Cell Division

Begg, K. J. and W. D. Donachie. 1985. Cell shape and division in *Escherichia coli*: Experiments with shape and division mutants. *J. Bacteriol.* **163**:615–622.

Begg, K. J., A. Takasuga, D. H. Edwards, S. J. Dewar, B. G. Spratt, H. Adachi, T. Ohta, H. Matsuzawa, and W. D. Donachie. 1990. The balance between different peptidoglycan precursors determines whether *Escherichia coli* cells will elongate or divide. *J. Bacteriol.* **172**:6697–6703.

Bi, E. G. and J. Lutkenhaus. 1990a. Analysis of *ftsZ* mutations that confer resistance to the cell division inhibitor of SulA (SfiA). *J. Bacteriol.* **172**:5602–5609.

Bi, E. B. and J. Lutkenhaus. 1990b. Interaction between the *min* locus and *ftsZ*. *J. Bacteriol.* **172**:5610–5616.

Bi, E. B. and J. Lutkenhaus. 1991. FtzZ ring structure associated with division in *Escherichia coli*. *Science* **354**:161–164.

Bupp, K. and J. van Heijenoort. 1993. The final step of peptidoglycan subunit assembly in *Escherichia coli* occurs in the cytoplasm. *J. Bacteriol.* **175**:1841–1843.

Cooper, S. 1989. The constrained hoop: An explanation of the overshoot in cell length during a shift-up of *Escherichia coli*. *J. Bacteriol.* **171**:5239–5243.

Cooper, S. 1990. Relationship between the acceptor-/donor radioactivity ratio and cross-linking in bacterial peptidoglycan: Application to surface synthesis during the division cycle. *J. Bacteriol.* **172**:5506–5510.

Cooper, S., 1991. *Bacterial Growth and Division: Biochemistry and Regulation of Prokaryotic and Eukaryotic Division Cycles*. Academic, San Diego.

Cooper, S. 1991. Synthesis of the cell surface during the division cycle of rod-shaped, gram-negative bacteria. *Microbiol. Rev.* **55**:649–674.

deBoer, P. A. J., R. E. Crossley, and L. I. Rothfield. 1992. Roles of MinC and MinD in the site-specific septation block mediated by the MinCDE system of *Escherichia coli. J. Bacteriol.* **174:** **63–70.**

deBoer, P., R. Crossley, and L. Rothfield. 1992. The essential bacterial cell-division protein FtsZ is a GTPase. *Nature (London)* **359:**254–256.

deBoer, P. A. J., W. R. Cook, and L. I. Rothfield. 1990. Bacterial cell division. *Annu. Rev. Genet.* **24:**249–274.

Donachie, W. D. 1993. The cell cycle of *Escherichia coli. Annu. Rev. Microbiol.* **47:**199–230.

Doyle, R. J., J. Chaloupka, and V. Vinter. 1988. Turnover of cell walls in microorganisms. *Microbiol. Rev.* **52:**554–567.

Gally, D., K. Bray, and S. Cooper. 1993. Synthesis of peptidoglycan and membrane during the division cycle of rod-shaped, gram-negative bacteria. *J. Bacteriol.* **175:**3121–3130.

Harold, F. M. 1990. To shape a cell: An inquiry into the causes of morphogenesis of microorganisms. *Microbiol. Rev.* **54:**381–431.

Helmstetter, C. E. 1987. Timing of synthetic activities in the cell cycle. *In* Escherichia coli *and* Salmonella typhimurium: *Cellular and Molecular Biology*, F. C. Neidhardt, J. S. Ingraham, K. B. Low, B. Magasanik, M. Schaechter, and H. E. Umbarger (eds.). American Society for Microbiology, Washington, DC, pp. 1594–1605.

Hiraga, S., T. Ogura, H. Niki, C. Ichinose, and H. Mori. 1990. Positioning of replicated chromosomes in *Escherichia coli. J. Bacteriol.* **172:**31–39.

Ikeda, M., M. Wachi, H. K. Jung, F. Ishino, and M. Matsuhashi. 1991. The *Escherichia coli mraY* gene encoding UDP-*N*-acetylmuramoyl-pentapeptide:Undecaprenyl-phosphate phospho-*N*-acetylmuramoyl-pentapeptide transferase. *J. Bacteriol.* **173:**1021–1026.

Jensen, K. F. and S. Pedersen. 1990. Metabolic growth rate control in *Escherichia coli* may be a consequence of subsaturation of the macromolecular biosynthetic apparatus with substrates and catalytic components. *Microbiol. Rev.* **54:**89–100.

Koch, A. L. 1988. Biophysics of bacterial walls viewed as stress-bearing fabric. *Microbiol. Rev.* **52:**337–353.

Koch, A. L. 1991. Effective growth by the simplest means: The bacterial way. *ASM News* **57:** 633–637.

Kubitschek, H. E. 1986. Increase in cell mass during the division cycle of *Escherichia coli* B/rA. *J. Bacteriol.* **168:**613–618.

MacAlister, T. J., W. R. Cook, R. Weigand, and L. I. Rothfield. 1987. Membrane-murein attachment at the leading edge of the division septum: A second membrane-murein structure associated with morphogenesis of the gram-negative bacterial division septum. *J. Bacteriol.* **169:**3945–3951.

Matsuhashi, M. 1994. Utilization of lipid-linked precursors and the formation of peptidoglycan in the process of cell growth and division: membrane enzymes involved in the final steps of peptidoglycan synthesis and the mechanisms of their regulation. In *Bacterial Cell Wall*, J.-M. Ghuysen and R. Hackenbeck (Eds.), Elsevier, Amsterdam, The Netherlands, pp. 55–71.

Mengin-Lecreulx, D., L. Texier, M. Rousseau, and J. van Heijenoort. 1991. The *murG* gene of *Escherichia coli* codes for the UDP-*N*-acetylglucosamine:*N*-acetylmuramyl-(pentapeptide) pyrophosphorylundecaprenol *N*-acetylglucosamine transferase involved in the membrane steps of peptidoglycan synthesis. *J. Bacteriol.* **173:**4625–4636.

Mulder, R., C. L. Woldringh, F. Tetart, and J.-P. Bouche. 1992. New *minC* mutations suggest different interactions of the same region of division inhibitor MinC with proteins specific for *minD* and *dicB* coinhibition pathways. *J. Bacteriol.* **174:**35–39.

Nanninga, N. and C. L. Woldringh. 1985. Cell Growth, Genome Duplication, and Cell Division. In *Molecular Cytology of* Escherichia coli, N. Nanninga (Ed.). Academic, London.

RayChaudhuri, D. and J. T. Park. 1991. *Escherichia coli* cell-division gene *ftsZ* encodes a novel GTP-binding protein. *Nature (London)* **369**:251–254.

Robinson, A. C., D. J. Ken, J. Sweeney, and W. D. Donachie. 1986. Further evidence for overlapping transcriptional units in an *Escherichia coli* cell envelope-cell division gene cluster: DNA sequence and transcriptional organization of the *ddl ftsQ* region. *J. Bacteriol.* **167**:809–817.

Shockman, G. D. and J.-V. Höltje. 1994. Microbial peptidoglycan (murein) hydrolases. In *Bacterial Cell Wall*. J.-M. Ghuysen and R. Hackenbeck (Eds.). Elsevier, Amsterdam, The Netherlands, pp. 131–166.

Walsh, C. T. 1989. Enzymes in the D-alanine branch of bacterial cell wall peptidoglycan assembly. *J. Biol. Chem.* **264**:2393–2396.

Zyskind, J. W. and D. W. Smith. 1992. DNA replication, the bacterial cell cycle, and cell growth. *Cell* **69**:5–8.

Osmoregulation

Csonka, L. N. 1989. Physiological and genetic responses of bacteria to osmotic stress. *Microbiol. Rev.* **53**:121–147.

Csonka, L. N. and A. D. Hanson. 1991. Prokaryotic osmoregulation: genetics and physiology. *Annu. Rev. Microbiol.* **45**:569–606.

Jebbar, M., R. Talibart, K. Gloux, T. Bernard, and C. Blanco. 1992. Osmoprotection of *Escherichia coli* by ectoine: Uptake and accumulation characteristics. *J. Bacteriol.* **174**:5027–5035.

Kaasen, I., P. Falkenburg, O. B. Styrvoid, and A. R. Strøm. 1992. Molecular cloning and physical mapping of the *otsBA* genes, which encode the osmoregulatory trehalose pathway of *Escherichia coli*: Evidence that transcription of activated by KatF (AppR). *J. Bacteriol.* **174**:889–898.

pH Regulation

Booth, I. R. 1985. Regulation of cytoplasmic pH in bacteria. *Microbiol. Rev.* **49**:359–378.

Foster, J. W. 1993. The acid tolerance response of *Salmonella typhimurium* involves transient synthesis of key acid shock proteins. *J. Bacteriol.* **175**:1981–1987.

Foster, J. W. 1992. Beyond pH homeostasis: The acid tolerance response of Salmonellae. *ASM News* **58**:266–270.

Slonczewski, J. L. 1992. pH-Regulated genes in enteric bacteria. *ASM News* **58**:140–144.

Jones, P. G., R. Krah, S. R. Tafuri, and A. P. Wolffe. 1992. DNA gyrase, CS7.4, and the cold shock response in *Escherichia coli*. *J. Bacteriol.* **174**:5798–5802.

Nutrient Starvation

Spector, M. P. and C. L. Cubitt. 1992. Starvation-inducible loci of *Salmonella typhimurium*: regulation and roles in starvation-survival. *Mol. Biol.* **6**:1467–1476.

Spector, M. P. and J. W. Foster. 1993. Starvation-stress response (SSR) of *Salmonella typhimurium*. Gene expression and survival during nutrient starvation. In *Starvation in Bacteria*. S. Kjelleberg (Ed.). Plenun Press, NY, pp. 201–224.

CHAPTER 12

ENDOSPORE FORMATION
(DIFFERENTIATION)

As shown in Table 12-1, six separate genera of bacteria produce **endospores**. Myxospores or microcysts, which sometimes develop within macroscopic fruiting bodies, are produced by *Myxococcus xanthus* and other members of the *Myxobacterales*, and by *Sporocytophaga*. *Azotobacter vinelandii* and other members of this genus develop **cysts**. Cyanobacteria produce specialized resting forms called **akinetes**. All of these resting forms exhibit greater resistance than vegetative cells to adverse environmental conditions but very few exhibit the extreme resistance displayed by the endospores produced by *Bacillaceae*. One function of all of these spores appears to be in the survival and dissemination of the species.

On a practical basis, the ability to produce resistant spores enables many organisms to survive autoclaving or other processes used in the preservation of foods or sterilization of materials used in medical procedures. Great care must be taken to establish the conditions required for their elimination. Concerted interest in the developmental and regulatory factors involved in sporulation also centers on the amenability of the process to study under controlled conditions. Thorough understanding of the metabolic, genetic, and regulatory factors governing sporulation may provide insights into other developmental processes in higher forms. Here we emphasize the comparative biology, physiology, and genetics of bacterial endospore formation and reentry of the spore into the vegetative growth cycle through activation, germination, and outgrowth.

Life Cycle of *Bacillus*

As shown in Figure 12-1, all spore-forming organisms undergo a life cycle that includes germination, outgrowth, multiplication, and sporulation. Germination represents the conversion of the dormant spore to a metabolically active cell capable of outgrowth. The cell can then proceed into the vegetative multiplication cycle, or, if conditions are unfavorable, proceed directly into sporulation (the **microcycle**). The key metabolic steps are the two-component sensing system involved in sporulation initiation and the transcriptional control mediated by the different σ factors synthesized at different times (see

549

TABLE 12-1. Characteristics of Endospore-Forming Bacteria[a,b]

Genus	G + C (mol%)	Shape	Distinguishing Metabolic Traits
Bacillus	33–66	Rods	**Catalase positive**. Most are strict aerobes
Sporolactobacillus	38–40	Rods	**Homolactic fermentation** Facultative anaerobe or microaerophilic
Clostridium	24–54	Rods or filaments	**Strict anaerobe**
Desulfotomaculum	37–50	Rods or filaments	**Sulfate reduction** and strict anaerobe
Sporosarcina	40–42	**Cocci in tetrads or packets**	Strict aerobe
Thermoactinomycetes	52–54.8	**Branched filaments**	Strict aerobe

[a]The major distinguishing factor that separates each spore former from the others is in bold type.
[b]From Slepecky, R. A. 1993. What is a *Bacillus*? In *Biology of Bacilli: Applications to Industry*. R. H. Doi and M. McGloughlin (Eds.). Butterworth-Heinemann, Boston, MA, pp. 1–21.

further discussion below). The sporulation process is, fundamentally, viewed as a modified prokaryotic cell division (**asymmetric division**).

Cytological Aspects of Bacterial Endospore Formation

The events in the sporulation process are divided into seven steps, as shown for *Bacillus* in Figure 12-1 and as outlined below:

Stage 0 (vegetative cells). Cells undergoing normal vegetative growth are defined as being in Stage 0 with regard to the sporulation cycle.

Stage I (axial filament). This stage is generally no longer recognized since it does not appear to be specific to sporulation and no mutants have been found to be arrested in this stage of development.

Stage II (septum formation). The first sign of sporulation is the formation of an asymmetrically sited division septum resulting in two distinct cells, the mother cell and the prespore. Each of the two cells receives one nuclear equivalent of DNA. Intermediary stages designated IIi, IIii, and IIiii are based on the isolation of mutants blocked at these points between Stage II and Stage III.

Stage IIii, IIiii (prespore engulfment). The prespore is engulfed by the mother cell to form a protoplast.

Stage III (forespore development). The engulfed protoplast is now referred to as the **forespore**. In most electron micrographs the forespore is rather amorphous in appearance, presumably because no peptidoglycan layer has, as yet, been formed to provide a defined shape.

Stage IV (cortex formation). Shortly after forespore engulfment, a dense, narrow band of peptidoglycan, the cortex layer, is formed between the inner and outer membranes of the forespore. At this point the beginnings of the spore coat can

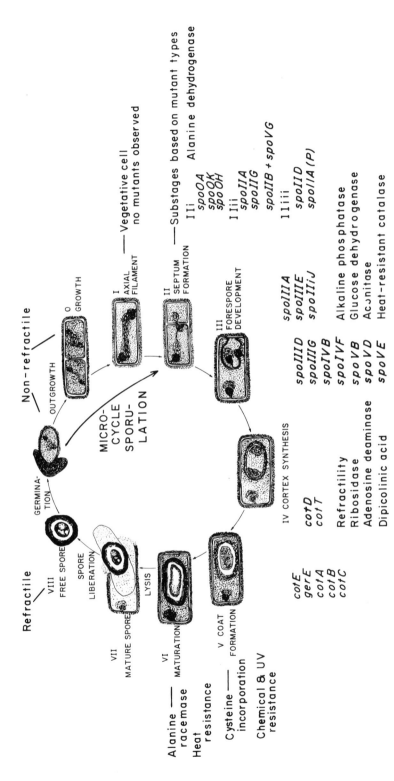

Fig. 12-1. Diagrammatic representation of the life cycle of *Bacillus subtilis*. As discussed in the text, the stages are designated by Roman numerals. Note that the stage originally designated as I is no longer recognized as a valid stage since mutants arrested at this stage have not been found. The diagram provides a representative sample of mutants blocked at the stage indicated. The approximate stage at which various enzymatic activities and other properties develop are also shown. The diagram represents a composite of diagrams from a number of reviews cited at the end of this chapter.

551

barely be discerned. As time progresses, the cortex becomes striated or multilayered.

Stage V (coat formation). Overlapping the cortex development process is the deposition of the multilayered spore coat. Discontinuous segments of coat material are observed. These segments coalesce into a continuous dense layer. Deposition of coat material at discontinuous points is more characteristic of coat formation in *Clostridium*. In *Bacillus*, deposition of coat material is a continuous process. However, there does not appear to be any fundamental difference in synthesis of the coat material in the two species.

Stage VI (maturation). Following the deposition of the cortex layers and formation of the spore coat, the final stage of maturation occurs. Observed as a whitening under phase contrast microscopy, this process is associated with the synthesis of dipicolinic acid and calcium uptake into the mature spore. The characteristic properties of resistance, dormancy, and germinability appear at this stage.

Stage VII (release of the mature spore). The mother cell is lysed and the mature spore is released as a dormant form.

The cytological events depicted in Figure 12-1 occur as an overlapping continuum rather than as separate and distinct events, as might possibly be inferred from the diagram presented. These morphological stages serve as useful reference points. Mutants that permit the sporulation process to proceed to one of these stages but prevent any further development beyond this point are designated spo0, spoII, spoIII, spoIV, and so on.

PHYSIOLOGICAL AND GENETIC ASPECTS OF SPORULATION

Sporulation Genes

The complex morphological and biochemical changes taking place during sporulation are governed by more than 100 gene products. The genes involved in sporulation are scattered around the chromosome and are interspersed with many genes that have no known role in sporulation. About 50 genetic loci, designated *spo*, have been identified and named according to the morphological stage at which the sporulation is blocked. A partial list of these genes is shown in Table 12-2. Other genes that participate in sporulation include those coding for a number of small acid-soluble proteins (SASPs) of the spore core, or for the proteins in the spore coat. Some of the genes involved in normal cell division, such as *ftsA* and *ftsZ*, are also involved in the asymmetric division process that initiates sporulation. Finally, those genes regulating spore germination (*ger*) are included since they must be incorporated into the spore genome during the sporulation process.

The temporal changes in gene expression that occur during sporulation are controlled by the sequential appearance or activation of new σ factors that bind to the core RNA polymerase (E) and confer on the holoenzyme (Eσ) the capacity to recognize new classes of sporulation-specific promoters. Known examples include: σ^H, which participates in the activation of early genes; σ^E and σ^F, which regulate early-intermediate genes; and σ^G and σ^K, which control the compartment-specific expression of gene sets activated at later times in the developing spore. Each class of promoters has a characteristic and

TABLE 12-2. A Partial List of Sporulation Genes in *B. subtilis*

Stage	Gene Designation	Function
IIi	*spo0A*	Response regulator; phosphorylation initiates sporulation via phosphorelay system
	spo0K	Regulates phosphorelay system
	spo0H	Encodes σ^H
IIii	*spoIIG*	Pro-σ^E and activating protease
	spoIIB + spoVG	Prespore engulfment; septum formation
IIiii	*spoIID*	Septal peptidoglycan hydrolysis
	spoIIA(P)	Processing of pro-σ^E
III	*spoIIIA*	Prespore engulfment
	spoIIIE	Controls σ^F expression
	spoIIiJ (kinA)	Histidine protein kinase; activates Spo0F
	spoIIID	Controls σ^E-dependent genes
	spoIIIG	Structural gene for σ^G
	spoIVB	Production of σ^K; signal peptide
	spoIVF (spoIIIF)	Processing of pro-σ^K
	spoVB	Cortex synthesis
	spoVD	Cortex synthesis
	spoVE	Cortex synthesis
IV	*cotD*	Coat synthesis
	cotT	Cortex synthesis
	gerE	Cortex synthesis
	cotA	Cortex synthesis
	cotB	Cortex synthesis
	cotC	Cortex synthesis

distinct polymerase consensus sequence located about 10 and 35 bp upstream of the transcription start site.

The *spoIIE* operon is a sporulation-specific transcription unit activated during the second hour of sporulation. Its promoter region does not conform to the consensus for any of the known sporulation-specific σ factors. Instead, it contains sequences that conform to the consensus sequence for vegetative promoters recognized by σ^A-associated RNA polymerase (Eσ^A).

Initiation. Bacterial endospore formation is initiated in response to starvation for carbon, nitrogen, or phosphorous. The very early stages of sporulation are reversible. Transfer to fresh culture medium will result in the resumption of vegetative growth. Thus, sporulation must be initiated by the accumulation of factors that inhibit vegetative growth and derepress the spore genome. How the cell senses this nutritional stress and transmits the information to the transcriptional apparatus has been the subject of a major portion of the research efforts on sporulation.

Two transcriptional regulators, σ^H and Spo0A, play key roles in initiation. The Spo0A protein contains an aspartyl residue in the N-terminal region that is the target of phosphorylation by a histidine protein kinase that is part of the **phosphorelay** system. This complex system of regulation of stationary-phase gene expression involves at least two histidine protein kinases, KinA and KinB. These protein kinases initiate the phosphorelay system by phosphorylating SpoOF, as shown in Figure 12-2. Other genes involved in this system include the *spo0B* operon that includes *obg*, a GTP-binding protein essential for vegetative growth. It has been suggested that the *obg* gene product may be the protein that senses the decrease in GTP levels that occurs following nutrient deprivation and provides the link to sporulation events via the phosphorelay system. The experimental results suggest that a threshold level of activated SpoOA (SpoOA-P) or of a component of the phosphorylation pathway must accumulate to induce sporulation gene expression and that most of the cells that are able to induce the expression of early genes that are directly activated by SpoOA-P go on to produce mature spores.

Initiation of the prespore and mother cell programs of gene expression in the correct compartments is crucial to the success of sporulation. Triggering of mother cell development is risky because it culminates in cell lysis. This is lethal unless it occurs in conjunction with a fully committed prespore program in the prespore. Thus, a sensitive but accurate signal from prespore to mother cell after septation would be important for determining cell fate during processing. A model system for this signal is shown in Figure 12-3. Neither σ^E nor σ^F are active after their synthesis. The σ^F activity in the prespore is inhibited by an anti σ protein (SpoIIAB). A third product of the *spoIIA* operon, SpoIIAA, is activated after completion of the spore septum allowing it to act as a substrate for the SpoIIAB kinase. The phosphorylated form of SpoIIAA forms a tight complex with SpoIIAB, thus releasing σ^F to initiate transcription of prespore-specific gene. Among these is an unknown gene whose product somehow acts vectorially across the spore septum to activate the SpoGA protease. This protease cleaves pro-σ^E to active form allowing it to initiate transcription of mother cell-specific genes.

The σ^H (SpoOH) controls many stationary phase genes. It is weakly expressed from a

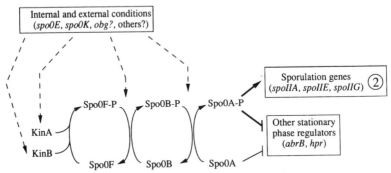

Fig. 12-2. Control of stationary-phase gene expression by the phosphorelay system. At least two histidine protein kinases, KinA and KinB, can transfer phosphoryl groups to the SpoOF protein in response to as yet unknown signals, perhaps internal or external nutritional conditions. The phosphoryl group is then tranferred via SpoOB to the SpoOA transcription factor. Phosphorylated SpoOA positively regulates the expression of several key early sporulation-specific genes (indicated by the line ending in an arrowhead). SpoOA-P also negatively regulates at least two genes controlling various stationary phase responses (indicated by the line ending in a bar). The nonphosphorylated form of SpoOA is also a weak negative regulator, as suggested by the thin line. Several other gene products are thought to influence the flux of phosphate through the phosporelay, as indicated by the dashed lines. Their precise targets in the pathway are not yet known, but in principle, they could act at any one of the four steps shown. From Errington, J. 1993. *Microbiol. Rev.* **57**:1–33.

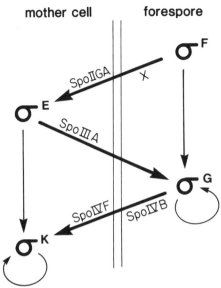

Fig. 12-3. Criss-cross regulation of compartmentalized σ factors. Proposed interrelationship of σ^F, σ^E, σ^G, and σ^K in the postseptation sporangium. The thin vertical lines indicate intracellular pathways of transcriptional control; the circular arrows indicate autoregulation. The heavy diagonal lines indicate intercellular pathways acting at the level of σ factor activity. From Losick, R. and P. Stragier. 1992. *Nature (London).* **355:**601–604.

σ^A promoter in midlog phase but is greatly increased after the initiation of sporulation. Among the various genes controlled by σ^H is one important sporulation-specific operon, *spoIIA*, which produces another σ factor. σ^F.

As discussed in Chapter 11, FtsZ is a Mg^{2+}-dependent GTPase that is recruited to the division site where it forms a ring that remains associated with the invaginating septum until septation is completed. Homologs of the cell division genes *ftsA* and *ftsZ* of *E. coli* have been identified in *B. subtilis.* The *ftsA* gene in *B. subtilis* was originally called *spoIIG* or *spoIIN*. Both *ftsA* and *ftsZ* are required for vegetative septation as well as the asymmetric septation that occurs in the initial stage of sporulation. The *ftsA* and *ftsZ* genes of *B. subtilis* constitute a simple operon expressed from promoter sequences immediately upstream of *ftsA*. Both *ftsA* and *ftsZ* are transcribed from a distinct promoter, p2 (*ftsAp2*), during sporulation but not during exponential growth. Transcription from p2 is dependent on RNA polymerase containing σ^H but does not require expression of a number of sporulation loci (*spo0A, spo0B, spo0E, spo0F,* or *spo0K*). Introduction of a *cat* cassette in the middle of promoter p2 does not affect vegetative growth but prevents postexponential symmetrical division and spore formation. Another mutant, designated *dds* (deficient in division and sporulation), located upstream of *ftsA*, does not show significant homology to other known genes. However, *dds* null mutants grow slowly, are filamentous, and exhibit a reduced level of sporulation.

Transition from Stage II to Stage III. Following the asymmetric septation, the chromosomes of the prespore and mother cell enter into markedly different programs of gene expression resulting from the segregation of σ factors σ^E and σ^F, as shown in Figure 12-

2. Many more genes are known to be transcribed by $E\sigma^E$ in the mother cell than by $E\sigma^F$ in the prespore. Engulfment of the prespore by the mother cell occurs at this stage. The engulfment process takes place by several discrete steps as indicated by the subdivision of the transition from Stage II to Stage III by the designations IIi, IIii, and IIiii, each requiring one or more specific gene products (Table 12-2). For example, *spoIIB* and *spoVG* double mutants fail to progress beyond the completion of the spore septum in Stage III.

Engulfment begins with peptidoglycan hydrolysis near the central part of the septal disk. This is accompanied by bulging of the two closely opposed membranes into the mother cell cytoplasm. Mutations in the σ^E-dependent *spoIID* gene cause a morphological block at intermediate Stage IIii. In Stage IIiii the edges of the septum begin to migrate toward the proximal pole of the cell. The edges of the septal membranes meet at the pole of the cell and fuse, completing the engulfment of the prespore within the mother cell cytoplasm (Stage III). This final step is prevented by the *spoIIAC*(P) mutations, indicating that a σ^F-dependent gene is necessary for the completion of engulfment.

Forespore Development. Once engulfment of the prespore is complete, it is referred to as a **forespore**. At this point, synthesis of **σ^G**, the product of the *spoIIIG* gene, begins with the resultant activation of the σ^G regulon. The σ^G regulon contains a large number of genes that orchestrate many changes in the properties of the developing spore. Regulation of σ^G occurs at the transcriptional and posttranscriptional level, however, the genes involved in this stage vary considerably with regard to their regulation. Mutations in *spoIIB* together with *spoVG*, and *spoIID* and in *spoIIM* block sporulation prior to engulfment. Mutations in *spoIIIA* and *spoIIIJ* allow engulfment to be completed but prevent further progression beyond Stage III. The onset of the late stages of forespore development are dependent on events occurring in the mother cell at the level of *spoIIIG* transcription and σ^G activity. Specific intercellular communication seems to be necessary to couple the activation of σ^G in the forespore to the program of gene expression in the mother cell (Fig. 12-3).

Three groups of σ^G-dependent genes are of major importance in the later stages of forespore development. These include the *ssp* genes encoding the SASPs found in the spore core, genes encoding sporulation-specific penicillin-binding proteins (PBPs), and the germination (*ger*) genes.

The SASPs are of two classes. The α/β-type SASPs are encoded by *sspA* through *sspD*. The γ-type SASPs are encoded by a single gene, *sspE*. All of the SASPs have a short conserved sequence recognized by a specific protease encoded by the *gpr* gene. The protease is synthesized as an inactive precursor that is activated only after germination. Protease cleavage of the SASPs during spore germination causes a rapid and complete degradation of the proteins to supply a source of amino acids for protein synthesis during outgrowth. In addition to the amino acid storage function, the α/β-type SASPs serve to protect the spore DNA from UV radiation damage by binding to the DNA in such a manner as to permit only the formation of thyminyl–thymine adducts (termed SP, for spore photoproducts). Repair of these thyminyl–thymine adducts is apparently much more error-free than repair of the cyclobutane (T:T) dimers normally produced by exposure to UV radiation. This SP-specific repair system is unique to spores and operates only during spore germination.

Of those genes concerned with spore germination, those at the *gerA* locus are the best characterized. Three genes at this locus, *gerAA*, *gerAB*, and *gerAC*, form an operon.

Mutants of any one of these genes are defective in the germination response to L-alanine. Mutations at the *gerB* locus, also comprised of three genes, cause a defective germination response to a mixture of asparagine, glucose, fructose, and KCl.

Final Stages of Sporulation.   The final stages of sporulation are controlled by the sporulation-specific σ factor σ^K. The synthesis of σ^K activates genes that contribute to the formation of the cortex and spore coat and other factors involved in maturation and release of the free spore. Regulation of σ^K synthesis must be timed so that it occurs only after the appearance of σ^E and sealing-off the forespore from the mother cell. A site-specific recombinase in the mother cell, encoded by the *spoIVCA* gene, brings together two partial coding sequences to form a circular DNA molecule termed *skin* (for *sigK* intervening element). Another level of regulation involves the processing of the inactive precursor, pro-σ^K, by proteolysis, to form σ^K.

Spore Cortex Synthesis.   The spore cortex synthesis involves the activity of several *spo* genes. The *spoVD* and *spoVE* genes map in a region of the chromosome containing a cluster of genes involved in cell wall synthesis and cell division. Both SpoVD and SpoVE proteins show amino acid sequence homology to PbpB (FtsI) and FtsW of *E. coli*. The *spoVD* gene is regulated independently of other genes in the cluster, is sporulation-specific, and the gene product is involved in cross-wall synthesis suggesting that its action may result in the production of a sporulation-specific peptidoglycan. The presence of σ^E-dependent promoter immediately upstream of *spoVE* allows expression during sporulation. The gene product appears to be similar to that of FtsW, suggesting a role in peptidoglycan synthesis. Several other genes, including *spoVB* (*spoIIIF*), *spoVG*, *gerJ*, and *gerM*, seem to be involved in cortex synthesis, but their role is less well defined.

Penicillin-binding proteins PBP 2B and PBP 3 increase during sporulation. The map location immediately upstream of *spoVD* suggests that the gene for PBP 2B may be a functional homolog of *pbpB* in *E. coli* and indicates a role in septum formation. The *dacB* gene product, PBP 5*, is the major sporulation-specific PBP. The 5* exhibits D-alanine carboxypeptidase activity and *dacB* transcription is associated with Eσ^E suggesting a role in synthesis of the peptidoglycan-like cortex layer of the spore. A null mutant of *dacA*, the structural gene for vegetative PBP 5, produces normal, heat-resistant spores, indicating that this PBP is not required for spore cortex synthesis.

Spore Coat Protein Synthesis.   The spore coat is composed of at least 15 polypeptides plus an insoluble protein fraction arranged in three morphological layers. Genes for at least seven coat proteins, designated *cotA–cotF* and *cotT*, have been identified. Most of the well-characterized *cot* genes have σ^K-dependent promoters even though they show considerable variation in their time of expression. The *cotE* gene has σ^E-dependent promoter that permits earlier expression but it also has a σ^K-dependent promoter allowing expression to continue once σ^K appears. The proteins CotA, CotB, and CotC are responsible for the formation of outer-coat proteins. The CotD, CotE, and CotT proteins are involved in the formation of inner-coat proteins. Mutations at *cotE* are deficient in proteins encoded by *cotA*, *cotB*, and *cotC* genes, indicating that the *cotE* gene product, CotE, is deposited on the outside surface of the inner coat, where it serves as a basement protein on which the other proteins assemble. Mutations in the *spoIVA* locus abolish cortex synthesis and interfere with the synthesis and assembly of the spore coat. The phenotypic properties of *spoIVA* mutants suggest a role for the gene product at an early stage in the morphogenesis of the spore outer layers. In the absence of SpoIVA, cortex synthesis is almost absent and coat proteins accumulate in swirls in the mother cell cytoplasm rather than on the surface of the outer-prespore membrane. The dual defects in both coat and cortex synthesis could be explained if coat deposition were dependent on completion of the cortex and SpoIVA was essential for cortex synthesis. Further study is needed to ascertain the role of SpoIVA in these final stages of sporulation.

ACTIVATION, GERMINATION, AND OUTGROWTH OF BACTERIAL ENDOSPORES

The bacterial spore evolved as a means of surviving periods of low nutrient supply and other harsh environmental conditions. At the same time, the dormant spore must possess the ability to respond quickly and specifically to conditions favorable to reentry into the vegetative cycle. Inherent in the sporulation process, then, is the incorporation of an efficient mechanism for germination and outgrowth. It is difficult to investigate the sporulation process without simultaneously considering the environmental factors that influence activation of the spore and its entry into the germinative stage.

ACTIVATION

The process of spore activation in natural environments is unknown. Under laboratory conditions, activation is most commonly achieved by heating (usually at 60–65 °C for 10 min or longer). Exposure to low pH, low temperature, reducing agents, or a number of chemical agents may be equally effective in activating bacterial endospores. Activation can be distinguished from germination by a number of criteria. Activated spores still retain their resistance to heat, are refractile, and resistant to staining, and still contain large amounts of dipicolinic acid (DPA). Activation is generally considered nonessential. However, the percentage of germinating spores is greatly increased by activation. Indeed, it has been found that the number of germinating spores is often very low unless some form of activation is employed. The mechanism of the activation process is not known with certainty. Some activating agents appear to increase the permeability of the spore to germinants. Activation is reversible, whereas spores are committed to germinate following brief exposure to an effective germinant.

GERMINATION

Germination occurs in response to specific germinants, which act as triggers that lead to rapid changes in the structure and physiology of the spore. As stated above, the spore is committed to germinate even if the germinant is removed. Commitment precedes the detectable changes in the spore: loss of heat resistance, ion fluxes, release of Ca^{2+} and DPA, hydrolysis of cortex peptidoglycan, rehydration of the core protoplast, and resumption of metabolic activity. During the early stages of germination, the light-scattering ability changes rapidly (within 6–8 min) from phase-bright to phase-gray to phase-dark.

Effective germinants include (1) **L-alanine** and alanine analogs; (2) **L-alanine plus inosine** or other ribonucleosides; (3) **sugars plus inorganic ions**; (4) **inorganic ions** alone, or (5) **asparagine, glucose, fructose, and KCl (AGFK).** Spores of different species of bacilli may respond to one or all of these combinations. D-Alanine competitively inhibits the germinant action of L-alanine. Alcohols and methyl anthranilate inhibit germinant triggering. Some inhibitors of germination are selective. For example, azide inhibits AGFK germination, whereas phenyl methyl sulfonyl fluoride inhibits only L-alanine germination. Inhibition of germination by protease inhibitors, such as tosyl arginine methyl ester, suggests a role for proteolytic activity in the germination process. Other inhibitors appear to be somewhat nonspecific in their action.

GERMINATION MONITORING

Qualitative tests for monitoring spore germination include loss of optical density, loss of heat resistance, and changes in appearance under the phase-contrast microscope. A plate test that is widely used to distinguish Ger^+ and Ger^- colonies involves detection of dehydrogenase-linked metabolism that is resumed upon germination by reduction of 2,3,5-triphenyltetrazolium chloride (a tetrazolium dye) added to the plating medium. Reduction of the tetrazolium dye stains the germinated colonies red, whereas ungerminated spores do not stain (appear white). A high correlation between the absence of or marked reduction in tetrazolium reduction and a measurable defect in spore germination aids in the selection of germination-defective mutants.

GERMINATION LOCI

Genetic analysis of spore germination in *B. subtilis* has identified a number of germination (*ger*) loci. Germination mutants are grouped according to their phenotype. Spores of *gerA* and *gerC* mutants fail to germinate in response to L-alanine but are able to respond to AGFK. Mutants that respond normally to L-alanine but are not stimulated in response to AGFK are designated *gerB*, *gerK* and *fruB*. Mutant spores that fail to germinate in response to either of these germinants have been designated *gerD* and *gerF*. The *gerE* gene may code for or regulate a protease required for the processing polypeptides in the spore coat during germination. These differences in response to germinants suggests the presence of at least two alternate systems for the detection and response to germinants. Mutations in a number of other loci (*gerE*, *gerJ*, *gerM*, *spoVIA*, *B*, *C*, *cotT*) affect both the germination properties of the spore and the overall structure of the cortex or coat.

MODELS OF SPORE GERMINATION

Several models for the mode of action of triggering germinants have been proposed. Models that include metabolism of the germinant infer that the germinant is converted to some intracellular metabolite. Models based on the premise that the germinant(s) are not metabolized, propose that the germinant acts allosterically on a receptor protein. Proponents of this latter model suggest that the germinant may alter membrane permeability or trigger proteolytic or cortex lytic enzyme activity.

METABOLIC CHANGES DURING GERMINATION

Dormant spores contain reduced levels of a number of important metabolites, particularly high-energy compounds, such as ATP, sugar phosphates, reduced pyridine nucleotides, ribonucleotide triphosphates, and tRNA. Many other compounds, such as adenine nucleotides, ribonucleotides, pyridine nucleotides, and other RNA species, are present at levels approximating those in growing cells. One potential high-energy phosphate donor, 3-phosphoglycerate (PGA), is present in the dormant spores of some, but not all, strains at very high concentrations. Accumulation of ATP occurs within a few minutes after germination begins (150-fold increase in the first 5 min). This initial increase in ATP occurs at the expense of endogenous energy sources, particularly the large PGA pool. Dormant spores contain PGA mutase, enolase, and pyruvate kinase, the three enzymes required for conversion of PGA to ATP. The dormant spore also contains sufficient enzymes of the glycolytic pathway and the hexosemonophosphate shunt since neither protein nor RNA synthesis are required for high-energy phosphate production through at least 40 min of germination. The mechanisms whereby these enzymes remain inactive in the dormant spore and are activated on germination are not known with certainty. After the exhaustion of the PGA pool, energy metabolism apparently occurs primarily via the hexose monophosphate shunt since many enzymes of the TCA cycle are absent in germinating spores. The glucose dehydrogenase gene (*gdh*) is expressed only during sporulation.

The synthesis of RNA begins by the second minute of germination in *B. megaterium* and constitutes a major use for the ATP generated at this stage. Synthesis of RNA during the first 20 min of germination can be completely accomplished from nucleotides stored in the dormant spore. Approximately 85% of the ribonucleotides required for RNA synthesis during the first 25–50 min of germination are derived from RNA degradation. This hydrolytic activity is necessary for rapid RNA synthesis early in germination since nucleotide biosynthesis is not demonstrable at this time.

Chromosome replication (DNA synthesis) does not begin until late in spore germination. Degradation of RNA supplies the nucleotides for DNA synthesis as well as for RNA synthesis. A temporal relationship between the increase in deoxyribonucleotide levels and the increased rate of DNA synthesis has been demonstrated. This finding suggests that DNA synthesis is regulated by the level of deoxyribonucleotides.

Protein synthesis begins within the first 2 or 3 min of germination and can be accomplished solely from stored nitrogen reserves through the first 60 min of germination. The free amino acid content of dormant spores is limited, but increases rapidly in germinating spores as a result of hydrolysis of spore protein. The proteins degraded in this process are a group of SASPs, which are formed during the sporulation process (see earlier discussion). In *B. megaterium* and *B. subtilis* a number of SASPs have been identified. Proteolytic activity is necessary for protein synthesis early in germination since biosynthesis of many amino acids is not demonstrable at this time.

Metabolism of small molecules during spore germination can be divided into two general stages: **Stage I**, or the **turnover stage**, from 0–70 min of germination; and **Stage II**, or the **synthesis Stage**, from 70 min and beyond. By the onset of Stage II, the germinating spore gains the capacity for de novo biosynthesis of all small molecules in the formation of DNA, RNA, and protein. Endogenous reserves of energy, amino acids, and nucleotides are exhausted by this time. During Stage I, small molecules are derived primarily from stored reserves. Stage I consists of three overlapping substages: Ia (0–15 min), Ib (0–20 min), and Ic (0–70 min), each representing the period when endogenous energy, nucleotide, and amino acid reserves, respectively, are sufficient for most of the metabolic and biosynthetic requirements.

OUTGROWTH

Although germinated spores lose the characteristics of the intact spore, they are by no means comparable to vegetative cells. Germinated spores have cytological structures typical of vegetative cells, are sensitive to environmental influences, and are metabolically active. However, they lack many of the activities and properties of vegetative cells. The transition from germinated spore to vegetative cell has been termed **outgrowth** and involves a number of metabolic changes. Generally, if a suitable nutritional environment is provided, the germinated spore will continue rapidly into the outgrowth process. If nutritional factors are lacking, the process may stop altogether, or the cells may proceed to sporulate and enter into **microcycle** sporulation. In the microcycle, asymmetric division, rather than symmetrical division, begins immediately after elongation of the cell is initiated and the cells proceed to sporulate. However, one round of DNA replication must occur prior to entry into the mcirocycle.

The isolation of mutants of *B. subtilis* that are temperature sensitive only during spore outgrowth suggests that there are functions that are specific to this stage of the life cycle. Studies of these mutants show that unique RNAs are synthesized during spore outgrowth. It has been possible to isolate these genes from a cloned library of *B. subtilis* DNA using hybridization with labeled RNA prepared from outgrowing spores in the presence of a large excess of competing vegetative RNA and the chromosomal locations of the transcribed sequences.

REFERENCES

Sporulation, Activation, Germination, and Outgrowth of Bacterial Endospores

Beall, B. and J. Lutkenhaus. 1992. FtsZ in *Bacillus subtilis* is required for vegetative septation and for asymmetric septation during sporulation. *Gene Dev.* **5**:447–455.

Beall, B. and J. Lutkenhaus. 1992. Impaired cell division and sporulation of a *Bacillus subtilis* strain with the *ftsA* gene deleted. *J. Bacteriol.* **174**:2398–2403.

Beall, B. and C. P. Moran, Jr. 1994. Cloning and characterization of *spoVR*, a gene from *Bacillus subtilis* in spore cortex formation. *J. Bacteriol.* **176**:2003–2012.

Buchanan, C. E. and M.-L. Ling. 1992. Isolation and sequence analysis of *dacB*, which encodes a sporulation-specific penicillin-binding protein in *Bacillus subtilis*. *J. Bacteriol.* **174**:1717–1725.

Buchanan, C. E. and A. Gustafson. 1992. Mutagenesis and mapping of the gene for a sporulation-specific penicillin-binding protein in *Bacillus subtilis*. *J. Bacteriol.* **174**:5430–5435.

Buchanan, C. E., A. O. Henriques, and P. J. Piggot. 1994. Cell wall changes during bacterial endospore formation. In *Bacterial Cell Wall*. J.-M. Ghuysen and R. Hackenbeck (Eds.). Elsevier, Amsterdam, The Netherlands, pp. 167–186.

Chung, J. D., G. Stephanopoulos, K. Ireton, and A. D. Grossman. 1994. Gene expression in single cells of *Bacillus subtilis*: Evidence that a threshhold mechanism controls the initiation of sporulation. *J. Bacteriol.* **176**:1977–1984.

Errington, J. 1993. *Bacillus subtilis* sporulation: Regulation of gene expression and control of morphogenesis. *Microbiol. Rev.* **57**:1–33.

Gholamhoseinian, A., A. Shen, J.-J. Wu, and P. Piggot. 1992. Regulation of transcription of the cell division gene *fstA* during sporulation of *Bacillus subtilis*. *J. Bacteriol.* **174**:4647–4656.

Losick, E. and P. Stragier. 1992. Crisscross regulation of cell-type-specific gene expression during development in *B. subtilis*. 1992. *Nature (London)* **355**:601–604.

Moir, A. 1990. The genetics of bacterial spore germination. *Annu. Rev. Microbiol.* **44**:531–553.

Setlow, P. 1992. I will survive: Protecting and repairing spore DNA. *J. Bacteriol.* **174**:2737–2741.

Slepecky, R. A. 1993. What is a *Bacillus*? In *Biology of Bacilli: Applications to Industry*. R. H. Doi and M. McGloughlin (Eds.). Butterworth-Heinemann, Boston, MA, pp. 1–21.

Smith, I., R. Slepecky, and P. Setlow. 1989. *Regulation of Procaryotic Development*. American Society for Microbiology, Washington, DC.

York, K., T. J. Kenney, S. Satola, C. P. Moran, Jr., H. Poth, and P. Youngman. 1992. Spo0A controls the σ^A-dependent activation of *Bacillus subtilis* sporulation-specific transcription unit *spo0IIE*. *J. Bacteriol.* **174**:2648–2658.

INDEX

Abnormal proteins, degradation of, 79
Abortive transduction, 115
Absidia glauca, mannitol utilization in, 347
Acetic acid:
    fermentation, 390–391
    production from glucose by *C. thermoaceticum*,
        381–382
Acetoacetate, pathway to, 13
Acetoin, pathways to, 13, 372–373, 384–385
Acetobacter aceti:
    production of acetic acid by, 390–391
    respiratory chain of, 390
Acetobacter xylinum, cellulose production by, 247,
    277, 278
Acetone, pathways to, 13, 379–380
Acetyl-CoA carboxylase, role of biotin in, 423
Acetylene reduction, correlation with nitrogen
    fixation, 439
N-Acetylglucosamine (NAG), 2
N-Acetylmuramic acid (NAM), 2
Acholeplasma, 24, 431
Achromobacter, free living nitrogen fixation, 438
Acidophilic bacteria, 26, 541
Acid tolerance response (ATR), 541
Acridine dyes (e.g., proflavin), intercalation, 136
Actinomyces, 26
Actinomycin D, DNA intercalating agent, 83
Acyl carrier protein (ACP), composition of, 423, 424
Addition (insertion) mutation, 134
Adenosine triphosphate (ATP), 14
    ATPase:
        F_1F_0 ATPase complex, 331, 332
        in DnaK protein, 186
        in energy production, 16

        Kdp ATPase, 338
        ATP-linked ion-motive pumps, 338
Adenylate cyclase, *cya* locus, 159
Adenylyltransferase (ATase; *glnE*), 194
Adsorption, bacteriophage, 207
Adventurous (A) motility, in gliding motility, 297
Aerobactin (siderophore), synthesis from lysine, 474
Aerobe, 24
Aerobic conditions:
    aerobic respiratory control (ArcA, ArcB), 189
    cytochrome oxidase production, 188
    fatty acid biosynthesis, 425–426
    metabolic pathways under, 188
Agaricus bisporous, mannitol accumulation in
    fruiting bodies, 347
Agrobacterium tumefaciens:
    plasmid in tumor induction, 97
    two-component system in, 180
Alanine dehydrogenases, 451, 453
Alarmone, 183
Alcaligenes eutrophus, autotrophic carbon dioxide
    assimilation in, 399
Alcohol dehydrogenase:
    isozymes of, 367
    yeast mutants deficient in, 368
Aldohexuronate metabolism, 348
Alfalfa (*Medicago sativa*), root infection and root
    nodulation in, 444
Algae, 237
Alk-1-enyl ethers (plasmalogens), 416
Alkaline phosphatase (PhoA), 196
Alkaliphilic bacteria, 26, 542
Alkyl ethers, long-chain alcohols bound to
    phospholipids, 416

Alkyl hydroperoxide reductase (*ahpC, ahpF*), 197

Alkylating agents, mutagenesis by, 144

Alleles, 10

Allolactose, 157

Amebae, 237

Amikacin, aminoglycoside antibiotic, 88

Amino acid(s):

biosynthesis, 15, 462–500

code, in protein synthesis, 58

decarboxylases, 452

deaminases, 452

general reactions of, 452

nonoxidative deaminases, 452

oxidases, 453

racemases, 457

transaminases, 456

Aminoacyl tRNA:

binding sites, 71

ligases, 57

δ-(L-α-Aminoadipyl)-L-cysteinyl-D-valine (ACV), in synthesis of penicillins and cephalosporins, 470

p-Aminobenzoate, biosynthesis, 491–492

Aminoglycosides, inhibition of translation, 88

resistance mechanisms (acetyltransferases, nucleotidyltransferases, phosphotransferases), 88

Aminoimidazole carboxamide ribonucleotide (AICAR), intermediate in purine biosynthesis, 503

Aminolevulinate synthesis, 463, 482, 483

Aminotransferases (amino acid transaminases), 456

Ammonia:

assimilation of, 450–452

commercial production by Haber–Bosch process, 439

Ammonia switch-off, in regulation of nitrogen fixation, 445

Amphipathic membrane lipids, 410

Amylases, 358

Amylopectin, structure of, 357

Amylose utilization, 355, 357

Amylosterum chailletti, fungal symbiont, cellulose degradation, 355

Anabaena:

free living nitrogen fixation in, 438

variabilis, arginine biosynthesis in, 465

Anaerobe, 24

Anaerobic bacteria, genera of, 26

Anaerobic conditions:

fatty acid biosynthesis under, 425

metabolic pathways, 188

terminal oxidoreductase production, 187, 188

Anaerobic respiration, 329

Anaerovibrio lipolytica, plasmalogens in, 416

Angolamycin, macrolide antibiotic, 87

Anapleurotic function, 25, 305, 319

Antibiotics:

affecting nucleic acid and protein synthesis, 80

affecting peptidoglycan synthesis, 254–255

agents affecting DNA metabolism, 80–85

intercalating agents, 80, 83

DNA gyrase inhibitors, 84

Anticodon, 9, 58

Antimicrobial agents, in peptidoglycan synthesis, 254, 255

Antiparallel, dsDNA structure, 29

Antiport transport systems, 337

Antisense RNA, plasmid copy number control, 98

Antiterminator loop, 167

Antiterminator pQ, 221

Aphanocapsa, free living nitrogen fixation, 438

Arabinose (*ara*) operon, 161–164

Archaebacteria (*Archaea*):

chromosomal DNA, nucleoproteins, 246

alkyl ethers in, 416, 417

unusual lipids in, 411

Archaebacterial lipids, 417

Arcyria cinerea (myxomycete), nucleus in, 237, 240

Arginine:

biosynthesis, 464–468

deiminase, 466

metabolism, organization and control in *N. crassa*, 467

regulon, *argR* locus, 171–173

Arogenate, precursor to phenylalanine and tyrosine, 489

Aromatic amino acids, pathways of biosynthesis, 484–487

Aromatic hydrocarbons, metabolism of, 360–384

Arthrobacter:

crystallopoietes, interpeptide bridge in peptidoglycan, 252

globiformis, glycine cleavage system in, 481

pyridinolis, malate requirement for growth on glucose, 319

Asparagine, biosynthesis, 471–472

Aspartase, 451, 454

Aspartate:

aminotransferase, 456

family of amino acid biosynthesis, 471–477

transcarbamylase, 167

Aspergillus niger, arginine biosynthesis in, 465

Asymmetric cell division, in sporulation, 550–551

Attenuation controls, in *trp* operon, 164–169

Atwater, root nodules and nitrogen fixation, 437

Autogenous translational repression, 216

Autolysins, 252

Autophosphorylation, in DnaK protein, 186

Autotrophs:

characteristics and metabolism of, 390–405

principal groups of, 392

Auxotroph, nutritional mutant, 133

Axoneme, in eukaryotic flagella, 281

Azomonas, free living nitrogen fixation in, 438

Azospirillum, nitrogen fixation in, 438, 445
Azotobacter croococcum:
  nitrogen fixation system in, 438, 441
  vinelandii:
    nitrogen fixation system in, 441
    cyst formation in, 549
Azotobacteriaceae, nitrogen fixing species, 538

Bacillus:
  acidocaldarius, pH homeostasis in, 542
  alcalophilus, optimal growth at alkaline pH, 542
  ammonia assimilation in, 451
  amylases in, 358
  arginine biosynthesis in, 465
  anthracis, 4, 271, 272, 275
  arginine biosynthesis in, 465
  branched-chain fatty acids in, 414
  cereus, glycerol teichoic acid in, 257
  endospore formation in, 19, 549–558
  firmus, optimum growth at alkaline pH, 542
  leicheniformis, glycerol teichoic acid in, 257
  polygalacturonate metabolism in, 352
  polymyxa, free living nitrogen fixation, 438
  pumilis, 97
    glutamine amidotransferase in, 456
    solvent-producing fermentations in, 377
  species, swarming in, 297
  sporulation cycle in, 549–552
  stearothermophilus, 419, 544
  subtilis, 75, 97, 113, 180
    glycerol and ribitol teichoic acids in, 257
    hexitol metabolism in, 246
    life cycle (sporulation cycle), 551
    S layer in, 249–251
    tryptophan supraoperon in, 487
Bacitracin, interference with peptidoglycan
  synthesis, 254–255
Bacteria, pH limits for growth, 27
Bacterial:
  cell, basic structural components, 2
  cell surface, general structure, 3, 261
  nucleoid (chromosome), 8, 239, 239–243
  viruses (c.f, bacteriophage), 202
Bacterial interspersed mosaic elements (BIME), 36
Bacteriochlorophylls, in photosynthetic bacteria, 391
Bacteriophage (bacterial viruses):
  attachment to cells and pili, 206
  capsid, 202
  characteristics, classification of, 202, 203
  core, 202
  enzyme activities in, 209
  genetics of, 202, 209
  growth curve, 204, 205
  infection cycles (lytic, virulent), 204
  lambda (λ), 202, 218–227
  Mu, 228–231
    in global regulatory systems, 183
    in transposition, 127

øX174, 202, 232–235
T2, T4, etc., 202, 205–217
temperate, 204
Bacteroid, in symbiotic nitrogen fixation, 444
Bacteroides,
  fatty acids in, 415
  guar gum (galactomannan) utilization, 348
  3-hydroxy fatty acids in, 415
  pectin utilization in, 349, 352
  plasmalogens in, 416
  polysaccharide utilization in, 357
Bactoprenol (C_{55}-isoprenol), 431
Basal body, 5
Base excision repair, 142
Basidiomycetes, lipids and fatty acids in, 414
Behavior, of flagella, 290
Beijerinkia, free living nitrogen fixation, 438
Benzer, fine-structure genetic map of lambda phage,
  212
Benzoate, anaerobic metabolism of, 363
Beta (β) sheet DNA binding proteins, 155
Beta (β) galactosidase, *lacZ* gene, 155
Betaine, in osmoregulation, 539–541
Binding proteins, β-sheet DNA, 155
Bifidobacterium, 26
  ring-containing fatty acids in, 415
Bioenergetics of pH homeostasis in acidophilic
  bacteria, 542
Biosynthetic pathways, 12, 14
Biotin, role in acetyl-CoA carboxylase system, 423
Bivalves, *Teredinidae*, symbiosis with marine
  bacteria, 438
Bordetella pertussis, 179
Borrelia burgdorferi, 95
Botrytis cinerea, pectin utilization in, 349
Bradyrhizobium:
  aminolevulinate synthesis in, 463
  nitrogen fixation in, 438, 440–444
Branched-chain amino acids, biosynthesis, 475–476
  regulation of, 479
Brucella, carbon dioxide requirement for growth, 25
Butanol, pathways to, 13, 378–379
Butyrate, pathway to, 13, 378–379
Butyribacterium retgeri, interpeptide bridge in, 249
Butyric acid, production of, 376
Butyrivibrio, solvent-producing fermentations in,
  376, 381
Butyrivibrio fibrisolvens, pectin utilization in, 349,
  352
Butyryl-Coenzyme A (CoA), formation in solvent
  fermentation, 377

Calothrix, free living nitrogen fixation, 438
Calvin-Bensen cycle, 23
Campbell-Evans cycle, 23
Campylobacter, carbon dioxide requirement for
  growth, 25
Candida albicans, arginine biosynthesis in, 465

Candida utilis, hexitol metabolism in, 237
Capsid, bacteriophage, 202
Capsomeres, of bacteriophage, 204
Capsule, 4, 270–278
Carbamoyl phosphate:
  energy yield from, 324
  precursor in arginine and pyrimidine
    biosynthesis, 465, 506
Carbamoyl phosphokinase, 466
Carbohydrate metabolism:
  energy production, 14
  isotope labelling patterns in, 370
  major pathways, 13, 305–409
Carbon and electron flow in *Clostridium
  acetobutylicum*, saccharide fermentation, 380
Carbon balance (carbon recovery), in fermentation,
  363
Carbon dioxide (C_1) balance, in fermentation, 364,
  365
Carbon dioxide:
  fixation in autotrophs, 394–395
  fixation in heterotrophs, 319
  nutritional requirement for, 25
Carbon/energy regulon, 160
Carbomycin, macrolide antibiotic, 87
Cardiolipin (diphosphatidylglycerol), 418
Carotenoid pigments, 393
Catabolite:
  activator protein, 159
  control, 158
  repression, 159, 344–345
Catalase (*katG*), 197, 373–374
Catechol, intermediate in metabolism of aromatic
  ring compounds, 360
Catechol siderophore (enterobactin, enterochelin),
  491
Cell:
  components, 238
  cycle, *Enterococcus hirae*, 521, 522
  division, *Bacillus subtilis*, 533–536
  division, *Escherichia coli*, 522–533
  division genes in the 2-min (*mra*) region of *E.
    coli*, 523, 534
  division genes in 133° region of *Bacillus subtilis*,
    534
  division, regulatory systems in, 530
  division, role of FtsZ in, 529
  mammalian, electron micrograph, 239
  membranes, 3
  structure and function, 1, 237–304
  surface, 1, 3
  wall, 2
Cellulose:
  degradation, 352–356
  in cell wall of eukaryotes, 247
Cellulosome, cellulase complex in, 353–355
Cellvibrio, cellulose utilization in, 353
Cephalosporins, synthesis from lysine, 470

Cephalosporium acremonium, biosynthesis of
  β-lactam antibiotics in, 470
Cephalosporium chrysogenum, hexitol metabolism
  in, 346
Chalcomycin, macrolide antibiotic, 87
Chaperone(s); chaperonins:
  in heat shock response, 186
  in protein folding, 73
  GroEL, GroES, DnaK, 73
Charging of tRNA, 57, 60
Chemical composition, bacterial cell, 11–12
Chemiosmotic driven transport porters, 337
Chemolithotrophs, 23, 332
Chemostat, continuous culture, 22
Chemotactic memory, methyl-accepting chemotaxis
  proteins, 289
Chemotaxis, 283–292
Chlorobium:
  carbon dioxide assimilation in, 395
  free living nitrogen fixation in, 438
Cholesterol, incorporation into cytoplasmic
  membranes, 411
Chorismic acid, in aromatic amino acid pathway,
  486–487
Chromatium vinosum, free living nitrogen fixation
  in, 438
Chi sequences, recombination, 118
Chitin, in cell walls of eukaryotes, 247
Chloramphenicol, inhibitor of translation, 86
Chlorobacteriaceae (green bacteria), 391
Chlorophyll *a*, 391
Chromosome, bacterial (nucleoid), 7–8
  partitioning, 47
  segregation, 8
  supercoiling, 49
Cilia, eukaryotic, 279, 280
Ciprofloxacin (quinolone), 85
Circular permutations, T4 phage genome, 212
Cis-trans test, 107, 212
Citrate:
  fermentation products from, 376
  synthetase, in citric acid (TCA) cycle, 316–319
  utilization by gram-negative bacteria, 385
Cladosporium cucumerinum, pectin utilization in,
  349
Clindamycin, inhibition of translation, 89, 90
Cloning vector, lambda phage, 227
Chlorophyll, in photosynthesis, 391–398
Chloroplasts, 392
Clostridium, 26
  arginine deiminase activity in, 466
  butylicum, nicotinamide adenine dinucleotide
    (NAD) biosynthesis, 497
  butyricum, free living nitrogen fixation in, 438
  pasteurianum, free living nitrogen fixation in,
    438, 441
  plasmalogens in, 416
  polygalacturonate metabolism in, 352

solvent-producing fermentations in, 376, 381
species, swarming in, 291
spore formation in, 19–20, 550
starch degradation in, 358
Stickland reaction in, 458
thermocellum, cellusome in, 353
thermohydrosulfuricum, cellobiose utilization in, 353
thermosaccharolyticum, cell wall, 251
Codons, in protein synthesis (nonsense, triplet), 57, 58
Coenzyme Q (benzoquinone), 23
Coiling, in DNA structure, 30
Cointegrate formation, 127
Colanic acid, M antigen, slime polysaccharide of *E. coli*, 276
Cold:
    effect on growth, 543–545
    shock response, 543
Colicin E1 (ColE1) plasmid, 96
Common pathway of aromatic amino acid biosynthesis, 484–486
Compactosome, 38
Compartmentalization of arginine and proline metabolism, 467–468
Competence:
    in *Bacillus subtilis*, 113
    in streptococci (competence activator protein), 111
Complementation (*cis/trans*) test, 107
Concatemers, in T4 phage infection, 212
Concerted feedback inhibition, 478
Conditional mutation, 134
Conjugation, bacterial, 94, 100, 104
    barriers to (surface exclusion, incompatibility), 104
    in enterococci, 108
Conjugative transposition in *Enterococcus faecalis*, 132
Consensus sequences, in promoters, 49
Constitutive gene expression, 158
Continuous culture (chemostat), 22
Control:
    negative, 152
    positive, 152
    transcriptional, 152
Core metabolites, 12
Core region, lipopolysaccharide, *Salmonella*, *Shigella*, 260
Corynebacterium:
    branched-chain fatty acids in, 414
    diphtheriae, lysogenic conversion in, 117
Corynemycolenic acid, 414
Corynolic acid, in *Corynebacterium* and *Mycobacterium*, 411
Cosmids, in lambda phage, 227
Cotransduction, 115
Coupling, between transcription and translation, 164

CRP-binding site (CAP), 156
Cyanobacteria:
    akinete formation, 549
    photosynthesis in, 391
Cyclic AMP (cAMP), in catabolite repression, 159–161
Cyclodextrins, in starch degradation, 359
Cycloheximide, inhibitor of translation, 86
Cyclopropane fatty acids, 415
Cycloserine, inhibition of peptidoglycan synthesis, 257
Cylindrospermum, free living nitrogen fixation, 438
Cysteine:
    biosynthesis, 481, 482
    desulfhydrase, 455
Cytochromes, 14, 17, 19
    oxidases, cytochrome *d* (*cyo*), cytochrome *o* (*cyo*), 187
    production of, in streptococci, 375
Cytology, cell structure, 237–304
Cytophagales, gliding motility in, 296
Cytoplasmic membrane, 3, 266–270
    transport systems in, 336

Dactinomycin (actinomycin D), 83
Dark reaction in photosynthesis, 394
Deaminases, amino acid, 453
Death phase, growth cycle, 21
Degrons, in degradation of normal proteins, 80
Deletion mutations, 134
Deletions, in transpositional recombination, 131
Denitrification, 447–449
Deoxycytidine monophosphate, precursor of hydroxymethylcytosine, 210
Deoxyribonucleic acid (DNA):
    A-form, 30
    B-form, 30
    bending, caused by CRP, 160
    binding proteins, 153
    cell-wall association, model of, 537
    double-stranded (dsDNA), 29
    gyrase, 33, 34, 40, 47, 84, 85
    ligase, 42, 43
    looping, 158, 161
    methylation, *dam* locus, 45
    polymerases (Pol I, Pol II, Pol III), 7, 38, 40, 42
        proofreading component in, DnaQ, 42
        T4 phage DNA polymerase, 214
    primase (*dnaG*), 38
    primosome, components of (DnaB, DnaC), 40
    repair systems, 139–143
    replication, 7, 36, 38–44
        initiation of, 44
        origin of, *oriC*, binding to membrane in *E. coli*, 43
        replicating fork, 39
        regulation, negative and positive control of, DnaA, DnaA boxes, IciA, 44

Deoxyribonucleic acid (DNA) (*Continued*)
termination of, terminus (*terA, B, C*), 40
semiconservative, Meselson and Stahl, 37, 38
T4 replication, 214
structure, 29
supercoiling, 36
topoisomerase, 47
Z-DNA, 30, 31
Derxia, nitrogen fixation in, 438
Desulfhydrase, cysteine, 455
Desulfovibrio, 26
desulfuricans, free living nitrogen fixation in, 438
gigas, free living nitrogen fixation in, 438
plasmalogens in, 416
sulfate reduction by, 399
Desulfotomaculum:
spore formation in, 550
sulfate reduction by, 400
Desulfurococcus, 419
Deuterium, labeling of NAD⁺ during reduction, 328
Dextran, production of, 273, 274
Dextrin, limit, 357
Diacetyl, metabolic pathway to, 13
Diadenosine tetraphosphate (ApppppA), 187
Diaminopimelic acid, in peptidoglycan, 251
Diatoms, marine, symbiotic association with
cyanobacteria, 438
Diauxic growth, 344
Dichotomous replication, 520
Differentiation, endospore formation, 549–562
Dihydrogen, production by nitrogenase, 440
Dihydroorotase, *pyrC*, 170
Dihydroxyuridine (DHU) loop, in tRNA, 58
Dimethylsulfoxide (DMSO), 187
Dinitrogen fixation, 439
Dinitrogenase, in nitrogen fixation, 439
Dismutation of pyruvate:
in fermentation by *S. faecalis*, 369
in yeast fermentation, 366
Division sites, positioning of, 531, 536
Downstream, region of DNA relative to polymerase
movement, 48
Drosophila, topoisomerases in, 33
Dynein (ATPase), in eukaryotic motility, 279, 281

Eclipse phase, bacteriophage growth, 204
Ectothiorhodospiras, photosynthetic, alkalophile,
542
Electrical potential (Δ ψ) and proton motive force,
325
Electrode potentials of important biological
systems, 329
Electron transport, respiratory pathways, 19,
187–189, 326–332
Elongation, of RNA, 48, 50
Elongation in polypeptide synthesis, 66
Elongation factor G, in protein synthesis, 71
Embden-Meyerhof (Embden-Meyerhof-Parnas,

EMP) pathway of glycolysis, 12, 13, 306–309
Endospores, 19, 549–562
endospore-forming bacteria, 550
stages of endospore formation, 550–552
Endotoxin, O antigen, lipopolysaccharide, 3
End-product inhibition and repression, 9, 10
Energy production, 14, 16, 322–334
Energy yield from glycolysis and TCA cycle, 332,
333
Engulfment, in sporulation cycle, 556
Enhancer sequences, in flagellar phase variation,
173
Enterochelin (Ent) iron transport system, 342
Entner-Doudoroff pathway, 13, 312–314
Enterobacter:
aerogenes, 23, 75
cloacae, free living nitrogen fixation, 438
Enterobacteriaceae: 75, 97, 100
autolysins, 252
pili, fimbriae, 297
porins in, 263
Enterobacterial common antigen (ECA), surface
glycolipid, 265
Enterobactin (enterochelin, catechol siderophore),
biosynthesis of, 491, 493
Enterococci, conjugation and pheromones in, 108
Enterococcus:
faecalis, 108
amino acid racemase in, 457
glycolipid in, 259
interpeptide bridge, 252
faecium (*E. hirae*) interpeptide bridge, 252
hirae (cf. *E. faecium*), 252
antibiotic synergy and, 257
growth of, 519–522
Entropy, 322
Environment, effect on growth, 536–546
Episomes, 95, 100
Ergosterol, biosynthesis of, 433
Erwinia:
Entner–Doudoroff pathway in, 313
Pectin utilization in, 348–352
Erythromycin, macrolide antibiotic, 87
Escherichia coli, 1, 7, 12, 17, 23, 26, 33, 35, 38,
44, 63, 65, 66, 73, 75, 90, 180, 185–189, 198,
207, 241, 242
aerobic and anaerobic metabolic pathways in,
188
aerobic and anaerobic respiratory pathways in,
190
arginine biosynthesis in, 465
aspartate aminotransferase, 456
electron transport chain, 326, 327
enterobactin (enterochelin) biosynthesis, 491, 493
fatty acid:
biosynthesis in, 421
composition of, 414
synthetase in, 424

flagellar structure, 282
flagellar, motility, and chemotaxis gene products, 283
folate biosynthesis, 491, 492
fucose utilization in, 349
gluconeogenesis in, 315
growth of, 522–533
interpeptide bridge, in peptidoglycan, 3, 4, 249
lipopolysaccharide (LPS) biosynthesis, 264
mannitol utilization in, 347
nicotinamide adenine dinucleotide (NAD) and NADP, biosynthesis, 496-497
outer membrane of, 249, 250
phages (bacteriophage) of, 203
phosphoglycerides in, 419
pili, fimbriae, 297
rhamnose utilization in, 348, 349
serine and threonine deaminases in, 455
transaminases in, 457
Ethanol, end product of fermentation, 13
Ethidium bromide, DNA intercalating agent, 83
Ethylating agents, adaptive response to, 144
Euascomycetes, lipids and fatty acids in, 414
Eubacterium, 26
Eukaryote, *Eukarya*:
    cell surface of lower eukaryotes, 247
    nucleus of lower eukaryotes, 237, 240
    topoisomerases in, 33
Evolutionary consequences of transposons, 132
Excision, control of in lambda phage infection, 223
Excision repair, by nucleotide excision, 139, 142, 143
Exponential phase, 21

Facilitated diffusion, in metabolite transport, 335
Facultative, 24
Farnesyl pyrophosphate, biosynthesis of, 433
Fatty acids, 23, 411–416
    biosynthesis of, 421–426
    branched-chain, 414
    degradation of (*fad* genes in), 429
    naturally occurring, 412, 413
    ring-containing (cyclopropane), 415
    straight-chain, 411
    synthetase, 424
    unsaturated, biosynthesis of, 425
Feedback control, in carbohydrate metabolism, 306
Feedback inhibition, 9, 10, 478
Feldman and Gunsalus, transaminases, 456
Fermentation, 14, 17
    acetone-butanol, 379
    balances, 363–365, 379
    balance, for *Lactobacillus pentoaceticus*, 365
    carbon and electron flow in *Clostridium acetobutylicum* fermentation, 380
    mixed acid, 187, 188, 382–385
    mixed solvent, 379
    pathways in specific groups of microorganisms, 362–390

products formed from pyruvate, 364
propionic acid, 385–390
yeast, 366
Ferridoxin:
    oxidation of reduced ferridoxin by hydrogenase, 377
    role in electron-transfer reactions, 382–383
    role in photosynthesis, 397–398
Ferrobacillus ferrooxidans, iron bacteria, 401
Fertility (F) factor, 95, 100
    F-prime (F′) formation, 106
    fertility inhibition, 105
    genetic map, 101
    Hfr formation, 105
    steps in F plasmid transfer, 103
Filament formation in cell division (ts) mutants, 527
Fine-structure map, T4 phage, 215
Fimbriae (pili), 5, 297–300
Flagella, 4, 5, 278–291
Flagellar gene system, 289
Flagellar phase variation, gene expression in, 171, 174, 288
Flagellar switch, model of, 287
Flavin adenine dinucleotide (FAD):
    function of, 328
    autooxidation of FADH and reduced flavoprotein, 372–374
Flexibacter polymorphus, gliding motility in, 296
Folic acid, biosynthesis of, 491–492, 502
Formate, in C_1 balance, 364–365
Forespore development, 556
Frameshift, resulting from base addition or deletion, 57
Frameshift mutation, 134
Fructose, fermentation products from, 376
Fructose bisphosphatase, in gluconeogenesis, 314
Fructose bisphosphate (FBP) aldolase, in Embden–Meyerhof–Parnas pathway of glycolysis, 306–309
Fruiting bodies, in *Myxobacterales*, 549
Fucose utilization, 348, 349
Fumarate (fumaric acid), 187
Fungi (yeasts and molds), 237
    lipids and fatty acids in, 414
Fusidic acid, inhibition of translation, 90
Fusobacterium, 26

Galactitol, catabolism of, 346–347
Galactose:
    (*gal*) operon, 161
    galactosidase, 155
    metabolism of, 345
Gallionella, iron bacteria, 401
Gap repair, postreplication daughter strand gap repair, 144
Gene, 10
    expression, phenotype, 10
    expression in bacteriophage infection, 209

Gene (*Continued*)
 splicing, T4 introns, 217
General amino acid control, 479
General secretory pathway (GSP), protein
  translocation, 74–76
Genetic regulation (repression), 9, 10
Genetic regulatory mechanisms, 154
Genetics:
 bacteriophage, 202
 of recombination, 120
 microbial, 10–11
Genomic map, T4 phage, 213
Genotype, 10
Gentamycins, inhibition of translation, 88, 89
Gibbs free energy (G), 322
Gleotheca, free living nitrogen fixation, 438
Gliding motility, 296–297
Global control networks, regulation at the whole
  cell level, 183–187, 537
Glucan, in cell wall of *S. cerevisiae*, 248
Glucitol, catabolism of, 346–347
Gluconeogenesis, hexose synthesis, 314–316
Gluconobacter:
 acetic acid production by, 390–391
 hexitol metabolism in, 347
Glucose:
 effect (catabolite repression), 344
 utilization in *Streptococcus mutans*, 375
Glucose-6-phosphatase, in gluconeogenesis, 314
Glucose-6-phosphate dehydrogenase (*zwf*), 198
Glutamate:
 biosynthesis, 463
 dehydrogenase (*gdh*), 193, 451
 family of amino acids, 462–470
 synthase, (*gltB*), 193, 451
Glutamine:
 biosynthesis, 463
 regulatory cascade, 193, 194
 synthetase (*glnA*, GlnS), in nitrogen assimilation,
  451
Glutathione:
 biosynthesis, 463
 reductase (*gorA*), 197
Glycerol:
 dissimilation of in *E. coli*, 385–386
 production from glucose in heterofermentative
  organisms, 375
 utilization by *Streptococcus faecalis*, 374
Glycerol phosphate, in teichoic acid, 257
Glycine, biosynthesis, 480
Glycine-betaine, in osmoregulation, 181
Glycine-serine interrelationships, 480
Glycogen:
 synthesis, 315
 utilization, 355
Glycolipids, 418–421
Glycolysis, glycolytic pathways, regulation,
  306–314

Glyoxylate cycle, 13, 319–321
Gram stain, 2
Gratuitous inducer, 157
Green bacteria (*Chlorobacteriaceae*), 391–392
Griffith, transformation in pneumococcus, 110
Griseofulvin, action on DNA replication, 84
Group-specific polysaccharides, in *Streptococcus
  pyogenes*, 258
Growth, 19
 curve, bacteriophage, one-step growth curve, 204
 cycle, 20–22, 518–519
 factors affecting, 22, 536–546
 gram-negative rods, 522–533
 gram-positive bacilli, 533–536
 gram-positive streptococci, 519–522
 regulation of, 518–548
 stages in cell cycle of gram-positive streptococci,
  520
Guanosine tetraphosphate, guanosine
  5′-diphosphate-3′-diphosphate (ppGpp); Magic
  Spot I:
 alarmone, in ribosome editing, 72
 in degradation of abnormal proteins, 79
 in regulation of *dnaA* operon and cell division,
  531
 in stringent control, 191
 RNA polymerase, 191
Guar gum (galactomannan) utilization, 348, 349
Gunsalus, and Feldman, transaminases, 456

Haber–Bosch, commercial ammonia production by,
  439
Haemophilus, 25, 110, 112
 influenzae:
  NAD biosynthesis in, 497
  phosphoglycerides (cardiolipin) in, 419
 parainfluenzae, cardiolipin in, 419
Halobacterium:
 halobium, 76
 salinarum, 246
Halotolerance, 540
Hansenula, methylotrophic yeast, 401
Haploid, 11
Harden–Young effect, in yeast fermentation, 366
Heat shock:
 proteins, 185
 response, 185–187
Helical structure, dsDNA, 30
Helix-turn-helix (HTH) DNA recognition motifs,
  153
 in lambda phage expression, 221, 222
Helreigel, root nodules and nitrogen fixation, 437
β-Hemolytic streptococci, group antigens, 272, 273
Heterofermentative lactic acid fermentations, 308,
  309, 368–372
Hexitols, metabolism of, 346–347
Hexose monophosphate pathways of glycolysis,
  308, 311

Hfr, interrupted matings, gene mapping, 105
High-energy phosphate, transfer of, 14
Histidine:
    biosynthesis of, 497–499, 502
    operons, 499
    permease, 341
    utilization (*hut*), 195
Histones, in eukaryotic cells, 243
Histone-like proteins (H, HLP, Hu), 35, 45, 46, 243, 245
Hofmann, discovery of lactobacillic acid, 415
Holiday intermediate, Chiasma formation, in recombination, 120
Homocysteine, as a methionine precursor, 474, 475
Homeostasis, pH, 334
Homofermentative lactic acid fermentations, 308, 309, 368–369
Homoserine, alternate pathways to methionine, 475
Hook-associated proteins (HAPS), in bacterial flagella, 280
Hostile environment, effect on growth, 538
Host killing, plasmid maintenance, *ccd* genes, 100
Host range, bacteriophage, 204
Hydrogen bacteria, 398
Hydrogen ion concentration (pH), 26, 27, 177
    effect on glucose fermentation by *S. faecalis*, 369
    effect on growth, 541–543
Hydrogen oxidation, 333
Hydrogen peroxide (H_2O_2), 197, 373–374
Hydrogen production during nitrogen fixation, 440
Hydrogenase:
    in hydrogen bacteria, 398
    in nitrogen fixation, 440
    oxidation of reduced ferridoxin, 377
Hydrostatic pressure, effect on growth, 546
Hydroxylamine, 448
Hydroxyl radicals, in oxidation stress, 197
Hydroxymethylcytosine (HMC), in T4 phage DNA, 205, 209
Hyphomicrobium, methylotrophic metabolism, 401

Incompatibility, of plasmids, 95
Induced mutations, 11
Inducer exclusion, 158, 344
Induction, of *lac* operon, 157
Initiation codon(s), in polypeptide synthesis, 65
Initiation of DNA replication, *oriC*, 43, 46
Initiation of polypeptide synthesis, 64, 65
Initiation of RNA transcription, 48, 49
Inorganic nitrogen, utilization for energy production, 334
Insertion sequences, in transposition, 127
Insertional inactivation, 183
Integrated suppression, 105
Integration:
    control of, in lambda phage infection, 223
    host factor (IHF), lambda phage, 223
Integration of F factor, 106

Integration of genetic markers in streptococci, *hex*, 111
Intercalating agents, agents affecting DNA metabolism, 80
Interpeptide bridges, in peptidoglycan, 249–252
Interphase nucleus, in myxomycete, 240
Intervening sequences in *Salmonella typhimurium*, 56
Introns, in T4 genes, 217
Inversions, in transpositional recombination, 131
Invertasome, in flagellar phase variation, 174
Ion gradients, establishment of, 338
Iron oxidation, chemolithotrophs, 334, 401
Iron–sulfur centers, 198
Iron transport, 341
Isocitrate lyase, in glyoxylate cycle, 319–320
Isoleucine, biosynthesis, 473, 475, 476
Isomaltose, 357
Isoprenoid structures, incorporation of mevalonic acid, 432–433
Isopropanol, pathways to, 378
Isopropyl-β-O-Thiogalactoside (IPTG), 157
Iteron, control of PI plasmid copy number, 99

Jacob, operon model of gene expression, 155

K antigens, *E. coli*, capsular polysaccharides, 275, 276
Kanamycins, aminoglycosides, inhibition of translation, 89
Kasugamycin, aminoglycoside, inhibition of translation, 89
2-Keto-3-deoxyoctulosonic acid (KDO), core region, LPS, 260, 263
Ketogluconate aldolase (Entner–Doudoroff) pathway of carbohydrate metabolism, 312–314
α-Ketoglutarate dehydrogenase complex in the TCA cycle, 316–317
Kinase, DnaK, heat activation of, 187
Klebsiella:
    aerogenes:
        aspartase, 451
        association with leaf nodulating plants, 438
    oxytoca, 75
    pneumoniae, 180, 195
        nitrogen fixation system in, 438, 441–447
Kujimycin, macrolide antibiotic, 87

Lactic acid producing fermentations, 367–376
Lacnospira multiparus, pectin utilization in, 349
β-Lactam antibiotics, biosynthesis of, from lysine, 470
Lactobacillic acid, 415
Lactobacillus, 26
    acidophilus, autolysins, 252
    arginine deiminase activity in, 466
    brevis, mannitol utilization in, 347

Lactobacillus (*Continued*)
  casei:
    bactoprenol, incorporation of mevalonic acid, 431
    interpeptide bridge, 252
  buchneri, Mg^{2+} binding by teichoic acid, 260
  fatty acids in, 415
  fermentation patterns in, 308
  fermenti, lipoteichoic acid in, 259
  fructosus, NAD biosynthesis in, 497
  plantarum, cell wall polymers, 257
  viridescens, interpeptide bridge, 252
Lactococcus:
  arginine deiminase activity in, 466
  fermentation patterns in, 308, 371–372
Lactose (*lac* genes):
  operon, 155–161
  permease, 155
  utilization, 344
Lag phase, 21
Lambda (λ) phage, 218–220
  in specialized transduction, 115
Lanosterol, biosynthesis of, 433
Latent period, bacteriophage infection, 204
Legumes, in nitrogen fixation, 438
Leloir pathway of carbohydrate metabolism, 346
Leptospira, isoleucine biosynthesis in, 477
Leptospiraceae, motility in, 296
Leptospirillum ferrooxidans, iron oxidation, 334
Leptothrix, iron bacteria, 401
Leucine, zipper DNA-binding domains, 155
Leucomycin, macrolide antibiotic, 87
Leuconostoc mesenteroides:
  fermentation pathways in, 308
  mannitol utilization in, 347
Leucosporidium, psychrophilic yeast, 544
Lichstein and Cohen, transaminases in bacteria, 456
Life cycle, sporulation cycle in *Bacillus*, 551
Ligase, DNA, 43
Lincomycin, inhibition of translation, 89, 90
Linkage map, *E. coli*, 45
Lipid A:
  biosynthesis of:
    genes and enzymes, 267
    pathway, 268
  in lipopolysaccharide of gram-negative bacteria, 261
Lipids, 410–435
  degradation of, 427–429
  in microorganisms, 411, 418
Lipoate, in pyruvate and α-ketoglutarate dehydrogenases, 317
Lipopolysaccharide, O antigen, endotoxin, 3, 260–366
Lipoprotein, 3, 264, 265
Lipoteichoic acid, 3, 257–260
Lithotroph, chemolithotroph, photolithotroph, 23
Locomotion, organs of, 278–297

Log phase, 21
Lon system: proteolytic control, 198
Long-chain fatty acids, biosynthesis of, 425
Lysins, role in cell division, 533
Lysine:
  biosynthesis in bacteria, 470, 473
  biosynthesis in fungi (yeasts and molds), 468, 469
Lysis inhibition, in bacteriophage infection, 212
Lysis-lysogeny decision, 219
Lysogenic:
  conversion, 116
  infection cycle, 204
Lysogeny, by lambda phage, 222
Lysophosphatidyl glycerides, 418
Lysozyme, in T4 phage, 212
Lytic infection cycle, bacteriophage, 204

Macrolides, inhibitors of translation, 87
Macromolecular synthesis (DNA, RNA, protein), 6, 28
Magasanik, catabolite repression, 344
Malate:
  alternative reactions in the TCA cycle, 316
  synthase, in glyoxylate cycle, 320–321
Maltose, metabolism of, 345–346
Maltotriose, 357
Mannan, in cell wall of *S. cerevisiae*, 248
Mannitol:
  catabolism of, 346
  metabolism of, 347
Mannitol-1-phosphate, conversion to ribulose-5-phosphate, 348
Mannose-sensitive adherence, type 1 pili, 297
Maturation of ribosomal RNAs (rRNAs), 64
Megalomycin, macrolide antibiotic, 87
Mecillinam (6-β-amidino-penicillanic acid), binding to PBP-2 of *E. coli*, 256
Megasphaera elsdenii, plasmalogens in, 416
Meister and Rudman, general transaminases, 456
Mellibiose metabolism, 348, 349
Membrane(s), 3
  cytoplasmic, 3
  mediated regulation, the Put system, 171
  nuclear, eukaryotic cell, 237
  outer membrane, gram-negative bacteria, 3
  outer membrane proteins, 196
Membranous organelles, 270
Menaquinone (vitamin K, methylnapthoquinone), biosynthesis of, 496
Merodiploid, partial diploid, 94
Meromictic, permanently stratified, lakes, 393
Meselson and Stahl, semiconservative replication, 36, 37
Mesophilic, 26
Mesosomes, 270
Messenger RNA (mRNA), 7, 28
Metabolic flow, from glucose, 12

Metabolic regulation in glycolysis and TCA cycle, 309

Meta-cleavage pathway, in aromatic ring metabolism, 360

Methanococcus voltae, cell surface of, 249, 250

Methanogens, methane-forming bacteria (*Methanobacterium, Methanobrevibacter, Methanococcus, Methanomicrobium, Methanogenicum, Methanospirillum, Methanosarcina*), 403–405
  free living nitrogen fixation in, 438
  unusual coenzymes participating in methanogenesis, 404

Methanol, oxidation by *Xanthobacter*, 399

Methanotrophs, 401

Methicillin, inhibition of peptidoglycan synthesis, 257

Methionine:
  aminopeptidase, 72
  biosynthesis, 473
    regulation of, 478
  deformylase, 72

Methylating agents, adaptive response to, 144

Methylation, of DNA, *dam* locus, 45

Methylobacterium, 401

Methylococcus, 401
  free living nitrogen fixation in, 438

Methylomonas, 401

Methylophilus, 401

Methylosinus, 401

Methylotrophs, 401, 402

Mevalonic acid, biosynthesis of, 430–432

Microbacterium, fermentation patterns in, 308

Microbial lipids, 418–419

Micrococcus:
  cryophilus, 545
  glutamicus, arginine biosynthesis in, 465
  luteus, interpeptide bridges in, 249
  roseus, interpeptide bridge in, 252

Microtubules, in eukaryotic cilia and flagella, 291

Microcycle sporulation, 549, 551

Mismatch repair, following recombination, 122, 142

Missense mutation, 133

Mitochondria, 6, 18, 19, 246, 247
  evolution from prokaryotes, 247

Mitomycins, inhibition of DNA, 83, 84

Mixed acid fermentations, balances, 383

Mixed-solvent fermentations, balances, 379

Modification enzymes, tRNA structure, 58

Modulon, 183

Molds:
  arginine biosynthesis in, 465
  nucleus, 237
  pH requirements for growth, 27

Monod:
  glucose effect (catabolite repression), 344
  operon model of gene expression, 155

Morphogenesis, steps of, in *E. coli*, 624

Motility, 4, 278–297

Mucor genevensis (fungus), fine structure of hyphal element, 240

Multivalent inhibition, regulation of amino acid biosynthesis, 478

Murein (peptidoglycan):
  synthesis of, 252, 253
  lipoprotein, major OMP in *E. coli*, 264

Mutagenesis, 10, 133, 134, 135

Mutant alleles, 10

Mutation:
  conditional, 134
  deletion, 134
  directed, 38
  frameshift, 134
  induced, 11
  insertion, 134
  rate, 134
  spontaneous, 10, 134
  suppressor, 136

Mutator phenotype, 142

Mycobacterium, 23
  branched-chain fatty acids in, 414
  glycolipids in, 420
  leprae, phenolic glycolipid in, 420–421
  tuberculosis, NAD biosynthesis in, 497

Mycoplasma, Mycoplasmataceae, 24, 76, 267
  arginine deiminase, 466
  glycolipids in, 410, 419
  sterol requirement for growth, 410, 431

Myxobacterales:
  gliding motility in, 296
  myxospores, in fruiting bodies of, 549

Myxococcus xanthus:
  gliding motility in, 296
  myxospores, 549

Myxomycete, nucleus in, 237, 240

N-Acetylmuramic acid, in peptidoglycan, 249

N-Acetylmuramyl:L-alanineamidase, autolysin, 252

Nalidixic acid, quinolone antibiotic, DNA gyrase inhibitor, 84, 85

Naphthoquinone (vitamin K), 23

Neisseria, 110, 112
  carbon dioxide requirement for growth, 25
  gonorrhoeae:
    arginine biosynthesis in, 465
    pilin variation in, 299–300
  meningitidis:
    maltose utilization in, 346
    pili, 300

N-end rule, N-degron, N-terminal protein residue, 80

Negative control, of transcription, 152, 154

Neomycins, aminoglycoside antibiotics, 88

Nerolidol pyrophosphate, biosynthesis of, 433

Neurospora, 23
  crassa, arginine biosynthesis in, 465

Neutralophilic, 26
Nicotinamide adenine dinucleotide (NAD), 14, 16, 17
  biosynthesis of, 496, 497
  NAD(P)H oxidase, 374
Nicotinamide adenine dinucleotide phosphate (NADP):
  biosynthesis of, 496, 497
  function in biosynthetic reactions, 17
Nitrification, 447–450
Nitrifying bacteria, 334, 399, 447–450
Nitrobacter winogradskyi, 399, 448
Nitrocystis oceanus, membranous organelle in, 270, 271
Nitrogen:
  assimilation, 16
  symbiotic association, 16, 193–196
  fixation, 193–196, 436–446
    genera of nitrogen fixing organisms, 438–439
    in *Rhizobium*, 18
  inorganic, metabolism, 446–452
  metabolism, 436–461
Nitrogenase (dinitrogen fixation), 195, 439–442
  regulation of, 445–447
Nitrosomonas europaea:
  denitrification, 399, 448
  membranous organelle, 270
Nocardia, free living nitrogen fixation, 438
Nodulation, and nitrogen fixation, 443–445
Nonconjugative plasmids, 95
Nonsense codons, in protein synthesis, 57
Nonsense mutations (amber, UAG; ochre, UGA; opal, UAA), 133
Novobiocin, DNA gyrase inhibitor, 84, 85
Nuclear membrane, eukaryotic cell, 237–239
Nucleic acid:
  core, in bacteriophage, 202
  interaction with protein, 31
Nucleoid, bacterial, 31, 239, 241, 242, 244
Nucleosomes, in eukaryotic cells, 243, 245
Nucleosome-like structures in prokaryotic cells, 243
Nucleotide excision repair, 139
Nucleotides, in DNA structure, 29
Nucleus:
  bacterial (c.f., nucleoid), 238, 239, 241–244
  eukaryotic, 237, 240
Nutrient deprivation, two-component systems, 177
Nutrition, microbial, 22–24, 545–546

O antigen, lipopolysaccharide, endotoxin, 3, 260
Octose, 2-keto-3-deoxyoctulosonic acid (KDO), 260–261
Okazaki fragments:
  in DNA replication, 7, 38
  in T4 phage replication, 214
Operator, 152
Operon, 153
  fusions, in *lacZ* search, 183

Origin of replication, 7
  oriC, binding to membrane in *E. coli*, 43
Organotroph, chemoorganotroph, photoorganotroph, 23, 24
Ornithine biosynthesis, 464
Ortho-cleavage pathway, in metabolism of aromatic compounds, 362
Osmium tetroxide, in Ryter–Kellenberger fixation technique, 241
Osmoregulation, 177, 182, 539–541
Osmotic control of gene expression, 179–183, 539–541
Outer membrane, gram-negative bacteria, 3, 260
  binding of origin of replication, *oriC*, 43
  porins, 263
  proteins (Omp), 179, 263
    osmoregulation of OmpF, OmpC, 182
    porin protein (PhoE), 196, 263
Oxidation-reduction:
  balance, in fermentations, 363
  in energy production, 15–16
  P_i assimilation, 16, 306
Oxidation stress, 197–198
Oxidation value, 363–365
Oxidative pentose phosphate cycle, 309, 312, 321, 391
Oxidative phosphorylation, 16, 325–332
Oxidized substrates, effect on fermentation products, 376
Oxolinic acid, DNA gyrase inhibitor, 84
Oxygen:
  growth conditions (aerobic, anaerobic, facultative), 24
  tolerance, 374
  toxicity, 373
  utilization by streptococci, 374–375

Panose, 357
p-Aminobenzoate (PAB) biosynthesis, 491
Paracoccus denitrificans, methylotrophic metabolism, 401
Paromomycins, aminoglycoside antibiotics, 88
Partitioning, of plasmids, 95
Pasteur effect, in yeast fermentation, 367
Pause sites, 73
Pectin (polymethylgalacturonate), pathways of utilization, 348, 351
Pediococcus, fermentation patterns in, 308
Penicillin:
  binding proteins, 253, 256
  synthesis from lysine, 470
  termination of cross-linking in peptidoglycan, 253
Penicillium:
  chrysogenum, β-lactam biosynthesis in, 470
  griseofulvum, 84
Pentaglycine, in peptidoglycan, 249
Pentose phosphate cycle, 13, 14, 391

Peptide bond formation, 70
Peptidoglycan, 2, 4, 5
  lipid-linked intermediates in, 525
  recycling of, 257
  structure and synthesis of, 248–257, 522–528
  switch between cell elongation and cell division, 528
Peptidyl tRNA hydrolase, 72
Peptidyltransferase, 69
Peptococcus, 26
  glycinophilus, glycine cleavage system in, 481
Peptostreptococcus, 26
  plasmalogens in, 416
Periplasm, periplasmic space, 3, 269
Periplasmic flagella, in spirochetes, 295
Periplasmic binding protein-dependent transport system, 335
Periseptal annuli, in cell division, 531
Permeability and transport, 269
Phage (c.f. bacteriophage), 202
Phase variation, flagellar, 171, 288
pH, hydrogen ion concentration:
  growth limits, 26, 27
  homeostasis (maintenance of intracellular pH), 334, 541
Phenotype, observed expression of genotype, 10
Phenylalanine:
  biosynthesis of, 484, 489–491
  deaminase, 455
Phenylalanyl tRNA, structure, 59, 61
Pheromones, in conjugation in enterococci, 108
Phosphate (P_i), assimilation of, 306
Phosphate starvation-controlled stimulon, 196–197
Phosphoenolpyruvate, in PEP transport systems (PTS), 339–340
Phosphoenolpyruvate (PEP) carboxykinase in gluconeogenesis, 314
Phosphoglycerides:
  alk-l-enyl ethers (plasmalogens), 416, 417
  alkyl ethers, 416
  biosynthesis of, 428
  in microorganisms, 418–419
Phosphoketolase pathway, 13, 308–311
Phospholipids, 3, 417
  biosynthesis of, 426
Phosphomycin (phosphonomycin; L-*cis*-1, 2-epoxypropylphosphonic acid), inhibition of peptidoglycan synthesis, 253
Phosphorelay system, in sporulation, 553–554
Phosphorylation:
  dephosphorylation, in regulation, 177
  oxidative, 14
  substrate level, 14, 306
Phosphoryl transfer potential (nucleoside triphosphates), 323
Phosphotransferase system, 160, 339, 342, 343
Photodimer, of thymine, in UV-irradiated DNA, 135
Photolithotroph, 23

Photoorganotroph, 23
Photophosphorylation, 398
Photoreactivation, 139
Photosynthetic bacteria and cyanobacteria, 391–398
Photosynthetic reaction center, 396
Phycobiliproteins, light-harvesting polypeptides, 397
Phycobilisomes, in *Fremyella diplosiphon*, 397
Phycomycetes, lipids and fatty acids in, 414
Pili, 5, 6, 297–300
Pilin variation in *Neisseria gonorrhoeae*, 299
Plasmalogens (alk-l-enyl ethers), 415, 417
Plasmids, 95
  replication control, 97
    in ColE1 plasmid, 98
    in P1 plasmid, iteron control, 99
  resistance genes (r-determinants) in, 96
Pneumococcus (*Streptococcus pneumoniae*):
  autolysin (N-acetylmuramyl:L-alanine amidase), 252
  capsule, 4, 271–273
  folate synthesis, 491
  transformation in, 110
Point mutation, single base change, 133
Polar mutations, 133
Polarity, in double-stranded DNA, 30
Poly(A) tails, in eukaryotes, 56
Polyamines (putrescine, spermidine, cadaverine):
  biosynthesis, 468
  in T4 phage, 205
Poly-β-hydroxybutyrate, 393
Polyene antibiotics (nystatin, amphotericin B), interference with sterol function, 410
Polymethyl galacturonate (pectin) catabolism, 352
Polycistronic message, from an operon, 153
Polymerase:
  DNA, 7, 40, 42, 44
  RNA (DNA-dependent RNA polymerase), 7
Polypeptide capsule, *B. anthracis*, *Y. pestis*, 275
Polypeptide synthesis, 64, 68, 69
  elongation, 66, 67
  initiation, 67
  peptide bond formation, 70
  ribosomal proteins, operons encoding, 65
  termination, 71
  translocation, 70
Polyphosphate, utilization as energy source, 387–388
Polyribosomes, 5
Polysialic acid (poly-N-acetylneuraminic acid), *E. coli*, 277
Polysomes, 5
Pores, in eukaryotic nuclear membrane, 237
Pores, porins, 3, 4, 179
Porfiromycin (mitomycin analog), 83
Porins, 263–264
Positioning of daughter chromosome, 532
Positive control, in transcription, 152, 154
Postreplication repair mechanisms, 145
Post-transcriptional modification, of tRNA, 58

Post-translational processing steps, 71
Potassium (K+), transport, 181, 341
Prephenate, precursor to phenylalanine and tyrosine, 489–490
Presecretion proteins, in general secretory system, 78
Prokaryotic gene expression, regulation of, 152
Proline:
  biosynthesis, 473, 464
  degradation pathway, 173
  osmoregulation, 539–541
  utilization, Put system, 171, 195
Propionibacterium, 26
  fatty acid biosynthesis in, 421
  plasmalogens in, 416
Propionic acid fermentation, 385–390
Propionispira, free living nitrogen fixation in, 438
Proteolytic control, Lon system, 198
Protocatechuate, intermediate in aromatic ring metabolism, 360
Proton motive force (PMF), 325
  conversion to usable energy, as ATP, 330–331
  in transport of metabolites, 331
Protoplast, 266, 267
Prototroph, wild type, 133
Pribnow's box (in transcription), 49
Primase (primer generating polymerase), 38
Primasome (RNA primer), 40
Proflavine, intercalating agent, 83
Prokaryote, cell surface of, 248
Promoters, consensus sequences, 49
Proofreading (by DNA polymerase, DNA Q), 42
Proteases, in lambda phage infection, 222
Protein:
  degradation of abnormal proteins, Lon protease, 79
  export, 74
  folding and chaperones (chaperonins), 73
  interaction with nucleic acid in regulation, 31
  synthesis, translation: 6–9, 56
    nonsense codons in, 57
    transfer RNA (tRNA) in, 57
  trafficking, *sec* gene products, 73–78
Proteus mirabilis, 6
  arginine biosynthesis in, 465
  capsular polysaccharide (N-acetylglucosamine), 275, 276
  pili and flagella of, 6
  swarming phenomenon, 291–295
Proteus vulgaris, swarming, 291–295
Proton motive force (PMF), 14, 330
Pseudomonas:
  acidovorans, conversion of chorismate to p-aminobenzoate, 491-493
  aeruginosa, 75, 181
    OmpP in, 263
  arginine biosynthesis in, 465
  arginine deiminase activity in, 466
  aromatic hydrocarbon metabolism in, 360
  branched-chain fatty acids in, 414
  fluorescens, 75

  arginine biosynthesis in, 465
  free living nitrogen fixation in, 438
  linderi (*Zymomonas mobilis*), Entner-Doudoroff pathway, 314
  maltophila, 415
  putida, 180
    aspartate aminotransferase, 456
Pseudomurein, in archaebacterial cell wall, 248
Pseudouridine, in Loop IV of tRNA, 58
Psychrophilic microorganisms, 26, 544–545
Psychrotolerant, psychrotropic microorganisms, 544
Pullulan, structure and utilization of, 357, 358
Puromycin, inhibition of translation, 90, 91
Purine:
  bases, abnormal and normal pairing, 137
  biosynthesis of, regulation, 500–503, 507–512
  fermentation of, 324
  interconversion, 512–513
  interconnecting pathways to riboflavin, folate, thiamine, histidine, 510
Purple bacteria, photosynthesis in, 393
Pyridoxal-5'-phosphate, role in enzymatic reactions of amino acids, 458
Pyrimidine:
  biosynthesis of, regulation, 500, 506–512
  interconversion of, 513–514
  pyrC locus, translational control of, 170
  pyrBI locus, in transcriptional attenuation, 167
Pyrodictium, 419
Pyrophosphate, inorganic, utilization of, 387–388
Pyruvate:
  alternate pathways of utilization, 330
  family of amino acid biosynthesis, 471
  formate lyase, 384
  major pathways of fermentation product formation from, 364

Quinolinic acid, precursor to NAD, 498
Quinolones, DNA gyrase inhibitors (ciprofloxacin), 85
Quinone (ubiquinone, hydroquinone), function of, 328

Racemases, amino acid, 457
Raffinose metabolism, 348, 350
Random coiling, in DNA structure, 31
Reaction center, in photosynthesis, 396
Receiver/regulatory proteins, 179
Recombination:
  genes, 122
  general recombination, 118
  integration of donor DNA into the genome, 117
  model for, 119
  recA-dependent, *recA*-independent recombination, 117, 121
  recombinational regulation of gene expression, 171
Reductive C$_4$-carboxylic acid cycle, 395
Reductive pentose phosphate (Calvin) pathway, 394
Regulation:
  aspartate family of amino acid biosynthesis, 476, 477

genetic (repression), 9–10, 154
gluconeogenesis, 315
lactose operon, 157
membrane-mediated, 171
metabolic (feedback inhibition), 9–10
multivalent inhibition, in aspartate family, 478
nitrogen metabolism (assimilation, fixation), 193–195
phosphate metabolism, 195–196
purine and pyrimidine biosynthesis, 508–512
Regulon, 153, 183
Replication of DNA, 6–7, 41
    initiation of, 43, 46
    of lambda (λ) phage, 225
    of T4 bacteriophage, 212
    origin of (oriC), 7, 40, 43
    replicating fork, 39
    termination, terminus (terA,B,C), 7, 40, 47
Replicon hypothesis and growth of gram-positive bacilli, 535
REP (repetitive extragenic palindromic) elements, 36
Repression (genetic regulation), 9, 10
    catabolite, 159
    permanent, 159
    transient, 159
Repressor proteins, 153
Resistance transfer factor (RTF), 96
Respiratory pathways, regulatory control, 190
Restriction and modification, 123, 124, 125, 210
Restriction endonucleases, 123
Retroregulation, in lambda phage infection, 225
Rhamnose utilization, 348, 349
Rhizopus, fermentation patterns in, 308
Rhizobiaceae:
    Rhizobium, 110, 180, 438
    in symbiotic nitrogen fixation, 442–446
Rhizoselenia, symbiotic nitrogen fixation, 438
Rho (rho;ρ) factor, 53
Rhodobacter:
    capsulatus, aminolevulinate synthesis in, 463
    spheroides, aminolevulinate synthesis in, 463
Rhodococcus erythropolis, hexitol utilization in, 347
Rhodocyclus, free living nitrogen fixation in, 438
Rhodomicrobium, free living nitrogen fixation in, 438
Rhodopseudomonas:
    capsulata, free living nitrogen fixation in, 438
    palustris, anaerobic benzoate metabolism in, 363
    sphaeroides, free liviing nitrogen fixation in, 438
Rhodospirillales, 26
Rhodospirillum:
    purpureus, glutamate dehydrogenase, 451
    rubrum:
        free living nitrogen fixation in, 438
        regulation of nitrogenase in, 446
Ribitol phosphate, in teichoic acid, 257
Riboflavin, biosynthesis of, 503–504
Ribonucleic acid (RNA):
    antisense (micRNA), 181
    messenger RNA (mRNA), 28

modifying enzymes, 56
polymerase, 7, 47, 48, 50, 51
    role of NusA and NusG, 51
    transcriptional factors, Class I, II, 51
processing, 52
ribonucleases:
    RNase H, 98
    RNase P, RNase III, RNase E, RNaseD, RNaseF, 54
    ribosomal (rRNA), 9, 28
    structure and processing, 55
synthesis (transcription), 47, 48, 49
rho-dependent terminators, 52, 53
rho-independent termination signal, 51
transfer RNA (tRNA), 9, 28, 53, 54, 57, 58, 59, 60
turnover, ribonucleases, ribozyme, 52, 54
Ribose-5-phosphate, pentose shunt pathway, 310
Ribosomes, 5, 61–66
    editing, 71
Ribosomal RNA (rRNA), 9, 28, 52, 61, 66, 176
Ribosomal proteins, 174
Ribosome editing, during translation, 72
Ribulose-5-phosphate, in pentose shunt pathway, 310
Richelia intracellularis, cyanobacteria, symbiotic nitrogen fixation, association with marine diatoms, 438
Rifampicin, rifamycin, inhibition of transcription, 48
Ristocetin, interference with peptidoglycan synthesis, 254
Root nodules, in soybean (Glycine max), 437
Rudman and Meister, general transaminases, 456
Ruminococcus, 26
    cellulose degradation in, 353
    plasmalogens in, 416

Saccharomyces carlsbergensis, maltose fermentation in, 346
Saccharomyces cerevisiae, 23
    ammonia assimilation in, 451
    aspartate aminotransferase in, 456
    cell structure of, 240
    fermentation in, 366
    lipids and fatty acids in, 414
    maltose fermentation in, 346
    serine biosynthesis in, 481
Salmonella typhimurium, 81–82, 171, 174, 179, 180, 184, 187, 191
    arginine biosynthesis in, 465
    core region, lipopolysaccharide, 260
    folate biosynthesis, 491
    gluconeogenesis in, 315
    glutamine amidotransferase in, 456
    growth of, 522–533
    lipopolysaccharide biosynthesis in, 264
    thiamine biosynthesis in, 502–505
Saturated fatty acids, biosynthesis of, 423–425
Secretion, sec gene, 74
Secretory pathways, gram-negative bacteria, 77

Secretory protein translocation system, 75
Segregation, of bacterial chromosome, 8
Selenocysteine, in formate dehydrogenase, 72
Selenomonas:
    plasmalogens in, 416
    ruminantium, plasmalogens in, 416
Semiconservative replication, 36, 37
Sensor proteins, 179
Sensory transduction system, 177–179
Sequential induction, in mandalate oxidation
    pathway, 362
Serine:
    deaminase, 452
    serine-glycine family of amino acid biosynthesis,
        480–482
Serratia marcescens:
    arginine biosynthesis in, 465
    swarming motility in, 291
Sex pheromones, in streptococci, 108, 109
Shape changes in *B. subtilis rodB* mutants, 538
Shigella, core region, lipopolysaccharide, 260
Shine-Dalgarno (S-D) sequence, 65
Shock-sensitive transport systems, 335, 336
Siderophores, iron-binding chelates, 341–342
Sigma (σ), subunit of RNA polymerase, 47, 49, 186
Sigma (σ), factors, in sporulation, 552–558
Signal peptidase (LepB), 77
Signal recognition particle (SRP) (*ffh*, *ffs*, *ftsY*
    genes), 77
Signal sequence, in proteins, 74
Signal transducing systems, 177–179
Sirex cyaneus (woodwasp), cellulose degradation in,
    355
Sisomycin, aminoglycoside antibiotic, 88
Slime layer, 270
Social (S) system, in gliding motility, 297
Solvent-producing fermentations, 13, 376–382
Sorbitan monooleate polyoxy-alkene, surfactant, 541
Sorbose, metabolism of, 347
SOS-inducible repair, 146–148
Specialized transduction, 115
Spectinomycins, aminoglycoside antibiotics, 88
Sphaerophorus, plasmalogens in, 416
Sphaerotilus, iron bacteria, 401
Spirillum atlanticum, 5
Spirillum linum, 5
Spirochaeta aurantia:
    chemotactic behavior in, 296
    branched-chain fatty acids in, 414
Spirochetes, motility in, 295–296
Spontaneous mutations, 134
Spore coat protein synthesis, 557
Spores (endospores): 19, 549–542
    activation, 558
    coat protein synthesis, 557
    cortex synthesis, 557
    germination, 558–561
    outgrowth, 561

sporulation cycle, *Bacillus*, 550–552
Sporocytophaga, 549
Sporolactobacillus, spore formation in, 550
Sporosarcina, spore formation in, 550
Sporulation:
    final stages of, 557
    forespore development, 556
    genes, 552–553
    initiation of, 553–555
    microcycle, 549
    phosphorelay system, 554
    regulation, 555
    transition from Stage II to Stage III, 555
Squalene, biosynthesis of, 430, 433–434
Stachyose metabolism, 348, 349
Stanier, mechanism of mandelate oxidation, 362
Staphylococcus aureus, 3, 87, 110
    fatty acids in, 414
    interpeptide bridge in, 3, 252
    mannitol production in, 347
    phosphoglycerides in, 419
    ribitol teichoic acids in, 257
Staphylococcus epidermidis, interpeptide bridge in,
    252
Starch utilization, 355
Starvation:
    effect on protein degradation, 79
    induced proteins, 545–546
Stationary phase, growth cycle, 21
Stem-loop structures, 166
Sterols, 410
    biosynthesis of, 430
    nutritional requirement for, 24
Stickland reaction, 459
Stimulon, 183
Streptococcus:
    arginine deiminase activity in, 466
    bovis, glutamate dehydrogeanse, 451
    faecalis, carbohydrate fermentation in, 71
    faecium (*hirae*), growth of, 519–520
    mutans:
        dextrans, levans, 273, 274
        glutamate dehydrogenase, 451
        mannitol production in, 347
    pneumoniae, 4
        autolysin (N-acetylmuramyl:L-alanine
            amidase), 252
        capsular polysaccharide structure, 4, 273
        folate biosynthesis, 491
        ribitol teichoic acids in, 257
    pyogenes:
        interpeptide bridge, 252
        M protein, 259
    salivarius, glutamate dehydrogenase, 451
    sanguis, glutamate dehydrogenase, 451
Streptolydigin, inhibition of transcription, 85
Streptomyces clavuligeris, β-lactam biosynthesis in,
    470

Streptovaricin, 85

Stress, environmental, effect on growth, 536–546

Stringent control, 189–193

Structure, bacteriophage, 209, 210

Structural gene, 152

Sugars (other than glucose), utilization of, 344–349

Sulfur bacteria, 393, 399

Supercoiling of DNA, 31, 32, 49

Sulfolobus solfataricus, 76
    aspartate aminotransferase in, 456
    phytanyl lipids in, 419

Superhelicity, of closed circular DNA, 32

Substrate level phosphorylation, 14, 322–324

Superoxide (anion), 25, 197, 373

Superoxide dismutase (*sodA*), 25, 197, 373–374

Super repressor, 158

Suppressor mutations, 136
    intragenic and extragenic suppressors, 138

Surface (S) layer, in eubacteria and archaebacteria, 248

Surface tension, effect on growth, 541

Svedberg unit (definition), 5

Swarming phenomenon, 291–295

Swivelase, GyrA component of DNA gyrase, 84

Swiveling, in DNA topoisomerase action, 33

Symbiosis (Sym) plasmid, 443

Symbiotic nitrogen fixation, 442–446

Symport transport systems, 337

Synergy, mecillinam and other penicillins and cephalosporins, 257

Tagatose (tagatose-6-phosphate) pathway, 344, 346

Target DNA, in transposons, 128

Teichoic acids, 257, 258

Temperate bacteriophage, 204

Temperature, effect on growth, 25, 26, 543–545

Terminase, in lambda phage infection, 227

Termination of polypeptide synthesis, 71

Termination of DNA replication (Tus), 7, 47

Termination of RNA synthesis, transcription, 48

Tetracyclines, inhibition of translation, 86, 87

Tetrapyrrole biosynthesis, 482–485

Thermoactinomycetes, spore formation, 550

Thermophilic microorganisms, 26, 356, 543–544

Thermotolerance, 186

Thermofilum, 419

Thermoplasma acidophilum (archaebacterium), 246

Thermoproteus, 419

Thermus thermophilus, 76

Thienamycin, interference with peptidoglycan synthesis, 257

Thiamine (vitamin B_1), biosynthesis of 502–506

Thiobacillus:
    acidophilus, pH homeostasis, 542
    iron bacteria, iron oxidation by, 334, 401

Thioredoxin, in deoxynucleotide synthesis, 506

Thiorhodaceae, sulfur bacteria, 399

Threonine:
    biosynthesis, 473
    deaminase, 455

Thylakoid (photosynthetic) membranes, red algae and cyanobacteria, 397

Thymidylate:
    synthase (*td* gene, in phage T4), 217
    in pyrimidine biosynthesis, 507

Tobramycins, aminoglycoside antibiotics, 88, 89

Toluene:
    oxidative ring fission by microorganisms, 361
    plasmids (TOL plasmids), 362

Topoisomerase(s), 32, 33, 49, 214, 217

Townsend, crop rotation, agricultural revolution, 436

Traffic ATPases, in metabolite transport, 336

Transcription, 6–9
    termination signal, 164

Transfer RNA (tRNA), 9, 58–61

Transformation, in streptococci, 110

Transaldolase, 13

Transaminases, 456–457

Transamination, 18

Transcription, 6, 8, 28, 47–50
    Transcriptional attenuation, 167–169
    Transcriptional control, 152, 479
    Coupled transcription and translation, 72
    Post-transcriptional modification, 58

Transduction, bacteriophage-mediated gene transfer, 94, 114
    low and high frequency transduction, 115, 116

Transfer RNA (tRNA), 9, 58–61, 28

Transfection, genetic transfer by bacteriophage DNA, 114

Transformation, bacterial, 94, 110, 112

Transglycosylase, cleavage of glycosidic bonds, 257

Transition mutations, 133

Transketolase, 13

Translation, 6, 7, 8, 9, 70
    translational control, *pyrC*, 170
    translational repression, 174–176

Translation factor, SelB (GTPase), 72

Translation control of *pyrC* expression, 170

Translational repression, 217

Translocation, 71
    inhibitors of, 86

Transport, metabolite, 335

Transposable elements, 126, 127, 128, 129

Transposition, in Mu phage, 228

Transposons, 127, 130

Transversion mutation, 133

Trehalose, in osmoregulation, 181

Treponema phagedenis,
    motility in, 296
    plasmalogens in, 416

Tricarboxylic acid (TCA, Krebs) cycle, 12, 13, 316–319
    function under anaerobic conditions, 319

Trichonympha sphaerica, (protozoon), cellulose digestion in, 355
Trimethyl-amine-N-oxide (TMAO), 187
Triplet codon, 9, 57
Tryptophan:
  biosynthetic pathway, 164, 484–488
  operon (*trp* operon), 164–168, 487–488
Tubules, in eukaryotic nuclear membrane, 237
Turgor, in osmoregulation, 181
Tweens, surfactants, effect on growth, 541
Two-component systems:
  adaptation to environment, 177
  properties of, 180
Tyrosine, biosynthesis, 487–491

Ubiquinol oxidase, in cytochrome *o* complex, 329
Ubiquinone:
  biosynthesis of, 494
  function of, 328
UDP-galactose, 161
UDP-glucose, 161
Underwinding, in DNA structure, 32
Uniport transport systems, 337
Unsaturated fatty acids, divergence of biosynthesis, 425
Upstream, regions of DNA relative to polymerase movement, 48
Uridine monophosphate (UMP), 167, 506, 507

Valine, biosynthesis, 475–476
Vancomycin, interference with peptidoglycan synthesis, 254–255
Vibrio: 26
  cholerae, 75, 177, 179
  parahaemolyticus, swarming, 291
Veillonella, 26
  parvula, 416
  plasmalogens in, 416
Viral receptors, on host cells, 204
Virion, bacteriophage, 204

Virulent or lytic infection cycle, of bacteriophage, 204
Vitamins, 23
  B vitamins (cobalamin, nicotinic acid, pantothenic acid, pyridoxin, riboflavin, thiamine), 23, 24
  K (menaquinone, methylnapthoquinone), 23
  biosynthesis of, 495

Water, effect on growth, 539
Weigle, reactivation and weigle mutagenesis, 149
Wild-type, normal allele expression, 10
Wilfarth, root nodules and nitrogen fixation, 437
Wood, H. G., and propionic acid fermentation, 388–389

Xanthobacter, 401
Xanthomonas, 110
Xenopus, topoisomerases in, 33
Xylulose-5-phosphate, 161, 310

Yeast phenylalanyl tRNA, structure, 61
Yeasts:
  arginine biosynthesis in, 465
  fermentation in, 366–367
  lipids and fatty acids in, 414
  nucleus in, 237
  pH requirements for growth, 26
  psychrophilic (*Leucosporidium*), 544
Yersinia pestis:
  polypeptide capsule, 275
  porins, 263

Z-DNA, 31
Zinc fingers, in DNA-binding proteins, 153
Zipper, leucine, DNA-binding domains, 155
Zymomonas mobilis (*Pseudomonas lindneri*), Entner-Doudoroff pathway in, 314
Zwischenferment, glucose-6-phosphate dehydrogenase, 308, 310